STATISTICS
for
SCIENTISTS and ENGINEERS

PRENTICE-HALL INTERNATIONAL, INC., *London*
PRENTICE-HALL OF AUSTRALIA, PTY, LTD., *Sydney*
PRENTICE-HALL OF CANADA, LTD., *Toronto*
PRENTICE-HALL OF INDIA (PRIVATE) LTD., *New Dehli*
PRENTICE-HALL OF JAPAN, INC., *Tokyo*

STATISTICS
for
SCIENTISTS *and* ENGINEERS

R. LOWELL WINE

Professor of Statistics
Hollins College

PRENTICE-HALL, INC., Englewood Cliffs, N. J.

Library of Congress Catalog Card Number 64–17363
Printed in the United States of America
84611 C

To Ruth

PREFACE

This book is designed as a beginning one year textbook in modern statistics with elementary calculus a prerequisite. It should be useful for *anyone* wanting to learn statistics starting with first principles. It is an expanded version of lecture notes used in a two-quarter course (three lectures per week) in statistics taught to juniors, seniors and first year graduate students in all areas of science and engineering at Virginia Polytechnic Institute and the University of Virginia. The material was also used several times with industrial research groups.

In writing this book I have tried to keep a balance between mathematical (theoretical) statistics and applied statistics. Many of the concepts are introduced by examples from applied statistics after which the concepts are formulated in mathematical terms and given a theoretical treatment. By presenting the material in this way, it is hoped that the reader will gain some real insight into the nature of statistics and at the same time learn how to apply the statistical procedures to actual experimental situations.

There are good statistical books for research workers in science and engineering, but generally speaking they are not very useful as a first introduction to statistics. In the first chapters of this book the reader is given an introduction and grounding in the foundations of statistics along with many examples and exercises. After this, some of those topics which engineers and scientists find most useful are introduced and developed. It is not intended that the topics selected for discussion be given an exhaustive treatment, but rather that they be developed to the point where they are useful to the practitioner of statistics. For the reader interested in special techniques or more sophisticated methods, some exercises and references have been given at the end of each chapter. (For example, very little is said about sampling techniques, quality control and acceptance sampling. There are already good textbooks in these areas.)

The examples are selected to illustrate the principles presented. That is, they are selected so that the principles of statistics can be understood without special knowledge of a particular subject matter field. These so-called common-sense examples refer to such things as heights, weights, tensile strengths, teaching methods and scores, and measures of objects which should be familiar to readers with at least two years of college.

Each chapter contains a copious supply of exercises which are applicable to many and widely different fields of science and engineering. Any reader should find numerous exercises of special interest. The exercises are designed to give the reader practice in applying the material presented, to extend both his practical and theoretical concepts of statistics, and to encourage him to look carefully at new concepts which are outlined only in exercises.

The book may be used as a text for either a theoretical or applied course in statistics. As a theoretical text emphasis should be given to Chaps. 3, 4 and 5, to those sections of the remaining chapters which deal primarily with the mathematical development, and to those exercises which stress proofs and new concepts. As an applied text the proofs and mathematical development may be cut considerably and the sections and exercises on application stressed.

This is a book on statistics. It should be useful to anyone learning the problems of numerical analysis in experimentation or planned investigations. The book should also be useful for anyone seeking some of the foundations of statistics and the way in which part of the statistical structure may be developed.

Many groups and individuals have helped in developing this book. First, I wish to gratefully acknowledge my thanks to Dr. John E. Freund for reading the manuscript and making numerous helpful suggestions, many of which have been included. I wish to acknowledge my gratitude to the several classes of students who read and made helpful suggestions on parts of the manuscript; to the United States Weather Bureau for data used in several exercises; to friends in the Celanese Corporation of America for data (coded) and advice on special problems; to friends in the West Virginia Pulp and Paper Company, United States Steel, and White Sands Missile Range for the opportunity to see applications to special problems; to the Statistics Department of Virginia Polytechnic Institute for typing several chapters of the first draft of the manuscript; to my colleagues in the Mathematics and Psychology Departments at Hollins College for valuable criticism and suggestions; to Miss Margaret Shinnick for reading most of the manuscript and commenting on style and content; to David and Suellen Wine, my son and daughter, for obtaining the data in Tables 5.2 and 5.5; and to many other persons who have published data which are reproduced and acknowledged at the appropriate places in the text. Further, I am indebted to the Danforth Foundation for a travel grant and to the personnel of the Hollins College library for their kind assistance on numerous occasions.

Finally, I wish to express my appreciation to Professor E.S. Pearson for his kind permission to reproduce Tables III, IV and VII from *Biometrika*, Table IX from *Biometrika Tables for Statisticians*, and Table X from *Tables of the Ordinates and Probability Integral of the Distribution of the Correlation Coefficient in Small Samples;* and to McGraw-Hill, Inc. for kind permission to reproduce Table VIII. I am indebted to Professor Sir Ronald A. Fisher, F.R.S., Cambridge and to Dr. Frank Yates, F.R.S., Rothamsted, also to Messrs. Oliver and Boyd Ltd., Edinburgh, for permission to reprint Table VI from their book *Statistical Tables for Biological, Agricultural and Medical Research.* Also, I wish to express my appreciation for permission to reproduce Tables 8.1, 8.2 and 10.9 from the *Journal of the American Statistical Association;* Tables 8.3 and 8.6 from books published by McGraw-Hill, Inc.; Tables 10.10 and 16.3 from *Biometrics;* Table 16.1 from a publication by the American Cyanamid Company; and Table 16.5 from the *Annals of Mathematical Statistics.*

R. LOWELL WINE

Roanoke, Virginia

CONTENTS

4 PROBABILITY

5 SAMPLING AND SAMPLING DISTRIBUTIONS— EXPECTATION AND ESTIMATION

13 AN INTRODUCTION TO FACTORIALS

14 REGRESSION AND RELATED TOPICS

1

INTRODUCTION

1.1. A DEFINITION OF STATISTICS

Modern statistics is a new and vigorous discipline. It is so new that some of the men who were most instrumental in establishing statistics as it is known today are still actively engaged in research and teaching. Its vigor can be attested from the fact that statistics is growing so rapidly that it is impossible to incorporate many of the latest techniques in a textbook, for by the time the last section is written the first chapters already need revision.

Statistics is playing an increasingly important role in research activities. For this reason it is necessary that special training in statistics be given as early as possible so that experimentation and scientific investigations do not suffer. The study of statistics should not be viewed as just another area of study which is merely desirable for the scientist and engineer; instead, statistics should be viewed as a very sensitive instrument which is capable of successfully coping with many of the most difficult problems posed by modern investigations. Ignoring the use of statistics in many of our research activities today should no more be tolerated than that of ignoring tractors and combines in the wheat fields of Kansas or of ignoring the latest drugs in the treatment of ailments.

The term "statistics" is old, but its present-day interpretation is very young. The term no longer simply refers to the collection and compilation of data; instead, *statistics* is often called *the science of decision-making in the face of uncertainty*. It has to do with both the deductive and the inductive process, that is, both mathematical and scientific procedures. Statistics currently deals with the theoretical development and application of methods suitable to numerical measurements.

1

Whenever data are collected, statistical methods may be used. In fact, anyone who attempts to work with data acts like or has occasion to act like a statistician. Statistics is a science, based upon mathematics, which deals with such problems as (1) planning a program or an experiment for obtaining data so that reliable conclusions can be drawn from the data, (2) tabulating and analyzing the data, (3) deciding what interpretations and conclusions can properly be drawn from the data, (4) determining to what extent the conclusions are reliable, and (5) justifying by mathematics the methods used in (1), (2), (3), and (4). *Statistical methods* are those procedures used in designing and planning experiments and in collecting, analyzing, and interpreting data. *Statistical theory* has to do with the mathematical development and justification of the methods used.

Statistical methods may be thought of as falling in two classes. Those methods which are used more meaningfully to describe a set of data but which do not involve generalizations are commonly called *descriptive statistical methods*. Those methods which are used on a relatively small set of data to generalize concerning the nature of a much larger set of possible data make up methods of *statistical inference*.

Descriptive statistical methods or, simply, *descriptive statistics*, include those methods which are used in making and describing such well-known objects of our everyday experience as graphs, charts, and tables. Such examples as the batting average of leading hitters, defense-spending graphs, airplane travel charts, stock market averages, census figures, production of automobiles by months, and the index of living costs represent only a few of the illustrations of descriptive statistics we see regularly. Thus, many of the results and techniques of descriptive statistics are known to most of us.

The methods of statistical inference are not so well-known, even though illustrations of their use are fairly common. We read, for example, that the Gallup poll makes a survey and predicts that Joe Brown will be elected governor instead of Sam Jones, or that Kinsey makes some inference about the sex habits of the American female, or that it has been proved that one brand of cigarettes contains less tar than other brands, or that a manufacturer claims that the average life span of a certain type of light bulb is 2500 hours. In each case, we read, and perhaps take issue with, the conclusion, but we know little or nothing about the methods used in arriving at these inferred statements.

The student is cautioned against thinking that methods of descriptive statistics and statistical inference are always distinct and clearly defined. As a matter of fact, most methods which are used in descriptive statistics are also applied in statistical inference. The two terms are generally used with reference to the *kinds of problems* we wish to consider, not with reference to particular formulas or series of formulas. For example, the "average number of defective parts" of a manufactured article might be used in either sense. The term "average" may *describe* the number of defective parts manufactured on a given day, or it may be used to *infer* the number of defective parts which will be produced per day during the remainder of the year.

Since statistics in some way touches so many of our daily activities, it is not possible to give a descriptive and short definition of the term. However, for our purposes it is probably adequate to think of *statistics as both a pure and an applied science which is involved in creating, developing, and applying procedures in such a way that the uncertainty of inferences may be evaluated in terms of probability.* It should be noted that deductive techniques, as used in mathematics, are required in developing the procedures.

1.2. SCIENTIFIC METHOD AND APPLIED STATISTICS

The student will soon realize that the selection and application of some statistical methods, particularly those used in the analysis and design of experiments, are similar to what the scientist and engineer do when setting up a hypothesis, planning and conducting an experiment, and testing the hypothesis by using the experimental data. In both the scientific and statistical disciplines one is concerned with such things as planning and analyzing an experiment in such a way as to establish a fact (the hypothesis) within the framework of a specific theory. In addition to this, the statistical discipline generally requires that a measure of the degree of uncertainty, in terms of probability, accompany the inference drawn from the experiment.

Even if the experimenter understands the theoretical framework and knows specifically what hypothesis he wishes to test, it is still not always obvious which collection of statistical techniques (methods) should be applied. For just as there are many ways to get from Denver to Boston, say, there are usually many ways to collect and use data statistically to justify a statement within a fixed theoretical framework. The investigator normally first looks for a relatively short and relatively simple procedure, but these are not the primary considerations. He wants to draw the correct conclusion within the framework of the experiment in the most efficient way, taking into account, among other things, time, money, and relative importance of the investigation. Thus, in addition to knowing several alternative statistical routes, the investigator must make a decision about the best one to select. In other words, he must decide which "statistical model," including a group of accompanying techniques, to use in the experiment, it being understood that *a model is an idealization of a particular experimental situation.*

There are many statistical models which may, or may not, be useful in solving real problems. For example, the normal curve may be used as a model in describing the distribution of the grades of all beginning mathematics students at a large engineering school. A model is not always so simple, nor is it always used in such a straightforward way, but the objective in using a model is always the same—to make the concept, analysis, and conclusion simpler to understand and to disseminate. Once the model is selected, the resulting conclusion can be relied on only to the extent that the model approximates the situation being studied.

Statistical models have not been constructed for many possible experimental situations. Thus, it may be desirable in a specific investigation to construct and develop the properties of a new model, and this is much easier when one already knows something about statistical models and the associated procedures.

1.3. A BRIEF HISTORY OF STATISTICS

Even though data have been compiled almost from the beginning of recorded time, the science of statistics has never been so broad in its scope as it is today. At first, statistics seems to have consisted in census taking. About 3050 B.C. the ancient Egyptians collected data concerning wealth and population before building the pyramids. There are two censuses of the Israelites recorded in the book of Numbers. In 594 B.C. a census was taken in Greece for the purpose of levying taxes. There are many other records of census taking in most countries of the world from early times to the present.

In addition to taking the census, the Romans prepared surveys of the entire country and kept records of births and deaths. After the Middle Ages, certain individuals as well as governments started keeping records on such things as wealth, armies, commerce, laws, and national resources.

About the middle of the eighteenth century, Gottfried Achenwall, a professor of philosophy in a German university, first used the word *statistik*, and the name *statistics* was introduced into England by E. A. W. Zimmerman about 1787. The Royal Statistical Society of London was founded in 1834 and the American Statistical Association in 1839; each of these societies holds meetings periodically and publishes papers of current interest. Thus, anyone who is interested may follow the growth of statistics over the last 125 years by looking at the records of these societies. In the first number of the *Journal of the Royal Statistical Society*, issued in 1838–39, we read, "Statistics may be said, in the words of the prospectus of this society, to be the ascertaining and bringing together of those facts which are calculated to illustrate the conditions and prospects of society."

The *theory of probability* and the *normal distribution* have been very important in the development of statistics, and they are now of primary importance in the theory and application of statistics. In the middle of the seventeenth century, some gambling experiences with a particular game of dice led Chevalier de Méré, a French nobleman, to consult the famous Blaise Pascal (1623–62) concerning the most advantageous way to bet. Pascal solved the problem, and de Méré posed another problem which Pascal investigated. This led to a private correspondence with the French lawyer-mathematician Pierre de Fermat (1601–65) and to the first foundations of the theory of probability. After becoming acquainted with the contents of this correspondence, Christiaan Huygens (1629–95) developed some new ideas and in 1654 published a first book on probability. Jacob Bernoulli (1654–1705), Abraham de Moivre (1667–1754), and Pierre Simon Laplace (1749–1827) made great contributions to the early theory and application of probability. Abraham de Moivre is responsible for the equation of the normal curve (1733). Much later, Laplace and Karl Friedrich Gauss (1777–1855) developed the same results independently of each other.

Laplace in his studies of the origin of comets seems to have been the first to attack problems relating to rules of "inductive behavior," that is, to the adjustment of our behavior to a limited number of observations. The geologist Charles Lyell (1797–1875), the biologist Charles Darwin (1809–82), and the monk Johann Gregor Mendel (1822–84) in his experimental breeding of plants based some of their work on statistical arguments. None of these men was a statistician, and they did not spend their energies in placing statistics on a firm foundation.

Karl Pearson (1857–1936), initially a mathematical physicist, after becoming interested in evolution spent nearly half a century in serious statistical research. He helped to found the journal *Biometrika* and gave the study of statistics its first great impetus. Sir Ronald Fisher (1890–1962) made many important contributions to statistics. Since the 1920's Fisher and his students have also stimulated great interest in applications of statistics in many fields, particularly agriculture, biology, and genetics. Some of the basic theory on hypothesis-testing was presented by J. Neyman (1894–) and E. S. Pearson (1895–) as late as 1936 and 1938.

Thus, it was not until early in this century that statistics started to be used to any large extent outside of census taking and other specialized areas such as genetics and astronomy. Since the late 1920's interest in the application of statistical methods to all types of problems has grown rapidly. It is interesting to note that prior to the 1920's applied statistics was predominately descriptive in nature, and that since the 1920's statistical inference has grown to constitute nearly all of statistics. Indeed, today purely descriptive methods play a very minor and almost incidental role in statistical applications.

It would be pointless to try to enumerate all the areas in which statistical methods are used. To mention only a few, statistics is becoming increasingly important in agriculture, astronomy, business, biology, economics, chemistry, engineering, industrial studies, insurance, medical research, meteorology, physics, psychology, sociology, and transportation. The use of statistical methods in each of these fields grew in a different way. In biology applications started early; in transportation late. In agriculture there is large variation, which is difficult to control; in the physical sciences, in many cases, the variation is small and relatively easy to control.

Statistics is not at the same stage of development in all of the different fields of study. Different aspects are considered important in different areas, and special techniques are required at the research level. However, the fundamental principles on which any of the special techniques are built are the same, and many are discussed in this book.

Fuller presentations of these and other topics are found in the references of this chapter. The student is urged to read further. In fact, the serious student should form a habit of reading some references with each chapter.

REFERENCES

1. Churchman, C. West, *Theory of Experimental Inference*. New York: The Macmillan Company, 1948.

2. Cox, Gertrude M., "Statistical Frontiers," *Journal of the American Statistical Association*, Vol. **52** (1957), pp. 1–12.

3. Fisher, R. A., *Statistical Methods for Research Workers*. London: Oliver and Boyd, 1938, Chap. 1.

4. Fisher, R. A., "The Expansion of Statistics," *American Scientist*, Vol. **42** (1954), pp. 275–82, 293.

5. Fitzpatrick, Paul J., "Leading American Statisticians in the Nineteenth Century," *Journal of the American Statistical Association*, Vol. **52** (1957), pp. 301–21.

6. Hoadley, Walter E., Jr., "Statisticians—Today and Tomorrow," *Journal of the American Statistical Association*, Vol. **54** (1959), pp. 1–11.

7. Hotelling, Harold, "The Impact of R. A. Fisher on Statistics," *Journal of the American Statistical Association*, Vol. **46** (1951) pp. 35–46.

8. Lachman, Roy, "The Model in Theory Construction," *Psychological Review*, Vol. **67** (1960) pp. 113–29.

9. Neyman, J., *First Course in Probability and Statistics*. New York: Holt, Rinehart & Winston, Inc., 1950, Chap. 1.

10. Pearson, E. S., "Karl Pearson, an Appreciation of some Aspects of his Life and Work, Part I: 1857–1906," *Biometrika*, Vol. **28** (1936), pp. 193–257.

11. Pearson, E. S., "Karl Pearson, an Appreciation of some Aspects of his Life and Work, Part II: 1906–1936," *Biometrika*, Vol. **29** (1938), pp. 161–248.

12. Walker, Helen M., "Bi-centenary of the Normal Curve," *Journal of the American Statistical Association*, Vol. **29** (1934), pp. 72–75.

13. Wolfowitz, J., "Abraham Wald, 1902–1950," *Annals of Mathematical Statistics*, Vol. **23** (1952), pp. 1–13.

14. Youden, W. J., "The Fisherian Revolution in Methods of Experimentation," *Journal of the American Statistical Association*, Vol. **46** (1951), pp. 47–50.

2

FREQUENCY DISTRIBUTIONS

The methods of statistics are concerned with the study of variation. The nature of the variation is shown by a frequency distribution. Several types of distributions for both continuous and discrete variables are considered. Characteristics of distributions such as the central value and amount of scatter are discussed.

2.1. DISCUSSION OF TERMS

We shall assume that you know the meaning of such terms as *data, object, individual, variable, collection* and *set*. Some terms in common usage, such as *observation* and *population*, have different or special meanings as they are applied in statistics and, thus, will need to be defined and discussed.

We shall think of an *observation* as a recording (usually numerical) of information. Scores, measurements, ranks, and categories are types of observations. A *score* is a numerical assessment of an individual or an object on a scale. Examples are the grade on a chemistry test, the points scored by a basketball team during a game, and an intelligence quotient. Scores are also used to indicate quality of such things as meat and butter. We usually think of a *measurement* as a numerical value indicating the extent or size of such things as the height of a tree, the length of a rod, the volume of a liquid, the weight of a chemical compound, the temperature of a room, the pressure of a gas, the intensity of sound, and the amount of electric current. The term *rank* is used to express the position of an object in regard to some stated quality relative to all other objects under consideration at the moment. For example, in a goodness-of-taste experiment involving three brands of grape juice, say A, B, C, one judge might give brand A rank 3, brand B rank 1 and brand C rank 2, where rank 1 indicates best taste, rank 2 second best, and rank 3 third best. Ranks are also used in beauty contests, academic achievement of the West Point graduating class, certain market reports, etc. *Categories* are used to indicate color of eye, type of tree, kind of metal used in making a machine part, degree of interest in a television program, etc. Illustrations of observations which we do not consider to fall in the above four types, to mention only a few, are the number of defective parts in a manufacturing process, the rate of germination of a seed, the proportion of cars exceeding the speed limit, and the number of students taking a course.

An observation is made on some characteristic of an *object*. For example, the object might be a human skull and the observation might be head width, the object a day and the observation maximum temperature, or the object a human female and the observation color of hair. In each case, the observation would differ from object to object. Thus, for example, we call head width a variable, since it differs from skull to skull. We shall always think of a *value of a variable* as being a number. Thus, observation and value of a variable are synonymous terms so long as the recording of information is numerical.

Variables may be continuous or discrete. A *continuous variable* is a variable which can assume any real value between two distinct numbers, and a *discrete variable* is one which can assume only isolated values. Examples of variables which are discrete are the number of heads in 100 tosses of a coin, and the number of mining accidents per year. Length of life of a light bulb, velocity of a jet airplane, height of an adult male, brightness of light, and amount of heat resulting from friction are examples of continuous variables. The reader is cautioned against thinking of height, say, as a discrete variable. It is true that any ordinary set of data from experience will assume only isolated values, and, thus, you might tend to think of these data as being discrete. But heights can be measured only to the accuracy of the instrument, say to the nearest ten-thousandth of an inch. This introduces a rounding-off error, so the recorded measurement only approximates a number which could fall anywhere between two distinct real numbers.

The statistician is not concerned with a single observation except as it relates to a collection. He is concerned with a collection having some common observable attribute, and he calls the collection of such observations a *population*. Sometimes the set of objects on which the observations are made is said to be a population, but the statistician works with the population of observations (or population of values of the variable). For example, one might think of a collection of similar rocks as being the population, but the statistician usually thinks of the population as being the collection of similar observations, say specific gravity, made on the rocks. Also, all cans of sliced pineapple prepared in the Hawaiian Islands during a given season might be thought of as a population, but the statistician usually thinks of the population as being the collection of similar observations, say, weight of sliced pineapple with the juice drained off, made on all these cans. Thus, when a statistician refers to a population, he usually thinks in terms of a collection of numbers and not a collection of objects. That is, he thinks in terms of the set of values of a variable.

It should be observed that, even though the statistician deals with numerical values, he does not forget the objects on which observations are made. Thus, he would not say, "Half the measurements are over 69 inches," when dealing with heights of adult human males in Roanoke City. He would say, instead, that half of the males in Roanoke City 21 years of age and over are more than 69 inches tall.

We have already observed that there is a population of observations corresponding to a population of objects. Several examples have been given of a population of values of a single variable. Such a population is called a *univariate population*. We may wish also to consider populations of values of two, three, or more variables. A population in two variables is called *bivariate*, one in three variables *trivariate*, and one in more than one variable *multivariate*. For example, corresponding to a population of adult males we may have a univariate population of heights, a bivariate population of heights and weights, or a trivariate population of heights, weights, and ages.

A population may consist of a finite or an infinite number of observations. If it is finite and contains N observations, we say that the *size of the population* is N; otherwise, it is called an *infinite population*. If N is sufficiently small, we can comprehend the nature of the population just by examining the set of observations. However, if the size of the population is finite and large or infinite, it should be reduced in some way so that we can more easily grasp its nature. This can be done

1. Tabularly or graphically by classifying the observations into classes of values of the variable.
2. Arithmetically by finding the numerical value of a few "features", *called parameters*, determined from the observations which practically characterize the population; that is, which bring out the most important aspects of the population.

It is with parameters that the statistician is mainly concerned. Since populations differ, they cannot all be described in the same way. However, there are certain features which most have in common, and it is these features (parameters) which we first examine in some detail.

In most cases the investigator has observations on not all of the objects of a population; he has only a *sample* which is some subcollection of the population of values. Whenever this happens, the best he can do is to estimate the nature of the population. This is done by determining the nature of the sample and using the properties of the sample to estimate the properties which characterize the population. The process by which we reach conclusions about a population from a sample taken from the population is known as *statistical inference*. Most of this book deals with problems of statistical inference, but in this chapter we wish to discuss 1 and 2 of the preceding paragraph along with some similar methods used to describe a sample. It will soon be observed that the graphical methods are the same for studying both population and sample, and that the arithmetic methods for population and sample differ in some few but important respects.

Not all types of observations are subject to the same statistical methods. In the first chapters only those methods which have broadest applications will be described; the more specialized methods will be left for later chapters. To illustrate, in most statistical studies the order in which observations are obtained is not important, but in certain weather and manufacturing studies, for example, order is very important. Thus, special methods are needed for dealing with observations falling in this category.

2.2. FREQUENCY DISTRIBUTIONS IN ONE VARIABLE

We shall now consider some methods for "picturing" the distribution of observations in a population or a sample. Typical examples are used to illustrate the methods which are generally applicable. (For more extensive treatment of the methods found in this section, see the references at the end of this chapter.)

2.2.1. Continuous Variable

If the population or sample is small, it can probably best be described by a simple listing of the values of the variable. For example, the weights of the three whooping cranes in captivity listed in increasing order might adequately describe the population. The weights would hardly need be reduced or summarized in any way.

Note. If the sole objective of the investigation is to describe a set of observations, then it makes no difference whether the set is thought of as being a population or as being a sample. However, when the main objective is to use the sample to make statistical inferences with regard to the population, the arithmetic methods for samples differ in some respects from those for populations, and, in this case, the description of the sample is considered an incidental part of the statistical procedure.

Some finite populations or samples are of such a size that the investigator might be in doubt as to what procedure to follow in order best to describe it. If the size of the population or sample were smaller, he would give a simple listing of the values; if the size were larger, he would definitely classify the observations in some way. The data in Table 2.1 might be considered to fall in this category. However, we shall use these *raw data*, that is, data listed in the order obtained, alphabetical order, or some other arbitrary way, to illustrate some of the tabular and graphic methods of classifying observations. It is not suggested that all the methods presented in connection with these data be used at one time or, in fact, that any need be used.

Table 2.1
Percentage of Humans in Each State 65 Years
Old and Over in 1950*

State	Per cent	State	Per cent
Alabama	6.5	Nebraska	9.8
Arizona	5.9	Nevada	6.9
Arkansas	7.8	New Hampshire	10.8
California	8.5	New Jersey	8.1
Colorado	8.7	New Mexico	4.9
Connecticut	8.8	New York	8.5
Delaware	8.3	North Carolina	5.5
Florida	8.6	North Dakota	7.8
Georgia	6.4	Ohio	8.9
Idaho	7.4	Oklahoma	8.7
Illinois	8.7	Oregon	8.7
Indiana	9.2	Pennsylvania	8.4
Iowa	10.4	Rhode Island	8.9
Kansas	10.2	South Carolina	5.4
Kentucky	8.0	South Dakota	8.5
Louisiana	6.6	Tennessee	7.1
Maine	10.2	Texas	6.7
Maryland	7.0	Utah	6.2
Massachusetts	10.0	Vermont	10.5
Michigan	7.2	Virginia	6.5
Minnesota	9.0	Washington	8.9
Mississippi	7.0	West Virginia	6.9
Missouri	10.3	Wisconsin	9.0
Montana	8.6	Wyoming	6.3

* Bureau of the Census, *County and City Data Book 1956* (A Statistical Abstract Supplement), page 2.

Generally, the first thing to do in order better to grasp the nature of the data is to arrange the numbers in increasing order of magnitude or to list each number and the corresponding frequency of its occurrence. Of course, in doing either our attention is focused on the numbers themselves and not on the objects associated with those numbers. Table 2.2 shows a rearrangement of the data of Table 2.1, listed in order of increasing percentage. A simple graphical representation of the data is given in the form of a *dot frequency diagram* in Fig. 2.1. In general, Table 2.2 and Fig. 2.1 are not used

Table 2.2
Percentages from Table 2.1 Arranged in Increasing Order

4.9	6.5	7.2	8.5	8.7	9.8
5.4	6.6	7.4	8.5	8.8	10.0
5.5	6.7	7.8	8.5	8.9	10.2
5.9	6.9	7.8	8.6	8.9	10.2
6.2	6.9	8.0	8.6	8.9	10.3
6.3	7.0	8.1	8.7	9.0	10.4
6.4	7.0	8.3	8.7	9.0	10.5
6.5	7.1	8.4	8.7	9.2	10.8

so much to show the nature of the population, but are most frequently used as intermediate steps to a more meaningful "picture" of the population illustrated below.

In passing, we observe some facts which are immediately evident from Table 2.2 or Fig. 2.1, but which require considerable time to obtain when determined from Table 2.1. The saving in time becomes greater as the number of observations increases. We observe that the highest percentage is 10.8 and the lowest is 4.9, with a *range* of $10.8 - 4.9 = 5.9$. We see what percentages are repeated most often and which percentages do not occur. Two-thirds of the states have percentages in the interval from 6.5 to 9.0, which is less than half the range. A large concentration of values is found in the interval from 8.5 to 9.0. In fact, 17 of the 48 states have percentages in this interval.

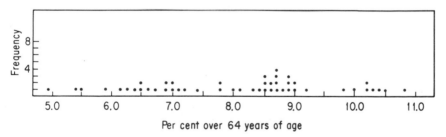

Fig. 2.1 Dot Frequency Diagram for Data of Table 2.1

The pattern of dots in Fig. 2.1 is so erratic that it is difficult to "picture" the set as a whole. Often a better picture of the observations is made possible by grouping them into *classes* (intervals or cells). This has a smoothing effect on the graph if the grouping is properly carried out. It is for this reason that we seek the answer to the following two questions: how many classes should be used, and what should be the length (see the next paragraph for definition) of each? There is no single answer to either question which holds in all cases. However, some general principles are given to guide the investigator in arriving at answers to these questions. The length of each class should be the same, unless there is very good reason to the contrary. Whenever there is a large number of observations, say more than 200, experience indicates that in most cases 10 to 20 classes are adequate. In our problem, assuming for the moment that grouping is desirable, we would expect to have fewer classes. In determining the number of classes in any problem, two things should be kept in mind. One is that the most natural or convenient classes be selected, and the other is that we select so that the "picture" of the population is as smooth as possible.

That numerical value which divides two successive classes is called a *class boundary*. The *class length* is the numerical difference in the two boundaries of that class. If the classes have equal length, then boundaries of the extreme classes are obtained by decreasing the smallest boundary value by the class length and by increasing the largest boundary value by the class length; otherwise, these extreme boundaries are usually assigned when the particular problem is specified.

In our illustration, one unit is the most convenient class length. Further, noting that the largest value is 10.8 and the smallest is 4.9 and that the range is 5.9, we decide to use seven classes. We still must determine whether to use intervals like 7.0 to 8.0 or like 7.5 to 8.5. For reasons which become clear at the end of the next paragraph, we select intervals of the latter type.

It is important that the class boundaries be selected so that there is no doubt concerning the class into which each observation falls. Thus, it is necessary that we select the boundaries so that they differ from any observation or give a rule which automatically places in a unique class any observation which is the same as a boundary. For example, with two successive intervals like 7.5–8.5 and 8.5–9.5 we must give a rule for placing the observation 8.5. If we wish to have an observation fall in a unique class automatically, we may usually select class boundaries halfway between two possible adjacent observations. For example, 7.55 is halfway between 7.5 and 7.6. There are computational advantages to selecting boundaries like 7.5, but the more appropriate method is to select boundaries like 7.55, since our observed values are rounded off according to the rule that 7.2, say, represents any number between 7.15 and 7.25. Thus, for illustrative purposes we select the class boundaries as 4.55, 5.55, 6.55, 7.55, 8.55, 9.55, 10.55, and 11.55.

The mid-value of a class is called the *class mark* and is used as the representative value for that class. The frequency with which observations fall in a class is called *class frequency*.

Once we have decided on the number of classes and the location of the class boundaries, a *frequency table* such as Table 2.3 may be prepared. (So long as a table contains a frequency column and a column for class boundaries or class marks, it is called a frequency table, no matter how many other columns it may have.) From the class frequency column of Table 2.3, we note that grouping produced a smoothing effect and that observations fell most often in the interval 8.55–9.55. Using this table, we are able to present the observations graphically in the form of frequency diagrams in several standard ways. It should be noted that all the columns of Table 2.3 are not required in a single study or in a single diagram.

Table 2.3
Frequency Table of Data in Table 2.1 (Grouped)

Class Boundaries	Class Mark x_i	Tabulation	Class Frequency f_i	Relative Frequency f_i/N	Cumulative Frequency f_i'	Cumulative Relative Frequency (f_i'/N)
4.55– 5.55	5.05	⫴	3	0.062	3	0.062
5.55– 6.55	6.05	⊬⊬⊬ ⏐	6	0.125	9	0.188
6.55– 7.55	7.05	⊬⊬⊬ ⫴⫴	9	0.188	18	0.375
7.55– 8.55	8.05	⊬⊬⊬ ⫴⫴	9	0.188	27	0.562
8.55– 9.55	9.05	⊬⊬⊬ ⊬⊬⊬ ⫴	13	0.271	40	0.833
9.55–10.55	10.05	⊬⊬⊬ ⫴	7	0.146	47	0.979
10.55–11.55	11.05	⏐	1	0.021	48	1.000

Perhaps the most useful frequency diagram is the *frequency histogram* (relative frequency histogram if the right-hand vertical axis is used) illustrated in Fig. 2.2. Putting the class boundaries and marks on the horizontal axis, x-axis, and the frequencies on the vertical axis, f-axis, of a rectangular co-ordinate system, we erect a rectangle above each interval whose area is proportional (most often equal) to the frequency of that class. Note that in our illustration the mark x_i of the ith class (i being any integer from 1

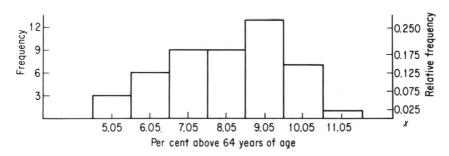

Fig. 2.2 Frequency Histogram for Data in Table 2.3

through k, the number of classes) is also the mid-point of the base of the ith rectangle making up the histogram, and that f_i represents the height of the ith rectangle as well as the frequency of the ith class. *Hence, the area of each rectangle is a number which is the same as the frequency for the corresponding interval, and the total area of the histogram, that is, the sum of the areas of the rectangles, is the total frequency.* In particular, the total frequency is 48 and the total area is $(48)\cdot(1) = 48$ square units. If some class length had not been equal to the others, it would have been necessary to adjust the height of its rectangle so as to express the area of this rectangle in units of all other rectangles. (Uses and limitations of the historgam will be discussed later.)

Histograms are normally used for grouped data. Grouping introduces approximations, and this limitation is reflected in the histogram, which may be used to indicate that each value in an interval appears with the same relative frequency. Thus, in the interval from 5.55 to 6.55 the histogram indicates that 5.6, 5.7, . . . , 6.5 occur with equal relative frequencies. This is clearly not the case, as we can see by looking at Table 2.2 This limitation of the histogram becomes less troublesome as the class length becomes smaller or the number of observations becomes larger.

The *frequency polygon* in Fig. 2.3 illustrates another method of representing frequency diagrams. A frequency polygon is constructed by plotting the points $(x_1, f_1), (x_2, f_2), \ldots, (x_k, f_k)$ on a rectangular co-ordinate system and then connecting these points by straight-line segments. In order to

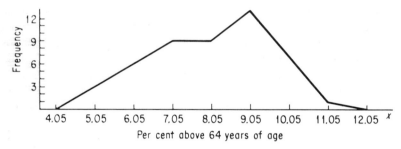

Fig. 2.3 Frequency Polygon for Data in Table 2.3

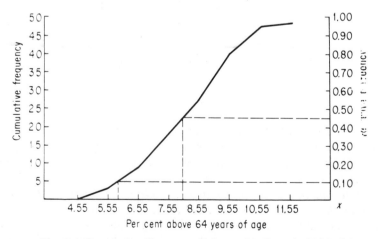

Fig. 2.4 Cumulative Frequency Polygon for Data in Table 2.3

complete the picture it is often desirable to add one class mark with zero frequency to each end of the diagram.

Frequency polygons can be misleading if they are incorrectly used. For example, the graph in Fig. 2.3 is for grouped data. Thus, caution must be used in "reading off" frequencies from the graph corresponding to values of x which are not class marks. A frequency polygon is used mainly to indicate the shape of a given set of data.

There is another type of frequency diagram, called the *cumulative frequency polygon* or *ogive*, which is often used. Figure 2.4 illustrates the graph of the ogive using data from Table 2.3. The main difference between a cumulative frequency polygon and the corresponding frequency polygon is that of locating the point (x_i', f_i') corresponding to (x_i, f_i) in the frequency polygon. The x_i' co-ordinate is either the upper or lower class boundary, depending on whether the cumulative frequency polygon is a "less than" or a "more than" type of diagram, and the f_i' is the cumulative frequency for the class containing x_i. Figure 2.4 is an ogive which gives the number of states with percentages less than the corresponding number which is given on the horizontal scale. Thus, in our example, we use the upper boundary of a class when we plot the polygon. (Some important uses of the ogive will be brought out later.) It should be observed at this time that in a cumulative frequency polygon the height above a particular point on the x-axis is the same numerical value as the area in the corresponding histogram to the left of that point. For example, the height of the cumulative polygon above $x = 8.55$ in Fig. 2.4 is 27, and the area to the left of $x = 8.55$ in Fig. 2.2 is also 27.

Figures 2.2, 2.3, and 2.4 illustrate methods of "picturing" distributions in terms of frequencies when the left-hand vertical axis is used. However, each of these graphs could have been drawn by using relative frequencies or proportions in place of frequencies, as is indicated by the right-hand vertical axis in Fig. 2.2 and 2.4. This would not change the shape of the diagrams, but the area of the histogram and the area under the polygons would be reduced in the ratio $1:48$. In general, the areas would be divided by N, where N denotes the total frequency. Again, it would be quite natural for us to call these diagrams *distributions*. In order to distinguish the diagrams, we might use the terms frequency histogram and relative frequency histogram, frequency polygon and relative frequency polygon, and cumulative frequency polygon and cumulative relative frequency polygon.

The points (vertices) of the polygons may be connected by smooth curves in some cases. We then call them *frequency* and *cumulative frequency curves* when frequencies are used, and *relative frequency* and *cumulative relative frequency curves* when relative frequencies are used. We shall see later that the word "density" is associated with one class of graphs and the word "distribution" is associated with another. For this reason we avoided using the term "distribution" in connection with any of the graphs mentioned above.

We now consider two practical uses of the ogive. From this graph the investigator may determine the number or relative frequency of observations less than a given value. Conversely, he may determine the value below which the observations fall with approximately a given relative frequency. Such values are called *quantiles* or *fractiles*. Either the frequency scale or the relative frequency scale may be useful. Generally, the relative frequency scale is employed, because the investigator is interested in comparing two sets of observations or in comparing one set of observations with a standard.

Special quantiles such as percentiles, deciles, and quartiles are normally used. Those x values of the observations corresponding to relative frequencies of 0.01, 0.02, ... , 0.99 or percentages of 1, 2, ... , 99 are called *first percentile, second percentile, ... , ninety-ninth percentile* and are denoted by P_1, P_2, ... , P_{99}; those x values corresponding to percentages of 10, 20, ... , 90 are called *first decile, second decile, ... , ninth decile* and are denoted by D_1, D_2, ... , D_9; and those x values corresponding to percentages of 25, 50, 75 are called *first quartile, second quartile, third quartile* and are denoted by Q_1, Q_2, Q_3. Clearly, the fiftieth percentile P_{50}, the fifth decile D_5, and the second quartile Q_2 represent the same x value. This particular x value is called the *median* of the distribution. In order to find $P_{10} = D_1$, say, from Fig. 2.4, we draw a line parallel to the x-axis passing through the 0.10 point of the right-hand vertical axis and a point on the cumulative polygon. From the point of intersection on the polygon we drop a perpendicular to the x-axis. The point where the perpendicular cuts the x-axis is the tenth percentile value and is $P_{10} = 5.85$. That is, 5.85 is the value (approximate) below which ten per cent of the observations fall. In order to find the *percentile rank* graphically for any observation P_k we reverse the steps given above and find the percentage point k (relative frequency $k/100$) on the right-hand vertical scale. For example, if $P_k = 8.00$, it follows that $k = 0.45$. Clearly, the cumulative polygon may be used to find the percentage (approximate) of the states with x values between two values x_1 and x_2. We must keep in mind that graphic methods introduce approximation.

The tables and diagrams presented so far are obviously for finite collections of data. The data may represent all the observations in a finite population or only a finite sample from an infinite population or larger finite population. In either case the methods presented may be thought of as describing or picturing the set of observations.

If the population is infinite in size, an investigator will never have all observations. Thus, he must always work with a sample which, at best, cannot indicate the true nature of the population in all details. Since the sample gives only an approximation to the population, he usually introduces some theoretical distribution, that is, mathematical model, which is thought to describe the population of possible or hypothetical observations. The statistician studies many such theoretical distributions which are thought to represent the true distribution of observations. The *normal distribution*, also called Gaussian distribution, is the most important continuous theoretical distribution and will be discussed in detail later along with others.

At this time we consider an illustration of a sample of observations taken from a population which may be thought of as an infinite population. Table 2.4 gives the acidity level expressed in *pH* units of 30,268 batches of Virginia soil tested in 1955. (The population is the set of *pH* values of all batches of soil in Virginia.) The maximum frequency is for the *pH* range 6.2–6.3 (class boundaries are 6.15 and 6.35); the distribution is smooth, and drops more rapidly for small *pH* values than for large values. Note that the first class has no fixed lower boundary and that the last class has no fixed upper boundary. Also, note that the class boundaries are not

Table 2.4
Acidity Level of Batches of Virginia Soil Tested in 1955*

pH Range	Frequency	pH Range	Frequency
Below 5.0	551	6.2–6.3	3548
5.0–5.1	1015	6.4–6.5	3055
5.2–5.3	2072	6.6–6.7	2497
5.4–5.5	2852	6.8–6.9	1755
5.6–5.7	3147	7.0–7.1	1222
5.8–5.9	3362	7.2–7.3	915
6.0–6.1	3521	Above 7.3	756
		Total	30,268

* Courtesy of Virginia Polytechnic Institute Soil Testing Laboratory.

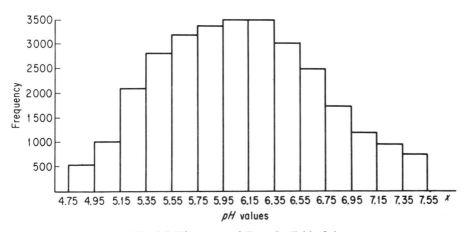

Fig. 2.5 Histogram of Data in Table 2.4

given and that the *limits* of the actual observed values are indicated in the table in place of boundaries.

The histogram in Fig. 2.5 contains more intervals than the one in Fig. 2.2 and has a smoother appearance. The first and last intervals are made the same length as the others and it is assumed that none of the batches had *pH* levels above 7.55 or below 4.75. We shall use Fig. 2.5 later in connection with a discussion on the theoretical distribution.

2.2.2. Discrete Variable

Many times the investigator obtains observations by counting such things as the number of petals on a daisy, the number of successes in *n* tosses of a coin, or the number of cars passing a certain point each day. Observations of this type are referred to as *counted data* or *enumeration data*. At this time we wish to consider methods of describing or picturing such data, and at the same time to compare these methods with those used in handling a continuous variable. This can best be done with the use of an illustration.

The data in Table 2.5, taken from *Genetics*, Vol. 38, show the frequency distribution of the number of bristles on the sixth abdominal sternite of female fruit flies (Drosophila pseudoobscura). The distribution is fairly regular, with a maximum frequency at 19 and near maximum frequency at 20. There is a steady decrease in frequency as the number of bristles increases or decreases from 19 and 20.

Table 2.5
Bristles on the Sixth Abdominal Sternite of Female Fruit Flies*

Number of Bristles	Frequency	Relative Frequency	Cumulative Frequency	Cumulative Rel. Freq.
12	2	0.002	2	0.002
13	0	0.000	2	0.002
14	4	0.005	6	0.007
15	21	0.026	27	0.033
16	47	0.058	74	0.092
17	104	0.129	178	0.220
18	123	0.152	301	0.373
19	148	0.183	449	0.556
20	146	0.181	597	0.739
21	91	0.113	686	0.849
22	79	0.098	765	0.947
23	27	0.033	792	0.980
24	8	0.010	800	0.990
25	7	0.009	807	0.999
26	1	0.001	808	1.000

* Bruce Wallace, J. C. King, Carol V. Madden, Bobbie Kaufmann, and E. C. Mc-Gunnigle, "An Analysis of Variability Arising Through Recombination," *Genetics*, Vol. **38** (1953), p. 272–307.

The frequency distribution of Table 2.5 is represented graphically in Fig. 2.6, in which the number of bristles is the abscissa and the corresponding frequency is the ordinate. This figure clearly shows that the variable takes only isolated values. For some purposes, however, rectangles are drawn with unit base and height equal to the frequency. In this case, the number of bristles is the mid-value of the base of each rectangle. This gives a histogram which has the appearance of those in Fig. 2.2 and 2.5, but it should be kept in mind that the variable is not continuous when a histogram is used for a discrete variable.

The cumulative frequency polygon for a discrete variable shown in

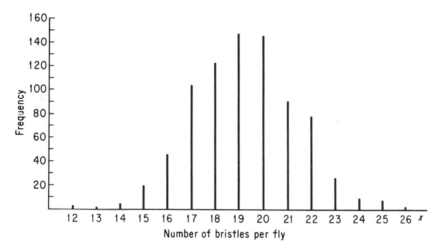

Fig. 2.6 Frequency Distribution of Data in Table 2.5

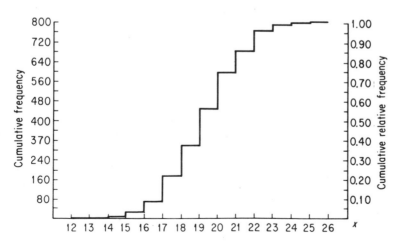

Fig. 2.7 Cumulative Frequency Polygon for Data in Table 2.5

Fig. 2.7 is different in nature from the one shown in Fig. 2.4. We might term the graph in Fig. 2.7 a step polygon, where the height of a step indicates the frequency corresponding to a given number of bristles. The ordinate represents the total number of flies with number of bristles equal to or less than the corresponding abscissa value. We could represent the cumulative frequency polygon by isolated ordinates above each abscissa value. The same applications can be made to both representations. From this graph it is possible to obtain percentiles and percentile ranks, as we did in the continuous case. However, it should be noted that corresponding to certain relative frequencies (percentile ranks) there is a whole interval of values which could be used as percentiles. For example, P_{22} may be any value between 17 and 18, including 17.

Just as in the continuous case, the set of data in a discrete experiment may represent all the observations or only a sample. If all the observations are known, the methods just described represent an accurate (in so far as the particular method is capable) picture of the distribution. If the observations in an experiment represent only a sample from either a finite or infinite population, then the methods just described give, at best, a good approximation to the true distribution. As in the continuous case, this leads to consideration of discrete theoretical distributions which are thought to represent true distributions of observations or populations. The *binomial distribution* and the *Poisson distribution* are perhaps the most useful discrete theoretical distributions—the binomial being finite and the Poisson infinite. (These distributions, along with certain others, will be discussed in detail later.)

The frequency distributions of the three illustrative examples as indicated by Fig. 2.2, 2.5, and 2.6 are roughly *bell-shaped;* that is, they are symmetrical, with most of the values falling somewhere near the middle of the range of values. Distributions of this type are the most common in practice, but many important variables have frequency distributions of a very different form. For example, if the variable is wealth of an American household or number of vehicles passing a certain point on a highway per ten second period, the frequency distribution looks like the right half of a bell-shaped distribution. A distribution of this type is said to be *skewed to the right,* a skewed distribution being one which lacks symmetry with respect to a vertical axis. Some variables are uniformly or approximately *uniformly distributed;* that is, each value of the variable occurs with equal or nearly equal frequency. For example, the frequency of occurrence of each number on the face of an honest die in 6000 tosses would be approximately 1000. Some distributions have maximum points at each end of the range of values and a minimum point near the middle. Such distributions are known as *U-shaped distributions.* Pearse [8] gives an example of an U-shaped distribution, showing the frequencies of estimated intensities of cloudiness at Greenwich during the years 1890–1904 (excluding 1901) for the month of July. (Further examples of the above types of distributions, along with other types of distributions, may be found in the exercises in Sect. 2.2.3.)

2.2.3. Exercises

Since many exercises in other sections refer to the data in these, these should be carefully prepared and saved for later use.

2.1. Give three examples of a bell-shaped distribution.

2.2. Give two examples of a uniform distribution.

2.3. Give two examples of a distribution skewed to the right.

2.4. Describe and illustrate a distribution which is not bell-shaped or uniform or skewed to the right.

2.5. Which of the distributions in Exercises 2.1 through 2.4 are for a continuous variable? A discontinuous variable?

2.6. Which of the distributions in Exercises 2.1 through 2.4 represents a population? A sample?

2.7. Draw a frequency histogram and a frequency polygon of the distribution in Table 2.6 of excess yardage of 100 denier acetate yarn over the specified minimum (99,000 yards per bobbin).

Table 2.6

Excess Yardage in Hundreds of Yards	Number of Bobbins	Excess Yardage in Hundreds of Yards	Number of Bobbins
0–1	2	16–17	19
1–2	0	17–18	47
2–3	0	18–19	57
3–4	0	19–20	48
4–5	0	20–21	42
5–6	3	21–22	18
6–7	0	22–23	29
7–8	0	23–24	24
8–9	8	24–25	10
9–10	8	25–26	2
10–11	7	26–27	2
11–12	13	27–28	1
12–13	26	28–29	3
13–14	31	29–30	1
14–15	26	30–31	3
15–16	21	Total	451

2.8. Draw a frequency histogram and cumulative relative frequency polygon of the following distribution of time required to reach target reaction temperature for hydrolysis of cellulose triacetate:

Table 2.7

Time in Seconds Required to Reach Temperature	Number of Observations	Time in Seconds Required to Reach Temperature	Number of Observations
10–14	2	55–59	30
15–19	32	60–64	21
20–24	196	65–69	13
25–29	936	70–74	5
30–34	2314	75–79	1
35–39	1757	80–84	4
40–44	678	85–89	0
45–49	234	90–94	1
50–54	66	95–99	2

From the cumulative polygon, determine graphically what proportion of the samples reach reaction temperature in less than 42 sec; between 20 sec and 48 sec.

2.9. The degree (arithmetic mean of 24 recordings in tenths) to which clouds covered the sky during 1957 at the weather station in Roanoke, Virginia are given in Table 2.8.

Table 2.8

Sky Cover from Midnight to Midnight (tenths)	Number of Days
0	15
1	19
2	25
3	23
4	28
5	41
6	37
7	42
8	41
9	39
10	55

(a) Draw a relative frequency histogram and cumulative polygon. (b) Determine the three quartiles and then half the difference between the first and third quartile. For what proportion of the days was the sky: completely covered with clouds? Without a trace of clouds?

2.10. Draw a frequency and relative frequency histogram on the same graph of the distribution of 1088 teak trees by three-inch-diameter classes listed in Table 2.9.*

Table 2.9

Diameter of Trees (inches)	Frequency
4.5– 7.5	8
7.5–10.5	26
10.5–13.5	50
13.5–16.5	120
16.5–19.5	181
19.5–22.5	215
22.5–25.5	213
25.5–28.5	145
28.5–31.5	76
31.5–34.5	36
34.5–37.5	18
	Total 1088

* A. L. Griffith and Bakshi Sant Ram, "The Silvicultural Research Code," Vol. **2,** *The Statistical Manual*, Office of the Geodetic Branch, Survey of India, Dehra Dun, India, viii, 214 pp., 1947.

2.11. (a) Draw a frequency and relative frequency histogram on the same graph of the distribution of precipitation each day at the weather station in Roanoke, Virginia, as shown in Table 2.10.

Table 2.10

Precipitation in Inches	Number of Days
Trace	48
0.01–0.09	55
0.10–0.19	22
0.20–0.29	12
0.30–0.39	11
0.40–0.49	5
0.50–0.59	7
0.60–0.69	3
0.70–0.79	7
0.80–0.89	2
0.90–0.99	2
1.00 and over	6

(b) Use the histogram in writing a short summary (around 100 words) of the record of rainfall at Roanoke in 1957.

2.12. The verbal scores of the 1958 Graduate Record Examination of 112 college seniors are as follows:

480	570	590	590	480	480	440
500	500	480	570	620	440	390
450	400	500	500	320	610	420
550	390	500	470	530	540	610
470	400	520	520	480	560	630
480	390	640	420	620	560	650
490	620	520	660	420	550	370
530	340	480	380	440	480	480
530	350	480	480	430	660	430
530	470	570	480	530	470	400
510	480	480	640	590	460	370
460	480	420	380	440	610	500
460	540	370	530	630	510	580
500	480	530	480	430	640	520
570	360	600	570	510	500	450
560	510	540	370	450	420	660

Use these data to construct a frequency table and a dot frequency diagram.

2.13. (a) Use the data in Exercise 2.12 to construct a frequency table with nine appropriately chosen classes of equal length. (b) Draw the frequency histogram and cumulative relative frequency polygon. (c) Determine what proportion of the students had scores less than 400; greater than 600. Find the median.

2.14. The body weights in grams of 62 male bobwhite quail trapped by Dr. Vince Schultz in Ohio in 1946–47 and 1947–48 are as follows:

210.0	198.5	230.3	198.5	202.0
203.8	198.5	173.6	198.5	194.9
189.4	198.5	180.7	184.3	187.8
196.4	174.7	183.3	184.3	233.9
179.6	177.0	192.5	191.4	202.0
195.1	177.0	201.3	191.4	212.6
217.9	198.5	176.6	194.9	198.5
199.3	181.6	174.3	222.2	212.6
217.8	198.5	163.4	198.5	233.9
215.6	194.8	199.7	177.2	184.3
220.2	184.3	191.8	184.3	
181.4	216.2	166.5	194.9	
198.5	212.6	198.5	202.0	

(a) Construct a frequency table with appropriately chosen classes (discuss with instructor). (b) Draw the histogram and the cumulative polygon for these data. (c) Use (a) and (b) to write a short summary of the weight distribution of these 62 quails.

2.15. Table 2.11 gives the maximum and minimum temperature in degrees Fahrenheit for each day of the summer of 1957 (starting with June 21 and ending with September 23) at the weather station in Roanoke, Virginia.

Table 2.11

Temperature (°F)		Temperature (°F)		Temperature (°F)	
Maximum	Minimum	Maximum	Minimum	Maximum	Minimum
83	64	92	70	78	53
76	58	83	63	70	58
68	56	83	56	82	62
75	59	82	57	87	58
87	57	84	63	86	60
88	62	84	63	89	61
78	54	88	63	96	66
81	55	92	63	95	65
87	54	94	67	93	64
92	58	89	67	92	65
82	67	94	62	89	66
84	57	95	63	88	70
90	55	86	71	87	62
93	64	83	62	72	66
91	71	81	53	77	63
87	59	85	53	70	63
92	62	90	57	69	63
94	63	91	62	81	65
85	69	89	70	85	62
85	63	88	68	88	69
88	54	93	66	91	64
93	63	85	67	85	67
93	65	85	68	83	63
95	72	85	71	77	65
94	67	91	72	68	64
89	70	77	63	66	60
80	70	65	61	66	60
75	68	68	62	81	64
88	66	80	62	87	66
95	61	83	56	88	65
98	67	85	57	73	53
97	71	80	60		

(a) Construct frequency tables for the maximum and minimum temperatures, letting the length of each class be five units. (b) Draw the cumulative relative frequency polygons on the same graph in order to compare the distributions of maximum and minimum temperatures.

2.16. (a) Determine the difference in the maximum and minimum temperature for each day in Exercise 2.15 and construct a frequency table of these differences. (b) Write a short summary of the distribution of ranges of temperatures for the data in Exercise 2.15.

2.17. Table 2.12 gives by states the personal income and the state debt per capita in 1955 (based on the estimated population on July 1, 1954, excluding armed forces overseas) and the expenditure per pupil in average daily attendance for public elementary and secondary day school in 1954.

Table 2.12

State*	Per Capita Debt	Income	Expenditure per Pupil	State	Per Capita Debt	Income	Expenditure per Pupil
Alabama	$ 22.97	$1181	$151	Nebraska	$ 2.49	$1540	$262
Arizona	4.76	1577	282	Nevada	6.80	2434	294
Arkansas	65.84	1062	139	New Hampshire	76.86	1732	256
California	68.82	2271	345	New Jersey	161.67	2311	333
Colorado	15.66	1764	280	New Mexico	36.56	1430	265
Connecticut	165.33	2499	297	New York	96.99	2263	362
Delaware	344.92	2513	325	North Carolina	70.29	1236	177
Florida	25.99	1654	229	North Dakota	34.41	1372	262
Georgia	64.56	1333	177	Ohio	57.05	2062	254
Idaho	3.92	1462	238	Oklahoma	89.40	1506	224
Illinois	33.26	2257	319	Oregon	108.40	1834	337
Indiana	75.71	1894	280	Pennsylvania	109.55	1902	299
Iowa	10.92	1577	274	Rhode Island	77.56	1957	268
Kansas	85.60	1647	264	South Carolina	91.84	1108	176
Kentucky	23.16	1238	153	South Dakota	.29	1245	275
Louisiana	79.58	1333	247	Tennessee	34.55	1256	166
Maine	131.50	1593	199	Texas	16.43	1614	249
Maryland	177.47	1991	268	Utah	5.92	1553	208
Massachusetts	167.39	2097	298	Vermont	19.98	1535	245
Michigan	73.75	2134	283	Virginia	31.36	1535	193
Minnesota	26.92	1691	287	Washington	92.01	1987	305
Mississippi	42.79	946	123	West Virginia	141.16	1288	186
Missouri	2.69	1800	233	Wisconsin	1.30	1774	293
Montana	70.09	1844	328	Wyoming	12.70	1753	330

* Bureau of the Census, *Statistical Abstract of the U. S.*, 1957.

(a) Group the data and construct a frequency table for the per capita debt and draw the frequency histogram. (b) Group the data and construct a frequency table for the per capita income and draw the frequency histogram. (c) Group the data and construct a frequency table for the expenditure per pupil and draw the frequency histogram.

2.18. (a) Draw a cumulative polygon for the per capita debt of Exercise 2.17 and write a short summary of the outstanding features of the distribution. (b) Draw a cumulative polygon for the per capita income of Exercise 2.17 and write a short summary of the outstanding features of the distribution. (c) Draw a cumulative polygon for the expenditure per pupil of Exercise 2.17 and write a short summary of the outstanding features of the distribution.

2.19. Toss simultaneously five similar coins 50 times, recording the number of heads resulting from each toss. Draw a graph which pictures this information.

2.20. (a) Combine the results from Exercise 2.19 of all the tosses of all the students in the class and draw a graph which pictures the information.

(b) What proportion of the tosses resulted in the appearance of five heads? Zero heads? Two heads? Three heads? Do these proportions seem reasonable? See if you can compute mathematically the proportion of times five heads would occur if all coins are fair.

2.21. Count the number of letters in each word of the first three paragraphs of Sect. 2.2.1 and construct a frequency table and frequency graph showing this information.

2.22. Count the number of steps you take in walking by the most direct route (keeping off the grass) from the main door of the campus post office to the main door of the administration building. Obtain this same information from 29 other students and construct a table and graph of your joint findings.

2.3. PARAMETERS

The tabular and graphic methods described in Sect. 2.2 give very little accurate quantitative information about distributions. For example, we might see that two histograms differ, but find it impossible to describe how much they differ. Thus, some sort of arithmetical description is desirable. For certain distributions found in practice, particularly the bell-shaped and uniform, such a description can be satisfactorily brought about by means of two parameters—one which measures central tendency and the other, dispersion. For other distributions the two additional descriptive measures of symmetry and peakedness are also important. Measures of symmetry indicate to what degree the data are balanced about some central value, and measures of peakedness indicate how flat or peaked the distribution may be.

In this section we think only of measures which characterize *finite populations and samples which are not to be used for purposes of statistical inference.* We consider measures which characterize infinite populations and samples used for statistical inference in later sections.

2.3.1. Measures of Central Tendency

Observations have a tendency to cluster at some particular location on the scale of measurement. We shall now consider some parameters which measure central tendency of a set of observations, also called *measures of location*, and at the same time introduce some notation which will be standard throughout the book.

The most important measure of location, both practically and theoretically, is the *arithmetic mean*, or *mean*, as it is usually called. It is defined as the sum of all the variable values divided by the number of such values. If we denote the mean by μ (lower-case Greek letter mu) and N observations have values denoted by $x_1, x_2, \ldots, x_N$, then, in symbols, the mean is defined by

$$\mu = \frac{x_1 + x_2 + \cdots + x_N}{N}$$

The mean can also be written as

$$\mu = \frac{\sum\limits_{i=1}^{N} x_i}{N} \qquad (2.1)$$

where the $\sum$ (capital Greek letter sigma) is the usual symbol for the sum of the values indicated; that is

$$\sum_{i=1}^{N} x_i = x_1 + x_2 + \cdots + x_N$$

For grouped data the mean may be approximated by using for computation the formula

$$\mu_g = \frac{\sum\limits_{i=1}^{k} f_i x_i}{N} \qquad (2.2)$$

where f_i is the frequency and x_i the mark of the ith class, k is the number of classes, and

$$N = \sum_{i=1}^{k} f_i$$

is the total number of observations. If the number of observations is large and the class lengths are sufficiently small, the approximation to the mean μ given by Eq. (2.2) is quite good. The saving in time of computation is considerable for a large set of data, especially if a frequency table is to be prepared anyway. If the observations are not grouped but a frequency table is prepared, as in Table 2.5, Formula (2.2) gives the exact value of the mean μ.

The data in Sect. 2.2.1. relating to the percentage of humans 65 years old and over in the United States in 1950 will be used to illustrate the applications of Formulas (2.1) and (2.2). Letting x_1 denote the percentage in Alabama, x_2 the percentage in Arizona, ..., x_{48} the percentage in Wyoming and using the percentages from Table 2.1 in Formula (2.1) gives

$$\mu = \frac{6.5 + 5.9 + \cdots + 6.3}{48} = \frac{387.2}{48}$$

or

$$\mu = 8.067 \text{ per cent}$$

Formula (2.2) may be applied to the arranged data in Table 2.2. Letting $x_1 = 4.9$, $x_2 = 5.4$, ..., $x_8 = 6.5$, ..., $x_{23} = 8.7$, ..., $x_{34} = 10.8$ so that

$f_1 = 1, f_2 = 1, \ldots, f_8 = 2, \ldots, f_{23} = 4, \ldots, f_{34} = 1$, we have on substituting in Formula (2.2)

$$\mu = [1(4.9) + 1(5.4) + \cdots + 2(6.5) + \cdots + 4(8.7) + \cdots + 1(10.8)]/48$$

$$= \frac{387.2}{48}$$

$$= 8.067 \text{ per cent}$$

Clearly, these two procedures lead to the same correct (except for rounding-off errors) arithmetic mean of the population of 48 percentages. Applying Formula (2.2) to the grouped data in Table 2.3 gives a slightly different value. Let $x_1 = 5.05$, $x_2 = 6.05$, $\ldots$, $x_7 = 11.05$; then $f_1 = 3, f_2 = 6, \ldots$, $f_7 = 1$, and by Formula (2.2)

$$\mu_g = \frac{3(5.05) + 6(6.05) + \cdots + 1(11.05)}{48}$$

$$= \frac{386.4}{48}$$

or

$$\mu_g = 8.050 \text{ per cent}$$

This example involving percentages was used only to illustrate the applications of Formulas (2.1) and (2.2). It is not necessarily intended that this is the most suitable measure of central tendency for this example. The nature of a distribution and the nature of the unit of measure of the variable, as we shall see later, determine to a large extent the type of descriptive measure which is most appropriate.

Geometrically, the mean is that point on the x-axis where a uniform and vertical sheet of metal in the shape of a histogram would balance on a pivot. Thus, we may think of the arithmetic mean as being located at the centroid or center of gravity of a distribution.

At this time, we give definitions of other measures of central tendency, for the most part leaving the treatment of their application and the discussion of some of the advantages and disadvantages of each for a more appropriate place. The three types of measures of location commonly used are the median, the mode, and the means (geometric and harmonic as well as arithmetic).

Arrange the N values of the variable x in increasing (or decreasing) order of magnitude and denote them by $x_{(1)}, x_{(2)}, \ldots, x_{(N)}$. Let μ_m denote the median of the N values. If N is odd, the median is the middle value; if N is even, it is taken as the mid-point of the middle pair of values. We have already learned that the median is the same as the second quartile, the fifth decile, or the fiftieth percentile. If the data are grouped, the median can be thought of geometrically as that point on the x-axis which has equal areas under the histogram on both sides of a vertical line through the point. The median is important in certain situations and ranks second to the mean in usefulness. For example, the median income is a more meaningful measure of income in the state of Delaware, say, than the arithmetic mean. This is so since a few very large incomes would not affect the typical measure of income, namely the median, but would increase the mean to the extent that it would be worthless as a typical measure of income.

The *mode*, M, is that value of x in a collection which occurs with maximum frequency. If there are q values of x which have the same maximum frequency, then there are q modes.

The *geometric mean*, G, is the Nth root of the product of the N values $x_1, x_2, \ldots, x_N$; that is

$$G = \sqrt[N]{x_1 x_2 \ldots x_N}$$

or

$$G = \sqrt[N]{\prod_{i=1}^{N} x_i} \tag{2.3}$$

where $\prod$ (capital Greek letter pi) denotes the product of N values. Note that the logarithm of the geometric mean is the arithmetic mean of the logarithms of the observations. The geometric mean is not used if any of the variable values are negative or zero. It is used chiefly in averaging rates or ratios rather than quantities, and this has rather restricted application.

The *harmonic mean*, H, of N values $x_1, x_2, \ldots, x_N$ (all different from zero) is the reciprocal of the arithmetic mean of their reciprocals; that is

$$H = \frac{N}{\sum_{i=1}^{N} \left(\frac{1}{x_i} \right)} \tag{2.4}$$

This measures the average time rates and is also very limited in its application.

Among other measures of location there is the *mid-range* which is the mid-value of the largest and smallest values of N; that is

$$\text{mid-range} = \frac{x_{(1)} + x_{(N)}}{2} \tag{2.5}$$

where $x_{(1)}$ and $x_{(N)}$ denote the smallest and largest values, respectively.

Experiments will be described later (Exercises 2.3.4) in which each of these measures of location is considered most appropriate. They are defined at this time to show that there is more than one measure of location, and to indicate that the experimenter does not automatically select the arithmetic mean when considering central tendency—he must have some basis for selecting a measure of location. However, the arithmetic mean is the only measure of central tendency which is readily amenable to theoretical treatment, and it is, as we shall see later, in a certain sense the "best measure of location" of some of the most widely used distributions.

2.3.2. Measures of Dispersion

We are likely to wonder how typical a particular measure of location really is, that is, how many variable values are near this single measure. Intuitively, we feel the answer must depend on the degree of cluster about the location parameter. Further, in order to compare two or more distributions with the same numerical value of some particular location parameter, we naturally turn to some numerical measure of the amount of scatter or

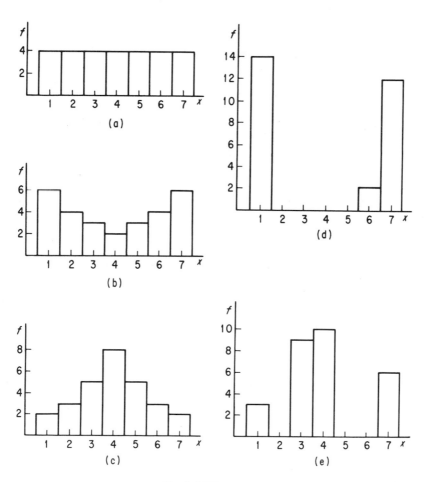

Fig. 2.8 Histograms

dispersion in the distributions. In particular, a buyer of metal bolts of a given diameter would rather deal with that manufacturer making the most uniform bolts, that is, bolts with diameters having the least amount of scatter. In order to be more concrete and to have material for illustration, we shall consider the five histograms in Fig. 2.8, which have means located at 4 or near 4, but which obviously have different amounts of scatter.

If we assume that a measure of dispersion is desirable, the question is how do we determine it? In order to answer this question as well as certain others, we set about defining and illustrating some of the most common measures, namely, range, mean deviation, variance, standard deviation, and coefficient of variation. It should be noted that all these measures would not be calculated in any given experiment. Generally, only one would be selected, that one being the measure which seems most appropriate for the particular experiment. The variance and standard deviation are of greatest importance, both theoretically and practically, and are the measures which we naturally turn to first in a given experimental situation.

The *range*, R, is given by

$$R = x_{(N)} - x_{(1)} \qquad (2.6)$$

where $x_{(N)}$ and $x_{(1)}$ are the largest and smallest values of N values of the variable, respectively. It is seldom used as a descriptive parameter of a population, since it indicates very little about the way the distribution appears inside the interval of values. For example, the range of each distribution in Fig. 2.8 is 7, but the distributions are obviously quite different. For a very small number of values the range might be considered satisfactory. However, this measure is mainly attractive because it is computationally convenient and because of the very small size of repeative samples obtained in certain industrial procedures.

The amount of scatter is clearly dependent on how much the set of values deviates from some central value. The greater the scatter, the larger the total deviation. Since the total deviation depends on the number of values as well as the amount of scatter, we think in terms of "average deviation" to avoid the difficulty of number of values.

The *mean deviation*, δ_m, is defined to be the arithmetic mean of the absolute value of the deviations of the observations from the median; that is

$$\delta_m = \frac{\sum_{i=1}^{N} |x_i - \mu_m|}{N} \qquad (2.7)$$

where μ_m denotes the median and $|x_i - \mu_m|$ is always positive. Until recently the mean deviation was almost always defined in terms of deviations from the mean μ, but deviations from the median are computationally and theoretically easier to use. Whenever the mean and median are the same, the mean deviation about the mean would be the same as δ_m; otherwise, they would differ. In any case, the mean deviation is very difficult, if not impossible, to use in certain important theoretical work and hence is not as useful as it might seem at first—the computation of absolute values limits its use in theory.

Since the absolute value causes trouble, it is natural that we try to avoid its use. For this and other reasons we define the variance which is computationally more difficult, but theoretically quite satisfactory.

The *variance*, σ^2 (lower-case Greek letter sigma squared), is defined to be the arithmetic mean of the squares of the deviations from the mean μ; that is

$$\sigma^2 = \frac{\sum_{i=1}^{N} (x_i - \mu)^2}{N} \tag{2.8}$$

The *standard deviation* is the positive square root of the variance and is denoted by σ. This gives a measure of dispersion which is expressed in terms of the unit of measure of the variable values. For grouped data the variance may be approximated by using the formula

$$\sigma_g^2 = \frac{\sum_{i=1}^{k} f_i(x_i - \mu)^2}{N} \tag{2.9}$$

where f_i is the frequency, x_i is the mark of the ith class, k is the number of classes, and N the total number of observations. If the observations are not grouped but a frequency table is prepared as in Table 2.5, Formula (2.9) gives the exact value of the variance σ^2.

To illustrate the applications of Formulas (2.8) and (2.9), we use the data relating to the percentage of humans over 64 years old in the United States in 1950. Referring to Table 2.1 and using the notation of Sect. 2.3.1 as well as the value $\mu = 8.067$ per cent, we have, from Formula (2.8)

$$\sigma^2 = \frac{(6.5 - 8.067)^2 + (5.9 - 8.067)^2 + \cdots + (6.3 - 8.067)^2}{48}$$

or

$$\sigma^2 = 2.13 \text{ square per cent}$$

If we use Table 2.2 and the notation of Sect. 2.3.1, Formula (2.9) may be applied to obtain

$$\sigma^2 = [1(4.9 - 8.067)^2 + 1(5.4 - 8.067)^2 + \cdots + 2(6.5 - 8.067)^2$$
$$+ \cdots + 4(8.7 - 8.067)^2 + \cdots + 1(10.8 - 8.067)^2]/48$$

or

$$\sigma^2 = 2.13 \text{ square per cent}$$

Applying Formula (2.9) to the grouped data in Table 2.3 and using the notation of Sect. 2.3.1, we have

$$\sigma_g^2 = \frac{3(5.05 - 8.067)^2 + 6(6.05 - 8.067)^2 + \cdots + 1(11.05 - 8.067)^2}{48}$$

or

$$\sigma_g^2 = 2.29 \text{ square per cent}$$

The standard deviation is

$$\sigma = \sqrt{2.13} = 1.46 \text{ per cent}$$

Form σ_g^2 we find that the standard deviation is approximately $\sigma_g = 1.51$ per cent. It is clear that "per cent" is a more meaningful measure than "square per cent" used for σ^2.

The *coefficient of variation*, v, of a set of observations is simply the standard deviation divided by the arithmetic mean; that is,

$$v = \frac{\sigma}{\mu} \tag{2.10}$$

This parameter is used mainly to bring out the degree of spread of the observations in terms of the mean. It is useful in comparing two distributions with widely differing means (see Exercise 2.43). For the above example

$$v = \frac{1.46}{8.067}$$

or

$$v = 0.181$$

which is a dimensionless measure.

Note. Often v is defined as the standard deviation expressed as a percentage of the arithmetic mean; that is, $v = 100\sigma/\mu$ per cent

The relative merits of these measures of dispersion will be discussed more fully later. They were introduced at this time to show that scatter can be measured in several meaningful ways, and to indicate that the investigator does not automatically select the standard deviation, say, when considering scatter—he must have some basis for selecting a measure of dispersion.

Perhaps it should be mentioned that the mean and variance or standard deviation, the median and mean deviation, and the mid-range and range are used together as a general rule. Furthermore, the mean and variance are used almost altogether for the large body of distributions which are roughly bell-shaped.

2.3.3. Use of Mean and Standard Deviation in Bell-Shaped Distributions

The standard deviation may be used to describe the frequency with which observations fall near the mean or in an interval about the mean. In particular, if the distribution is bell-shaped and the number of observations in the population is reasonably large, then approximately $\frac{2}{3}$ of the observations fall in the interval from $\mu - \sigma$ to $\mu + \sigma$, $\frac{19}{20}$ of the observations fall in the interval from $\mu - 2\sigma$ to $\mu + 2\sigma$, and 997/1000 of the observations fall in the interval from $\mu - 3\sigma$ to $\mu + 3\sigma$. This property is easily illustrated by using the data of Tables 2.2 and 2.5

Example 2.1. Find the proportion of observations in Table 2.2 falling in the three intervals defined above.

We know that $\mu = 8.07$ and $\sigma = 1.46$. Thus, the interval from $\mu - \sigma$ to $\mu + \sigma$ becomes 6.61–9.53. By actual count 30 observations, or $\frac{30}{48}$ of the observations, fall in this interval; that is, the relative frequency of observations in this interval is 0.625. Also, 47 of 48 observations fall in the interval from $\mu - 2\sigma = 5.15$ to $\mu + 2\sigma = 10.99$, and 48 observations fall in the interval from $\mu - 3\sigma = 3.69$ to $\mu + 3\sigma = 12.46$. That is, the relative frequencies of observations in these intervals are 0.979 and 1.000. Even though the size of the population is not very large, the proportion of observations falling in the intervals is fairly close to those (0.667, 0.950, 0.997) given above.

For the grouped data, $\mu_g = 8.05$ and $\sigma_g = 1.51$. Thus, the intervals are 6.54–9.56, 5.03–11.07 and 3.52–12.58 with the relative frequencies of observations falling in these intervals being 0.646, 0.958, and 1.000, respectively.

For either the grouped or the ungrouped data we can see that roughly $\frac{2}{3}$ of the states have percentages falling within one standard deviation, 1.5, of the mean, 8.1. Thus, without counting we might have felt reasonably sure that approximately 32 states have percentages from 6.6 to 9.6. For large masses of observations this use of the mean and standard deviation furnishes an easy way to describe numerically the nature of the observations, and also saves time in determining the proportion of observations between two boundaries.

The fruit fly data found in Table 2.5 may be used to illustrate the above methods for a large population. It is found in Sect. 2.4 that $\mu = 19.21$ and $\sigma = 2.11$. Thus, we would expect roughly $\frac{2}{3}$ of the observations or 539 observations to fall in the interval from 17.10–21.32, that is, that roughly 539 female fruit flies have 18, 19, 20, or 21 bristles on the sixth abdominal sternite. By counting, we find 508 observations or 62.9 per cent of the observations with 18, 19, 20, or 21 bristles. The estimated number of observations, 539, differs from the actual number, 508, more than we might wish. This difficulty can generally be overcome by using a histogram and thinking momentarily of the number of bristles possible as being uniformly distributed over the continuum of values for each class. In this case, the actual number would be 533.2, which is fairly close to the estimated value, 539.

2.3.4. Exercises

In this set of exercises, as well as others, we give some problems which are designed to amplify the concepts already introduced and some which are designed to bring out other concepts defined or explained for the first time in the exercise. The references at the end of the chapter may be consulted if further, more detailed information relating to these concepts is desired.

In Exercises 2.23 through 2.31, use the values indicated in Fig. 2.8. In order that the calculations be exact, think of all the variable values in a class as falling at the class mark unless otherwise indicated. Thus, for example, in Fig. 2.8e, the ordered values are 1, 1, 1, 3, 3, . . . , 7.

2.23. Determine the median mid-range and mode for the values in each figure.

2.24. Determine the mean for the values in each figure.

2.25. Compute the variance and standard deviation for the values in Figs. 2.8a, b, and c.

2.26. Compute the variance and standard deviation for Figs. 2.8d and e.

2.27. Compute the mean deviation for the values in Figs. 2.8a, b, and c.

2.28. Compute the mean deviation for the values in Figs. 2.8d and e.

2.29. Let

$$\delta'_m = \frac{\sum\limits_{i=1}^{N} |x_i - \mu|}{N} \tag{2.11}$$

be the definition of the *mean deviation about the mean*. Compute δ'_m for the values in Figs. 2.8d and e. Compare these values with δ_m found in Exercise 2.28.

2.30. Compute the coefficient of variation for the values in each of the figures.

2.31. Bring the measures of dispersion computed together in one table and write a summary comparing these measures with each other and relating them to the amount of spread in the figures.

2.32. Assuming the values within each class to be distributed uniformly, that is, all values in the class to occur with the same frequency or relative frequency, determine what proportion of the observations fall between $\mu - \sigma$ and $\mu + \sigma$ for Figs. 2.8a, b, and c.

Answer. 0.57, 0.51, and 0.60.

Hint. Let the base of the ith rectangle in a histogram be c_i and the area be f_i, where f_i is the frequency of the class with boundaries x_i and $x'_i = x_i + c_i$. Then the height of the rectangle is f_i/c_i, and the area f_o of that part of the rectangle to the left of the line perpendicular to the x-axis through any point x_o between x_i and $x_i + c_i$ is given by

$$f_o = \frac{(x_o - x_i)f_i}{c_i} \tag{2.12}$$

The area of that part of the rectangle to the right of x_o is given by

$$f_i - f_o \quad \text{or} \quad \frac{(x_i' - x_o)f_i}{c_i}$$

Thus, f_o is the frequency of those values from x_i through x_o, and $f_i - f_o$ is the frequency of those values from x_o through x_i'.

2.33. Using the assumption and method of Exercise 2.32, determine what proportion of the observations fall between $\mu_m - \delta_m$ and $\mu + \delta_m$ in Figs. 2.8a, b, and c.

2.34. (a) Show that when a histogram is used the number of observations between $\mu - \sigma$ and $\mu + \sigma$ in the fruit fly example (last paragraph of Sect. 2.3.3.) is actually 533.2. (b) Find the proportion of observations between $\mu - 2\sigma$ and $\mu + 2\sigma$.   *Answer.* 0.958.

2.35. (a) Find the arithmetic, geometric, and harmonic means of the numbers 2 and 8. (b) Find the arithmetic, geometric, and harmonic means of the numbers 5, 8, and 25. (c) Compare these three means.

Note. For positive numbers the harmonic mean is always the smallest, and the arithmetic mean is the largest.

2.36. A student on successive tests answered correctly 8 out of 40 questions, 36 out of 60 questions, and 18 out of 20 questions. For example, on the second test he did three times as well as he did on the first. Find the average rate of improvement (geometric mean) between successive tests.
Answer. $G = 3\sqrt{2}/2 = 2.12$.

Note. Suppose we try to use the arithmetic mean (which is $\frac{9}{4}$) as the correct average rate to determine the third grade, assuming we do not know any grade except the first (which is 20, based on 100 points). This would give $\frac{20}{100}(\frac{9}{4})(\frac{9}{4}) = \frac{81}{80} = 101$, which is an impossible grade. On the other hand, using the geometric mean, we obtain

$$\frac{20}{100}\left(\frac{3\sqrt{2}}{2}\right)\left(\frac{3\sqrt{2}}{2}\right) = 90$$

which is the correct grade.

2.37. A car travels 40 miles per hour for 50 miles in one direction and 60 miles per hour on the return trip. Find the average (harmonic mean) rate per hour for the round trip.

Note. The arithmetic mean of 50 miles per hour is not appropriate, as can be checked by determining the time required to travel the 100 miles. Observe that the term "average" might refer to either the arithmetic, geometric, or harmonic mean and for this reason is not often used in statistics.

2.38. Table 2.13 gives the foul-shooting record of the five regular basketball players on a university team:

Table 2.13

Name	Number of Shots Attempted	Percentage of Foul Shots Made
Jim	140	50
Tom	300	60
Harley	120	70
Clay	100	40
Dewitt	240	80

Find the percentage of foul shots made by the team (that is, the five regular players).

Hint. In problems of this type the measurements should be weighted according to the number of objects having each measurement, the relative importance of each measurement, etc. Thus, the *weighted arithmetic mean* μ_w of a set of values $x_1, x_2, \ldots, x_k$ which have weights $w_1, w_2, \ldots, w_k$, respectively, is given by

$$\mu_w = \frac{\sum w_k x_k}{\sum w_k} \tag{2.13}$$

It should be observed that the weights need not be integers.

2.39. (a) The median and mean for the values indicated in Fig. 2.8c are each 4. Change only one value so that a 7 becomes 21 and then compute the median and mean. (b) Compare these measures of location with 4, noting the effect that only one large (or small) value has on the mean and that the mean changes in the direction in which the distribution is skewed.

Note. In a symmetrical distribution the arithmetic mean, the median, and the mode (antimode, as in Fig. 2.8b) are the same, but for skewed distributions they differ, the median falling between the arithmetic mean and mode. In fact, for moderately skewed unimodal distributions these three measures are approximately related by the formula

$$\mu - M = 3(\mu - \mu_m) \tag{2.14}$$

according to empirical evidence and a mathematical explanation by Doodson [4].

2.40. (a) Prove that $\sigma = R/2$ when $N = 2$. (b) Let three values ordered from highest to lowest be denoted by x_1, x_2, x_3. Then

$$R = x_1 - x_3 = (x_1 - x_2) + (x_2 - x_3) = k_1 R + k_2 R$$

where $k_1 = (x_1 - x_2)/(x_1 - x_3)$ and $k_2 = (x_2 - x_3)/(x_1 - x_3)$. Prove that

$$\sigma = \frac{R\sqrt{2(1 - k_1 k_2)}}{3} \tag{2.15}$$

(c) Show that Formula (2.15) becomes $\sigma = R\sqrt{2}/3$ when $x_1 = x_2$ and $\sigma = R\sqrt{6}/6$ when x_2 is halfway between x_1 and x_3. (d) Find σ for the values 1, 2, 7 using the definition; using Formula (2.15).

2.41. (a) Prove that $\sum(x_i - \mu) = 0$. (b) Illustrate that this formula is true for values 1, 2, 2, 3, 4, 6, 10. (c) Find $\sum(x_i - \mu_m)$ for the values in (b).

2.42. (a) Verify that $\delta_m < \delta'_m$ for the values in Exercise 2.41(b). (b) Show that for N values $x_1, x_2, \ldots, x_N$ the mean deviation about an arbitrary value x_0 can never be less than δ_m.

2.43. Is there more variation in the weights of female albino rats or in the weights of female humans? Use the following samples to answer this question:

Weights of rats in grams: 205, 190, 180, 230, 215, 195, 210, 170, 215, 190.
Weights of humans in pounds: 105, 130, 110, 130, 120, 140, 105, 95, 130, 115, 135, 125.

Hint. Use the coefficient of variation to compare these two samples, realizing that the conclusion might not be the same for the populations.

2.44. (a) Use the distribution in Exercise 2.39 to compute skewness by the following two formulas:

$$SK = \frac{\mu - M}{\sigma} \tag{2.16}$$

$$\alpha_3 = \frac{\dfrac{\sum(x_i - \mu)^3}{N}}{\sigma^3} \tag{2.17}$$

(b) Changing both values from 7 to 21 in Exercise 2.39, compute SK and α_3 and compare with the measures obtained in (a).

Note. SK is called the *Pearsonian coefficient of skewness*, since he introduced it, realizing that μ and M differ in case the distribution is skewed. From the definition it is clear that SK is positive for a distribution skewed to the right and negative for a distribution skewed to the left, since the mean tends to go in the direction in which a distribution is skewed. The most important and widely used measure of skewness is α_3 (lowercase Greek letter alpha). This is based on moments of the distribution.

As a distribution becomes more skewed both SK and α_3 become numerically larger, but not at the same rate. Actually these two measures are used more generally to compare asymmetry of two distributions. There are several other measures of skewness or asymmetry, but the two given here should be sufficient.

From Fig. 2.8e we see that SK is zero, since $M = 4 = \mu$. Thus, if the measure of skewness (for example, SK) is zero, it does not necessarily follow that the distribution is symmetric.

2.45. Karl Pearson in 1906 introduced for the purpose of measuring *peakedness* or *kurtosis* of a unimodal distribution the moment-ratio given by

$$\alpha_4 = \frac{1}{N}\frac{\sum(x_i - \mu)^4}{\sigma^4} \tag{2.18}$$

Using the values from Fig. 2.8a, b, and c, compute α_4 and compare in view of the note below.

Note. For the normal distribution, which is a special bell-shaped distribution, α_4 is 3, and this is the standard for comparison. Distributions with a ratio less than, equal to, or greater than 3 are known as *platykurtic, mesokurtic,* and *leptokurtic,* respectively.

2.4. NUMERICAL CALCULATIONS

There are several useful devices for shortening calculations involving the parameters already introduced. We shall consider at this time some of the most common procedures along with some illustrative examples. (The explanations are for the finite discrete populations. Other cases are considered in Chap. 5.)

Theorem 2.1. *The variance of N values* $x_1, x_2, \ldots, x_N$ *is given by*

$$\sigma^2 = \frac{\sum\limits_{i=1}^{N} x_i^2 - \left(\sum\limits_{i=1}^{N} x_i\right)^2 / N}{N} \tag{2.19}$$

or

$$\sigma^2 = \frac{N \sum\limits_{i=1}^{N} x_i^2 - \left(\sum\limits_{i=1}^{N} x_i\right)^2}{N^2} \tag{2.20}$$

Proof. From the definition of variance given in Formula (2.8), we obtain

$$N\sigma^2 = \sum_{i=1}^{N} (x_i - \mu)^2$$

Expanding the right-hand side of this equation, using the definition of μ, and collecting terms, we obtain the following

$$N\sigma^2 = \sum_{i=1}^{N} (x_i^2 - 2\mu x_i + \mu^2)$$
$$= (x_1^2 - 2\mu x_1 + \mu^2) + (x_2^2 - 2\mu x_2 + \mu^2) + \cdots + (x_N^2 - 2\mu x_N + \mu^2)$$
$$= (x_1^2 + x_2^2 + \cdots + x_N^2) - 2\mu(x_1 + x_2 + \cdots + x_N) + N\mu^2$$
$$= \sum_{i=1}^{N} x_i^2 - 2 \frac{\sum\limits_{i=1}^{N} x_i}{N} \cdot \sum_{i=1}^{N} x_i + N \left(\frac{\sum\limits_{i=1}^{N} x_i}{N}\right)^2$$
$$= \sum_{i=1}^{N} x_i^2 - \frac{\left(\sum\limits_{i=1}^{N} x_i\right)^2}{N}$$

or

$$N\sigma^2 = \frac{N \sum\limits_{i=1}^{N} x_i^2 - \left(\sum\limits_{i=1}^{N} x_i\right)^2}{N}$$

Formulas (2.19) and (2.20) are obtained from these last two equations, respectively, by dividing each by N.

By a similar argument it can be shown that Formula (2.9) for grouped values reduces to

$$\sigma_g^2 = \frac{\sum_{i=1}^{k} f_i x_i^2 - \left(\sum_{i=1}^{k} f_i x_i\right)^2 / N}{N} \tag{2.21}$$

or

$$\sigma_g^2 = \frac{N \sum_{i=1}^{k} f_i x_i^2 - \left(\sum_{i=1}^{k} f_i x_i\right)^2}{N^2} \tag{2.21a}$$

where

$$N = \sum_{i=1}^{k} f_i$$

Theorem 2.1 is especially useful when a calculating machine is available, Formula (2.20) being better than Formula (2.19) for most purposes, since fewer recordings and one less division are required. It is also useful whenever μ is a repeating decimal. These are illustrated in Example 2.2.

Example 2.2. Find σ^2 for the values 2, 3, 5.
Letting $x_1 = 2$, $x_2 = 3$, and $x_3 = 5$, we find that

$$\sum_{i=1}^{3} x_i = 10 \quad \text{and} \quad \sum_{i=1}^{3} x_i^2 - 38$$

Hence, by Formula (2.20),

$$\sigma^2 = \frac{(3)(38) - (10)^2}{3 \cdot 3} = \frac{14}{9} = 1.555 \ldots \doteq 1.556$$

where $\doteq$ means "approximately equal to." However, if we find μ first, the following procedure is used

$$\mu = \tfrac{10}{3} = 3.3$$

and

$$\sigma^2 = \frac{\sum_{i=1}^{3} (x_i - \mu)^2}{N} = \frac{(2 - 3.3)^2 + (3 - 3.3)^2 + (5 - 3.3)^2}{3}$$

$$= \frac{4.67}{3} = 1.5566 \ldots \doteq 1.557$$

The second method introduces a rounding-off error immediately, but the rounding-off error need not be introduced in the first method unless we wish to divide as a final step.

Theorem 2.2. *Adding a constant (positive or negative) to each of N values adds the same constant to the mean but does not change the variance or standard deviation.*

Proof. Let $y_i = x_i + k$ $(i = 1, 2, \ldots, N)$, where k is any constant. Let μ_y and μ_x denote the means of the $N\,y$ values and the $N\,x$ values, respectively. Then

$$
\begin{aligned}
\mu_y &= \frac{\displaystyle\sum_{i=1}^{N} y_i}{N} = \frac{\displaystyle\sum_{i=1}^{N} (x_i + k)}{N} \\[2mm]
&= \frac{(x_1 + k) + (x_2 + k) + \cdots + (x_N + k)}{N} \\[2mm]
&= \frac{(x_1 + x_2 + \cdots + x_N) + Nk}{N} \\[2mm]
&= \frac{\displaystyle\sum_{i=1}^{N} x_i + Nk}{N} = \frac{\displaystyle\sum_{i=1}^{N} x_i}{N} + k \\[2mm]
&= \mu_x + k
\end{aligned}
$$

Letting σ_y^2 and σ_x^2 denote the variances for the y values and x values, respectively, we obtain

$$
\begin{aligned}
\sigma_y^2 &= \frac{\displaystyle\sum_{i=1}^{N} (y_i - \mu_y)^2}{N} \\[2mm]
&= \frac{\displaystyle\sum_{i=1}^{N} [(x_i + k) - (\mu_x + k)]^2}{N} \\[2mm]
&= \frac{\displaystyle\sum_{i=1}^{N} (x_i - \mu_x)^2}{N} = \sigma_x^2
\end{aligned}
$$

and

$$
\sigma_y = \sigma_x
$$

The reader may wish to show that Theorem 2.2 is true when the values are grouped.

Example 2.3. Use Theorems 2.1 and 2.2 to determine μ and σ^2 for the values in Table 2.2.

The values are $x_1 = 4.9$, $x_2 = 5.4$, $\ldots$, $x_{21} = 8.0$, $\ldots$, $x_{48} = 10.8$. If

we let $k = -8.0$, it follows that $y_1 = -3.1$, $y_2 = -2.6$, ..., $y_{21} = 0.0$, ...,
$y_{48} = 2.8$

$$\sum_{i=1}^{48} y_i = 3.2$$

and

$$\sum_{i=1}^{48} y_i^2 = 102.44$$

Thus, by definition

$$\mu_y = \frac{3.2}{48} = 0.067$$

and by Formula (2.20)

$$\sigma_y^2 = \frac{(48)(102.44) - (3.2)^2}{(48)^2}$$

$$= \frac{4917.12}{2304} = 2.13$$

By Theorem 2.2

$$\mu_x = \mu_y - k = 0.067 - (-8.0) = 8.067$$

and

$$\sigma_x^2 - 2.13$$

Theorem 2.3. *Multiplying each of N values by the same constant also multiplies the mean by this constant, the standard deviation by this constant, and the variance by this constant squared.*

Proof. Let

$$y_i = kx_i \qquad (i = 1, 2, \ldots, N)$$

where k is a constant. Then, by definition and substitution

$$\mu_y = \frac{\sum y_i}{N} = \frac{\sum (kx_i)}{N} = k \cdot \frac{\sum x_i}{N} = k \cdot \mu_x$$

$$\sigma_y^2 = \frac{\sum (y_i - \mu_y)^2}{N} = \frac{\sum (kx_i - k\mu_x)^2}{N}$$

$$= \frac{k^2 \sum (x_i - \mu_x)^2}{N} = k^2 \cdot \sigma_x^2$$

and

$$\sigma_y = k\sigma_x$$

The proofs for grouped data are similar.

Note. In the above proof the limits for the summation were dropped. That is, $\sum$ replaced

$$\sum_{i=1}^{N}$$

From this point on, the limits of summation will be dropped when it does not lead to confusion, that is, when it is clear from the context of the problem what the limits are.

Theorem 2.4. *If* $y_i = (x_i - x_o)/K$, *where* x_o *is any constant and* K *is any constant different from zero, then* $\mu_y = (\mu_x - x_o)/K$, $\sigma_y^2 = \sigma_x^2/K^2$, *and* $\sigma_y = \sigma_x/K$.

The proof follows immediately from Theorems 2.2 and 2.3.

Example 2.4. Apply Theorem 2.4 to the data of Table 2.2.

Let x_i be defined as in Example 2.3; let $x_0 = 8.0$ and $K = \frac{1}{10} = 0.1$. Then

$$y_1 = \frac{4.9 - 8.0}{0.1} = \frac{-3.1}{0.1} = -31, \qquad y_2 = -26, \ldots, y_{48} = 28$$

$$\sum y_i = 32$$

and

$$\sum y_i^2 = 10244$$

Thus, by definition

$$\mu_y = \tfrac{32}{48} = 0.67$$

and by Formula (2.20)

$$\sigma_y^2 = \frac{(48)(10244) - (32)^2}{(48)^2} = 213$$

From Theorem 2.4, substituting the above values, we have

$$\mu_x = K\mu_y + x_0 = \tfrac{1}{10}(0.67) + 8.0 = 8.067$$

$$\sigma_x^2 = K^2\sigma_y^2 = (\tfrac{1}{10})^2 \cdot 213 = 2.13$$

and

$$\sigma_x = K\sigma_y = \tfrac{1}{10}\sqrt{213} = 1.46$$

Thus, the mean, variance, and standard deviation obtained by this method are the same as those obtained from the definition. As the numbers get larger and the frequency increases, transformations of this kind become more important as time-saving devices and aids to limiting errors in computation.

Theorem 2.5. *The mean deviation from the median is the sum of the values in the higher half minus the sum of the values in the lower half divided by N; that is*

$$\delta_m = \frac{\sum \text{higher half of values} - \sum \text{lower half}}{N} \tag{2.22}$$

the median being excluded if N is an odd number.

Proof. Let the values be arranged in decreasing order of magnitude and denoted by

$$x_{(1)}, x_{(2)}, \ldots, x_{(N)}$$

where $x_{(1)}$ is the largest value and $x_{(N)}$ is the smallest value. Let $x_{(u)}$ be the smallest value equal to or greater than μ_m and $x_{(l)}$ be the largest value equal to or less than μ_m. Then the first half of values $x_{(1)}, x_{(2)}, \ldots, x_{(u)}$ represent μ_m or points to the right of μ_m and the second half of values $x_{(l)}, \ldots, x_{(N)}$ represent μ_m or points to the left of μ_m. The numerical distance between two points on opposite sides of μ_m is the value to the right minus the value to the left; it is also the distance between the point to the left and μ_m plus the distance between the point to the right and μ_m. In particular

$$x_{(1)} - x_{(N)} = |x_{(1)} - \mu_m| + |x_{(N)} - \mu_m|, \text{ etc.} \tag{2.23}$$

Now, from the definition of mean deviation, we may write

$$\delta_m = \frac{\sum |x_{(i)} - \mu_m|}{N}$$

or, rearranging terms, we have

$$\delta_m = [(|x_{(1)} - \mu_m| + |x_{(N)} - \mu_m|) + (|x_{(2)} - \mu_m| + |x_{(N-1)} - \mu_m|) \\ + \cdots + (|x_{(u)} - \mu_m| + |x_{(l)} - \mu_m|)]/N \tag{2.24}$$

Substituting Eq. (2.23) in Eq. (2.24) and collecting terms gives

$$\delta_m = \frac{(x_{(1)} + x_{(2)} + \cdots + x_{(u)}) - (x_{(N)} + x_{(N-1)} + \cdots + x_{(l)})}{N}$$

or

$$\delta_m = \frac{\sum_{i=1}^{u} x_{(i)} - \sum_{i=l}^{N} x_{(i)}}{N} \tag{2.25}$$

Example 2.5. Find the mean deviation for the data of Table 2.2.

Using the notation of Example 2.3 and the definition, we find the median to be

$$\mu_m = \frac{x_{24} + x_{25}}{2} = \frac{8.4 + 8.5}{2} = 8.45$$

From the definition of mean deviation we have

$$\delta_m = \frac{|4.9 - 8.45| + |5.4 - 8.45| + \cdots + |10.8 - 8.45|}{48}$$

$$= \frac{3.55 + 3.05 + \cdots + 2.35}{48}$$

$$= \frac{57.6}{48} = 1.20 \text{ per cent}$$

Using Theorem 2.5 and the fact that the sum of the smallest 24 values is 164.8 and the sum of the largest 24 values is 222.4, we obtain

$$\delta_m = \frac{222.4 - 164.8}{48}$$

$$= \frac{57.6}{48} = 1.20 \text{ per cent}$$

Clearly, computing δ_m by the second method has several advantages, one of them being that it is not necessary that μ_m be found. It should be noted that the mean deviation is 82 per cent of the standard deviation for this population. In general, for a bell-shaped distribution the mean deviation is about 80 per cent of the standard deviation. It can be shown that this percentage is 79 for the fruit fly distribution.

Any good computational technique will include check devices, the best techniques having checks at each important step of the calculations. A device which is used to check the sum and the sum of squares of a set of values is given in Theorem 2.6.

Theorem 2.6. *If $x_1, x_2, \ldots, x_N$ is any set of N values, then*

$$\sum (x_i + 1)^2 = \sum x_i^2 + 2 \sum x_i + N \qquad (2.26)$$

Proof. We have immediately that

$$\sum (x_i + 1)^2 = \sum (x_i^2 + 2x_i + 1)$$
$$= \sum x_i^2 + 2 \sum x_i + N$$

The formula for grouped data is

$$\sum_{i=1}^{k} f_i(x_i + 1)^2 = \sum_{i=1}^{k} f_i x_i^2 + 2 \sum_{i=1}^{k} f_i x_i + \sum_{i=1}^{k} f_i \qquad (2.26a)$$

where k is the number of groups.

Example 2.6. Adding one to each value x_i of Table 2.2, verify that Theorem 2.6 holds.

We find directly that

$$\sum x_i = 387.2 \quad \text{and} \quad \sum x_i^2 = 3225.64$$

Adding one to each value x_i gives $x_1 + 1 = 5.9$, $x_2 + 1 = 6.4, \ldots,$ $x_{48} + 1 = 11.8$, and

$$\sum (x_i + 1)^2 = (5.9)^2 + (6.4)^2 + \cdots + (11.8)^2$$
$$= 4048.04$$

But

$$\sum x_i^2 + 2 \sum x_i + N = 3225.64 + 2(387.2) + 48 = 4048.04$$

Hence, we have verified that Theorem 2.6 holds, and we have increased our confidence that $\sum x_i$ and $\sum x_i^2$ are correct.

2.5.　APPROXIMATIONS IN CALCULATIONS

In all calculations we should bear in mind that the usual rules for rounding off numbers are not always adequate. If only one or two mathematical operations are required the "rounded-off" answer is usually satisfactory, but when long series of operations are performed there is obviously more opportunity to make sizable errors. Thus, we should look more closely at the real nature of the "rounding-off" methods.

With this in mind we proceed to define range numbers, that is, approximate numbers written in a special form. An *approximate value of a number*, or, more briefly, *approximate number*, is a number which differs from the true value by some restricted small amount. Thus, if x denotes the true value and x' an approximate value, then the error, ϵ or Δx, is the absolute value of the difference; that is, $\epsilon = |x - x'|$.

If ϵ were known, the true value could be determined by using an approximate value, and, in this case, we would not need any rules for rounding off. However, ϵ is not known, but is often specified to be less than or equal to some small positive quantity, say η. Thus, we could say, knowing x', that the true value x is in the range from $x' - \eta$ to $x' + \eta$. For example, when heights are recorded to the nearest inch, η is 0.5 in., and a recording of 68 in. is intended to represent any number from $68 - 0.5 = 67.5$ to $68 + 0.5 = 68.5$ in. Thus, when we say that 68 in. is an approximate number with two significant figures, called the *significant figures form*, we understand that the true value falls in the range of values from 67.5 to 68.5 in. We might have indicated this by writing 68 ± 0.5 in., or, in a more compact notation, 68 (0.5). Dwyer [5] refers to numbers written in this form as *approximate-error numbers*, since 68 is approximate and 0.5 is the maximum error.

An approximate number may be expressed in many ways. It is important that we understand that, regardless of the way we write it, *an approximate number represents a range within which the true value of the number is located.* For some purposes the form

$$\begin{bmatrix} x_h \\ x_l \end{bmatrix}$$

where x_h is the largest possible value and x_l is the smallest possible value of x in a range, is the best way to write an approximate number. An approximate number expressed in this form is called a *range number*, and the two values x_h and x_l are the *components of the range number*.

In the special case in which the maximum error η is one-half unit in the last decimal position, the significant figures, approximate-error, and range number forms can all be used to express an approximate number. For example, if $b = 2.1$ is an approximate number with two significant figures, we understand that $\eta = 0.05$ and may write

$$2.1 = 2.1(0.05) = \begin{bmatrix} 2.15 \\ 2.05 \end{bmatrix}$$

However, if η is not one-half unit, the significant figures form cannot be used. Other disadvantages of the usual significant figures form will become apparent in Examples 2.7 and 2.9. We must first indicate how to operate (add, subtract, multiply, etc.) with range numbers.

Let a and b be two approximate numbers written in significant figures form. Then

$$\begin{bmatrix} a_h \\ a_l \end{bmatrix} \quad \text{and} \quad \begin{bmatrix} b_h \\ b_l \end{bmatrix}$$

are the range forms of a and b, respectively. In order to obtain the sum $a + b$ of a and b in range form, first express a and b in terms of range numbers; then the sum is given by

$$\begin{bmatrix} a_h \\ a_l \end{bmatrix} + \begin{bmatrix} b_h \\ b_l \end{bmatrix} = \begin{bmatrix} (a + b)_h \\ (a + b)_l \end{bmatrix} = \begin{bmatrix} a_h + b_h \\ a_l + b_l \end{bmatrix} \tag{2.27}$$

where $(a + b)_h = a_h + b_h$ is the largest value which could be obtained by adding a value in the range from a_l to a_h to a value in the range from b_l to b_h, and $(a + b)_l = a_l + b_l$ is the smallest value which could be obtained by adding a value in the range from a_l to a_h to a value in the range from b_l to b_h. It should be noted that Formula (2.27) gives the sum of two approximate numbers even though they are not given in terms of significant figures. The difference $a - b$ in range form is given by

$$\begin{bmatrix} a_h \\ a_l \end{bmatrix} - \begin{bmatrix} b_h \\ b_l \end{bmatrix} = \begin{bmatrix} (a-b)_h \\ (a-b)_l \end{bmatrix} = \begin{bmatrix} a_h - b_l \\ a_l - b_h \end{bmatrix} \tag{2.28}$$

provided we let

$$-\begin{bmatrix} a_h \\ a_l \end{bmatrix} = \begin{bmatrix} -a_l \\ -a_h \end{bmatrix} \tag{2.29}$$

Other operations are indicated in Example 2.7.

Example 2.7. Let 2.1, 1.7, 1.3, and 3.2 be approximate numbers with two significant figures. Determine range numbers for each of the following

    a. 2.1×1.3

    b. $2.1 + 1.7 + 1.3 + 3.2$

    c. $(1.7 \times 3.2) - (2.1 \times 1.3)$

    d. $3.2 \div 1.3$

    e. $(1.3)^8$

We have, for a, b, c, d, e, respectively

$$2.1 \times 1.3 = \begin{bmatrix} 2.15 \\ 2.05 \end{bmatrix} \times \begin{bmatrix} 1.35 \\ 1.25 \end{bmatrix} = \begin{bmatrix} 2.9025 \\ 2.5625 \end{bmatrix}$$

$$2.1 + 1.7 + 1.3 + 3.2 = \begin{bmatrix} 2.15 \\ 2.05 \end{bmatrix} + \begin{bmatrix} 1.75 \\ 1.65 \end{bmatrix} + \begin{bmatrix} 1.35 \\ 1.25 \end{bmatrix} + \begin{bmatrix} 3.25 \\ 3.15 \end{bmatrix}$$

$$= \begin{bmatrix} 8.50 \\ 8.10 \end{bmatrix}$$

$$(1.7 \times 3.2) - (2.1 \times 1.3) = \begin{bmatrix} 1.75 \\ 1.65 \end{bmatrix} \times \begin{bmatrix} 3.25 \\ 3.15 \end{bmatrix} - \begin{bmatrix} 2.15 \\ 2.05 \end{bmatrix} \times \begin{bmatrix} 1.35 \\ 1.25 \end{bmatrix}$$

$$= \begin{bmatrix} 5.6875 \\ 5.1975 \end{bmatrix} - \begin{bmatrix} 2.9025 \\ 2.5625 \end{bmatrix} = \begin{bmatrix} 3.1250 \\ 2.2950 \end{bmatrix}$$

$$3.2 \div 1.3 = \begin{bmatrix} 3.25 \\ 3.15 \end{bmatrix} \div \begin{bmatrix} 1.35 \\ 1.25 \end{bmatrix} = \begin{bmatrix} \dfrac{3.25}{1.25} \\ \dfrac{3.15}{1.35} \end{bmatrix}$$

$$= \begin{bmatrix} 2.60 \\ 2.33 \end{bmatrix}$$

and

$$(1.3)^8 = (1.3 \times 1.3)^4 = \left(\begin{bmatrix} 1.35 \\ 1.25 \end{bmatrix} \times \begin{bmatrix} 1.35 \\ 1.25 \end{bmatrix} \right)^4$$

$$= \left(\begin{bmatrix} 1.8225 \\ 1.5625 \end{bmatrix} \times \begin{bmatrix} 1.8225 \\ 1.5625 \end{bmatrix} \right)^2 = \begin{bmatrix} 3.32150625 \\ 2.44140625 \end{bmatrix}^2$$

$$= \begin{bmatrix} 11.0324037687890625 \\ 5.9604644775390625 \end{bmatrix} = \begin{bmatrix} 11.04 \\ 5.96 \end{bmatrix}$$

where $3.15/1.35 \doteq 2.33$ is rounded off so that the lower component becomes smaller, and the components of $(1.3)^8$ are rounded off so that they define a range of values including the exact components which appear in the next-to-the-last step of the calculations.

Example 2.8. Compute the values in Example 2.7 by the usual rounding-off methods.

We have, for a, b, c, d, e, respectively

$$2.1 \times 1.3 = 2.73 \doteq 2.7$$

$$2.1 + 1.7 + 1.3 + 3.2 = 8.3$$

$$(1.7 \times 3.2) - (2.1 \times 1.3) = 5.44 - 2.73 = 2.71 \doteq 2.7$$

$$3.2 \div 1.3 \doteq 2.46154 \doteq 2.5$$

$$(1.3)^8 = (1.69)^4 = (2.8561)^2 = 8.15730721 \doteq 8.2$$

Example 2.9. Compare the results in Examples 2.7 and 2.8.

In order to compare the solutions, we write the answers in Example 2.8, expressed in significant figures form, as range numbers. Thus

$$2.7 = \begin{bmatrix} 2.75 \\ 2.65 \end{bmatrix}, \quad 8.3 = \begin{bmatrix} 8.35 \\ 8.25 \end{bmatrix}, \quad 2.5 = \begin{bmatrix} 2.55 \\ 2.45 \end{bmatrix}, \quad \text{and} \quad 8.2 = \begin{bmatrix} 8.25 \\ 8.15 \end{bmatrix}$$

It is clear that

$$\begin{bmatrix} 2.9025 \\ 2.5625 \end{bmatrix} \quad \text{and} \quad \begin{bmatrix} 2.75 \\ 2.65 \end{bmatrix}$$

do not tell the same story. The first range number tells us that the product is some value in the range from 2.5625 to 2.9025 and the second that the product is some value in the range from 2.65 to 2.75. Clearly, the first represents all the values, and the second fails to represent many possible values of the product. Similar and more exaggerated statements can be made concerning the comparison of the other values. Compare, for example

$$(1.3)^8 = \begin{bmatrix} 11.04 \\ 5.96 \end{bmatrix} \quad \text{and} \quad (1.3)^8 = \begin{bmatrix} 8.25 \\ 8.15 \end{bmatrix}$$

If we attempt to compare the computed values by using significant figures, we run into some interesting complications. For the product 2.1×1.3 we have only one significant figure, namely 3, when using range numbers, but the usual rule gives two, namely, 2.7. For the quantity $(1.7 \times 3.2) - (2.1 \times 1.3)$ the usual rule gives 2.7 with two significant figures, but the range number method fails to indicate that any single number can be used. Thus, it appears that in some cases the rules normally used in determining significant figures are altogether misleading. For further discussion of range numbers and approximate error numbers, see Dwyer [5].

The above discussion involving range numbers was used to emphasize some of the reasons why the rounding-off rules should be applied with caution. We will not at this time get into an exhaustive discussion of the pros and cons of the rounding-off methods, but we will point out that the usual argument in favor of these methods is that numbers which overestimate and numbers which underestimate a set of measurements behave in such a way that the errors tend to cancel (balance out) each other in a series of calculations.

2.6. EXERCISES

The first exercises are intended to lead to a better understanding of the numerical methods described in Sects. 2.4 and 2.5; the last refer to all topics of this chapter as well as to certain related topics.

2.46. Compute the mean for the weights of the humans in Exercise 2.43, using (a) Theorem 2.2, (b) Theorem 2.3, and (c) Theorem 2.4.

2.47. Compute the mean deviation for the weights of the rats in Exercise 2.43, using (a) the definition, (b) Theorem 2.5, and (c) Formula (2.11).

2.48. Compute the mean deviation for the weights of the humans in Exercise 2.43, using (a) the definition, (b) Theorem 2.5, and (c) Formula (2.11).

2.49. Consider weights of the rats in grams in Exercise 2.43. (a) Use Formula (2.20) to find the variance of the weights. (b) Subtract 200 grams from each weight and use Theorem 2.2 to find the variance of the weights. (c) Let $x_0 = 200$ and $K = 5$ and use Theorem 2.4 to find the variance of the weights. (d) The variance has been obtained by four methods if we include the method [definition in Formula (2.8)] used in Exercise 2.43. Compare the amount of computation required in obtaining these answers; the accuracy of the answers.

2.50. Increase the weight of one rat in Exercise 2.43 by five grams and answer all the parts in Exercise 2.49.

2.51. Compute the variance for the weights of the humans in Exercise 2.43, using any method in Sect. 2.4.

2.52. Illustrate Theorem 2.6, using (a) the weights of the rats in Exercise 2.43, and (b) the weights of the humans in Exercise 2.43.

2.53. (a) Find the variance σ_g^2 for the grouped data in Table 2.3. (b) Find $\sigma_c^2 = \sigma_g^2 - c^2/12$, where c is the common class length. (c) In Sect. 2.3.2 we found that $\sigma^2 = 2.13$, using the data in Table 2.2. Compare σ^2, σ_c^2, and σ_g^2.

Note. When parameters are computed from a grouped frequency distribution, certain errors are introduced as a result of assuming the variable values are concentrated at the mid-point of the class intervals. Sheppard [9], Wold [10, 11], Craig [2] and others have suggested corrections for specified parameters for certain types of distributions. The corrected variance σ_c^2, defined in Exercise 2.53(b) and known as Sheppard's correction, is an improvement over σ_g^2 for bell-shaped distributions with sufficiently small class lengths, as is illustrated in the above case. However, for certain grouped frequency distributions σ_c^2 is not necessarily an improvement over σ_g^2 and should be used with caution.

2.54. (a) Find the mean, variance, and standard deviation for the maximum temperatures for the grouped data in Exercise 2.15(a). (b) Find the mean, variance, and standard deviation for the minimum temperature for the grouped data in Exercise 2.15(a). (c) Use the mean and standard deviation computed in (b) to describe the frequency distribution.

2.55. (a) Find the mean, variance, and standard deviation for the differences (maximum temperature minus minimum temperature) for the grouped data in Exercise 2.16(a). (b) Use the mean and standard deviation found in (a) to describe the frequency distribution of differences.

2.56. (a) Find the mean, variance, and standard deviation for the grouped data (verbal scores) in Exercise 2.13(a). (b) Describe the frequency distribution of verbal scores in terms of the mean and standard deviation.

2.57. (a) Find the median and mean deviation δ_m for the differences (maximum temperature minus minimum temperature) for the grouped data in Exercise 2.16(a). (b) Find the median and mean deviation δ_m for the verbal scores in Exercise 2.13(a). (c) Using σ_g found in Exercises 2.55(a) and 2.56(a), compute δ_m/σ_g in each case and compare considering the note below.

Note. The mean deviation is about 80 per cent of the standard deviation for unimodal curves which are symmetric or nearly symmetric. In particular, for a "normal" distribution the ratio is $\sqrt{2/\pi} \doteq 0.798$.

2.58. Show that $\delta_m/\sigma = 0.791$ for the fruit fly data in Table 2.5.

2.59. Use range numbers to find the following

(a) $(1.2) \times (1.7)$ (e) $1.2 - 1.7$

(b) $(1.2)^3$ (f) $2.3 \div 1.2$

(c) $(1.2)^4$ (g) $(1.2)^3 - (1.7) \times (0.7)^2$

(d) $1.7 - 1.2$ (h) $(2.13 \div 1.2) - (1.7 \times 0.7)$

2.60. Prove Formula (2.28).

2.61. (a) Express each of the following as range numbers

202(3)   2475(50)

2.73(0.07)   2.222(0.022)

(b) Express each of the following as approximate-error numbers

$$\begin{bmatrix} 3.75 \\ 3.71 \end{bmatrix} \quad \begin{bmatrix} 42.1 \\ 40.1 \end{bmatrix} \quad \begin{bmatrix} 7.63 \\ 7.58 \end{bmatrix}$$

2.62. Evaluate and express results in range form

(a) $\begin{bmatrix} 2.79 \\ 2.73 \end{bmatrix} - \begin{bmatrix} 0.56 \\ 0.37 \end{bmatrix} + \begin{bmatrix} -1.17 \\ -1.26 \end{bmatrix}$

(b) $\begin{bmatrix} 2.79 \\ 2.73 \end{bmatrix} \div \begin{bmatrix} 0.56 \\ 0.37 \end{bmatrix}$

(c) $\begin{bmatrix} 0.27 \\ -0.04 \end{bmatrix} \div \begin{bmatrix} -1.17 \\ -1.26 \end{bmatrix}$

2.63. Evaluate and express results in approximate-error form

(a) $44.7(0.3) + 22.6(0.8) - 31.2(1.3)$

(b) $\dfrac{2.76(0.3)}{0.465(0.095)}$

(c) $\sqrt{22.3(1.1)}$

(d) $\dfrac{6}{2.3(0.2)}$

2.64. The heights in inches of 20 men are

64.0	72.5	69.0	69.5
68.5	70.5	69.5	69.0
71.5	70.0	70.5	70.0
73.5	71.0	71.5	70.5
69.5	69.5	71.0	72.0

(a) Find the mean height of these men. (b) Write each height as two significant figures, always increasing numbers ending in 5 by 0.5, and find the mean. (c) Write each height as two significant figures, always changing numbers ending in 5 to the nearest even integer, and find the mean. (d) Compare the results obtained in (b) and (c) with (a).

Note. We shall always use the method in (c).

2.65. Prove that δ'_m is not greater than σ.

2.66. The frequency of each of the values $x = 1, 2, \ldots, n$ is b. Find the mean and variance of this distribution.

2.67. The frequency of each of the values $x = 2, 4, \ldots, 2n$ is c. Find the mean and variance of this distribution.

2.68. (a) Prove that the mean of the distribution whose relative frequencies at $x = 0, 1, 2, \ldots, r, \ldots$ are

$$e^{-\mu}, \; e^{-\mu}\mu/1!, \; e^{-\mu}\mu^2/2!, \; \ldots, \; e^{-\mu}\mu^r/r, \; \ldots \text{ is } \mu.$$

(b) Prove that the variance of the distribution in (a) is also μ.

2.69. The relative frequency of the values $x = 0, 1, 2, \ldots, n$ are the successive terms in the expansion of $(\frac{1}{2} + \frac{1}{2})^n$. Find the mean and variance of this distribution.

2.70. The relative frequency of the values $x = 0, 1, 2, \ldots, n$ are the successive terms in the expansion of $(p + q)^n$, where $p + q = 1$. Show that $\mu = np$ and $\sigma^2 = npq$.

REFERENCES

1. Bennett, C. A. and N. L. Franklin, *Statistical Analysis in Chemistry and the Chemical Industry*. New York: John Wiley & Sons, Inc., 1954, Chaps. 1 and 2.

2. Craig, C. C., "Sheppard's Corrections for a Discrete Variable," *Annals of Mathematical Statistics*, Vol. **7** (1936), pp. 55–61.

3. Dixon, W. J. and F. J. Massey, *An Introduction to Statistical Analysis*. New York: McGraw-Hill, Inc., 1957, Chaps. 1, 2, and 3.

4. Doodson, A. T., "Relation of Mode, Median and Mean in Frequency Curves," *Biometrika*, Vol. **11** (1917), pp. 425–29.

5. Dwyer, P. S., *Linear Computations*. New York: John Wiley & Sons, Inc., 1951, Chaps. 1 and 2.

6. Freund, J., *Modern Elementary Statistics*. Englewood Cliffs, N. J.: Prentice-Hall, Inc., 1952, Chaps. 2, 4, and 5.

7. Kendall, M. G. and A. Stuart, *The Advanced Theory of Statistics*, Vol. **1**. New York: Hafner Publishing Co., Inc., 1958, Chaps. 1 and 2.

8. Pearse, G. E., "On Corrections for the Moment Coefficient of Frequency-Distributions when there are Infinite Ordinates at one or both Terminals of the Range," *Biometrika*, Vol. **20**, (1928), pp. 314–55.

9. Sheppard, W. F., "On the Calculation of the Most Probable Values of Frequency Constants for Data Arranged According to Equidistant Divisions of a Scale," *Proceedings of the London Mathematical Society*, Vol. **29** (1898), p. 353.

10. Wold, H., "Sulle Correzione di Sheppard," *Giorn. Ist. Ital. Att.*, Vol. **4** (1934), p. 304.

11. Wold, H., "Sheppard's Correction Formulae in Several Variables," *Skand. Akt.*, Vol. **17** (1934), p. 248.

3

THEORETICAL
DISTRIBUTIONS

The few distributions which are used as models for most of the work done in statistics are defined and a few simple properties of each are enumerated. The density and distribution functions are defined and compared.

3.1. INTRODUCTION

We have discussed the terms frequency distribution and relative frequency distribution, using tables and graphs to describe the "shape" of a distribution. Now we wish to present mathematical functions which "describe" these distributions and which serve as models for most collections of data. This is what we do, for example, when we use points, lines, and planes in geometry for models. We think of these terms as being conceptual idealizations of real-life objects, and we use the models because they lend themselves so easily to mathematical manipulations. For example, the draftsman in preparing blue prints for an apartment house uses lines (images of geometrical lines) to represent walls, windows, doors, etc.

In Chap. 2 distributions were discussed in terms of frequencies and relative frequencies. Now, as our treatment of distributions becomes more formal and more theoretical with the introduction of models and theoretical distributions, we use the term *probability* in place of relative frequency. We think of probability as a sort of idealization of such terms as "relative frequency," "proportion," and "part of," and introduce formal properties of probability as they are required in the development of theoretical distributions. In Chap. 4 we define *probability of an event* as the limit of the relative frequency with which it occurs, and demonstrate how such a definition is useful in estimating probability or in arriving at meaningful assignments of probability. Even though the axiomatic definition and resulting properties of probability introduced in this chapter are desirable for a formal development of theoretical distributions, they are of no real value when it comes to assigning probabilities to events or to numerical values of events.

3.2. DISCRETE DISTRIBUTIONS OF ONE VARIABLE

The essential properties of discrete density and distribution functions have already been suggested in Sect. 2.2.2. Thus, we need only introduce notation in order carefully to describe the general case.

For our purposes we think of probability as a measure associated with a real number which has been assigned to an event in an experiment. Thus, probability is a measure associated with an event in an experiment. For example, we may associate the probability measure 0.25 with the occurrence of two heads (event) in two tosses of a balanced coin (experiment). If the real number assigned to the event is "the number of heads," then we may say, for this experiment, that the probability of 2 is 0.25. In this chapter we are concerned with properties of two important probability functions and not with assignments of probability measures to events (or to real numbers of events). Thus, for our immediate purposes the formal axiomatic approach to probability is sufficient.

Assume that x may take on either a finite number (see Exercises 2.66 and 2.67) or a countably infinite number (see Exercise 2.68) of discrete real values denoted by $x_1, x_2, \ldots, x_k$, where k may be either finite or countably infinite. Let

$$S_p = \{x_{p_1}, x_{p_2}, \ldots, x_{p_{k'}}\}$$

be the pth subset of the set

$$S = \{x_1, x_2, \ldots, x_k\} \tag{3.1}$$

[It is to be understood that p is either a positive integer or countably infinite, that $k' \leq k$ and k' may be countably infinite, and that x_{p_j}, the jth ($j = 1$, $2, \ldots, k'$) element in subset p, is an element from S.] Let $P[S_p]$ be a nonnegative real number assigned to S_p. Whenever a subset S_p contains the single number $x_{p_1} = x_i$, we may write $P[x = x_i]$ or $f(x_i)$ in place of $P[S_p]$. Then the discrete variable x is said to have a *probability density function f*, provided that the real number $f(x_i)$ assigned to x_i ($i = 1, 2, \ldots, k$) satisfies the following conditions

(a) $f(x_i)$ is a positive real number for each x_i

(b) $\sum_{i=1}^{k} f(x_i) = 1$ (3.2)

(c) $P[S_p] = \sum_{j=1}^{k'} f(x_{p_j})$ for each S_p

We also call f a *density function* or *probability function*. Assuming that it will lead to no confusion in the future, we use the traditional notation $f(x)$ to denote density function. [For the discrete case, the reader should note that $f(x_i)$ denotes "function value" and $f(x)$ denotes "function."] When $f(x)$ is a density function, $f(x_i) = P[x = x_i]$ gives the probability that x has the value x_i; that is, $f(x_i)$ is the probability measure of x_i.

We are usually interested in relations such as $x < x_i$, $x \geq x_i$, or $x_a \leq x \leq x_b$, where a and b are any two positive integers such that $a \leq b$ or a is a positive integer and b is countably infinite. In this context, we may replace (c) in Eq. (3.2) by the following restricted but useful form

(c′) $\begin{cases} P[x_a \leq x \leq x_b] = \sum_{i=a}^{b} f(x_i) \text{ where the real numbers in } S \text{ defined} \\ \text{in Eq. (3.1) are ordered; that is,} \\ x_1 < x_2 < \cdots < x_k. \end{cases}$

The reader should note that in (c′) it makes no difference whether we think of x in the relation $x_a \leq x \leq x_b$ as assuming only values in S or as assuming all real numbers between x_a and x_b inclusive, it being tacitly assumed that $f(x)$ is zero for all values of x not in S.

Example 3.1. Think of the values of x as being 1, 2, 3, 4, 5, 6 and as representing the numbers on the faces of a die. If the die is properly balanced, it seems reasonable to let the probability assigned to each value be $\frac{1}{6}$, i.e., $f(1) = f(2) = \cdots = f(6) = \frac{1}{6}$. Clearly, $f(x_i)$ is a density function, for Conditions (3.2a) and (3.2b) hold by definition, and Condition (3.2c) follows immediately. A graphical description of this density function in the form of a line diagram is shown in Fig. 3.1. Each height (length of line segment) represents the probability of the corresponding number. As an illustration

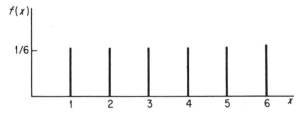

Fig. 3.1 Theoretical Density Function for Die

of Condition (3.2c) we note that the probability of the occurrence of an even face or of the set $\{2, 4, 6\}$ is $f(2) + f(4) + f(6) = \frac{1}{2}$.

We think of the distribution function as being an idealization of a cumulative relative frequency distribution, just as the density function is an idealization of a relative frequency distribution. In many problems we are interested in the probability that the value of a real number is equal to or less than some specified value x. It is customary to define such a cumulative probability for all real values of x from $-\infty$ to ∞. If we let

$$x_1 < x_2 < \cdots < x_k$$

the *cumulative probability function F* is defined by the rule

$$F(x) = \begin{cases} 0, & \text{when } x < x_1 \\ \sum_{i=1}^{r} f(x_i), & \text{when } x_r \leq x < x_{r+1} \quad (r = 1, 2, \ldots, k-1) \\ 1, & \text{when } x \geq x_k \end{cases} \quad (3.3)$$

This is also called the distribution function of the discrete random variable x. [The reader should, just as with $f(x)$, note that $F(x)$ will be used in the traditional sense; that is, $F(x)$ may denote "the value at x" or "function of x." The context in which $F(x)$ is used should make its meaning clear.] It follows from Definition (3.3) that the distribution function has the following two important properties

$$F(x) \text{ is a nondecreasing function of the continuous variable } x. \quad (3.4)$$

$$0 \leq F(x) \leq 1 \quad \text{for each value of } x. \quad (3.5)$$

Example 3.2. The distribution function associated with the density function in Example 3.1 is given by

$$F(x) = \begin{cases} 0, & \text{when } x < 1 \\ \frac{1}{6}, & \text{when } 1 \leq x < 2 \\ \frac{2}{6}, & \text{when } 2 \leq x < 3 \\ \frac{3}{6}, & \text{when } 3 \leq x < 4 \\ \frac{4}{6}, & \text{when } 4 \leq x < 5 \\ \frac{5}{6}, & \text{when } 5 \leq x < 6 \\ 1, & \text{when } x \geq 6 \end{cases} \quad (3.6)$$

The graph of Eq. (3.6) is shown in Fig. 3.2. It should be observed that $F(x)$ is a nondecreasing function of x with a finite number of jumps (or steps or saltuses) and that the ith jump is $f(x_i)$.

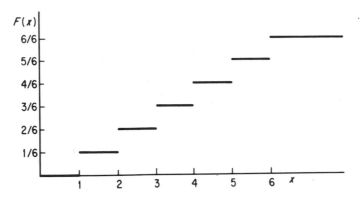

Fig. 3.2 Theoretical Distribution Function for Die

The mean μ and variance σ^2 of the discrete variable x with density function $f(x)$ are given by

$$\mu = \sum_{i=1}^{k} x_i f(x_i) \tag{3.7}$$

and

$$\sigma^2 = \sum_{i=1}^{k} (x_i - \mu)^2 f(x_i) \tag{3.8}$$

respectively. In case N is finite it is easy to show that Formulas (2.2) and (2.9) are special cases of Eqs. (3.7) and (3.8). If we assume each of the N observations to have probability $1/N$, it follows that x_i has probability $f(x_i) = f_i/N$, since $\sum_{i=1}^{k} f_i = N$. Thus

$$\mu = \sum_{i=1}^{k} x_i f(x_i) = \sum_{i=1}^{k} x_i \frac{f_i}{N} = \frac{\sum_{i=1}^{k} x_i f_i}{N}$$

and

$$\sigma^2 = \sum_{i=1}^{k} (x_i - \mu)^2 f(x_i) = \sum_{i=1}^{k} (x_i - \mu)^2 \frac{f_i}{N} = \frac{\sum_{i=1}^{k} (x_i - \mu)^2 f_i}{N}$$

Theorem 3.1. *If $f(x)$ is the density function of the discrete variable x_i ($i = 1, 2, \ldots, k$), then*

$$\sum_{i=1}^{k} x_i^2 f(x_i) = \sigma^2 + \mu^2 \tag{3.9}$$

Proof. From Eq. (3.8) we have

$$\sigma^2 = \sum (x_i^2 - 2\mu x_i + \mu^2) f(x_i)$$
$$= \sum x_i^2 f(x_i) - 2\mu \sum x_i f(x_i) + \mu^2 \sum f(x_i)$$
$$= \sum x_i^2 f(x_i) - 2\mu \cdot \mu + \mu^2 \cdot 1 \qquad \text{by Eqs. (3.7) and (3.2)}$$

or

$$\sigma^2 = \sum x_i^2 f(x_i) - \mu^2$$

Solving this last equation for $\sum x_i^2 f(x_i)$ gives Eq.(3.9).

Example 3.3. Find the mean and variance of the discrete variable with density function given in Example 3.1.

By Definition (3.7)

$$\mu = 1 \cdot \tfrac{1}{6} + 2 \cdot \tfrac{1}{6} + \cdots + 6 \cdot \tfrac{1}{6} = \tfrac{21}{6} = \tfrac{7}{2} = 3.5$$

and, by Definition (3.8)

$$\sigma^2 = (1 - 3.5)^2 \cdot \frac{1}{6} + (2 - 3.5)^2 \cdot \frac{1}{6} + \cdots + (6 - 3.5)^2 \cdot \frac{1}{6}$$

$$= \frac{17.50}{6} = \frac{8.75}{3} \doteq 2.92$$

Also, since $\sum x_i^2 f(x_i) = 1^2 \cdot \tfrac{1}{6} + 2^2 \cdot \tfrac{1}{6} + \cdots + 6^2 \cdot \tfrac{1}{6} = \tfrac{91}{6}$, we have, by Theorem 3.1 and Definition (3.7)

$$\sigma^2 = \tfrac{91}{6} - (\tfrac{7}{2})^2 = \tfrac{35}{12} \doteq 2.92$$

Example 3.4. Find the probability of values 2, 3, 4, and 5 in Example 3.1. From Eq.(3.2c′) we obtain

$$P[2 \leq x \leq 5] = f(2) + f(3) + f(4) + f(5)$$

$$= \tfrac{1}{6} + \tfrac{1}{6} + \tfrac{1}{6} + \tfrac{1}{6}$$

$$= \tfrac{2}{3}$$

Note. Since $F(x_5) = \sum_{i=1}^{5} f(x_i)$ and $F(x_1) = f(x_1)$, we have

$$P[2 \leq x \leq 5] = \sum_{i=1}^{5} f(x_i) - \sum_{i=1}^{1} f(x_i) = F(x_5) - F(x_1)$$

$$= \tfrac{5}{6} - \tfrac{1}{6} = \tfrac{2}{3}$$

It can be shown that for any two positive integers a and b such that $a \leq b$ that

$$F(x_b) - F(x_{a-1}) = \sum_{i=a}^{b} f(x_i) \tag{3.10}$$

The six values resulting from the toss of a well-balanced die have been used to illustrate the meaning of theoretical density and distribution functions. At this time we present two discrete distributions which are of great importance in practice as well as theory. They are also used frequently in the later chapters of this book.

3.2.1. Dichotomous Distribution

The simplest of all distributions and one which occurs so often in practice results from our tendency to place each observation from a population in one or the other of two categories (or classes or groups), mutually exclusive and together exhaustive. For example, we divide a population of some manufactured product into defective and nondefective items, a population of students into male and female, a population of responses to a question into yes and no, and a population of diseased people into those recovering and those who do not. Generally, the variable is one we would consider to be qualitative, but since only two categories are involved, it is easy to think of the variable as being quantitative. It is customary to call an observation a "success" or a "failure" depending on which category it falls into, and for mathematical reasons to denote failure by 0 and success by 1. It does not make any difference, so far as the mathematics is concerned, whether we denote success by 1 or by 0. We use 1, following convention.

Letting p denote the probability with which 1 (success) occurs and $q = 1 - p$ the probability with which 0 (failure) occurs, we may write the *density function for a dichotomous variable $d(x; p)$* as

$$f(x) = d(x; p) = \begin{cases} q, & \text{when } x = 0 \\ p, & \text{when } x = 1 \end{cases} \tag{3.11}$$

Once p is known, $d(x; p)$ is uniquely determined. We say that $d(x; p)$ represents a one-parameter family of distributions, since only one parameter is required in order to determine the distributions. For $p = 0.3$, say, we have the particular dichotomous distribution with density function

$$d(x; 0.3) = \begin{cases} 0.7, & \text{when } x = 0 \\ 0.3, & \text{when } x = 1 \end{cases}$$

The distribution function is

$$F(x) = \begin{cases} 0, & \text{when } x < 0 \\ q, & \text{when } 0 \le x < 1 \\ 1, & \text{when } x \ge 1 \end{cases} \tag{3.12}$$

If we use Eqs.(3.7) and (3.8), it follows that the mean and variance of a dichotomous variable with density function defined by Eq.(3.11) are

$$\mu = 0 \cdot q + 1 \cdot p = p \tag{3.13}$$

and

$$\sigma^2 = (0 - p)^2 \cdot q + (1 - p)^2 \cdot p$$
$$= p^2 \cdot q + q^2 \cdot p$$
$$= pq(p + q)$$

or

$$\sigma^2 = pq \tag{3.14}$$

The student should be cautioned concerning the dual use of the symbol p. By definition, p denotes the probability of $x = 1$. On the other hand, p also is that value of the variable x which is the mean. Figure 3.3 should make the distinction clear.

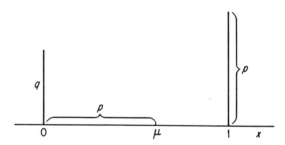

Fig. 3.3 Dichotomous Distribution

The dichotomous population just considered is sometimes called a *binomial population*. We refrain from using the term binomial distribution at this time, because it might lead to confusion when we discuss sampling distributions determined from dichotomous distributions, as discussed in Sect. 5.3.

3.2.2. Poisson Distribution

We have already considered in some detail the nature of two discrete distributions with finitely many possible values of the variable x. Now we describe a discrete distribution called the Poisson distribution with 0, 1, 2, 3, . . . being the possible values of x.

The *Poisson density function $p(x; m)$* is given by

$$f(x) = p(x; m) = \frac{e^{-m} m^x}{x!}, \qquad x = 0, 1, 2, 3, \ldots \tag{3.15}$$

where m is a positive real constant and e is the base of the natural logarithms, or

$$e = \lim_{u \to \infty} \left(1 + \frac{1}{u}\right)^u$$

It should be recalled that

$$e = 1 + \frac{1}{1!} + \frac{1}{2!} + \frac{1}{3!} + \cdots \doteq 2.71828$$

It should be observed that the variable x in Eq.(3.15) needs no subscript, since its possible values are the nonnegative integers. The Poisson distribution is uniquely determined by the single parameter m. Thus, we sometimes say that the Poisson is a one-parameter family of distributions and use the notation $f(x; m)$ or $p(x; m)$ to denote this. The notation $p(x; 1.0)$ denotes the Poisson density function with $m = 1.0$ and represents one member of the family.

For any positive real number m and any nonnegative integer x, we see that $e^{-m} m^x/x!$ is always positive. Further

$$\sum_{x=0}^{\infty} f(x) = \sum_{x=0}^{\infty} \frac{e^{-m} m^x}{x!} = e^{-m} \sum_{x=0}^{\infty} \frac{m^x}{x!}$$

$$= e^{-m} \left(1 + \frac{m}{1!} + \frac{m^2}{2!} + \cdots \right) = e^{-m} \cdot e^m = 1$$

The three conditions in Eq.(3.2) hold, if we assume that the probability of any set S_p is given by Condition (3.2c). Hence, we were justified in calling Eq. (3.15) a density function.

It can be shown (see Exercise 2.68) that both the mean and variance of a discrete variable with density function given by Eq. (3.15) are equal to m, i.e.

$$\mu = \sigma^2 = m \quad \text{for a Poisson distribution} \tag{3.16}$$

Solution of many problems found in practice are, as we shall see later, much easier to obtain due to this rare property.

In any applied problem we would not expect x to be infinitely large. Thus, the Poisson distribution can serve only as a model and an approximation to real-life situations. However, the limitation is not as restrictive as it might seem at first.

In order to get some idea of how little we are restricted, since, in fact, x may get infinitely large, let us consider Table 3.1, which lists $p(x; m)$ correct to four decimal places for selected values of m. From this table it is clear that the cumulated probabilities from some small value of x upwards is zero if only four significant figures are required. Thus, for example

$$\sum_{x=8}^{\infty} p(x; 1.0) = 0.0000$$

since

$$\sum_{x=0}^{7} p(x; 1.0) = 1.0000$$

Table 3.1

Poisson Density Function;* $p(x; m) = e^{-m} m^x / x!$

(Entries in the table are values of $p(x; m)$ for indicated values of x and m)

x \ m	0.5	1.0	1.5	2.0	2.5	3.0	4.0	5.0
0	0.6065	0.3679	0.2231	0.1353	0.0821	0.0498	0.0183	0.0067
1	0.3033	0.3679	0.3347	0.2707	0.2052	0.1494	0.0733	0.0337
2	0.0758	0.1839	0.2510	0.2707	0.2565	0.2240	0.1465	0.0842
3	0.0126	0.0613	0.1255	0.1804	0.2138	0.2240	0.1954	0.1404
4	0.0016	0.0153	0.0471	0.0902	0.1336	0.1680	0.1954	0.1755
5	0.0002	0.0031	0.0141	0.0361	0.0668	0.1008	0.1563	0.1755
6	0.0000	0.0005	0.0035	0.0120	0.0278	0.0504	0.1042	0.1465
7		0.0001	0.0008	0.0034	0.0099	0.0216	0.0595	0.1044
8		0.0000	0.0001	0.0009	0.0031	0.0081	0.0298	0.0653
9			0.0000	0.0002	0.0009	0.0027	0.0132	0.0363
10				0.0000	0.0002	0.0008	0.0053	0.0181
11					0.0000	0.0002	0.0019	0.0082
12						0.0001	0.0006	0.0034
13						0.0000	0.0002	0.0013
14							0.0001	0.0005
15							0.0000	0.0002
16								0.0000

* For extensive tables of $p(x; m)$ see Burington, R. S. and D. C. May, *Handbook of Probability and Statistics with Tables*. Sandusky, Ohio: Handbook Publishers, Inc., 1953; or E. C. Molina, *Poisson's Exponential Binomial Limit*. New York: D. Van Nostrand, 1942.

Example 3.5. Suppose a certain population is assumed to have the Poisson distribution with mean of 1.5; determine the theoretical frequencies for a representative sample of 100 observations taken from this population.

We simply multiply each entry in the column headed by $m = 1.5$ in Table 3.1 by 100 to obtain

x	0	1	2	3	4	5	6	7	8
$100 \cdot p(x)$	22.31	33.47	25.10	12.55	4.71	1.41	0.35	0.08	0.01

Rounding off these theoretical frequencies, we would expect the sample to have 22 observations with value 0, 33 with value 1, 25 with value 2, etc.

In practice, the value x usually represents the number of objects which possess a rare property. For example, x may denote the number of people killed by automobile accidents per day in a large city, the number of counts recorded by a Geiger counter per two minute interval, the number of blowholes in each of n castings, or the number of white blood corpuscles on slides of a given size. We discuss these and other applications in the exercise sections and after the properties of samples have been given in Chaps. 5 and 6.

3.2.3. Exercises

3.1. A balanced die, without dots, is colored red on one side, green on two sides, and blue on the remaining sides. (a) Define density and distribution functions which describe the distribution of colors. (b) Give an illustration in which the density function of (a) serves as a model.

3.2. (a) Find the density and distribution functions of the sum of two numbers (dots on two dice) which appear on two balanced dice. (b) Find the mean and variance of the sums. (c) Graph the density and distribution functions, indicating the location of the mean on the density graph.

3.3. (a) The *discrete uniform distribution* has the density function $f(x) = 1/m$, where $x = 1, 2, \ldots, m$. Find the distribution function as well as the mean and variance of the variable x. (b) Give an illustration of a particular uniform distribution.

3.4. (a) Given $f(x) = c(x - 2)^2$, where $x = 1, 2, \ldots, 6$, find c so that $f(x)$ is a density function. (b) Find the distribution function and graph it. (c) Find the mean and variance of this variable x.

3.5. Given $f(x) = c(x - 2)$, where $x = 1, 2, \ldots, 6$, is it possible to find a constant c so that $f(x)$ is a density function? Why?

3.6. If 1000 observations closely approximate a Poisson distribution with mean 2.5, how many of them would have x values of 2, 4, 6, 8, 10?

3.7. The variance of a population known to have the Poisson distribution is 2.0. (a) Determine the distribution function of this population. (b) Use Table 3.1 to find what proportion of the x values are less than 3; between 2 and 5; between 2 and 5 inclusive.

3.8. (a) Find c so that $f(x) = c/x!$, where $x = 0, 1, 2, \ldots$ is a density function. (b) Find $P[x < 2]$; $P[x \leq 2]$ for the density function in (a).

3.9. Assume that the number of telephone calls x coming to a certain switchboard during a period of one minute is approximately distributed as a Poisson variable with mean 5. (a) For what proportion of one minute intervals will there be more than ten calls coming to the switchboard per one minute interval? (b) If the switchboard can handle, at most, 12 calls per minute, what proportion of one minute intervals will the switchboard be overtaxed?

3.10. The number x of white blood corpuscles on slides of a fixed size is assumed to be distributed as a Poisson variable with mean 4. If more than ten indicates a dangerous surplus of white corpuscles, what proportion of slides fall in this category?

3.11. Let

$$f(x) = (\tfrac{1}{2})^x$$

where $x = 1, 2, \ldots$. (a) Is $f(x)$ a density function? (b) If $f(x)$ is a density function, find the mean of this distribution. (Strictly speaking, we should say "mean of the variable x," but this is a common expression which we sometimes use.)

Hint. Differentiate both sides of the identity $1 + a + a^2 + \cdots = 1/(1 - a)$ with respect to a; let $a = \tfrac{1}{2}$, etc.

(c) This function serves as a model for a balanced coin which is tossed until a head appears for the first time on the xth toss. Give another illustration in which $f(x) = (\frac{1}{2})^x$ serves as a model.

3.3. CONTINUOUS DISTRIBUTIONS OF ONE VARIABLE

The regular forms of the histograms and cumulative polygons discussed in Chap. 2 suggest that in favorable cases data are approximations to distributions which can be represented by smooth curves and be given simple mathematical expressions. Now we consider how this can be accomplished.

Think of the case where the area of the histogram is made equal to one. From the discussion in Chap. 2 it is clear that the sum of the areas of neighboring rectangles is equal to the relative frequency with which (proportion of times) the value of x falls in the intervals which make up the bases of those rectangles. Now suppose we think of subdividing each interval into two, then four, then eight, etc. smaller intervals. (For a finite collection of observations the histograms would look smoother for a while as the number of rectangles increased, but eventually the resulting histograms would become more and more irregular.) Conceptually, in an infinite population we would rarely have irregular relative frequencies as the number of subdivisions increases. Since the property of smoothness would continue to hold as the number of subdivisions increases indefinitely, the area under the limiting or idealized curve between any two given values of x should be equal to the relative frequency with which x would lie in the interval determined by those two values of x.

The function $f(x)$ whose graph is the limiting curve of the series of histograms just described is considered the mathematical model and is called the density function of the variable x. Formally, we say the continuous variable x has a *probability density function* $f(x)$ if it satisfies the following properties

(a) $f(x)$ is a single-valued nonnegative real number for all real values of x.

(b) $\int_{-\infty}^{\infty} f(x)\,dx = 1$ $\qquad\qquad\qquad\qquad\qquad\qquad$ (3.17)

(c) $\int_{a}^{b} f(x)\,dx = P[a < x < b]$

where $P[a < x < b]$ denotes the probability with which x falls between any two real values a and b for which $a < b$.

We also call $f(x)$ or f a *density function* or *probability function* of the continuous variable x.

The curve in Fig. 3.4 represents a smooth density function which goes from c to $+\infty$. Since it is a density function, the three properties of Eq.

(3.17) must be satisfied. For example, the geometric representation of property (c) is the area under the curve and above the x axis which is between the two vertical lines $x = a$ and $x = b$. This is a visual image of the probability with which x falls in the interval between a and b.

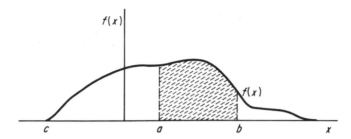

Fig. 3.4 Theoretical Density Function for Continuous Variable x

Comparing the definitions of discrete and continuous density functions as given by Eqs. (3.2) and (3.17), respectively, we see that they satisfy the same kinds of properties. However, there is one obvious difference which may prove to be annoying if it is not discussed. It is the way in which we think geometrically of probabilities. In the discrete case the probability of a set of (discrete) points in an interval is the sum of lengths of line segments above the points; in the continuous case the probability of the set of (all) points in an interval is an area above the interval. In one case, we see probability measured in terms of lineal units (lengths) and in the other case it is measured in terms of square units (area).

Since the area over a point is zero this leads us, in the continuous case, to define *the probability of a single value x_o to be zero*. In case there are those who think this is unreasonable, consider the following argument: let d be a small positive number. Then the probability of the set of points in the interval from $x_o - d$ to $x_o + d$ is given by

$$A = \int_{x_o-d}^{x_o+d} f(x) \, dx$$

It is possible to find a value of x, say x_1, in the interval $x_o - d$ to $x_o + d$ such that the area of the rectangle of height $f(x_1)$ above the interval is $A = 2df(x_1)$. Now for a given x_1, $f(x_1)$ is a fixed value, since $f(x)$ is assumed to be a single-valued continuous function. Thus

$$\lim_{d \to 0} 2d f(x_1) = 0$$

and so the probability of the reduced interval, a single point x_o, is zero. Hence, in the continuous case it turns out that the density function value $f(x_o)$ is not the probability for x_o, whereas in the discrete case the value $f(x_o)$

is the probability for x_o. For these reasons we are led to the use of $f(x)\ dx$, called the *probability element* or *density element*, when we are discussing probabilities in the continuous case. Here dx is the usual differential element defined in the calculus.

The interval $a < x < b$ was used in Definition (3.17) when we gave the third property of a density function. In view of the above presentation, it is clear that we could have used $a \leq x \leq b$ or $a < x \leq b$ or $a \leq x < b$ without changing the probability associated with the intervals. Thus, if we had used the interval $a \leq x \leq b$ in Eq. (3.17), property (c) in Definitions (3.2) and (3.17) of a density function would have appeared the same.

The distribution function $F(x)$ of the continuous variable x which is the idealization of the cumulative polygon is defined by

$$F(x) = \int_{-\infty}^{x} f(x)\ dx \left[\text{or} \quad \int_{-\infty}^{x} f(t)\ dt \right] \tag{3.18}$$

It should be noted that the variable of integration is a "dummy" variable and that the distribution function is a function of the upper limit of the integral expression. It follows from Definition (3.18) that

$$\begin{cases} F(x) \text{ is a nondecreasing function of } x \text{ and} \\ 0 \leq F(x) \leq 1. \end{cases} \tag{3.19}$$

From Eq.(3.18) it is clear how the distribution function can be obtained from the density function. Conversely, $f(x)$ can be found from $F(x)$ by differentiation; that is

$$f(x) = \frac{dF(x)}{dx} \qquad [\text{or} \quad f(x)\ dx = dF(x)]$$

The mean μ and variance σ^2 of a continuous variable x with density function $f(x)$ are defined by

$$\mu = \int_{-\infty}^{\infty} x f(x)\ dx \tag{3.20}$$

and

$$\sigma^2 = \int_{-\infty}^{\infty} (x - \mu)^2 f(x)\ dx \tag{3.21}$$

Theorem 3.2. *If $f(x)$ is the density function of the continuous variable x, then*

$$\int_{-\infty}^{\infty} x^2 f(x)\ dx = \sigma^2 + \mu^2 \tag{3.22}$$

Proof. Starting with Eq. (3.21) and using Eqs. (3.20) and (3.17), we have

$$\sigma^2 = \int_{-\infty}^{\infty} (x^2 - 2\mu x + \mu^2) f(x)\, dx$$

$$= \int_{-\infty}^{\infty} x^2 f(x)\, dx - 2\mu \int_{-\infty}^{\infty} x f(x)\, dx + \mu^2 \int_{-\infty}^{\infty} f(x)\, dx$$

or

$$\sigma^2 = \int_{-\infty}^{\infty} x^2 f(x)\, dx - \mu^2$$

Solving this last equation for

$$\int_{-\infty}^{\infty} x^2 f(x)\, dx$$

gives Eq. (3.22).

Theorem 3.3. *If x has distribution function F(x), then the probability of x falling between any two points x = a and x = b (a < b) is given by F(b) − F(a).*

Proof. The result follows immediately from Definition (3.18). Thus

$$F(b) - F(a) = \int_{-\infty}^{b} f(x)\, dx - \int_{-\infty}^{a} f(x)\, dx$$

$$= \int_{-\infty}^{a} f(x)\, dx + \int_{a}^{b} f(x)\, dx - \int_{-\infty}^{a} f(x)\, dx$$

$$= \int_{a}^{b} f(x)\, dx$$

Even though there are real and important differences between discrete and continuous distributions, it is clear that the forms used in defining density and distribution functions as well as means and variances are very similar. Thus, in much of the theoretical presentation in the remaining chapters we shall treat only the continuous case in detail and leave the discrete case for the reader. Generally, this requires only changing $\int dx$ to $\sum$ and going through the same steps.

To avoid stating all definitions and theorems twice, some authors in mathematical studies use a type of integral due to Stieltjes. The Stieltjes integral in one summation process includes the summation denoted by $\sum$ and the usual Riemann integral (infinite summation) denoted by $\int$. We shall not use this type of integral, since it is beyond the scope of this book.

In order to illustrate the definitions of this section we shall now examine two very important continuous distributions. We consider the simplest distribution first.

3.3.1. Uniform Distribution

If the density function of a variable x is constant over some region (domain) and zero elsewhere, the variable x is said to be *uniformly distributed* over that region. Thus, a continuous variable with density function

$$f(x) = u(x; c, d) = \begin{cases} \dfrac{1}{d-c}, & \text{for } c \le x \le d \\ 0, & \text{otherwise} \end{cases} \tag{3.23}$$

is uniformly distributed over the interval from c to d. This is also referred to as a *rectangular distribution*. The parameters are c and d. Thus, the uniform distribution belongs to a two-parameter family of distributions. The graph of a typical uniform distribution is shown in Fig. 3.5.

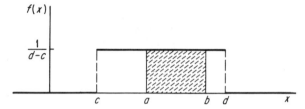

Fig. 3.5 A Typical Uniform Density Function

The distribution function is given by

$$F(x) = \begin{cases} 0, & \text{when } x < c \\ \dfrac{x-c}{d-c}, & \text{when } c \le x \le d \\ 1, & \text{when } x > d \end{cases} \tag{3.24}$$

and the typical graph is shown in Fig. 3.6.

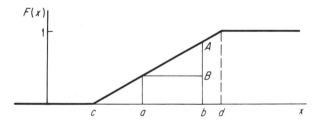

Fig. 3.6 A Typical Uniform Distribution Function

According to Theorem 3.3 the shaded area in Fig. 3.5 is numerically the same as the length of the line segment AB in Fig. 3.6. That is, the numerical value of the length of AB is also the probability of x falling in the interval from a to b. Using Eq. (3.10), we can find a similar line segment in the discrete case which measures the probability of discrete values falling in the closed interval from c to d. (By closed interval we mean that the end points c and d are included in the interval.) The close similarity between discrete and continuous distribution functions, as indicated by this property, is one of the reasons why $F(x)$ is at the heart of mathematical studies in statistics.

Since the density function is symmetrical about $x = (c + d)/2$, the center of the interval from c to d, we would expect the mean to be this value. Using Eq. (3.20), we find that

$$\mu = \int_{-\infty}^{\infty} x f(x) \, dx$$

$$= \int_{-\infty}^{c} x \cdot 0 \, dx + \int_{c}^{d} x \left(\frac{1}{d-c}\right) dx + \int_{d}^{\infty} x \cdot 0 \, dx$$

$$= \frac{d^2 - c^2}{2(d-c)}$$

or

$$\mu = \frac{d+c}{2} \tag{3.25}$$

which verifies our expectation. It can be shown by using Theorem 3.2 and Formula (3.25) that the variance is

$$\sigma^2 = \frac{(d-c)^2}{12} \tag{3.26}$$

This distribution is of great importance both practically and theoretically. In applied work it is useful in studying rounding errors in measurements which are made within a specified accuracy. For example, weights of humans are usually made and recorded to the nearest pound. It is assumed that the difference in the recorded and true weight is some number between -0.5 and 0.5 and that the error is uniformly distributed over this interval. For this study the density function would be $f(x) = 1$ when $-0.5 \leq x \leq 0.5$ and $f(x) = 0$ otherwise. The mean would be 0 and the variance $\frac{1}{12}$.

The uniform distribution is of greater importance theoretically due to the following very important theorem, which we state without proof.

Theorem 3.4. Let $f(x)$ be any density function of a continuous variable x and $F(x)$ be its distribution function. Then $f(x)$ may be transformed to the uniform density function

$$g(u) = 1 \qquad 0 \leq u \leq 1$$

by letting $u = F(x)$.

It is clear that u must range from 0 to 1, since this is the range of $F(x)$.

It is possible with the use of this theorem to exhibit many properties of continuous distributions in general by proving them for this particular uniform distribution. It follows from this theorem that there is at least one transformation which transforms any continuous distribution into any other continuous distribution. One transformation obtained is by combining the transformation which transforms the first distribution into the uniform distribution with the inverse (reverse) of the transformation which transforms the second distribution into the uniform distribution.

3.3.2. Normal Distribution

The most important distribution in statistics is the normal distribution. The *normal density function* $n(x; \mu, \sigma)$ is given by

$$f(x) = n(x; \mu, \sigma) = \frac{1}{\sqrt{2\pi}\,\sigma} e^{-\frac{1}{2}\left(\frac{x-\mu}{\sigma}\right)^2}, \qquad -\infty < x < \infty \qquad (3.27)$$

where μ and σ are parameters which also happen to be the mean and standard deviation of the continuous variable x which is normally distributed. The normal density function is also called the *Gaussian function* and the *error function*.

The graph of a typical normal density is given in Fig. 3.7. It is clear that the curve is symmetrical about the line $x = \mu$ and hence by symmetry the mean must be μ. The function is defined for all real values of x and has points of inflection at $x = \mu \pm \sigma$.

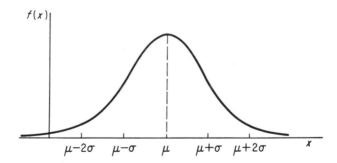

Fig. 3.7 Typical Normal Density Curve

Sometimes we refer to μ as the *location parameter* and σ as the *scale parameter* of this two-parameter family of distributions. For a fixed σ, if we change μ the resulting curves keep the same shape but have different locations along the x-axis. However, if μ is held fixed and σ is allowed to vary, the curves have different spreads. Figure 3.8 shows three curves with $\mu = 5$ and $\sigma = 1$, 2, and 3, respectively. The area under each curve is one, but there are proportionately more values of x further away from the mean when σ gets larger and larger.

Since we defined $n(x; \mu, \sigma)$ to be a density function, it must be true that

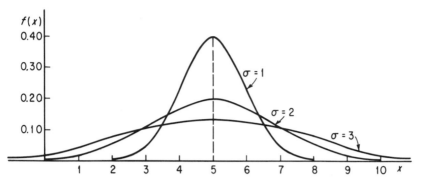

Fig. 3.8 Normal Density Curves with $\sigma = 1, 2, 3$

$$\int_{-\infty}^{\infty} n(x; \mu, \sigma)\, dx = 1 \qquad (3.28)$$

This fact is not obvious. Since $n(x)$ is such an important function, we shall take time to verify Eq. (3.28) and to become familiar with some of the troublesome problems involved in manipulating $n(x)$. Let the area under the curve of $n(x)$, as shown in Fig. 3.7, be A. Then we have

$$A = \frac{1}{\sqrt{2\pi}\,\sigma} \int_{-\infty}^{\infty} e^{-\frac{1}{2}\left(\frac{x-\mu}{\sigma}\right)^2} dx$$

Letting $u = (x - \mu)/\sigma$, we obtain $dx = \sigma\, du$ and

$$A = \frac{1}{\sqrt{2\pi}} \int_{-\infty}^{\infty} e^{-u^2/2}\, du$$

We could evaluate this integral by using a book of tables of integrals. However, we use the following method to determine A:
Since u is a dummy variable, we may write

$$A = \frac{1}{\sqrt{2\pi}} \int_{-\infty}^{\infty} e^{-v^2/2}\, dv$$

Thus

$$A^2 = \frac{1}{\sqrt{2\pi}} \int_{-\infty}^{\infty} e^{-u^2/2}\, du \cdot \frac{1}{\sqrt{2\pi}} \int_{-\infty}^{\infty} e^{-v^2/2}\, dv$$

or

$$A^2 = \frac{1}{2\pi} \int_{-\infty}^{\infty} \int_{-\infty}^{\infty} e^{-\frac{1}{2}(u^2 + v^2)}\, du\, dv$$

Letting $u = r \sin \theta$ and $v = r \cos \theta$, we find that $du\, dv = r\, d\theta\, dr$ and hence

$$A^2 = \frac{1}{2\pi} \int_0^\infty \int_0^{2\pi} r e^{-r^2/2} \, d\theta \, dr$$

$$= \int_0^\infty r e^{-r^2/2} \, dr$$

$$= [-e^{-r^2/2}]_0^\infty = 1$$

Since $n(x)$ is always nonnegative, A is positive and hence $A = 1$.

The *normal distribution function* $N(x; \mu, \sigma)$ is

$$F(x) = N(x; \mu, \sigma) = \frac{1}{\sqrt{2\pi}\,\sigma} \int_{-\infty}^x e^{-\frac{1}{2}\left(\frac{y-\mu}{\sigma}\right)^2} \, dy \qquad (3.29)$$

The integral in Eq. (3.29) can not be expressed so as to have a simple functional form, but it can be computed by numerical methods. Since Eq. (3.29) changes whenever either μ or σ changes, we consider a particular member of this two-parameter family of distributions. Letting $\mu = 0$ and $\sigma = 1$ and replacing x by t in Eq. (3.27), we obtain

$$n(t; 0, 1) = \frac{1}{\sqrt{2\pi}} e^{-t^2/2} \qquad (3.30)$$

which is called the *standard normal density function*. Hence, the *standard normal distribution function* is given by

$$N(t; 0, 1) = \frac{1}{\sqrt{2\pi}} \int_{-\infty}^t e^{-u^2/2} \, du \qquad (3.31)$$

Extensive tables of $n(t)$ and $N(t)$ (to 15 decimal places) have been compiled by the New York Mathematical Tables Project [11] and others [4, 12]. For our purposes the abbreviated Tables I and II in the back of this book are adequate. Graphs of the standard normal density and distribution functions are given in Figs. 3.9 and 3.10, respectively. Note that the shaded area in Fig. 3.9 is numerically the same as the ordinate $N(t_0)$ in Fig. 3.10. This shows the type of values given in Table II.

Before we consider uses of these tables, it should be observed that they may be used to find ordinates and areas for any member of the family of normal distributions. For if we let

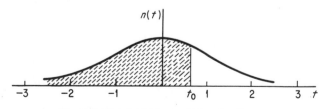

Fig. 3.9 Standard Normal Density Curve

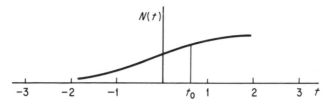

Fig. 3.10 Standard Normal Distribution Curve

$$t = \frac{x - \mu}{\sigma} \tag{3.32}$$

in the probability element $n(x; \mu, \sigma)\, dx$, we obtain

$$\frac{1}{\sqrt{2\pi}\,\sigma}\, e^{-(1/2)t^2}\, \sigma\, dt = \frac{1}{\sqrt{2\pi}}\, e^{-t^2/2}\, dt = n(t; 0, 1)\, dt$$

where

$$n(t; 0, 1) = n\left(\frac{x - \mu}{\sigma}; 0, 1\right)$$

is the same function as the one given in Eq.(3.30). From this point on, when $n(t)$ and $N(t)$ are used, it will be understood that the standard normal distribution is being referred to.

Due to the symmetry of the normal distribution, only nonnegative values of t are given in Tables I and II. Thus, to facilitate the use of Table I we need the relation

$$n(-t) = n(t) \tag{3.33}$$

and to facilitate the use of Table II we need the relations

$$\int_0^t n(u)\, du = N(t) - 0.5 \tag{3.34}$$

$$\int_{-t}^t n(u)\, du = 2N(t) - 1 \tag{3.35}$$

$$N(-t) = 1 - N(t) \tag{3.36}$$

The proofs of these relationships will be left to the student.

3.3.3. Illustrations for the Normal Distribution

Three intervals of importance in studies of bell-shaped distributions were given in Sect. 2.3.3. Using Table II, we can now determine the proportion of t values falling in any interval of a normally distributed variable.

Example 3.6. For the standard normal distribution find the probability of values falling within one standard deviation of the mean; that is, find the proportion of standard normal values falling between $t = -1$ and $t = 1$.

Also find the proportion falling between $t = -2$ and $t = 2$; $t = -3$ and $t = 3$.

From Table II, using Eq. (3.35), we obtain

$$N(t = 1) - N(t = -1) = 2(0.8413) - 1 = 0.6826 = 68.26 \text{ per cent}$$

$$N(t = 2) - N(t = -2) = 2(0.9772) - 1 = 0.9544 = 95.44 \text{ per cent}$$

$$N(t = 3) - N(t = -3) = 2(0.9987) - 1 = 0.9974 = 99.74 \text{ per cent}$$

Example 3.7. What symmetric interval about the mean of a standard normal distribution contains (a) 90 per cent, (b) 95 per cent, (c) 99 per cent of the t values?

For (a) we must find t such that $2N(t) - 1 = 0.9000$. or $N(t) = 0.9500$. Using linear interpolation in Table II, we get $t = 1.645$. Thus, the interval is from -1.645 to 1.645. Similarly, we find the 95 per cent interval to reach from -1.960 to 1.960 and the 99 per cent interval to reach from -2.575 to 2.575. (The three values 1.645, 1.96, and 2.575 are used so often that they should be memorized.)

In general, if t_α denotes that value of the standard normal distribution for which

$$\int_{t_\alpha}^{\infty} n(t; 0, 1)\, dt = \alpha$$

where α is a positive number less than 1, we say the $100(1 - 2\alpha)$ per cent symmetric interval of x about μ reaches from $\mu - t_\alpha \sigma$ to $\mu + t_\alpha \sigma$. That is, $\mu \pm t_\alpha \sigma$ are the limits of a $100(1 - 2\alpha)$ per cent symmetric interval about μ.

Example 3.8 For the normal distribution with mean 5 and variance 4, find what proportion of the x values fall between $x_1 = 1$ and $x_2 = 7$.

The area under the normal density curve $n(x; 5, 2)$ from 1 to 7 is the same as the area under the standard normal density curve from

$$t_1 = \frac{x_1 - \mu}{\sigma} = \frac{1 - 5}{2} = -2$$

to

$$t_2 = \frac{x_2 - \mu}{\sigma} = \frac{7 - 5}{2} = 1$$

Thus, using Table II and Eq. (3.36), we obtain

$$N(x = 7; 5, 2) - N(x = 1; 5, 2) = N(t = 1; 0, 1) - N(t = -2; 0, 1)$$

$$= N(t = 1) - [1 - N(t = 2)]$$

$$= 0.8413 - (1 - 0.9772)$$

$$= 0.8185$$

The relation between the x-scale and t-scale is brought out in Fig. 3.11, where the appropriate scales are placed under a normal curve. From Table I we find that $n(t = 0) = 0.3989 \doteq 0.40$, $n(t = 1) = 0.2420 \doteq 0.24$, and $n(t = 2) = 0.0540 \doteq 0.05$. Hence, $n(x = 5) = n(t = 0)/2 \doteq 0.20$, $n(x = 7) \doteq 0.12$, and $n(x = 9) \doteq 0.03$. (The curve in Fig. 3.11 is not a "true picture" of either density function, since the lengths of the units of measurements on the horizontal and vertical axes are different. It should be understood that the two vertical scales differ also.)

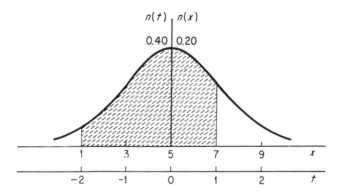

Fig. 3.11 Comparison of Scales for Normal Density Curve

Example 3.9. The teachers in a certain department of a large American university assign grades to the beginning class by means of the normal distribution. Determine how many of 500 students will receive each of the grades A, B, C, D, and F if F is given to a student whose grade falls in the interval $(-\infty, \mu - 1.5\sigma)$, D is given if the grade falls in the interval $(\mu - 1.5\sigma, \mu - 0.5\sigma)$, etc.

The t values which determine the intervals are -1.5, -0.5, 0.5, and 1.5. From Table II we find that the areas above these five intervals are 0.0668, 0.2417, 0.3830, 0.2417, and 0.0668, respectively. Then the number of students who will be assigned each grade is as follows: $500(0.0668) = 33.4 \doteq 33$ will receive the grade F, 121 the grade D, 192 the grade C, 121 the grade B, and 33 the grade A.

Actually, once the percentages 7, 24, 38, 24, 7 have been determined, the department would give the grade of A to the top 7 per cent, the grade of B to the next 24 per cent, etc. It should be observed that, even though the extreme theoretical intervals go to infinity, the practical limits are $\mu - 2.5\sigma$ and $\mu + 2.5\sigma$ in this case. Thus all the practical intervals have length σ in this case.

Theoretical distributions were introduced in an effort to find simple functions which closely approximate real data and which serve as models for collections of measurements. The next example is given to illustrate how the model density curve can be fitted to the histogram. Later this example will be used to determine if the fit is satisfactory.

Example 3.10. Fit a normal density curve to the teak tree data of Exercise 2.10, assuming that we have reason to believe that this is a representative sample from a normal distribution with mean μ and variance σ^2.

This fitted curve serves as an approximation to the population curve under the assumption that the sample size is large enough so that the sample mean $\hat{\mu} = 21.69$ and sample variance $\hat{\sigma}^2 = 34.5156$ are close to the population mean and variance. Table 3.2 illustrates how the normal curve with mean $\hat{\mu}$ and variance $\hat{\sigma}^2$ is fitted to the data. (The circumflex symbol ^ above the symbol for a parameter denotes estimate of the parameter.) Columns 5 and 6 of Table 3.2 may be used to compare the area under the fitted curve and the histogram for the various class intervals. In this problem there appears to be good agreement for all intervals except possibly for the class interval 22.5–25.5.

Table 3.2
Fitting of Normal Curve to Teak Tree Histogram

Upper Class Boundaries x	$\dfrac{x - 21.69}{5.875}$ t	Area to Left of t A	Area over Interval to Left of t ΔA	Theoretical Frequency $1088\,\Delta A$	Observed Frequency f	Length of Ordinate at Class Mark $1088\,n(x)$
7.5	−2.42	.0078	.0078	8.5	8	2.1
10.5	−1.90	.0287	.0209	22.7	26	7.2
13.5	−1.39	.0823	.0536	58.3	50	18.9
16.5	−0.88	.1894	.1071	116.5	120	38.6
19.5	−0.37	.3557	.1663	180.9	181	60.6
22.5	0.14	.5557	.2000	217.6	215	73.4
25.5	0.65	.7422	.1865	202.9	213	68.5
28.5	1.16	.8770	.1348	146.7	145	49.3
31.5	1.67	.9525	.0755	82.1	76	27.3
34.5	2.18	.9854	.0329	35.8	36	11.5
37.5	2.69	.9964	.0110	12.0	18	3.8

It is not generally necessary actually to graph the fitted normal curve when it is being compared with the frequency histogram. However, if for any reason a graph seems desirable, we multiply the density function $n(x;$ 21.69, 5.875) by the sample size 1088 so that the area under the resulting curve is the same as that of the histogram. Using Table I, we obtain the ordinates of the fitted curve shown in Table 3.2, where the ordinate of the class mark is given by $1088n(x) = 1088n(t)/5.875 = 185.2n(t)$. Figure 3.12 pictures the goodness of fit. In Sect. 15.2 we present a statistical procedure designed to "test the goodness of fit."

There is a special kind of graph paper, called *normal probability* (or *cumulative normal*) *graph paper*, which may be used to determine whether a

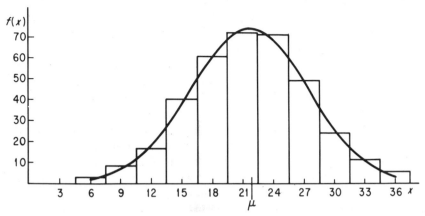

Fig. 3.12 Normal Curve Fitted to a Frequency Histogram

set of observations may have been drawn from a normal distribution. This graph paper is prepared by transforming the vertical scale (as is indicated in Example 3.11) so that a cumulative normal distribution curve such as the one shown in Fig. 3.10 becomes a straight line. Since observations are so often assumed to be drawn from a normal distribution, this graph paper is very useful in practice. Figure 3.13 shows normal probability graph paper as it is used in Example 3.11. The reader should note that Theorem 3.4 can be used to transform any continuous distribution function so that the graph of the transformed distribution function appears as a straight line.

Example 3.11. Plot the teak tree data of Exercise 2.10 on normal probability graph paper.

First, we find the cumulative relative frequencies below the class bound-

Table 3.3
Cumulative Relative Frequencies for the Teak Tree Data

Upper Class Boundaries	Cumulative Frequencies	Cum. Relative Frequencies
7.5	8	0.007
10.5	34	0.031
13.5	84	0.077
16.5	204	0.188
19.5	385	0.354
22.5	600	0.551
25.5	813	0.747
28.5	958	0.881
31.5	1034	0.950
34.5	1070	0.983
37.5	1088	1.000

aries shown in Table 3.3. Then we place 7.5, 10.5, . . . , 37.5 at equal intervals on the horizontal scale of normal probability graph paper, using as much of the scale as is convenient. Along the right-hand vertical scale, which is already marked, we locate the cumulative relative frequencies and plot points with class boundaries as abscissas and cumulative relative frequencies as ordinates, as indicated in Fig. 3.13. Using a straight edge, we draw the best-fitting line by sight. In this case, the points either fall on or very near the line. Hence, we conclude that the observations could have come from a normal population. This is the method commonly used to determine whether the frequent assumption that the observations are drawn from a normal population is tenable. Note that the left-hand vertical scale in Fig. 3.13 is used when we obtain cumulative relative frequencies above class boundaries.

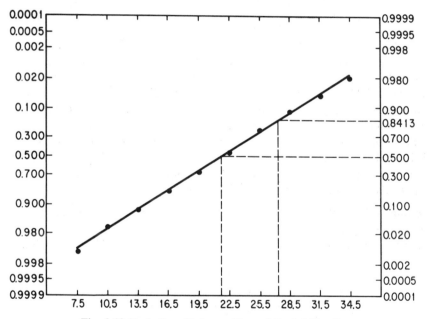

Fig. 3.13 Teak Tree Data on Normal Probability Paper

Furthermore, when the fit is good we can estimate the mean and variance of the normal distribution. The .50 point on the vertical scale corresponds to the estimated mean $\hat{\mu}$ on the horizontal scale, and the .8413 point (see Table II or Example 3.6) corresponds to $\hat{\mu} + \hat{\sigma}$. Thus, $\hat{\sigma}$ can be obtained by subtraction. In particular, we find that $\hat{\mu} = 21.6$ and $\hat{\mu} + \hat{\sigma} = 27.4$. Thus $\hat{\sigma} = 27.4 - 21.6 = 5.8$. We see that the mean and standard deviation obtained by graphic methods are very near those obtained by numerical methods in Example 3.10.

The reader should be cautious in drawing conclusions when fitting normal curves to data. Even though the fitted curve may very nearly fit the data, as illustrated by Examples 3.10 and 3.11, it does not necessarily follow that the sample was drawn from the indicated normal population.

The above examples were given to familiarize the reader with the normal distribution and to indicate some of its many uses. As the theory develops, we shall consider other uses of this distribution. At this time we give some reasons why the normal distribution is so important in the theory and application of statistics. The reasons are

1. Distributions encountered in practice are frequently believed to be normal or approximately normal.
2. The mathematics is highly developed and relatively simple.
3. It is important as a limiting distribution.

3.3.4. Exercises

3.12. Let

$$f(x) = \begin{cases} c, & \text{when } -2 \leq x \leq 3 \\ 0, & \text{otherwise} \end{cases}$$

Determine c so that $f(x)$ is a density function. Then find the distribution function and graph both functions.

3.13. Let $f(x; \theta) = 1/\theta$, $0 \leq x \leq \theta$, be a one-parameter family of distributions. Graph the density curves for $\theta = \frac{1}{2}; 1; 2$.

3.14. (a) The variable x is uniformly distributed over the interval $0 \leq x \leq 12$. Determine the density and distribution functions of x and graph both. (b) Determine the mean and variance of the variable x. (c) Assume that x represents the position of the minute hand on the face of a clock in a jewelry shop which strikes the hour every time the hand reaches 12. Every day at a time determined by chance an individual aimlessly walks about town and stops in this shop. On what proportion of his trips would he arrive at least 20 minutes before the clock strikes?

3.15. Let $f(x; \theta) = c/2$, $(\theta - 2) \leq x \leq (\theta + 2)$, be a one-parameter family of distributions. (a) Determine c so that $f(x; \theta)$ is a density function. (b) Find the mean and variance of x. (c) Graph two density curves from this family of distributions.

3.16. Prove Formula (3.26), using Theorem 3.2 and Formula (3.25).

3.17. The variable x has density function $f(x) = 1$, $(k - \frac{1}{2}) \leq x \leq (k + \frac{1}{2})$, where k is an arbitrary constant. For $k = 70$ in. with what probability would we obtain the particular sample 70.4, 69.7, 70.1? Explain.

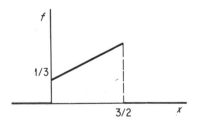

3.18. It is known that x is distributed as

is indicated by the wide-line graph at the bottom of page 83. (a) Determine the density and distribution functions for x. (b) Find the mean and variance of this distribution. (c) Find the proportion of x values between $\frac{1}{2}$ and 1.

3.19. The *triangular density function* is given by

$$f(x; a, b) = \frac{1}{a}\left(1 - \frac{|x - b|}{a}\right) \qquad \text{for } |x - b| < a$$

(a) Find the mean and variance of this distribution. (b) Graph the density curve for $a = 2$ and $b = 0$; for $a = 1$ and $b = 1$. (c) Find the distribution function and graph for one particular curve; say, for $a = b = 1$. (d) When $a = b = 1$, find the probability that x falls between 0.8 and 1.3.

3.20. Find c so that $f(x) = cx$, $0 \le x \le 1$, is a density function, and then find the mean and variance of the distribution of x.

3.21. Use Theorem 3.4 to find the transformation which transforms the distribution in Exercise 3.20 into the uniform density function $f(u) = 1$, $0 \le u \le 1$.

3.22. (a) Use Theorem 3.4 to transform the triangular distribution in Exercise 3.19 into the unit uniform density function. (b) What values of u for the transformed distribution correspond to the values $x = -1, -0.7, -0.3, 0, 0.3, 0.7, 1$, when the particular triangular density function $f(x; 1, 1)$ is used?

3.23. Assume x to be normally distributed with mean 4 and standard deviation 3. Find
(a) the proportion of values greater than $x = 7$, i.e., $P[x > 7]$
(b) $P[1 \le x < 7]$
(c) $P[-2 \le x \le 7]$
(d) $P[0 < x < 5]$

3.24. Prove each of the following: (a) Formula (3.34), (b) Formula (3.35), (c) Formula (3.36).

3.25. If the mean lifetime of a certain kind of battery is 600 days, with a standard deviation of 50 days, what percentage of this type of battery can be expected to last anywhere from 500 to 800 days? Assume that the lifetimes are normally distributed.

3.26. The bacteria content of a canned food product must be less than 70 to be acceptable. Long experience indicates that the mean bacteria content is 68 with a standard deviation of 0.9. What proportion of the cans must be declared not acceptable, if the bacteria content of the cans is assumed to be normally distributed?

3.27. A certain school with high standards requires that a student have a verbal score on a given college board examination in excess of 540 in order to be considered for admission. It is known that the scores are approximately normally distributed with mean 490 and standard deviation 80. What percentage of students taking the college board examination would not be considered for admission because of a low score?

3.28. A one in. bolt manufactured at a certain plant is rejected unless its diameter measures between 0.98 and 1.02 in. If the diameters of bolts made at this plant are approximately normally distributed with mean 1.000 in. and variance 0.000049 sq. in. what percentage of bolts are rejected due to improper size?

3.29. If x is normally distributed with mean μ and variance $\frac{1}{4}$, find

(a) $P[x \geq \mu + \frac{1}{4}]$

(b) $P[(\mu - 1) < x < (\mu + 1)]$

(c) $P[x > (\mu + 1) \quad \text{or} \quad x < (\mu - 2)]$

3.30. If x is normally distributed with mean 10 and variance 4, find a number x_0 such that

(a) $P[x < x_0] = 0.05$

(b) $P[12 < x < x_0] = 0.10$

(c) $P[x > |x_0|] = 0.20$

3.31. If x is normally distributed with mean 10 and variance 9, find the limits of a symmetric interval about the mean such that the area above the interval and below the normal density curve is (a) 0.90, (b) 0.95, (c) 0.98.

3.32. If the grades in your class in statistics were assigned by using the normal distribution, how many would receive each of the grades A, B, C, D, F?

3.33. (a) The grades in a certain school are A, B+, B, C+, C, D, F. What proportion of the students in a class would receive each grade if they were assigned by means of the normal distribution? (b) What proportion of students in a school giving only the grades "excellent," "good," "passing," and "failing" would receive each grade if they were assigned by means of the normal distribution?

3.34. Use data in Exercise 2.7 to prepare a new frequency table with intervals 0–4, 4–8, 8–12, etc. (a) Compute the mean $\hat{\mu}$ and variance $\hat{\sigma}^2$ for this data.
$$\text{Answer. } \hat{\mu} = 17.74; \quad \hat{\sigma}^2 = 19.1977.$$
(b) Fit a normal density curve to this data. How good is the fit? (c) Plot this data on normal probability graph paper. (d) Use (c) to estimate graphically the mean and standard deviation of the fitted normal distribution. (e) Estimate graphically the standard deviation of the fitted normal distribution in (c) by finding $\hat{\mu} - 2\hat{\sigma}$ and $\hat{\mu} + 2\hat{\sigma}$.

3.35. Use the data in Exercise 2.8 to answer (a), (b), (c), (d), and (e) of Exercise 3.34. $\text{Answer to (a). } \hat{\mu} = 34.50, \hat{\sigma}^2 = 45.8082.$

3.36. Use the data in Table 2.4 to answer (a), (b), (c), (d), and (e) of Exercise 3.34. $\text{Answer to (a). } \hat{\mu} = 6.256, \hat{\sigma}^2 = 0.3776.$

Hint. In computing the mean and variance, let 4.85 and 7.45 be the class marks of the lower and upper classes respectively.

3.37. (a) Find c so that $f(x) = ce^{-x}$, $x \geq 0$, is a density function. (b) Find the mean and variance of x.

3.38. (a) Find c so that $f(x) = cx^\alpha e^{-x}$, $x \geq 0$, α a positive integer, is a density function.

Hint. Use the fact that

$$\int_0^\infty x^\alpha e^{-x}\, dx = \alpha!$$

when α is a positive integer.

(b) Find the mean and variance of x.

3.39. (a) Show that the area under the *Cauchy density* function $f(x; m) = \{\pi[1 + (x - m)^2]\}^{-1}$, $-\infty < x < \infty$, is one. (b) Graph $f(x; 0)$ and the standard normal density curves using the same axes.

Note. It can be shown that the moments above the first, defined in Exercise 3.45, all turn out to be infinite and the mean is defined in a restricted sense.

3.40. (a) Show that $f(x; \theta) = (1 + \theta)x^\theta$, $\theta > 0$, $0 \leq x \leq 1$, is a density function. (b) Graph $f(x; 1)$, $f(x; 2)$, and $f(x; 3)$. (c) Find the mean and variance of x with density function $(1 + \theta)x^\theta$, $\theta > 0$, $0 \leq x \leq 1$.

3.41. A density function of x is defined by

$$f(x) = \begin{cases} \dfrac{\sin x}{2}, & \text{when } 0 \leq x \leq \pi \\ 0, & \text{otherwise} \end{cases}$$

(a) Find the distribution function and sketch its graph. (b) Compute $P[x < \pi/3]$ and $P[\pi/3 < x < 2\pi/3]$. (c) Determine x_0 such that $P[x < x_0] = 0.05$. (d) Determine x_0 such that $P[(\mu - x_0) < x < (\mu + x_0)] = 0.95$.

3.42. We said after giving Eq. (3.29) that $N(x; \mu, \sigma)$ [or $N(t; 0, 1)$] can be computed by numerical methods. When t is small, the standard normal distribution function $N(t; 0, 1) = N(t)$ may be evaluated by using the series expansion

$$N(t) = \frac{1}{2} + \frac{1}{\sqrt{2\pi}}\left(t - \frac{t^3}{2 \cdot 3} + \frac{1}{2!} \cdot \frac{t^5}{2^2 \cdot 5} - \cdots\right) \qquad (3.37)$$

(a) Evaluate $N(0.1)$ and $N(0.2)$, using the first three terms in the parentheses and compare with the values found in Table II. (b) Prove Eq. (3.37), using the series expansion of $e^{-t^2/2}$.

3.43. For large values of t, Eq. (3.37) converges too slowly to be of practical use. For large values of t the following series may be used:

$$N(t) = 1 - n(t) \cdot \frac{1}{t}\left(1 - \frac{1}{t^2} + \frac{3}{t^4} - \frac{3 \cdot 5}{t^6} + \cdots \pm R\right) \qquad (3.38)$$

where R is numerically less than the last term considered. (a) Evaluate $N(2)$ and $N(3)$, using the first four terms in the parentheses and compare with the values found in Table II. Use R to determine the maximum error in $N(2)$ and $N(3)$ accurate to four decimal places. (b) Prove Eq. (3. 38) by repeatedly integrating by parts.

3.44. A more useful method of computing $N(t)$ for all values of t is given by a continued fraction expression due to Laplace. The expression is

$$N(t) = 1 - n(t) \cdot \left(\cfrac{1}{t + \cfrac{1}{t + \cfrac{2}{t} + \cdots}} \right) \tag{3.39}$$

Use Eq. (3.39) to compute $N(0.2)$ and $N(2)$. Compare these results with those found in Table II.

Note. There are other numerical methods for computing tables of the standard normal distribution function. These methods are particularly useful when high-speed computing machines are available.

3.45. Generally, moments of the first and second order are adequate for most practical and theoretical work. However, occasionally moments of higher order are useful. Thus, the kth *moment about the origin* of a continuous variable x with density function $f(x)$ is defined by

$$\mu'_k = \int_{-\infty}^{\infty} x^k f(x)\, dx \tag{3.40}$$

where k is a positive integer. Sometimes it is desirable to calculate the moments of a continuous function $h(x)$ of x. Thus, the kth *moment of a continuous function* $h(x) = h$ [of x which has density function $f(x)$] is defined by

$$\mu'_{k:h} = \int_{-\infty}^{\infty} h^k(x) f(x)\, dx \tag{3.41}$$

One particular function of importance is $h(x) = x - \mu$. Hence, substituting $h = x - \mu$ in Eq. (3.41) and replacing $\mu'_{k:x-\mu}$ by μ_k, we have the formula which gives the kth *moment about the mean* μ; that is,

$$\mu_k = \int_{-\infty}^{\infty} (x - \mu)^k f(x)\, dx \tag{3.42}$$

(a) Prove that $\mu_2 = \mu'_2 - (\mu'_1)^2$. Note that $\mu_2 = \sigma^2$ and $\mu'_1 = \mu$. (b) Derive a formula for calculating μ_k in terms of $\mu'_k, \mu'_{k-1}, \ldots, \mu'_2, \mu'_1$. (c) Use the formula found in (b) to obtain $\mu_2, \mu_3,$ and μ_4.

3.46. Find the third and fourth moments about the mean, μ_3, and μ_4, for the (a) uniform variable with density function given by Eq. (3.23), (b) triangular variable with density function $f(x; 1, 1)$ given in Exercise 3.19, (c) variable with density function determined in Exercise 3.20, (d) variable with density function determined in Exercise 3.38.

3.47. For many distributions, moments of higher order than the first or second are difficult to determine directly from the definition. An indirect method using what is known as the *moment generating function* (MGF) is often employed. The MGF is also very desirable to have for theoretical considerations. Thus, *the MGF, $M_x(t)$, of a continuous variable x with density function $f(x)$ is defined by*

$$M_x(t) = \int_{-\infty}^{\infty} e^{tx} f(x)\, dx \tag{3.43}$$

The MGF is a function of t only—the subscript x being used to indicate the variable of the distribution. We assume that $f(x)$ is a density function such that $M_x(t)$ converges for some values of t. Expanding e^{tx} in a power series, substituting this in Eq. (3.43), and evaluating gives

$$M_x(t) = 1 + \mu_1' \cdot t + \mu_2' \cdot \frac{t^2}{2!} + \mu_3' \cdot \frac{t^3}{3!} + \cdots \qquad (3.44)$$

The coefficient of $t^k/k!$ in this expansion is the kth moment about the origin. Thus, if the MGF is known or can be found and can be expanded into a power series which is convergent for some values of t, then the kth moment μ_k' can be obtained by inspection. For certain density functions $f(x)$ the MGF is of such a form that it is more convenient to find the kth moment given by

$$\mu_k' = \frac{d^k M}{dt^k}\bigg|_{t=0} \qquad (3.45)$$

where

$$\frac{d^k M}{dt^k}\bigg|_{t=0}$$

denotes the kth derivative of the MGF with respect to t evaluated at $t=0$.

Note. It should be pointed out that the MGF does not always exist. However, for most density functions used in practice, it does exist.

(a) Prove Formula (3.44). (b) Prove Formula (3.45).

3.48. (a) Show that the MGF of the uniform density function defined in Eq. (3.23) is

$$\frac{e^{dt} - e^{ct}}{(d-c)t}$$

(b) Use Eq. (3.44) to show that the kth moment about the origin is

$$\frac{d^{k+1} - c^{k+1}}{d - c} \cdot \frac{1}{k+1}$$

(c)   Use Eq. (3.45) to find the kth moment μ_k'.
(d) Find the MGF and kth moment μ_k' for the unit uniform density function $f(x) = u(x; 0, 1)$.
(e) Find μ_2, μ_3, and μ_4.

3.49. (a) Let u have the standard normal distribution with density function

$$\frac{1}{\sqrt{2\pi}} e^{-u^2/2}$$

Show that the MGF of u, $M_u(t)$, is $e^{t^2/2}$.

Hint. Complete the square in the exponent of the integrand of $M_u(t)$.

(b) Find the moments of the standard normal distribution.

3.50. (a) Find the MGF of the variable x with density function given in Exercise 3.37. (b) Find μ_k' from the MGF.

3.51. (a) Show that the MGF of the variable x with density function

$$f(x) = \frac{x^\alpha e^{-x}}{\alpha!}$$

where α is a positive integer and $x \geq 0$, is $(1 - t)^{-(\alpha+1)}$ (b) Find μ'_k and σ^2. Compare with results of Exercise 3.38(b).

3.52. In order to generate moments of the type given by Eq. (3.41), we generalize the definition of MGF given in Exercise 3.47. Thus, to generate moments of $h(x)$, an arbitrary function of x, we replace x by $h(x)$ in Eq. (3.43), giving

$$M_{h(x)}(t) = \int_{-\infty}^{\infty} e^{h(x) \cdot t} f(x)\, dx \qquad (3.46)$$

If c is an arbitrary constant and $h(x)$ is any function of x for which the MGF exists, then prove that

$$M_{ch(x)}(t) = M_{h(x)}(ct) \qquad (3.47)$$

and

$$M_{h(x)+c}(t) = e^{ct} M_{h(x)}(t) \qquad (3.48)$$

Note. The MGF and the generalized MGF along with the properties of Eqs. (3.47) and (3.48) are very useful in the derivation of numerous theorems. We shall use these relations in Chap. 5 to great advantage.

3.53. (a) Show that the variable x with normal density function $n(x; \mu, \sigma)$ has the MGF given by

$$M_x(t) = e^{\mu t + \sigma^2 \cdot t^2 / 2} \qquad (3.49)$$

Hint. Use Exercise 3.49 and properties (3.47) and (3.48).

(b) Use Eq. (3.49) to find μ'_1, μ'_2, μ'_3.
(c) Find μ_2 and μ_3.

3.4. MULTIVARIATE DISTRIBUTIONS

The definitions and theorems of this chapter can be generalized to more than one variable. Thus, we denote a density function in two variables x and y by $f(x, y)$ and one in n variables $x_1, x_2, \ldots, x_n$ by $f(x_1, x_2, \ldots, x_n)$. In the continuous case, $f(x, y)$ represents a surface in three dimensions and $f(x_1, x_2, \ldots, x_n)$ represents a hypersurface in $n + 1$ dimensions. The volume under the surface $f(x, y)$ and above the rectangular region determined by $a < x < b$ and $c < y < d$ gives the probability that the pair of variables x and y fall in this rectangle. Generalizing Eq. (3.17), we say that the n continuous variables $x_1, x_2, \ldots, x_n$ have a *multivariate (joint) density function* $f(x_1, x_2, \ldots, x_n) = f$ if the following conditions are satisfied

(a) $f(x_1, x_2, \ldots, x_n)$ is a single-valued nonnegative real number for all real values of $x_1, x_2, \ldots, x_n$

(b) $\displaystyle\int_{-\infty}^{\infty} \cdots \int_{-\infty}^{\infty} f(x_1, \ldots, x_n)\, dx_1 \ldots dx_n = 1 \qquad (3.50)$

(c) $\displaystyle\int_{a_n}^{b_n} \cdots \int_{a_1}^{b_1} f(x_1, \ldots, x_n)\, dx_1 \ldots dx_n$
$= P[a_1 < x_1 < b_1, \ldots, a_n < x_n < b_n]$

where $P[a_1 < x_1 < b_1, \ldots, a_n < x_n < b_n]$ denotes the probability with which x_1 falls between any two real values a_1 and b_1, x_2 falls between any two real values a_2 and b_2, $\ldots$, x_n falls between any two real values a_n and b_n simultaneously.

The student will not have any difficulty giving the defining properties for a multivariate density function $f(x_1, x_2, \ldots, x_n)$ for n discrete variables.

The *multivariate (joint) distribution function* $F(x_1, \ldots, x_n)$ of the n continuous variables $x_1, \ldots, x_n$ is given by

$$F(x_1, \ldots, x_n) = \int_{-\infty}^{x_n} \cdots \int_{-\infty}^{x_1} f(t_1, \ldots, t_n) \, dt_1 \ldots dt_n \qquad (3.51)$$

where the dummy variables $t_1, \ldots, t_n$ replace $x_1, \ldots, x_n$ in the integrand, F being a function of the upper limits $x_1, \ldots, x_n$. It follows that

$$0 \leq F(x_1, \ldots, x_n) \leq 1 \qquad (3.52)$$

for all values for which $f(x_1, \ldots, x_n)$ is defined.

The mean μ_i and variance σ_i^2 of the variable x_i $(i = 1, 2, \ldots, n)$ are given by

$$\mu_i = \int_{-\infty}^{\infty} \cdots \int_{-\infty}^{\infty} x_i f(x_1, \ldots, x_i, \ldots, x_n) \, dx_1 \ldots dx_i \ldots dx_n \qquad (3.53)$$

and

$$\sigma_i^2 = \int_{-\infty}^{\infty} \cdots \int_{-\infty}^{\infty} (x_i - \mu_i)^2 f(x_1, \ldots, x_i, \ldots, x_n) \, dx_1 \ldots dx_i \ldots dx_n \qquad (3.54)$$

A new parameter σ_{ij} involving a pair of variables x_i and $x_j (i \neq j)$, called the *covariance of x_i and x_j* $(i, j = 1, 2, \ldots, n)$, is defined by

$$\sigma_{x_i x_j} = \sigma_{ij} = \int_{-\infty}^{\infty} \cdots \int_{-\infty}^{\infty} (x_i - \mu_i)(x_j - \mu_j) f \, dx_1 \ldots dx_n \qquad (3.55)$$

When the variables x and y are uncorrelated in a certain sense which is discussed at length in Chap. 4, we say that the variables are independently distributed. For the moment, we give the following definition:

The variables $x_1, \ldots, x_n$ are *independently distributed* if and only if their joint density function $f(x_1, \ldots, x_n)$ can be expressed as a product of the marginal density functions $f_1(x_1)$, $\ldots, f_n(x_n)$ for all values of $x_1, \ldots, x_n$ for which f is defined. (3.56) The *marginal density function* $f_i(x_i)$ for x_i $(i = 1, \ldots, n)$ is given by integrating $f(x_1, \ldots, x_n)$ with respect to all other variables $(x_1, \ldots, x_{i-1}, x_{i+1}, \ldots, x_n)$ between the limits $-\infty$ and ∞.

It should be observed that the variables $x_1, \ldots, x_n$ in the function $f(x_1,$

$\ldots$, x_n) are always *independent* in the usual mathematical sense, but they are not necessarily *independently distributed*. Usually we say that variables which satisfy Definition (3.56) are *independent in the probability sense*. The illustrations and exercises should clarify the concept.

Now we give brief descriptions of two bivariate distributions. Further properties of joint distributions will be discussed and illustrated in later chapters.

3.4.1. The Bivariate Normal Distribution

The bivariate normal distribution is just as important (or even more important) among bivariate distributions as the univariate normal distribution is among univariate distributions. The *bivariate normal density function* $n(x, y; \mu_x, \mu_y, \sigma_x, \sigma_y, \rho) = n(x, y)$ is given by

$$f(x, y) = n(x, y)$$

$$= \frac{1}{2\pi\sigma_x\sigma_y\sqrt{1 - \rho^2}}\, e^{\frac{-1}{2(1-\rho^2)}\left[\left(\frac{x-\mu_x}{\sigma_x}\right)^2 - 2\rho\left(\frac{x-\mu_x}{\sigma_x}\right)\left(\frac{y-\mu_y}{\sigma_y}\right) + \left(\frac{y-\mu_y}{\sigma_y}\right)^2\right]} \tag{3.57}$$

where the parameters μ_x, μ_y, σ_x^2, σ_y^2 are the means and variances of x and y, and ρ, called the *correlation coefficient*, is defined by

$$\rho = \frac{\sigma_{xy}}{\sigma_x\sigma_y} \tag{3.58}$$

If we defined a new function $n^*(x, y)$ obtained from Eq. (3.57) by replacing μ_x, μ_y, σ_x, σ_y, ρ and $1/(2\pi\sigma_x\sigma_y\sqrt{1 - \rho^2})$ with a, b, c, d, e, and f respectively, it can be shown that, in order for $n^*(x, y)$ to be a density function satisfying Eqs. (3.53), (3.54), (3.55), and (3.58), the constants a, b, c, d, e, f must, in fact, have the specific values given in Eq. (3.57). This will be left as an exercise for the student.

It is informative to study the nature of the density surface. We note that the density function $n(x, y)$ is constant when the exponent of e in Eq. (3.57) is a positive constant k, that is, when

$$\left(\frac{x - \mu_x}{\sigma_x}\right)^2 - 2\rho\left(\frac{x - \mu_x}{\sigma_x}\right)\left(\frac{y - \mu_y}{\sigma_y}\right) + \left(\frac{y - \mu_y}{\sigma_y}\right)^2 = 2(1 - \rho^2)k \tag{3.59}$$

The points satisfying Condition (3.59) lie on an ellipse with center at the point (μ_x, μ_y). The location of the major axis of the ellipse depends on σ_x, σ_y, and ρ. The major axis has positive slope when $\rho > 0$ and negative slope when $\rho < 0$. In case $\sigma_x = \sigma_y$ and $\rho \neq 0$, the slope is either 1 or -1. When $\rho = 0$, the major axis is parallel to the x axis if $\sigma_x > \sigma_y$ and is parallel to the y aixs if $\sigma_x < \sigma_y$. When $\sigma_x = \sigma_y$ and $\rho = 0$, the ellipse reduces to a circle. The ellipse (3.59) is known as a contour ellipse. It indicates the type of scatter of points (x, y) taken from a bivariate normal population.

For a given bivariate normal density function, as k becomes larger, the corresponding contour ellipse becomes larger, but the value of $n(x, y)$ decreases. Figure 3.14 shows the normal density surface for $\rho > 0$ along with two contour curves, indicating the relation of k to the surface. The contour ellipses for the bivariate normal distribution are analogous to the limits $\mu \pm t_\alpha \sigma$ for the univariate normal distribution. Thus, it is possible to find a $k_{2\alpha}$ such that the volume under the surface and inside the contour ellipse is $100(1 - 2\alpha)$ per cent of the total volume.

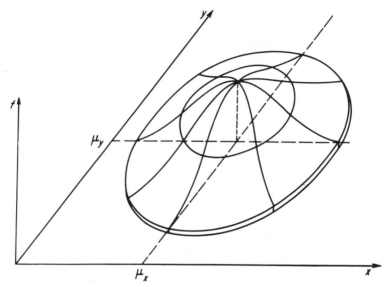

Fig. 3.14 Bivariate Normal Density Surface with Contour Ellipses

Letting

$$u = \frac{x - \mu_x}{\sigma_x} \quad \text{and} \quad v = \frac{y - \mu_y}{\sigma_y}$$

we find that Eq. (3.57) reduces to

$$n(u, v) = \frac{1}{2\pi\sqrt{1 - \rho^2}}\, e^{-\frac{1}{2(1-\rho^2)}(u^2 - 2\rho uv + v^2)} \tag{3.60}$$

since $dx\, dy = \sigma_x \sigma_y\, du\, dv$. This is called the *standardized bivariate normal density function*. It involves one variable parameter ρ and has been extensively tabulated by Owen [17].

Observe that when $\rho = 0$ the standard normal density function $f(u, v)$

factors into the product $f(u) \cdot f(v)$ of two standard density functions; that is

$$\frac{1}{2\pi} e^{-\frac{1}{2}(u^2 + v^2)} = \frac{1}{\sqrt{2\pi}} e^{-u^2/2} \cdot \frac{1}{\sqrt{2\pi}} e^{-v^2/2}$$

The student should be cautioned at this point. Even though the variables in a joint normal distribution are independent when $\rho = 0$ this is not generally true. However, if variables are independently distributed, it necessarily follows that $\rho = 0$ (see Exercise 3.65). The distribution of u for any fixed v is the same, no matter what value of v is selected; that is, the distribution of u is independent of v. Likewise, we see that the distribution of v is independent of u. Thus, it is natural to say that u and v are independently distributed.

3.4.2. A Discrete Bivariate Distribution

We discuss a discrete distribution in order to show how the properties in Sect. 3.4 for a continuous variable may be used as guides in the discrete case. Consider the following discrete bivariate function

$$f(x, y) = \begin{cases} f(1, 1) = 0.1 \\ f(1, 2) = 0.1 \\ f(1, 3) = 0.2 \\ f(2, 1) = 0.1 \\ f(2, 2) = 0.2 \\ f(2, 3) = 0.3 \\ \quad 0 \text{ for all other pairs of values of } x \text{ and } y. \end{cases} \tag{3.61}$$

First, we note that this is a density function, since (1) $f(x, y)$ is single-valued and nonnegative, (2)

$$\sum_{x=1}^{2} \sum_{y=1}^{3} f(x, y) = 1$$

and (3) the sum of any subset of these six functional values gives the probability of obtaining the corresponding pairs (x, y). For example

$$\sum_{x=1}^{1} \sum_{y=2}^{3} f(x, y) = f(1, 2) + f(1, 3) = 0.1 + 0.2 = 0.3$$

which is the probability of obtaining the pair $(1, 2)$ or $(1, 3)$.

We may think of the bivariate distribution function $F(x_r, y_s)$ $(r = 1, \ldots, k; s = 1, \ldots, l; x_r < x_{r+1}; y_s < y_{s+1})$ for the six discrete points of Eq. (3.61) as given by

$$F(x_r, y_s) = \begin{cases} F(x_1 = 1, y_1 = 1) = f(1, 1) & = 0.1 \\ F(1, 2) & = f(1, 1) + f(1, 2) & = 0.2 \\ F(1, 3) & = f(1, 1) + f(1, 2) + f(1, 3) & = 0.4 \\ F(2, 1) & = f(1, 1) + f(2, 1) & = 0.2 \\ F(2, 2) & = f(1, 1) + f(1, 2) + f(2, 1) + f(2, 2) = 0.5 \\ F(2, 3) & = 1.0 \end{cases}$$

$$(3.62)$$

However, in mathematical considerations it is customary to define the distribution function $F(x, y)$ for all pairs of real values of x and y. Thus, in general, we define the *bivariate distribution function* $F(x, y)$ as follows

$$F(x, y) = \begin{cases} 0, & \text{when } x < x_1 \quad \text{or} \quad y < y_1 \\ F(x_r, y_s), & \text{when } x_r \leq x < x_{r+1} \text{ and } y_s \leq y < y_{s+1}, \text{ it} \\ & \text{being understood } x_{k+1} = \infty \text{ and } y_{l+1} = \infty \\ & \text{and that } x_k \text{ and } y_l \text{ do not appear together} \\ 1, & \text{when } x \geq x_k \quad \text{and} \quad y \geq y_l \end{cases} \quad (3.63)$$

For example, using the density function (3.61), we obtain

$$F(x, y) = \begin{cases} 0, & \text{when } x < 1 \quad \text{or} \quad y < 1 \\ F(1, 1), & \text{when } 1 \leq x < 2 \quad \text{and} \quad 1 \leq y < 2 \\ F(1, 3), & \text{when } 1 \leq x < 2 \quad \text{and} \quad 3 \leq y < \infty \\ \text{etc.} \\ 1, & \text{when } x \geq 2 \quad \text{and} \quad y \geq 3 \end{cases} \quad (3.64)$$

The means, variances, covariance, and correlation coefficient for the density function (3.61) are obtained as follows

$$\mu_x = \sum_{x=1}^{2} \sum_{y=1}^{3} x f(x, y)$$

$$= 1f(1, 1) + 1f(1, 2) + 1f(1, 3) + 2f(2, 1) + 2f(2, 2) + 2f(2, 3)$$

$$= 1.6$$

$$\mu_y = \cdots = 2.3$$

$$\sigma_x^2 = \sum_{x=1}^{2} \sum_{y=1}^{3} (x - 1.6)^2 f(x, y) = \cdots = 0.240$$

$$\sigma_y^2 = \cdots = 0.610$$

$$\sigma_{xy} = \sum_{x=1}^{2} \sum_{y=1}^{3} (x - 1.6)(y - 2.3)f(x, y)$$

$$= (-0.6)(-1.3)f(1, 1) + (-0.6)(-0.3)f(1, 2) + (-0.6)(0.7)f(1, 3)$$
$$+ (0.4)(-1.3)f(2, 1) + (0.4)(-0.3)f(2, 2) + (0.4)(0.7)f(2, 3)$$
$$= 0.020$$

and

$$\rho = \frac{0.020}{\sqrt{(0.240)(0.610)}} = 0.052$$

Thus, it follows that the variables x and y are not independently distributed since $\rho \neq 0$.

3.4.3. Exercises

The first exercises are intended to lead to a better understanding of the ideas presented in Sect. 3.4; the last refer to all topics of this chapter, as well as to certain related topics. It is to be understood that $f(x, y) = 0$ for all pairs of x and y values not mentioned in the exercises.

3.54. Let the joint distribution of the two variables $0 \leq x \leq 1$ and $0 \leq y \leq 1$ have the joint density function $f(x, y)$ given by $f(x, y) = 1$. (a) Find the joint distribution function $F(x, y)$. (b) Find the means, variances, and covariance. (c) Calculate $F(1, 1)$; $P[x < 1]$. (d) Calculate $P[(x + y) < 1]$; $P[2x > y]$; $P[(x^2 + y^2) < 4]$.

3.55. Let $f(x, y) = ce^{-(x+y)}$, $x \geq, y \geq 0$. (a) Find c so that $f(x, y)$ is a density function. (b) Find $F(x, y)$. (c) Calculate $F(1, 1)$; $P[x < 1]$. (d) Evaluate $P[(x + y) < 1]$; $P[x > y]$. (e) Find the means, variances, and covariance. Note that x and y are independently distributed, since $f(x, y) = f(x) \cdot f(y)$, where $f(x) = e^{-x}$ and $f(y) = e^{-y}$ are marginal density functions.

3.56. Let $f(x, y) = c, 0 \leq x \leq 1, 0 \leq y \leq x$. (a) Find c so that $f(x, y)$ is a density function. (b) Determine the means and variances. (c) Find $F(x, y)$.

3.57. Let $f(x, y) = 10,000, 0.49 < x < 0.50, 0.50 < y < 0.51$. Suppose we interpret x to be the diameter in inches of shafts made by one machine and y to be the inside diameter in inches of bushings made by another machine. If we assume that the bushing fits the shaft satisfactorily when the difference in their diameters is between 0.0016 and 0.0064 in., what proportion of the shafts and bushings will fit?

3.58. Give an illustration in which the joint density function in Exercise 3.55 might be used as a model.

3.59. (a) Find c so that $f(x, y) = cxy$, $0 \leq x \leq 1$, $0 \leq y \leq x$, is a joint density function. (b) Find $F(x, y)$. (c) Compute the means and variances. (d) Calculate $F(1, 1)$; $P[x > \frac{1}{2}, y < \frac{1}{2}]$; $P[x > \frac{1}{2}]$.

3.60. Let the joint density function of the two variables $x \geq 0$ and $y \geq 0$ be given by $f(x, y) = cxye^{-(x+y)}$. (a) Find the value of c. (b) Calculate $P[x < 1, y < 1]$, $P[x < 1]$. (c) Find $F(x, y)$ if possible.

3.61. Consider the following discrete bivariate distribution

$$f(x, y) = \begin{cases} 2c, & \text{when } x = 1 \quad \text{and} \quad y = 1, 2, 3 \\ c, & \text{when } x = 2 \quad \text{and} \quad y = 1, 3 \\ 4c, & \text{when } x = 2 \quad \text{and} \quad y = 2 \end{cases}$$

(a) Find c so that $f(x, y)$ is a density function. (b) Compute the means, variances, and correlation coefficient.

Note. Marginal density functions $f(x)$ and $f(y)$ do not exist such that $f(x, y) = f(x) \cdot f(y)$, even though $\rho = 0$.

(c) Determine $F(x, y)$ and find $F(1.5, 2.5)$.

3.62. Consider the following discrete joint distribution

$$f(1, 1, 1) = f(1, 2, 1) = f(1, 2, 2) = c,$$
$$f(1, 1, 2) = f(2, 1, 2) = f(2, 2, 2) = 2c,$$
$$f(2, 1, 1) = f(2, 2, 1) = 3c$$

(a) Find c so that $f(x_1, x_2, x_3)$ is a density function. (b) Compute the means and variances. (c) Compute the covariances and correlation coefficients. (d) Find $F(1, 1, 2)$; $F(1, 2, 1)$; $F(2, 2, 1)$; $F(1, 1.5, 2)$.

3.63. Let

$$f(x, y, z) = ce^{-(x+y+z)}, \qquad x \geq 0, \quad y \geq 0, \quad z \geq 0.$$

(a) Find c so that $f(x, y, z)$ is a density function. (b) Find $F(x, y, z)$. (c) Find means and variances. (d) Determine the covariances.

3.64. Assume that the heights x and weights y of adult human males are distributed with joint normal density function $n(x, y; 68, 160, 3, 20, 0.5)$. (a) Write the expression for the joint density function. (b) Determine the function of y obtained by letting $x = 68$. Is this a normal function? Is it a normal density function? (c) Find the major axis of the contour ellipse.

3.65. If x and y are independently distributed with density function $f(x, y)$, prove that $\rho = 0$.

3.66. The definitions and properties of moments and moment generating functions for discrete variables are similar to those given for the continuous case in Exercises 3.45, 3.47, and 3.52. (a) Show that the MGF of a variable having the Poisson distribution with mean μ is given by $M_x(t) = e^{\mu(e^t - 1)}$. (b) Using Eq. (3.45), show that the mean and variance are both equal to μ.

3.67. A function which provides a generalization of the factorial and which is used extensively in distribution theory is given by the definite integral

$$\Gamma(\alpha) = \int_0^\infty x^{a-1} e^{-x} \, dx, \qquad \alpha > 0 \qquad (3.65)$$

This is called the *gamma function*. (a) Show that $\Gamma(\alpha + 1) = \alpha\Gamma(\alpha)$.

Hint. Use integration by parts.

(b) If k is a positive integer less than α, show that $\Gamma(\alpha + 1) = \alpha(\alpha - 1)$ $\ldots (\alpha - k) \Gamma(\alpha - k)$. Further, if $\alpha = n$ is an integer, it follows that $\Gamma(n + 1) = n!$ (c) In statistical application α is usually an integer or a multiple of one-half. Thus, if we knew $\Gamma(\frac{1}{2})$, we could compute $\Gamma(\alpha + 1)$ for almost any value needed. Show that $\Gamma(\frac{1}{2}) = \sqrt{\pi}$. (d) Find $\Gamma(1), \Gamma(\frac{3}{2})$, $\Gamma(2), \Gamma(\frac{5}{2}), \Gamma(3)$. (e) Computing $\Gamma(\alpha + 1)$ by using the above relations is impractical for large values of α. Thus, $\log \Gamma(n + 1)$ has been extensively tabulated. For very large n, the following approximation, known as *Sterling's formula*, can be used: $\Gamma(\alpha + 1) = \alpha! \doteq \sqrt{2\pi} \, \alpha^{\alpha + (1/2)} e^{-\alpha}$. The approximation gets better as α increases. Use logarithms and Stirling's formula to find $\Gamma(101)$ and $\Gamma(\frac{99}{2})$.

3.68. The density function $f(x; \alpha, \beta)$ for the *gamma distribution* is given by

$$f(x; \alpha, \beta) = \frac{1}{\alpha! \, \beta^{\alpha+1}} x^\alpha e^{-x/\beta}, \qquad x > 0, \quad \beta > 0, \quad \alpha > -1 \qquad (3.66)$$

Changing β changes the scale of the axes. Thus, letting $y = x/\beta$, we obtain

$$f(y; \alpha) = \frac{1}{\alpha!} y^\alpha e^{-y}, \qquad y > 0, \quad \alpha > -1 \qquad (3.66a)$$

which may be used as the gamma density function. (a) Prove that $f(x; \alpha, \beta)$ is a density function. (b) Show that the MGF of the variable x with density function (3.66) is given by $M_x(t) = (1 - \beta t)^{-(\alpha+1)}$, provided $t < 1/\beta$. (c) Show that $\mu = (\alpha + 1)\beta$ and $\sigma^2 = (\alpha + 1)\beta^2$ (d) Show that the distribution function $F(x; \alpha)$ is given by

$$F(x; \alpha) = \frac{\Gamma(x; \alpha)}{\Gamma(\alpha)}$$

where

$$\Gamma(x; \alpha) = \int_0^x y^{\alpha+1} e^{-y} \, dy$$

The function $F(x; \alpha) \equiv \Gamma(x; \alpha)$ is called the *incomplete gamma function* and has been extensively tabulated by Karl Pearson [22]. We shall show later (in Chap. 7) that the very useful chi-square distribution is an incomplete gamma distribution.

3.69. A function which is important in distribution theory is defined by the definite integral

$$B(\alpha, \beta) = \int_0^1 x^{\alpha-1} (1 - x)^{\beta-1} \, dx \qquad (3.67)$$

where α and β are constants greater than minus one. This is called the *beta function*. (a) Prove that $B(\alpha, \beta) = B(\beta, \alpha)$. (b) Prove that

$$B(\alpha, \beta) = \frac{\Gamma(\alpha)\Gamma(\beta)}{\Gamma(\alpha + \beta)}$$

Hint. Let $x = y^2$ in Eq. (3.65), write $\Gamma(\alpha) \cdot \Gamma(\beta)$ as a double integral, and then transform to polar co-ordinates, following the pattern of evaluating A^2 in Sect. 3.2.2.

3.70. The density function $f(x; \alpha, \beta)$ for the *beta distribution* is given by

$$f(x; \alpha, \beta) = \frac{1}{B(\alpha + 1, \beta + 1)} x^\alpha (1 - x)^\beta$$

$$0 < x < 1, \quad \alpha > -1, \quad \beta > -1$$

(3.68)

(a) Prove that $f(x; \alpha, \beta)$ is a density function. (b) The MGF of the beta distribution does not have a simple form, but the moments μ_k can be found directly. Show that

$$\mu_k' = \frac{\Gamma(\alpha + \beta)\Gamma(\alpha + k + 1)}{\Gamma(\alpha + \beta + k + 2)\Gamma(\alpha + 1)}$$

Thus, determine μ and σ^2. (c) Show that the distribution function $F(x; \alpha, \beta)$ is given by

$$F(x; \alpha, \beta) = \frac{B(x; \alpha, \beta)}{B(\alpha, \beta)}$$

where

$$B(x; \alpha, \beta) = \int_0^x t^\alpha (1 - t)^\alpha \, dt, \quad 0 < x < 1$$

The function $F(x; \alpha, \beta) \equiv I(x; \alpha, \beta)$ is called the *incomplete beta function* and has also been extensively tabulated by Karl Pearson [21]. We shall show later (in Chap. 9) that the very useful F distribution is an incomplete beta distribution.

3.71. It is desirable to have a reasonably simple mathematical expression which can generate models for most of the distributions found in applied statistics. Karl Pearson [18, 19, 20] has proposed the differential equation

$$\frac{df}{dx} = \frac{(x + a)f}{b + cx + dx^2}$$

(3.69)

for this purpose. This equation does have a large family of solutions which includes many distributions found in practice. By letting the constants a, b, c, d assume different relations to each other, Pearson classified the solutions of Eq. (3.69) into 12 families of curves, those of one family being called Type I curves, those of a second family being called Type II curves, etc. The beta distributions represent Type I and Type II (when $\alpha = \beta$) curves, the gamma distributions are Type III curves, and the normal distributions are Type VII curves. See Craig [8], Elderton [10], and Kendall [14] for complete accounts of the Pearson types. (a) Show that when $a = -\mu$, $b = \sigma^2$, and $c = d = 0$, the solution of Eq. (3.69) is $f = n(x; \mu, \sigma)$. (b) Determine the solution of Eq. (3.69) when $a = b = d = 0$. (c) Determine the solution of Eq. (3.69) when $a = b = c = 0$.

Note 1. It can be shown that the constants a, b, c, d in Eq. (3.69) can be expressed in terms of the first four moments. Thus, only the first four moments are required in order to specify any density function satisfying Eq. (3.69). Thus, an approximation to any density function in the family may be obtained by computing estimates of the first four moments from a sample of observations.

Note 2. Other methods for expressing a large family of distributions in simple form have been presented. Among those the best-known method is called the *Gram-Charlier series*, which states that under fairly general conditions a wide class of density functions $f(x)$ may be expressed as an infinite series of terms made up of the normal density function and its derivations. That is, $f(x) = a_0 n_0(t) + a_1 n_1(t) + a_2 n_2(t) + \ldots$, where a_i are constants, $t = (x - \mu)/\sigma$, and $n_i(t)$ is the ith derivative of the standard normal density $n(t; 0, 1) = n_0(t)$. For more on the Gram-Charlier series see Refs. [2, 5, 6, 7, 8, 15, 16].

REFERENCES

1. Anderson, R. L. and T. A. Bancroft, *Statistical Therory in Research*. New York: McGraw-Hill, Inc., 1952, Chaps. 3, 4, and 5.
2. Aroian, L. A., "The Type B Gram-Charlier Series," *Annals of Mathematical Statistics*, Vol. **8** (1937), p. 183.
3. Bennett, C. A. and N. L. Franklin, *Statistical Analysis in Chemistry and the Chemical Industry*. New York: John Wiley & Sons, Inc., 1954, Chap. 4.
4. Burington, R. S. and D. C. May, *Handbook of Probability and Statistics with Tables*. Sandusky, Ohio: Handbook Publishers, Inc., 1953, Tables VII, VIII, and IX.
5. Charlier, C. V. L., *Researches into the Theory of Probability*. Lund, 1906.
6. ———, "A New Form of the Frequency Function," *Meddelande from Lunds Astronomiska Observat*, Series II, No. 51 (1928).
7. ———, *Applications à l'astronomie* (one of the series in Borel's *Traite du Calcul des Probabilities*, Gauthier-Villars, Paris), 1931.
8. Craig, C. C., "A New Exposition and Chart for the Pearson System of Frequency Curves," *Annals of Mathematical Statistics*, Vol. **7** (1936), pp. 16–28.
9. Dixon, W. J. and F. J. Massey, *An Introduction to Statistical Analysis*. New York: McGraw-Hill, Inc., 1957, Chap. 5.
10. Elderton, W. P., *Frequency Curves and Correlation*. London: Cambridge University Press, 1938.
11. *Tables of Probability Functions, Vol. II.*, Federal Works Agency, Mathematical Tables Project (sponsored by National Bureau of Standards), New York, 1942.
12. *Handbook of Chemistry and Physics*, Chemical Rubber Publishing Co., Cleveland, Ohio, 1959.
13. Hoel, P. G., *Introduction to Mathematical Statistics*, 2nd ed. New York: John Wiley & Sons, Inc., 1954, Chaps. 2 and 5.
14. Kendall, M. G. and A. Stuart, *The Advanced Theory of Statistics*, Vol. **1**. New York: Hafner Publishing Co., Inc., 1958, Chaps. 1, 3, 5, and 6.
15. Kenney, J. F., *Mathematics of Statistics*, Part 2. New York: D. Van Nostrand Co., Inc., 1947, Chaps. 2, 3.

16. Mood, A. M., *Introduction to the Theory of Statistics*. New York: McGraw-Hill, Inc., 1950, Chaps. 3, 4, 5, and 6.
17. Owen, D. B., *The Bivariate Normal Probability Distribution*, Office of Technical Services, Department of Commerce, Washington, D. C., 1957.
18. Pearson, K., "Contributions to the Mathematical Theory of Evolution," *Philosophical Transactions of the Royal Society of London*, Series A, Vol. **185** (1894), pp. 71–110.
19. Pearson, K., "Contributions to the Mathematical Theory of Evolution, II: Skew Variation in Homogeneous Material," *Philosophical Transactions of the Royal Society of London*, Series A, Vol. **186** (1895), pp. 343–414.
20. Pearson, K., "On the Systematic Fitting of Curves to Observations and Measurements," *Biometrika*, Vol. **1** (1902), pp. 265–303; Vol. **2** (1902), pp. 1–23.
21. Pearson, K., *Tables of the Incomplete Beta Function*. London: Cambridge University Press, 1932.
22. Pearson, K., *Tables of the Incomplete Gamma Function*. London: Cambridge University Press, 1922.
23. Wilks, S. S., *Mathematical Statistics*. Princeton, N. J.: Princeton University Press, 1943, Chaps. 2 and 3.

4

PROBABILITY

The concept of probability is treated from the relative-frequency point of view. Some rules for computing probabilities are discussed.

4.1. INTRODUCTION

In the earlier chapters we have been concerned with statistical distributions from a descriptive point of view. We have indicated that distributions found in practice may be represented by simple mathematical forms which can be characterized by a few parameters. So long as these parameters are known and the nature of the distribution is known, the mathematical expression of the distribution can be explicitly given. However, these parameters are seldom known, and we are faced with the question, "How can we determine which mathematical expression is correct (or most appropriate) when only a sample of the population is available?" Clearly, we would not expect, on the basis of the information obtained from a sample, to make any statement concerning the nature of the distribution of the population with the same certainty as we could if we had the whole population. But it is possible to make statements with less certainty in terms of probability.

4.2. DEFINITIONS OF PROBABILITY

In Chap. 3 we introduced formal properties of "mathematical probability" (formerly called "probability") of *real numbers* assigned to events, outcomes of experiments. There we did not discuss probabilities of *events* or the problem of *assigning* a meaningful probability. For example, in the axiomatic definition of probability given in Eq. (3.2), each of the assignments

101

Table 4.1
Assigned Probabilities

Case	Number on a Balanced Die					
	1	2	3	4	5	6
I	.9	.09	.009	.0009	.00009	.00001
II	.95	.01	.01	.01	.01	.01
III	.1	.1	.1	.1	.3	.3
IV	$\frac{1}{6}$	$\frac{1}{6}$	$\frac{1}{6}$	$\frac{1}{6}$	$\frac{1}{6}$	$\frac{1}{6}$

(and millions more) of probability in Table 4.1 to a number on the face of a balanced die is satisfactory. However, anyone who has to use the results of experimentation to take action knows that the four cases are not equally satisfactory. Further, most investigators would prefer to use results of previous experiments along with certain assumptions to estimate and assign probabilities to events. Thus, in this chapter we give an operational definition of probability, as contrasted with mathematical probability of Chap. 3, which is very useful in spite of its obvious shortcomings.

It can be shown that the basic properties suggested by and derived from the frequency approach to probability are like those obtained from the axiomatic approach to mathematical probability. Furthermore, such an approach allows for an interpretation of probability which is familiar to the experimenter, and, at the same time, satisfies the properties of mathematical probability. From this point on we shall think of the relative-frequency concept of *probability of events* as being a special meaningful interpretation of mathematical probability assigned to real numbers associated with events.

We discuss the relative-frequency concept of probability in terms of *events*, understanding that the x used in Chap. 3 is a real-valued function defined over all the events of an experiment. Hence, the relative-frequency approach to probability is easily expressed in terms of x, that is, in terms of numbers assigned to events. An investigator in a particular experimental situation usually knows what number to assign to an event. Hence, he would have no difficulty in changing probability statements about events to probability statements about real numbers x.

It is not easy to give the term probability a precise meaning which would satisfy everyone. Some would want the term to be characterized so as to include all the ideas generally associated with the word probability. Most statisticians are concerned with the term as it is used in a particular sense. For example, they want to know how sample means or sample standard deviations vary from sample to sample, if it is assumed that samples of the same size are drawn from the same population. These facts, and others, lead to disagreement as to how best to characterize the term. We shall take the somewhat restricted point of view of many statisticians and think of probability of an event as the "limit of the relative frequency with which it occurs."

In order to arrive at a clear understanding of what is meant by "limit of the relative frequency," we first consider two illustrations and a very restricted intermediate interpretation of probability.

Example 4.1. Consider the simple experiment of tossing a coin. We assume that the only two meaningful outcomes are the occurrence of a head H or the occurrence of a tail T, and that they are *mutually exclusive,* since both sides of the coin cannot turn up simultaneously. Of course, after a toss it is conceivable that the coin might stand on edge or vanish, but we speak of the occurrence of a head or a tail as the *only two possible outcomes.* Furthermore, we assume the coin to be well balanced and to be tossed fairly. Thus, on a given toss we would say that there is nothing to favor the occurrence of a head more or less than the occurrence of a tail (that is, the two possible alternatives are *equally likely*) and that the chance of either occurring is $\frac{1}{2}$. If obtaining a head is called event E, we say that the chance of event E is $\frac{1}{2}$.

Example 4.2. Consider the chance of drawing a certain type of card from a pack of 52 playing cards. We assume that we have the same chance of drawing any one of the 52 cards and that we cannot draw more than one card in one draw; that is, the 52 possible alternatives are equally likely and mutually exclusive. Thus, if the pack is well shuffled and the selection of a card is random, then the chance of getting one particular card is $\frac{1}{52}$. The chance of getting an ace in one draw is $\frac{4}{52} = \frac{1}{13}$, since there are four such cards in a pack of 52. Furthermore, the chance of getting a heart in one draw is $\frac{13}{52} = \frac{1}{4}$. If obtaining a heart in one draw is called event E, then the chance of event E is $\frac{1}{4}$.

These two illustrations lead us to the following *classical concept of probability of events* for simple experiments:

> If an experiment can result in N mutually exclusive and equally likely possible outcomes, n of which correspond to the occurrence of an event E, then the *probability of the event E* is n/N, or briefly $P[E] = n/N$.

Note that, even though the term "equally likely" is a primitive or indefinable term applied to an experiment, this approach to probability may be criticized as being circular. In such case, the reader may prefer to think of "equally likely" as being synonymous with the term "equally probable," and, hence, think of *probability of an event E* as being expressed as the sum of n equal probabilities $1/N$ *assigned* to a set of N mutually exclusive and exhaustive experimental outcomes.

The classical interpretation certainly agrees with the usual concept of probability as applied in simple experiments, but it leaves something to be desired in many more complex experiments (or investigations). For example, this approach does not apply when it is impossible to count the number of equally likely outcomes, such as the number of cars on the Merritt Parkway tomorrow or the number of people who will die from cancer in the twentieth century. Thus, the classical concept is inadequate in determining the probability of a light bulb's being defective or of an unborn baby's being female or of an American's living more than one hundred years. Furthermore, it may be that the outcomes in an experiment are not equally likely. For example, a coin may be biased in favor of "heads" or a die may be "loaded." Hence, we sense the need for another practical concept of probability—one that does not depend on a complete *a priori* analysis. This suggests an empirical approach.

Suppose that in a sequence of n trials of a specified experiment an event E occurs n_1 times. The ratio n_1/n is called the relative frequency of the event E and is denoted by $R[E]$. *Then the probability $P[E]$ of the event E is the limit approached by $R[E]$ as n increases indefinitely*, it being assumed a limit value exists. We call this the *relative-frequency concept of probability*. Such a definition is an *operational definition*, a type often used in science and engineering, and is useful because it allows empirical estimates of probability to be obtained from relative frequencies. (It should be noted for this interpretation of probability that an operational limit is not the same as the usual mathematical limit. The following three examples and other material developed later should make the distinction increasingly clear.)

It is clear that the relative-frequency concept of probability is based on observational evidence, and that the classical concept is not. For example, according to the classical concept we say that the probability of obtaining a head in a toss of an ideal coin is $\frac{1}{2}$ without even tossing the coin, but according to the relative-frequency approach we would need to toss the coin many times before arriving at a satisfactory estimate of the probability. With this fact in mind, we use the results obtained from several trials to take a closer look at the meaning of probability as given in the relative-frequency interpretation.

Example 4.3. A quarter, believed to be ideal, was tossed 200 times with the following results (reading across rows):

```
H H T  H H H H T  H H T  H H H H H T  T  H T  H T  H H T  H H T  H T
T  T  T  H H T  H T  H H H T  T  H T  T  H H H H H H H H T  H T  H T  H
T  T  T  T  T  H H T  H T  T  T  T  H H T  T  H T  H H T  T  T  H H H H H H
H H T  H T  H T  H H H H T  H T  T  T  T  T  H T  H H T  H H H H H T  T  H
T  T  H T  H H H T  T  T  H T  H H T  H H H T  H H T  H T  T  T  T  T  T  T
H T  T  H H T  T  T  H T  H H H T  T  T  H H H T  T  H T  T  H T  T  T  T  H
H T  T  H H H H T  T  T  T  H H H T  H T  T  T  T  H
```

We will not say whether 200 tosses represents "many trials." Perhaps it does not. If 200 tosses do represent many trials, then, according to the relative-frequency concept, the probability of event E, occurrence of a head on a single toss, is nearly $\frac{104}{200} = 0.52$. To illustrate the effect that increasing n has on the relative frequencies $R[E]$, we compute the ratio of heads after the first 20 tosses, the first 40 tosses, etc. to obtain $R[E] = \frac{14}{20} = 0.70$, 0.63, 0.63, 0.56, 0.59, 0.57, 0.57, 0.54, 0.53, 0.52. Thus, we see that as the number of trials increases the corresponding ratios fluctuate less and less and usually get closer and closer to 0.50, which is the assumed correct probability.

Note. We computed ratios after each 20 new trials in order to see the nature of the fluctuations in the sequence without having to look through so many terms. Selecting ratios at regular intervals could lead to wrong conclusions, for it might happen that we would get a sequence of large (or small or unusual) ratios.

There is another sense in which we may think of the relative-frequency interpretation and which gets us closer to the way the statistician usually thinks of probability.

Example 4.4. Suppose we toss a quarter and record the number of heads (a) every 20 tosses for 15 times, (b) every 200 tosses for 15 times, and (c) every 2000 tosses for 15 times and obtain the results shown in Table 4.2.

Table 4.2
Number of Heads per Group

Group	Number of Tosses		
	20	200	2000
1	14	104	1010
2	11	91	990
3	13	99	1012
4	7	96	986
5	14	99	991
6	10	108	988
7	11	101	1004
8	6	101	1002
9	9	101	976
10	9	110	1018
11	9	108	1021
12	6	103	1009
13	6	98	1000
14	10	101	998
15	13	109	988

The percentages of heads are shown in Table 4.3. In the groups with 20 tosses the percentage of heads varies (at random) between 30 and 70; in the groups with 200 tosses it varies between 45.5 and 55.0; and in the groups with

Table 4.3
Percentage of Heads per Group

Group	Number of Tosses		
	20	200	2000
1	70	52.0	50.50
2	55	45.5	49.50
3	65	49.5	50.60
4	35	48.0	49.30
5	70	49.5	49.55
6	50	54.0	49.40
7	55	50.5	50.20
8	30	50.5	50.10
9	45	50.5	48.80
10	45	55.0	50.90
11	45	54.0	51.05
12	30	51.5	50.45
13	30	49.0	50.00
14	50	50.5	49.90
15	65	54.5	49.40

2000 tosses between 48.80 and 51.05. In Fig. 4.1 these percentages are plotted as dot frequency diagrams. The third diagram suggests that the percentages cluster about a fixed value of approximately 50 percent, and that the variation about this value gets smaller when the percentages are calculated from larger groups. The fixed value divided by 100 is, by definition, the probability of obtaining a head on a single toss of the coin.

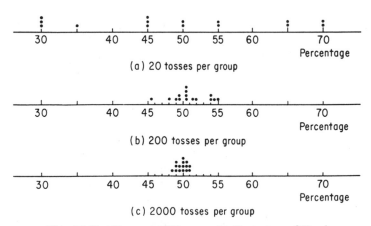

Fig. 4.1 Dot Frequency Diagrams for Percentage of Heads

Note. It may take many more than 15 groups before the fixed value can be approximated with sufficient accuracy. Fewer groups are required as the number of tosses per group increases.

We have observed with the coin-tossing experiment that relative frequencies may be used to estimate probabilities from observational results. The relative-frequency concept of probability may also be used with more realistic data. For example, if we found that 2540 peas out of 10,000 of a certain strain were pale green or yellow, we would estimate the probability as $0.254 \doteq 0.25$ that peas of this strain are pale green or yellow; if a worker examined 8250 machine parts during the day and found 174 defective, we would estimate the probability as 0.021 that a part which the factory produces is defective.

In each of the above illustrations the variable is discrete. Now we consider an example in which the variable is continuous.

Example 4.5. Assume that we have a roulette wheel with a pointer which is equally likely to stop at any position (point) of the wheel. (Other circular objects, such as the face of a clock with the minute hand as pointer, could be used.) Using the classical concept of probability, we would say that the *probability that the pointer falls within a given interval* on the scale of the wheel is equal to the ratio of the length of the interval to the length of the whole scale. Now if we think of the length of the interval converging to zero, then, according to the relative-frequency concept, *the probability of the pointer's stopping at a given point selected in advance of the experiment is zero.* Thus, we see that even though the probability is zero that the pointer will stop at any single point, selected in advance, the event if regarded as accurately definable is not an impossible event.

This example indicates that the classical concept of probability can be applied to a certain type of problem with a continuous variable, particularly when each possible outcome is assumed to be equally likely. However, there are some difficulties in relating empirical results to true probabilities. For we have that old problem, common to all mathematical theories of physical phenomena, of passing through an infinite number of steps. In addition to this, we have another problem peculiar to probability theory. In the relative-frequency interpretation, we do not say

$$\lim_{n \to \infty} R[E] = P[E]$$

in the mathematical sense of limit; that is, we do not say that for an arbitrarily small positive real number ϵ there exists a large n, say n', such that $|R[E] - P[E]| < \epsilon$ for every $n > n'$. Instead, we say, operationally, that "the probability $P[E]$ is the limit approached by $R[E]$ as n increases indefinitely." Such a statement allows an element of *uncertainty* and thus, in a sense, is defined within the framework of the concept itself. Thus, even though the usual desirable properties of probability hold for the relative-frequency interpretation, we cannot rigorously derive them from such a definition.

The relatively simple interpretations of probability given in this section, even though they have some obvious limitations, are adequate for most practical problems. In situations in which these interpretations are not broad enough to give probability of an event, we simply assume that it is some fixed but unknown value which obeys the usual properties of probability given in Sect. 4.3 (and Chap. 3).

There are other ways of interpreting probability, but ours has the advantage of being simple. For other more complete discussions of probability see the following books devoted to its theory [1, 2, 3, 4, 5, 6, 8, 9, 10, 11, 14, 16, 17]. Halmos [7] gives an elementary discussion of a mathematical concept of probability, and Carnap [2] lists (on 15 pages) a selected bibliography.

4.2.1. Exercises

4.1. What are the extreme values of the probability of an event? Give an illustration of each.

4.2. A balanced die is fairly tossed. Let the event A denote the occurrence of an odd number, B denote the occurrence of an even number, C denote the occurrence of a number greater than 3, D denote the occurrence of a number less than 3, and E denote the occurrence of the number 3. (a) Find $P[A]$, $P[B]$, $P[D]$. (b) Which of these events are equally likely? (c) Which of these events are mutually exclusive? (d) Find the probability of event C or event D. (e) Find the probability of event B or event C.

4.3. A card is drawn from a well-shuffled deck of playing cards. (a) What is the probability of obtaining a jack or queen? (b) What is the probability of not obtaining a heart? (c) What is the probability of obtaining a heart or an ace?

4.4. Two balanced coins are tossed. (a) List the equally likely occurrences. (b) What is the probability of obtaining two heads? (c) Is the occurrence of a head mutually exclusive of the occurrence of a tail?

4.5. How would you estimate the probability that the acidity level of a sample of Virginia soil selected at random exceeds a pH level of 6.0, assuming Table 2.4 to be an accurate distribution? What is this probability?

4.6. Using Table 2.5, explain how you would estimate the probability that a female fruit fly (of this strain) caught at random (see Chap. 5 for more about this term) has more than 20 bristles on the sixth abdominal sternite.

4.7. Estimate the probability that the excess yardage of 100 denier acetate yarn of a bobbin selected at random is less than 20 yards, assuming that the table in Exercise 2.7 is an accurate distribution.

4.8. How would you estimate experimentally the probability of getting an appointment within two weeks with a certain busy medical doctor?

4.9. A coin may be used to connect the repeated-trials concept of probability to the equally likely concept of probability. Give an illustration in which the equally likely concept of probability seems most appropriate; the repeated-trials concept seems most appropriate.

4.3. PROPERTIES OF PROBABILITY

We now present rules for computing probabilities of compound or related events in terms of probabilities of simple events. For simplicity, we consider properties of probability involving only two events associated with an experiment, properties of probability involving more than two events being straightforward generalizations of properties of probability for two events. Furthermore, these properties are discussed and illustrated in terms of the classical interpretation, even though they hold in much more general situations.

Let I denote every possible outcome of an experiment, and call it the *universal event*. If A is an arbitrary event of an experiment, then the event "not A," denoted by $\bar{A}$, is that event of the experiment in which A does not occur. $\bar{A}$ is sometimes called the *complementary event of A*. We now define other events of an experiment in terms of simple events A and B, using the connectives "and" and "or," which are denoted by $\cap$ and $\cup$, respectively.

Definition 4.1. Let A and B be any two arbitrary events of an experiment.

(a) The event "A and B," denoted by $A \cap B$ or $B \cap A$, is that event in which event A and event B occur together.

(b) The event "A or B," denoted by $A \cup B$ or $B \cup A$, is that event in which either A occurs or B occurs or both A and B occur. Note that the connective "or" is used in the inclusive sense of either or both.

Example 4.6. Let ten similar circular disks in a box be marked by the numbers 1, 2, 3, . . . , 10, respectively. A single disk is selected at random. Let A denote the event that the number is a positive odd integer less than 11; that is, event A occurs if either 1, 3, 5, 7, or 9 is drawn. Let B denote the event that the number is a positive integer less than 4; that is, event B occurs if either 1, 2, or 3 is drawn. Use these two simple events to illustrate the above definitions.

Since the possible outcomes of the experiment are 1, 2, 3, . . . , 10, the universal event I occurs if either 1, 2, 3, . . . , or 10 is drawn. Thus, the complementary event $\bar{A}$ occurs if either 2, 4, 6, 8, or 10 is drawn, and the complementary event $\bar{B}$ occurs if either 4, 5, . . . , or 10 is drawn. Clearly, $A \cup \bar{A} = I$, and $B \cup \bar{B} = I$.

Since events A and B have only the numbers 1 and 3 in common, $A \cap B$ is that event in which either 1 or 3 occurs. Further, the event $A \cup B$ occurs if either 1, 2, 3, 5, 7, or 9 is drawn.

Events can conveniently be expressed in "set" notation. For example, we may think of A as a collection of numbers 1, 3, 5, 7, 9 and write

$$A = \{1, 3, 5, 7, 9\}$$

understanding that A is a "set" whose "elements" are the positive integers 1, 3, 5, 7, 9. In a similar manner we write

$$I = \{1, 2, 3, \ldots, 10\} \quad \text{and} \quad B = \{1, 2, 3\}$$

So, in set symbolism, it follows that

$$\bar{A} = \{2, 4, 6, 8, 10\}$$
$$\bar{B} = \{4, 5, 6, 7, 8, 9, 10\}$$
$$A \cap B = \{1, 3\}$$
$$A \cup B = \{1, 2, 3, 5, 7, 9\}$$

The events "A and B" and "A or B" are called the *intersection of A and B* and the *union of A and B*, respectively. (We may think of "set A" as the mathematical counterpart of the real-world "event A" and use them interchangeably.)

In practical applications the probability of a simple event $P[A]$, say, is usually taken to be an estimate obtained by computing the "appropriate" relative frequency from observation data, or, in certain cases, it is a value obtained from the classical interpretation by taking into account a priori considerations. Since $\bar{A}, \bar{B}, A \cap B$, and $A \cup B$ are events of the universal event I with simple events A and B, the probabilities $P[\bar{A}]$, $P[\bar{B}]$, $P[A \cap B]$, and $P[A \cup B]$ may be determined by the method used in determining $P[A]$ $P[B]$. However, it is usually desirable to compute such probabilities by applying properties like those given by Eqs. (4.2), (4.3), (4.3a), (4.4), and (4.4a).

In Sect. 4.2 we discussed the probability of an event in terms of all possible outcomes in an experiment. Now we consider the probability of an event in terms of some subset of all the possible outcomes, or, as is commonly stated, subject to the condition that we restrict the number of possible outcomes. This concept is illustrated in the following example.

Example 4.7. Use the events defined in Example 4.6 to find the probability that the number drawn is a positive odd integer on the condition that the number must be less than 4; that is, find the *probability of A given that B has occurred.*

When B occurs, the disk must be marked with 1, 2, or 3. In order for A to occur, if it is known that B has occurred, the disk must be marked with 1 or 3. Thus, the required probability is $\frac{2}{3}$.

Definition 4.2. If an experiment can result in N mutually exclusive and equally likely possible outcomes, $n(B) \neq 0$ of which correspond to the occurrence of the event B and $n(A \cap B)$ of which correspond to the occurrence of the event A if it is given that event B has occurred, then the *probability* *of event A given that event B has occurred,* denoted $P[A \mid B]$, is given by $n(A \cap B)/n(B)$, or, briefly,

$$P[A \mid B] = \frac{n[A \cap B]}{n[B]}$$

More generally, for any two events A and B such that $P[B] \neq 0$,

$$P[A \mid B] = \frac{P[A \cap B]}{P[B]}$$

Note. The symbol | in $P[A \mid B]$ is read "given." Thus, we read $P[A \mid B]$ as "probability of event A given event B" and call it the *conditional probability of event A.* It should be obvious that the restricted definition is like the classical interpretation of probability.

An understanding of terms like "mutually exclusive," "exhaustive," and "independent events" is useful. Definition 4.3 and Example 4.8 should make clear the meanings of such terms.

Definition 4.3. (a) Two or more events E_1, E_2, ... , E_k are *mutually exclusive* if the occurrence of any one precludes the occurrence of each of the others.

(b) A collection of events is *exhaustive* if it includes every possible occurrence in the experiment.

(c) The event E_1 *is independent of* E_2 when

$$P[E_1 \mid E_2] = P[E_1]$$

Note that independent events are defined in terms of probability.

Example 4.8. In addition to the two events A and B defined in Example 4.6, we introduce the three new events

$$C = \{4, 5, 6, 7, 8\}$$
$$D = \{7, 8, 9, 10\}$$
$$F = \{3, 6, 9\}$$

all being events associated with the 10 disks in a box. Use these five events (subsets of I) to illustrate Definition 4.3.

Events B and C are mutually exclusive. Events D and F are not mutually exclusive, since they occur simultaneously when the number is 9.

The collection of events B, C, and D is exhaustive. Also, events $\bar{B}$, $\bar{C}$, and $\bar{F}$ make up an exhaustive collection. The probability of the occurrence of some event in a collection which is exhaustive is 1.

The event C is independent of event D, since the probability of the occurrence of event C is the same regardless of whether we are restricted to the numbers of event D or not; that is,

$$P[C \mid D] = \frac{n(C \cap D)}{n(D)} = \frac{2}{4} = 0.5$$

and

$$P[C] = \frac{n(C)}{10} = \frac{5}{10} = 0.5$$

Further, we see that event D is independent of event C, since $P[D \mid C] = \frac{2}{5}$ and $P[D] = \frac{4}{10}$ are the same. In general, when an event E_1 is independent of E_2, it follows, as will be shown later in the proof of Eq. (4.2), that E_2 is independent of E_1 and thus we say that E_1 *and* E_2 *are independent events*. Sometimes we say that E_1 and E_2 are *independent in the probability sense*. Note that A is *not independent* of B, for $P[A \mid B] = \frac{2}{3}$ and $P[A] = \frac{5}{10}$ are different. Note also that B is not independent of A, for $P[B \mid A] = \frac{2}{5}$ and $P[B] = \frac{3}{10}$ are different. Events which are not independent are said to be *dependent events*.

Since expressions involving simple events connected with "not," "and," or "or" are also events, we could use the classical or the relative-frequency concept to determine directly the probability of any such compound event. However, it is usually simpler to determine the probability of compound or related events in terms of probabilities of simple events with the aid of one or more of the rules stated below. (These properties of probabilities of events A and B are listed together for easy reference, and they hold for any of the numerical interpretations of probability.)

$$0 \leq P[A] \leq 1 \tag{4.1}$$

$$P[\bar{A}] = 1 - P[A] \tag{4.2}$$

$$P[A \cap B] = P[A] \cdot P[B \mid A] = P[B] \cdot P[A \mid B] \tag{4.3}$$

$$P[A \cap B] = P[A] \cdot P[B] \qquad \text{if } A \text{ and } B \text{ are independent} \tag{4.3a}$$

$$P[A \cup B] = P[A] + P[B] - P[A \cap B] \tag{4.4}$$

$$P[A \cup B] = P[A] + P[B] \qquad \text{if } A \text{ and } B \text{ are mutually exclusive} \tag{4.4a}$$

After illustrating these properties, we shall show how they follow directly from the classical concept of probability.

Example 4.9. Use the events defined in Examples 4.6 and 4.8 to illustrate properties (4.2), (4.3), (4.3a), (4.4), and (4.4a).

The event B can occur in three mutually exclusive and equally likely ways and the event $\bar{B}$ in seven. Thus $P[B] = \frac{3}{10}$, $P[\bar{B}] = \frac{7}{10}$, and $1 - P[B] = \frac{7}{10}$. So Eq. (4.2) is verified.

According to Example 4.8, events A and B are not independent. Clearly,

$P[A] \cdot P[B \mid A] = (\frac{5}{10}) \cdot (\frac{2}{5}) = 0.2$ and $P[B] \cdot P[A \mid B] = (\frac{3}{10}) \cdot (\frac{2}{3}) = 0.2$. But $A \cap B$ is that event in which the number is 1 or 3, and hence by the classical interpretation, $P[A \cap B] = \frac{2}{10} = 0.2$. This verifies Eq. (4.3). Note that $P[A] \cdot P[B] = (\frac{5}{10}) \cdot (\frac{3}{10}) = 0.15 \neq P[A \cap B]$. Further, since C and D are independent events with $P[C] = \frac{5}{10}$, $P[D] = \frac{4}{10}$, and $P[C \cap D] = \frac{2}{10}$, it follows that Eq. (4.3a) holds (in this special case).

The events A and B are not mutually exclusive. Thus, $P[A] + P[B] - P[A \cap B] = \frac{5}{10} + \frac{3}{10} - \frac{2}{10} = \frac{6}{10}$. But $P[A \cup B] = \frac{6}{10}$, since the event $A \cup B$ can occur if the number is 1, 2, 3, 5, 7, or 9. Hence Eq. (4.4) is verified. The events B and C are mutually exclusive, and $B \cup C$ can occur if the number is 1, 2, 3, 4, 5, 6, 7, or 8. Thus, $P[B \cup C] = \frac{8}{10}$ and $P[B] + P[C] = \frac{3}{10} + \frac{5}{10} = \frac{8}{10}$, so Eq. (4.4a) is verified. Note that C and D are not mutually exclusive even though they are independent events.

Property (4.1) follows directly from the classical interpretation. To prove Eq. (4.2), let n denote the number of mutually exclusive and equally likely possible outcomes of an experiment, $n(A)$ of which correspond to the occurrence of the event A; then $n - n(A)$ of the outcomes correspond to the occurrence of the event $\bar{A}$. Thus, by the classical interpretation and substitution, $P[\bar{A}] = \dfrac{n - n(A)}{n} = 1 - \dfrac{n(A)}{n} = 1 - P[A]$.

By Definition 4.2 we have $P[A \mid B] = n(A \cap B)/n(B)$ when $n(B) \neq 0$. On dividing both numerator and denominator by n, we have

$$\frac{n(A \cap B)/n}{n(B)/n} = \frac{P[A \cap B]}{P[B]}$$

Thus $P[A \mid B] = P[A \cap B]/P[B]$, or $P[A \cap B] = P[B] \cdot P[A \mid B]$ when $P[B] \neq 0$. If $P[B] = 0$, we define $P[A \cap B]$ to be zero, so the relation holds. The proof of $P[A \cap B] = P[A] \cdot P[B \mid A]$ follows by interchanging A and B in the above proof. In case A is independent of B, we have $P[A \mid B] = P[A]$ and on substitution in Eq. (4.3) this gives $P[A \cap B] = P[B] \cdot P[A]$. If B is independent of A, we obtain $P[A \cap B] = P[A] \cdot P[B]$. Since $P[B] \cdot P[A] = P[A] \cdot P[B]$, we see that the statements "B is independent of A" and "A is independent of B" lead to the same result. Thus, we feel justified in saying A and B are independent events.

If the events A and B are mutually exclusive, then the event $A \cup B$ can result in $n(A) + n(B)$ possible occurrences. Hence

$$P[A \cup B] = \frac{n(A) + n(B)}{n}$$

$$= \frac{n(A)}{n} + \frac{n(B)}{n}$$

$$= P[A] + P[B]$$

If A and B are not mutually exclusive, then they have $n(A \cap B)$ of the possible occurrences in common. Thus, the event $A \cup B$ can result in $n(A) + n(B) - n(A \cap B)$ possible occurrences. In this case it follows that

$$P[A \cup B] = \frac{n(A) + n(B) - n(A \cap B)}{n}$$

$$= \frac{n(A)}{n} + \frac{n(B)}{n} - \frac{n(A \cap B)}{n}$$

$$= P[A] + P[B] - P[A \cap B]$$

Properties (4.3), (4.3a), (4.4), and (4.4a) are easily generalized to more than two events. See the exercises in Sect. 4.6 for formulas and the examples below for illustrations.

Example 4.10. What is the probability of obtaining any sequence of heads H and tails T resulting from five fair tosses of a well-balanced coin?

By the classical interpretation the probability $P[H]$ of obtaining heads on a given toss is $P[H] = p = \frac{1}{2}$ and the probability $P[T]$ of obtaining tails on a given toss is $P[T] = P[\text{not } H] = 1 - p = q = \frac{1}{2}$. Since any two tosses are considered independent, the probability of obtaining the particular ordering $HHTHT$ is $p \cdot p \cdot q \cdot p \cdot q = \frac{1}{2} \cdot \frac{1}{2} \cdot \frac{1}{2} \cdot \frac{1}{2} \cdot \frac{1}{2} = \frac{1}{32}$. This is a generalization of Property (4.3a). Since $p = q = \frac{1}{2}$ this is the probability of obtaining any one of the 32 mutually exclusive sequences. Thus, the 32 possible sequences are mutually exclusive and equally likely.

Example 4.11. Thirty pumpkin seeds of the same size and shape, 9 being of strain A and 21 of strain B, are placed in a box and thoroughly mixed. One seed is drawn and the strain recorded and then a second seed is drawn and the strain recorded. What is the probability that (a) both seeds are from strain A? (b) One seed is from strain A and the other from strain B?

Let A_1 denote a seed of strain A on the first draw and A_2 denote a seed of strain A on the second draw. In (a) we wish to find P[seed of strain A on the first draw and seed of strain A on the second draw] or $P[A_1 \cap A_2]$. A_1 and A_2 are not independent, since the probability of A_2 depends on the first draw. Thus, by Eq. (4.3) we have

$$P[A_1 \cap A_2] = P[A_1] \cdot P[A_2 \mid A_1] = \tfrac{9}{30} \cdot \tfrac{8}{29} = \tfrac{12}{145} = 0.083$$

since on the second draw there are only 8 seeds of strain A and 29 seeds altogether.

To solve (b) we must find P[seed of strain A and seed of strain B] or $P[(A_1 \cap \bar{A}_2) \cup (\bar{A}_1 \cap A_2)]$; that is, we first draw a seed of strain A and then one of strain B or we first draw a seed of strain B and then one of strain A. Since $A_1 \cap \bar{A}_2$ and $\bar{A}_1 \cap A_2$ are events which are mutually exclusive, we may use Eqs. (4.4a) and (4.3) to write

$$P[(A_1 \cap \bar{A}_2) \cup (\bar{A}_1 \cap A_2)] = P[A_1 \cap \bar{A}_2] + P[\bar{A}_1 \cap A_2]$$
$$= P[A_1] \cdot P[\bar{A}_2 | A_1] + P[\bar{A}_1] \cdot P[A_2 | \bar{A}_1]$$
$$= \tfrac{9}{30} \cdot \tfrac{21}{29} + \tfrac{21}{30} \cdot \tfrac{9}{29} = \tfrac{63}{145} \doteq 0.434$$

The method used to solve Example 4.11 illustrates how Formulas (4.3) and (4.4a) are applied in fairly complicated situations. However, in this case it is a reasonably simple matter to find these probabilities directly from the classical interpretation. To do this we must enumerate the total number of possible equally likely outcomes and the number of outcomes which are favorable to the occurrence of the event. The total number of possible outcomes is $30 \cdot 29$, since there are 30 possibilities on the first draw and with each of these there are 29 possibilities on the second draw. The total number of outcomes favorable to event $A_1 \cap A_2$ is $9 \cdot 8$, since there are only 9 seeds of strain A on the first draw and 8 of strain A on the second draw. Thus,

$$P[A_1 \cap A_2] = \frac{9 \cdot 8}{30 \cdot 29} \doteq 0.083$$

Further, the number of outcomes favorable to event $(A_1 \cap \bar{A}_2) \cup (\bar{A}_1 \cap A_2)$ is $9 \cdot 21 + 21 \cdot 9 = 2 \cdot 9 \cdot 21$, since when a seed of strain A is drawn first, there are 9 possibilities and with each of these there are 21 seeds of strain B possible on the second draw, and when a seed of strain B is drawn first, there are 21 possibilities and with each of these there are 9 seeds of strain A possible on the second draw. Thus,

$$P[(A_1 \cap \bar{A}_2) \cup (\bar{A}_1 \cap A_2)] = \frac{2 \cdot 9 \cdot 21}{30 \cdot 29} \doteq 0.434$$

Example 4.12. Make a frequency table of the number of heads occurring in the sequences of Example 4.10.

We could list the 32 possible sequences and count the number of sequences with 0, 1, 2, 3, 4, and 5 heads. However, the work can be shortened some if we classify sequences according to number of heads, ignoring order, and then count the number of orders possible, that is, by counting the distinct orders of $TTTTT$, $HTTTT$, $HHTTT$, $HHHTT$, $HHHHT$, $HHHHH$. Clearly, there is only one order for the first (or sixth) set of five symbols. For the second (or fifth) set of five symbols there are five orders, since H(or T) can appear in any one of five positions. For the third (or fourth) set of five symbols there are

$$\frac{32 - (1 + 1 + 5 + 5)}{2} = 10 \text{ orders}$$

Since $1 + 1 + 5 + 5$ of the 32 sequences have already been identified, this leaves 20 sets of five symbols with either two heads or two tails, there being

10 of each due to symmetry. (A direct method of computing this frequency will be given in Sect. 4.4.) Thus, we have the following frequency table for the number of heads appearing on five fair tosses of a balanced coin:

Number of Heads	0	1	2	3	4	5
Frequency	1	5	10	10	5	1

Example 4.13. Use the events defined in Examples 4.6 and 4.8 to find $P[\bar{A} \cup (C \cap D)]$.

We find this probability by two methods. First, identify the event $\bar{A} \cup (C \cap D)$ and use the classical interpretation of probability. The event $\bar{A}$ denotes a disk marked 2, 4, 6, 8, or 10, and the event $C \cap D$ denotes a disk marked 7 or 8. Thus, event $\bar{A} \cup (C \cap D)$ denotes a disk marked 2, 4, 6, 7, 8, or 10. Thus,

$$P[\bar{A} \cup (C \cap D)] = \tfrac{6}{10} = 0.6$$

since there are 10 disks and six are favorable to the occurrence of event $\bar{A} \cup (C \cap D)$.

The second method is repeatedly to apply the formulas on p. 112 so as to reduce $P[\bar{A} \cup (C \cap D)]$ to an expression involving probabilities of single events and then use the classical interpretation of probability and conditional probability. Thus, we have

$$
\begin{aligned}
P[\bar{A} \cup (C \cap D)] &= P[\bar{A}] + P[C \cap D] - P[\bar{A} \cap (C \cap D)] \\
&= P[\bar{A}] + P[C \cap D] - P[C \cap D] \cdot P[\bar{A} \mid (C \cap D)] \\
&= 1 - P[A] + P[C] \cdot P[D \mid C](1 - P[\bar{A} \mid (C \cap D)]) \\
&= 1 - \tfrac{5}{10} + \tfrac{5}{10} \cdot \tfrac{2}{5}(1 - \tfrac{1}{2}) \\
&= \tfrac{6}{10}
\end{aligned}
$$

4.4. FORMULAS FOR COUNTING

Examples 4.11 and 4.12 indicate that some systematic method of enumerating the possible ways events can occur would be desirable in order to determine the probability of simple or compound events. We now present some methods which aid in the problem of counting.

4.4.1. Permutations

Consider a set of n *different* objects, such as books on a shelf. Let $x \leq n$ of the objects be selected and arranged in a line from left to right, such as x books on an empty second shelf. Any arrangement of any x objects is called a *permutation* of x objects selected from n.

A different permutation is obtained by interchanging any two objects of a given permutation. Clearly, there are many permutations when x is large. We wish to count the number of different permutations of n distinct objects taken x at a time. For example, if the objects are denoted by a, b, c, and $x = 2$, the six permutations ab, ba, ac, ca, bc, cb represent all permutations of three objects taken two at a time.

In the general case, think of x positions on a line as being fixed and determine the number of ways in which x objects can be selected from n and placed on the line. The position on the extreme left, that is, the first position, can be filled in n ways. Once this position is filled, there remains $(n - 1)$ objects to place in the second position. Thus, for each choice in the first position there are $(n - 1)$ choices for the second position. Hence, there are $n(n - 1)$ choices for the first two positions. Repeating this argument, we find there are $n - x + 1$ objects left with which to fill the xth position. Hence, there is total of $n(n - 1) \ldots (n - x + 1)$ ways in which to place n objects in x positions. Letting $_nP_x$ denote the number of permutations of n objects taken x at a time, we may write

$$_nP_x = n(n - 1) \ldots (n - x + 1) \qquad (4.5)$$

For example, eight out of a class of 20 students may occupy the eight front seats in a classroom in $_{20}P_8 = 20 \cdot 19 \ldots (20 - 8 + 1) = 20 \cdot 19 \ldots 13$ ways.

Theorem 4.1. *The number of ways of permuting x objects selected from n distinct objects is given by*

$$_nP_x = \frac{n!}{(n - x)!} \qquad (4.6)$$

Proof. Since $_nP_x = n(n - 1) \ldots (n - x + 1)$, we may write

$$_nP_x = [n(n - 1) \ldots (n - x + 1)] \cdot \frac{[(n - x) \ldots 2 \cdot 1]}{[(n - x) \ldots 2 \cdot 1]}$$

$$= \frac{n!}{(n - x)!}$$

When $x = n$, Eq. (4.5) reduce to

$$_nP_n = n(n - 1) \cdots 1 = n! \qquad (4.7)$$

If we let $x = n$ in Eq. (4.6), 0! appears in the denominator. We define 0! to be 1, noting that this definition is consistent with the property that $n!/n = (n - 1)!$ when $n = 1$.

Sometimes we do not have n different objects. Instead, we may have n_1 objects of one kind, n_2 objects of a second kind, ..., n_k objects of a kth kind so that $n_1 + n_2 + \ldots + n_k = n$. Let the number of permutations of these n objects be denoted by $_nP(n_1, \ldots, n_k)$. Obviously, $_nP(n_1, \ldots, n_k)$ is less than $_nP_n$ when some n_i ($i = 1, 2, \ldots, k$) is different from one. For the moment suppose the n_1 like objects of the first kind are made different by markings. Thus they can be arranged into $n_1!$ different orders. In a similar way the n_2 like objects of the second kind may be made different momentarily to give $n_2!$ times as many permutations as before. Continuing this procedure, the total number of permutations after all like objects have momentarily been made different is $n_1!n_2! \ldots n_k!$ times as large as the number of permutations $_nP(n_1, \ldots, n_k)$. But the number of permutations in this case is $_nP_n$; that is,

$$n_1! \ldots n_k! \, _nP(n_1, \ldots, n_k) = \, _nP_n$$

Thus it follows that

$$_nP(n_1, \ldots, n_k) = \frac{_nP_n}{n_1! \ldots! n_k!}$$

or

$$_nP(n_1, \ldots, n_k) = \frac{n!}{n_1! \ldots! n_k!} \tag{4.8}$$

For example, the number of permutations of 3 dimes, 2 nickels and 1 penny is $_6P(3, 2, 1) = 6!/(3! \, 2! \, 1!) = 60$.

4.4.2. Combinations

Next, suppose that we have n distinct objects from which we wish to select x without regard to their arrangement. Such an unordered selection is called a *combination*. Thus, if two students are selected from the front row which contains three, say, Bob, Joe, and Ted, the selection of Bob and Joe is the same combination as Joe and Bob. The total number of combinations of two students is three in this case, namely, Joe and Bob, Bob and Ted, and Ted and Joe.

In order to determine the number of combinations in the general case, let $\binom{n}{x}$ denote the total number of combinations possible in selecting x objects from n different objects. The number of permutations of any given selection (combination) of x objects is $x!$ .Thus, the total number of permutations of n different objects selected x at a time is

$$x! \binom{n}{x} = \, _nP_x = \frac{n!}{(n-x)!}$$

Hence, it follows that

$$\binom{n}{x} = \frac{n!}{x! \, (n-x)!} \tag{4.9}$$

Note that $_nP(x, n-x) = \binom{n}{x}$.

Example 4.14. In how many ways can two students be selected from three?

By Eq. (4.9) we obtain

$$\binom{3}{2} = \frac{3!}{2!\,(3-2)!} = 3$$

which is the same value obtained earlier by listing all combinations.

Example 4.15. A balanced coin is fairly tossed five times. What is the probability of obtaining three heads and two tails?

From Example 4.10 we found that the probability of obtaining any particular order of five simple events is $\frac{1}{32}$. The number of possible orderings of three heads and two tails is given by

$$\binom{5}{3} = \frac{5!}{3!\,2!} = 10$$

(see Example 4.12 for another method). But these ten orderings represent mutually exclusive ways in which the compound event of three heads and two tails can occur. Thus, the desired probability, if we use a generalization of Eq. (4.4a), is given by

$$P[3 \text{ heads and 2 tails}] = \overbrace{\tfrac{1}{32} + \cdots + \tfrac{1}{32}}^{10 \text{ times}} = \tfrac{10}{32}$$

Example 4.16. Consider a dichotomous type of experiment in which a success (event E, say) occurs with probability p, and a failure (event $\bar{E}$) occurs with probability $1 - p = q$ in a single trial. Find the probability of exactly x successes in n independent trials of the experiment.

Consider a particular sequence of x consecutive successes followed by $n - x$ failures. The probability of this particular sequence, if we use a generalization of Eq. (4.3a), since all trials are independent, is given by

$$\overbrace{p \cdot p \cdot \,\cdots\, \cdot p}^{x \text{ times}} \cdot \overbrace{q \cdot q \cdot \,\cdots\, \cdot q}^{(n-x) \text{ times}} = p^x q^{n-x} \tag{4.10}$$

The probability of obtaining x successes and $n - x$ failures in any other sequence is the same as Eq. (4.10), since the p's and q's are simply rearranged to correspond to the new sequence of successes and failures.

It is necessary that we count the number of different sequences in order to find the required probability. The number of sequences is simply the number of permutations of n objects, x of which are of one kind (p), and $n - x$ of which are of a second kind (q). Using Eq. (4.8), we find this to be

$$\frac{n!}{x!\,(n-x)!} \tag{4.11}$$

Now these $n!/x!(n-x)!$ sequences represent mutually exclusive compound events, each with a probability of $p^x q^{n-x}$. Thus, by a generalization of Eq. (4.4a), the probability of one or the other of these sequences is the sum of $n!/x!(n-x)!$ probabilities $p^x q^{n-x}$; that is, the required probability is the product of the quantities in Eqs. (4.10) and (4.11). Since x may have any one of the values $0, 1, \ldots, n$, we think of the probability of obtaining x successes in n independent trials of an event which occurs with probability p in a single trial as a function of x given by

$$b(x;n,p) = \frac{n!}{x!(n-x)!} p^x (1-p)^{n-x} \tag{4.12}$$

where $x = 0, 1, \ldots, n$.

The function (4.12) is known as the *binomial density function* or *Bernoulli density function*. The name *binomial* comes from the fact that Eq. (4.12) is a term in the expansion of the binomial $[(1-p)+p]^n$ or $(q+p)^n$. We see that $b(x;n,p)$ is a density function $f(x) = b(x;n,p)$ since

$$\sum_{x=0}^{n} f(x) = \sum_{x=0}^{n} \frac{n!}{x!(n-x)!} (1-p)^{n-x} p^x$$

$$= (1-p)^n + n(1-p)^{n-1}q + \frac{n(n-1)}{2}(1-p)^{n-2}q^2 + \cdots + p^n$$

$$= [(1-p)+p]^n = 1$$

The binomial distribution can be applied to many practical problems. We consider one use of the binomial now, leaving others to be given later.

Example 4.17. It is known from long experience that a certain manufacturing process produces one per cent defective units. What is the probability that there will be more than one defective unit in a random selection of 100 units?

We wish to find $b(2) + b(3) + \cdots + b(100)$. In this case, it is easier to find $b(0; 100, 0.01)$ and $b(1; 100, 0.01)$ first and then find the required probability by subtraction; that is

$$P[x \geq 2] = 1 - b(0) - b(1)$$

$$= 1 - \frac{100!}{0!\,100!}(0.01)^0(0.99)^{100} - \frac{100!}{1!\,99!}(0.01)^1(0.99)^{99}$$

$$= 1 - (1.99)(0.99)^{99}$$

$$= 1 - 0.736 = 0.264$$

since $\log(0.99)^{99} = 99(-1 + 0.99564) = 9.56836 - 10$ so that $(0.99)^{99} = 0.370$ and $(1.99)(0.99)^{99} = 0.736$.

From Example 4.17 it is clear that obtaining certain probabilities or cumulative probabilities for a binomial distribution involves lengthy calculations. To aid in solving problems in which the binomial distribution is used, tables have been presented in Ref. [12, 13, 15].

4.5. DISTRIBUTION, PROBABILITY, AND RANDOM VARIABLE

In this chapter we have largely restricted our discussion of probability of events to the discrete case. This restriction was not necessary, since the relative-frequency interpretation of probability of events does not exclude the continuous case, and since we assume that the basic properties in this chapter are also valid for the continuous case.

In Chap. 3 we discussed mathematical properties of probability density and distribution functions defined over all or part of the real numbers x. We have not discussed the problem of assigning values of x to the events in an experiment, since this is usually obvious in a particular experiment. The only requirement is that x be a *real-valued function* defined over *all possible occurrences* (component or elementary events) in an experiment. This means that exactly one real number must be assigned to each occurrence and that each possible occurrence has such an assignment. Such a real-valued function, x, is also called a *random variable* (or *variate* or *chance variable*), indicating that its dependent variable is a special kind of variable. The random variable is *discrete* if it is capable of assuming only a finite or countably infinite number of distinct values and is *continuous* in an interval if it can assume any value lying in that interval.

In this chapter we have discussed the probability of events of an experiment; in Chap. 3 we discussed the probability density and distribution functions of a real-valued function x, called a random variable, defined over all the elementary events of an experiment. Thus, once the values of the random variable x are assigned to all the elementary events of an experiment, we can relate the properties of probability discussed in this chapter to those in Chap. 3. Changing the statements about probabilities of events to statements about probabilities of random variables, we see how the properties of distributions given in Chap. 3 are rooted in the more practical approach to probability given in this chapter.

By thinking of an actual experiment it may be made clearer to say that a random variable assumes values with associated probabilities. Before a particular experiment is run, the outcome is an uncertain value, and, so far as we know, it can be any value in a range. It is in this sense that the variable depends on chance, and it is in this sense that we shall often think of the variable.

4.6. EXERCISES

4.10. A well-balanced coin is tossed four times. Find the probability of obtaining exactly zero heads; exactly one head; exactly two heads.

4.11. Two fuses in a box of ten are defective. If three fuses are selected at the same time, what is the probability that at least one is defective?

4.12. (a) In how many ways may seven people be ordered on a bench? (b) In how many ways may seven people be ordered around a circular table?

4.13. Compare the probability of rolling a 10 with two dice with that of rolling a 15 with three dice.

4.14. A box contains eight similar circular disks, each bearing exactly one of the symbols 1, 2, 3, ..., 8. If two disks are drawn at the same time, what is the probability that (a) 1 and 2 are selected? (b) 1 or 2 is selected? (c) Neither 1 nor 2 is selected? (d) Both 1 and 2 are not selected?

4.15. Three similar coins are tossed. What is the probability of getting exactly two faces alike?

4.16. How many numbers can be formed by rearranging the digits in the number 4130131 when numbers beginning with 0 are excluded?

4.17. Box A contains three white and four black balls. Box B contains two white and three black balls. The boxes are alike and all balls are the same size. (a) If two balls are chosen from each box, what is the probability that they will be the same color? (b) If a box is selected (at random) and two balls drawn from it, what is the probability that they will be the same color?

4.18. A box contains 12 similar circular disks, each bearing exactly one of the symbols 2, 3, 4, ..., 13 so that the universal event is made up of the numbers 2, 3, 4, ..., 13. Let A denote the event that a number is a factor of 36, B the event that a number is a multiple of 3, C the event that a number is a remainder on division by 9, D the event that a number is a positive odd integer, and E the event that a number is of the form $4k + 2$, where $k = 0, 1, 2$. (a) Which pairs of events are mutually exclusive? (b) Which pairs of events are independent? Explain. (c) Which three events, if any, are independent? Explain. (d) An event is made up of the numbers 3, 5, 6, 7. Express this as a compound event in terms of some of the simple events A, B, C, D, and E. (e) If possible, express the event made up of the numbers 11, 12, 13 in terms of some of the simple events A, B, C, D, and E.

4.19. If n and x are nonnegative integers such that $(n - 1) \geq x$, prove that

$$\binom{n}{x} = \binom{n - 1}{x - 1} + \binom{n - 1}{x}$$

4.20. Let A, B, and C denote any three events. (a) Prove that

$$P[A \cap B \cap C] = P[A] \cdot P[B \mid A] \cdot P[C \mid A \cap B] \qquad (4.13)$$

(b) Write $P[A \cap B \cap C]$ as the product of three factors in five other ways. (c) Write an expression similar to Eq. (4.13) for n events.

4.21. Let A, B and C denote any three events. (a) Prove that

$$P[A \cup B \cup C] = P[A] + P[B] + P[C]$$
$$- P[A \cap B] - P[A \cap C] - P[B \cap C] + P[A \cap B \cap C] \qquad (4.14)$$

(b) Write an expression similar to Eq. (4.14) for n events. What is the form of this expression when the n events are mutually exclusive?

4.22. A well-balanced die is tossed and the number n on the top face observed. Then n well-balanced dice are tossed together. Find the probability that a total of exactly nine points will show.

4.23. Six cards of the same size have identical backs. Three cards have a red face, two cards a white face, and one card a blue face. (a) If the first two cards drawn have red faces, what is the probability that the third card is red? (b) If three consecutive cards are drawn, what is the probability that they all have different colors? (c) Compare the probability of drawing the sequence red, white, white with the probability of drawing the sequence white, red, red. (d) If three consecutive cards are drawn, what is the probability that the last (third) card is red? What is the probability that the last card is blue?

4.24. The probabilities of joint (or compound) events are often easier to study with the aid of a table, particularly when the simple events are not mutually exclusive. We consider the case of two simple events A and B. On a single trial of an experiment one and only one of the following events occur: $A \cup B$, $A \cap \bar{B}$, $\bar{A} \cap B$, $\bar{A} \cap \bar{B}$. Letting a, b, c, d denote the number of cases favorable to the occurrence of the mutually exclusive events $A \cap B$, $A \cap \bar{B}$, $\bar{A} \cap B$, $\bar{A} \cap \bar{B}$, respectively, we may present the information as shown in Table 4.4. We see, for example, that there are $a + b$ cases favorable to the occurrence of A, $b + d$ cases favorable to the occurrence of "not B," and n possible cases.

Table 4.4
Mutually Exclusive Joint Events in Terms of Two Simple Events

	B	$\bar{B}$	*Total*
A	a	b	$a + b$
$\bar{A}$	c	d	$c + d$
Total	$a + c$	$b + d$	$a + b + c + d = n$

In a particular experiment there are 30 balls of the same size and weight. Suppose that 11 are painted blue and marked by x, 7 are painted red and marked by x, 9 are blue and not marked, and 3 are red and not marked. Make a table like Table 4.4 and use it to determine (a) the probability of a ball's being red or marked, (b) the probability of a red ball being marked, (c) if color is independent of marking.

4.25. The events and number of cases shown in Table 4.4 may also be presented in a *Venn diagram*, shown in Fig. 4.2. Use such a diagram to answer the questions in Exercises 4.17 and 4.24.

4.26. The Venn diagram is particularly useful in problems with three events

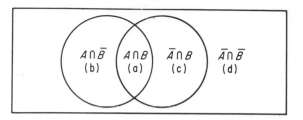

Fig. 4.2 Venn Diagram of Table 4.4

A, B, C which are not mutually exclusive. Such a diagram is made by drawing three intersecting circles such that eight regions, like those in Figure 4.3, are formed. These regions represent the mutually exclusive joint events $A \cap B \cap C$, $A \cap B \cap \bar{C}$, $A \cap \bar{B} \cap C$, $A \cap \bar{B} \cap \bar{C}$, $\bar{A} \cap B \cap C$, $\bar{A} \cap B \cap \bar{C}$, $\bar{A} \cap \bar{B} \cap C$, $\bar{A} \cap \bar{B} \cap \bar{C}$. (It should be noted that the Venn diagram is restricted to cases with two or three simple events, that is, to two or three circles.)

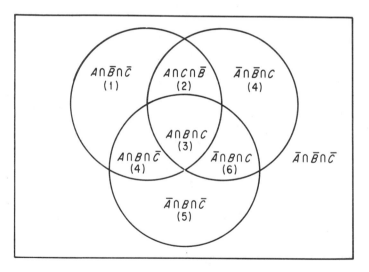

Fig. 4.3 Venn Diagram for Events A, B and C

In Fig. 4.3 we may think of events A, B, and C as referring to the sets of students enrolled in French, German, and Russian, respectively. Thus, we may think of 30 students, majoring in a foreign language at a school, as being distributed as indicated in the diagram. That is, three students are enrolled in the three languages, four students are enrolled in French and German only, two students are enrolled in French and Russian only, five students are enrolled in German only, etc. Use Fig. 4.3 to answer the following questions: (a) How many of the 30 students are not enrolled in any of the three languages? How many are enrolled in French? In German? In Russian? (b) How many are enrolled in both French and German? Both French and Russian? Both German and Russian? (c) If the proportion of students in the various categories is assumed to be constant from year to year, what is the probability that a student enrolled in Russian will also be enrolled in German? (d) If the proportion of students in the various categories is assumed to be constant, what is the probability that a student enrolled in both French and German will also be enrolled in Russian?

4.27. If $P[A \cap B] = P[A] \cdot P[B]$, prove that $P[A \cap \bar{B}] = P[A] \cdot P[\bar{B}]$, $P[\bar{A} \cap B] = P[\bar{A}] \cdot P[B]$, and $P[\bar{A} \cap \bar{B}] = P[\bar{A}] \cdot P[\bar{B}]$. Note that A and B are independent when $P[A \cap B] = P[A] \cdot P[B]$.

4.28. It is possible to have A and B independent, A and C independent, and B and C independent and still to have $A \cap B$ and C, say, dependent or to have A, B, and C dependent. That is, it is possible for three events A, B, and C to be *pairwise independent* and not be *mutually independent*. For this reason we say events A, B, and C are *mutually independent* if

$$\left\{ \begin{array}{l} P[A \cap B] = P[A] \cdot P[B] \\ P[A \cap C] - P[A] \cdot P[C] \\ P[B \cap C] = P[B] \cdot P[C] \\ P[A \cap B \cap C] = P[A] \cdot P[B] \cdot P[C] \end{array} \right. \qquad (4.15)$$

hold. (a) Let A, B, C be events such that $P[A \cap B \cap C] = P[A \cap \bar{B} \cap \bar{C}] = P[\bar{A} \cap B \cap \bar{C}] = P[\bar{A} \cap \bar{B} \cap C] = 0$ and $P[A \cap B \cap \bar{C}] = P[A \cap \bar{B} \cap C] = P[\bar{A} \cap B \cap C] = P[\bar{A} \cap \bar{B} \cap \bar{C}] = \frac{1}{4}$. Show, using a Venn diagram, that the events are not independent. (b) Show that the events illustrated in Fig. 4.3 are not independent. (c) Write the set of 11 probability statements, corresponding to Eq. (4.15), which are sufficient for the mutual independence of events A, B, C, D. Write the conditions which are sufficient for the mutual independence of events A_1, A_2, ..., A_n. (d) In applications it appears that practically all, if not all, events which are pairwise independent are also mutually independent. Thus, the distinction between pairwise and mutual independence is largely of theoretical interest. The reader might wish to try to give a practical application (interpretation) of (a) in Exercise 4.28 or to describe any other application.

    Hint. See the discussion in Ref. [4].

4.29. In the definition for mutual independence of three events given in Exercise 4.28, the system of equations (4.15) may be replaced by the system of eight (2^3) equations of the form

$$P[A' \cap B' \cap C'] = P[A'] \cdot P[B'] \cdot P[C'] \qquad (4.16)$$

where A' denotes either A or $\bar{A}$, B' either B or $\bar{B}$, and C' either C or $\bar{C}$. (a) Show that Eq. (4.16) follows from Eq. (4.15). (b) Use Eq. (4.16) to solve (a) and (b) of Exerscie 4.28. (c) Write the conditions corresponding to Eq. (4.16) which are sufficient for the mutual independence of events $A_1, A_2, \ldots, A_n$.

4.30. Prove that

$$\sum_{x=0}^{n} \binom{n}{x} = 2^n, \qquad n = 1, 2, 3, \ldots, \text{ any positive integer} \qquad (4.17)$$

4.31. Ten pennies and a dime are in box A, and 20 pennies are in box B. Five coins are taken from box A and placed in box B, and then the 25 coins of box B are thoroughly mixed. Then five coins are taken from box B and placed in box A. What is the probability that the dime is in box A?

4.32. In a bridge game, North and South were dealt 11 trumps. What is the probability that East and West were each dealt one trump?

4.33. For purposes of bidding in the game of bridge, we commonly let an ace count four points, a king three points, a queen two points, a jack one point and other cards zero points. If the maximum point count for partners is assumed to be 40, what is the probability of partners holding exactly 38 points between them?

4.34. Definition 4.2 should usually be applied in the computation of simple conditional probabilities. However, there are times when the very controversial (Refs. [3, 10, 17]) Bayes' formula [see Eq. (4.18)] should be applied. According to Bayes' theorem, if $B_1, \ldots, B_n$ are mutually exclusive and exhaustive events and if A can only occur in combination with one of the n events $B_1, \ldots, B_n$, then

$$P[B_i \,|\, A] = \frac{P[B_i] \cdot P[A \,|\, B_i]}{P[B_1] \cdot P[A \,|\, B_1] + \cdots + P[B_n] \cdot P[A \,|\, B_n]},$$
$$i = 1, \ldots, n \qquad (4.18)$$

Prove Eq. (4.18).

4.35. Suppose three machines M_1, M_2, and M_3 in a factory make exactly the same parts and that they are packed in the same type of boxes. Suppose machine M_i ($i = 1, 2, 3$) makes n_i boxes per day, of which p_i per cent, on the average, are declared defective. If from all boxes produced by the three machines in a single day we select (at random) a box and from this box we select (at random) a part which proves to be defective, what is the probability that we have chosen a box made by M_2?

4.36. Prove that

$$P[A_1 \cup A_2 \cup \cdots \cup A_n] + P[\bar{A}_1] \cdot P[\bar{A}_2] \cdots P[\bar{A}_n] = 1$$

when $A_1, \ldots, A_n$ are independent events.

4.37. We have generally illustrated the properties of probability in terms of number of favorable occurrences of discrete events. However, there are problems in which probability is easiest to determine in terms of admissible regions in space, that is, in terms of line segments in one dimension, areas in two dimensions, and volumes in three dimensions. In such situations any finite portion of the space, no matter how small, contains an infinite number of points, making the usual method of counting the number of favorable occurrences meaningless. In the geometrical problem it is usually assumed that the probability (or *probability measure*) of any subdivision of the admissible region is directly proportional to the size of the subdivision, even though it is possible to give a geometrical treatment to problems in cases where probability of a subdivision of a fixed size varies with location. For example, when the admissible region is a square of area A, the probability measure of any section (of the square) with area a is a/A. We now illustrate geometrical probabilities with the *Buffon needle problem*.

Consider an infinite set of ruled parallel lines such that any two adjacent lines are d units apart. If a needle of length l ($l < d$) is randomly tossed on the set of lines, what is the probability that the needle will intersect one of the lines?

Let y denote the distance from the mid-point of the needle to the nearest line, and let x denote the angle the needle makes with the perpendicular. The admissible values of x and y are given by $-(\pi/2) \le x \le (\pi/2)$ and $0 \le y \le (d/2)$, respectively. By constructing a figure, as Fig. 4.4, it is clear that $y < (l/2) \cos x$ when the needle intersects the nearest line; otherwise, $y \ge (l/2) \cos x$. Thus, the curve defined by $y = (l/2) \cos x$ is the boundary between the region of intersection and nonintersection. In the rectangular co-ordinate system of x and y, as in Fig. 4.5, the admissible region for the experiment is represented by a rectangle

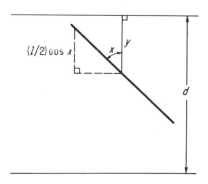

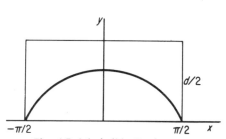

Fig. 4.4 Ruled Board   **Fig. 4.5** Admissible Regions

π units long and $d/2$ units high, and the admissible region for a single intersection of the needle and a line is represented by the area of the rectangle which is under the curve. Thus, the probability that the needle will intersect a line one time is given by

$$\frac{\int_{-\pi/2}^{\pi/2} \frac{l}{2} \cos x \, dx}{\frac{d}{2}\pi} = \frac{2l}{\pi d}$$

For the case where $l > d$, show that the probability of the needle intersecting at least one line is

$$\frac{2l}{\pi d}\left(1 - \frac{\sqrt{l^2 - d^2}}{l}\right) + \frac{2}{\pi} \text{ arc cos } \frac{d}{l}$$

4.38. A triangular figure is tossed on a board ruled as in Buffon's needle problem. If the longest side of the triangle has length less than d, what is the probability that some part of the triangular figure will cover any portion of a line?

REFERENCES

1. Arley, N. and K. R. Buch, *Introduction to the Theory of Probability and Statistics*. New York: John Wiley & Sons, Inc., 1950.

2. Carnap, R., *Logical Foundations of Probability*. Chicago: University of Chicago Press, 1950.

3. Cramer, H., *The Elements of Probability Theory and Some of Its Applications*. New York: John Wiley & Sons, Inc., 1955.

4. Feller, W., *An Introduction to Probability Theory and its Applications*. New York: John Wiley & Sons, Inc., 1950.

5. Fry, T. C., *Probability and its Engineering Uses*. New York: The Macmillan Company, 1928.

6. Goldberg, S., *Probability: An Introduction*. Englewood Cliffs, N. J.: Prentice-Hall, Inc., 1960.

7. Halmos, P. R., "The Foundations of Probability," *The American Mathematical Monthly*, Vol. **51** (1944), pp. 493–510.

8. Kerrich, J. E., *An Experimental Introduction to the Theory of Probability*. Copenhagen: Ejnar Munksgaard, 1946.

9. Kolmogoroff, A., *Foundations of the Theory of Probability*. New York: Chelsea Publishing Company, 1950.

10. Munroe, M. E., *Theory of Probability*. New York: McGraw-Hill, Inc., 1951.

11. Nagel, E., *Principles of the Theory of Probability*, International Encyclopedia of Unified Science, Vol. **1**, No. 6 (1939), Chicago.

12. *Tables of the Binomial Probability Distribution*, Applied Mathematical Series 6, National Bureau of Standards, Washington 25, D. C., 1950, 387 pp.

13. Pearson, Karl, *Tables of the Incomplete Beta-Function*. London: Cambridge, University Press, 1934, 494 pp.

14. Reichenbach, H., *The Theory of Probability—An Inquiry into the Logical and Mathematical Foundations of the Calculus of Probability*. Berkeley, Calif.: University of California Press, 1949.

15. Romig, H. G., 50–100 *Binomial Tables*. New York: John Wiley & Sons, Inc., 1952, 172 pp.

16. Uspensky, J. V., *Introduction to Mathematical Probability*. New York: McGraw-Hill, Inc., 1937.

17. Wadsworth, G. P. and J. G. Bryan, *Introduction to Probability and Random Variables*. New York: McGraw-Hill, Inc., 1960.

5

SAMPLING

AND SAMPLING DISTRIBUTIONS—

EXPECTATION AND ESTIMATION

The value of a given statistic can be expected to vary from sample to sample. This variation is studied in terms of sampling distributions (of means, proportions, totals, variances, etc.) determined from both finite and infinite distributions. Some desirable properties of statistics are discussed. The method of maximum likelihood for estimating parameters is described.

5.1. INTRODUCTION

A portion of a population of values, namely a sample, is normally used to study the population or the characteristics of the population. Since many possible samples can be drawn from a given population, many possible impressions or estimates of the nature of the population can be inferred. Thus, it is desirable that we select a measure to characterize the sample which corresponds to a parameter in the population and which shows relatively little variation from sample to sample. This leads us to a study of fluctuations of the measure or measures selected to characterize the sample and to considerations of desirable properties for this measure (or these measures) to possess.

5.2. SAMPLE AND STATISTIC

Let $x_1, x_2, \ldots, x_n$ denote the set of numerical values of n observations selected from a larger set, the set being either a discrete one or a continuum. We refer to the smaller set as a *sample of size n* drawn from a univariate population. A numerical value determined from some or all of the values which make up the sample is called a *statistic*. One or more statistics may be used in describing or characterizing a sample of observations. We now consider some of the simple statistics used in drawing inferences concerning the larger number of similar observations which might have been obtained.

Since the primary function of a statistic is to lead to a clear understanding of the nature of the population, and since the population is usually studied by looking at the parameters, it is natural that we name and attempt to define statistics as we did parameters. The notation for a parameter and the corresponding statistic usually differ, particularly if the statistic is one which is often used. Following the custom, we usually denote parameters by Greek letters and the corresponding statistics by the corresponding Latin letters. Thus, the standard deviation for a population is denoted by σ and the standard deviation for a sample by s. Similarly, μ and m are used to denote the mean of a population and a sample, respectively. Generally, we use $\bar{x}$ in place of m in this book, since this is common practice and since m is used for other purposes.

The sample mean $\bar{x}$ of a sample of size n is given by

$$\bar{x} = \frac{\sum_{i=1}^{n} x_i}{n} \tag{5.1}$$

The sample median is denoted by x_m and is defined to be the middle numerical value if n is odd and the average (arithmetic mean) of the two middle values if n is even. Corresponding to each parameter which measures central tendency in the population, we define a statistic which measures central tendency in the sample in exactly the same way. The same notation is used to denote both parameter and statistic for all measures of central tendency except the mean and median, since there is little occasion to confuse them.

Three measures of dispersion in a sample are of interest at this time. The *range* is defined and denoted as in Sect. 2.3.2. The *mean deviation* about the sample median, d_m, is defined by

$$d_m = \frac{\sum |x_i - x_m|}{n} \tag{5.2}$$

and the *variance s^2* of a sample is defined by

$$s^2 = \frac{\sum (x_i - \bar{x})^2}{n - 1} \tag{5.3}$$

It should be observed that d_m, x_m, s^2, $\bar{x}$, and $(n-1)$ replace δ_m, μ_m, σ^2, μ, and N, respectively. In case it seems more natural to use n as the divisor in defining s^2, we point out that $n-1$ is used in order that s^2 have one of the very desirable properties of statistics to be discussed in Sect. 5.6.2, namely, that the expected value (mean) of all statistics of a given kind computed from samples of size n be equal to the corresponding parameter.

It should be observed that a sample is finite and that the values of a sample constitute a discrete set, no matter whether they have been drawn from a discrete or a continuous distribution. When our only interest is in describing the sample, it makes no difference whether the values have been drawn from a discrete or continuous distribution. However, when our interest is in making a statement about the population from which the sample was drawn, we must distinguish between the two cases. Thus, we consider separately sampling from a finite and discrete population and sampling from an infinite and discrete or continuous population.

5.3. SAMPLING DISTRIBUTIONS DERIVED FROM A FINITE POPULATION

We may examine variation in statistics of one kind (say, means of samples of size n drawn by the same method from one population) form either the experimental or the mathematical point of view. Both approaches are in common use, and something is to be gained by studying each. We illustrate both methods in a simple case in order to compare them. Afterwards, we devote most of our discussion to mathematical sampling.

There are many methods by which we may draw a sample of n objects from a population. In this section we consider the two most useful methods.

Let the observations in the finite population be denoted by x_1, x_2, ..., x_N Some of these values may be the same numerically. We say that an observation is *randomly selected* or is a *random observation* if each of the N observations in the population has an equal chance of being (is equally likely to be) selected.

Let S denote a sample of n observations with x_1' denoting the first observation selected, x_2' the second observation selected, ..., x_n' the nth observation selected. Assume that each observation x_i' $(i = 1, ..., n)$ is randomly selected. We say that S is a *random sample with replacement or a simple random sample* if x_1' is returned to the population before x_2' is selected, x_2' is returned to the population before x_3' is selected, ..., x_{n-1}' is returned to the population before x_n' is selected. On the other hand, if, for each i $(i = 1, 2, \ldots, n-1)$, x_i' is not replaced before x_{i+1}' is drawn, we say that S is a *random sample without replacement*.

It should be noted that the probability of selecting each member in a simple random sample remains the same, but the probability of selecting x_i' in a random sample without replacement changes according to how many observations have already been drawn. That is, in obtaining a sample of size n, the population remains the same for each selection in the first case, but the population changes for each selection in the second case.

5.3.1. Sampling Without Replacement

First, consider sampling without replacement. Let five disks of the same kind marked 1, 3, 5, 7, and 9, respectively, be placed in a bowl. The distribution of this population is shown in Table 5.1. After the disks have been thoroughly mixed, two are drawn one after another (or simultaneously), the numbers recorded, and the disks then returned to the bowl. Then the disks are thoroughly mixed again, two are drawn, and the numbers are recorded. This process can be repeated as many times as we wish. This is an illustration of *experimental sampling without replacement from a finite population of five different numbers.*

Table 5.1
Discrete Uniform Population of Disks

Number on Disk	Frequency
1	1
3	1
5	1
7	1
9	1

Note that drawing one disk after another for two draws may be thought of as leading to exactly the same sample as drawing two simultaneously, provided the disks are not mixed between the first and second draws. However, if the disks are mixed between draws, we might expect that on a given trial the two methods would lead to different samples, but it is assumed that in the "long run" the two methods of drawing two disks lead to approximately the same collection of samples. Verification of this assumption is left to the student.

Example 5.1. Draw 100 experimental random samples of size two without replacement from the population in Table 5.1 and compute the mean $\bar{x}$ of each sample. Construct a table showing the distribution of $\bar{x}$.

One set of 100 random samples without replacement is shown in Table 5.2. The frequency and relative frequency distributions of $\bar{x}$ are shown in Table 5.3 and represent an *approximation to the sampling distribution of $\bar{x}$.* We call this *an experimental or empirical sampling distribution.*

If another 100 samples were selected by the same method and a relative frequency table of their $\bar{x}$'s constructed, we would expect to get a slightly different approximation to the *sampling distribution of $\bar{x}$.* If two such distributions are noticeably very different, we could obtain a more stable experi-

Table 5.2

Random Samples Without Replacement Drawn
from the Population in Table 5.1

No.	Sample	No.	Sample	No.	Sample	No.	Sample
1	1, 3	26	9, 1	51	3, 7	76	7, 9
2	3, 5	27	7, 3	52	3, 1	77	9, 3
3	9, 5	28	7, 1	53	5, 9	78	5, 9
4	3, 1	29	9, 1	54	7, 3	79	5, 1
5	7, 5	30	7, 5	55	7, 5	80	9, 3
6	9, 3	31	5, 1	56	5, 9	81	7, 1
7	7, 9	32	9, 5	57	7, 3	82	1, 5
8	5, 9	33	7, 1	58	9, 5	83	1, 3
9	7, 3	34	1, 3	59	7, 3	84	5, 7
10	1, 5	35	7, 9	60	5, 1	85	7, 5
11	9, 7	36	9, 1	61	5, 1	86	1, 9
12	3, 1	37	7, 5	62	7, 5	87	9, 5
13	3, 1	38	7, 5	63	9, 1	88	9, 1
14	3, 7	39	7, 3	64	3, 5	89	3, 7
15	5, 7	40	7, 3	65	7, 9	90	1, 7
16	9, 7	41	9, 3	66	1, 7	91	1, 5
17	3, 1	42	7, 3	67	5, 7	92	1, 7
18	5, 9	43	9, 5	68	7, 1	93	5, 3
19	7, 9	44	1, 9	69	7, 1	94	5, 3
20	9, 5	45	7, 5	70	9, 3	95	1, 7
21	9, 7	46	3, 5	71	5, 9	96	3, 1
22	7, 5	47	5, 1	72	9, 5	97	9, 1
23	9, 1	48	3, 7	73	7, 3	98	9, 1
24	9, 5	49	9, 3	74	9, 7	99	7, 3
25	9, 3	50	7, 5	75	7, 9	100	7, 9

Table 5.3

Experimental Sampling Distribution of $\bar{x}$ Obtained from Table 5.2

$\bar{x}$	Frequency	Rel. Freq.
2	9	0.09
3	8	0.08
4	13	0.13
5	24	0.24
6	20	0.20
7	14	0.14
8	12	0.12

mental sampling distribution by increasing the number of samples to, say, 500; that is, we would expect less variation from one experimental sampling distribution of 500 $\bar{x}$'s to another than we would for experimental sampling distributions determined from 100 samples. In any case, it seems reasonably clear that we should be able to choose enough samples so as to be satisfied that the resulting experimental sampling distribution gives a good idea (approximation) of how means of samples of size two change.

The theoretical sampling distribution or, for short, *sampling distribution* obtained by mathematical methods is the model for an experimental sampling distribution. The mathematical method is illustrated in Example 5.2. After discussing the theoretical sampling distribution, we show how this distribution can be used to make predictions about what actually happens in experimental sampling from a finite population.

Example 5.2. Find the *sampling distribution of means* of samples of size two drawn without replacement from the finite population (disks marked 1, 3, 5, 7, and 9, respectively) given in Table 5.1 and Example 5.1.

All possible samples which can be drawn from this population without replacement are

$$1, 3 \quad 1, 5 \quad 1, 7 \quad 1, 9 \quad 3, 5$$
$$3, 7 \quad 3, 9 \quad 5, 7 \quad 5, 9 \quad 7, 9$$

These ten samples are equally likely events. The means of these ten samples are distributed as shown in Table 5.4, and this is the *sampling distribution of means* required.

Table 5.4
Theoretical Sampling Distribution Without Replacement
of $\bar{x}$ Obtained from Table 5.1

$\bar{x}$	Frequency	Rel. Freq.
2	1	0.10
3	1	0.10
4	2	0.20
5	2	0.20
6	2	0.20
7	1	0.10
8	1	0.10

The distribution of Table 5.4 is clearly a probability distribution; that is, it gives probabilities. The third column gives the probability of the corresponding mean in the first column. Further, the probability of obtaining any subset of means is found by adding the appropriate probabilities in the third column.

Example 5.3. What is the probability of obtaining a mean of 3, 5, or 7 (that is, a value which is found in the original population) if a sample of size two is drawn without replacement from the population of Table 5.1?

It follows that $P[\bar{x} = 3$ or 5 or $7] = 0.10 + 0.20 + 0.10 = 0.40$.

The distribution of Table 5.4 also serves as a limiting distribution (or model) of an experimental sampling distribution one would get by drawing more and more samples of size two. Comparing Tables 5.3 and 5.4, we see that some relative frequencies in Table 5.3 are near (in value to) those in Table 5.4, whereas others are not so near. However, if we found an experimental sampling distribution of means by drawing 500 samples, we could be "almost certain" that this distribution of means would be closer to the theoretical sampling distribution than an experimental sampling distribution of 100 means. As the number of samples of size two gets larger and larger, we can be "almost sure" that the corresponding experimental sampling distribution gets nearer and nearer the sampling distribution. Thus, it is in this sense that the theoretical sampling distribution is a limiting distribution for the experimental sampling distribution. Hence, by discussing the theoretical sampling distribution alone, we can get a clear picture of the nature of the variability in means computed from random samples drawn without replacement.

We used sample means to compare experimental and theoretical sampling distributions, even though the comparison could have been made with the use of any other statistic. Since the mean is a fairly simple statistic, and since it plays such a central role in applications, we shall continue to use it in much of our discussion on sampling.

By comparing the third columns of Tables 5.3 and 5.4, we have already noted differences in the relative frequencies of the experimental sampling distribution of 100 means and the theoretical sampling distribution. In Example 5.4 we examine differences in their means and variances.

Example 5.4. Compute the mean and variance for the experimental sampling distribution of Table 5.3; for the sampling distribution of Table 5.4. Compare these means and variances.

Denoting the mean and variance of the experimental sampling distribution by $\bar{\bar{x}}$ and $s_{\bar{x}}^2$, we find that

$$\bar{\bar{x}} = \frac{\sum f\bar{x}}{100} = \frac{528}{100} = 5.28$$

and

$$s_{\bar{x}}^2 = \frac{\sum f\bar{x}^2 - \frac{(\sum f\bar{x})^2}{100}}{99} = \frac{3090 - \frac{(528)^2}{100}}{99} = 3.05$$

Note that only a sample of values (means) is used to compute the mean and variance. Thus the notation and formulas for samples are required. However, for the sampling distribution shown in Table 5.4 all the values (means) are known, and thus the mean and variance are parameters of a population and require the notation $\mu_{\bar{x}}$ and $\sigma_{\bar{x}}^2$. We find that

$$\mu_{\bar{x}} = \frac{\sum f\bar{x}}{10} = \frac{50}{10} = 5.00$$

and

$$\sigma_{\bar{x}}^2 = \frac{\sum f\bar{x}^2 - \frac{(\sum f\bar{x})^2}{10}}{10} = \frac{280 - \frac{(50)^2}{10}}{10} = 3.00$$

Both the mean and variance for the experimental sampling distribution of 100 means are larger than they are for the theoretical sampling distribution. Such would not always be true. Actually, we expect $\bar{x}$ to be smaller than $\mu_{\bar{x}}$ about as often as it is larger, and we expect the same thing to be true of the variances. Furthermore, we expect $\bar{\bar{x}}$ to get closer to $\mu_{\bar{x}}$ "almost always" and $s_{\bar{x}}^2$ to get closer to $\sigma_{\bar{x}}^2$ "almost always" as the experimental sampling distribution gets larger.

So far, we have compared two sampling distributions, empirical and theoretical, with the sample size fixed at two. It is more important to study the relationship between the population and the theoretical sampling distribution of some statistic, while allowing the sample size n to vary. In terms of parameters, we wish to find the relation between $\mu_{\bar{x}}$ and μ (or μ_x) and $\sigma_{\bar{x}}^2$ and σ^2 (or σ_x^2), where μ and σ^2 denote the mean and variance of the population from which the sampling distribution of $\bar{x}$ was determined.

Example 5.5. Compute μ and σ^2 for the population given in Table 5.1. Use $\mu_{\bar{x}}$ and $\sigma_{\bar{x}}^2$ found in Example 5.4 to compare μ and $\mu_{\bar{x}}$; σ^2 and $\sigma_{\bar{x}}^2$.
Using Table 5.1, we find that $\mu = \frac{25}{5} = 5.00$ and

$$\sigma^2 = \frac{165 - \frac{(25)^2}{5}}{5} = 8.00$$

Thus, for a particular population, a particular statistic, and a particular sample size, we see that the means, μ and $\mu_{\bar{x}}$, of the two distributions are the same and that the variance $\sigma_{\bar{x}}^2$ of the sampling distribution is considerably smaller than the variance $\sigma^2 = 8.00$ of the population. As a matter of fact, σ^2 and $\sigma_{\bar{x}}^2$ satisfy the equation

$$\sigma_{\bar{x}}^2 = \frac{\sigma^2}{n} \cdot \frac{N-n}{N-1}$$

where N denotes the population size and n the sample size, for

$$\frac{\sigma^2}{n} \cdot \frac{N-n}{N-1} = \frac{8}{2} \cdot \frac{5-2}{5-1} = 3 = \sigma_{\bar{x}}^2$$

Note. Since the sampling distribution of means $\bar{x}$ is also a *population of $\bar{x}$'s*, the original population from which the samples are drawn is sometimes called the *parent population of x's.*

The relationships pointed out in Example 5.5 for a particular case are true in the general case. The general statement is found in Theorem 5.1.

The standard deviation of the sampling distribution of means is sometimes called the *standard error of the mean.* In general, the standard deviation of the sampling distribution of any statistic t is called the standard error of that statistic and denoted by σ_t.

Theorem 5.1. *If N denotes the size of any finite population and n the size of a sample selected without replacement, then for all possible samples of size n the mean of the means $\mu_{\bar{x}}$ is equal to the population mean μ; the variance of means $\sigma_{\bar{x}}^2$ is equal to the population variance σ^2 times a factor $(N - n)/[n(N - 1)]$; that is*

$$\mu_{\bar{x}} = \mu \tag{5.4}$$

and

$$\sigma_{\bar{x}}^2 = \frac{\sigma^2}{n} \cdot \frac{N - n}{N - 1} \tag{5.5}$$

Proof. If the population values are denoted by

$$x_1, x_2, \ldots, x_N$$

then all possible samples of size n may be indicated by $(x_1, x_2, \ldots, x_{n-1}, x_n)$, $(x_1, x_2, \ldots, x_{n-1}, x_{n+1}), \ldots, (x_{N-n+1}, x_{N-n+2}, \ldots, x_{N-1}, x_N)$. There are $\binom{N}{n}$ such samples, each having the same probability of $p_i = 1 / \binom{N}{n}$, where $i = 1, 2, \ldots, \binom{N}{n}$. Now $\mu_{\bar{x}}$ is the mean of the mean of all these samples; that is

$$\mu_{\bar{x}} = \bar{x}_1 p_1 + \bar{x}_2 p_2 + \cdots + \bar{x}_{\binom{N}{n}} \cdot p_{\binom{N}{n}}$$

or

$$\mu_{\bar{x}} = \left[\frac{x_1 + x_2 + \cdots + x_{n-1} + x_n}{n} + \frac{x_1 + x_2 + \cdots + x_{n-1} + x_{n+1}}{n} \right. $$
$$\left. + \cdots + \frac{x_{N-n+1} + x_{N-n+2} + \cdots x_{N-1} + x_N}{n} \right] \Big/ \binom{N}{n}$$

We must now count the number of times each x_i occurs. For a particular x_i which occurs in a sample of size n, there are $(n - 1)$ other x's which may be selected from $N - 1$ values; that is, there are $\binom{N-1}{n-1}$ ways of selecting the other $(n - 1)$ x's. Since each x_i occurs the same number of times, we may write

$$\mu_{\bar{x}} = \frac{\binom{N-1}{n-1} \cdot x_1 + \binom{N-1}{n-1} \cdot x_2 + \cdots + \binom{N-1}{n-1} \cdot x_N}{n\binom{N}{n}}$$

$$= \frac{\binom{N-1}{n-1}}{n\binom{N}{n}} \cdot \sum_{j=1}^{N} x_j$$

$$= \frac{1}{N} \cdot \sum_{j=1}^{N} x_j = \mu_x = \mu$$

since

$$\frac{\binom{N-1}{n-1}}{n\binom{N}{n}} = \frac{\dfrac{(N-1)!}{(n-1)!\,(N-n)!}}{n \cdot \dfrac{N!}{n!\,(N-n)!}} = \frac{1}{N}$$

The derivation of Eq. (5.5) involves a little more algebra, but is straightforward. Using Theorem 3.1, we have

$$\sigma_{\bar{x}}^2 = \bar{x}_1^2 \cdot p_1 + \bar{x}_2^2 \cdot p_2 + \cdots \bar{x}_{\binom{N}{n}}^2 \cdot p_{\binom{N}{n}} - \mu_{\bar{x}}^2$$

$$= \frac{[(x_1 + \cdots + x_n)^2 + (x_1 + \cdots + x_{n+1})^2 + \cdots + (x_{N-n+1} + \cdots + x_N)^2]}{n^2\binom{N}{n}}$$

$$- \frac{(x_1 + \cdots + x_N)^2}{N^2}$$

Two types of terms are involved in squaring these sums, namely, x_i^2 and $2x_i x_j$ $(i \neq j)$. We know that each x_i^2 in the square brackets occurs $\binom{N-1}{n-1}$ times by using an argument similar to that used in obtaining Eq. (5.4). By the same kind of argument we find that a particular product of $x_i x_j$ in the square brackets occurs $\binom{N-2}{n-2}$ times. Since this is true for every pair x_i and x_j $(i \neq j)$, we may write

$$\sigma_{\bar{x}}^2 = \frac{\binom{N-1}{n-1} \cdot (x_1^2 + \cdots + x_N^2)}{n^2\binom{N}{n}} - \frac{x_1^2 + \cdots + x_N^2}{N^2}$$

$$+ \frac{2\binom{N-2}{n-2}(x_1 x_2 + \cdots + x_{N-1} x_N)}{n^2\binom{N}{n}} - \frac{2(x_1 x_2 + \cdots + x_{N-1} x_N)}{N^2}$$

(5.6)

Since

$$\frac{\binom{N-1}{n-1}}{n^2\binom{N}{n}} - \frac{1}{N^2} = \frac{1}{nN} - \frac{1}{N^2} = \frac{N-n}{nN^2} \qquad (5.7)$$

and

$$\frac{2\binom{N-2}{n-2}}{n^2\binom{N}{n}} - \frac{2}{N^2} = 2\left[\frac{n-1}{nN(N-1)} - \frac{1}{N^2}\right] = \frac{-2(N-n)}{nN^2(N-1)} \qquad (5.8)$$

we may, after substituting Eqs. (5.7) and (5.8) in Eq. (5.6), write

$$\sigma_{\bar{x}}^2 = \left(\frac{N-n}{nN^2}\right) \cdot (x_1^2 + \cdots + x_N^2) - 2\left[\frac{N-n}{nN^2(N-1)}\right] \cdot (x_1 x_2 + \cdots + x_{N-1} x_N)$$

or

$$\sigma_{\bar{x}}^2 = \frac{N-n}{n(N-1)}\left[\frac{N-1}{N^2}(x_1^2 + \cdots + x_N^2) - \frac{2}{N^2}(x_1 x_2 + \cdots + x_{N-1} x_N)\right]$$

$$(5.9)$$

But

$$\sigma^2 = \frac{N\sum\limits_{i=1}^{N} x_i^2 - \left(\sum\limits_{i=1}^{N} x_i\right)^2}{N^2} = \frac{x_1^2 + \cdots + x_N^2}{N} - \left(\frac{x_1 + \cdots + x_N}{N}\right)^2$$

or

$$\sigma^2 = \left(\frac{1}{N} - \frac{1}{N^2}\right) \cdot (x_1^2 + \cdots + x_N^2) - \frac{2}{N^2}(x_1 x_2 + \cdots + x_{N-1} x_N) \qquad (5.10)$$

Hence, on substituting Eq. (5.10) in Eq. (5.9), we get

$$\sigma_{\bar{x}}^2 = \frac{N-n}{n(N-1)} \sigma^2$$

It should be noted that nothing was said about the form of the parent population, and that $\sigma_{\bar{x}}^2 < \sigma^2$ when the sample size is greater than one. Further, note that when N is large when compared to n, the factor $(N-n)/(N-1)$ is nearly one, but always less than one. Thus

$$\sigma_{\bar{x}}^2 \doteq \frac{\sigma^2}{n} \qquad (5.11)$$

is a good approximation when n is small when compared to N; that is, the approximation in Eq. (5.11) is good when $(N-n)/(N-1)$ is near one.

5.3.2. Sampling with Replacement

We have considered in some detail sampling from a finite population without replacement. Next we consider the problem of sampling from a finite population with replacement. This method is not as useful in a practical sense as the one already discussed. However, when it comes to sampling from an infinite population, sampling with replacement is the method most commonly used—at least in the theoretical sampling distribution. We introduce this method of sampling now, using a finite population in order to make clear the technique and to have a connecting link between sampling without replacement from a finite population and sampling with replacement from an infinite population. Example 5.6 illustrates experimental sampling with replacement.

Example 5.6. Draw 100 random samples of size two with replacement from the population in Table 5.1 and compute the mean of each sample. Construct a table showing the experimental sampling distribution of 100 $\bar{x}$'s.

Place five disks marked 1, 3, 5, 7, and 9, respectively, in a bowl and mix thoroughly. Draw one disk, record its marking, replace it in the bowl, and mix again. Draw a second disk, record its marking, and replace it in the bowl. This gives a random sample of size two with replacement. Repeating this process 99 times, we obtain the 100 samples shown in Table 5.5, and after computing the 100 means we construct the sampling distribution of means shown in Table 5.6. If another 100 such sample means were determined

Table 5.5
Random Samples With Replacement Drawn
from the Population in Table 5.1

No.	Sample	No.	Sample	No.	Sample	No.	Sample	No.	Sample
1	3, 3	21	1, 9	41	5, 3	61	5, 1	81	3, 7
2	5, 3	22	3, 7	42	9, 3	62	3, 5	82	1, 7
3	1, 1	23	9, 3	43	5, 9	63	3, 1	83	5, 7
4	7, 3	24	9, 1	44	7, 3	64	7, 7	84	7, 7
5	1, 5	25	5, 9	45	5, 5	65	1, 1	85	1, 9
6	3, 5	26	5, 3	46	9, 9	66	9, 7	86	3, 3
7	5, 5	27	1, 9	47	9, 5	67	1, 3	87	3, 7
8	5, 7	28	9, 5	48	9, 7	68	9, 5	88	3, 1
9	9, 3	29	1, 9	49	7, 3	69	3, 5	89	1, 1
10	3, 3	30	5, 5	50	3, 7	70	9, 7	90	1, 7
11	5, 7	31	9, 3	51	3, 1	71	9, 7	91	1, 5
12	7, 3	32	1, 1	52	5, 5	72	1, 3	92	9, 1
13	3, 7	33	3, 3	53	9, 1	73	1, 5	93	7, 7
14	3, 3	34	1, 3	54	5, 9	74	7, 1	94	7, 3
15	1, 7	35	5, 1	55	5, 9	75	3, 5	95	5, 9
16	5, 9	36	1, 5	56	9, 1	76	5, 5	96	3, 5
17	9, 1	37	1, 5	57	3, 1	77	3, 5	97	9, 7
18	7, 5	38	7, 1	58	7, 1	78	9, 5	98	5, 7
19	7, 3	39	7, 1	59	7, 7	79	7, 1	99	5, 1
20	3, 9	40	3, 5	60	7, 9	80	9, 5	100	1, 3

Table 5.6
Experimental Sampling Distribution of $\bar{x}$
Obtained from Table 5.5

$\bar{x}$	Frequency	Rel. Freq.
1	4	0.04
2	8	0.08
3	13	0.13
4	18	0.18
5	25	0.25
6	10	0.10
7	15	0.15
8	6	0.06
9	1	0.01

and the relative-frequency table constructed, we would expect to get a different experimental sampling distribution of $\bar{x}$. However, just as in the illustration of sampling without replacement we can get a very good idea of the theoretical sampling distribution (model) by taking n sufficiently large.

The method of obtaining a theoretical sampling distribution with replacement from a finite population is illustrated in Example 5.7.

Example 5.7. Find the sampling distribution of means of samples of size two drawn with replacement from the finite population (disks marked 1, 3, 5, 7, and 9, respectively) given in Table 5.1 and Example 5.1.

All possible samples which can be drawn from this population with replacement are

$$1, 1 \quad 1, 3 \quad 1, 5 \quad 1, 7 \quad 1, 9$$
$$3, 1 \quad 3, 3 \quad 3, 5 \quad 3, 7 \quad 3, 9$$
$$5, 1 \quad 5, 3 \quad 5, 5 \quad 5, 7 \quad 5, 9$$
$$7, 1 \quad 7, 3 \quad 7, 5 \quad 7, 7 \quad 7, 9$$
$$9, 1 \quad 9, 3 \quad 9, 5 \quad 9, 7 \quad 9, 9$$

It should be noted that the same value can appear on each draw. For example, 1, 1 and 7, 7 are possible samples. Further, 1, 3 and 3, 1 represent different samples (ordered samples), even though they contain the same values. In the first sample, 1 was drawn first and then 3, but in the second sample the reverse is true. These 25 samples are mutually exclusive and equally likely and thus have the same probability. Computing the mean of each sample, we obtain the sampling distribution of means shown in Table 5.7. This distribution serves as a model for the distribution shown in Table 5.6.

Example 5.8. Compare the means and variances of the distributions found in Tables 5.6 and 5.7.

The mean and variance of the experimental sampling distribution of 100

Table 5.7
Theoretical Sampling Distribution with Replacement
of $\bar{x}$ Obtained from Table 5.1

$\bar{x}$	Frequency	Rel. Freq.
1	1	0.04
2	2	0.08
3	3	0.12
4	4	0.16
5	5	0.20
6	4	0.16
7	3	0.12
8	2	0.08
9	1	0.04

means shown in Table 5.6 are

$$\bar{\bar{x}} = \tfrac{478}{100} = 4.78$$

and

$$s_{\bar{x}}^2 = \frac{2626 - \dfrac{(478)^2}{100}}{99} = 3.45$$

The mean and variance of the sampling distribution of Table 5.7 are

$$\mu_{\bar{x}} = \tfrac{125}{25} = 5.00$$

and

$$\sigma_{\bar{x}}^2 = \frac{725 - \dfrac{(125)^2}{25}}{25} = 4.00$$

Even though the mean $\bar{\bar{x}}$ and variance $s_{\bar{x}}^2$ of the experimental sampling distribution are not the same as the mean $\mu_{\bar{x}}$ and variance $\sigma_{\bar{x}}^2$ of the theoretical sampling distribution, it can be shown that $\bar{\bar{x}}$ and $s_{\bar{x}}^2$ get closer and closer to $\mu_{\bar{x}}$ and $\sigma_{\bar{x}}^2$ "almost always" as the number of samples drawn goes up.

It should be observed that the variance obtained when one is sampling with replacement is larger than the variance obtained when sampling without replacement; that is, $4.00 > 3.00$. Thus, we would not expect Eq. (5.5) to give the relation between σ^2 and $\sigma_{\bar{x}}^2$ when sampling with replacement. Theorem 5.2 relates $\mu_{\bar{x}}$ and $\sigma_{\bar{x}}^2$ to μ and σ^2 in this case.

Theorem 5.2. *If N denotes the size of any finite population and n the size of a sample selected with replacement, then for all possible samples of size n the mean of the means $\mu_{\bar{x}}$ is equal to the population mean μ, and the variance of means $\sigma_{\bar{x}}^2$ is equal to the population variance σ^2 times a factor $1/n$; that is*

$$\mu_{\bar{x}} = \mu \qquad (5.12)$$

and

$$\sigma_{\bar{x}}^2 = \frac{\sigma^2}{n} \qquad (5.13)$$

Proof. Let the population values be denoted by

$$x_1, x_2, \ldots, x_N$$

Then all samples of size n may be indicated by $(x_1, x_1, \ldots, x_1, x_1), (x_1, x_1, \ldots, x_1, x_2), \ldots, (x_1, x_1, \ldots, x_1, x_N), \ldots, (x_N, x_N, \ldots, x_N, x_1), \ldots, (x_N, x_N, \ldots, x_N, x_N)$. There are N^n such samples, for the first observation may be any one of N values, the second observation any one of N values, ..., the nth observation any one of N values. Now each sample has the same probability $p_k = 1/N^n$ $(k = 1, \ldots, N^n)$. Hence, the mean of the mean of all these samples is given by

$$\mu_{\bar{x}} = \bar{x}_1 p_1 + \bar{x}_2 p_2 + \cdots + \bar{x}_{N^n} \cdot p_{N^n}$$

or

$$\mu_{\bar{x}} = \frac{\dfrac{x_1 + \cdots + x_1}{n} + \cdots + \dfrac{x_N + \cdots + x_N}{n}}{N^n} \qquad (5.14)$$

In order to simplify this expression further, we count the number of times each x occurs in Eq. (5.14). There are N^n samples, each with n values. Therefore, Eq. (5.14) contains $n \cdot N^n$ x-terms in all. Since there are N values in the population, and since each x obviously occurs the same number of times, then any particular x appears $n \cdot N^n/N = n \cdot N^{n-1}$ times. Thus,

$$\mu_{\bar{x}} = \frac{n \cdot N^{n-1}(x_1 + \cdots + x_N)}{n \cdot N^n} = \frac{x_1 + \cdots + x_N}{N} = \mu$$

The derivation of Eq. (5.13) follows a similar pattern, and will be left as an exercise for the student. Using the variances computed in Examples 5.5 and 5.8, we see that

$$\frac{\sigma^2}{n} = \frac{8.00}{2} = 4.00 = \sigma_{\bar{x}}^2$$

which verifies Eq. (5.13) in a special case.

Example 5.9. The probability that a sample mean falls between two limits depends on the method of sampling, the size of the sample, and the population from which the sample is drawn. For the parent population given in Table 5.1 and a sample of size two determine the probability that a sample mean $\bar{x}$ falls between 3 and 7 inclusive when the sample is randomly drawn (a) without replacement; (b) with replacement.

Using Table 5.4, we find for (a) that

$$P[3 \leq \bar{x} \leq 7] = P[\bar{x} = 3] + \cdots + P[\bar{x} = 7] = 0.10 + \cdots + 0.10 = 0.80$$

For (b), using Table 5.7, we obtain $P[3 \leq \bar{x} \leq 7] = 0.76$. Clearly, we expect the probability to be smaller for the sampling distribution with replacement, since the means are more disperse.

Example 5.10. What is the probability that a random sample of size two drawn from the population of Table 5.1 has a mean falling between $\mu_x - 2\sigma_x$ and $\mu_x + 2\sigma_x$ when the sample is drawn (a) without replacement; (b) with replacement?

According to Example 5.4 the mean μ_x and variance σ_x^2 for the sampling distribution without replacement given in Table 5.4 are 5.00 and 3.00, respectively. Thus $\sigma_x = 1.73$, and the limits for (a) are $5.00 \pm 2(1.73)$, or 1.54 and 8.46. Using Table 5.4, we find the probability to be $P[1.54 < \bar{x} < 8.46] = 1.00$. The mean and variance of the sampling distribution with replacement are 5.00 and 4.00, respectively. Thus $\sigma_x = 2.00$ and the required probability, when Table 5.7 is used, is $P[5.00 - 2(2.00) < \bar{x} < 5.00 + 2(2.00)] = P[1 < \bar{x} < 9] = 0.92$.

Example 5.11. What size sample must be drawn from a population of size N in order for the standard deviation of the sampling distribution of means to be half the standard deviation of the parent population?

Let $\sigma = \sigma_0$. Then $\sigma_x = \sigma_0/2$. The sample size depends on the method of sampling. When sampling with replacement, we have $\sigma_x^2 = \sigma^2/n$. Thus

$$n = \frac{\sigma^2}{\sigma_x^2} = \frac{\sigma_0^2}{\left(\dfrac{\sigma_0}{2}\right)^2} = 4$$

When sampling without replacement, we have

$$\sigma_x^2 = \frac{\sigma^2}{n} \cdot \frac{N-n}{N-1}$$

Thus, in our problem

$$\left(\frac{\sigma_0}{2}\right)^2 = \frac{\sigma_0^2}{n}\left(\frac{N-n}{N-1}\right)$$

or

$$(N-1)n = 4(N-n)$$

or

$$n = \frac{4N}{N+3}$$

It is very interesting to note that the sample size does not depend on the population size when one is sampling with replacement, but it does depend on the population size when one is sampling without replacement. For when $N = 9$, in Example 5.11, the sample size must be four when we are sampling with replacement and three when we are sampling without replacement; when $N = 3$, the sample sizes are four and two, respectively.

Further, note that the variance of the sampling distribution of means can be made as small as we wish simply by choosing n large enough.

So far, the sampling distributions of means determined from samples drawn from finite populations are the only particular sampling distributions discussed. Sampling distributions of other statistics are considered in the next set of exercises, and those determined from infinite populations are discussed in Sect. 5.4. Sampling distributions derived from particular populations are discussed in later sections.

5.3.3. Exercises

5.1. The observations of a sample of size seven are arranged in increasing order of magnitude and denoted by $x_1, x_2, \ldots, x_7$, respectively. (a) Which, if any, of the following are statistics?

(i) $(x_1 + x_7)/2$

(ii) $(x_2 + \cdots + x_6)/5$

(iii) $x_6 - x_2$

(iv) $[(x_6 + x_7) - (x_1 + x_2)]/2$

(v) $7(x_1 + x_7)/2$

(vi) $(2x_1 + x_3 + x_5 + 2x_7)/6$

(vii) x_4

(viii) $(1/x_1 + \cdots + 1/x_7)/7$

(ix) $(x_2^2 + x_6^2)/2$

(x) $(x_2 + x_4 + x_6)/3$

(b) Which, if any, of the values in (a) may be used as measures of central tendency? As measures of dispersion? (c) Give a generalized expression for each of the values in (a).

Hint. Let $x_1, x_2, \ldots, x_n$ denote an ordered sample and consider cases where n is odd and even.

5.2. The five disks of a finite population are marked 1, 4, 7, 10, 13. List all samples of size three which can be drawn from this population without replacement and use in the calculations of (a), (b), (c), (d), (e), (f), and (g): (a) Find the sampling distribution of means. (b) Find the sampling distributions of totals. (c) Find the sampling distribution of medians. (d) Find the sampling distribution of ranges. (e) Find the sampling distribution of variances. (f) Find the sampling distribution of standard deviations. (g) Find the sampling distribution of $(x_1 + x_3)/2$, where x_1 and x_3 denote the smallest and largest values of a sample. (h) Find the means of the sampling distributions in (a), (b), (c), (d), (e), (f), and (g). (i) Compute the mean, median, range, variance, and standard deviation of the parent population and compare each with the appropriate values in (h). (j) Find the variances of the sampling distributions in (a), (b), and (g). Find the variance of the parent population and compare with each of the variances in (a), (b), and (g).

5.3. The four disks of a finite population are marked 1, 4, 7, and 10. List all samples of size three which can be drawn from this population with replacement and use to answer (a), (b), (c), (d), (e), (f), (g), (h), (i), and (j) of Exercise 5.2.

5.4. The six disks of a finite population are marked 1, 1, 1, 3, 3, and 5. List all samples of size two which can be drawn from this population without replacement and use to answer (a), (b), (c), (d), (e), (f), (g), (h), (i), and (j) of Exercise 5.2.

5.5. Work Exercise 5.4 if the samples are drawn with replacement.

5.6. Prove Eq. (5.13) of Theorem 5.2.

5.7. Use Exercise 5.2 to find the probability that a sample drawn at random from the parent population has the following statistic within two standard deviations (in units of the statistic) of the mean of the sampling distribution: (a) mean, (b) median, (c) $(x_1 + x_2)/2$.

5.8. Use Exercise 5.3 to answer (a), (b), and (c) of Exercise 5.7.

5.9. Use Exercise 5.4 to answer (a), (b), and (c) of Exercise 5.7.

5.10. What size sample must be drawn without replacement from a population of size N in order for the standard deviation of the sampling distribution of means to be $1/k$th the standard deviation of the parent population?

5.11. Assume that three finite populations of sizes 1000, 10,000 and 100,000, respectively, have the same variance. A random sample of size 50 is drawn without replacement from each of these populations. Compute $\sigma_{\bar{x}}$ for each and compare their magnitudes.

5.12. The discrete uniform distribution has the density function $f(x) = 1/m$, where $x = 1, 2, \ldots, m$. Samples of size n are drawn with replacement. Determine the sampling distribution of means.

5.13. Work Exercise 5.12 if samples of size n are drawn without replacement.

5.14. Determine the sampling distribution of ranges of samples of size n drawn with replacement from the population of Exercise 5.12.

5.15. Determine the sampling distribution of medians of samples of size n drawn without replacement from the population of Exercise 5.12.

5.16. (a) Find the mean and variance of the sampling distribution of means in Exercise 5.12. (b) What proportion of sample means lie within one standard deviation of the population mean?

5.17. Use Exercise 5.14 to determine what proportion of ranges of samples of size n are greater than $m/2$; less than $m/4$.

5.18. A discrete distribution has density function $f(x) = (x - 2)^2/31$ when $x = 1, 2, \ldots, 6$. Samples of size n are drawn with replacement. (a) Determine the sampling distribution of totals. (b) Determine the sampling distribution of medians. (c) Find the mean and variance of the sampling distribution in (a); in (b). (d) Find an interval which contains 90 per cent of the means of samples of size n drawn from the parent population. What are the limits of the shortest such interval?

5.19. If samples of size n are drawn with replacement from the dichotomous distribution whose density function is given in Eq. (3.11) and the total number of successes is denoted by T, then the density function of the sampling distribution of T, according to Formula (4.12), is

$$f(T) = \frac{n!}{T!(n-T)!} p^T q^{n-T} \qquad (5.15)$$

where

$$T = 0, 1, \ldots, n$$

Thus, it is clear that the binomial distribution may be considered a sampling distribution of totals obtained from n drawings from a dichotomous distribution. Show that the mean μ_T and variance σ_T^2 of the binomial distribution are given by

$$\mu_T = np \quad \text{and} \quad \sigma_T^2 = npq = np(1-p) \qquad (5.16)$$

5.20. The moment generating function (MGF), $M_x(t)$, of the variable x with discrete density function $f(x)$ is defined by

$$M_x(t) = \sum e^{tx} f(x) \qquad (5.17)$$

where the summation is over all values of x for which $f(x) \neq 0$. [See Formula (3.43) of Exercise 3.47 for further information.] Use the moment generating function for the binomial distribution in Exercise 5.19 to find μ_T and σ_T^2.

5.21. Let samples of size n be drawn without replacement from a finite dichotomous population of size N in which there are Np "successes" and Nq "failures" $(p + q = 1)$. Then prove that the relative frequency of x successes and $n - x$ failures in a sample of size n (i. e., the *hypergeometric* density function of x) is given by

$$f(x) = h(x, n - x; N, p, q) = \frac{\binom{Np}{x} \cdot \binom{Nq}{n-x}}{\binom{N}{n}} \qquad (5.18)$$

Note. The distribution derives its name from the fact that the values $f(x)$ can be expressed as successive terms of a hypergeometric series.

5.22. Show that for the hypergeometric distribution

$$\mu_x = np \quad \text{and} \quad \sigma_x^2 = np(1-p) \cdot \frac{N-n}{N-1} \qquad (5.19)$$

The mean is the same as for the binomial distribution, but the variance is less. It can be shown that when Np and N approach infinity in such a manner that the ratio $p = Np/N$ becomes fixed, the hypergeometric distribution reduces to the binomial distribution. Further, if $n/N < 0.1$,

$p < 0.1$, and N is fairly large, the Poisson distribution may be used as a satisfactory approximation in most cases.

5.23. A dichotomous distribution of size eight contains three defective and five nondefective parts. (a) Use Eq. (5.18) to find the frequency distribution of the number of defective parts in a random sample (without replacement) of size five. (b) Compute the mean and variance for this distribution and show that the results obtained agree with those obtained by using Eq. (5.19). In the past the hypergeometric distribution has most often been used in connection with *acceptance sampling*, but this is by no means the only area of application.

5.24. There are many experiments in which we require more than two categories of classification. For example, in response to a question one might obtain answers such as "yes," "no," "don't know," and "no answer;" in tossing a die there are six categories. Thus, we consider generalizations of the dichotomous type distributions of Exercises 5.19, 5.20, 5.21, 5.22, and 5.23.

Suppose that there are k mutually exclusive and exhaustive possible outcomes $O_1, O_2, \ldots, O_k$ of an experiment for which the probabilities are $p_1, p_2, \ldots, p_k$, respectively, and

$$\sum_{i=1}^{k} p_i = 1$$

Let samples of size n be drawn with replacement from such a population. Let x_i denote the number of times outcome O_i occurs ($i = 1, \ldots, k$). Prove that the joint density function $f(x_1, \ldots, x_k)$ for the random variables $x_1, \ldots, x_k$ is

$$f(x_1, \ldots, x_k) = \frac{n!}{x_1! \ldots x_k!} p_1^{x_1} \ldots p_k^{x_k} \tag{5.20}$$

where each x_i may range from zero to n inclusive and

$$\sum_{i=1}^{k} x_i = n$$

Since the terms in the expansion of

$$(p_1 + p_2 + \cdots + p_k)^n = 1$$

are those given by Eq. (5.20), we call this distribution the *multinomial distribution*.

It should be noted that Eq. (5.20) is a joint density function of only $k - 1$ random variables, since the kth x is exactly determined by the relation

$$\sum_{i=1}^{k} x_i = n$$

when the other $k - 1$ x's are specified. Since the binomial density was written as $f(x)$ rather than $f(x, n - x)$, we may write the multinomial density as $f(x_1, \ldots, x_{k-1})$ in place of $f(x_1, \ldots, x_{k-1}, x_k)$ or

$$f\left(x_1, \ldots, x_{k-1}, n - \sum_{i=1}^{k-1} x_i\right)$$

5.25. Show that for the multinomial distribution

$$\begin{cases} \mu_{x_i} = np_i \\ \sigma^2_{x_i} = np_i(1 - p_i) \quad \text{and} \\ \text{cov}\,(x_i, x_j) = \sigma_{x_i x_j} = -np_i p_j \quad (i \neq j; i, j = 1, \ldots, k) \end{cases} \qquad (5.21)$$

Hint. The MGF, $M = M_{x_1 \ldots x_{k-1}}(t_1, \ldots, t_{k-1})$, given by

$$M = \sum e^{t_1 x_1 + \ldots + t_{k-1} x_{k-1}} f(x_1, \ldots, x_{k-1}) \qquad (5.22)$$

may prove useful. It is to be understood that the summation is over all values of $x_1, \ldots, x_{k-1}$ for which $f(x_1, \ldots, x_{k-1}) \neq 0$.

5.26. Let samples of size n be drawn without replacement from the parent distribution with k categories defined in Exercise 5.24. Let x_i denote the number of times outcome O_i occurs $(i = 1, \ldots, k)$. Prove that the probability of density function $f(x_1, \ldots, x_k) = h(x_1, \ldots, x_k; N, p_1, \ldots, p_k, n) \equiv h$ is given by

$$h = \frac{\binom{Np_1}{x_1} \cdots \binom{Np_k}{x_k}}{\binom{N}{n}} \qquad (5.23)$$

where each x_i may range from zero to n inclusive and

$$\sum_{i=1}^{k} x_i = n$$

This expression, Eq. (5.23), is an extension of Eq. (5.18), the density function for the hypergeometric distribution.

5.27. In a human population of size ten it is known that four answer "no," five answer "yes," and one answers "don't know" in response to a certain question. (a) Use Eq. (5.23) to find the frequency distribution of answers if a sample of size four is drawn at random (without replacement) from this population. (b) Find the mean and variance for the individuals answering "yes." What is the covariance of the "yes" and "no" answers?

5.4. SAMPLING FROM AN INFINITE POPULATION

It is possible to discuss both experimental and theoretical sampling from an infinite population just as we did with a finite population. We may think of the sampling distribution of some statistic determined from experimental sampling as approximating the sampling distribution of the same statistic determined from theoretical sampling—the theoretical sampling distribution being thought of as a model. Since experimental sampling from an infinite population is carried on as it is for finite populations, we limit most of our discussion in this chapter to theoretical sampling distributions.

Before we continue, it should be observed that infinite populations fall into three classes. A population may take on (1) every value on a continuum, (2) a countable discrete set of values, or (3) infinitely many values, only k of them being different. The normal distribution is an illustration of (1), the Poisson distribution is an example of (2), and a die in which every number may be thought of as occurring infinitely many times illustrates (3).

In this section we discuss sampling distributions in terms of means, just as we did when sampling from finite populations. In particular, we first observe how Formula (5.5) in Theorem 5.1, relating variances of a finite parent population and the corresponding theoretical sampling distribution of means, enables us to extend the relationship to populations with infinitely may discrete values [types (2) and (3) in the above paragraph]. For as N becomes infinitely large and we sample without replacement, we have $\sigma_{\bar{x}}^2 = \sigma^2/n$, since $(N - n)/(N - 1)$ has 1 as a limit. This is the relationship stated by Eq. (5.13) of Theorem 5.2. Thus, when sampling without replacement from an infinite discrete population, we are led to the same result as sampling with replacement from a finite population. Further, it will be shown that this relationship between two variances exists when one is sampling with or without replacement from any infinite population, including type (1) above.

It is of interest to consider the connection between random sampling with and without replacement. Conceptually, drawing a random sample without replacement from a bowl containing indefinitely many disks marked 1, indefinitely many marked 2, ... , indefinitely many marked k is the same as drawing a random sample with replacement from a bowl containing k disks marked $1, 2, \ldots, k$, respectively. In particular, drawing a random sample of two objects with replacement from a finite population is the same as drawing a sample of two objects without replacement from an infinite population.

In a continuous population the probability of drawing a particular value is zero. Hence, the probability of drawing a second value is not affected by the first draw. It follows that random drawings from an infinite population are independent of each other. It is in this sense that the two methods of sampling already discussed amount to the same thing when one is sampling from an infinite population. So in the remainder of this book, unless it is otherwise indicated, when we speak of sampling at random from an infinite population, we may mean either sampling with or without replacement.

We have already noted that for infinitely discrete populations Eq. (5.13) of Theorem 5.2 can be obtained from Eq. (5.5) of Theorem 5.1 by letting N become infinitely large. However, Eq. (5.13) along with Eq. (5.12) can be obtained by another method. This method indicates how the corresponding relations may be derived for the continuous population—that is, the forthcoming method serves as a bridge between finite discrete and continuous cases and illustrates a method which is very important to the mathematics of theoretical sampling.

Let an infinitely large discrete population have the following distribution, $f(x)$ being the density function

x	x_1	x_2	$\ldots$	x_k
$f(x)$	$f(x_1)$	$f(x_2)$	$\ldots$	$f(x_k)$

We wish to determine the mean $\mu_{\bar{x}}$ and variance $\sigma_{\bar{x}}^2$ of the theoretical sampling distribution of means of all random samples of size two. Let x_i and x_j $(i, j = 1, 2, \ldots, k)$ denote the values resulting from the first and second draws of the random variable x. Since $f(x_i)$ and $f(x_j)$ denote the probabilities of the occurrence of x_i and x_j, respectively, and since x_i and x_j are independent, the probability $f(x_i, x_j)$ of the joint occurrence of x_i and x_j is $f(x_i) \cdot f(x_j)$, by Eq. (4.3a). Thus, the probability $f[(x_i + x_j)/2]$ of obtaining the mean $(x_i + x_j)/2$ computed from this sample is $f(x_i) \cdot f(x_j)$. If we use Eq. (3.7), the mean of all possible means is given by

$$\mu_{\bar{x}} = \sum^{k^2} \left(\frac{x_i + x_j}{2} \right) f\left(\frac{x_i + x_j}{2} \right) = \sum_{i=1}^{k} \sum_{j=1}^{k} \left[\frac{x_i + x_j}{2} \right] \cdot f(x_i) \cdot f(x_j)$$

$$= \tfrac{1}{2} \sum_{i=1}^{k} \sum_{j=1}^{k} x_i f(x_i) \cdot f(x_j) + \tfrac{1}{2} \sum_{j=1}^{k} \sum_{i=1}^{k} x_j f(x_i) f(x_j)$$

$$= \tfrac{1}{2} \sum_{i=1}^{k} x_i f(x_i) \cdot \sum_{j=1}^{k} f(x_j) + \tfrac{1}{2} \sum_{j=1}^{k} x_j f(x_j) \cdot \sum_{i=1}^{k} f(x_i)$$

or

$$\mu_{\bar{x}} = \tfrac{1}{2} \sum_{i=1}^{k} x_i f(x_i) + \tfrac{1}{2} \sum_{j=1}^{k} x_j f(x_j) \tag{5.24}$$

since

$$\sum_{i=1}^{k} f(x_i) = \sum_{j=1}^{k} f(x_j) = 1 \tag{5.25}$$

By definition, the mean μ of the given distribution is

$$\mu = \sum_{i=1}^{k} x_i f(x_i) \quad \text{or} \quad \mu = \sum_{j=1}^{k} x_j f(x_j) \tag{5.26}$$

Substituting Eq. (5.26) in Eq. (5.24) gives

$$\mu_{\bar{x}} = \tfrac{1}{2}\mu + \tfrac{1}{2}\mu = \mu \tag{5.27}$$

Further, by Formula (3.9), the variance of the theoretical sampling distribution of means is given by

$$\sigma_{\bar{x}}^2 = \sum^{k^2} \left(\frac{x_i + x_j}{2}\right)^2 f\left(\frac{x_i + x_j}{2}\right) - \mu_{\bar{x}}^2$$

$$= \sum_{i=1}^{k} \sum_{j=1}^{k} \left(\frac{x_i^2 + 2x_i x_j + x_j^2}{4}\right) f(x_i) f(x_j) - \mu_{\bar{x}}^2$$

or

$$\sigma_{\bar{x}}^2 = \tfrac{1}{4}\left[\sum_{i=1}^{k} x_i^2 f(x_i) \cdot \sum_{j=1}^{k} f(x_j) + 2 \sum_{i=1}^{k} x_i f(x_i) \cdot \sum_{j=1}^{k} x_j f(x_j)\right.$$
$$\left. + \sum_{j=1}^{k} x_j^2 f(x_j) \cdot \sum_{i=1}^{k} f(x_i)\right] - \mu_{\bar{x}}^2 \tag{5.28}$$

Substituting Eqs. (3.9), (5.25), and (5.26) in Eq. (5.28) gives

$$\sigma_{\bar{x}}^2 = \tfrac{1}{4}[(\sigma^2 + \mu^2)\cdot 1 + 2\mu\cdot\mu + (\sigma^2 + \mu^2)\cdot 1] - \mu^2$$

or, on reduction

$$\sigma_{\bar{x}}^2 = \frac{\sigma^2}{2} \tag{5.29}$$

By a similar method it can be shown that for samples of size n

$$\mu_{\bar{x}} = \mu \tag{5.30}$$

and

$$\sigma_{\bar{x}}^2 = \frac{\sigma^2}{n} \tag{5.31}$$

when samples are drawn with or without replacement from an infinitely large discrete population.

Next we determine the mean and variance of the theoretical sampling distribution of means of all random samples of size two drawn from an infinitely large population with random variable x ($-\infty \leq x \leq \infty$) having continuous density function $f(x)$. Denote the first and second draws of a random sample by x_1 and x_2, respectively. Since x_1 is a random variable, it can take any value of x in the interval ($-\infty, \infty$), and it has density function $f(x_1)$, where $f(x_1)$ is of the same form as $f(x)$. Thus, $f(x_1)\Delta x_1$ is the probability that x_1 falls between the limits of any interval Δx_1. Similarly, x_2 is a random variable with density function $f(x_2)$, having the same form as $f(x)$, and $f(x_2)$. Δx_2 is the probability that x_2 falls between the limits of any interval Δx_2. Since x_1 and x_2 are independent, the probability $f(x_1, x_2) \Delta x_1 \Delta x_2$ that a random sample will simultaneously have the first value between the limits of Δx_1 and the second value between the limits of Δx_2 is $f(x_1) \Delta x_1 \cdot f(x_2) \Delta x_2$, by Eq. (4.3a). Thus, the probability $f[(x_1 + x_2)/2]$ of the mean $(x_1 + x_2)/2$, computed from the random sample x_1 and x_2, falling in the interval $\Delta[(x_1 + x_2)/2]$ determined from the joint intervals Δx_1 and Δx_2 (area geometrically) is

$$f\left(\frac{x_1 + x_2}{2}\right) \Delta\left(\frac{x_1 + x_2}{2}\right) = f(x_1)\,\Delta x_1 \cdot f(x_2)\,\Delta x_2$$

From this argument, if we use Eq. (3.20), if follows that the mean of the means is give by

$$\mu_{\bar{x}} = \int_{-\infty}^{\infty} \bar{x} f(\bar{x})\,d\bar{x}$$

$$= \int_{-\infty}^{\infty} \left(\frac{x_1 + x_2}{2}\right) f\left(\frac{x_1 + x_2}{2}\right) d\left(\frac{x_1 + x_2}{2}\right)$$

$$= \int_{-\infty}^{\infty} \int_{-\infty}^{\infty} \left(\frac{x_1 + x_2}{2}\right) f(x_1) f(x_2)\,dx_1\,dx_2$$

$$= \tfrac{1}{2}\left[\int_{-\infty}^{\infty}\int_{-\infty}^{\infty} x_1 f(x_1) f(x_2)\,dx_2\,dx_1 + \int_{-\infty}^{\infty}\int_{-\infty}^{\infty} x_2 f(x_1) f(x_2)\,dx_1\,dx_2\right]$$

$$= \tfrac{1}{2}\left\{\int_{-\infty}^{\infty} x_1 f(x_1)\left[\int_{-\infty}^{\infty} f(x_2)\,dx_2\right]dx_1 + \int_{-\infty}^{\infty} x_2 f(x_2)\left[\int_{-\infty}^{\infty} f(x_1)\,dx_1\right]dx_2\right\}$$

$$= \tfrac{1}{2}\left[\int_{-\infty}^{\infty} x_1 f(x_1)\,dx_1 + \int_{-\infty}^{\infty} x_2 f(x_2)\,dx_2\right]$$

or

$$\mu_{\bar{x}} = \mu$$

since

$$\int_{-\infty}^{\infty} f(x_1)\,dx_1 = \int_{-\infty}^{\infty} f(x_2)\,dx_2 = 1$$

and

$$\int_{-\infty}^{\infty} x_1 f(x_1)\,dx_1 = \int_{-\infty}^{\infty} x_2 f(x_2)\,dx_2 = \mu$$

The variance of the theoretical sampling distribution of means of size two, if we use Eqs. (3.22), (3.17), and (3.20), is

$$\sigma_{\bar{x}}^2 = \int_{-\infty}^{\infty} \bar{x}^2 f(\bar{x})\,d\bar{x} - \mu_{\bar{x}}^2$$

$$= \tfrac{1}{4}\int_{-\infty}^{\infty}\int_{-\infty}^{\infty} (x_1^2 + 2x_1 x_2 + x_2^2) f(x_1) f(x_2)\,dx_1\,dx_2 - \mu^2$$

$$= \tfrac{1}{4}\left\{\int_{-\infty}^{\infty} x_1^2 f(x_1)\left[\int_{-\infty}^{\infty} f(x_2)\,dx_2\right]dx_1 + 2\int_{-\infty}^{\infty} x_1 f(x_1)\left[\int_{-\infty}^{\infty} x_2 f(x_2)\,dx_2\right]dx_1\right.$$

$$\left. + \int_{-\infty}^{\infty} x_2^2 f(x_2)\left[\int_{-\infty}^{\infty} f(x_1)\,dx_1\right]dx_2\right\} - \mu^2$$

$$= \tfrac{1}{4}\left[\int_{-\infty}^{\infty} x_1^2 f(x_1)\,dx_1 + 2\mu\int_{-\infty}^{\infty} x_1 f(x_1)\,dx_1 + \int_{-\infty}^{\infty} x_2^2 f(x_2)\,dx_2\right] - \mu^2$$

$$= \tfrac{1}{4}[(\sigma^2 + \mu^2) + 2\mu\cdot\mu + (\sigma^2 + \mu^2)] - \mu^2$$

or

$$\sigma_{\bar{x}}^2 = \frac{\sigma^2}{2}$$

The reader should check each step in these last two derivations to be sure the reasons are clear. The proofs are given in great detail so that a comparison may be made with the proofs involving $\sum$ symbols.

The general proof follows a similar argument leading to the results of Theorem 5.3.

Theorem 5.3. *Let x be a continuous random variable distributed with mean* μ, *variance* σ^2 *and density function* $f(x)$. *If random samples of size n are drawn from this distribution, then the sampling distribution of means has mean* $\mu_{\bar{x}}$ *equal to the population mean and variance* $\sigma_{\bar{x}}^2$ *equal to the population variance times a factor of 1/n; that is,*

$$\mu_{\bar{x}} = \mu$$

and

$$\sigma_{\bar{x}}^2 = \frac{\sigma^2}{n}$$

Thus, sampling at random with replacement from a finite population or sampling at random from an infinite population leads to the same results concerning $\sigma_{\bar{x}}^2$. That is, the variance of the sample mean is equal to the population variance divided by the sample size, provided the population variance is finite.

This is an extremely important property in the application of statistics, since it means that the distribution of the sample mean becomes more and more concentrated about the population mean (since $\mu_x = \mu_{\bar{x}}$). Thus, "as the sample size increases, we become more certain that the sample mean is a good estimate of the population mean." This is a rough statement of the *law of large numbers* in terms of $\bar{x}$. One form of a more precise statement is that

$$P[|\bar{x} - \mu| > k] \le \frac{\sigma^2}{nk^2} = \frac{\sigma_{\bar{x}}^2}{k^2} \tag{5.32}$$

where k is any positive real number. Formula (5.32) is known as *Chebyshev's inequality*. [For other forms of Eq. (5.32) and more discussion, see the next set of exercises.] It should be observed at this time that the random variable need not be $\bar{x}$.

Without knowing the form of the distribution of the parent population, it is possible, if the variance is known, to determine a sample size n such that any desired proportion of the sample means falls within k units of the true mean. The following example illustrates this.

Example 5.12. Determine the smallest random sample for which the probability is at least 0.90 that the difference between the average number of defectives drawn from a dichotomous population and the true proportion defective p unknown is not greater than 0.1.

From Eqs. (3.13) and (3.14) we have $\mu = p$ and $\sigma^2 = p(1 - p)$. Now n must be large enough so that the following inequality holds

$$P[|\bar{x} - \mu| \leq 0.1] > 0.90$$

But from Eq. (5.32) we obtain

$$P[|\bar{x} - \mu| \leq k] > 1 - \frac{\sigma^2}{nk^2}$$

or

$$P[|\bar{x} - \mu| \leq 0.1] > 1 - \frac{p(1 - p)}{n(0.1)^2}$$

Hence we must find n such that

$$1 - \frac{p(1 - p)}{n(0.1)^2} = 0.90$$

or

$$n = 1000\, p(1 - p)$$

We must find n such that the inequality (5.32) is satisfied no matter what the value of p. It is clear that $p = \frac{1}{2}$ makes $p(1 - p) = \frac{1}{4}$ a maximum and gives the required value

$$n = 250$$

Later we shall find n by taking into account the sampling distribution of proportion of defective. Using this knowledge shows that a much smaller sample size would suffice. But Chebyshev's inequality has the advantage that it does not depend on the distribution of the random variable (so long as the random variable has finite mean and variance).

If we restrict our attention to the sampling distribution of the mean (or total), there is a property which is much more noteworthy and useful than the Chebyshev form of the statement of the law of large numbers. It is the most important theorem in statistics; it is the theorem which gives the normal distribution such a central place in both the theory and application of statistics. The following statement of this theorem, called the *central-limit theorem*, is one of many forms in which it is written.

Theorem 5.4 (Central-limit Theorem). *Let the random variable x be distributed with mean μ and variance σ^2 (but with density function unknown). Then the distribution of the sample mean $\bar{x}$ is very closely approximated by the normal distribution with mean μ and variance σ^2/n when n is large.*

A general proof of this theorem is beyond the scope of this book. However, a restricted proof is outlined in Exercise 5.69. Among the mathematicians who have worked on various phases of this fundamental theorem are DeMoivre, Laplace, Gauss, Chebyshev, Liapounoff, Lindeberg, Lévy, Feller, and Cramér. Laplace [20] first stated the theorem in 1812. In 1901 Liapounoff [23, 24] gave a rigorous proof under fairly general conditions. Feller, Khintchine, Lévy, and others [11, 12, 19, 22] found most general conditions under which the theorem is valid.

It is an amazing fact that nothing is required of the form of the distribution function except that the variance be finite. In application, the requirement of finite variance is no real restriction, since in almost any practical problem the range of the variate is finite. This implies that the variance is necessarily finite.

One using this theorem would like to know how large n must be before the normal approximation is adequate. There is no simple answer. The size of n depends on the shape of the parent distribution. Unless the parent distribution is unusual in shape, a sample no smaller than 30 should furnish a reasonable approximation. Of course, in cases where it is known that the parent distribution is normal or approximately normal, fewer than 30 observations might be used with discretion.

We have already observed the piling-up effect of sample means in the neighborhood of the population mean in Table 5.4 and Table 5.7. In these places there is even a hint that the limiting distribution might be approximately normal. Assuming that the distribution of samples of size sixteen drawn from the population (1, 3, 5, 7, 9) given in Table 5.1 yields a fair approximation to the normal distribution, we consider the following example.

Example 5.13. The mean of the population of Table 5.1 is $\mu = 5$. If means are computed for samples of size sixteen randomly drawn with replacement, use Theorem 5.4 to determine a symmetrical interval about μ in which 90 per cent of the means fall.

From Example 5.5 we find the population variance to be $\sigma^2 = 8$. According to Theorem 5.4, $\bar{x}$ is approximately normally distributed with $\mu_{\bar{x}} = 5$ and $\sigma_{\bar{x}} = \sqrt{\frac{8}{16}} \doteq 0.707$. For the standard normal distribution the interval from $t_1 = -1.645$ to $t_2 = 1.645$ contains 90 per cent of the t values (see Example 3.7). We wish to find two values $\bar{x}_1$ and $\bar{x}_2$ such that $P[\bar{x}_1 \leq \bar{x} \leq \bar{x}_2] = 0.90$. That is, $\bar{x}_1$ and $\bar{x}_2$ must be determined from

$$\frac{\bar{x} - \mu_{\bar{x}}}{\sigma_{\bar{x}}} = \pm 1.645$$

or

$$\bar{x} = \pm 1.645 \cdot \sigma_{\bar{x}} + \mu_{\bar{x}}$$

Thus, the required interval is $3.84 \leq \bar{x} \leq 6.16$.

The student could make some judgment as to the goodness of the normal approximation by drawing, say, 100 samples of size sixteen and computing their means. If roughly 90 of these means fall in the interval $3.84 \leq \bar{x} \leq 6.16$, he might think the approximation is adequate. (For some students it might require 500 or 1000 or more samples to reach a satisfactory decision.)

Note. Usually the sampling distribution of means approaches with increasing n the normal curve faster in the neighborhood of the mean than in regions some distance from the mean. Thus, the further a point is from the population mean, the more slowly the sampling distribution of means can be expected to approach the normal curve.

In Sect. 2.4 we learned, for finite populations, that if $y = kx$, k being a nonzero constant, then $\mu_y = k\mu_x$ and $\sigma_y^2 = k^2\sigma_x^2$. These relations can be shown to hold for discrete infinite and continuous populations. Using these relations, we may prove Theorem 5.5.

Theorem 5.5. *Let the random variable x be distributed with mean μ, variance σ^2, and density function $f(x)$. The theoretical sampling distribution of totals $T = \sum x_i$ of random samples of size n drawn from the parent population has mean μ_T and variance σ_T^2 given by*

$$\mu_T = n\mu \tag{5.33}$$

and

$$\sigma_T^2 = \begin{cases} n\sigma^2 \cdot \dfrac{N-n}{N-1}, & \text{when sampling from a finite population} \\ & \text{without replacement} \\ n\sigma^2, & \text{otherwise} \end{cases} \tag{5.34}$$

Proof. We have immediately that

$$\mu_T = \mu_{nx} = n\mu_x = n\mu$$

and

$$\sigma_T^2 = \sigma_{nx}^2 = n^2\sigma_x^2$$

It should be clear from the theorems of this and the last section that we may work with either the sampling distribution of means or the sampling distribution of totals, whichever seems most appropriate in a given problem. We now illustrate theorems and formulas of this section.

Example 5.14. A random sample of size two is drawn from the uniform distribution with density function

$$f(x) = \begin{cases} 1, & \text{when } -\tfrac{1}{2} \leq x \leq \tfrac{1}{2} \\ 0, & \text{otherwise} \end{cases}$$

Find (a) the mean and variance of the sampling distribution of means, (b) the density function of the sampling distribution of means, (c) the probability that a sample mean will fall between $-\frac{1}{4}$ and $\frac{1}{4}$, and (d) the probability that a sample mean will fall within two standard errors of the population mean.

For (a), use Formulas (3.25) and (3.26) to obtain $\mu = 0$ and $\sigma^2 = \frac{1}{12}$. Then from Theorem 5.3 it follows that $\mu_{\bar{x}} = 0$ and $\sigma_{\bar{x}}^2 = \frac{1}{12}/2 = \frac{1}{24}$.

To solve (b), let x_1 denote the first observation and x_2 the second observation in the sample. Since the sample is random, the observations are independently distributed. Thus, if $f(x_1) = f(x)$ is the density function of x_1 and $f(x_2) = f(x)$ is the density function of x_2, the joint density function $f(x_1, x_2)$ of x_1 and x_2 is given by $f(x_1, x_2) = f(x_1) \cdot f(x_2)$, or

$$f(x_1, x_2) = \begin{cases} 1, & \text{when } -\tfrac{1}{2} \le x_1 \le \tfrac{1}{2} \text{ and } -\tfrac{1}{2} \le x_2 \le \tfrac{1}{2} \\ 0, & \text{otherwise} \end{cases} \tag{5.35}$$

The graph of this function is shown in Fig. 5.1.

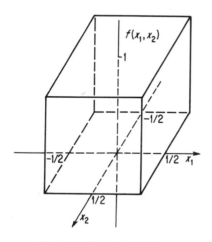

Fig. 5.1 Bivariate Uniform Distribution

We require the density function of $\bar{x}$. This may be found from the joint density function $f(x_1, \bar{x})$ of x_1 and $\bar{x}$. The function $f(x_1, \bar{x})$ must satisfy the relation

$$f(x_1, x_2)\, dx_1\, dx_2 = f(x_1, \bar{x})\, dx_1\, d\bar{x}$$

or

$$f(x_1, x_2)\, dx_2 = f(x_1, \bar{x})\, d\bar{x}$$

Since $\bar{x} = (x_1 + x_2)/2$, it follows that $x_2 = 2\bar{x} - x_1$ and $dx_2 = 2\, d\bar{x}$. Thus, at a fixed value of x_1

$$f(x_1, x_2)\, dx_2 = f(x_1, 2\bar{x} - x_1) \cdot 2\, d\bar{x}$$

and

$$f(x_1, \bar{x}) = f(x_1, 2\bar{x} - x_1) \cdot 2$$

or

$$f(x_1, \bar{x}) = \begin{cases} 2, & \text{when } -\tfrac{1}{2} \le x_1 \le \tfrac{1}{2} \text{ and } \dfrac{2x_1 - 1}{4} \le \bar{x} \le \dfrac{2x_1 + 1}{4} \\ 0, & \text{otherwise} \end{cases} \tag{5.36}$$

since $-\frac{1}{2} \le x_2 \le \frac{1}{2}$ implies $-\frac{1}{2} \le 2\bar{x} - x_1 \le \frac{1}{2}$ implies

$$\frac{x_1 - \tfrac{1}{2}}{2} \le \bar{x} \le \frac{x_1 + \tfrac{1}{2}}{2}$$

implies

$$\frac{2x_1 - 1}{4} \leq \bar{x} \leq \frac{2x_1 + 1}{4}$$

The graph of $f(x_1, \bar{x})$ is shown in Fig. 5.2, from which it should be noted that the volume of the parallelepiped is one and that the base is not a rectangle. Since $f(x_1, \bar{x})$ is a density function

$$\int_{-\infty}^{\infty} \int_{-\infty}^{\infty} f(x_1\ \bar{x})\ dx_1\ d\bar{x} = 1$$

or

$$\int_{-1/2}^{1/2} \left(\int_{-1/2}^{2\bar{x} + (1/2)} 2\ dx_1 + \int_{2\bar{x} - (1/2)}^{1/2} 2\ dx_1 \right) d\bar{x} = 1$$

Let $I = \int_{-1/2}^{2\bar{x} + (1/2)} 2\ dx_1 + \int_{2\bar{x} - (1/2)}^{1/2} 2\ dx_1$. Clearly, I is a nonnegative function for all real numbers in the domain $-\frac{1}{2} \leq \bar{x} \leq \frac{1}{2}$. Further

$$\int_{-1/2}^{1/2} I\ d\bar{x} = 1 \quad \text{and} \quad 0 \leq \int_{a}^{b} I\ d\bar{x} \leq 1$$

for any two real numbers a and b for which $-\frac{1}{2} \leq a < b \leq \frac{1}{2}$. Thus, I

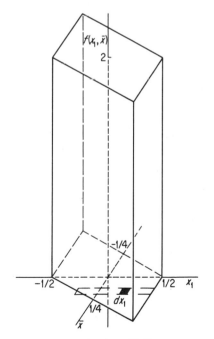

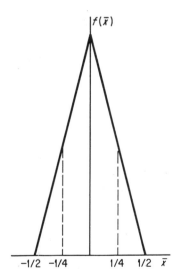

Fig. 5.2 Bivariate Uniform Distribution

Fig. 5.3 Graph of Eq. 5.37, the Marginal Distribution of $\bar{x}$ Obtained from Eq. 5.35

satisfies the conditions of a density function of $\bar{x}$. Letting this function be denoted by $f(\bar{x})$, we may write

$$
f(\bar{x}) = \begin{cases} \int_{2x-(1/2)}^{1/2} 2\,dx_1, & \text{when } 0 \le \bar{x} \le \tfrac{1}{2} \\ \int_{-1/2}^{2x+(1/2)} 2\,dx_1, & \text{when } -\tfrac{1}{2} \le \bar{x} < 0 \\ 0, & \text{otherwise} \end{cases}
$$

or, on evaluation

$$
f(\bar{x}) = \begin{cases} 2 - 4\bar{x}, & \text{when } 0 \le \bar{x} \le \tfrac{1}{2} \\ 2 + 4\bar{x}, & \text{when } -\tfrac{1}{2} \le \bar{x} < 0 \\ 0, & \text{otherwise} \end{cases} \tag{5.37}
$$

The graph of Eq. (5.37) is shown in Fig. 5.3. The function $f(\bar{x})$ is called the *marginal density* function of $\bar{x}$. A more complete discussion of density functions obtained from joint density functions is found in the exercises which follow.

The above method illustrates in a simple case how density functions of statistics may be found. Another method, the method using moment generating functions, is widely used and, for most of the sampling distributions required in this course, is the simpler method. This method is discussed in a later set of exercises.

To find the probability required in Example 5.14(c), use Eq. (5.37) and a property of symmetry to write

$$
\begin{aligned}
P[-\tfrac{1}{4} \le \bar{x} \le \tfrac{1}{4}] &= 2 \cdot P[0 \le \bar{x} \le \tfrac{1}{4}] \\
&= 2 \cdot \int_0^{1/4} (2 - 4\bar{x})\,d\bar{x} \\
&= 2[2\bar{x} - 2\bar{x}^2]_0^{1/4} \\
&= \tfrac{3}{4}
\end{aligned}
$$

This probability is easily verified by looking at Fig. 5.3.

Since the density function of $\bar{x}$ is known, we find from the definition of mean and variance that

$$
\begin{aligned}
\mu_{\bar{x}} &= \int_{-\infty}^{\infty} \bar{x} f(\bar{x})\,d\bar{x} \\
&= \int_{-1/2}^{0} \bar{x}(2 + 4\bar{x})\,d\bar{x} + \int_0^{1/2} \bar{x}(2 - 4\bar{x})\,d\bar{x} \\
&= 0
\end{aligned}
$$

and, using Formula (3.22), we obtain

$$\sigma_{\bar{x}}^2 = \int_{-\infty}^{\infty} \bar{x}^2 f(\bar{x}) \, d\bar{x} - \mu_{\bar{x}}^2$$

$$= \int_{-1/2}^{0} \bar{x}^2 (2 + 4\bar{x}) \, d\bar{x} + \int_{0}^{1/2} \bar{x}^2 (2 - 4\bar{x}) \, d\bar{x} - 0$$

$$= \tfrac{1}{24}$$

These values check with those found earlier without the density function of $\bar{x}$ being known. This points up the importance of Theorem 5.3.

The standard error of the mean is $\sigma_{\bar{x}} = \sqrt{6}/12 \doteq 0.204$. Since $\mu = \mu_{\bar{x}}$, we may use the sampling distribution of the mean to find the probability that a sample mean will fall within two standard errors of the population mean. Thus

$$P[\mu_{\bar{x}} - 2\sigma_{\bar{x}} \le \bar{x} \le \mu_{\bar{x}} + 2\sigma_{\bar{x}}] = P\left[-\frac{\sqrt{6}}{6} \le \bar{x} \le \frac{\sqrt{6}}{6}\right]$$

$$= 2 \cdot P\left[0 \le \bar{x} \le \frac{\sqrt{6}}{6}\right]$$

$$= \int_{0}^{\sqrt{6}/6} (2 - 4\bar{x}) \, d\bar{x}$$

$$= \tfrac{2}{3}[\sqrt{6} - 1] \doteq 0.966$$

Since 96.6 per cent of the means of samples of size two fall between -0.408 and 0.408, 96.6 per cent of the totals fall between -0.816 and 0.816 when random samples are drawn from the uniform distribution from $-\tfrac{1}{2}$ to $\tfrac{1}{2}$.

5.4.1. Some Properties of Linear Combinations of Means

In Sect. 5.4 we discussed some properties of sampling distributions of a single sample mean. In many problems the experimenter is interested in comparing several means. In particular, to compare the mean yield of process A with the mean yield of process B he must know the nature of the distribution of $\bar{x}_A - \bar{x}_B$, the difference in the means of samples taken from the two processes.

Now let us determine the nature of the mean and variance of the sampling distribution of a linear combination of random variables.

Theorem 5.6. *Let*

$$l = \sum_{i=1}^{p} a_i y_i$$

where a_i are real constants, and the y_i are random variables with means $\mu_{y_i} = \mu_i$, variances $\sigma_{y_i}^2 = \sigma_i^2$ and covariances $\mathrm{cov}\,(y_i, y_j) = \sigma_{ij}\ (i, j = 1, \ldots, p;\ i \neq j)$. Then

$$\mu_l = \sum a_i \mu_i \qquad\qquad (5.38)$$

and

$$\sigma_l^2 = \sum a_i^2 \sigma_i^2 + 2 \sum_{i<j} a_i a_j \sigma_{ij} \tag{5.39}$$

Proof. Consider the case where $p = 2$. Let $f(y_1, y_2)$ be the joint density function of y_1 and y_2. Then by a generalization of Definition (3.41) with the use of Formulas (3.53), (3.54), and (3.55), we obtain

$$\mu_l = \int_{-\infty}^{\infty} \int_{-\infty}^{\infty} lf(y_1, y_2)\, dy, dy_2 = \int_{-\infty}^{\infty} \int_{-\infty}^{\infty} (a_1 y_1 + a_2 y_2) f(y_1, y_2)\, dy, dy_2$$

$$= a_1 \int_{-\infty}^{\infty} \int_{-\infty}^{\infty} y_1 f(y_1, y_2)\, dy_1\, dy_2 + a_2 \int_{-\infty}^{\infty} \int_{-\infty}^{\infty} y_2 f(y_1, y_2)\, dy_1\, dy_2$$

$$= a_1 \mu_1 + a_2 \mu_2$$

and

$$\sigma_l^2 = \int_{-\infty}^{\infty} \int_{-\infty}^{\infty} (l - \mu_l)^2 f(y_1, y_2)\, dy_1\, dy_2$$

$$= \int_{-\infty}^{\infty} \int_{-\infty}^{\infty} [a_1(y_1 - \mu_1) + a_2(y_2 - \mu_2)]^2 f(y_1, y_2)\, dy_1\, dy_2$$

$$= a_1^2 \int_{-\infty}^{\infty} \int_{-\infty}^{\infty} (y_1 - \mu_1)^2 f(y_1, y_2)\, dy_1\, dy_2$$

$$+ a_2^2 \int_{-\infty}^{\infty} \int_{-\infty}^{\infty} (y_2 - \mu_2)^2 f(y_1, y_2)\, dy_1\, dy_2$$

$$+ 2a_1 a_2 \int_{-\infty}^{\infty} \int_{-\infty}^{\infty} (y_1 - \mu_1)(y_2 - \mu_2) f(y_1, y_2)\, dy_1\, dy_2$$

$$= a_1^2 \sigma_1^2 + a_2^2 \sigma_2^2 + 2a_1 a_2 \sigma_{12}$$

The proof for the case of p random variables is left as an exercise for the student. For finite populations the proofs are analogous to those for the infinite populations.

Corollary 5.1. *If, in addition to the conditions of Theorem* **5.6,** *we assume that the* y_i *are mutually independent, then*

$$\sigma_l^2 = \sum a_i^2 \sigma_i^2 \tag{5.40}$$

but Eq. (5.38) *remains the same.*

Proof. If y_i is independent of y_j, then cov $(y_i, y_j) = 0$, and the second term of Eq. (5.39) vanishes.

Corollary 5.2. *Let* $\bar{x}_i$ ($i = 1, 2$) *be the mean of a random sample of size* n_i *drawn from a population with mean* $\mu_{x_i} = \mu_i$ *and variance* $\sigma_{x_i}^2 = \sigma_i^2$. *If* $\bar{x}_1$ *and* $\bar{x}_2$ *are independently distributed, then*

$$\mu_{x_1-x_2} = \mu_1 - \mu_2 \tag{5.41}$$

and

$$\sigma^2_{\bar{x}_1-\bar{x}_2} = \frac{\sigma^2_1}{n_1} \cdot \frac{N_1 - n_1}{N_1 - 1} + \frac{\sigma^2_2}{n_2} \cdot \frac{N_2 - n_2}{N_2 - 1} \qquad (5.42a)$$

when random samples are drawn without replacement from finite populations of sizes N_1 and N_2, or

$$\sigma^2_{\bar{x}_1-\bar{x}_2} = \frac{\sigma^2_1}{n_1} + \frac{\sigma^2_2}{n_2} \qquad (5.42b)$$

otherwise (i.e., when random samples are drawn with replacement from finite populations or with or without replacement from infinite populations).

Proof. Let $y_1 = \bar{x}_1$ and $y_2 = \bar{x}_2$ so that $l = a_1 y_1 + a_2 y_2 = \bar{x}_1 - \bar{x}_2$, where $a_1 = 1$ and $a_2 = -1$. Then on substituting these values in Eqs. (5.38) and (5.40) and applying Theorem 5.1 or 5.2 or 5.3, we obtain Eqs. (5.41), (5.42a), and (5.42b).

Note that $\mu_{\bar{x}_1+\bar{x}_2} = \mu_1 + \mu_2 \neq \mu_{\bar{x}_1-\bar{x}_2}$, but $\sigma^2_{\bar{x}_1+\bar{x}_2} = \sigma^2_{\bar{x}_1-\bar{x}_2}$ when $\bar{x}_1$ and $\bar{x}_2$ are independently distributed.

Corollary 5.3. *If, in addition to the conditions of Corollary* **5.2,** *we assume that $n_1 = n_2 = n$, $\sigma^2_1 = \sigma^2_2 = \sigma^2$, and $N_1 = N_2 = N$ when the populations are finite, then*

$$\mu_{\bar{x}_1-\bar{x}_2} = \mu_1 - \mu_2$$

and

$$\sigma^2_{\bar{x}_1-\bar{x}_2} = \frac{2\sigma^2}{n} \cdot \frac{N - n}{N - 1}$$

when samples are drawn without replacement from finite populations, or

$$\sigma^2_{\bar{x}_1-\bar{x}_2} = \frac{2\sigma^2}{n}$$

otherwise.

5.4.2. Exercises

5.28. Prove Theorem 5.3 for the case where $n = 3$, using the method prior to the statement of the theorem.

5.29. On the average, $\frac{3}{4}$ of the seeds of a certain strain germinate. (a) If 80 seeds are planted, how many on the average (can be expected to) will germinate? Find the variance of the number that will germinate.

Hint. Use Eq. (5.16) of Exercise 5.19.

(b) What is the probability that fewer than the expected number minus 5 will germinate? Use Chebyshev's inequality. (c) Under what conditions can the results of (a) and (b) be expected to be valid?

5.30. Long experience with testing the tensile strength of a certain manufactured fiber indicates that, for the process under standard conditions, the mean is $\mu = 3$ lb and the standard deviation is $\sigma = 0.2$ lb. (a) Use Chebyshev's inequality to determine the smallest random sample for which the probability is at least 0.90 that the sample mean will not differ from the population mean by more than 0.05 lb. (b) Use the central-limit theorem (i.e., the normal approximation) to find n in (a). (c) The mean of a random sample of size twenty-five was found to be 2.90 lb. Use Chebyshev's inequality to determine the probability of a sample mean's being this small or smaller. (d) Use the normal approximation to answer (c).

5.31. Work Example 5.14 after replacing "sampling distribution of means" by "sampling distribution of totals" and "sample mean" by "sample total."

5.32. Let a random sample of size n_i be drawn from an infinite population with mean μ_i and variance σ_i^2 ($i = 1, 2, 3, 4$). Let $\bar{x}_i$ denote the mean of the sample drawn from the ith population. Assume that the sample means $\bar{x}_1, \bar{x}_2, \bar{x}_3$, and $\bar{x}_4$ are independently distributed. (a) Determine the mean and variance of $\bar{x}_1 + 2\bar{x}_2 - 3\bar{x}_3$. (b) Determine the mean and variance of $\bar{x}_1 + \bar{x}_2 - \bar{x}_3 - \bar{x}_4$. (c) Determine the covariance of x_{11} and $\bar{x}_1$; i. e., find cov $(x_{11}, \bar{x}_1)$. (d) Answer (a), (b), and (c) when $n_i = n$, $\mu_i = \mu$, and $\sigma_i^2 = \sigma^2$ ($i = 1, 2, 3, 4$).

5.33. In Example 5.14 we introduced the term "marginal density function of $\bar{x}$." Let us consider the more general problem, where $f(x, y)$ is the joint density function of two continuous variates (for the discrete case use $\sum$ in place of $\int dx$). Suppose that we are interested in finding a density function $f_1(x)$ of the single variate x. Since $f_1(x)$ is to be a density function, this means, among other things, that $f_1(x)$ must satisfy the condition

$$P[a < x < b] = \int_a^b f_1(x)\, dx$$

for any pair of constants a and b ($a < b$). Now in terms of the joint density function $f(x, y)$ we must have

$$P[a < x < b] = \int_a^b \int_{-\infty}^{\infty} f(x, y)\, dy\, dx$$

Thus, the required marginal density function is given by

$$f_1(x) = \int_{-\infty}^{\infty} f(x, y)\, dy \tag{5.43}$$

since this definition allows the Conditions (3.2) of the density function to be satisfied. Further, the cumulative marginal distribution function of x, for example, is defined by

$$F_1(x) = \int_{-\infty}^{-x} \int_{-\infty}^{\infty} f(x, y)\, dy\, dx = \int_{-\infty}^{x} f_1(x)\, dx = F(x, \infty) \tag{5.44}$$

Similarly, the marginal density function of y is given by

$$f_2(y) = \int_{-\infty}^{\infty} f(x, y)\, dx$$

and the cumulative marginal distribution of y by $F_2(y) = F(\infty, y)$.

In general, if $f(x_1, x_2, \ldots, x_k)$ is a joint density function of the variates $x_1, \ldots, x_k$, the marginal density function of any subset of p variates $x_{i_1}, \ldots, x_{i_p}$ is given by integrating $f(x_1, \ldots, x_k)$ with respect to all other variates between the limits $-\infty$ and $+\infty$.

Find the marginal density functions and cumulative marginal distribution functions for the joint density function found in (a) Exercise 3.55, (b) Exercise 3.59, (c) Sect. 3.4.2. (d) Find the marginal density functions $f(x, y)$ and $f(x)$, using the density function $f(x, y, z)$ found in Exercise 3.63. (e) Find the marginal density function of the variates $x_1, \ldots, x_{k-2}$ for the density function given in Eq. (5.20). (f) Find the marginal density function of the variates x_1, x_2 for the density function given in Eq. (5.23) when $k = 5$.

5.5. PROPERTIES OF EXPECTATIONS

Often professional gamblers pose questions which relate to how much, in the long run, one might *expect* to win in a game of chance. In fact, most of us pose similar questions relating to decisions we must make in our day-by-day activities. [We have already used the term "expected number" in Exercise 5.29(b)] Mathematicians and statisticians answer the question in terms of "expected value."

There are many places in statistics where it is desirable to use the term "expected value of ———," where the blank space might be filled in with any statistic, a function of one or more statistics such as $\bar{x}_1 + 2\bar{x}_2 - 3\bar{x}_3$, e^{tx}, etc. As a matter of fact, means, variances, moments, and moment generating functions may all be expressed in terms of the "expected value" notation. Further, once the properties of expectation are developed, we may use the same notation for both the discrete and continuous cases.

By definition, the mean or the *expectation* or the *expected value*, $E(x)$, of a random variable x with density function $f(x)$ is

$$E(x) = \mu_x = \begin{cases} \sum xf(x), & \text{for the discrete case} \\ \int_{-\infty}^{\infty} xf(x)\, dx, & \text{for the continuous case} \end{cases} \tag{5.45}$$

In developing the properties of expectation, we consider only the continuous case. It is assumed that the student can supply the notation and argument for the discrete case.

Before we introduce a discussion of some desirable properties of statistics, desirable from the point of view of estimating parameters, it will be useful to introduce some properties of expected values which are based on continuous functions of random variables. With this in mind, let $h(x)$ be any continuous function of x and let x be a continuous random variable with density function $f(x)$. Then the expectation or expected value $E[h(x)] = E(h)$ of $h(x)$ is given by

$$E[h(x)] = \int_{-\infty}^{\infty} h(x)f(x)\,dx \tag{5.46}$$

In defining kth moments and the moment generating functions in Exercises 3.45 and 3.47, we gave special forms of Eq. (5.46). For example, when $h(x) = x^k$, Formula (5.46) becomes

$$E(x^k) = \int_{-\infty}^{\infty} x^k f(x)\,dx \tag{5.47}$$

which is the kth moment about the origin defined in Eq. (3.40). Also, when $h(x) = e^{tx}$, Formula (5.46) becomes

$$E(e^{tx}) = \int_{-\infty}^{\infty} e^{tx} f(x)\,dx \tag{5.48}$$

which is the moment generating function (MGF) defined in Eq. (3.43).

In general, if $h(x_1, \ldots, x_k)$ is a continuous function of $x_1, \ldots, x_k$, and if $x_1, \ldots, x_k$ are continuous random variables with joint density function $f(x_1, \ldots, x_k)$, then the expectation, $E[h(x_1, \ldots, x_k)] = E(h)$, of $h(x_1, \ldots, x_k) = h$ is given by

$$E[h(x_1, \ldots, x_k)] = \int_{-\infty}^{\infty} \cdots \int_{-\infty}^{\infty} h(x_1, \ldots, x_k)f(x_1, \ldots, x_k)\,dx_1 \ldots dx_k \tag{5.49}$$

In defining the mean, variance, and covariance of x_i and x_j for multivariate distributions given in Sect. 3.4, we gave special forms of Eq. (5.49). For example, when

$$h(x_1, \ldots, x_k) = (x_i - \mu_i)(x_j - \mu_j)$$

Formula (5.49) becomes

$$E[(x_i - \mu_i)(x_j - \mu_j)]$$
$$= \int_{-\infty}^{\infty} \cdots \int_{-\infty}^{\infty} (x_i - \mu_i)(x_j - \mu_j)f(x_1, \ldots, x_k)\,dx_1 \ldots dx_k \tag{5.50}$$

which is the covariance, $\sigma_{ij} = \text{cov}\,(x_i, x_j)$ of x_i and x_j defined in Eq. (3.55). Note that when $i = j$ Formula (5.50) becomes the definition of the variance of x_i; that is

$$\sigma_i^2 = V(x_i) = E[(x_i - \mu_i)^2] = E\{[x_i - E(x_i)]^2\} \tag{5.51}$$

Also, many of the expressions in Sect. 5.4.1 relating to linear combinations of random variables may be considered special cases of Eq. (5.49).

We now list together some basic properties of expectation for easy reference. It is to be understood that a denotes any constant, and that h_m is an abbreviation for $h_m (x_1, \ldots, x_k)$, which is the mth ($m = 1, 2$) continuous function of $x_1, \ldots, x_k$

$$E(a) = a \tag{5.52}$$

$$E(ah_i) = aE(h_i) \tag{5.53}$$

$$E(h_1 + h_2) = E(h_1) + E(h_2) \tag{5.54}$$

$$E(x_1 + \cdots + x_k) = E(x_1) + \cdots + E(x_k) \tag{5.55}$$

$$E(x_i^2) = V(x_i) + [E(x_i)]^2 = \sigma_i^2 + \mu_i^2 \tag{5.56}$$

$$E(x_i \cdot x_j) = E(x_i) \cdot E(x_j) \tag{5.57}$$

when x_i and x_j ($i \neq j$) are independently distributed.

Note also that the formulas for moment generating functions as given in Eqs. (3.43), (3.46), (3.47), and (3.48) can be expressed in terms of expectations. These properties are not difficult to prove with the aid of the definitions of expectation given in Eqs. (5.46) and (5.49), the definitions of marginal density [for example, see Eq. (5.43)], and other properties given earlier.

5.5.1. Proofs of Some Properties of Expectation

We illustrate the nature of the proofs for $k = 2$ by proving Eqs. (5.55) and (5.57). By Definition (5.49) and properties of integral calculus, we may write

$$E(x_1 + x_2) = \int_{-\infty}^{\infty} \int_{-\infty}^{\infty} (x_1 + x_2) f(x_1, x_2) \, dx_1 \, dx_2$$

$$= \int_{-\infty}^{\infty} x_1 \left[\int_{-\infty}^{\infty} f(x_1, x_2) \, dx_2 \right] dx_1 + \int_{-\infty}^{\infty} x_2 \left[\int_{-\infty}^{\infty} f(x_1, x_2) \, dx_1 \right] dx_2$$

The marginal density functions, according to Eq. (5.43), are

$$g_1(x_1) = \int_{-\infty}^{\infty} f(x_1, x_2) \, dx_2 \quad \text{and} \quad g_2(x_2) = \int_{-\infty}^{\infty} f(x_1, x_2) \, dx_1$$

Thus, by substitution we find that

$$E(x_1 + x_2) = \int_{-\infty}^{\infty} x_1 g_1(x_1) \, dx_1 + \int_{-\infty}^{\infty} x_2 g_2(x_2) \, dx_2$$

Since $g_1(x_1)$ and $g_2(x_2)$ are density functions, it follows from the definition of expectation that

$$E(x_1 + x_2) = E(x_1) + E(x_2)$$

Also

$$E(x_1 \cdot x_2) = \int_{-\infty}^{\infty} \int_{-\infty}^{\infty} x_1 \cdot x_2 f(x_1, x_2)\, dx_1\, dx_2$$

Since x_1 and x_2 are independently distributed

$$f(x_1, x_2) = f_1(x_1) \cdot f_2(x_2)$$

and

$$
\begin{aligned}
E(x_1 \cdot x_2) &= \int_{-\infty}^{\infty} \int_{-\infty}^{\infty} x_1 \cdot x_2 f_1(x_1) \cdot f_2(x_2)\, dx_1\, dx_2 \\
&= \int_{-\infty}^{\infty} x_1 f_1(x_1)\, dx_1 \cdot \int_{-\infty}^{\infty} x_2 f_2(x_2)\, dx_2 \\
&= E(x_1) \cdot E(x_2)
\end{aligned}
$$

It should be noted that Eq. (5.57) requires the independence assumption, but Eq. (5.55) does not.

The properties of expectation are particularly useful in proving properties of variance. Four important formulas are

$$V(x + a) = V(x) \tag{5.58}$$

$$V(ax) = a^2 V(x) \tag{5.59}$$

$$V(x_1 + x_2) = V(x_1) + V(x_2) \tag{5.60}$$

if x_1 and x_2 are independent

$$V(a_1 x_1 + \cdots + a_k x_k) = \sum_{i=1}^{k} a_i^2 V(x_i) \tag{5.61}$$

if the x_i are mutually independent and at least one constant $a_i \neq 0$ (see Corollary 5.1).

To prove Eq. (5.58) we write, using Eqs. (5.51), (5.55), and (5.52)

$$
\begin{aligned}
V(x + a) &= E\{[x + a - E(x + a)]^2\} \\
&= E\{[x + a - E(x) - a]^2\} \\
&= E\{[x - E(x)]^2\} \\
&= V(x)
\end{aligned}
$$

The proof of Eq. (5.60) is slightly more involved. Using the above properties and the fact that $E(x_1) = \mu_1$ and $E(x_2) = \mu_2$ are constants, we may write

$$
\begin{aligned}
V(x_1 + x_2) &= E\{[x_1 + x_2 - E(x_1 + x_2)]^2\} \\
&= E\{[x_1 - E(x_1) + x_2 - E(x_2)]^2\} \\
&= E[(x_1 - \mu_1)^2 + 2(x_1 - \mu_1)(x_2 - \mu_2) + (x_2 - \mu_2)^2] \\
&= E[(x_1 - \mu_1)^2] + 2E[(x_1 - \mu_1)(x_2 - \mu_2)] + E[(x_2 - \mu_2)^2] \\
&= V(x_1) + V(x_2)
\end{aligned}
$$

since

$$E[(x_1 - \mu_1)(x_2 - \mu_2)] = E(x_1 \cdot x_2 - \mu_1 x_2 - \mu_2 x_1 + \mu_1 \mu_2)$$
$$= E(x_1 \cdot x_2) - E(\mu_1 x_2) - E(\mu_2 x_1) + E(\mu_1 \mu_2)$$
$$= E(x_1) \cdot E(x_2) - \mu_1 E(x_2) - \mu_2 E(x_1) + \mu_1 \mu_2$$
$$= 0$$

5.5.2. Conditional Expectations

Conditional expectations are useful when one is working with bivariate (or multivariate) distributions. Let $f(x_1, x_2)$ denote the bivariate density function of the continuous random variables x_1 and x_2, and let $g(x_1)$ and $g(x_2)$ denote the marginal density functions of x_1 and x_2, respectively. Then, using our definition of conditional probability, we define the *conditional density function of x_1 for a given x_2* by

$$f(x_1 \mid x_2) = \frac{f(x_1, x_2)}{g(x_2)} \tag{5.62}$$

for each x_2 for which $g(x_2) \neq 0$, and the *conditional density function of x_2 for a given x_1* by

$$f(x_2 \mid x_1) = \frac{f(x_1, x_2)}{g(x_1)} \tag{5.63}$$

for each x_1 for which $g(x_1) \neq 0$. Hence, the corresponding conditional expectations are defined by

$$E(x_1 \mid x_2) = \int_{-\infty}^{\infty} x_1 f(x_1 \mid x_2) \, dx_1 \tag{5.64}$$

and

$$E(x_2 \mid x_1) = \int_{-\infty}^{\infty} x_2 f(x_2 \mid x_1) \, dx_2 \tag{5.65}$$

respectively.

When x_2 is fixed, $E(x_1 \mid x_2)$ is a mean which is fixed. But when x_2 is allowed to vary, Eq. (5.64) is a function of x_2, in which case it can be shown that

$$E[E(x_1 \mid x_2)] = E(x_1) \tag{5.66}$$

Likewise, when $E(x_2 \mid x_1)$ is considered a function of x_1, it can be shown that

$$E[E(x_2 \mid x_1)] = E(x_2) \tag{5.67}$$

In the particular case where x_1 and x_2 are independent variables, it follows that

$$E(x_1 \mid x_2) = E(x_1) \tag{5.68}$$

and

$$E(x_2 \mid x_1) = E(x_2) \qquad (5.69)$$

The proofs of Eqs. (5.66), (5.67), (5.68), and (5.69) are left as an exercise for the student. In listing the properties of conditional expectation we considered only the continuous case. It is assumed that the student can supply the notation and argument for the discrete case.

The properties of this section are used extensively in the next section and in later chapters. Thus, the reader is encouraged to work several of the exercises on expectation at the end of this chapter.

5.6. ESTIMATION

We have already indicated in an intuitive way that statistics are used to estimate parameters which in turn help to characterize populations. We have actually worked a few problems involving statistics. However, much of our time has been spent in the study of sampling distributions of statistics, that is, in the study of a statistic as a random variable with a sampling distribution. Now, we wish to assess the merits of statistics in terms of the properties of the sampling distributions.

We have used the sample mean $\bar{x}$ and sample median x_m to estimate the population mean μ; we have used the sample variance s^2 to estimate the population variance σ^2. In a careful study of the nature of these and other estimates, it is natural that we consider properties which a "good" estimate should possess, and attempt to determine which, if any, estimate is "best" for a given purpose.

Recall that a statistic was defined (see Sect. 5.2) as a numerical value determined from some or all of the values which make up a sample. Often the statistic is computed from the sample values in exactly the same way as the corresponding parameter is determined from the population values. Thus, when the sample is representative of the population, we would normally expect the statistic to be representative of the parameter. For this reason we require that the observations of the sample be randomly selected; otherwise, we could reasonably infer very little about the parameter.

In studying properties of estimation it is customary that a distinction be made between the *rule* which defines a statistic and the particular *value* which results from applying the rule to a particular sample. Thus, for example, when we consider $x_1, \ldots, x_n$ as representing particular values of a random sample, the mean $\bar{x}$ is a value which is called an *estimate* of the population mean. When, before a particular sample is actually taken, we think of x_1, $\ldots, x_n$ as representing n random observations, then the mean $\bar{x}$ is a random variable and is called an *estimator* of the population mean. The distinction between *estimator* and *estimate* is the same as that between a function $f(x)$ and a functional value $f(c)$. $f(x)$ is a variable defined in some domain of x, and $f(c)$ is a constant corresponding to a specified value of x equal to constant c.

In general, let θ denote an unknown parameter of a distribution with density function $f(x)$. Let $x_1, \ldots, x_n$ be a random sample of size n taken from this distribution. If we think of the ith $(i = 1, \ldots, n)$ sample observation x_i before a particular value is drawn, it is a random variable which can take on any value of x. Let

$$t_n \equiv t(x_1, \ldots, x_n) \tag{5.70}$$

a function of the n sample observations, denote any statistic corresponding to θ. When t_n is thought of as a function of n random variables, it is called an *estimator* of θ. After the observations have been made, that is, after a particular sample has been drawn, the statistic t_n is a particular value which is called an *estimate* of θ. Our problem now is to study *estimators*, that is, properties of sampling distributions of statistics, not estimates.

5.6.1. Consistent Estimators

We observe that an estimator should probably not be considered bad simply because it can assume a value which deviates considerably from the true value θ. However, if the bulk of the values of t_n deviated considerably from θ, we might consider t_n a bad estimator of θ, particularly if this is the case as n becomes large. Thus, one of the first desirable properties we might require is that there be high probability that the estimator be near the parameter it is intended to estimate when the sample size is large. This leads us to the following definition.

Definition. An estimator t_n is a *consistent* estimator of θ if, for any positive numbers δ and ϵ, no matter how small, there exists an integer n' such that the probability that $|t_n - \theta| < \epsilon$ is greater than $1 - \delta$ for all $n > n'$; that is

$$P[|t_n - \theta| < \epsilon] > 1 - \delta \qquad \text{for all } n > n'$$

This definition is obviously similar to the definition of convergence in the mathematical sense, except that here we say that, given any small ϵ, we can find a sample size large enough so that, for all larger sample sizes, the *probability* that t_n differs from the true value θ more than ϵ is as small as we please. In this case we say t_n *converges in probability* to θ. So convergence in probability means that t_n is a consistent estimator of θ.

It is not difficult to show that $\bar{x}$ is a consistent estimator of μ, if it is assumed that the population variance σ^2 is finite. Since, in this case, the variance σ^2/n of the sampling distribution of $\bar{x}$ approaches zero as n approaches infinity, and $E(\bar{x}) = \mu$ for any n, it follows that the sampling distribution of $\bar{x}$ must become concentrated at μ for large n. In fact, if the population is normal, so that the true median and mode are the same as the mean μ, it follows that $\bar{x}$ is a consistent estimator of the population median and mode. Also, the population variance and many other parameters have consistent estimators.

Not all estimators are consistent. For example, the estimator x_1, the first observation of a random sample of size n, of μ is not consistent. Further, if $\bar{x}$ is the mean of a random sample drawn from a Cauchy distribution with density function

$$\frac{1}{\pi} \cdot \frac{1}{[1 + (x - \mu)^2]}, \qquad -\infty < x < \infty \qquad (5.71)$$

and mean μ, it can be shown that $\bar{x}$ is not consistent. Note that the population variance does not exist. In both these cases the sampling distribution of the estimator is the same as the original distribution. Thus, the estimator does not increase in accuracy as n increases.

The criterion of consistency is not very practical, since it has to do with a limiting property. We point out two reasons why this is so. First, samples have a finite number of observations, and the definition requires an infinite number. Second, when there is one consistent estimator t_n of θ, it is possible to define infinitely many. For example, when t_n is consistent, so is

$$\frac{n + a}{n + b} t_n \qquad (5.72)$$

for all fixed real numbers $a \neq -n$ and $b \neq -n$.

5.6.2. Unbiased Estimator

Next, we give a property of a good estimator which is designed to restrict the number of possible consistent estimators. When t_n is a consistent estimator of θ, its sampling distribution must have practically all of its values in a neighborhood of θ when n is large. Thus, the measure of central tendency of t_n must be at or very near θ when n is large. If we *remove the restriction* "*for large n,*" we may select from among all consistent estimators a much smaller class by applying the following definition.

Definition. An estimator t_n is an *unbiased estimator* of θ if

$$E(t_n) = \theta$$

This definition applies for all n and θ, and requires that the *mean* of the sampling distribution of any statistic equals the parameter which the statistic is supposed to estimate. We might have required that the *median* (or any other measure of central tendency) of the sampling distribution of t_n be equal to θ. But we did not; we defined an unbiased estimator in the above way because of its mathematical convenience, and we shall use the term "unbiased" in this way.

Let $\bar{x}$ and s^2 denote the mean and variance of a random sample of n observations drawn from the same infinite population with mean μ and variance σ^2. We have already stated that the mean of the sampling distribution of $\bar{x}$ is equal to the population mean. This follows directly from Eqs. (5.53) and (5.55) and the definition of μ. That is

$$E(\bar{x}) = E\left(\frac{x_1 + \cdots + x_n}{n}\right)$$

$$= \frac{1}{n} E(x_1 + \cdots + x_n)$$

$$= \frac{1}{n} [E(x_1) + \cdots + E(x_n)]$$

$$= \frac{1}{n} (\mu + \cdots + \mu)$$

$$= \mu$$

Thus, $\bar{x}$ is an unbiased estimator of μ. If we use the properties indicated, it follows that

$$E(s^2) = E\left[\frac{\sum_{i=1}^{n} x_i^2 - \dfrac{\left(\sum_{i=1}^{n} x_i\right)^2}{n}}{n-1}\right]$$

$$= \frac{1}{n-1} E\left(\sum_{i=1}^{n} x_i^2 - n\bar{x}^2\right)$$

$$= \frac{1}{n-1}\left[\sum_{i=1}^{n} E(x_i^2) - nE(\bar{x}^2)\right]$$

$$= \frac{1}{n-1}\left[\sum_{i=1}^{n} (\sigma^2 + \mu^2) - n(\sigma_{\bar{x}}^2 + \mu_{\bar{x}}^2)\right]$$

$$= \frac{1}{n-1}\left[n(\sigma^2 + \mu^2) - n\left(\frac{\sigma^2}{n} + \mu^2\right)\right]$$

or

$$E(s^2) = \sigma^2 \tag{5.73}$$

Thus, s^2 is an unbiased estimator of σ^2. This is the primary reason we defined the sample variance s^2 as we did.

If we had defined the sample variance as

$$(s')^2 = \frac{\sum_{i}^{n} (x_i - \bar{x})^2}{n} \tag{5.74}$$

then we would have

$$E\left[\frac{\sum (x_i - \bar{x})^2}{n}\right] = E\left[\frac{n-1}{n}\frac{\sum (x_i - \bar{x})^2}{n-1}\right]$$

$$= \frac{n-1}{n}E(s^2)$$

$$= \frac{n-1}{n}\sigma^2$$

or

$$E(s'^2) = \sigma^2 - \frac{\sigma^2}{n} \tag{5.75}$$

Thus, we would have a biased estimator s'^2 of σ^2 which is too small on the average. For small sample sizes the degree of bias is great, but as n approaches infinity the bias vanishes.

It can be shown (see Exercise 5.52) that s^2 is a consistent estimator. Thus, the biased estimator

$$s'^2 = \left(\frac{n-1}{n}\right) s^2$$

is a consistent estimator. We know that the first observation x_1 in a random sample of n observations is not a consistent estimator of μ. Since $E(x_1) = \mu$, x_1 is an unbiased estimator of μ. Therefore, unbiased estimators may not be consistent. Thus, neither property implies the other. However, a consistent estimator t_n with finite $E(t_n)$ must tend to be unbiased for large n.

5.6.3. Best Estimators

There may be several consistent and unbiased estimators of a parameter from which we wish to select the best according to some criterion. It seems fairly reasonable that we should select that estimator which is consistently closest to the parameter being estimated. But what should the measure of closeness be? It is customary that the variance of the sampling distribution be used for this purpose.

In view of the preceding discussion, we say that any estimator which has minimum variance among all possible estimators of θ is an *efficient estimator*. There may be several estimators of θ which are efficient. In comparing any estimator t'_n with variance $V(t'_n)$ against an efficient estimator t_n with variance $V(t_n)$, we use the ratio

$$\frac{V(t_n)}{V(t'_n)} \tag{5.76}$$

called the *efficiency of t'_n*.

For example, consider the comparison of two estimators, the mean and the median, computed from a random sample of size n drawn from a normal population with mean μ and variance σ^2. Let $t_n = \bar{x}$ and $t'_n = x_m$. It can be shown that t_n is normally distributed with mean μ and variance $V(t_n) = \sigma^2/n$, and, for large n, that t'_n is approximately normally distributed with mean μ and variance

$$V(t'_n) = \frac{\pi}{2} \cdot \frac{\sigma^2}{n}$$

Both estimators are consistent and unbiased, but the variance of the mean is less than the variance of the median. Therefore, in this case, the mean is considered a better estimator of μ than the median when n is large. For small n the variance of the median must be determined for each n. Since it appears that $V(t_n) < V(t'_n)$ always, we say that $\bar{x}$ is a better estimator of μ than x_m if samples are from a normal population. The reader is cautioned that this is not necessarily the case for all distributions.

For samples of size n randomly drawn from a normal population, it can be shown that the minimum variance of all unbiased estimators of μ is σ^2/n. Since $V(\bar{x}) = \sigma^2/n$, $\bar{x}$ is an efficient estimator, and x_m has an efficiency of $2/\pi$ when n is large.

For fixed n, it is possible for one estimator t_n to be biased and another estimator t'_n to be unbiased and still to have t_n consistently closer to θ than t'_n. In this case the measure of variability of t_n about θ is not the variance. It is

$$E[(t_n - \theta)^2] \tag{5.77}$$

which is called the *second moment of t_n about θ*. When θ is equal to the mean of t_n, then Eq. (5.77) reduces to the variance of t_n; that is, when $\theta = E(t_n)$,

$$E[(t_n - \theta)^2] = V(t_n)$$

With this in mind, we give the following more general definition as a basis for selecting good estimators.

Definition. An estimator t_n is called a *best estimator* of the parameter θ if t_n is such that

$$E[(t_n - \theta)^2] \leq E[(t'_n - \theta)^2] \tag{5.78}$$

where t'_n is any estimator of θ.

Note that this definition does not require either the consistency or unbiasedness property. However, for consistent estimators, as n approaches infinity, we see that the unbiased property necessarily follows. This definition of a best estimator has certain disadvantages, but so do other definitions which could be substituted. In any case, our definition of best estimator has proved to be very useful in both theory and application.

Example 5.15. Let $x_1, x_2, \ldots, x_n$ be a random sample from a population with mean μ and variance σ^2. Let

$$t'_n = \frac{x_1 + x_n}{n}$$

$$t''_n = \frac{3x_1 - x_2 - x_{n-1} + 4x_n}{5}$$

$$t'''_n = \tfrac{1}{2} x_1 + \tfrac{1}{3} x_2 + \tfrac{1}{6} x_n$$

(a) Which, if any, of the estimators t'_n, t''_n, t'''_n are unbiased estimators of μ? Show why. (b) Which of these three estimators is best? (c) Of all possible linear unbiased estimators of μ, which is the best? Show why.

According to expectation properties (5.53) and (5.55), we have

$$E(t'_n) = \frac{1}{n} E(x_1 + x_n)$$

$$= \frac{1}{n} [E(x_1) + E(x_n)]$$

Since x_i $(i = 1, \ldots, n)$ is a random variable distributed like the parent population, it follows that $E(x_i) = \mu$ and $V(x_i) = \sigma^2$ for each i. Thus

$$E(t'_n) = \frac{1}{n} (\mu + \mu) = \frac{2\mu}{n}$$

Following a similar argument, we find that

$$E(t''_n) = \frac{3E(x_1) - E(x_2) - E[x_{n-1}] + 4E(x_n)}{5} = \mu$$

and

$$E(t'''_n) = \tfrac{1}{2} E(x_1) + \tfrac{1}{3} E(x_2) + \tfrac{1}{6} E(x_n) = \mu$$

Thus t''_n and t'''_n are unbiased estimators of μ when n is equal to or greater than 4 and 3, respectively. The estimator t'_n underestimates μ except when $n = 2$, in which case it is an unbiased estimator of μ.

In order to determine which of the three estimators is best, we must compute $E[(t_n - \mu)^2]$ for each. When $E(t_n) \neq \theta$, we have

$$E[(t_n - \theta)^2] = E[\{ [t_n - E(t_n)] + [E(t_n) - \theta] \}^2]$$
$$= E\{[t_n - E(t_n)]^2\} + 2E\{[t_n - E(t_n)][E(t_n) - \theta]\}$$
$$+ E\{[E(t_n) - \theta]^2\}$$

or

$$E[(t_n - \theta)^2] = V(t_n) + [E(t_n) - \theta]^2 \tag{5.79}$$

since

$$E\{[t_n - E(t_n)][E(t_n) - \theta]\} = [E(t_n) - \theta]E[t_n - E(t_n)]$$
$$= [E(t_n) - \theta][E(t_n) - E(t_n)]$$
$$= 0$$

Thus, for the first estimator of μ we have, on substituting in Eq. (5.79)

$$E[(t'_n - \mu)^2] = V(t'_n) + \left(\frac{2\mu}{n} - \mu\right)^2$$

Since, when Eq. (5.61) is applied

$$V(t'_n) = V\left(\frac{x_1 + x_n}{n}\right) = \frac{1}{n^2} V(x_1) + \frac{1}{n^2} V(x_n) = \frac{2\sigma^2}{n^2}$$

the second moment of t'_n about μ becomes

$$E[(t'_n - \mu)^2] = \frac{2\sigma^2}{n^2} + \frac{\mu^2}{n^2}(2 - n)^2$$

Because the other two estimators of μ are unbiased, the last term of Eq. (5.79) vanishes, and we have

$$E[(t''_n - \mu)^2] = V(t''_n) = \tfrac{9}{25} V(x_1) + \tfrac{1}{25} V(x_2) + \tfrac{1}{25} V(x_{n-1}) + \tfrac{16}{25} V(x_n)$$

$$= \tfrac{27}{25} \sigma^2$$

and

$$E[(t'''_n - \mu)^2] = V(t'''_n) = \tfrac{1}{4} V(x_1) + \tfrac{1}{9} V(x_2) + \tfrac{1}{36} V(x_n)$$

$$= \tfrac{7}{18} \sigma^2$$

Hence, t'''_n is always a better estimator than t''_n, since $V(t'''_n) < V(t''_n)$. The comparison of t'_n and t'''_n depends on n and μ. First, we note that t'''_n is not defined unless $n \geq 3$. Now $V(t'_n) < V(t'''_n)$ when

$$\frac{2\sigma^2}{n^2} + \frac{\mu^2(2 - n)^2}{n^2} < \frac{7}{18}\sigma^2$$

or

$$\frac{\mu^2(n - 2)^2}{n^2} < \sigma^2\left(\frac{7}{18} - \frac{2}{n^2}\right)$$

or

$$\frac{18\mu^2(n - 2)^2}{7n^2 - 36} < \sigma^2 \qquad (5.80)$$

Finally, t'_n is the best estimator of μ when Eq. (5.80) and $n \geq 4$ are satisfied; otherwise, t'''_n is the best estimator when $n \geq 4$.

Any linear estimator of μ is of the form

$$t_n = a_1 x_1 + \cdots + a_n x_n$$

where $a_1, \ldots, a_n$ are real numbers with at least one a_i ($i = 1, \ldots, n$) being different from zero. Now t_n is unbiased when $E(t_n) = \mu$. Since $E(t_n) = a_1 E(x_1) + \cdots + a_n E(x_n) = (a_1 + \cdots + a_n)\mu$, t_n is an unbiased estimator of μ when $a_1 + \cdots + a_n = 1$. This restriction need not be stated if we write t_n as

$$t_n = \frac{b_1 x_1 + \cdots + b_n x_n}{b_1 + \cdots + b_n} \tag{5.81}$$

where $b_1, \ldots, b_n$ are real numbers not all of which are zero. Since Eq. (5.81) is an unbiased estimator of μ, the second moment of t_n about μ is also the variance of t_n.

On applying Eq. (5.61) we find the variance of t_n to be

$$V(t_n) = \frac{\sum b_i^2}{(\sum b_i)^2} \sigma^2 \tag{5.82}$$

We require values of $b_1, \ldots, b_n$ which make $V(t_n)$ a minimum. From the calculus, we know that such values must satisfy the n equations

$$\frac{\partial V(t_n)}{\partial b_1} = 0, \quad \cdots, \quad \frac{\partial V(t_n)}{\partial b_n} = 0$$

simultaneously. The partial derivative of $V(t_n)$ with respect to b_j is

$$\frac{\partial V(t_n)}{\partial b_j} = \frac{(\sum b_i)^2 (2 b_j) - (\sum b_i^2)(2 \sum b_i)}{(\sum b_i)^4} \quad (j = 1, \ldots, n)$$

Letting each of these n partial derivatives equal zero leads to the unique solution

$$b_j = \frac{\sum b_i^2}{\sum b_i} = \frac{b_1^2 + \cdots + b_n^2}{b_1 + \cdots + b_n} \quad (j = 1, \ldots, n)$$

That is, $V(t_n)$ is a minimum when $b_1 = \ldots = b_n = b$. Hence, the coefficient of x_i in Eq. (5.81) is

$$\frac{b_i}{b_1 + \cdots + b_n} = \frac{b}{nb} = \frac{1}{n}$$

so that $t_n = \bar{x}$. Therefore, no linear combination of the observations can have a smaller variance than $\bar{x}$; that is, $\bar{x}$ is the best of the linear unbiased estimates of μ.

As was pointed out earlier, when the random sample is drawn from a normal population, the minimum variance of all unbiased estimators of μ is σ^2/n. Since $V(\bar{x}) = \sigma^2/n$, it follows that $\bar{x}$ is not only the best linear unbiased estimator of μ, but is the best function of any kind which can be used to estimate μ.

5.7. THE METHOD OF MAXIMUM LIKELIHOOD

By this time the reader must have asked, "Is it possible to describe a method which gives estimators with desirable properties, or must one always apply some aribitrary method when introducing an estimator of a parameter?" Fortunately, it is possible to describe a method with many desirable properties. There are actually several methods [3, 5, 18], all of which have some merit.

We describe a very popular method, the method of maximum likelihood described in general by Fisher [13, 14]. This method is attractive because it gives, under fairly general conditions, estimators which are often best or nearly so and which are often quite easy to obtain. There are cases where a maximum likelihood estimator is very poor, but in most applications the estimators have desirable properties. Another important feature of the method of maximum likelihood is that it yields estimators which are approximately normally distributed for large samples. For fixed n the maximum likelihood estimators are often biased. Fortunately, in cases where an unbiased estimator is desirable, it is often possible to multiply the maximum likelihood estimator by a coefficient involving only n so that the resulting estimator is unbiased.

Let $f(x; \theta)$ denote the density function of a random variable x, where θ is the parameter to be estimated. The form of the function is assumed to be known, but θ is unknown. Let $x_1, \ldots, x_n$ denote a random sample of n observations. Then there are n random variables, each with a density function of the form $f(x_i; \theta)$ $(i = 1, \ldots, n)$. The joint density function, $f(x_1, \ldots, x_n; \theta)$, of the random variables $x_1, \ldots, x_n$ is given by

$$f(x_1, \ldots, x_n; \theta) = f(x_1; \theta) \ldots f(x_n; \theta)$$

since $x_1, \ldots, x_n$ are mutually independent. If we think of $x_1, \ldots, x_n$ as values for a particular (fixed) sample, then $f(x_1, \ldots, x_n; \theta)$ is a function of θ only. Such a function is called the likelihood function and is indicated by $L(\theta)$. Thus, for a fixed set of sample values $x_1, \ldots, x_n$

$$L(\theta) = \prod_{i=1}^{n} f(x_i; \theta) \tag{5.83}$$

is the *likelihood function of θ*. That value of θ, denoted by $\hat{\theta}$, for which $L(\theta)$ is a maximum is called the *maximum likelihood estimator of θ*. Definition (5.83) applies for both discrete and continuous cases.

The generalization of Eq. (5.83) to a univariate density function with more than one parameter is straightforward. Let $f(x; \theta_1, \ldots, \theta_k)$ denote the density function of a random variable x, where $\theta_1, \ldots, \theta_k$ represent k parameters to be estimated. Then for a fixed sample of values $x_1, \ldots, x_n$ the *likelihood function of $\theta_1, \ldots, \theta_k$* is

$$L(\theta_1, \ldots, \theta_k) = \prod_{i=1}^{n} f(x_i; \theta_1, \ldots, \theta_k) \qquad (5.84)$$

Those values of the k parameters, denoted by $\hat{\theta}_1, \ldots, \hat{\theta}_k$, for which $L(\theta_1, \ldots, \theta_k)$ is a maximum are called the maximum *likelihood estimators of* $\theta_1, \ldots, \theta_k$, respectively.

Example 5.16. A random sample of size n is drawn from a normal distribution with mean μ and variance σ^2.

(a) Find the maximum likelihood estimator of μ when $\sigma^2 = 1$. (b) Find the maximum likelihood estimators of μ and σ^2. For (a) the univariate density function is

$$f(x; \mu) = \frac{e^{-\frac{1}{2}(x-\mu)^2}}{\sqrt{2\pi}}$$

and the likelihood function is

$$L(\mu) = \frac{e^{-\frac{1}{2}(x_1-\mu)^2}}{\sqrt{2\pi}} \cdots \frac{e^{-\frac{1}{2}(x_n-\mu)^2}}{\sqrt{2\pi}}$$

or

$$L(\mu) = \frac{e^{-\frac{1}{2}\Sigma(x_i-\mu)^2}}{(2\pi)^{n/2}}$$

That value of μ which maximizes $L(\mu)$ will also maximize $\log_e L(\mu) \equiv l(\mu)$. Now

$$l(\mu) = -\frac{1}{2}\sum(x_i - \mu)^2 - \frac{n}{2}\log_e(2\pi)$$

Therefore

$$\frac{dl(\mu)}{d\mu} = -\frac{1}{2}\sum 2(x_i - \mu)(-1)$$

When

$$\frac{dl(\mu)}{d\mu} = 0$$

it follows that $\sum(x_i - \hat{\mu}) = 0$, or

$$\hat{\mu} = \frac{\sum x_i}{n} = \bar{x}$$

Thus, the maximum likelihood estimator is the sample mean which is unbiased.

For (b) the univariate density function is

$$f(x\,;\mu,\sigma^2) = \frac{e^{-(x-\mu)^2/2\sigma^2}}{\sqrt{2\pi}\cdot\sigma}$$

and the likelihood function is

$$L(\mu,\sigma^2) = \frac{e^{-\Sigma\,(x_i-\mu)^2/(2\sigma^2)}}{(2\pi)^{n/2}\sigma^n}$$

Those values of μ and σ^2 which maximize $L(\mu,\sigma^2)$ will also maximize $\log_e L(\mu,\sigma^2) \equiv l(\mu,\sigma^2)$. Now

$$l(\mu,\sigma^2) = -\frac{\Sigma\,(x_i-\mu)^2}{2\sigma^2} - \frac{n}{2}\log_e\sigma^2 - \frac{n}{2}\log_e(2\pi)$$

Therefore

$$\frac{\partial l(\mu,\sigma^2)}{\partial\mu} = \frac{-\Sigma\,2(x_i-\mu)(-1)}{2\sigma^2}$$

and

$$\frac{\partial l(\mu,\sigma^2)}{\partial\sigma^2} = \frac{-\Sigma\,(x_i-\mu)^2(-1)}{2(\sigma^2)^2} - \frac{n}{2}\cdot\frac{1}{\sigma^2}$$

When

$$\frac{\partial l(\mu,\sigma^2)}{\partial\mu} = 0 \quad\text{and}\quad \frac{\partial l(\mu,\sigma^2)}{\partial\sigma^2} = 0$$

it follows that

$$\frac{\Sigma\,(x_i-\hat\mu)}{\hat\sigma^2} = 0$$

and

$$\frac{\Sigma\,(x_i-\hat\mu)^2}{\hat\sigma^4} - \frac{n}{\hat\sigma^2} = 0$$

Solving these last two equations simultaneously gives

$$\hat\mu = \bar x \quad\text{and}\quad \hat\sigma^2 = \frac{\Sigma\,(x_i-\hat\mu)^2}{n} = \frac{\Sigma\,(x_i-\bar x)^2}{n}$$

Clearly, $\hat\mu$ is an unbiased estimator of μ, but $\hat\sigma^2$ is a biased estimator of σ^2. However, as we have already noted, $n\hat\sigma^2/(n-1) = s^2$ is an unbiased estimator of σ^2.

Example 5.17. A random sample of size n is drawn from a dichotomous population with density function

$$d(x\,;p) = \begin{cases} 1-p, & \text{when } x = 0 \\ p, & \text{when } x = 1 \end{cases}$$

(a) Find the maximum likelihood estimator of p if the successful event

$(x = 1)$ occurs n_1 times where $0 < n_1 < n$. (b) Find the maximum likelihood estimate if it is known that $p = 0.2, 0.4, 0.6,$ or 0.8, $n = 6$, and $n_1 = 2$.

The likelihood function is the product, in some order, of n_1 probabilities p and $n - n_1$ probabilities $1 - p$. That is

$$L(p) = p^{n_1}(1 - p)^{n-n_1} \tag{5.85}$$

Even though x is a discrete variable, the likelihood function may be considered a continuous function of p for $0 < p < 1$. Now

$$l = \log_e L(p) = n_1 \log_e p + (n - n_1) \log_e (1 - p)$$

and

$$\frac{dl}{dp} = \frac{n_1}{p} - \frac{n - n_1}{1 - p}$$

When $dl/dp = 0$, then

$$n_1(1 - \hat{p}) - (n - n_1)\hat{p} = 0$$

or

$$\hat{p} = \frac{n_1}{n}$$

For (b), Eq. (5.85) is a function of a discrete variable p. To find the maximum likelihood estimate of p, we simply substitute each possible value of p in Eq. (5.85) and observe which value makes $L(p)$ a maximum. Now

$$L(p = 0.2) = (0.2)^2(0.8)^4 = 0.016384$$
$$L(p = 0.4) = (0.4)^2(0.6)^4 = 0.020736$$
$$L(p = 0.6) = (0.6)^2(0.4)^4 = 0.009216$$
$$L(p = 0.8) = (0.8)^2(0.2)^4 = 0.001024$$

Thus $\hat{p} = 0.4$ is the maximum likelihood estimate of p.

If n and n_1 were not specified, we would expect the estimator to be that fraction closest to n_1/n. In the above case $n_1/n = \frac{2}{6} \doteq 0.33$ is closer to 0.4 than 0.2. Thus, we would expect 0.4 to be the estimate of p.

5.8. EXERCISES

5.34. Let x_1, x_2, x_3 be independent random variables with $E(x_1) = 3$, $E(x_2) = 1$, $E(x_3) = -1$, $V(x_1) = 1$, $V(x_2) = 2$, and $V(x_3) = 3$. Find: (a) $E(x_1 + 2x_2 - 3x_3)$, (b) $E[(x_1 - x_2) \cdot (x_2 - x_3)]$, (c) $V(x_1 + 2x_2 - 3x_3)$. (d) Which, if any, of the answers in (a), (b), and (c) depend on the independence of the random variables? (e) Find cov $(x_1, \bar{x})$, where $\bar{x} = (x_1 + x_2 + x_3)/3$.

5.35. Prove Eqs. (5.52), (5.53), and (5.54), properties of expectation.

5.36. Prove Eq. (5.55) for $k = 3$; in general.

5.37. Prove Eq. (5.56) for a bivariate distribution.

5.38. Prove Eqs. (5.59) and (5.61).

5.39. Two dice are tossed and the sum x on the two faces observed. Find $E(x)$ and $V(x)$.

5.40. Work Exercise 5.39 for n dice.

5.41. A well-balanced coin is tossed until a head appears. What is the expected number of tosses?

5.42. Prove Theorem 5.3.

5.43. Prove Theorem 5.6.

5.44. The random variable x is normally distributed with mean μ and variance σ^2. Find $E[(x - \mu)^4]$ and $E(x^3)$.

5.45. A seasonal item brings a net profit of P dollars for each item sold and a net loss of L dollars for each item not sold at the end of the season. Suppose the number of customer orders, x, during any season is a random variable which has the Poisson density function [Eq. (3.15)] with $m = 5$. How many items should be stocked so that the expected value of the profit is a maximum?

5.46. Prove properties (5.66) and (5.68) on conditional expectation.

5.47. The joint density function $f(x_1, x_2)$ of two discrete random variables x_1 and x_2 is given in the table. Find the conditional density function of x_1 and the expectation $E(x_1 | x_2)$ for each x_2. Use Eq. (5.66) to find $E(x_1)$.

Table 5.8

x_1 \ x_2	1	2	3
0	0.2	0.1	0.0
1	0.0	0.1	0.3
2	0.1	0.2	0.0

5.48. If $f(x_1, x_2) = 2$, $0 \leq x_1 \leq x_2$, $0 \leq x_2 \leq 1$, find the conditional density functions, $E(x_1 | x_2)$ and $E(x_2 | x_1)$ and $E(x_2)$.

5.49. A box contains five similar circular disks, each bearing exactly one of the marks 1, 2, 3, 4, 5. If two disks are drawn without replacement, what is the expected value of the sum of the two numbers?

5.50. If $f(x_1, x_2) = x_1 + x_2$, $0 < x_1 < 1$, $0 < x_2 < 1$, find $E(x_2 | x_1)$ and $E(x_2)$.

5.51. If t_n is an unbiased estimator of θ, can it be expected, on repeated sampling, to underestimate the true parameter θ half the time? Explain.

5.52. If the variance of a random sample of size n is defined as

$$s'^2 = \frac{\sum (x_i - x)^2}{n}$$

it can be shown that

$$V(s'^2) = \frac{\mu_4 - \sigma^4}{n}$$

where

$$\mu_4 = E[(x - \mu)^4]$$

(a) Show that s'^2 is a consistent estimator. Show that s^2 is also a consistent estimator. (b) Find $V(s'^2)$ when the sample is drawn from a normal population.

5.53. For random samples of size $2n$ let

$$\bar{x}_1 = \frac{\sum\limits_{i=1}^{2n} x_i}{2n} \quad \text{and} \quad \bar{x}_2 = \frac{\sum\limits_{i=1}^{n} x_i}{n}$$

denote two estimators of the population mean μ. Which is the better estimator of μ? Explain why.

5.54. Let $x_1, \ldots, x_7$ be a random sample from a population with mean μ and variance σ^2. Let

$$t_1 = \frac{x_2 + \cdots + x_6}{5} \qquad t_2 = x_6 - x_2$$

$$t_3 = \frac{(x_6 + x_7) - (x_1 + x_2)}{2} \qquad t_4 = \frac{7(x_1 + x_7)}{2}$$

$$t_5 = \frac{2x_1 + x_3 + x_5 + 2x_7}{6}$$

denote estimators of μ. (a) Which, if any, of the estimators of μ are unbiased? (b) Which of the five estimators is best? (c) Let

$$t_6 = \frac{x_2^2 + x_6^2 - \dfrac{x_2 + x_6}{2}}{2}$$

Is t_6 an unbiased estimator of σ^2? Explain. (d) Are any of the estimators of μ consistent? Explain.

5.55. Three random and independent samples of sizes $n_1 = 12$, $n_2 = 8$, and $n_3 = 4$ drawn from a population with variance σ^2 have unbiased variances s_1^2, s_2^2, and s_3^2, respectively. (a) Prove that

$$\frac{12s_1^2 + 8s_2^2 + 4s_3^2}{24}$$

is an unbiased estimator of σ^2. (b) Construct another linear function of the three sample variances which is an unbiased estimator of σ^2.

5.56. Let $x_1, \ldots, x_k$ be random variables which are independently distributed with means 0 and variances $\sigma_1^2, \ldots, \sigma_k^2$. Prove that

$$t = \frac{b_1 x_1 + \cdots + b_k x_k}{b_1 + \cdots + b_k}$$

has minimum variance when the weight b_i is inversely proportional to σ_i^2 $(i = 1, \ldots, k)$.

5.57. (a) Find the variance of t in Exercise 5.56 when $b_i = 1/k$ $(i = 1, \ldots, k)$. (b) Assign values, not all equal, to the σ^2's in Exercise 5.56 and compare the variance of t found in (a) with that found in Exercise 5.56.

5.58. A random sample of size n is drawn from a normal population with mean zero and variance σ^2. Find the maximum likelihood estimator of σ^2.

5.59. Find the maximum likelihood estimator of θ for the density function

$$f(x\,;\theta) = \theta\, e^{-\theta x}, \qquad x \geq 0$$

5.60. Find the maximum likelihood estimator of m for the Poisson density function.

5.61. Two observations drawn at random from a population with density function

$$f(x\,;m) = 1 + m(x - \tfrac{1}{2}), \qquad 0 < x < 1$$

are to be used to find an estimator of m. (a) Find the maximum likelihood estimator of m. (b) If the observations are 0.65 and 0.90, what is the maximum likelihood estimate of m? (c) Discuss the properties of this estimator.

5.62. Suppose the random variable x is uniformly distributed over an interval of length l with mean μ, that is center point, unknown. (a) Find the maximum likelihood estimator of a random sample of size n taken from this distribution. (b) Find the maximum likelihood estimate of μ if the random observations are $3.2, 2.7, 3.0, 2.2$, and 3.4, and it is known that $l = 2$. (c) Use the results of (b) to select the best estimate of μ and to define the best density function.

5.63. Let $p_1, p_2, \ldots, p_k$ denote the probabilities that k mutually exclusive and exhaustive possible outcomes $O_1, O_2, \ldots, O_k$, respectively, will occur on a single trial of an experiment. A random sample of size n is drawn with replacement from such a distribution. Find the maximum likelihood estimators $\hat{p}_i$ $(i = 1, \ldots, k)$, if n_i denotes the observed frequency in the ith category O_i and $n_1 + \cdots + n_k = n$ and $p_1 + \cdots + p_k = 1$.

5.64. For a certain large group of people, suppose $p^2 + 2pr$, $q^2 + 2qr$, $2pq$, and r^2 denote the probabilities of the four blood groups A, B, AB, and O, respectively, where $p + q + r = 1$. (It is to be understood that p, q, and r represent the probabilities in the population of the genes for A, B, and O, respectively.) (a) In a random sample of $n = n_1 + n_2 + n_3 + n_4$ people from the population in question it is found that n_1, n_2, n_3, and n_4 have blood groups A, B, AB, and O, respectively. Find the maximum likelihood estimators of p, q, and r. (b) Find the maximum likelihood estimates of p, q, and r when the actual frequencies for the four blood groups are $n_1 = 192$, $n_2 = 119$, $n_3 = 62$, and $n_4 = 127$.

5.65. Since the likelihood function is a function of parameters, the maximum likelihood estimators for multivariate distributions are found by the method used with univariate distributions. Let $(x_1, y_1), (x_2, y_2), \ldots,$ (x_n, y_n) denote a random sample of n pairs taken from a bivariate normal distribution with parameters $\mu_x, \mu_y, \sigma_x^2, \sigma_y^2$ and ρ. (a) Find the maximum likelihood estimator of ρ when $\mu_x = \mu_y = 0$ and $\sigma_x^2 = \sigma_y^2 = 1$. (b) Find the maximum likelihood estimators of $\mu_x, \mu_y,$ and ρ when $\sigma_x^2 = \sigma_y^2 = 1$. (c) Find the maximum likelihood estimators of $\sigma_x^2, \sigma_y^2,$ and ρ when $\mu_x = \mu_y = 0$. (d) Find the maximum likelihood estimators of $\mu_x, \mu_y, \sigma_x^2, \sigma_y^2,$ and ρ.

5.66. In Sect. 5.4.1 we considered only linear combinations of random variables. Sometimes it is desirable that we find the mean and variance of an arbitrary function $y = h(x_1, \ldots, x_k)$ of k variables $x_1, \ldots, x_k$. It is usually difficult to find the exact values of the mean and variance of y. However, if $y = h(x_1, \ldots, x_k)$ is approximately linear over practically the whole range of variation of $(x_1, \ldots, x_k)$ then y can be adequately represented by the linear terms of its Taylor series expansion. Then the mean and variance can be found by the methods already described.

For the case where $k = 2$ let the means, variances, and covariance of x_1 and x_2 be denoted by $\mu_1, \mu_2, \sigma_1^2, \sigma_2^2$ and σ_{12}. On expanding $y = h(x_1, x_2)$ about the point (μ_1, μ_2) and neglecting terms of degree higher than one in $(x_1 - \mu_1)$ and $(x_2 - \mu_2)$, we have

$$y \doteq h(\mu_1, \mu_2) + h_1(\mu_1, \mu_2)(x_1 - \mu_1) + h_2(\mu_1, \mu_2)(x_2 - \mu_2)$$

where

$$h_i(\mu_1, \mu_2) = \left.\frac{\partial y}{\partial x_i}\right|_{\substack{x_1 = \mu_1 \\ x_2 = \mu_2}} \qquad (i = 1, 2)$$

are partial derivatives of h with respect to x_1 and x_2 evaluated at (μ_1, μ_2). Hence, using the properties of expectation and Theorem 5.6, we have immediately

$$E(y) \doteq h(\mu_1, \mu_2) \tag{5.86}$$

and

$$V(y) \doteq h_1^2(\mu_1, \mu_2)\sigma_1^2 + h_2^2(\mu_1, \mu_2)\sigma_2^2 + 2h_1(\mu_1, \mu_2) \cdot h_2(\mu_1, \mu_2)\sigma_{12} \tag{5.87}$$

since

$$E[h(\mu_1, \mu_2)] = h(\mu_1, \mu_2)$$
$$E[h_i(\mu_1, \mu_2)(x_i - \mu_i)] = h_i(\mu_1, \mu_2)E(x_i - \mu_i) = 0$$

and $V[y - h(\mu_1, \mu_2)] = V(y)$. If x_1 and x_2 are independent, the last term of Eq. (5.87) vanishes. The derivation of expressions for $E(y)$ and $V(y)$ in the general case is left as another exercise for the student.

Note. As a rule, it appears that Formulas (5.86) and (5.87) are adequate when the standard deviation of any variate is not greater than

20 per cent of its mean. (This is just an empirical rule and should be used with caution.)

(a) Find the approximate mean and variance of $y = (x_1 - x_2)/x_1$. (b) In case x_1 and x_2 are independent, an example of (a) is given in *Statistical Analysis in Chemistry and the Chemical Industry*, by Bennett and Franklin (pp. 52–54), in which y is the percentage of moisture in a sample of coal, x_1 is the original weight, and x_2 is the final weight after heating. Study this reference. Other examples appear in *Statistical Theory with Engineering Applications*, by Hald (pp. 246–51), and *Statistical Methods in Research and Production*, by Davies (pp. 48–50).

5.67. Often, as in Example 5.14, we make transformations of variables. The evaluation of single and multiple integrals is often simplified by applying the appropriate transformation. Further, we are often required to obtain the density function (or distribution function) of some statistic $g \equiv g(x_1, \ldots, x_m)$ defined in terms of the variates $x_1, \ldots, x_n$ from an n-fold integral in which the integrand is the joint density function $f(x_1, \ldots, x_n)$. For these reasons we state and discuss an important theorem involving transformations of double integrals (the ideas are easily extended to n-fold integrals). It is assumed that the reader is already familiar with transformations of integrals in the univariate case.

If the transformation $x_1 = t_1(y_1, y_2)$, $x_2 = t_2(y_1, y_2)$ represents a continuous one-to-one mapping of the closed region R of the $x_1 x_2$-plane on the region R' of the $y_1 y_2$-plane, and if the functions t_1 and t_2 have continuous first derivatives and their Jacobian

$$J\left(\frac{x_1, x_2}{y_1, y_2}\right) = \begin{vmatrix} \dfrac{\partial x_1}{\partial y_1} & \dfrac{\partial x_1}{\partial y_2} \\ \dfrac{\partial x_2}{\partial y_1} & \dfrac{\partial x_2}{\partial y_2} \end{vmatrix} = \frac{\partial x_1}{\partial y_1} \cdot \frac{\partial x_2}{\partial y_2} - \frac{\partial x_2}{\partial y_1} \cdot \frac{\partial x_1}{\partial y_2}$$

is either everywhere positive or everywhere negative, then

$$\int_R \int f(x_1, x_2)\, dx_1\, dx_2 = \int_{R'} \int f[t_1(y_1, y_2), t_2(y_1, y_2)]\left| J\left(\frac{x_1, x_2}{y_1, y_2}\right) \right| dy_1\, dy_2$$

$$(5.88)$$

where

$$\left| J\left(\frac{x_1, x_2}{y_1, y_2},\right) \right|$$

denotes the absolute value of the Jacobian. To avoid having to solve for the inverse functions in a set of equations, it is convenient to know that, under very general conditions, the Jacobian of the inverse system of functions is the reciprocal of the Jacobian of the original system; i.e.

$$J\left(\frac{y_1, y_2}{x_1, x_2}\right) = J^{-1}\left(\frac{x_1, x_2}{y_1, y_2}\right)$$

$$(5.89)$$

Thus, in Example 5.14, if we let $x_1 = x_1$ and

$$\bar{x} = \frac{x_1}{2} + \frac{x_2}{2}$$

then

$$J\left(\frac{x_1, \bar{x}}{x_1, x_2}\right) = \begin{vmatrix} 1 & 0 \\ \frac{1}{2} & \frac{1}{2} \end{vmatrix} = \frac{1}{2}$$

so that

$$J\left(\frac{x_1, x_2}{x_1, \bar{x}}\right) = 2$$

Hence, from the fact that

$$\int_{-1/2}^{1/2} \int_{-1/2}^{1/2} 1 \, dx_1 \, dx_2 = 1$$

and the above theorems, we may write

$$\int_{-1/2}^{1/2} \left[\int_{-1/2}^{2\bar{x}+(1/2)} 2 \, dx_1 + \int_{2\bar{x}-(1/2)}^{1/2} 2 \, dx_1 \right] d\bar{x} = 1$$

(From Exercise 5.33 we know that the function of $\bar{x}$ inside the brackets is a marginal density function.)

(a) Find the density function of the sampling distribution of means of samples of size three drawn at random from the parent population of Example 5.14.

(b) Find the density function of the sampling distribution of means of samples of size n drawn at random from the distribution with density function $f(x) = e^{-x}$, $x \geq 0$.

(c) Solve (a) if the population has density function $f(x; \theta) = (1 + \theta)x^\theta$, $\theta > 0$, $0 \leq x \leq 1$.

(d) Find the density function of the sampling distributions of variances of samples of size two drawn at random from the parent population of Example 5.14.

5.68. The Chebyshev inequality applies to any random variable [for example, the sample mean $\bar{x}$ considered in Eq. (5.32), the range, the median], provided it has a mean and variance. Now, we state and prove the general theorem: Let y be a random variable with mean $\mu = E(y)$ and variance $\sigma^2 = V(y)$. Then if k is any positive number

$$P[|y - \mu| \geq k] \leq \frac{\sigma^2}{k^2} \qquad (5.90)$$

Proof. Let $f(y)$ be the density function of y and note that

$$\sigma^2 = \int_{-\infty}^{\infty} (y - \mu)^2 f(y) \, dy$$

may be written as

$$\sigma^2 = \int_{-\infty}^{\mu-k} (y - \mu)^2 f(y)\, dy + \int_{\mu-k}^{\mu+k} (y - \mu)^2 f(y)\, dy$$
$$+ \int_{\mu+k}^{\infty} (y - \mu)^2 f(y)\, dy$$

(5.91)

Since the second integral on the right-hand side of Eq. (5.91) is non-negative, we have

$$\sigma^2 \geq \int_{-\infty}^{\mu-k} (y - \mu)^2 f(y)\, dy + \int_{\mu+k}^{\infty} (y - \mu)^2 f(y)\, dy$$

or

$$\sigma^2 \geq \int_{|y-\mu|\geq k} (y - \mu)^2 f(y)\, dy$$

In this last expression the integrand is always at least as large as $k^2 f(y)$ over the range of integration. Thus, we may write

$$\sigma^2 \geq k^2 \int_{|y-\mu|\geq k} f(y)\, dy = k^2 P[|y - \mu| \geq k]$$

and it follows that Eq. (5.90) holds.

(a) Use Chebyshev's inequality in the proof of the following important property known as the *weak law of large numbers*. Let $y_1, y_2, \ldots$ be an arbitrary sequence of random variables with expectations $E(y_1), E(y_2), \ldots$. Assume that

$$\sum_{i=1}^{n} y_i$$

has a variance for each positive integer n. If

$$V\left(\sum_{i=1}^{n} \frac{y_i}{n}\right)$$

approaches zero as n approaches ∞, and if k is a positive number, then

$$P\left\{\left|\frac{1}{n}\sum_{i=1}^{n} [y_i - E(y_i)]\right| \geq k\right\} \quad \text{approaches 0 as } n \text{ approaches } \infty \quad (5.92)$$

or

$$P\left\{\left|\frac{1}{n}\sum_{i=1}^{n} [y_i - E(y_i)]\right| < k\right\} \to 1 \quad \text{as } n \to \infty \quad (5.93)$$

where the symbol $\to$ denotes "approaches."

(b) Prove the following corollary to Eqs. (5.92) and (5.93). Let $\bar{x}$ be the sample mean of a random sample of size n drawn from a population with mean μ and variance σ^2. If $k > 0$, then

$$P[|\bar{x} - \mu| \geq k] \to 0 \quad \text{as } n \to \infty \quad (5.94)$$

or

$$P[|\bar{x} - \mu| < k] \to 1 \quad \text{as } n \to \infty \tag{5.95}$$

Note. See Feller [10] and Munroe [29] for further discussions of the topics in this exercise as well as discussions on the strong law of large numbers.

5.69. We said earlier that the general proof of the central-limit theorem is beyond the scope of this book, but stated that we would give a restricted proof later. Now we give this proof for distributions which have moment generating functions. (It can be shown that a distribution function is uniquely determined by its MGF.) The proof consists in showing that the MGF for the sample mean approaches the MGF for the normal distribution and in applying a theorem which states that "two random variables which have the same MGF have the same density functions, except possibly at points of discontinuity." (Actually, the MGF is of more importance in determining the distribution of sample statistics than it is in finding moments.)

Proof. Let $f(x)$ be the density function of any random variable x with mean μ and variance σ^2 which has a MGF. The MGF, $M_y(t)$, of the standardized variate $y = (x - \mu)/\sigma$ is given by

$$M_y(t) = \int_{-\infty}^{\infty} e^{[(x-\mu)/\sigma]t} f(x)\, dx \tag{5.96}$$

By Theorem 5.3, the mean and variance of the random variable $\bar{x}$ are $\mu_{\bar{x}} = \mu$ and $\sigma_{\bar{x}}^2 = \sigma^2/n$. So the MGF, $M_z(t)$, for $z = (\bar{x} - \mu)/(\sigma/\sqrt{n})$ may be written, using the expectation notation, as

$$M_z(t) = E\left(e^{\frac{x-\mu}{\sigma/\sqrt{n}} \cdot t}\right) = E\left(e^{\frac{t}{n} \cdot \Sigma \frac{x_i - \mu}{\sigma/\sqrt{n}}}\right)$$

Since $x_1, \ldots, x_n$ is a random sample with x_i having density functions $f(x_i) = f(x)$ $(i = 1, \ldots, n)$, the joint density function of the set $(x_1, \ldots, x_n)$ of variates is given by

$$f(x_1, \ldots, x_n) = \prod_{i=1}^{n} f(x_i)$$

Thus, we have

$$M_z(t) = \int_{-\infty}^{\infty} \cdots \int_{-\infty}^{\infty} e^{\frac{t}{\sqrt{n}} \cdot \Sigma \frac{x_i - \mu}{\sigma}} \prod_{i=1}^{n} f(x_i)\, dx_i$$

$$= \prod_{i=1}^{n} \left[\int_{-\infty}^{\infty} e^{\frac{t}{\sqrt{n}}\left(\frac{x_i - \mu}{\sigma}\right)} f(x_i)\, dx_i \right]$$

or

$$M_z(t) = \prod_{i=1}^{n} \left[M_y\left(\frac{t}{\sqrt{n}}\right) \right] = \left[M_y\left(\frac{t}{\sqrt{n}}\right) \right]^n$$

since by Eq. (5.96) each factor has MGF $M_y(t/\sqrt{n})$. Using Eq. (3.44), we may write

$$M_y\left(\frac{t}{\sqrt{n}}\right) = 1 + \mu'_{1:y} \cdot \frac{t}{\sqrt{n}} + \frac{\mu'_{2:y}}{2!}\left(\frac{t}{\sqrt{n}}\right)^2 + \frac{\mu'_{3:y}}{3!}\left(\frac{t}{\sqrt{n}}\right)^3 + \cdots$$

$$(5.97)$$

Since $y = (x - \mu)/\sigma$, $\mu'_{1:y} = \mu_y = (\mu_x - \mu)/\sigma = 0$, $V(y) = [V(x)]/\sigma^2 = 1$ and $\mu'_{2:y} - (\mu'_{1:y})^2 = \mu'_{2:y} = 1$. Thus Eq. (5.97) may be written as

$$M_y\left(\frac{t}{\sqrt{n}}\right) = 1 + \frac{1}{n}\left(\frac{t^2}{n} + \frac{1}{3!\sqrt{n}}t^3\mu'_{3:y} + \cdots\right)$$

Remembering that, by definition

$$\lim_{n\to\infty}\left(1 + \frac{x}{n}\right)^n = e^x$$

and observing that

$$\left[1 + \frac{1}{n}\left(\frac{t^2}{2} + \frac{1}{3!\sqrt{n}}t^3\mu'_{3:y} + \cdots\right)\right]^n$$

is of the form of $[1 + (x/n)]^n$, we may write

$$\lim_{n\to\infty} M_z(t) = e^{t^2/2}$$

$$(5.98)$$

since

$$\left(\frac{1}{3!\sqrt{n}}t^3\mu'_{3:y} + \cdots\right)$$

vanishes as $n \to \infty$. According to Exercise 3.49(a), the moment generating function of the standard normal distribution is

$$M_u(t) = e^{t^2/2}$$

Since in the limit the MGF of z is the same as the MGF of u, we conclude that in the limit z must be normally distributed with mean zero and variance 1, no matter what the distribution of x (so long as it has a MGF). Thus in the limit $\bar{x}$ is normally distributed with mean μ and variance σ^2/n.

5.70. Sampling from a normal population is so important that we devote the next four chapters to this topic. However, in order to illustrate some of the principles of this chapter the student should prove that "if x is normally distributed with mean μ and variance σ^2, then the sampling distribution of the means of random samples of size n is normally distributed with mean μ and variance σ^2/n."

Outline of the derivation. Write the joint density function $x_1, \ldots, x_n$ as the product of n univariate density functions; let

$$x_n = n\bar{x} - \sum_{i=1}^{n-1} x_i$$

so as to obtain a joint density function of $x_1, \ldots, x_{n-1}, \bar{x}$, and then find the marginal density function of $\bar{x}$ by integrating over $x_1, \ldots, x_{n-1}$. The marginal density function is the required normal density function.

REFERENCES

1. Anderson, R. L. and T. A. Bancroft, *Statistical Theory in Research*. New York: McGraw-Hill, Inc., 1952, Chaps. 4, 5, and 6.

2. Bennett, C. A. and N. L. Franklin, *Statistical Analysis in Chemistry and the Chemical Industry*., New York: John Wiley & Sons, Inc., 1954, Chaps. 3 and 4.

3. Brunk, H. D., *An Introduction to Mathematical Statistics*. Boston: Ginn & Company, 1960, Chaps. 5, 6, 7, and 9.

4. Cochran, W. G., *Sampling Techniques*. New York: John Wiley & Sons, Inc., 1953, Chap. 4.

5. Cramér, H., *Mathematical Methods of Statistics*.Princeton, N. J.: Princeton University Press, 1946, Chaps. 15, 16, and 17.

6. ———, *Random Variables and Probability Distributions*, Cambridge Tracts in Mathematics, No. 36, London: Cambridge University Press, 1937.

7. ———, "On the Composition of Elementary Errors," *Skandinavisk Aktvarietidskrift*, 1928, pp. 13 and 141.

8. Davies, O. L., *Statistical Methods in Research and Production*, 3rd ed. London: Oliver and Boyd, 1958, Chap. 3.

9. Dixon, W. J. and F. J. Massey, *An Introduction to Statistical Analysis*, 2nd ed. New York: McGraw-Hill, Inc., 1957, Chaps. 4 and 16.

10. Feller, W., *An Introduction to Probability Theory and Its Applications*. New York: John Wiley & Sons, Inc., 1958.

11. ———, "Über den Zentralen Grenzwertsatz der Wahrscheinlichkeits-rechnung," *Math. Zeitschr.*, Vol. **40** (1935), p. 521.

12. ———, "Über den Zentralen Grenzwertsatz der Wahrscheinlichkeits-rechnung, II," *Math. Zeitschr.*, Vol. **42** (1937), p. 301.

13. Fisher, R. A., "On the Mathematical Foundation of Theoretical Statistics," *Philosophical Transactions of the Royal Society*, Series *A*, Vol. **222** (1922), pp. 309–368.

14. ———, "Theory of Statistical Estimation," *Proceedings of the Cambridge Philosophical Society*, Vol. **22** (1925), pp. 700–725.

15. Fraser, D. A. S., *Statistics; An Introduction*. New York: John Wiley & Sons, Inc., 1958, Chaps. 6, 7, and 8.

16. Hald, A., *Statistical Theory with Engineering Applications*. New York: John Wiley & Sons, Inc., 1952, Chaps. 8 and 9.

17. Hoel, P. G., *Introduction to Mathematical Statistics*, 2nd ed. New York: John Wiley & Sons, Inc., 1954, Chaps. 5 and 6.

18. Kendall, M. G., *The Advanced Theory of Statistics*, Vol. 1. New York: Hafner Publishing Co., Inc., 1952, Chaps. 8 and 10.

19. Khintchine, A., "Sul Dominio di Attrazione della Legge di Gauss," *Giorn. Ins. Italiano d. Attvari*, Vol. **6** (1935), pp. 378–.

20. Laplace, P. S., *Théorie Analytique des Probabilités*, 1st ed. Paris, 1812.

21. Lévy, P., *Calcul des Probabilités*. Paris: Gauthier-Villars et Cie, 1925.

22. ———, "Propriétés Asymptotiques des Sommes de Variables Aléatoires

Indépendantes ou Enchainées," *Journ. Math. Pures Appl.*, Vol. **14** (1935), pp. 347–.

23. Liapounoff, A., "Sur Une Proposition de la Théorie des Probabilitiés," *Bull. Acad. Sc.*, St-Pétersbourg, Vol. **13** (1900), pp. 359–.

24. ———, "Nouvelle Forme du Théorème Sur la Limite de Probabilité," *Mém. Acad. Sc.*, St-Pétersbourg, Vol. **12,** No. 5 (1901).

25. Lindeberg, J. W., "Eine Neve Herleitung des Exponentialgesetzes in der Wahrscheinlichkeitsrechnung," *Math. Zeitschr.*, Vol. **15** (1922), pp. 211–.

26. Mises, R. von, "Deux Nouveaux Théorèmes de Limite dans le Calcul des Probabilités," *Revue Fac. Sc.*, Istanbul, Vol. **1** (1935), pp. 61–.

27. ———, "Les Lois de Probabilité Pour les Fonctions Statistiques," *Ann Inst.*, Henri Poincaré, Vol. **6** (1936), pp. 185–.

28. Mood, A. M., *Introduction to the Theory of Statistics.* New York: McGraw-Hill, Inc., 1950, Chaps. 3, 5, 7, and 10.

29. Munroe, M. E., *Theory of Probability.* New York: McGraw-Hill, Inc., 1951.

30. Wilks, S. S., *Elementary Statistical Analysis.* Princeton, N. J.: Princeton University Press, 1951, Chap. 9.

6

SAMPLING
FROM NORMAL POPULATIONS

In many applications it is reasonable to assume that the parent population is approximately normally distributed, or that the data can be transformed so as to be approximately normally distributed. In either case sampling distributions derived from a parent normal distribution are of prime importance in problems of estimation and tests of hypotheses. We state and illustrate some very important theorems concerning sampling distributions. (A proof or an outline of a proof of each of these theorems is found among the exercises at the end of the chapter.)

6.1. INTRODUCTION

The concept of a sampling distribution of *any statistic* determined from *any population* should be clear by now. In Chap. 5, we showed how the mean and variance of sampling distributions of means, totals, and linear functions of random variables are related to the means and variances of the parent populations. In the exercises, sampling distributions of certain other statistics, such as median and range, determined from samples drawn from particular populations were considered. In a special case, we illustrated how the density function of the sample mean $\bar{x}$ might be found by using the marginal density function. But the emphasis was generally on understanding the nature of the relationship between parameters in the parent population and parameters in the sampling distributions.

In this chapter we study the sampling distributions from the point of view of *density functions obtained from a parent normal density function.* Such functions enable us to compute tables useful in obtaining confidence intervals and in testing hypotheses. In order to point up the usefulness of the theorems of this chapter, we introduce *methods* for finding confidence intervals and testing hypotheses without actually theoretically justifying the techniques.

In Chap. 5 we discussed properties of estimators. We learned that, for a particular sample, the estimator takes on a particular value which may be used to estimate the unknown value of a parameter. Such an estimate is sometimes called a *point estimate.* Unfortunately, a point estimate, even a *best point estimate,* may deviate so much from the parameter that a single value may not be considered satisfactory. Thus, it is customary to estimate a parameter θ by any value in an *interval.*

Let $x_1, \ldots, x_n$ be a random sample from a distribution with density function $f(x)$ and parameter θ. Let $t \equiv t(x_1, \ldots, x_n)$, a function of the n sample observations, be a statistic corresponding to θ. Let $f(t)$ be the density function of the random variable t. If it is possible to determine two values, t_1 and t_2, of the statistic t such that, for the parameter being estimated

$$P[t_1 \leq \theta \leq t_2] = \gamma \qquad (6.1)$$

where γ is some fixed probability, the set of values between t_1 and t_2 inclusive is called a *confidence interval of θ.* Thus, any value in the interval is considered a possible value of the parameter θ. The values t_1 and t_2 are called the *confidence limits.* The measure of probability associated with the confidence interval is called the *confidence level or* the *confidence coefficient.* References [3, 7, 9] give theoretical developments of confidence intervals.

Sometimes we refer to the confidence interval as the 100γ per cent confidence interval. It should be noted that for a particular sample, limits are found which either *do* or *do not* include the true value of θ. Thus, when we say that θ lies in a 100γ per cent confidence interval, we mean that this is true, on the average, 100γ per cent of the time. That is, if an experiment is repeated a large number of times and a 100γ per cent interval is computed each time, then in approximately 100γ per cent of the experiments, the limits include the true value of θ.

In general, we say that a *statistical hypothesis* is an assumption concerning the density function of a random variable (or a set of random variables), and a *test of hypothesis* is a procedure for deciding whether to reserve judgment or reject the hypothesis. The definition of hypothesis is broad enough to cover many different types found in the study of statistics. In a restricted case, we suppose the form of the density function to be known, so that the statistical hypothesis specifies the value (or values) of each parameter of the density function. That is, the hypothesis specifies some one member (or subset of members) of a family of density functions. In particular, in the simplest cases of a statistical hypothesis, only *one parameter is assumed to have a single value*, the form of the density function along with all other parameters being known. Any hypothesis which assumes, for a given density function, that each parameter has a single fixed value is called a *simple hypothesis;* otherwise, it is called a *composite hypothesis.* Many hypotheses will be amply illustrated in the remaining sections of this book.

Associated with each hypothesis there are often several test procedures. In this chapter, we use that procedure which best illustrates the theorem under consideration. To apply a test in the simplest case, select an estimator of the parameter of the hypothesis. Then any rule which divides the set of all possible values of an estimator into two sets, one being the region of rejection, called the *critical region*, and the other the region of indecision, is called *a test procedure* for the statistical hypothesis. If, as a result of an experiment, the value of an estimator falls in the critical region, we *reject* the hypothesis; otherwise, we *fail to reject* the hypothesis. When the hypothesis is rejected, we usually accept some alternative value (or values) of the parameter associated with the choice of the critical region. The critical region, as we shall see shortly, may be composed of two or more disjoint sets. The probability that a value of the estimator falls in the critical region is called the *significance level of the test* and is denoted by α. Since the main emphasis of the test procedure is to reject the hypothesis, we generally refer to the statistical hypothesis under test as the *null hypothesis.* To illustrate the meaning of the above terms, consider the following example.

Example 6.1. Use the teak tree data of Exercise 2.10: (a) Find a point estimate of the population mean of teak trees. (b) Find a 95 per cent confidence interval for the true mean diameter. (c) Test the hypothesis that the true mean diameter is 22 in. when the significance level is five per cent.

Suppose the 1088 diameters represent a random sample of a large population of teak trees, all of the same age and grown in the same general area and environment of India. Use the sample mean to make statements about the population mean. Since the sample size is so large, we may actually assume the sample mean to be approximately normally distributed. According to Example 3.10, the mean and variance of the sample are $\bar{x} = 21.69$ and $s^2 = 34.5156$. Thus, 21.69 is not only a point estimate but may be considered a best point estimate of the population mean μ.

In order to solve (b) and (c), we assume that the 1088 diameters represent a random sample from a single population and that the sample size is large enough so that the sampling distribution of means is closely approximated by the normal distribution. Finally, we assume that $n = 1088$ is large enough so that we may use 34.5156 as the population variance. Thus, the standard deviation of the sampling distribution of $\bar{x}$, that is, the standard error of $\bar{x}$, is given by

$$\sigma_{\bar{x}} = \sqrt{\frac{34.5156}{1088}} = 0.1781$$

It can be shown that 95 per cent confidence limits of μ are given by $\bar{x} - t_{.025}\sigma_{\bar{x}}$ and $\bar{x} + t_{.025}\sigma_{\bar{x}}$ where $t_{.025}$ is the upper 2.5 per cent value of the standard normal variate t. Thus, in our example the confidence limits are

$$\bar{x} \pm t_{.025}\sigma_{\bar{x}} = 21.69 \pm (1.960)(0.1781) = 21.69 \pm 0.35$$

or 21.34 and 22.04.

Hence, the 95 per cent confidence *interval* for μ, symmetric about $\bar{x}$, is given by

$$21.34 \leq \mu \leq 22.04 \tag{6.2}$$

The interval (6.2) is determined from a single sample of n values. Another random sample would usually lead to a different interval. That is, the limits vary from sample to sample. In fact, in a long series of random samples of size n drawn from a normal population with mean μ and variance $\sigma^2 = 34.5156$, we would expect about 95 per cent of the resulting intervals to include the true fixed value of μ. It is for this reason that we call the typical interval

$$\bar{x} - t_{.025}\frac{\sigma}{\sqrt{n}} \leq \mu \leq \bar{x} + t_{.025}\frac{\sigma}{\sqrt{n}} \tag{6.3}$$

a 95 per cent confidence interval. Since the sampling distribution of $\bar{x}$ is symmetric and the limits are equidistant from a sample mean, Relation (6.3) is sometimes called a *symmetric* 95 *per cent confidence interval*. If Relation (6.3) were used to find intervals for each of 100 random samples of size n, we would expect about 95 of the intervals to include μ, not knowing which of about five would fail.

When we use a 95 per cent confidence interval, we feel that our chance of covering the true mean is fairly good. If we wish to be more confident that the interval covers μ, we should use a higher confidence level, say 98 or 99 per cent. To be completely certain that the interval covers μ, we could take as our interval all values from $-\infty$ to $+\infty$. But such a statement does not really tell us anything about the mean of the population. In fact, the more confident we wish to be that an interval includes the true value of the parameter, the less we have to be sure of.

To find a nonsymmetric 95 per cent confidence interval, we let $\bar{x} - t_{.01}\sigma_{\bar{x}}$ and $\bar{x} + t_{.04}\sigma_{\bar{x}}$ be the lower and upper limits, respectively. Then, since $t_{.01} = 2.326$ and $t_{.04} = 1.751$, the limits become $\bar{x}_1 = 21.69 - (2.326)(0.1781) = 21.28$ and $\bar{x}_2 = 21.69 + (1.751)(0.1781) = 22.00$, so the 95 per cent confidence interval is

$$21.28 \leq \mu \leq 22.00$$

In this case, it is clear that the symmetric 95 per cent confidence interval is shorter than the nonsymmetric 95 per cent confidence interval. It can be shown that this is always true for a fixed confidence level and a fixed sample size when $\bar{x}$ is symmetrically distributed about its mean.

In the test of a null hypothesis the choice of a critical region depends on what we expect the parameter under test to be if the hypothesized value is not true. Lehmann [8] and others [3, 7, 9, 15, 16] give some principles which are useful in determining the most desirable critical region in a given experimental situation. The intuitive method we give should generally be adequate for the relatively simple cases considered in this book.

Let us suppose, in our example, that the true mean is equal to or less than 22. Since the null hypothesis is $\mu = 22$ in., the alternative values for μ are those less than 22 in. That is, the so-called *alternative hypothesis* is that $\mu < 22$ in. We require a critical region, defined in terms of $\bar{x}$, which allows us to accept the alternative hypothesis when the null hypothesis is rejected. It seems reasonable to suppose that values of $\bar{x}$ somewhat less than 22 are the only values which would make us want to reject the null hypothesis in favor of the alternative hypothesis. Thus, we define the critical region to be that set of values of $\bar{x}$ less than $\bar{x}_0$ for which $P[\bar{x} \leq \bar{x}_0] = \alpha$ when the null hypothesis is true. That is, $\bar{x}_0$ is selected so that

$$P[\bar{x} \leq \bar{x}_0; \ \mu_{\bar{x}} = 22, \ \sigma_{\bar{x}} = 0.178] = 0.05$$

or

$$P\left[t \leq \frac{\bar{x}_0 - 22}{0.178}\right] = 0.05$$

Since

$$P[t \leq -1.645] = 0.05$$

it is clear that $\bar{x}_0$ must satisfy the equation

$$\frac{\bar{x}_0 - 22}{0.178} = -1.645$$

Thus, $\bar{x}_0 = 22 - 1.645(0.178) = 21.61$, and the critical region is made up of all values of $\bar{x}$ for which $\bar{x} \leq 21.61$. Since our sample mean $\bar{x} = 21.69$ falls in the region of indecision rather than the critical region, we fail to reject the hypothesis that $\mu = 22$ in. This does not imply that the mean is 22 in.; it simply means that we do not have enough evidence to say the true mean is less than 22 in.

Some points in connection with the above problem will be discussed in detail later on. At this time we wish to examine some properties of sampling distributions of measures of central tendency derived from normal parent populations.

6.2. SOME PROPERTIES OF LINEAR FUNCTIONS OF RANDOM VARIABLES

If the parent population is normal, there is no better estimator of the population mean than the sample mean. In spite of this, other statistics which measure central tendency are sometimes used in place of the sample mean. Generally, these statistics are used because they are already available or are easy and relatively inexpensive to obtain. For example, the census bureau and many governmental agencies give most of their summary data in the form of medians. Mid-ranges may be computed in certain industrial processes almost as fast as the data are collected. Actually, if data are easy and relatively inexpensive to collect, it may be possible to place more reliance on medians or mid-ranges of samples of size sixty than on means of samples of size thirty. It may also be possible to make the statistical analysis in less time.

As a matter of convenience for the reader, we now give in one group statements of seven important theorems relating to measures of central tendency and the normal distribution. Following the statements illustrations of applications are presented.

Theorem 6.1. *Let x_i be the ith observation in a random sample of size n with x_i drawn from a normal population with mean μ_i and variance σ_i^2. Let a linear combination l of these random variables be defined by*

$$l = \sum_{i=1}^{n} a_i x_i$$

where a_i are real constants not all zero. Then the random variable l is normally distributed with mean

$$\mu_l = \sum_{i=1}^{n} a_i \mu_i$$

and variance

$$\sigma_l^2 = \sum_{i=1}^{n} a_i^2 \sigma_i^2$$

Symbolically, the theorem may be stated as *if $x_i \sim n_i(\mu_i, \sigma_i^2)$, x_i's are independently distributed and $l = \sum a_i x_i$, some $a_i \neq 0$; then $l \sim n(\mu_l = \sum a_i \mu_i, \sigma_l^2 = \sum a_i^2 \sigma_i^2)$.*

Theorem 6.2. *If x is normally distributed with mean μ and variance σ^2, then the sampling distribution of the means of random samples of size n is normally distributed with mean μ and variance σ^2/n. Symbolically, if $x \sim n(\mu, \sigma^2)$, then $\bar{x} \sim n(\mu_{\bar{x}} = \mu, \sigma_{\bar{x}}^2 = \sigma^2/n)$.*

Theorem 6.3. *If x is normally distributed with mean μ and variance σ^2, then the sampling distribution of the median x_m of random samples of size n approaches the normal distribution with mean μ and variance $\pi\sigma^2/2n$ as n be-*

comes large. Symbolically, if $x \sim n(\mu, \sigma^2)$, *then* $x_m \overset{\cdot}{\sim} n(\mu_{x_m} = \mu, \sigma_{x_m} = \pi\sigma^2/2n)$ *as n becomes large.*

Theorem 6.4. *If x represents the number of successes in n independent trials of an event for which p is the probability of success in a single trial, then*

$$y = \frac{x - np}{\sqrt{np(1 - p)}}$$

has a distribution that approaches the normal distribution with mean 0 and variance 1 as the number of trials becomes increasingly large. Symbolically, if $x \sim b(x; n, p)$, *then* $y \overset{\cdot}{\sim} n(\mu = 0, \sigma^2 = 1)$ *if n is large.* [For a fixed p, n is considered sufficiently large if $np \geq 5$ and $n(1 - p) \geq 5$.]

Theorem 6.5. *The proportion of successes x/n, where x and n are defined as in Theorem 6.4, is approximately normally distributed with mean p and variance $p(1 - p)/n$ when n is sufficiently large.*

Theorem 6.6. *If* $\bar{x}_1$ *and* $\bar{x}_2$ *are normally and independently distributed with means* μ_1 *and* μ_2 *and variances* σ_1^2/n_1 *and* σ_2^2/n_2, *respectively, then* $d \equiv \bar{x}_1 - \bar{x}_2$ *is normally distributed with mean* $\delta = \mu_d = \mu_1 - \mu_2$ *and variance*

$$\sigma_d^2 = \frac{\sigma_1^2}{n_1} + \frac{\sigma_2^2}{n_2}$$

It is understood that n_1 *and* n_2 *denote the sample sizes. Symbolically, if* $\bar{x}_1$ *and* $\bar{x}_2$ *determined from random samples of sizes* n_1 *and* n_2, *respectively, and are independently distributed with* $\bar{x}_i \sim n(\mu_i, \sigma_i^2/n_i)$ $(i = 1, 2)$, *then*

$$d = \bar{x}_1 - \bar{x}_2 \sim n\left(\mu_d = \mu_1 - \mu_2, \sigma_d^2 = \frac{\sigma_1^2}{n_1} + \frac{\sigma_2^2}{n_2}\right)$$

Theorem 6.7. *Let* $x_i (i = 1, 2)$ *represent the number of successes in* n_i *independent trials of an event for which* p_i *is the probability of success in a single trial. Let* $p_i' = (x_i/n_i)$ *denote the sample proportions of success. Then, when the number of trials* n_1 *and* n_2 *are sufficiently large, the difference* $d = p_1' - p_2'$ *in the sample proportions is approximately normally distributed with mean* $\mu_d = p_1 - p_2$ *and variance*

$$\sigma_d^2 = \frac{p_1(1 - p_1)}{n_1} + \frac{p_2(1 - p_2)}{n_2}$$

Symbolically, if $x_i \sim b(x_i; n_i, p_i)$ $(i = 1, 2)$, and x_1 and x_2 are independently distributed, then

$$d = \frac{x_1}{n_1} - \frac{x_2}{n_2} \overset{\cdot}{\sim} n\left[\mu_d = p_1 - p_2, \sigma_d^2 = \frac{p_1(1 - p_1)}{n_1} + \frac{p_2(1 - p_2)}{n_2}\right]$$

The proofs of Theorems 6.1 and 6.4 follow from properties of moment generating functions and are outlined in the next set of exercises. Theorems 6.2, 6.5, 6.6, and 6.7 are actually corollaries of Theorems 6.1 and 6.4. For example, if we let $a_i = 1/n$ in Theorem 6.1, we have Theorem 6.2. Further, if in Theorem 6.1 the two random variables are $\bar{x}_1$ and $\bar{x}_2$, $a_1 = 1$, $a_2 = -1$, and the variances are $\sigma_{\bar{x}_1}^2$ and $\sigma_{\bar{x}_2}^2$, then we have Theorem 6.6. The proof of Theorem 6.3 is the most difficult.

6.3. APPLICATIONS

We now give a variety of applications of the theorems of Sect. 6.2. From Theorem 6.2 it is clear that when the parent population is normal the distribution of sample means is normal, no matter how small the sample size.

Example 6.2. A manufacturer of bolts specifies that the diameter of a certain type should be 2.5 cm. The standard process makes bolts with diameters approximately normally distributed with a standard deviation of 0.1 cm. Adjustments are regularly required in order that the bolts be not too small or too large. Determine if the machine needs adjustment if a random sample of nine measurements has a mean diameter of $\bar{x} = 2.62$ cm.

We wish to test the null hypothesis that $\mu = 2.5$ cm, which is often denoted by

$$H_0 : \mu = 2.5 \text{ cm}$$

In order to protect against making bolts which are too small or too large, it is desirable that adjustments be made if the sample mean is small enough to be below a fixed small $\bar{x}_1$ or large enough to be above a fixed large $\bar{x}_2$. Thus, the problem is to determine $\bar{x}_1$ and $\bar{x}_2$. We make the somewhat arbitrary decision to find values so that, when $\mu = 2.5$ and $\sigma = 0.1$

$$P[\bar{x} \leq \bar{x}_1] = 0.025 \quad \text{and} \quad P[\bar{x} \geq \bar{x}_2] = 0.025$$

or

$$P\left[t \leq \frac{\bar{x}_1 - 2.5}{0.1/\sqrt{9}}\right] = 0.025 \quad \text{and} \quad P\left[t \geq \frac{\bar{x}_2 - 2.5}{0.1/\sqrt{9}}\right] = 0.025$$

Since

$$P[t \leq -1.960] = 0.025 \quad \text{and} \quad P[t \geq 1.960] = 0.025$$

we have

$$\frac{\bar{x} - 2.5}{0.1/\sqrt{9}} = \pm 1.960$$

or

$$\bar{x}_1 = 2.435 \quad \text{and} \quad \bar{x}_2 = 2.565$$

So the critical region is made up of all those values of $\bar{x}$ for which

$$\bar{x} < 2.435 \quad \text{or} \quad \bar{x} > 2.565$$

The values $\bar{x}_1$ and $\bar{x}_2$ are called *critical points* and represent boundary points between the region of indecision and the critical regions. Since the sample mean $\bar{x} = 2.62$ is greater than 2.565, we reject the hypothesis that the population mean is 2.5 cm. That is, we accept the alternate hypothesis that the mean is different from 2.5 cm. Actually, in this case, we would say that the mean is larger than 2.5 cm, realizing that there is a very small (less than 0.025) chance of being wrong. Thus, on the basis of the sample, the machine needs to be adjusted to make smaller bolts.

Above we defined the critical region in terms of $\bar{x}_1$ and $\bar{x}_2$ and used the sample mean to make a decision. However, we could reach a conclusion just as well if the values $\bar{x}_1$, $\bar{x}_2$ and sample mean were expressed in terms of standard normal deviates. In this case we find directly from the table $t_1 = -1.960$ and $t_2 = 1.960$, so that the critical region is composed of all those values of t for which $t < -1.96$ or $t > 1.96$. Further, from the sample mean we find

$$t = \frac{\bar{x} - \mu}{\sigma / \sqrt{n}} = \frac{2.62 - 2.5}{0.1 / \sqrt{9}} = \frac{(0.12)3}{0.1} = 3.6$$

Table 6.1

Probabilities for the Binomial Distribution

$$b(x; n, p) = \binom{n}{x} p^x (1 - p)^{n-x}$$

Number of Successes x	Density(a) n = 10 p = 0.2	Density(b) n = 10 p = 0.4	Density(c) n = 20 p = 0.2	Density(d) n = 20 p = 0.4	Normal Approximation for(d)
0	.1074	.0060	.0115	.0000	.0003
1	.2684	.0403	.0576	.0005	.0012
2	.3020	.1209	.1369	.0031	.0045
3	.2013	.2150	.2054	.0123	.0140
4	.0881	.2508	.2182	.0350	.0351
5	.0264	.2007	.1746	.0746	.0718
6	.0055	.1115	.1091	.1244	.1197
7	.0008	.0425	.0545	.1659	.1632
8	.0001	.0106	.0222	.1797	.1803
9	.0000	.0016	.0074	.1597	.1632
10	.0000	.0001	.0020	.1171	.1197
11			.0005	.0710	.0718
12			.0001	.0355	.0351
13			.0000	.0146	.0140
14			.0000	.0049	.0045
15			.0000	.0013	.0012
16			.0000	.0003	.0003
17			.0000	.0000	.0001
18			.0000	.0000	.0000
19			.0000	.0000	.0000
20			.0000	.0000	.0000

Since 3.6 is greater than $t_2 = 1.96$, we reject H_0 and make the same conclusion as before. Since there is usually less computation this way, we usually solve problems in terms of standard values.

Example 6.3. Graph particular binomial distributions in the form of histograms and compare with the corresponding normal approximations.

From Table I in *Handbook of Probability and Statistics with Tables* by Burington and May we obtained probabilities for the binomial distribution which are shown in columns 2,3,4, and 5 of Table 6.1. Using these values, we construct the histograms shown in Fig. 6.1. Note that corresponding to each discrete value x_i for the binomial density functions we have a unit interval from $x_i - \frac{1}{2}$ to $x_i + \frac{1}{2}$ for the histogram. The height of each rectangle in the histogram is the same as the density function value. Hence, summing any subset of density function values $f(n)$ is the same numerically as summing the areas of corresponding rectangles. In each case we obtain probabilities.

A normal curve obviously would not fit Figs. 6.1a, b, and c well. However, as shown in Fig. 6.2, a normal curve does fit Fig. 6.1d reasonably well.

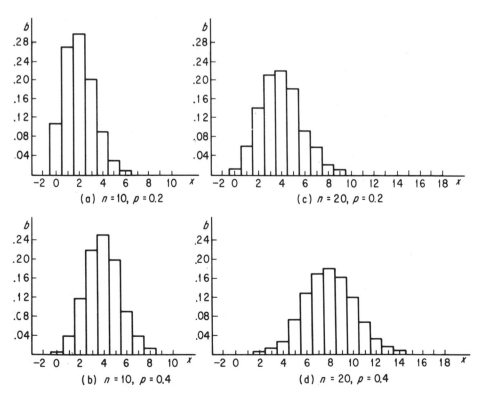

Fig. 6.1 Binomial Histograms

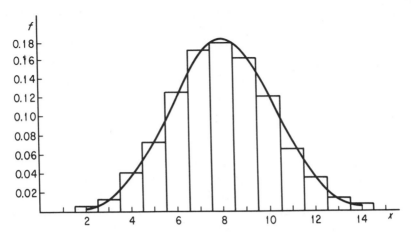

Fig. 6.2 Normal Curve Approximation to the Binomial

The normal distribution has mean $\mu = np = 8$ and standard deviation $\sigma = \sqrt{np(1-p)} = 2.19$. The area of the histogram and the area under the normal curve are both 1. Further, for the histogram the area above the interval from $x_i - \frac{1}{2}$ to $x_i + \frac{1}{2}$ is $f(x_i)$, and the corresponding area for the normal distribution is

$$\int_{x_i - (1/2)}^{x_i + (1/2)} n(x; np, npq)\, dx$$

Since the normal curve reaches from $-\infty$ to $+\infty$, it is customary to let the interval for $x = 0$ reach from $-\infty$ to $\frac{1}{2}$ and the interval for $x = n$ reach from $n - \frac{1}{2}$ to $+\infty$. The last column of Table 6.1 shows how good the normal approximation is for each value of x for the density function (d). (The values were found by using the methods of Example 3.10.)

Example 6.4. The occurrence of a one or two in an honest toss of a well-balanced die is considered a success. Use the normal distribution to find the approximate probability of being successful (a) in more than 15 of 50 tosses and (b) exactly 15 times in 50 tosses. (c) Check (a) and (b) using either Pearson's [11] or Romig's [13] binomial tables.

In (a) and (b) we find the probabilities for the discrete binomial function, using the continuous normal approximation as described in Example 6.3. Thus, in order to find $b(16) + \cdots + b(50)$, we first determine the standard normal t corresponding to $x - \frac{1}{2} = 15.5$; that is

$$t = \frac{x - \frac{1}{2} - \mu}{\sigma} = \frac{x - \frac{1}{2} - np}{\sqrt{np(1-p)}} = -0.35 \tag{6.4}$$

Hence

$b(16) + \cdots + b(50) = P[x = 16, \ldots, 50; \text{ binomial with } n = 50, \ p = \tfrac{1}{3}]$

$\qquad\qquad\qquad \doteq P[x \geq 15.5; \text{ normal with } \mu = \tfrac{50}{3}, \ \sigma^2 = \tfrac{100}{9}]$

$\qquad\qquad\qquad = P[t \geq -0.35; \text{ normal with } \mu = 0, \ \sigma^2 = 1]$

$\qquad\qquad\qquad = 0.6368$

For (b), using the same method, we obtain

$b(15; n = 50, p = \tfrac{1}{3}) \doteq P[14.5 \leq x \leq 15.5; \text{ normal with } \mu = \tfrac{50}{3}, \sigma^2 = \tfrac{100}{9}]$

$\qquad\qquad\qquad = P[-0.65 \leq t \leq -0.35; \text{ normal with } \mu = 0, \sigma^2 = 1]$

$\qquad\qquad\qquad = P[t \geq -0.65] - P[t \geq -0.35]$

$\qquad\qquad\qquad = 0.7422 - 0.6368$

$\qquad\qquad\qquad = 0.1054$

Using Romig's tables, we must interpolate. Since

$\qquad P[x = 16, \ldots, 50; \text{ binomial with } n = 50, \ p = 0.34] = 0.6679$

and

$\qquad P[x = 16, \ldots, 50; \text{ binomial with } n = 50, \ p = 0.33] = 0.6120$

we have, by linear interpolation

$\qquad P[x = 16, \ldots, 50; \text{ binomial with } n = 50, \ p = \tfrac{1}{3}] = 0.6120 + \tfrac{1}{3}(0.0559)$

$\qquad\qquad\qquad\qquad\qquad\qquad\qquad\qquad = 0.6306$

Further, by linear interpolation

$\qquad P[x = 15; \text{ binomial with } n = 50, \ p = \tfrac{1}{3}] = 0.1103 - \tfrac{1}{3}(0.1103 - 0.1020)$

$\qquad\qquad\qquad\qquad\qquad\qquad\qquad\qquad = 0.1075$

From this we see that the approximations agree to two significant figures. This is quite adequate for many purposes.

Example 6.5. Find the shortest 90 per cent confidence interval for the difference between unknown population means μ_1 and μ_2 if it is known that the two populations are normal with variances $\sigma_1^2 = 70$ and $\sigma_2^2 = 180$, respectively. Suppose we find that the means of random samples of sizes $n_1 = 10$ and $n_2 = 20$ are $\bar{x}_1 = 23$ and $\bar{x}_2 = 28$. (The reader can furnish his own interpretation for such a problem. For example, a manufacturer may wish to compare the means of measurements made on articles produced by the oldest and newest machines in the plant; a scientist may wish to make such a comparison using measurements obtained by two different methods; an engineer may wish to test tensile strengths of two different metals or concretes, etc.)

Theorem 6.6 is used to find symmetric 90 per cent limits, since they give the shortest interval. Since the standard normal t is given by

$$t = \frac{\bar{x}_1 - \bar{x}_2 - (\mu_1 - \mu_2)}{\sqrt{\dfrac{\sigma_1^2}{n_1} + \dfrac{\sigma_2^2}{n_2}}} \tag{6.5}$$

the limts for $\mu_1 - \mu_2$ are given by

$$\bar{x}_1 - \bar{x}_2 \pm t_{.05}\left(\sqrt{\frac{\sigma_1^2}{n_1} + \frac{\sigma_2^2}{n_2}}\right) \tag{6.6}$$

where $t_{.05}$ is the upper five per cent value obtained from tables of the normal distribution. In particular, the limits are

$$23 - 28 \pm 1.645(\sqrt{\tfrac{70}{10} + \tfrac{180}{20}})$$

or

$$-11.58 \quad \text{and} \quad 1.58$$

It appears that μ_1 is very likely to be less than μ_2, but μ_1 could be equal to or slightly greater than μ_2.

Example 6.6. We may use Theorem 6.4 and the number of successes in a random sample of size n to find approximate confidence limits for p when n is large. (a) Find approximate $100 (1 - \alpha)$ per cent confidence limits for p when n is large. (b) Use the results obtained in (a) to determine the 95 per cent confidence limits when $n = 100$ if, from a random sample, the total number of successes is $x = 20$.

Again, it is left for the reader to supply his own interpretations. For example, an inspector may wish to know the minimum and maximum proportion of defectives to expect if a random sample has x/n defectives; a biologist may wish to determine the survival rate of a certain type of insect under a given set of conditions; one sampling public opinion may use this method to determine what proportions of a population can be expected to vote a certain way or select a certain article, etc. It should be understood that in each case the experimenter is at liberty to set the degree of confidence with which he wishes to work.

The reader, no doubt, has already noticed, in the special case where $\theta = \mu$ and the statistic is a mean $\bar{x}$ which is normally distributed, that the statement

$$P[t_1 \leq \theta \leq t_2] = \gamma = 1 - \alpha$$

becomes

$$P[\bar{x} - t_{\alpha/2}\cdot\sigma_{\bar{x}} \leq \mu \leq \bar{x} + t_{\alpha/2}\cdot\sigma_{\bar{x}}] = 1 - \alpha \tag{6.7}$$

where $t_{\alpha/2}$ is the standard normal value defined by $P[t \geq t_{\alpha/2}] = \alpha/2$. Further, it should be clear that

$$P\left[-t_{\alpha/2} \le \frac{\bar{x} - \mu}{\sigma_{\bar{x}}} \le t_{\alpha/2}\right] = 1 - \alpha \tag{6.8}$$

is equivalent to Eq. (6.7), since

$$\bar{x} - t_{\alpha/2} \cdot \sigma_{\bar{x}} \le \mu \le \bar{x} + t_{\alpha/2} \cdot \sigma_{\bar{x}}$$

implies that

$$-t_{\alpha/2} \le \frac{\bar{x} - \mu}{\sigma_{\bar{x}}} \quad \text{and} \quad \frac{\bar{x} - \mu}{\sigma_{\bar{x}}} \le t_{\alpha/2}$$

implies that

$$-t_{\alpha/2} \le \frac{\bar{x} - \mu}{\sigma_{\bar{x}}} \le t_{\alpha/2}$$

μ and $\sigma_{\bar{x}}$ being fixed parameters and $\bar{x}$ being a sample value of the random variable $\bar{x}$.

Now in case Theorem 6.4 applies, x, the number of successes in n independent trials, may replace $\bar{x}$ in Eq. (6.8), so that we may write

$$P\left[-t_{\alpha/2} \le \frac{x - np}{\sqrt{np(1 - p)}} \le t_{\alpha/2}\right] = 1 - \alpha \tag{6.9}$$

For a given value of x in a sample of size n with fixed $t_{\alpha/2} \equiv t_0$, there is a range of values of p which satisfy the inequality

$$-t_0 \le \frac{x - np}{\sqrt{np(1 -\!\!- p)}} \le t_0$$

We can find the limits of the range of values of p by solving the two equations

$$\frac{x - np}{\sqrt{np(1 - p)}} = \pm t_0 \tag{6.10}$$

(Actually, a correction for continuity should be used. That is, we should write

$$\frac{x - \frac{1}{2} - np}{\sqrt{np(1 - p)}} = -t_0 \quad \text{and} \quad \frac{x + \frac{1}{2} - np}{\sqrt{np(1 - p)}} = +t_0$$

But this leads to further complications in solving for p. Anyway, for large values of n the correction makes very little difference in the limits. For small values of n the correction probably should be taken into account.) On squaring both sides, the two equations in (6.10) become the same. Then the resulting quadratic in p is

$$(x - np)^2 = t_0^2[np(1 - p)]$$

or

$$(n^2 + nt_0^2)p - (2nx + nt_0^2)p + x^2 = 0$$

The $(1 - \alpha)$ confidence limits of p obtained by solving this equation by the quadratic formula are

$$p = \frac{2nx + nt_0^2 \pm t_0\sqrt{(-4x^2 + 4nx + nt_0^2)n}}{2(n^2 + nt_0^2)} \tag{6.11}$$

To solve (b) we let $n = 100$, $x = 20$, and $t_{.025} = t_0 = 1.96$ in Eq. (6.11). This gives the limits

$$p_1 = 0.133 \quad \text{and} \quad p_2 = 0.289$$

Formula (6.11) may be used to obtain good approximate limits to the true proportion whenever the sample size is sufficiently large. However, the calculations involved are lengthy and unnecessary, in most cases, since charts have already been prepared. Table III may be used to find limits for p for confidence coefficients 0.95 and 0.99. In the rare case when other confidence coefficients are required, Formula (6.11) may be used. In order to find the 95 per cent confidence limits of p for (b), use the chart in Table III. Locate $\frac{20}{100} = 0.20$ on the bottom horizontal line, and follow the vertical line above 0.20 until it intersects the bottom curved line for $n = 100$. With a straightedge falling on this point of intersection and placed in a horizontal position, locate a point where the straightedge intersects the first vertical line, the scale for p. We read the value $p_1 = 0.125$. In a similar way we find $p_2 = 0.290$ by using the top curved line for $n = 100$. These values compare favorably, to two significant figures, with those obtained by the formula. Since the limits, at best, are only approximations, we probably should not use more than two significant figures in any event.

Example 6.7. It is informative to use an approximate method which is less accurate than either of the above methods, but which has the advantage of being short. Replacing $\bar{x}$ in Eq. (6.7) by $p' = x/n$, the proportion of successes in n independent trials, and remembering (see Theorem 6.5) that $\mu_{p'} = p$ and $\sigma_{p'}^2 = p(1 - p)/n$ for sample proportions, we obtain the following inequality with approximate limits

$$p' - t_0\sqrt{\frac{p(1 - p)}{n}} \le p \le p' + t_0\sqrt{\frac{p(1 - p)}{n}}$$

Allowing for the correction for continuity, we may write

$$p' - \frac{1}{2n} - t_0\sqrt{\frac{p(1 - p)}{n}} \le p \le p' + \frac{1}{2n} + t_0\sqrt{\frac{p(1 - p)}{n}}$$

But the population proportion p which is unknown appears on both sides of the inequalities. However, if n is large enough, we may replace p with $p' = x/n$, the sample joint estimate of the true proportion, to obtain

$$p' - \frac{1}{2n} - t_0 \sqrt{\frac{p'(1-p')}{n}} \leq p \leq p' + \frac{1}{2n} + t_0 \sqrt{\frac{p'(1-p')}{n}} \qquad (6.12)$$

Using Eq. (6.12) and the sample values of Example 6.6(b), we find the limits

$$p' \pm \left[\frac{1}{2n} + t_0 \sqrt{\frac{p'(1-p')}{n}} \right] = 0.2 \pm \left[0.005 + 1.96 \sqrt{\frac{(0.2)(0.8)}{100}} \right]$$

$$= 0.200 \pm 0.083$$

or

$$p_1 = 0.117 \quad \text{and} \quad p_2 = 0.283$$

Ignoring the correction $1/2n = 0.005$, we find $p_1 = 0.122$ and $p_2 = 0.278$. Thus, to two decimal places, we obtain limits 0.12 and 0.28 in both cases, and they are both lower than the more exact limits of 0.13 and 0.29.

Example 6.8. In a particular city, in response to a certain question 100 of 400 women answered "yes," and 150 of 500 men answered "yes." Use these data to establish 90 per cent confidence limits for the true difference in the proportion of women and men responding "yes." Assume that the samples were randomly drawn.

If it is assumed that Theorem 6.7 applies, approximate confidence limits of $p_1 - p_2$ are given by

$$p_1' - p_2' \pm t_{.05} \sqrt{\frac{p_1(1-p_1)}{n_1} + \frac{p_2(1-p_2)}{n_2}}$$

where the subscript $_1$ denotes women and the subscript $_2$ men. Assuming the samples to be large enough so that the sample proportions may be used as good point estimates of the true proportions, we introduce a further approximation in the confidence limits by using

$$p_1' - p_2' \pm t_{.05} \sqrt{\frac{p_1'(1-p_1')}{n_1} + \frac{p_2'(1-p_2')}{n_2}} \qquad (6.13)$$

Substituting the given values in Eq. (6.13) along with $t_{.05} = 1.645$ gives the limits -0.099 and -0.001. That is, the 90 per cent confidence interval is

$$-0.099 \leq p_1 - p_2 \leq -0.001$$

Thus, we conclude that a larger proportion of men in the city will vote "yes" on the question, understanding that we may be wrong (but it is not likely).

It is informative to note at this time that the standard normal value

$t_{\alpha/2}$ used to obtain the symmetric $(1 - \alpha)$ 100 per cent confidence interval of μ_s is the same value used to make the corresponding symmetric α level test of the null hypothesis $\mu_s = k$, where k is a real number and s is any statistic which is normally distributed with mean μ_s and variance σ_s^2. However, the confidence limits and the corresponding critical points in terms of the statistic s are not the same. Actually, they would be the same only in very rare cases.

Consider Example 6.2. For the five per cent level test of the null hypothesis $\mu = 2.5$ cm the symmetric 2.5 per cent critical points are $t_1 = -1.96$ and $t_2 = 1.96$ in terms of the standard statistic t, and $\bar{x}_1 = 2.435$ and $\bar{x}_2 = 2.565$ in terms of the sample statistic $\bar{x}$. But the confidence limits are

$$\bar{x} \pm t_{.025}\sigma_{\bar{x}} = 2.62 \pm 1.96 \left(\frac{0.1}{\sqrt{9}}\right) = 2.620 \pm 0.065$$

or

$$\bar{x}_1 = 2.555 \quad \text{and} \quad \bar{x}_2 = 2.685$$

which are quite different from the critical points 2.434 and 2.565.

For the information given in Example 6.2 the 95 per cent confidence interval for the diameter of the average bolt manufactured is

$$2.555 \leq \mu \leq 2.685$$

In this case, it might be observed that the hypothesized mean $\mu = 2.5$ cm does not fall in the confidence interval. This agrees with our earlier conclusion, where we rejected the null hypothesis and accepted the fact that the mean is larger than 2.5 cm. As a matter of fact, the confidence interval indicates what values the parameter is likely to take on. Thus, in this case at least, we see that a statement at the five per cent level can be made about the null hypothesis if we have the 95 per cent confidence interval. The reader should be cautioned on this point lest he think this is always the case. Actually, in most cases the symmetric confidence interval is used since it is shortest, whereas the critical region, depending on the alternative hypothesis, is often not symmetric, in which case the confidence limits cannot be used to reach a decision regarding the null hypothesis. Thus we may state the following rule: If, in problems in which Theorems 6.1 through 6.7 apply, the critical region is divided into two parts, each with equal probability and reaching to infinity, then the symmetric confidence interval can be used to indicate a decision on the null hypothesis $\mu = k$, the alternative hypothesis being $\mu \neq k$, provided the same statistic is appropriate in each case.

This brings us to the second point. In some cases, such as in Example 6.8, different statistics are used in determining confidence limits and critical points. In Example 6.8 the statistic used to find the confidence limits was

$$t \doteq \frac{(p_1' - p_2') - (p_1 - p_2)}{\sqrt{\dfrac{p_1'(1 - p_1')}{n_1} + \dfrac{p_2'(1 - p_2')}{n_2}}} \tag{6.14}$$

However, in order to test the corresponding null hypothesis $p_1 - p_2 = 0$ (or $p_1 = p_2$), symmetric critical regions being used, the most appropriate statistic is

$$t \doteq \frac{(p_1' - p_2') - 0}{\sqrt{p'(1 - p') \left(\dfrac{1}{n_1} + \dfrac{1}{n_2} \right)}} \tag{6.15}$$

where

$$p' \equiv \frac{x_1 + x_2}{n_1 + n_2}$$

For under the assumption of the null hypothesis $p_1 = p_2$, we may let p be the common true proportion of successes for both populations, and use the pooled sample to obtain a single estimate p' of the true proportion. Since more observations go into finding the estimate p', we would expect p' to be better than either p_1' or p_2' in the sense that the sampling distribution of p' has less spread than either of the sampling distributions of p_1' or p_2'.

Example 6.9. Use the data of Example 6.8 to test at the ten per cent level the null hypothesis $p_1 = p_2$ against the alternative hypothesis $p_1 \neq p_2$.

If it is assumed that Theorem 6.7 applies, the critical region, in terms of the standard normal variate, is the set of values for which $t < -1.645$ or $t > 1.645$. The standard normal deviate for our samples is given by

$$t \doteq \frac{\frac{100}{400} - \frac{150}{500}}{\sqrt{\frac{250}{900} \cdot \frac{650}{900} \left(\frac{1}{400} + \frac{1}{500} \right)}} = -1.66$$

Since the sample value of the statistic falls in the negative critical region, we reject the null hypothesis and conclude that the true mean proportion of women answering "yes" will be less than for men. The student should satisfy himself that the use of Eq. (6.14) could lead to a different conclusion.

Example 6.10. Compare the mean and median for small sample sizes.

We indicated earlier that the mean is generally preferred to the median for a given sample size. The reason for this is indicated in Theorem 6.3, since so many of the populations encountered in practice are approximately normally distributed. According to the theorem

$$\sigma_{x_m}^2 = \frac{\pi}{2} \cdot \frac{\sigma^2}{n} = \frac{\pi}{2} \cdot (\sigma_{\bar{x}}^2)$$

when n is large. That is, the *standard error of the median* is $\pi/2 \doteq 1.57$ times as large as the *standard error of the mean* for large samples.

For small n it can be shown [5, 10] that the relationship $\sigma_{\bar{x}_m}^2 > \sigma_{\bar{x}}^2$ still holds except in the case where $n = 2$. Table 6.2 shows how the variance of the median changes for small samples drawn from a normal population. Thus, when estimating the population mean μ from a random sample of size $n > 2$, we prefer the sample mean to the sample median, because the sample mean deviates less from μ, on the average.

Table 6.2

Variance of the Median when $\sigma_{\bar{x}}^2 = 1$ and Samples are Drawn from a Normal Population.

n	2	3	4	5	6	7	10
$\sigma_{\bar{x}_m}^2$	1.00	1.35	1.19	1.44	1.29	1.47	1.39

There is another sense in which we may relate the sample mean and sample median. Suppose we require that the variance of the statistic (mean or median) which estimates the population mean not to exceed a fixed value σ_0^2. Then the sample sizes n and n' of the mean and median, respectively, must be large enough so that

$$\frac{\sigma^2}{n} \leq \sigma_0^2 \quad \text{and} \quad \frac{\pi}{2} \cdot \frac{\sigma^2}{n'} \leq \sigma_0^2$$

Thus, when the standard errors of the mean and median equal σ_0^2

$$\frac{\sigma^2}{n} = \frac{\pi}{2} \cdot \frac{\sigma^2}{n'}$$

or

$$n' = \frac{\pi}{2} n \doteq 1.57\, n \quad \text{for large } n \tag{6.16}$$

Thus, when one is sampling from a normal population, the sampling distribution of the median of samples of size $(\pi/2)\, n$ is the same as the sampling distribution of the mean of samples of size n. Hence, an inference statement based upon a sample mean of size n can be expected to be just as dependable as one based upon a sample median of size $(\pi/2)n$. In particular, if the population variance is known, the length of a 100 $(1 - \alpha)$ per cent confidence interval for the population mean as determined from the mean of a sample of size 40 is the same as the length of a 100 $(1 - \alpha)$ per cent confidence interval determined from the median of a sample of size $(\pi/2)(40) \doteq 63$. The interval determined from the median of a sample larger than 63 would be shorter.

Example 6.11. In industrial work and in other areas, the statistical control of quality in a continuing process has proven to be very important. Actual control situations are varied and are studied with the aid of many kinds of statistics. We consider a special case involving sample means computed from random samples drawn from a normal population with mean μ and variance σ^2.

In a hypothetical investigation, suppose that one determination of raw pulp viscosity is made on each work day (hour, or any constant period of time) at roughly the same time for 30 weeks. Further, suppose the mean raw pulp viscosity per week for each of the 30 weeks is as follows.

Table 6.3

Week	Mean	Week	Mean	Week	Mean
1	140	11	135	21	135
2	135	12	155	22	125
3	155	13	135	23	150
4	145	14	140	24	160
5	135	15	140	25	140
6	135	16	140	26	195
7	160	17	145	27	190
8	140	18	145	28	180
9	145	19	150	29	190
10	125	20	155	30	165

(a) Prepare a *control chart* for the 30 means. (b) Discuss the use of the control chart in (a). (c) Give a short discussion of the term *quality control*.

In constructing the control chart in Fig. 6.3, we assume the observations to be randomly drawn from a normal population with mean $\mu = 145$ and variance $\sigma^2 = 256$. Therefore, the sample means $\bar{x}$ are normally distributed with mean $\mu_{\bar{x}} = 145$ and variance $\sigma_{\bar{x}}^2 = \frac{256}{5} = 51.2$. The sample means were based on samples of size five, there being five work days. Each

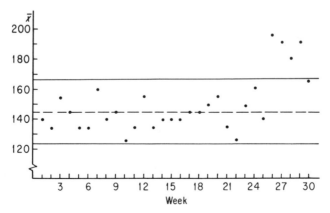

Fig. 6.3 Control Chart for Means

point on the control chart was located by letting the week number be the abscissa and the corresponding mean be the ordinate.

The three lines parallel to the axis of abscissas are for means $\mu - 3\sigma/\sqrt{5} = 123.5$, $\mu = 145$, and $\mu + 3\sigma/\sqrt{5} = 166.5$. The top line is called the *upper control limit*, the bottom line the *lower control limit* and $\mu = 145$ the *center line*. Under the assumptions, the probability of a sample mean's falling above the line $\bar{x} = 166.5$ is 0.00135, and the probability of a sample mean's falling below the line $\bar{x} = 123.5$ is 0.00135. Thus, so long as the assumptions hold, the sample means vary randomly about the center line $\bar{x} = \mu = 45$ with a probability of 0.00270 of falling outside the two limits. Since the chance of a sample mean's falling outside the limits is so small, we say that the process is *out of control* whenever this happens.

In our example the process went out of control during the twenty-sixth week, and actually stayed out of control for the next three weeks. With such strong evidence that all is not right, the production process should be investigated. Thus, such things as raw material, machines, ways in which the machines are adjusted and operated, and various other aspects of the process should be examined and corrected when necessary.

The control limits in this particular example are symmetrical about the population mean and located three standard errors away. Sometimes the limits may not be symmetrically placed. In fact, only one limit may be required. For example, in many tensile-strength investigations, say, we require only a lower control limit since the object is to eliminate production of weak material.

Control limits may be located any number of units on either side of the population mean, say. However, it is customary to locate them in such a way that whenever a point falls outside of the control region the investigator (process engineer or trouble-shooter) can expect to look for serious trouble, and not waste time looking for minor or nonexistent trouble spots. Actually, control limits should be placed so as to strike an economical balance between looking for trouble that does not exist and failing to look for trouble that does exist.

Since the population mean and variance are not often known, the center and control limit lines are usually located by applying estimates obtained from several of the first samples. This involves problems which can be handled better in later chapters.

Statistical control of quality has widespread applications. A large body of methods in quality control have developed since W. A. Shewhart first introduced the techniques in industry in 1926. The control chart, even though it can be used with almost any statistic, represents only one of the methods. For the student interested in reading more on the use of control charts and other methods in statistical control of quality, there are many books and articles available. A few references [1, 2, 4, 12, 14] are given at the end of this chapter.

In almost any kind of repeated-measures investigation we expect a certain amount of variation. We think of part of the variation as resulting from chance (or random) causes and part as resulting from assignable (or systematic or controllable) causes. If, in a process, only random causes account for the variation, we say that the process is in *statistical control.* Otherwise, the process is out of statistical control, and our job is to measure and correct (or adjust) the assignable causes. It is doubtful that a process can ever be brought in complete control, but it can be brought close enough for practical purposes.

There are many quality characteristics which are measurable, for example, the tensile strength of wire, life of a light bulb, number of defective units in a lot, and number of rough spots on a surface. Terms like *variable* and *attribute* are often used to distinguish between the continuous and discrete cases in statistical quality control. (The above concepts of quality control, along with others, are discussed in other chapters and in the exercises of this chapter.)

The reader should recognize that the control chart provides a graphic way of repeatedly testing the same hypothesis relative to the quality of a product. The purpose of the test is to determine repeatedly, on the basis of a small sample, whether a desired standard of quality is actually being met. The hypothesis subject to test is that the "process is in control." (In Example 6.11, the statement "process is in control" means "a mean of samples of size five is not significantly different from 145.") If for any sample one fails to reject the hypothesis, it is *presumed* that the process is operating satisfactorily. If for any sample the hypothesis is rejected, it is presumed that the process is not operating satisfactorily, that is, is out of control, and a search is made to detect the source of assignable variation.

6.4. ERROR TYPES AND OPERATING CHARACTERISTIC CURVES

Once the null and alternative hypotheses, the sample size, and the significance level of the test are decided on, a sample (usually random) is selected from a population under consideration, and the sample is used to make a decision concerning the null hypothesis. Unfortunately, in reaching a decision, we cannot be sure that a mistake will not be made. In fact, there are two types of mistakes which are possible. It may happen that the null hypothesis is true and we conclude that it is false. In this case, we say the mistake is a *type* 1 *error.* The probability of making this error is sometimes called the *size of the type* 1 *error* and is the same as the significance level of the test. It may happen that the null hypothesis is false and we fail to reject it. This mistake is called a *type* 2 *error.* The frequency with which type 1 and type 2 errors are made is very important to the experimenter. We shall see that the frequency of errors can be controlled to some extent.

The meaning of the terms just introduced, as well as others already discussed in hypothesis testing, can best be brought out by an illustration.

Example 6.12. We discuss the average breaking strength of a certain type of yarn as measured in ounces, assuming the population of breaking strength of yarn, x, manufactured by Bob to be normally distributed with variance $\sigma^2 = 100$. Manufacturer Bob contends that the mean μ is at least 1000 oz. Thus, our null hypothesis H_0 is that $\mu \geq 1000$; that is, $H_0: \mu \geq 1000$. The distributer, Joe, however, wants some kind of statistical verification of this hypothesized value of the mean breaking strength against the alternative hypothesis that $\mu < 1000$; that is $H_a: \mu < 1000$. Suppose that Joe and Bob agree on selecting at random 25 pieces of yarn and that the breaking strength of each piece be tested by some standard procedure. It should be observed that Joe would have no further worry if $\bar{x}$ were greater than 1000 oz. However, with an $\bar{x}$ below 1000 Joe would wonder how often he could expect to get such a small sample mean due to chance causes alone. He may doubt the claim of Bob. In any case, it is decided that the statistical test should be carried out. This requires that Joe and Bob agree on the maximum size of the type 1 error, that is, that they agree on how often they are willing to reject H_0 on the basis of the results of one sample when H_0 is actually true.

In order to see how they might arrive at a mutually satisfactory answer, we shall consider the sampling distribution of $\bar{x}$ for samples of size 25. From previous theorems we know that $\bar{x}$ is normally distributed with mean $\mu_{\bar{x}} = \mu = 1000$ and variance $\sigma_{\bar{x}}^2 = \sigma^2/n = \frac{100}{25} = 4$ under the assumptions that the null hupothesis is true. We have used the most meaningful μ of H_0, that is, $\mu = 1000$. The graph of this distribution is shown in Fig. 6.4.

The distributor will maintain that if $\bar{x}$ is less than some fixed value, say, $\bar{x}_2$ (see Fig. 6.4), the hypothesis must be rejected and the manufacturing process improved. On the other hand, the manufacturer will propose a smaller value of $\bar{x}$, say, $\bar{x}_1$. However, the manufacturer and distributor must agree

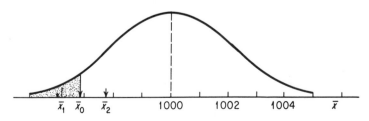

Fig. 6.4 Normal Density

on a value, say, $\bar{x}_0$, before proceeding with the test. The value $\bar{x}_0$ is the critical point, and it will usually lie between $\bar{x}_1$ and $\bar{x}_2$. The region to the left of the critical point is the critical region, wherein we reject the hypothesis, and the area above the critical region and below the density curve which we denote by α is the significance level of the test. This value α gives the probability of rejecting the null hypothesis when it is true; that is, it gives the probability of making the type 1 error. The student should note that once $\bar{x}_0$ is agreed on, α can be found, or if agreement is reached on the size of α, then $\bar{x}_0$ can be computed. The usual value of α (also called the "producer's risk") is five per cent, although, in general, this is a matter of compromise. The region to the right of the critical point is the region of indecision, or region of nonrejection, and if our sample $\bar{x}$ falls in this region, we fail to reject the hypothesis. Actually, if no changes are made in the manufacturing process, we act as though we accept the hypothesis tentatively. It should be noted that even though the statistician may be able to reserve judgment when the sample mean falls in the noncritical region, the man who is responsible can take no such stand.

Returning to our original problem, suppose it is decided that $\alpha = 5$ per cent. Then, since $\bar{x} \sim n(1000, 4)$ under the null hypothesis, we can use Table II to find our critical point, which turns out to be 996.71, since $t_{.05} = -1.645 = (\bar{x}_0 - 1000)/2$ implies $\bar{x}_0 = -3.290 + 1000$. The critical region is that region for which $\bar{x} < 996.7$. Since the mean of the random sample of 25 pieces of yarn actually turned out to be 990 oz, and since $990 < 996.7 = \bar{x}_0$, we reject the hypothesis that the average breaking strength of yarn made by Bob is equal to or greater than 1000 oz and conclude that the average breaking strength is less than 1000 oz, understanding that there is a chance of making an error. Actually, since 990 is five standard deviations away from the mean, there is less than a 0.0000003 chance of being wrong.

We make a few observations at this time. *First, it should be observed that the sampling distribution of $\bar{x}$ is a normal distribution.* For a large sample size, say, $n \geq 30$, the mean $\bar{x}$ can be expected to be normally distributed regardless of the distribution of x, if it is assumed that x has finite variance. For small samples, we must assume that x is normally distributed before the above argument holds. *Second, we used the particular value $\mu = 1000$ of the null hypothesis to draw the curve in Fig. 6.4 and to locate the critical point and critical region for the test.* The reason for this seems obvious. For example, suppose we had used the value $\mu = 1010$ to draw the curve; then the critical point would have been 1006.71, and the critical region would have been all those $\bar{x}$ values for which $\bar{x} < 1006.71$. Thus, if a particular sample had a mean of 1000, we would find ourselves in a position of rejecting the hypothesis that $\mu \geq 1000$, which is ridiculous. *Third, the critical point would be different for a different sample size n or a different value of* *α or a different alternative hypothesis.* Clearly, these three things should be decided on before proceeding with the test. *Fourth, the location of the critical region depends on the nature of the distribution as well as the statements of the hypotheses.* For example, if the hypothesis had been that $\mu \leq 1000$, then only very large values of the sample mean $\bar{x}$ would make the experimenter want to reject the hypothesis. Thus, the critical region would be made up of all those means which are greater than some critical value $\bar{x}_0$ which is larger than 1000; that is, $\bar{x} > \bar{x}_0 > 1000$. If the null hypothesis had been that $\mu = 1000$, with the alternative hypothesis $\mu \neq 1000$, the critical region would be made up of two parts, as has been illustrated in Examples 6.2 and 6.9. For a symmetric distribution the two parts are located so that $\alpha/2$ of the area of the density function is above each part, and these regions are located at the ends of the scale of measurements. Such tests are called *two-tailed tests.* If the critical region is located at one end of the scale of measurements the test is called a *one-tailed test.*

In the above illustration of the test of the hypothesis that the population mean of the breaking strength of yarn manufactured by Bob is equal to or greater than 1000 oz, we have proceeded in a definite way. We now set down a general procedure for testing hypotheses, and in other examples in this and later chapters we follow this procedure.

1. State the null and the alternative hypotheses.
2. State the assumptions and specify n and α. (Ideally, α and β, the probability of making the type 2 error, should be specified.)
3. Specify the statistic to be used.
4. Determine the critical region from the required table.
5. Compute the statistic from the sample.
6. Write the conclusion in terms of the statement of the hypothesis.

Sometimes we fail to reject the null hypothesis when it is not true. In such a case we make a type 2 error. Denoting the probability of such an error by β, we determine its value for specified values of the alternative hypothesis by

$$\beta = P[\text{failing to reject } H_0; H_a \text{ is true}] \qquad (6.17)$$

The right-hand side of Eq. (6.17) is read as "the probability of failing to reject the null hypothesis, given that the alternative hypothesis is true."

Example 6.13. In the yarn experiment, assume that the true mean is $\mu = 998$ oz. Thus, the probability of making a type 2 error is given in Fig. 6.5 by the area under the curve with mean $\mu = 998$ and above the region of nonrejection, that set of values of x for which $x \geq 996.7$. In particular, we have

$$\beta = P[\text{failing to reject } \mu_0 = 1000; \text{normal with } \mu = 998 \text{ and } \sigma_{\bar{x}}^2 = 4]$$

$$= \int_{996.7}^{\infty} \frac{1}{2\sqrt{2\pi}} e^{-\frac{1}{2}\left(\frac{\bar{x}-998}{2}\right)^2} d\bar{x}$$

$$= \int_{-.65}^{\infty} \frac{1}{\sqrt{2\pi}} e^{-t^2/2} dt \quad \text{when} \quad t = \frac{\bar{x}-998}{2}$$

$$= 1 - F(-0.65) = 1 - 0.26 = 0.74$$

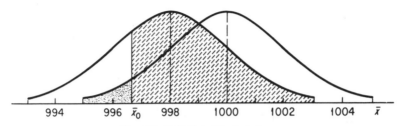

Fig. 6.5 β (998) is Shaded Area

Since the true value of μ is not usually known, we graph $\beta = \beta(\mu)$ for different values of μ, and this graph is called the *operating characteristic curve* (abbreviated O. C. curve). It gives us at a glance the probability of making a type 2 error for any alternate true value of μ.

The general formula for computing β when σ^2 is known is derived below for the null hypothesis that $\mu \geq \mu_0$, where the alternative hypothesis is $H_a: \mu < \mu_0$. We have

$$\beta(\mu) = P[\text{failing to reject } \mu_0; \ \mu \text{ is true}]$$

or

$$\beta(\mu) = \frac{\sqrt{n}}{\sqrt{2\pi}\,\sigma} \int_{x_0}^{\infty} e^{-\frac{1}{2}\left(\frac{\bar{x}-\mu}{\sigma/\sqrt{n}}\right)^2} d\bar{x} \tag{6.18}$$

where $\bar{x}_0 = \mu_0 + t_a \cdot \sigma/\sqrt{n}$. If we let

$$t = \frac{\bar{x}-\mu}{\dfrac{\sigma}{\sqrt{n}}}$$

Eq. (6.18) becomes

$$\beta(\mu) = \frac{1}{\sqrt{2\pi}} \int_{t_0}^{\infty} e^{-t^2/2} dt \tag{6.19}$$

where

$$t_0 = (\bar{x}_0 - \mu)\frac{\sqrt{n}}{\sigma} = \left(\mu_0 + \frac{t_a\sigma}{\sqrt{n}} - \mu\right)\cdot\frac{\sqrt{n}}{\sigma} = t_\alpha + \frac{\mu_0 - \mu}{\dfrac{\sigma}{\sqrt{n}}}$$

Thus, we may write

$$\beta(\mu) = 1 - \int_{-\infty}^{t_0} \frac{1}{\sqrt{2\pi}} e^{-t^2/2} \, dt = 1 - F(t_0)$$

or

$$\beta(\mu) = 1 - F\left(t_\alpha + \frac{\mu_0 - \mu}{\frac{\sigma}{\sqrt{n}}}\right) \tag{6.20}$$

Example 6.14. In Examples 6.12 and 6.13 we have $\sigma = 10$, $n = 25$, $\mu_0 = 1000$, $\alpha = 0.05$, and $t_{.05} = -1.645$. Thus Eq. (6.20) becomes

$$\beta(\mu) = 1 - F\left(\frac{1000 - \mu}{2} - 1.645\right) \tag{6.21}$$

In order to plot the curve $\beta(\mu)$ shown in Fig. 6.6, we determine the following points from Eq. (6.21)

$$\beta(999) = 1 - F(0.5 - 1.645) = 1 - F(-1.145) = 1 - 0.126 = 0.874$$

$$\beta(998) = 1 - F(1 - 1.645) = 1 - F(-0.645) = 1 - 0.259 = 0.741$$

$$\beta(997) = 1 - F(-0.145) = 0.558$$

$$\beta(996) = 1 - F(0.355) = 0.361$$

$$\beta(995) = 1 - F(0.855) = 0.196$$

$$\beta(994) = 1 - F(1.355) = 0.088$$

$$\beta(993) = 1 - F(1.855) = 0.032$$

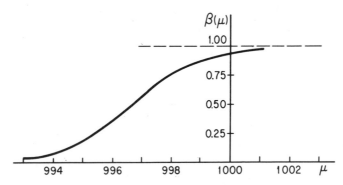

Fig. 6.6 Operating Characteristic Curve

Suppose $x \sim n(\mu, \sigma^2)$ and σ^2 is known; then the general formula for computing β for the hypothesis that $\mu \le \mu_0$ against the alternative hypothesis $H_a: \mu > \mu_0$ is given by

$$\beta(\mu) = F\left(t_\alpha + \frac{\mu_0 - \mu}{\frac{\sigma}{\sqrt{n}}}\right) \tag{6.22}$$

If the null hypothesis is $\mu = \mu_0$ and the alternative hypothesis is $\mu \neq \mu_0$, then for symmetric critical regions

$$\beta(\mu) = F\left(t_{1-(\alpha/2)} + \frac{\mu_0 - \mu}{\frac{\sigma}{\sqrt{n}}}\right) - F\left(t_{\alpha/2} + \frac{\mu_0 - \mu}{\frac{\sigma}{\sqrt{n}}}\right) \tag{6.23}$$

In theoretical work the function $p(\mu)$ given by

$$p(\mu) = 1 - \beta(\mu) \tag{6.24}$$

called the *power function* of the test of a hypothesis involving μ, is very useful in comparing tests of means. It is clear that the power function of a test gives the probability of rejecting the null hypothesis when the null hypothesis is false; that is

$$p(\mu) = P[\text{reject } H_0; H_a \text{ is true}] \tag{6.25}$$

or

$$p(\mu) = P[\text{reject } H_0; H_0 \text{ is false}]$$

Thus, the power is greatest when the probability of an error of the second kind is smallest. A power function has an advantage over the operating characteristic function in that it is associated with the critical region just as α is. That is

$$p(\mu) = 1 - P[\text{sample mean falling in noncritical region}; H_a \text{ true}]$$
$$= P[\text{sample mean falling in critical region}; H_a \text{ true}]$$

and

$$\alpha = P[\text{sample mean falling in critical region}; H_0 \text{ true}]$$

For the same sample size and the same significance level, test one is said to be *more powerful* than test two, at a specified value of μ, if the power of test one is larger than test two, or if the probability of committing a type 2 error is smaller for test one. In general, we prefer that test which is more powerful.

6.5. EXERCISES

6.1. Of the first 300 babies born in November in a certain city, 180 were boys and 120 girls. If it is assumed that these 300 babies represent a random sample from a population, estimate by a 95 per cent confidence interval the proportion of male births in the population.

6.2. Of two machines used to test hardness in a laboratory, it is desired to determine whether machines A and B are consistent with each other. The variance of the readings from each machine is known to be 0.16. A standard block was used, and eight determinations made on each machine showed sample means of $\bar{x}_A = 66.7$ and $\bar{x}_B = 67.3$. Test the appropriate hypothesis at the five per cent level, using the general test procedure for testing hypotheses outlined in six steps on p. 218.

6.3. Two sets of 100 students each were taught to read by two different methods, respectively. After instruction was over, a reading test gave the following sample results: $\bar{x} = 73.4$, $\bar{y} = 70.3$, $s_x = 8$, and $s_y = 10$. Assume that the samples are large enough so that s_x and s_y may be used in place of σ_x and σ_y. (a) Determine a 90 per cent confidence interval for $\mu_x - \mu_y$. (b) Determine how large an equal-size sample from each group should have been used if it is desired to estimate $\mu_x - \mu_y$ to within one unit with a probability of 0.95.

6.4. In a poll taken among college students, 46 out of 200 fraternity men favored a certain proposition, and 51 out of 300 nonfraternity men favored it. (a) Find a 95 per cent confidence interval for the difference in proportions favoring the proposition. (b) Test the hypothesis that the true proportions favoring the proposition are equal.

6.5. Draw the operating characteristic curve for the test in Example 6.2. Use the curve and graphic methods to find the probability of making the type 2 error when the true mean is 2.44; when it is 2.60.

6.6. The bacteria content of a food product must be less than 65.0 to be acceptable. A sample of 16 cans from a lot of the product has a mean content of 65.4. Long experience indicates that the standard deviation of the bacteria content is 0.4. (a) If the probability of false rejection of the lot is to be 0.05, should the lot be rejected on the basis of the sample evidence? (b) Draw the operating characteristic curve for this test. What is the chance of nonrejection of a lot which has mean bacteria content of 66.0? (c) Draw on the same graph with (b) the operating characteristic curve for the above test when samples of size 64 are used in place of samples of size 16. Suppose we require that the probability of the type 2 error not be greater than 0.10. Find the smallest bacteria count for which $\alpha = 0.05$, $\beta = 0.10$, and $n = 16$; for which $\alpha = 0.05$, $\beta = 0.10$, and $n = 64$. Use this information to make a statement about the effect of sample size on the test. (d) Suppose the specification that the true mean bacteria content be 65.0 allows the true mean content to be as large as 65.05 without rejection of the lot. For the five per cent level test that the mean bacteria content not be larger than 65.0, determine the sample size so that for a sample mean of $\bar{x} = 65.05$ the probability of the type 2 error is at most 0.10; that is, $\beta = 0.10$. (The reader should observe here that we wish to find the sample size which insures that the sample mean not be larger than the maximum allowable specification when the sizes of the type 1 and type 2 errors are fixed and the true mean actually does not exceed 65.0.)

6.7. (a) Consider the test of H_0: $\mu \leq \mu_0$ against H_a: $\mu > \mu_0$, where μ_0 is some constant. Let $\bar{x}$ denote a sample mean and let $d \equiv \bar{x} - \mu_0$. Find the smallest sample size n which guarantees that for a fixed d the probabilities of the type 1 and type 2 errors will not exceed specific values of α and β, respectively. It is to be understood that the parent population is normally distributed with variance σ^2 and that the critical point is denoted by $\bar{x}_0$. [See Exercise 6.6(d) for a numerical approach to the problem.] (b) Use the expression derived in (a), the standard normal table and specified values of $D \equiv d/\sigma$, α, and β to make a table of minimum sample sizes. Compute at least eight values, letting $\alpha = 0.05$ and 0.01, $\beta = 0.10$ and 0.05, and D equal to values of interest to the student.

6.8. (a) A lot of rolls of paper is acceptable for making bags for grocery stores if its mean breaking strength is not less than 40 lb. A random sample of 25 pieces of paper from the lot had a mean breaking strength of 39 lb. Long experience indicates that the standard deviation of the breaking strength is 2 lb. Should the lot be rejected if $\alpha = 0.1$? (b) Draw the operating characteristic curve for the test in (a). What is the chance of nonrejection of a lot which has true mean breaking strength of 39 lb? How does this curve compare with the one drawn in Exercise 6.6(b)? (c) Draw on the same graph with (b) the operating characteristic curve for the above test when $n = 64$ replaces $n = 25$. Use these two curves to make a statement about the effect of the sample size on the test if $\beta \leq 0.05$. (d) Suppose the specification that the true mean breaking strength be 40 lb allows the true mean breaking strength to be as small as 39.5 lb without rejection of the lot. For the one per cent level test that the mean breaking strength not be smaller than 40 lb, determine the sample size so that for a sample mean of $\bar{x} = 39.5$ the probability of the type 2 error is at most 0.05. (e) Graph the power functions for the two tests discussed in (a) and (c). Use these power functions to determine graphically the probability of rejecting H_0 when $\mu = 39$ lb; when $\mu = 38$ lb. Notice that the test using a sample of size 64 is more powerful in both cases than the test using $n = 25$.

6.9. (a) Consider the test of H_0: $\mu > \mu_0$ against H_a: $\mu < \mu_0$, where μ_0 is some constant. Let $\bar{x}$ denote a sample mean and let $d \equiv \bar{x} - \mu_0$. Find the smallest sample size n which guarantees that for a fixed d the probabilities of the type 1 and type 2 errors will not exceed α and β, respectively. It is understood that the parent population is normally distributed with variance σ^2. [See Exercise 6.8(d) for a numerical approach and Exercise 6.7(a) for an analogous problem.] (b) Use the expression derived in (a), the standard normal table, and specified values of $D \equiv d/\sigma$, α, and β to make a table of minimum sample sizes. Compute at least eight values, letting $\alpha = 0.05$ and 0.01, $\beta = 0.10$ and 0.05, and D equal to values of interest to the student.

Note. Even though Exercises 6.7 and 6.9 are for different tests, the derivations and calculations of one are like those for the other.

6.10. (a) Consider the test of H_0: $\mu = \mu_0$ against H_a: $\mu \neq \mu_0$, where μ_0 is some constant. Let $\bar{x}$ denote a sample mean and let $d \equiv |\bar{x} - \mu_0|$. Find the smallest sample size n which guarantees that for a fixed d the probabilities of the type 1 and type 2 errors will not exceed α and β, respectively. It is to be understood that the parent population is normally distributed with variance σ^2. (b) Use the expression derived in (a), the standard normal table, and specified values of $D \equiv d/\sigma$, α, and β to make a table of minimum sample sizes.

6.11. (a) Find the power function of the test of Exercise 6.2. (b) Find the power function for the test resulting from changing the sample size to 25 and the significance level of the test to 0.01. (c) Graph these two functions on the same graph. Give the co-ordinates of the points where the curves intersect.

6.12. According to the Mendelian inheritance theory, certain crosses of peas should give smooth and wrinkled seeds in the ratio of 3 to 1; that is, the relative frequency of wrinkled seeds is equal to 0.25. In a random sample of 720 seeds 500 were smooth. (a) What is the probability of getting exactly 500 smooth seeds? Use both the binomial density functions and the normal approximation to compute the probability and compare results. (b) Use the given information to test the Mendelian inheritance theory at the five per cent level. (c) Use the normal curve approximation to draw a power curve for the test in (b). From this curve determine power of the test if the true proportion of wrinkled seeds is actually .020 or 0.30. (d) Indicate how one could obtain a power curve using the binomial distribution without applying the normal approximation.

6.13. A random sample of size 49 was drawn from a normal distribution with a variance of 25. The mean and median of the sample were 21.3 and 22.1, respectively. Assume that 49 is a large sample size. (a) Find a 95 per cent confidence interval for the population mean, using the sample mean; using the sample median. (b) Compare the two intervals found in (a), writing a short summary statement. (c) What sample size would be required in order that the 95 per cent confidence interval for the population mean, as determined from a sample median, be one-half the length of the interval determined from the sample mean in (a)? Answer the same question if the term "one-half" is replaced by "k times," where k is any positive real number. (d) Find the confidence coefficient which gives a confidence interval based on the sample median which has the same length as the 95 per cent confidence interval based on the sample mean.

6.14. The ith observation of a random sample of size n is drawn from a normal distribution with mean $\mu_x = \mu_i = \alpha + \beta(k_i - \bar{k})$ and variance $\sigma_{x_i}^2 = \sigma_i^2 = \sigma^2$, where α, β, k_i, and

$$\bar{k} \equiv \sum_{i=1}^{n} \frac{k_i}{n}$$

are real constants. Let

$$b = \frac{\sum (k_i - \bar{k})(x_i - \bar{x})}{\sum (k_i - \bar{k})^2}$$

Find the sampling distribution of b, giving the mean and variance in simplest form.

6.15. Assuming that Theorems 6.1 and 6.4 hold, prove Theorems 6.2 and 6.5.

6.16. Assuming that Theorems 6.1 through 6.5 hold, prove Theorems 6.6 and 6.7.

6.17. Let $f(x_1, \ldots, x_n)$ denote the density function of the random variables $x_1, \ldots, x_n$ and let $h \equiv h(x_1, \ldots, x_n)$ be any single-valued function of these variates. Then the kth moment about the origin of h is defined by

$$\mu'_{k:h} = \int_{-\infty}^{\infty} \cdots \int_{-\infty}^{\infty} h^k(x_1, \ldots, x_n) f(x_1, \ldots, x_n) \, dx_1 \ldots dx_n \qquad (6.26)$$

and the moment generating function of h is defined by

$$M_h(t) = \int_{-\infty}^{\infty} \cdots \int_{-\infty}^{\infty} e^{ht} f(x_1, \ldots, x_n) \, dx_1 \ldots dx_n \qquad (6.27)$$

(a) Prove that if the variates $x_1, \ldots, x_n$ are independently distributed and

$$h \equiv a_1 x_1 + \cdots + a_n x_n$$

where $a_1, \ldots, a_n$ are real numbers not all zero, then

$$M_h(t) = M_{x_1}(a_1 t) \cdot M_{x_2}(a_2 t) \ldots M_{x_n}(a_n t) \qquad (6.28)$$

(b) Further, prove that when $h = \bar{x}$, and $x_1 = x$ is normally distributed with mean μ and variance σ^2, then

$$M_{\bar{x}}(t) = M_x^n \left(\frac{t}{n} \right) \qquad (6.29)$$

(c) In Exercise 3.49 we showed that when x_i is normally distributed with mean μ and variance σ^2, then $u = (x - \mu)/\sigma$ has moment generating function $M_u(t) = e^{t^2/2}$. Use the MGF for u to find the MGF of $\bar{x}$, where

$$\bar{x} = \frac{\sigma}{\sqrt{n}} u + \mu$$

(d) Use the MGF of $\bar{x}$ obtained in (c) along with the two important properties of MGF's mentioned in Exercise 5.69 to show that Theorem 6.2 holds. Note that in this way Theorem 6.2 can be proved in its own right. However, it is interesting to prove Theorem 6.1 directly and then show, as in Exercise 6.15, that Theorem 6.2 is a corollary. (e) Prove Theorem 6.1, using the two properties of MGF's mentioned in Exercise 5.69 along with Eq. (6.28).

6.18. In Exercise 5.69 we proved the central-limit theorem in case the MGF exists. That is, we proved that when x is distributed with mean μ and variance σ^2, then $\bar{x} \sim n(\mu, \sigma^2/n)$ for large n [or $y = (\bar{x} - \mu)/(\sigma/\sqrt{n}) \sim n$ $(0, 1)$ if n is sufficiently large]. Referring to Theorem 6.4, we note that the x which represents the number of successes in n independent trials is actually an $\bar{x}$ in a sample of size n when we think of each success as having the variate value 1 and each failure the variate value 0. Further, since np is the mean and $np(1 - p)$ the variance of the binomial variate x, we may write $(x - np)/[\sqrt{np(1 - p)}]$ in place of $(\bar{x} - \mu)/(\sigma/\sqrt{n})$ to complete the proof of Theorem 6.4.

There is a proof of Theorem 6.4 which does not depend on the central-limit theorem, but does depend on MGF's. Let x be distributed as a binomial. Then

$$t' = \frac{x - \mu}{\sigma} = \frac{x - np}{\sqrt{np(1 - p)}}$$

may be considered a standardized binomial variate. Since the MGF of x, according to Exercise 5.20, can be shown to be

$$M_x(t) = [(1 - p) + pe^t)^n \qquad (6.30)$$

it follows that

$$M_{t'}(t) = \epsilon^{(-\mu t)/\sigma}(1 - p) + pe^{t/\sigma})^n \qquad (6.31)$$

Further, using the appropriate series expansion, we find that

$$M_{t'}(t) = \frac{t^2}{2} + \text{terms in } t^k, \qquad k = 3, 4, \ldots \qquad (6.32)$$

The proof of the theorem is complete on showing that

$$\lim_{n \to \infty} M_{t'}(t) = M_u(t) = e^{t^2/2} \qquad (6.33)$$

The student should prove Theorem 6.4 by first proving Eqs. (6.31), (6.32), and (6.33).

6.19. The median is a particular quantile (Sect. 2.2). Let x_1 denote any sample quantile. Then it can be shown [6] that for large n the variance of the quantile x_1 is given by

$$V(x_1) = \sigma_{x_1}^2 = \frac{p(1 - p)}{nf_1^2}$$

where n denotes sample size, $1 - p$ the proportion of variate values below x_1 and $f_1 = f(x_1)$ the ordinate of the quantile x_1 in the parent distribution. (a) Show that when x_1 is the median and the parent population is normal, the standard error of the median is approximately $(\pi/2) \cdot (\sigma/\sqrt{n})$, σ being the standard deviation of the normal distribution. (b) Find the standard error of the first quantile of a sample of size n drawn from a normal distribution with variance σ^2.

6.20. A control chart for the mean of samples of size five is to be constructed.

Suppose that the process is in control when $\mu = 20$ and $\sigma^2 = 40$. (a) The upper and lower control limits for sample means are to be three standard deviations from the center line. Determine these limits. (b) The symmetric control limits for sample means are to be located so that the probability of a sample mean's falling outside is 0.006. What are the limits?

6.21. Suppose that a process is in control when $\mu = 10$ and $\sigma^2 = 25$. For random samples of size four drawn each hour for 40 consecutive hours, the means are as follows

Table 6.4

Sample	Mean	Sample	Mean	Sample	Mean	Sample	Mean
1	9.0	11	8.0	21	10.5	31	11.7
2	6.7	12	11.0	22	18.7	32	10.7
3	13.0	13	10.5	23	7.7	33	13.3
4	12.3	14	12.0	24	9.7	34	9.3
5	7.5	15	11.0	25	9.0	35	11.7
6	9.7	16	10.7	26	10.7	36	10.7
7	10.3	17	11.5	27	12.0	37	10.7
8	10.3	18	6.7	28	9.7	38	10.5
9	13.0	19	9.5	29	12.5	39	11.0
10	14.5	20	7.5	30	11.5	40	13.3

(a) Plot the 40 means on a control chart with upper and lower control lines three standard deviations from the center line. Does it appear that the process is out of control at any point? (b) Assume for the moment that the true population mean is unknown but that the variance is 25. Use the first 25 sample means to estimate the position of the center line, and to locate the three-standard deviation control lines. Plot the last 15 data values on such a control chart and determine if the process is in control after the twenty-fifth hour. (If the center value is not known, the usual practice is to take the first 25 sample values to establish the standard and then use the standard until there is evidence that it should be changed.)

Note. The reader should supply his own interpretation of the data in Exercises 6.21 and 6.22. The means might represent (in coded form) such things as life of an electric lamp, tensile strength of thread, diameter of ball bearings, life of a washer, number of surface defects, chemical composition of steel, and blowout time of a fuse. The proportion of defective units per sample (or lot) might refer to cigarette lighters, light bulbs for Christmas trees, transistors, spools of yarn, errors on a typed page, and ill students.

6.22. The procedures for preparing a control chart of *proportion (or number) of defective units* in repeated samples is similar to that used for sample means. [Theorem 6.5(or 6.4) may be used for this purpose. See the note at the end of Exercise 6.21 for possible interpretations.]

Suppose that a process is in control when the proportion of defective units is $p = 0.04$ or less. Suppose that samples of size 100 are taken at regular intervals and the number of defective units determined. (a) Let the lower control limit for the number of defective units be zero. Find the upper control limit for which the probability is 0.005 that a sample number defective will exceed it. (b) Make a control chart with the limits of (a), plot the following 40 sample "number of defective units" on this chart, and comment on the state of control. The number of defective units in 40 consecutive samples of 100 are (read across)

3	5	2	6	3	3	4	1	5	2
4	3	8	4	5	0	1	6	5	3
4	3	5	3	9	1	5	3	4	5
0	6	3	4	1	2	6	12	2	5

(c) Use the first 25 samples to estimate the true number of defective units. Use this estimate to work (a) and (b).

REFERENCES

1. Bennett, C. A. and N. L. Franklin, *Statistical Analysis in Chemistry and the Chemical Industry.* New York: John Wiley & Sons, Inc., 1954, Chaps. 10 and 11.

2. Cowden, Dudley J., *Statistical Methods in Quality Control.* Englewood Cliffs, N. J.: Prentice-Hall, Inc., 1957.

3. Cramér, H., *Mathematical Methods of Statistics.* Princeton, N. J.: Princeton University Press, 1946, Chaps. 31–35.

4. Grant, E. L., *Statistical Quality Control*, 2nd ed. New York: McGraw-Hill, Inc., 1952.

5. Hojo, T., "Distribution of the Median, Quartile and Interquartile Distance in Samples from a Normal Population," *Biometrika*, Vol. **23** (1931), pp. 315–360.

6. Kendall, M. G. and A. Stuart, *The Advanced Theory of Statistics*, Vol. 1. New York: Hafner Publishing Co., Inc., 1958, Chap. 10.

7. ———, *The Advanced Theory of Statistics*, Vol. 2, New York: Hafner Publishing Co., Inc., 1961, Chaps. 21, 21, 22, 23, and 24.

8. Lehmann, E. L., *Testing Statistical Hypotheses.* New York: John Wiley & Sons, Inc., 1959.

9. Mood, A. M., *Introduction to the Theory of Statistics.* New York: McGraw-Hill , Inc., 1950, Chaps. 11 and 12.

10. Pearson, Karl, "Appendix to a Paper by Professor Tokishige Hojo. On the Standard Error of the Median to a Third Approximation, etc.," *Biometrika*, Vol. **23** (1931), pp. 361–363.

11. ———, *Tables of the Incomplete Beta-Function.* London: Cambridge University Press, 1934, 494 pp.

12. Rice, W. G., *Control Charts in Factory Management.* New York: John Wiley & Sons, Inc., 1947.

13. Romig, H. G., *50–100 Binomial Tables*. New York: John Wiley & Sons, Inc., 1952, 172 pp.

14. Shewhart, W. A., *The Economic Control of Quality of a Manufactured Product*. New York: D. Van Nostrand Co., Inc., 1931.

15. Wald, A., "Contributions to the Theory of Statistical Estimation and Testing Hypotheses," *Annals of Mathematical Statistics*, Vol. **10** (1939) pp. 299–326.

16. Wald, A., *Statistical Decision Functions*. New York: John Wiley & Sons, Inc., 1950.

7

SAMPLING
FROM NORMAL POPULATIONS—
THE CHI-SQUARE DISTRIBUTION

We have considered (Chaps. 5 and 6) the nature of sampling distributions of means and medians. In this chapter we study properties and important applications of the sampling distribution of a measure of dispersion, namely, the chi-square distribution. Power curves are studied aş they relate to tests of hypotheses and to the determination of sample size.

7.1. INTRODUCTION

There are problems in which our first interest is in making some statement about a measure of the population dispersion. In case the population is normally distributed, we use the sample variance to make such a statement, since other statistics measuring dispersion are not as reliable. The chi-square distribution is introduced for problems involving the sample variance.

For convenience for the reader we now give in one group statements of four important theorems relating to measures of dispersion of samples drawn randomly from normal distributions. Following these theorems we give illustrations of applications and, in the set of exercises, outlines of proofs of key theorems. (Proofs of one or more of the four theorems may be found in each of the ten references at the end of the chapter.)

Theorem 7.1. *Let*

$$w = \sum_{i=1}^{n} \left(\frac{x_i - \mu_i}{\sigma_i} \right)^2 = \sum_{i=1}^{n} t_i^2$$

where the x_i are normally and independently distributed with means μ_i and variances σ_i^2 and $t_i = (x_i - \mu_i)/\sigma_i$. Then w is distributed with density function

$$f(w; n) = \frac{1}{2^{n/2} \Gamma\left(\frac{n}{2}\right)} e^{-w/2} w^{(n/2)-1}, \qquad w > 0 \tag{7.1}$$

mean n, and variance 2n. The gamma function $\Gamma(n/2)$ is defined in Exercise 3.67.

Note. The distribution of the random variable w is a special case of the gamma distribution and is known as the *chi-square distribution* with n *degrees of freedom,* n being the parameter which is the mean. Thus, it is natural that we use the square of the Greek letter chi, χ^2, in place of w.

Theorem 7.2. *The sum of k independent random variables $\chi_1^2, \ldots, \chi_k^2$ having chi-square distributions with $\nu_1, \ldots, \nu_k$ degrees of freedom, respectively, is distributed as chi-square with $\nu_1 + \ldots + \nu_k$ degrees of freedom.*

Theorem 7.3. *The sample mean and variance are independent random variables when one is sampling (randomly) from a normal population.*

Theorem 7.4. *If a random sample $x_1, \ldots, x_n$ is drawn from a normal population with mean μ and variance σ^2, then the statistic*

$$\frac{\sum (x_i - \bar{x})^2}{\sigma^2}$$

which is a random variable, has the χ^2 distribution with $n - 1$ degrees of freedom.

The χ^2 distribution is one of the most important sampling distributions in statistics. The illustrations given in this chapter represent only a small portion of the many applications of this distribution. Before we present applications, let us be sure that we understand the theorems.

First, the reader should note that the term "chi square" is used in two ways—it denotes the variate and is also the name of the distribution. This may cause some difficulty in the beginning, since this is not the case with other distributions mentioned so far. For example, the variate of the standard normal distribution is called t; it is not called "normal."

7.2. THE CHI-SQUARE DISTRIBUTION—PROPERTIES

Due to convention, the variate w and parameter n are usually replaced by χ^2 and v, respectively, and the density function (7.1) is written as

$$f(\chi^2 ; v) = \frac{1}{2^{v/2}\Gamma\left(\frac{v}{2}\right)} e^{-\chi^2/2}(\chi^2)^{(v/2)-1}, \qquad \chi^2 > 0 \tag{7.2}$$

The Greek letter v, nu, is used in place of μ and refers to degrees of freedom as well as to population mean. The density function (7.2) represents a one-parameter family of distributions. The graph of three members of this family is shown in Fig. 7.1, which illustrates the fact that the "shape" of the distribution changes with the degrees of freedom. This being the

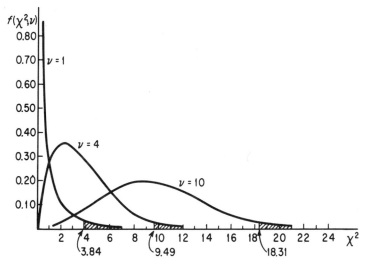

Fig. 7.1 χ^2 Distribution for $v = 1$, 4, and 10.
$\chi^2_{.05}$ Points are Indicated

case, we cannot use a single simple standardization process as with the normal distribution (where it was necessary to table only values for the standard normal variate t). Thus, it is necessary that the proportion (percentage) α of the area under the χ^2 curves to the right of the point χ^2_α be computed for each degree of freedom. That is, for any α, χ^2_α is that value such that

$$P[\chi^2 > \chi^2_\alpha] = \alpha \tag{7.3}$$

Values of χ^2_α corresponding to selected values of α are shown in Table IV.

The proportion of the area to the right of a χ^2 point is used instead of that to the left, since this is normally the way the χ^2 distribution is applied. Since the χ^2 distribution is nonsymmetric, it is necessary that the upper and lower α level points be computed separately.

Often we assume that a sample is randomly drawn from a single normal population with mean μ and variance σ^2. In this case, w in Theorem 7.1 becomes

$$w = \frac{\sum (x_i - \mu)^2}{\sigma^2}$$

But since the mean of a population is usually unknown, we use Theorem 7.4 rather than Theorem 7.1 in problems involving the population variance σ^2. According to these theorems χ^2 is distributed with one fewer degree of freedom when $\bar{x}$ is used instead of μ. This is the reason we sometimes say that one degree of freedom is used up in finding the estimate $\bar{x}$.

The useful statistic of Theorem 7.4 may also be written as

$$\frac{SSx}{\sigma^2} = \frac{(n-1)s^2}{\sigma^2} = \chi^2 \tag{7.4}$$

with $n - 1$ degrees of freedom, where SSx denotes sum of squares of x deviates and s^2 denotes the sample variance. Since the distribution of χ^2 depends on the number of degrees of freedom, we sometimes write "$\chi^2(n-1)$" in place of "χ^2 with $n - 1$ degrees of freedom." From Eq. (7.4) it is clear that the sampling distribution of s^2 is easily obtained from the distribution of $\chi^2(n-1)$ and that

$$s^2 = \frac{\sigma^2 \chi^2(n-1)}{n-1} \tag{7.5}$$

It should be apparent that Eq. (7.4) shows a relation between χ^2 and s^2 in very much the same way that $t = (\bar{x} - \mu)/(\sigma/\sqrt{n})$ relates t and $\bar{x}$. Hence, we sometimes refer to the statistic χ^2 as a standardized sample variance in much the same way we think of the statistic t as a standardized sample mean.

The relation in Eq. (7.4) may be used to establish a confidence interval with confidence coefficient $\gamma = 1 - \alpha$. This may be done by first finding two values of χ^2, say χ_1^2 and χ_2^2, such that

$$P[\chi_1^2 < \chi^2 < \chi_2^2] = \int_{\chi_1^2}^{\chi_2^2} f(\chi^2; \nu)\, d\chi^2 = \gamma \tag{7.6}$$

and then changing the inequalities, using Eq. (7.4), to obtain

$$P\left[\frac{SSx}{\chi_2^2} < \sigma^2 < \frac{SSx}{\chi_1^2}\right] = \gamma \tag{7.7}$$

The length of the 100γ per cent confidence interval given by Eq. (7.7) is

$$SSx\left[\frac{1}{\chi_1^2} - \frac{1}{\chi_2^2}\right]$$

and is shortest when

$$\left[\frac{1}{\chi_1^2} - \frac{1}{\chi_2^2}\right]$$

is a minimum, SSx being fixed for a particular sample. If appropriate tables were available, χ_1^2 and χ_2^2 could be found by inspection so as to minimize $[(1/\chi_1^2) - (1/\chi_2^2)]$, but in their absence the required computation using Eq. (7.6) is usually considered too long to be practical. Instead, in setting up a $100(1 - \alpha)$ per cent confidence interval, we usually choose $\chi_1^2 = \chi_{1-(\alpha/2)}^2$ and $\chi_2^2 = \chi_{(\alpha/2)}^2$; that is, χ_1^2 and χ_2^2 are selected so that $\alpha/2$ of the area is in each tail of the distribution. Such a choice gives an interval with length very near the minimum unless v is small. Sometimes in practice zero is taken as the lower bound of the interval and the upper limit is selected so that $\chi_1^2 = \chi_{1-\alpha}^2$.

7.3. APPLICATIONS OF THE CHI-SQUARE DISTRIBUTION

Example 7.1. Machine A is to be compared with a standard for the precision with which it cuts off pieces. The variance for the standard machine B is 0.030. A random sample of ten pieces cut off by machine A has a variance of 0.058. (a) Find a 95 per cent confidence interval for the variance σ_A^2 of machine A. (b) Is there a real difference between the variances of machines A and B?

Now $SSx = (10 - 1)(0.058) = 0.522$, and from Table IV, using nine degrees of freedom, we find $\chi_1^2 = \chi_{.975}^2 = 2.700$ and $\chi_2^2 = \chi_{.025}^2 = 19.02$. Thus, according to Eq. (7.7) a 95 per cent confidence interval of σ_A^2 is given by

$$\frac{0.522}{19.02} < \sigma_A^2 < \frac{0.522}{2.700}$$

or

$$0.027 < \sigma_A^2 < 0.193$$

Using $\chi_1^2 = \chi_{.95}^2 = 3.325$, we obtain $SSx/\chi_1^2 = 0.157$ and a 95 per cent confidence interval

$$0 < \sigma_A^2 < 0.157$$

The second interval is shorter than the first interval, but neither is the shortest 95 per cent confidence interval.

Earlier we noted that point estimates are not very useful. However, it should be mentioned here that $s^2 = 0.058$ is the best single estimate of the variance σ_A^2 in the sense that the mean of the sampling distribution of s^2 is σ_A^2, and the variance is smaller than that for other measures of dispersion commonly used. These statements are based upon the assumption that the parent population is normal.

The (b) part of Example 7.1 is answered by following the general procedure for testing hypotheses given on p. 218.

1. $H_o: \sigma^2 = 0.030$ and $H_a: \sigma^2 \neq 0.030$.

2. Assume that the ten sample values were independently obtained from the same normal population. Let the significance level be $\alpha = 0.02$.

3. The statistic to be used is

$$\chi^2 = \frac{(n-1)s^2}{\sigma^2} = \frac{9s^2}{0.030}$$

χ^2 is generally easier to use than s^2, since critical values can be read directly from the tables.

4. Since the alternative hypothesis includes values of σ^2 both larger and smaller than 0.030, we use a two-tailed test with one per cent of the area in each tail. Thus, the critical region includes all those values of χ^2 for which $\chi^2 < 2.088$ or $\chi^2 > 21.666$.

5. The sample statistic is

$$\chi^2 = \frac{9(0.058)}{0.030} = 17.4$$

6. Since 17.4 does not fall in the critical region, we fail to reject the null hypothesis. That is, we do not have enough evidence to say that the population variance is different from 0.030. (As a practical matter, if the assumptions of step 2 are considered realistic and the person responsible for action agrees to the test procedure, then machine A would probably be allowed to go on cutting pieces with the provision that it be checked regularly.)

The reader should realize that Example 7.1 was presented for illustrative purposes. An investigator would not normally solve both (a) and (b) with one set of data, nor would he be likely to use a 95 per cent confidence interval in one part and a two per cent significance level in the other. Further, he would be unwise to let α have such a small value without taking into account the size of the type 2 error and the consequences of making such an error. We illustrate this problem in Example 7.4.

On letting $n = 1$ in Theorem 7.1, we see that $w = t_1^2 \equiv t^2$ is distributed as χ^2 with one degree of freedom, provided t is distributed normally with zero mean and unit variance. The connection between these distributions is made clearer in the following example.

Example 7.2. Relate intervals of the t variate with values of the $t^2 = \chi^2$ variate.

Consider the symmetric interval $-t_0 < t < t_0$, where t_0 is a positive real number. The square of any value between 0 and t_0 is a value between 0 and t_0^2, and the square of any value between $-t_0$ and 0 is a value between $(-t_0)^2 = t_0^2$ and 0. Thus, if t is a value in the interval $-t_0 < t < t_0$, then t^2 is a value in the interval $0 < t < t_0^2$ and

$$P[-t_0 < t < t_0] = P[0 < t^2 < t_0^2]$$

In particular, if $t_0 = 0.126$, $t_0^2 = 0.0159$ and, according to Tables II and IV, we see that

$$P[-0.126 < t < 0.126] = 0.10$$

and

$$P[0 < t^2 < 0.0159] = 0.10$$

That is, that portion of the normal distribution which is close about the mean becomes the left tail of the χ^2 distribution.

If $t_0 = 1.96$, then

$$P[t > 1.96] = 0.025 = P[t < -1.96]$$

so that

$$P[t^2 > (1.96)^2] = P[\chi^2 > 3.84] = 0.05$$

That is, five per cent of the χ^2 values are greater than 3.84 when t falls in the intervals $t < -1.96$ or $t > 1.96$. So that portion of the normal distribution which is in two symmetric tails becomes the right tail of the χ^2 distribution. From this it should be clear that any two-tailed α level test using symmetric tails of the normal distribution as a critical region is the same test as a one-tailed test using the right tail of a χ^2 distribution with one degree of freedom as the critical region. (However, no χ^2 test can replace a one-tailed normal test, nor is it reasonable to replace a two-tailed χ^2 test with a three-interval normal test. The student should be cautioned that the above statements apply only for the χ^2 distribution with one degree of freedom.)

7.4. DEGREES OF FREEDOM

The χ^2 distribution is represented by a family of curves. There is one curve for each value of the parameter ν. The particular value of the parameter which determines a curve is also the mean of the distribution or the number of degrees of freedom of the distribution which the curve represents. Since the distribution of the statistic SSx $[= \sum (x_i - \bar{x})^2 = (n - 1)s^2]$ for sample size n depends on the χ^2 distribution with $\nu = n - 1$ degrees of freedom, and, since we say, "χ^2 is distributed as χ^2 with $n - 1$ degrees of freedom," it is only natural that we say "SSx is distributed [as $\sigma^2\chi^2$] with $n - 1$ degrees of freedom." Thus, we say that the *estimator* s^2 of σ^2 as well as the statistic $(n - 1)s^2$ *has* $n - 1$ *degrees of freedom*, $n - 1$ being a constant.

So far we have considered the number of degrees of freedom associated with the variance estimator obtained from a sample of size n drawn from a single normal population with variance σ^2. However, independent samples of sizes n_1 and n_2 obtained from two populations with a common variance σ^2 may be used to estimate σ^2. In this case, we say the variance *estimator* s_p^2 *has* $\nu = (n_1 - 1) + (n_2 - 1)$ *degrees of freedom*. One argument for saying this goes as follows. If the statistics $SSx_1 = (n_1 - 1)s_1^2$ and $SSx_2 = (n_2 - 1)s_2^2$ are computed from independent samples, they are independent random variables. Thus

$$\chi_1^2 = \frac{(n_1 - 1)s_1^2}{\sigma^2} \quad \text{and} \quad \chi_2^2 = \frac{(n_2 - 1)s_2^2}{\sigma^2}$$

are also independent random variables distributed with $\nu_1 = n_1 - 1$ and $\nu_2 = n_2 - 1$ degrees of freedom, respectively, and, according to Theorem 7.2

$$\chi_1^2 + \chi_2^2 = \frac{(n_1 - 1)s_1^2 + (n_2 - 1)s_2^2}{\sigma^2} \tag{7.8}$$

is distributed with $\nu = \nu_1 + \nu_2 = n_1 + n_2 - 2$ degrees of freedom. Hence, we say that

$$SSx_1 + SSx_2 = (n_1 - 1)s_1^2 + (n_2 - 1)s_2^2 = \sigma^2(\chi_1^2 + \chi_2^2)$$

is distributed [as $\sigma^2(\chi_1^2 + \chi_2^2) = \sigma^2\chi^2$] with $n_1 + n_2 - 2$ degrees of freedom and that

$$s_p^2 = \frac{SSx_1 + SSx_2}{n_1 + n_2 - 2}$$

has $n_1 + n_2 - 2$ degrees of freedom. In general, when $SSx_1 = (n_1 - 1)s_1^2$, $\ldots$, $SSx_k = (n_k - 1)s_k^2$ are computed from independent samples drawn from k normal populations with a common variance σ^2, we say that $SSx_1 + \cdots + SSx_k$ has $(n_1 - 1) + \cdots + (n_k - 1)$ degrees of freedom and that the estimator

$$s_p^2 = \frac{SSx_1 + \cdots + SSx_k}{(n_1 - 1) + \cdots + (n_k - 1)}$$

has $n_1 + \cdots + n_k - k$ degrees of freedom.

The above discussion indicates how the number of degrees of freedom of a variance estimator associated with a χ^2 distribution may be found, but it does not necessarily give the student an insight into a real understanding of "degrees of freedom." The term was introduced by Fisher, who apparently had in mind the degrees of freedom of a dynamical system, that is, the number of independent co-ordinate values necessary to determine the system. At the present time the term is used in different senses, as discussed below.

In selecting a random sample of n values from a population, we select a value for each of n variables, each variable ranging over the population of values. In this case the selection involves n degrees of freedom of choice, since the n values can be arbitrarily obtained within the specification of the system.

Considerations of statistics obtained from samples may lead to restrictions in the form of linear relations among sample values. For example, if the mean $\bar{x}$ of a sample of n values $x_1, \ldots, x_n$ is fixed, and if $n - 1$ of the values of x are selected arbitrarily, the remaining value is determined, and we say that there are $n - 1$ degrees of freedom of choice. In general, if r independent linear relations [9] are imposed on the n values in a sample, only $n - r$ of the values can be selected arbitrarily, and we say that there are $n - r$ degrees of freedom of choice.

Further, if a sample of size n is grouped into k intervals and the number of values in $k - 1$ intervals is arbitrarily assigned, then the number of values falling in the kth interval is determined. In this case we say there are $k - 1$ degrees of freedom.

We refer most often to the number of degrees of freedom associated with quadratic forms. In the expression $SSx = \sum (x_i - \bar{x})^2$ there are only $n - 1$ arbitrary choices possible on the deviates $e_i \equiv x_i - \bar{x}$ when the mean $\bar{x}$ is fixed (that is, when there is only one linear relation restricting the x values). In general, if k independent linear relations are imposed on the n values in a quadratic form, then it is possible by a suitable nonsingular linear transformation on the variables to express the quadratic form as the sum of squares of exactly $n - k$ of the new independent variables. The number of degrees of freedom is $n - k$, being the same as the *rank* of the original quadratic form [2, 6]. For example, in the simple case where there is only one linear restriction of the form $\sum x_i = n\bar{x}$ we may write, when $n = 2$

$$SSx = (x_1 - \bar{x})^2 + (x_2 - \bar{x})^2 = \left(\frac{x_1 - x_2}{\sqrt{2}}\right)^2 \tag{7.9}$$

or

$$e_1^2 + e_2^2 = q_1^2$$

where $q_1 = (x_1 - x_2)/\sqrt{2}$, and, when $n = 3$

$$SSx = (x_1 - \bar{x})^2 + (x_2 - \bar{x})^2 + (x_3 - \bar{x})^2$$

$$= \left(\frac{x_1 - x_2}{\sqrt{2}}\right)^2 + \left(\frac{x_1 + x_2 - 2x_3}{\sqrt{6}}\right)^2 \tag{7.10}$$

or

$$e_1^2 + e_2^2 + e_3^2 = q_1^2 + q_2^2$$

where $q_2 = (x_1 + x_2 - 2x_3)/\sqrt{6}$.

When SSx is fixed, the degrees of freedom for Eqs. (7.9) and (7.10) are obviously $2 - 1 = 1$ and $3 - 1 = 2$, respectively.

The expression "degrees of freedom" is also used to denote the number of independent comparisons which can be made between members of a sample. This is discussed in Sect. 10.7.

Different members of the family of χ^2 distributions with ν degrees of freedom have different means. Sometimes it is desirable to make a transformation so that all members of the transformed family have the same mean. Since the mean of

$$\chi^2 = \frac{SSx}{\sigma^2} = \frac{(n-1)s^2}{\sigma^2}$$

is $\nu = n - 1$, the mean of

$$\frac{\chi^2}{n-1} = \frac{s^2}{\sigma^2}$$

when a generalization of Theorem 2.3 is used, is 1. We say that the ratio s^2/σ^2 is distributed as "χ^2 per degree of freedom with ν degrees of freedom" and denote this by "χ^2/ν with ν d.f." Graphs of three members of the family of χ^2/ν distributions are shown in Fig. 7.2 and values, χ_α^2/ν, of χ^2/ν corresponding to selected values of α are shown in Table V. The χ^2/ν with ν d.f. distribution makes the comparison of two variances with unequal sample sizes easier.

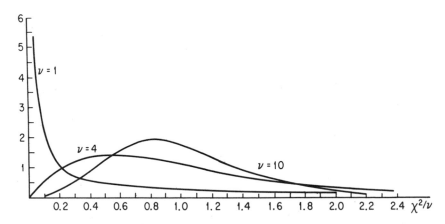

Fig. 7.2 χ^2/ν Distribution for $\nu = 1$, 4, and 10

7.5. POWER CURVES FOR CHI-SQUARE TESTS

In Example 7.1 we mentioned that an experimenter would be unwise to select the size of the type 1 error without taking into account the size of the type 2 error. Some problems involving β or $1 - \beta$ in a one-tailed α-level χ^2 test are discussed in the following examples.

Example 7.3. A random sample of size eleven is drawn from a normal population with unknown variance σ^2. For null hypothesis $\sigma^2 = 10$ and alternative hypothesis $\sigma^2 > 10$ (a) describe the appropriate one-sided five per cent level test, (b) determine the power curve and (c) use this curve to find the size of the type 2 error when the true variance is actually 25.

Using the statistic $\chi^2 = (11 - 1)s^2/10$ with ten degrees of freedom, we find the critical point $\chi^2_{.05} = 18.31$. Thus, the critical region is made up of all values of χ^2 such that $\chi^2 > 18.31$. If $\chi^2 = \chi_0^2$ for a particular sample falls in the critical region, we reject the null hypothesis and conclude that $\sigma^2 > 10$; otherwise, we fail to reject the null hypothesis and conclude that we do not have enough evidence to say $\sigma^2 > 10$.

The power of the test described above is given by

$$p(\sigma^2) = P[\text{sample } \chi_0^2 \text{ falling in critical region; } H_a \text{ true}]$$
$$= P[\chi^2 > 18.31; \sigma^2 > 10]$$
$$= P[(11 - 1)s^2 > 10(18.31); \sigma^2 > 10]$$
$$= P\left[\frac{10s^2}{\sigma^2} > \frac{10}{\sigma^2}(18.31); \sigma^2 > 10\right]$$

or

$$p(\sigma^2) = P\left[\chi^2 > \frac{18.31}{\lambda^2}; \sigma^2 > 10\right] \tag{7.11}$$

where $\lambda^2 = \dfrac{\sigma^2}{10}$.

From Eq. (7.11) it is clear that the power increases with λ^2 or σ^2, since $18.31/\lambda^2$ decreases as λ^2 increases. Further, since

$$P[\chi^2 > \chi^2_{1-\beta}] = 1 - \beta(\sigma^2) = p(\sigma^2)$$

it follows that

$$\chi^2_{1-\beta} = \frac{18.31}{\lambda^2}$$

or

$$\lambda^2 = \frac{18.31}{\chi^2_{1-\beta}} = \frac{\chi^2_\alpha}{\chi^2_{1-\beta}} \tag{7.12}$$

The power curve may be graphed as a function of λ^2 as well as a function of σ^2. Corresponding to selected values of the power function $p(\lambda^2) = 1 - \beta$ (λ^2), we use Table IV to find $\chi^2_{1-\beta}$, which is used in Eq. (7.12) to determine λ^2. Thus, for ten degrees of freedom we find the following

Table 7.1

β	.90	.75	.50	.25	.10	.05	.01
$\chi^2_{1-\beta}$	15.99	12.55	9.34	6.74	4.87	3.94	2.56
λ^2	1.14	1.47	1.96	2.72	3.76	4.65	7.15

Plotting the points $(\lambda^2, 1 - \beta)$ or $(\lambda^2, p(\lambda^2))$ and connecting them by a smooth curve, we obtain the power curve shown in Fig. 7.3. (The variances σ^2 corresponding to selected values of λ^2 are indicated for the case where the hypothesized variance is $\sigma_0^2 = 10$.)

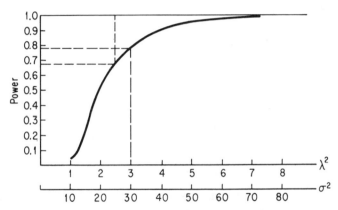

Fig. 7.3 Power Curve for the One-Sided χ^2 Test of $\sigma^2 = \sigma_0^2$
When $\nu = 10$ and $\alpha = .05$. $\sigma^2 = \sigma_0^2 \lambda^2 = 10\lambda^2$

Using the graph, we see that the power of the test of the null hypothesis $\sigma^2 = \sigma_0^2$ is 0.78 when the true variance σ^2 is three times larger than σ_0^2. In particular, if $\sigma_0^2 = 10$, then the probability of rejecting the null hypothesis $\sigma^2 = 10$ when $\sigma^2 = 30$ is 0.78. Hence, the probability of failing to reject the null hypothesis is $1 - 0.78 = 0.22$; that is, the probability of the type 2 error is 0.22.

We may find $\beta(\sigma^2)$ directly from the graph as $1 - p(\sigma^2)$. Thus, for $\sigma^2 = 25$ we find $\beta(25) = 1 - p(25) = 1 - 0.675 = 0.325$. That is, in roughly one-third of the cases where the true variance is 25 or greater, we would conclude that the variance is not significantly larger than ten, and this could be very serious in many experiments.

If it is required in a five per cent level one-sided test of the null hypothesis $\sigma^2 = \sigma_0^2$ that β be not greater than 0.10 when the true variance is $2.5\sigma_0^2$, then the sample size must be considerably larger than 11. Actually, we must

Table 7.2

Ratio $\lambda^2 = \chi^2_{.05}/\chi^2_{1-\beta}$ for the one-sided five per cent level χ^2 test of the null hypothesis $\sigma^2 = \sigma_0^2$ against the alternative $\sigma^2 > \sigma_0^2$. (Useful in finding power, size of type 2 error, and sample size.)

ν	$\chi^2_{.05}$	λ^2					
		$\beta = .01$	.05	.10	.25	.50	.80
2	5.99	294.3	58.4	28.4	10.4	4.32	1.80
3	7.81	67.9	22.2	13.4	6.44	3.30	1.68
4	9.49	32.0	13.3	8.92	4.94	2.83	1.58
5	11.07	20.0	9.66	6.88	4.15	2.54	1.52
6	12.59	14.4	7.70	5.71	3.64	2.35	1.47
8	15.51	9.42	5.67	4.44	3.06	2.11	1.41
10	18.31	7.16	4.65	3.76	2.72	1.96	1.37
12	21.03	5.89	4.02	3.34	2.49	1.85	1.33
14	23.68	5.08	3.60	3.04	2.33	1.78	1.30
16	26.30	4.53	3.30	2.82	2.21	1.71	1.28
18	28.87	4.11	3.07	2.66	2.11	1.67	1.27
20	31.41	3.80	2.89	2.52	2.03	1.62	1.26
25	37.65	3.27	2.58	2.29	1.89	1.55	1.23
30	43.77	2.93	2.37	2.13	1.79	1.49	1.21
40	55.76	2.52	2.10	1.92	1.66	1.42	1.18
50	67.50	2.27	1.94	1.79	1.59	1.37	1.16
60	79.08	2.11	1.83	1.70	1.51	1.33	1.15
70	90.53	1.99	1.75	1.64	1.47	1.31	1.14
80	101.9	1.90	1.68	1.59	1.43	1.28	1.13
90	113.1	1.83	1.64	1.54	1.40	1.27	1.12
100	124.3	1.77	1.60	1.51	1.38	1.25	1.11

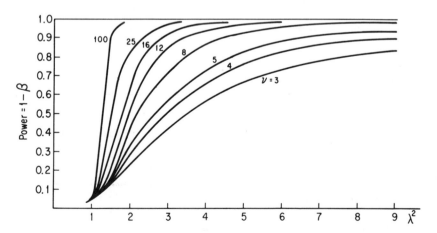

Fig. 7.4 Power Curves for the One-Sided .05 Level χ^2 Test of $\sigma^2 = \sigma_0^2$ Against $\sigma^2 > \sigma_0^2$. $\lambda^2 = \sigma^2/\sigma_0^2$

find $n - 1$ such that the ratio $\lambda^2 = \chi^2_{.05}/\chi^2_{.90}$ is not greater than 2.5. Table 7.2 may be used to determine the sample size. For $\nu = 20$, $\lambda^2 = 2.52$, and for $\nu = 25$, $\lambda^2 = 2.29$. Using linear interpolation, we find $\lambda^2 = 2.48$ when $\nu = 21$. Thus, we feel safe in saying that a sample of size $n = \nu + 1 = 22$ will give the required protection against errors of types 1 and 2, even though linear interpolation is not very good for ν less than 30.

Table 7.2 may also be used in constructing a family of power or operating characteristic curves which are adequate for most purposes. Eight power curves are shown in Fig. 7.4. These curves may be of use in finding the approximate probability of rejecting the null hypothesis $\sigma^2 = \sigma_0^2$ when the true variance has a value σ_1^2 such that $\sigma_1^2 > \sigma_0^2$.

7.6. SAMPLE SIZE FOR A TEST

The curves in Fig. 7.4 may also be used to find the minimum sample size in a five per cent level test which insures that β does not exceed a specified value for a particular value of σ^2. This is illustrated in the following example.

Example 7.4. In Example 7.1, suppose the null hypothesis is $\sigma^2 = 0.0009$ and the alternative hypothesis is $\sigma^2 > 0.0009$. Further, suppose a random sample is to be taken from machine A to determine whether the company should go to the expense of buying a new machine. (a) Determine what sample size should be drawn in order that $\alpha = 0.05$ and $\beta = 0.05$ when $\sigma^2 = 0.0027$. (b) Discuss the consequences of making the type 1 or type 2 error.

When $\sigma^2 = 0.0027$ and $\sigma_0^2 = 0.0009$, $\lambda^2 = 3$. Since $\beta = 0.05$, the power is to be 0.95 when $\sigma^2 = 0.0027$. Thus, with the aid of Fig. 7.4 we find, by interpolation, that $\nu = n - 1 = 20$. This means that when a random sample of size 21 is taken from machine A which actually has true variance 0.0027, and a five per cent level one-sided χ^2 test is applied, then the chance of failing to reject the hypothesis $\sigma^2 = 0.0009$ is not more than 0.05. That is, the probability of rejecting the null hypothesis is 0.95. For the one-sided five per cent level test, the above can be stated in more general terms as "for a sample of size 21 the chance of making a type 2 error is not more than 0.05 when σ^2 is not less than 0.0027." As a matter of fact, the null hypothesis $\sigma^2 = \sigma_0^2$ could also be stated as $\sigma^2 \leq \sigma_0^2$ (for reasons similar to those given following Example 6.10. which involved $\mu \geq 1000$). The sample size can be found directly from Table 7.2 for fixed α, β, and σ^2.

In answering (b), note that as a result of the test of the null hypothesis one of two decisions *must* be made. Either decision can be correct or in error. Associated with each decision is some action. If the decision is wrong, the action is wrong. Thus, if machine A actually does cut pieces with the precision of the standard, and the decision resulting from the test is that it does not, a new machine will be unnecessarily purchased at considerable expense. On the other hand, if the decision is that the null hypothesis cannot be rejected, and machine A actually does cut pieces with a variance greater than 0.0027, then the company produces substandard parts and is in a position to lose customers at considerable expense to the company.

It should be observed at this point that the decision concerning the sizes of α and β is not one the consulting statistician should be called upon to make. Indeed, he has no rule for fixing the maximum sizes of type 1 and type 2 errors. This is in the hands of the investigator or supervisor. For example, in the problem about the machines which cut pieces, someone in a responsible position concerning money matters probably should determine the sizes of α and β. If the expense to the company in making a type 1 or type 2 error is about the same, then the size of α and β should be about the same. But the decision as to whether the size of these errors should be 0.01, 0.05, 0.10, or something else is not the prerogative of the statistician. Naturally, the common size of α and β should be as small as possible. However, there is a nonzero limit to the size. For the smaller the probability of making an error, the larger the sample must be, and the greater the expense to the company in time and money, among other things. Clearly, someone other than the consulting statistician must make a decision on what values to assign α and β in order to keep a "reasonable balance" within the company.

In case the unnecessary expense incurred by making an error of type 2 is decidedly greater than that incurred in making a type 1 error, the size of β should be smaller than the size of α. The question is, how much smaller? Again, this is a problem for someone responsible for taking action resulting from the statistical test procedure. Even the problem of estimating how expensive each type of error will be for the company is a complicated problem in accounting.

7.7. SUMMARY REMARKS

We have shown how to determine the appropriate sample size for assigned values of α, β, and λ^2. It is often the case that β must be determined when α, λ^2, and n are known—this is straightforward. In fact, it should be clear by now that, when three of the four variables α, β, n, and λ^2 are arbitrarily selected, the fourth is uniquely determined, it being assumed that random samples are drawn from a single normal population. For example, when β, λ^2, and n are specified, α may be uniquely determined. We may use Formula (7.12) to find $\chi_\alpha^2 = \lambda^2 \cdot \chi_{1-\beta}^2$ and then look in Table IV with $\nu = n - 1$ degrees of freedom for this value.

The discussion so far has centered around the one-sided right-tailed

χ^2 test. It is assumed that the student can give similar arguments for an α level χ^2 test of the null hypothesis $\sigma^2 \doteq \sigma_0^2$ against the alternative hypothesis $\sigma^2 < \sigma_0^2$. For example, the power function for such a test is given by

$$p(\sigma^2) = P\left[\chi^2(\nu) < \frac{\chi^2_{1-\alpha}(\nu)}{\lambda^2}; \ \sigma^2 < \sigma_0^2\right] \tag{7.13}$$

where $\chi^2(\nu)$ denotes a random variable distributed as chi-square with $\nu = n - 1$ degrees of freedom and $\lambda^2 = \sigma^2/\sigma_0^2$. Using Eq. (7.13), we may construct a table similar to Table 7.2 and a family of curves similar to those of Fig. 7.4. For a specified variance $\sigma_1^2 < \sigma_0^2$ and fixed error sizes α and β the sample size n may be determined from

$$\lambda_1^2 = \frac{\sigma_1^2}{\sigma_0^2} = \frac{\chi^2_{1-\alpha}(n-1)}{\chi^2_\beta(n-1)} \tag{7.14}$$

For a two-sided α level χ^2 test of the null hypothesis $\sigma^2 = \sigma_0^2$ against the alternative hypothesis $\sigma^2 \neq \sigma_0^2$ the methods are straightforward. Generally, the two critical points χ_1^2 and χ_2^2 are taken as the $\chi^2_{1-(\alpha/2)}(\nu)$ and $\chi^2_{(\alpha/2)}(\nu)$ points, respectively, of a chi-square distribution with $\nu = n - 1$ degrees of freedom. Then the power function for the test is given by

$$p(\sigma^2) = P\left[\chi^2(\nu) < \frac{\chi^2_{1-(\alpha/2)}(\nu)}{\lambda^2}\right] + P\left[\chi^2(\nu) > \frac{\chi^2_{(\alpha/2)}(\nu)}{\lambda^2}\right] \tag{7.15}$$

where $\lambda^2 = \sigma^2/\sigma_0^2$. If α and β are fixed and the specified variance σ_1^2 is smaller than σ_0^2 but not near σ_0^2, the sample size n may be determined approximately from

$$\frac{\sigma_1^2}{\sigma_0^2} = \frac{\chi^2_{1-(\alpha/2)}(\nu)}{\chi^2_\beta(\nu)} \tag{7.16}$$

If α and β are fixed and the specified variance σ_1^2 is larger than σ_0^2 but not near σ_0^2, the sample size n may be determined approximately from

$$\frac{\sigma_1^2}{\sigma_0^2} = \frac{\chi^2_{(\alpha/2)}(\nu)}{\chi^2_{1-\beta}(\nu)} \tag{7.17}$$

Table 7.2 and Fig. 7.4 were constructed for discussions involving a type 1 error of size 0.05 in one-sided tests with critical region in the right tail. But these constructions are inadequate for tests of this type. Some experiments would require that Fig. 7.4 include many more curves and that other figures be drawn for values of α such as 0.005, 0.01, 0.10, 0.20. There is a similar need for one-sided tests with critical regions in the left tail and for two-sided tests. The student can find further information on tests, power and operating characteristic curves, and size of sample in Refs. [1, 4, 6, 9, 10] at the end of this chapter.

7.8. EXERCISES

7.1. For a sample of size ten, $s^2 = 0.002285$. Find 95 per cent confidence intervals for σ^2 and σ.

7.2. Graph the curve for the chi-square distribution when $\nu = 2$; when $\nu = 3$.

7.3. Graph the curve for the χ^2/ν distribution when $\nu = 2$; when $\nu = 3$.

7.4. A random sample of size 16 has variance 2.23. Use a two-sided five per cent level test to determine whether the true variance is different from 1.5.

7.5. The seven random observations 40, 53, 34, 48, 49, 35, 53 are from a normal population with mean 50 and standard deviation 10. In order to acquaint the student with errors made in testing hypotheses, pretend that the population variance is unknown, and use a five per cent level two-sided test for the null hypothesis that the population variance is (a) 100, (b) 99, (c) 101, (d) 20, and (e) 500. In each case state whether the test conclusion is correct, or whether a type 1 error or type 2 error is made.

7.6. Prove Eqs. (7.9) and (7.10).

7.7. The lengths, in inches, of a random sample of eight parts cut off by a certain machine are

0.823	0.793	0.809	0.781
0.790	0.813	0.802	0.797

The specified standard deviation is not to exceed 0.005 in. Test, at the five per cent level, the hypothesis that the population standard deviation is 0.005 in.

7.8. The tensile strengths, in pounds, of a random sample of ten pieces of a certain type of yarn are

440	524	644	500	570
482	578	578	410	474

The specified standard deviation is not to exceed 60 lb. (a) Use a five per cent level test to test the hypothesis that the population standard deviation is 60 lb. (b) Find a 95 per cent confidence interval for the true variance, letting the lower limit be zero. (c) Find the power curve for the hypothesis in (a). (d) Use the curve found in (c) to determine the size of the type 2 error when the true variance is actually 7500. (e) Determine what sample size should be drawn in order that $\alpha = 0.05$ and $\beta = 0.10$ when $\sigma^2 = 9000$.

7.9. (a) Draw the power curve for the test in Exercise 7.7. Find σ^2 when $\beta = 0.10$. (b) Determine what sample size should be drawn in order that $\alpha = 0.01$ and $\beta = 0.05$ when $\sigma^2 = 0.004$.

7.10. (a) Draw the power curve for the test in Exercise 7.4 and use this curve to find σ^2 when $\beta = 0.05$. (b) Find the smallest sample size for the test in Exercise 7.4 for which $\alpha = 0.05$ and $\beta = 0.01$ when $\sigma^2 = 2.0$.

7.11. Suppose that x_i $(i = 1, 2, 3)$ is normally distributed with mean μ and variance σ^2. Let q_1 and q_2 be the linear combinations defined following Eqs. (7.9) and (7.10), respectively. (a) Find $E(q_j)$ $(j = 1.2)$. (b) Show that q_1 and q_2 are independently distributed.

Hint. Show that the covariance of q_1 and q_2 is zero. (According to Sect. 3.4.1, we know that the random variables u and v in a bivariate normal distribution are independently distributed when the covariance of u and v is zero.)

7.12. Use Eqs. (7.9) and (7.10) as guides in expressing

$$\sum_{i=1}^{4} (x_i - \bar{x})^2$$

as the sum of squares of three independent variables, q_1, q_2, and q_3. Prove that cov $(q_1, q_3) = $ cov $(q_2, q_3) = 0$.

Hint. See Exercise 7.11.

7.13. Use Eqs. (7.9) and (7.10) and Exercise 7.12 as guides in expressing

$$\sum_{i=1}^{n} (x_i - \bar{x})^2$$

as the sum of squares of $n - 1$ independent variables, $q_1, q_2, \ldots, q_{n-1}$. Prove that cov $(q_{n-2}, q_{n-1}) = 0$. If x_i $(i = 1, \ldots, n)$ is normally distributed with mean μ and variance σ^2, what can you say about the joint density function of $q_1, \ldots, q_{n-1}$?

7.14. Find the length of the shortest 95 per cent confidence interval for σ^2 when $\nu = 2$. What proportion of the χ^2 values are greater than the upper value χ_2^2?

7.15. (a) Show that the moment generating function of χ^2, $M_{\chi^2}(\theta)$, is $(1 - 2\theta)^{-\nu/2}$. (Note that the t used in earlier chapters is replaced by θ. This is done to avoid confusion with the standard normal t.) (b) Show that the moment generating function of

$$w = \sum_{i=1}^{n} t_i^2$$

given in Theorem 7.1, is $(1 - 2\theta)^{-n/2}$. (c) Use (a) and (b) to prove Theorem 7.1.

7.16. Use moment generating functions to prove Theorem 7.2.

7.17. Let x_i $(i = 1, \ldots, n)$ be normally distributed with mean μ and variance σ^2. Let

$$\sum_{i=1}^{n} (x_i - \bar{x})^2 = q_1^2 + \cdots + q_{n-1}^2$$

where the q's are defined in Sect. 7.4 and Exercise 7.13. (a) Prove that

$E(q_j) = 0$ and $V(q_j) = \sigma^2$ $(j = 1, \ldots, n - 1)$. (b) Prove that cov $(q_j,$ $q_{j'}) = 0$ for $j \neq j'(j' = 2, \ldots, n - 1)$. (c) Prove Theorem 7.4. (d) Prove cov $(\bar{x}, q_j) = 0$. (e) Prove Theorem 7.3.

Hint. Since $\bar{x}$ is independent of each q_j, it is independent of

$$\frac{q_1^2 + \cdots + q_{n-1}^2}{n - 1}$$

7.18. Prove

$$\sum_{i=1}^{n} (x_i - \mu)^2 = \sum_{i=1}^{n} (x_i - \bar{x})^2 + n(\bar{x} - \mu)^2 \tag{7.18}$$

Then use the identity (7.18) and Theorems 7.1, 7.2, and 7.3 to prove Theorem 7.4. (For further study of distributions of quadratic forms see Refs. [1, 5, 7, 10,].)

REFERENCES

1. Alexander, H. W., *Elements of Mathematical Statistics*. New York: John Wiley & Sons, Inc., 1961, Chaps. 2 and 5.

2. Anderson, T. W., *An Introduction to Multivariate Statistical Analysis*. New York: John Wiley & Sons, Inc., 1958, Chap. 7.

3. Anderson, R. L. and T. A. Bancroft, *Statistical Theory in Research*. New York: McGraw-Hill, Inc., 1952, Chap. 7.

4. Bennett, C. A. and N. L. Franklin, *Statistical Analysis in Chemistry and the Chemical Industry*. New York: John Wiley & Sons, Inc., 1954, Chap. 4.

5. Cochran, W. G., "The Distribution of Quadratic Forms in a Normal System," *Proceedings of the Cambridge Philosophical Society*, Vol. 30 (1934), pp. 178–191.

6. Cramér, H., *Mathematical Methods of Statistics*. Princeton, N. J.: Princeton University Press, 1946, Chap. 29.

7. Fisher, R. A., "Applications of Student's Distribution," *Metron*, Vol. 5, No. 3 (1925), pp. 90–104.

8. Fraser, D. A. S., *Statistics; An Introduction*. New York: John Wiley & Sons, Inc., 1958, Chap. 8.

9. Hald, A., *Statistical Theory with Engineering Applications*. New York: John Wiley & Sons, Inc., 1952, Chap. 10.

10. Hogg, R. V. and A. T. Craig, *Introduction to Mathematical Statistics*. New York: The Macmillan Company, 1959, Chaps. 2, 6, and 10.

8

SAMPLING
FROM NORMAL POPULATIONS—
THE STUDENT t DISTRIBUTION

The Student t distribution is used to establish confidence intervals and test hypotheses when the population variance is not known and the sample size is small. Properties and uses of the power function are given. Tolerance intervals are introduced and compared with confidence intervals. The t distribution is used in a paired-observations experiment and is compared with the sign test. Model and observational equations are discussed.

8.1. INTRODUCTION

When the sample mean was used in Sect. 6.3 to obtain a confidence interval or to test a hypothesis about the population mean, it was necessary that the population variance be known or that the sample be large. In many experiments the population variance is not known and the sample size is so small that the sample variance cannot be used in place of the population variance. Fortunately, due to the Student t distribution, the sample variance along with the sample mean may be used in making statements about the population mean.

We now give statements of four important theorems relating to the Student t distribution. Following these theorems we give illustrations of applications and, in the set of exercises, outlines of proofs of these key theorems.

Theorem 8.1. *If u has a normal distribution with mean 0 and variance 1, w has a χ^2-distribution with v degrees of freedom, and u and $\sqrt{w}$ are independently distributed, then the random variable*

$$t = \frac{u}{\sqrt{w/v}}$$

has density function

$$f(t; v) = \frac{\Gamma[(v+1)/2]}{\sqrt{\pi v}\,\Gamma(v/2)}\left(1 + \frac{t^2}{v}\right)^{-(v+1)/2}, \qquad -\infty < t < \infty \qquad (8.1)$$

with mean 0 and variance $v/(v-2)$ for $v > 2$. The statistic t is said to have the Student t distribution, or simply the t distribution, with v degrees of freedom.

Theorem 8.2. *If a random sample $x_1, x_2, \ldots, x_n$ is drawn from a normal population with mean μ and variance σ^2, then the statistic*

$$t = \frac{\bar{x} - \mu}{\sqrt{\dfrac{s^2}{n}}}$$

where $s^2 = \sum (x_i - \bar{x})^2/(n-1)$ and $\bar{x} = \sum x_i/n$, is distributed as the Student t distribution with $n-1$ degrees of freedom.

Theorem 8.3. *As the number of degrees of freedom of s^2 approaches infinity, the Student t distribution approaches the normal distribution with mean 0 and variance 1.*

Theorem 8.4. *If two populations are normal and have the same variance, then the statistic*

$$t = \frac{(\bar{x}_1 - \bar{x}_2) - (\mu_1 - \mu_2)}{\sqrt{s_p^2\left(\dfrac{1}{n_1} + \dfrac{1}{n_2}\right)}} \qquad (8.2)$$

has the Student t distribution with $n_1 + n_2 - 2$ degrees of freedom. Random samples $x_{11}, \ldots, x_{1n_1}$ and $x_{21}, \ldots, x_{2n_2}$ of sizes n_1 and n_2 are independently drawn from populations 1 and 2 with true means μ_1 and μ_2, respectively. The sample means are

$$\bar{x}_1 = \frac{\displaystyle\sum_{j=1}^{n_1} x_{1j}}{n_1} \quad \text{and} \quad \bar{x}_2 = \frac{\displaystyle\sum_{j=1}^{n_2} x_{2j}}{n_2}$$

and the pooled sample variance is

$$s_p^2 = \frac{\displaystyle\sum_{j=1}^{n_1}(x_{1j} - \bar{x}_1)^2 + \sum_{j=1}^{n_2}(x_{2j} - \bar{x}_2)^2}{n_1 + n_2 - 2} = \frac{\displaystyle\sum_{i=1}^{2}\sum_{j=1}^{n_i}(x_{ij} - \bar{x}_i)^2}{n_1 + n_2 - 2} \qquad (8.3)$$

8.2. PROPERTIES OF THE t DISTRIBUTION

Before considering applications of Theorems 8.1 through 8.4, we illustrate certain properties of the t distribution.

The density function (8.1) represents a one-parameter family of distributions. The graphs of three members of this family are shown in Fig. 8.1, which illustrates the fact that the "shape" of the distribution changes with the degrees of freedom. This being the case, we cannot tabulate appropriate values using only a single simple standardization process as with the normal distribution. Thus, it is necessary that the probability that a value of t fall in any particular interval be computed for each degree of freedom. Since the t distribution is symmetric, and since we usually require probabilities associated with the tail of the distribution, we let $t_\alpha(\nu)$ be that positive value of t with ν degrees of freedom such that

$$P[t > t_\alpha(\nu)] = \alpha$$

where α is any number between 0 and 0.5. Values of t_α corresponding to selected values of α are shown in Table VI. The probability that t is less than $-t_\alpha(\nu)$, a value in the left tail, is given by

$$P[t < -t_\alpha(\nu)] = P[t > t_\alpha(\nu)]$$

From Fig. 8.1 and Theorem 8.3, it is clear that the t distribution approaches the normal distribution as ν approaches ∞. It is for this reason that we sometimes call the normal distribution a t distribution with an infinite number of degrees of freedom and include probability points

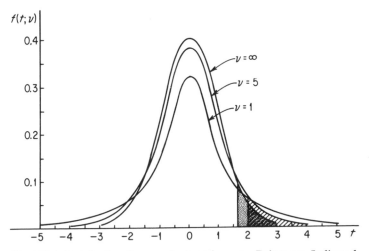

Fig. 8.1 t Distribution for $\nu = 1$, 5 and ∞. $t_{.05}$ Points are Indicated

$t_\alpha = t(\infty)$ in Table VI. Further, as the number of degrees of freedom approaches one, the t distribution is slower and slower in its rate of approach to the t axis. That is, for any point $t_\alpha(\nu)$ on the positive t axis, the probability that t is greater than $t_\alpha(\nu)$ gets larger as ν approaches one.

The dispersion of the t distribution increases as the number of degrees of freedom decreases. Since, according to Theorem 8.1, the variance is $\nu/(\nu - 2)$, we observe that, as ν changes from ∞ to 5 to 3, the variance changes from 1 to $\frac{5}{3}$ to 3. For $\nu = 1$, the density function for the t distribution of Eq. 8.1 reduces to

$$f(t; 1) = \frac{1}{\pi(1 + t^2)} \tag{8.4}$$

since $\Gamma(1) = 1$ and $\Gamma(\frac{1}{2}) = \sqrt{\pi}$ according to Exercise 3.67. The density function in Eq. (8.4) is a special case of the Cauchy density function. It can be shown that the Cauchy distribution does not have finite mean and variance. In fact, no moment of the Cauchy distribution is finite. The case where $\nu = 2$ is considered in the exercises.

Theorem 8.2 is actually a corollary of Theorem 8.1. By Theorem 7.3 we know that for a sample of size n the sample mean $\bar{x}$ and the sample variance $s^2 = SSx/(n - 1)$ are independently distributed. Thus

$$\frac{\bar{x} - \mu}{\sigma/\sqrt{n}} \quad \text{and} \quad \frac{(n - 1)s^2}{\sigma^2}$$

are independently distributed, μ, σ, and n being constants. But these random variables are distributed as u and w, respectively, in Theorem 8.1, where $\nu = n - 1$. Hence

$$t = \frac{u}{\sqrt{\dfrac{w}{n - 1}}} = \frac{\dfrac{\bar{x} - \mu}{\sigma/\sqrt{n}}}{\sqrt{\dfrac{(n - 1)s^2}{\sigma^2} \Big/ (n - 1)}} = \frac{\bar{x} - \mu}{\sqrt{\dfrac{s^2}{n}}} \tag{8.5}$$

is distributed as the Student t variable with $n - 1$ degrees of freedom. The reader should note that the degrees of freedom associated with t is the same as the number of degrees of freedom associated with s^2.

The two statistics $\bar{x}$ and s^2 used to compute t may or may not be determined from a single sample. If they are computed from samples of different sizes, the number of degrees of freedom for the t statistic is the same as that of s^2.

8.3. APPLICATIONS OF THE t DISTRIBUTION

The t statistic is used in very much the same way as the normal u, with the exception that the sample variance is required instead of the population variance. This leads to extra computations and more elaborate tables. The following examples bring out the differences in the applications of the t and normal distributions.

Example 8.1. A certain type of rat shows a mean gain of 65 grams during the first three months of life. Twelve rats were fed a particular diet from birth until age three months and the following weight gains in grams were observed: 55, 62, 54, 57, 65, 64, 60, 63, 58, 67, 63, and 61. (a) Is there reason to believe that the diet causes a change in the amount of weight gained? (b) Find a 95 per cent confidence interval for the mean weight gained.

We answer (a) by using the general procedure for testing hypotheses given on p. 218.

1. $H_0: \mu = 65$ grams and $H_a: \mu \neq 65$ grams, since an increase in weight less than or greater than 65 grams indicates that the diet has some effect.
2. Assume that the twelve sample values were independently obtained from the same normal population. Let the significance level be $\alpha = 0.05$.
3. The statistic to be used is

$$t = \frac{\bar{x} - 65}{\dfrac{s}{\sqrt{12}}}$$

since t values can be read directly from the tables.
4. For an experimenter who is equally concerned with detecting a gain in weight which is less than or greater than 65 grams, we take as the critical region all those values of t for which

$$t < t_{.975}(11) = -2.20 \quad \text{or} \quad t > t_{.025}(11) = 2.20$$

5. Since $\bar{x} = \frac{729}{12} = 60.75$ and

$$s^2 = \frac{12(44467) - (729)^2}{(11)(12)} = 16.38$$

the calculated t statistic is

$$t_c = \frac{60.75 - 65}{\sqrt{\dfrac{16.38}{12}}} = \frac{-4.25}{1.17} = -3.63$$

6. Since the sample statistic $t_c = -3.63$ falls in the left critical region,

we reject the hypothesis that $\mu = 65$ grams and conclude that the diet does cause a decrease in the amount of weight gained, understanding that there is a five per cent chance of making a type 1 error. Actually, the probability of saying that there is a decrease in weight gained when H_0 is true is only 0.025.

The symmetric 95 per cent confidence limits are given by $\bar{x} \pm t_{.025} \cdot s_{\bar{x}}$ where $s_{\bar{x}} = \sqrt{s^2/n}$. From steps 4 and 5 above, we have $\bar{x} = 60.75$, $t_{.025} = 2.20$, and $s_{\bar{x}} = 1.17$. Thus, the limits for the true weight gained are 58.17 and 63.32. That is

$$58.17 < \mu < 63.32 \tag{8.6}$$

Of course, the true mean either falls in the interval (8.6), or it does not. The 95 per cent simply indicates our degree of confidence that the interval in (8.6) includes the true mean.

If the alternative hypothesis in Example 8.1 (a) were $\mu < 65$ grams (or $\mu > 65$ grams), a one-tailed test should be used. Thus, if the significance level of the test remains the same ($\alpha = 0.05$), the critical region is made up of those values of t for which $t < -1.80$ (or $t > 1.80$). Hence, for a sample value of t falling in the interval $-2.20 < t < -1.80$ or the interval $t > 2.20$, the two tests (1) $H_0: \mu = 65$ grams against $H_a: \mu \neq 65$ grams and (2) $H_0: \mu = 65$ grams against $H_a: \mu < 65$ grams lead to different conclusions. This points up the importance of stating the correct alternative hypothesis for an α level test. For example, if the diet actually does lead to a decrease in mean weight gained, the second test would detect this more often.

8.4. POWER FOR THE t TEST

Example 8.2. Find the power function for the five per cent level test of the null hypothesis $H_0: \mu = 65$ grams against the alternative hypothesis $H_a: \mu < 65$ grams, when twelve random observations are taken from a normal population with unknown variance.

Clearly, the statistic to use is

$$t = \frac{\bar{x} - \mu}{\dfrac{s}{\sqrt{n}}} \tag{8.7}$$

and the critical region for the test is $t < -1.80 = t_{.95}(11) = t_0$ or $\bar{x} < \bar{x}_0$, where $\bar{x}_0 = t_0 \cdot s/\sqrt{12} + 65$. Following an argument similar to the one used in deriving Eq. (6.20), we have

$$p(\mu; \mu < 65) = P[\bar{x} < \bar{x}_0; \mu < 65]$$

$$= P\left[\frac{\bar{x} - \mu}{\frac{s}{\sqrt{12}}} < \frac{\bar{x}_0 - \mu}{\frac{s}{\sqrt{12}}}; \mu < 65\right]$$

$$= P\left[t \cdot \frac{s}{\sqrt{12}} + \mu < \bar{x}_0; \mu < 65\right]$$

$$= P\left[\frac{t \cdot \frac{s}{\sqrt{12}} + \mu - 65}{\frac{s}{\sqrt{12}}} < \frac{\bar{x}_0 - 65}{\frac{s}{\sqrt{12}}}; \mu < 65\right]$$

or

$$p(\mu; \mu < 65) = P\left[t + \frac{\mu - 65}{\frac{s}{\sqrt{12}}} < t_0; \mu < 65\right] \tag{8.8}$$

or

$$\begin{cases} p(\mu; \mu < 65) = P\left[t + \lambda\sqrt{\frac{11}{\chi^2}} < t_0; \mu < 65\right] & \text{where} \\ \\ \lambda = \frac{\mu - 65}{\frac{\sigma}{\sqrt{12}}} \quad \text{and} \quad \frac{\chi^2}{11} = \frac{s^2}{\sigma^2} \end{cases} \tag{8.9}$$

From Eq. (8.9) we see that the power function depends on the two random variables t and χ^2, each with eleven degrees of freedom.

In general, we say that the random variable

$$t' = t + \frac{\mu - \mu_0}{\frac{\sigma}{\sqrt{n}}} \sqrt{\frac{n-1}{\chi^2}} \tag{8.10}$$

is distributed as the *noncentral t with* $n - 1$ *degrees of freedom*. Useful tables for finding power are given in *Tables of the Non-Central t Distribution* by Resnikoff and Lieberman, Stanford University Press, along with Refs. [8, 9, 10]. [It is interesting to note that the power function obtained from Eq. (6.20) may be written as

$$p(\mu; \mu < \mu_0) = 1 - \beta(\mu) = P\left[u < u_\alpha - \frac{\mu - \mu_0}{\frac{\sigma}{\sqrt{n}}}; \mu < \mu_0\right]$$

or

$$p(\mu; \mu < \mu_0) = P\left[u + \frac{\mu - \mu_0}{\dfrac{\sigma}{\sqrt{n}}} < u_\alpha; \mu < \mu_0\right]$$

and that

$$u' \equiv u + \frac{\mu - \mu_0}{\dfrac{\sigma}{\sqrt{n}}}$$

is normally distributed with mean

$$\frac{\mu - \mu_0}{\dfrac{\sigma}{\sqrt{n}}}$$

and variance one.]

In general, the power function for the one-sided α level test of the hypothesis $\mu = \mu_0$ against the alternative hypothesis $\mu < \mu_0$ with critical region $t < t_0 = t_{1-\alpha}(n-1)$ is given by

$$p(\mu; \mu < \mu_0) = P[t' < t_0; \mu < \mu_0] \qquad (8.11)$$

where t and t' are defined by Eqs. (8.7) and (8.10), respectively, and

$$\lambda = \frac{\mu - \mu_0}{\sigma} \cdot \sqrt{n} = \Delta\sqrt{n} \qquad (8.12)$$

with $\Delta = (\mu - \mu_0)/\sigma$. If the alternative hypothesis is $\mu > \mu_0$ with critical region $t > t_0 = t_\alpha(n-1)$, the power function is

$$p(\mu; \mu > \mu_0) = P[t' > t_0; \mu > \mu_0] \qquad (8.13)$$

If the alternative hypothesis is $\mu \neq \mu_0$, and the critical region is defined by

$$t < t_{1-\alpha/2}(n-1) = t_1 \quad \text{and} \quad t > t_{\alpha/2}(n-1) = t_2$$

then the power function of the two-sided α level test of the hypothesis $\mu = \mu_0$ is given by

$$p(\mu; \mu \neq \mu_0) = P[t' < t_1] + P[t' > t_2] \qquad (8.14)$$

where t' is distributed as the noncentral t with $n - 1$ degrees of freedom given in Eq. (8.10). To illustrate the nature of power curves along with other properties associated with power, Tables 8.1 and 8.2 for right-sided one-tailed tests are given.

Example 8.3. Graph the power function for the five per cent level test in Example 8.2 and use the resulting curve to determine the power and probability of making the type 2 error when the true mean is 61 and the true variance is 16.

Use Table 8.1 with eleven degrees of freedom. Since the critical region for the test is in the left tail of the t distribution, the values of λ are negative. The required power curve is shown in Fig. 8.2.

Table 8.1

Values of $\lambda = (\mu - \mu_0)\sqrt{n}/\sigma$ for Specified Degrees of Freedom ν and Power p. One-sided Test with $\alpha = 0.05$.*

ν \\ p	0.10	0.20	0.30	0.40	0.50	0.60	0.70	0.80	0.90	0.95	0.99
1	0.64	1.60	2.46	3.35	4.31	5.38	6.63	8.19	10.51	12.53	16.46
2	0.50	1.15	1.63	2.07	2.49	2.92	3.40	3.98	4.81	5.52	6.88
3	0.45	1.02	1.43	1.79	2.13	2.48	2.85	3.30	3.93	4.46	5.47
4	0.43	0.96	1.34	1.67	1.99	2.30	2.64	3.04	3.60	4.07	4.95
5	0.42	0.92	1.29	1.61	1.91	2.21	2.53	2.90	3.43	3.87	4.70
6	0.41	0.90	1.26	1.57	1.86	2.15	2.46	2.82	3.33	3.75	4.55
7	0.40	0.89	1.24	1.54	1.82	2.11	2.41	2.77	3.26	3.67	4.45
8	0.40	0.88	1.22	1.52	1.80	2.08	2.38	2.73	3.21	3.62	4.38
9	0.39	0.87	1.21	1.50	1.78	2.06	2.35	2.70	3.18	3.58	4.32
10	0.39	0.86	1.20	1.49	1.77	2.04	2.33	2.67	3.15	3.54	4.28
11	0.39	0.86	1.19	1.48	1.75	2.02	2.32	2.66	3.13	3.52	4.25
12	0.38	0.85	1.19	1.47	1.74	2.01	2.30	2.64	3.11	3.50	4.22
13	0.38	0.85	1.18	1.47	1.74	2.00	2.29	2.63	3.09	3.48	4.20
14	0.38	0.84	1.18	1.46	1.73	2.00	2.28	2.62	3.08	3.46	4.18
15	0.38	0.84	1.17	1.46	1.72	1.99	2.27	2.61	3.07	3.45	4.17
16	0.38	0.84	1.17	1.45	1.72	1.98	2.27	2.60	3.06	3.44	4.16
17	0.38	0.84	1.17	1.45	1.71	1.98	2.26	2.59	3.05	3.43	4.14
18	0.38	0.83	1.16	1.45	1.71	1.97	2.26	2.59	3.04	3.42	4.13
19	0.38	0.83	1.16	1.44	1.71	1.97	2.25	2.58	3.04	3.41	4.12
20	0.38	0.83	1.16	1.44	1.70	1.97	2.25	2.58	3.03	3.41	4.12
25	0.37	0.83	1.15	1.43	1.69	1.95	2.23	2.56	3.01	3.38	4.09
30	0.37	0.82	1.15	1.42	1.68	1.94	2.22	2.54	2.99	3.37	4.06
∞	0.36	0.80	1.12	1.39	1.64	1.90	2.17	2.49	2.93	3.29	3.97

* This table is reproduced from J. Neyman and B. Tokarska, "Errors of the Second Kind in Testing 'Student's' Hypothesis," *Journal of American Statistical Association*, Vol. **31**, p. 322, Table I, with the permission of the editor of the journal.

When $\mu = 61$, $\sigma^2 = 16$, $n = 12$, and $\mu_0 = 65$, we find that

$$\lambda = \frac{(61 - 65)\sqrt{12}}{4} = -3.46$$

From Fig. 8.2 we observe that the power of the test is

$$p(\mu = 61; \mu_0 = 65) = p(\lambda = -3.46) = 0.94$$

Thus the probability of making the type 2 error is

$$\beta(\mu = 61; \mu_0 = 65) = 1 - 0.94 = 0.06$$

Table 8.2

Values of $\lambda = (\mu - \mu_0)\sqrt{n}/\sigma$ for Specified Degrees of Freedom ν and Power p. One-sided Test with $\alpha = 0.01$.*

ν \ p	0.10	0.20	0.30	0.40	0.50	0.60	0.70	0.80	0.90	0.95	0.99
1	4.00	8.07	12.27	16.70	21.47	26.79	33.00	40.80	52.37	62.40	82.00
2	2.08	3.20	4.12	4.98	5.83	6.73	7.73	8.96	10.74	12.26	15.22
3	1.66	2.44	3.03	3.56	4.07	4.59	5.17	5.87	6.87	7.75	9.34
4	1.48	2.14	2.63	3.06	3.47	3.88	4.34	4.88	5.64	6.28	7.52
5	1.38	1.98	2.42	2.81	3.17	3.54	3.93	4.41	5.07	5.62	6.68
6	1.32	1.88	2.30	2.66	2.99	3.33	3.70	4.13	4.74	5.25	6.21
7	1.27	1.82	2.22	2.56	2.88	3.20	3.55	3.96	4.53	5.01	5.91
8	1.24	1.77	2.16	2.49	2.80	3.11	3.44	3.84	4.39	4.84	5.71
9	1.22	1.74	2.11	2.43	2.74	3.04	3.36	3.75	4.28	4.72	5.56
10	1.20	1.71	2.08	2.39	2.69	2.99	3.31	3.68	4.20	4.63	5.45
11	1.18	1.69	2.05	2.36	2.65	2.94	3.26	3.63	4.14	4.56	5.36
12	1.17	1.67	2.03	2.33	2.62	2.91	3.22	3.58	4.09	4.50	5.29
13	1.16	1.65	2.01	2.31	2.60	2.88	3.19	3.55	4.04	4.46	5.23
14	1.15	1.64	1.99	2.29	2.57	2.86	3.16	3.51	4.01	4.42	5.18
15	1.14	1.63	1.98	2.28	2.56	2.84	3.14	3.49	3.98	4.38	5.14
16	1.14	1.62	1.97	2.26	2.54	2.82	3.12	3.47	3.95	4.35	5.11
17	1.13	1.61	1.96	2.25	2.53	2.80	3.10	3.45	3.93	4.33	5.08
18	1.13	1.60	1.95	2.24	2.51	2.79	3.08	3.43	3.91	4.31	5.05
19	1.12	1.60	1.94	2.23	2.50	2.78	3.07	3.41	3.89	4.29	5.03
20	1.12	1.59	1.93	2.22	2.50	2.77	3.06	3.40	3.88	4.27	5.01
25	1.10	1.57	1.90	2.19	2.46	2.73	3.02	3.35	3.82	4.20	4.93
30	1.09	1.55	1.89	2.17	2.44	2.70	2.99	3.32	3.78	4.16	4.88
∞	1.04	1.48	1.80	2.07	2.33	2.58	2.85	3.17	3.61	3.97	4.65

* This table is reproduced from J. Neyman and B. Tokarska, "Errors of the Second Kind in Testing 'Student's' Hypothesis," *Journal of American Statistical Association*, Vol. **31**, p. 322 , Table II with the permission of the editior of the journal.

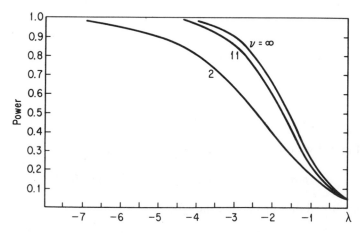

Fig. 8.2 Power Curve for the Left-Tailed One-Sided 5% Level t Test of $\mu = \mu_0$ When $\nu = 2$, 11 and ∞. $\lambda = (\mu - \mu_0)\sqrt{n}/\sigma$

This means that for a normal population with true mean 61 and true standard deviation of 4 we would fail to reject the null hypothesis $\mu = 65$ only six per cent of the time in repeated experiments using random samples of size twelve.

The term "fail to reject the null hypothesis" may seem strange to the reader. It might appear that when a sample value falls in the noncritical region the term "accept the null hypothesis" is more appropriate, since we "accept the alternative hypothesis" otherwise. Two points may be made in this connection. First, when we accept the alternative hypothesis, we know (or can find) the probability of making an error, but if we "accepted the null hypothesis" we would not know the probability of making an error without first knowing the true mean and true variance in Example 8.3, say. (But if we knew the true mean, we would have no need to test a hypothesis about the mean in the first place.) Second, the test procedure concerns the "rejection" of the null hypothesis. Thus, if one fails to reject the null hypothesis, this simply means that there is not enough evidence, as a result of the experiment, to say what the true value really is—it could be the value assumed in the null hypothesis or some other value. That is, the statistical test may lead to the conclusion "reserve judgment."

But anyone who performs experiments knows that, due to practical considerations, one cannot always reserve judgment when a sample value falls in the noncritical region. There are situations in which the statistical conclusion "fail to reject the null hypothesis" actually means that the experimenter "accepts the null hypothesis." That is, the experimenter concludes that the difference in the sample statistic and the hypothesized value of the parameter is small enough to be of no "practical significance." Thus, when an experimenter accepts the null hypothesis, he does so out of practical considerations rather than statistical considerations.

Tables 8.1 and 8.2 may be used to find power curves or operating characteristic curves for either left-tailed or right-tailed one-sided five per cent level or one per cent level tests. Curves for two-tailed tests may be obtained by using Eq.(8.14) and Ref. [11]. Also, these tables may be used to find the minimum sample size in an α level test which insures that β does not exceed a specified value for a particular value of λ. Table E on pp. 606–607 of *Design and Analysis of Industrial Experiments* by Davies, Hafner Publishing Co., may be used to find the sample size directly. This is illustrated in the following example.

Example 8.4. In the rat diet experiment (Example 8.1), suppose H_0: $\mu = 65$ grams and $H_a: \mu \neq 65$ grams. Further, suppose that it is known from past experience that $\sigma = 4$ grams and that a mean weight gain differing less than three grams from 65 grams does not appreciably effect future performance of the rats. What sample size should be taken if the probabilities of error of the first and second kind are not to exceed 0.05?

Letting δ denote the difference it is important to detect, we have

$\Delta = \delta/\sigma = \frac{3}{4} = 0.75$. Thus, for a two-sided test with $\alpha = \beta = 0.05$, we find that $n = 26$.

8.5. TOLERANCE LIMITS

Earlier we learned how to find two confidence limits which determine a confidence interval for a population parameter. That is, we learned how to find limits such that in repeated experiments of the same size a certain proportion of intervals determined by these two limits cover a population parameter. In some experiments it is not enough to find an interval which 100γ per cent of the time covers a parameter value. Instead, we require an interval which covers a *fixed portion of the population of values* with a specified confidence, say γ. Such intervals are called *tolerance intervals*, and their endpoints are called *tolerance limits*.

For example, in the rat diet experiment of Example 8.1 we may wish to find a 95 per cent tolerance interval for 90 per cent of the weights closest to the mean of a normal population of weights. It is obvious that the 95 per cent tolerance interval which covers 90 per cent of the values specified must be longer than the 95 per cent confidence interval which covers the single value μ. Since the 95 per cent confidence limits of μ may be expressed as

$$\bar{x} \pm t_{.025} \cdot s_{\bar{x}} = \bar{x} \pm k \cdot s$$

where

$$k = \frac{t_{.025}}{\sqrt{n}}$$

we determine tolerance limits from

$$\bar{x} \pm K \cdot s \qquad (8.15)$$

where K is determined so that the interval covers 90 per cent (P) of the population of values with confidence 95 per cent (γ). Values of K are given in Table 8.3. Thus, for $\gamma = 0.95$, $P = 0.90$, and $n = 12$, we find $K = 2.655$. Since $\bar{x} = 60.25$ and $s = \sqrt{16.38} = 4.05$, the 95 per cent tolerance limits are $60.25 \pm (2.655)(4.05)$, or 49.50 and 71.00. Clearly, the interval from 49.50 to 71.00 is considerably longer than the 95 per cent confidence interval given by Eq. (8.6.) Further, note that 100 per cent of the weights in the sample of Example 8.1 fall in the 95 per cent tolerance interval.

8.6. COMPARISON OF MEANS OF TWO SETS OF INDEPENDENT OBSERVATIONS

Problems often occur in practice which involve the comparison of two population means. If random samples $x_{11}, \ldots, x_{1n_1}$ and $x_{21}, \ldots, x_{2n_2}$

Table 8.3

Tolerance Factors for Normal Distributions*

n	$\gamma = 0.75$					$\gamma = 0.90$				
P	0.75	0.90	0.95	0.99	0.999	0.75	0.90	0.95	0.99	0.999
2	4.498	6.301	7.414	9.531	11.920	11.407	15.978	18.800	24.167	30.227
3	2.501	3.538	4.187	5.431	6.844	4.132	5.847	6.919	8.974	11.309
4	2.035	2.892	3.431	4.471	5.657	2.932	4.166	4.943	6.440	8.149
5	1.825	2.599	3.088	4.033	5.117	2.454	3.494	4.152	5.423	6.879
6	1.704	2.429	2.889	3.779	4.802	2.196	3.131	3.723	4.870	6.188
7	1.624	2.318	2.757	3.611	4.593	2.034	2.902	3.452	4.521	5.750
8	1.568	2.238	2.663	3.491	4.444	1.921	2.743	3.264	4.278	5.446
9	1.525	2.178	2.593	3.400	4.330	1.839	2.626	3.125	4.098	5.220
10	1.492	2.131	2.537	3.328	4.241	1.775	2.535	3.018	3.959	5.046
11	1.465	2.093	2.493	3.271	4.169	1.724	2.463	2.933	3.849	4.906
12	1.443	2.062	2.456	3.223	4.110	1.683	2.404	2.863	3.758	4.792
13	1.425	2.036	2.424	3.183	4.059	1.648	2.355	2.805	3.682	4.697
14	1.409	2.013	2.398	3.148	4.016	1.619	2.314	2.756	3.618	4.615
15	1.395	1.994	2.375	3.118	3.979	1.594	2.278	2.713	3.562	4.545
16	1.383	1.977	2.355	3.092	3.946	1.572	2.246	2.676	3.514	4.484
17	1.372	1.962	2.337	3.069	3.917	1.552	2.219	2.643	3.471	4.430
18	1.363	1.948	2.321	3.048	3.891	1.535	2.194	2.614	3.433	4.382
19	1.355	1.936	2.307	3.030	3.867	1.520	2.172	2.588	3.399	4.339
20	1.347	1.925	2.294	3.013	3.846	1.506	2.152	2.564	3.368	4.300
21	1.340	1.915	2.282	2.998	3.827	1.493	2.135	2.543	3.340	4.264
22	1.334	1.906	2.271	2.984	3.809	1.482	2.118	2.524	3.315	4.232
23	1.328	1.898	2.261	2.971	3.793	1.471	2.103	2.506	3.292	4.203
24	1.322	1.891	2.252	2.959	3.778	1.462	2.089	2.489	3.270	4.176
25	1.317	1.883	2.244	2.948	3.764	1.453	2.077	2.474	3.251	4.151
26	1.313	1.877	2.236	2.938	3.751	1.444	2.065	2.460	3.232	4.127
27	1.309	1.871	2.229	2.929	3.740	1.437	2.054	2.447	3.215	4.106
30	1.297	1.855	2.210	2.904	3.708	1.417	2.025	2.413	3.170	4.049
35	1.283	1.834	2.185	2.871	3.667	1.390	1.988	2.368	3.112	3.974
40	1.271	1.818	2.166	2.846	3.635	1.370	1.959	2.334	3.066	3.917
45	1.262	1.805	2.150	2.826	3.609	1.354	1.935	2.306	3.030	3.871
50	1.255	1.794	2.138	2.809	3.588	1.340	1.916	2.284	3.001	3.833
55	1.249	1.785	2.127	2.795	3.571	1.329	1.901	2.265	2.976	3.801
60	1.243	1.778	2.118	2.784	3.556	1.320	1.887	2.248	2.955	3.774
65	1.239	1.771	2.110	2.773	3.543	1.312	1.875	2.235	2.937	3.751
70	1.235	1.765	2.104	2.764	3.531	1.304	1.865	2.222	2.920	3.370
75	1.231	1.760	2.098	2.757	3.521	1.298	1.856	2.211	2.906	3.712
80	1.228	1.756	2.092	2.749	3.512	1.292	1.848	2.202	2.894	3.696
85	1.225	1.752	2.087	2.743	3.504	1.287	1.841	2.193	2.882	3.682
90	1.223	1.748	2.083	2.737	3.497	1.283	1.834	2.185	2.872	3.669
95	1.220	1.745	2.079	2.732	3.490	1.278	1.828	2.178	2.863	3.657
100	1.218	1.742	2.075	2.727	3.484	1.275	1.822	2.172	2.854	3.646
110	1.214	1.736	2.069	2.719	3.473	1.268	1.813	2.160	2.839	3.626
120	1.211	1.732	2.063	2.712	3.464	1.262	1.804	2.150	2.826	3.610
130	1.208	1.728	2.059	2.705	3.456	1.257	1.797	2.141	2.814	3.595
140	1.206	1.724	2.054	2.700	3.449	1.252	1.791	2.134	2.804	3.582
150	1.204	1.721	2.051	2.695	3.443	1.248	1.785	2.127	2.795	3.571
160	1.202	1.718	2.047	2.691	3.437	1.245	1.780	2.121	2.787	3.561
170	1.200	1.716	2.044	2.687	3.432	1.242	1.775	2.116	2.780	3.552
180	1.198	1.713	2.042	2.683	3.427	1.239	1.771	2.111	2.774	3.543
190	1.197	1.711	2.039	2.680	3.423	1.236	1.767	2.106	2.768	3.536
200	1.195	1.709	2.037	2.677	3.419	1.234	1.764	2.102	2.762	3.529
250	1.190	1.702	2.028	2.665	3.404	1.224	1.750	2.085	2.740	3.501
300	1.186	1.696	2.021	2.656	3.393	1.217	1.740	2.073	2.725	3.481
400	1.181	1.688	2.012	2.644	3.378	1.207	1.726	2.057	2.703	3.453
500	1.177	1.683	2.006	2.636	3.368	1.201	1.717	2.046	2.689	3.434
600	1.175	1.680	2.002	2.631	3.360	1.196	1.710	2.038	2.678	3.421
700	1.173	1.677	1.998	2.626	3.355	1.192	1.705	2.032	2.670	3.411
800	1.171	1.675	1.996	2.623	3.350	1.189	1.701	2.027	2.663	3.402
900	1.170	1.673	1.993	2.620	3.347	1.187	1.697	2.023	2.658	3.396
1000	1.169	1.671	1.992	2.617	3.344	1.185	1.695	2.019	2.654	3.390
∞	1.150	1.645	1.960	2.576	3.291	1.150	1.645	1.960	2.576	3.291

* Reproduced from C. Eisenhart, M. W. Hastay, and W. A. Wallis, *Techniques of Statistical Analysis*. New York: McGraw-Hill, Inc. Chap. 2, Table 2.1, pp. 102–107, with permission of the publisher.

Table 8.3

Tolerance Factors for Normal Distributions* (cont.)

n \ P	$\gamma = 0.95$					$\gamma = 0.99$				
	0.75	0.90	0.95	0.99	0.999	0.75	0.90	0.95	0.99	0.999
2	22.858	32.019	37.674	48.430	60.573	114.363	160.193	188.491	242.300	303.054
3	5.922	8.380	9.916	12.861	16.208	13.378	18.930	22.401	29.055	36.616
4	3.779	5.369	6.370	8.299	10.502	6.614	9.398	11.150	14.527	18.383
5	3.002	4.275	5.079	6.634	8.415	4.643	6.612	7.855	10.260	13.015
6	2.604	3.712	4.414	5.775	7.337	3.743	5.337	6.345	8.301	10.548
7	2.361	3.369	4.007	5.248	6.676	3.233	4.613	5.488	7.187	9.142
8	2.197	3.136	3.732	4.891	6.226	2.905	4.147	4.936	6.468	8.234
9	2.078	2.967	3.532	4.631	5.899	2.677	3.822	4.550	5.966	7.600
10	1.987	2.839	3.379	4.433	5.649	2.508	3.582	4.265	5.594	7.129
11	1.916	2.737	3.259	4.277	5.452	2.378	3.397	4.045	5.308	6.766
12	1.858	2.655	3.162	4.150	5.291	2.274	3.250	3.870	5.079	6.477
13	1.810	2.587	3.081	4.044	5.158	2.190	3.130	3.727	4.893	6.240
14	1.770	2.529	3.012	3.955	5.045	2.120	3.029	3.608	4.737	6.043
15	1.735	2.480	2.954	3.878	4.949	2.060	2.945	3.507	4.605	5.876
16	1.705	2.437	2.903	3.812	4.865	2.009	2.872	3.421	4.492	5.732
17	1.679	2.400	2.858	3.754	4.791	1.965	2.808	3.345	4.393	5.607
18	1.655	2.366	2.819	3.702	4.725	1.926	2.753	3.279	4.307	5.497
19	1.635	2.337	2.784	3.656	4.667	1.891	2.703	3.221	4.230	5.399
20	1.616	2.310	2.752	3.615	4.614	1.860	2.659	3.168	4.161	5.312
21	1.599	2.286	2.723	3.577	4.567	1.833	2.620	3.121	4.100	5.234
22	1.584	2.264	2.697	3.543	4.523	1.808	2.584	3.087	4.044	5.163
23	1.570	2.244	2.673	3.512	4.484	1.785	2.551	3.040	3.993	5.098
24	1.557	2.225	2.651	3.483	4.447	1.764	2.522	3.004	3.947	5.039
25	1.545	2.208	2.631	3.457	4.413	1.745	2.494	2.972	3.904	4.985
26	1.534	2.193	2.612	3.432	4.382	1.727	2.469	2.941	3.865	4.935
27	1.523	2.178	2.595	3.409	4.353	1.711	2.446	2.914	3.828	4.888
30	1.497	2.140	2.549	3.350	4.278	1.668	2.385	2.841	3.733	4.768
35	1.462	2.090	2.490	3.272	4.179	1.613	2.306	2.748	3.611	4.611
40	1.435	2.052	2.445	3.213	4.104	1.571	2.247	2.677	3.518	4.493
45	1.414	2.021	2.408	3.165	4.042	1.539	2.200	2.621	3.444	4.399
50	1.396	1.996	2.379	3.126	3.993	1.512	2.162	2.576	3.385	4.323
55	1.382	1.976	2.354	3.094	3.951	1.490	2.130	2.538	3.335	4.260
60	1.369	1.958	2.333	3.066	3.916	1.471	2.103	2.506	3.293	4.206
65	1.359	1.943	2.315	3.042	3.886	1.455	2.080	2.478	3.257	4.160
70	1.349	1.929	2.299	3.021	3.859	1.440	2.060	2.454	3.225	4.120
75	1.341	1.917	2.285	3.002	3.835	1.428	2.042	2.433	3.197	4.084
80	1.334	1.907	2.272	2.986	3.814	1.417	2.026	2.414	3.173	4.053
85	1.327	1.897	2.261	2.971	3.795	1.407	2.012	2.397	3.150	4.024
90	1.321	1.889	2.251	2.958	3.778	1.398	1.999	2.382	3.130	3.999
95	1.315	1.881	2.241	2.945	3.763	1.390	1.987	2.368	3.112	3.976
100	1.311	1.874	2.233	2.934	3.748	1.383	1.977	2.355	3.096	3.954
110	1.302	1.861	2.218	2.915	3.723	1.369	1.958	2.333	3.066	3.917
120	1.294	1.850	2.205	2.898	3.702	1.358	1.942	2.314	3.041	3.885
130	1.288	1.841	2.194	2.883	3.683	1.349	1.928	2.298	3.019	3.857
140	1.282	1.833	2.184	2.870	3.666	1.340	1.916	2.283	3.000	3.833
150	1.277	1.825	2.175	2.859	3.652	1.332	1.905	2.270	2.983	3.811
160	1.272	1.819	2.167	2.848	3.638	1.326	1.896	2.259	2.968	3.792
170	1.268	1.813	2.160	2.839	3.627	1.320	1.887	2.248	2.955	3.774
180	1.264	1.808	2.154	2.831	3.616	1.314	1.879	2.239	2.942	3.759
190	1.261	1.803	2.148	2.823	3.606	1.309	1.872	2.230	2.931	3.744
200	1.258	1.798	2.143	2.816	3.579	1.304	1.865	2.222	2.921	3.731
250	1.245	1.780	2.121	2.788	3.561	1.286	1.839	2.191	2.880	3.678
300	1.236	1.767	2.106	2.767	3.535	1.273	1.820	2.169	2.850	3.641
400	1.223	1.749	2.084	2.739	3.499	1.255	1.794	2.138	2.809	3.589
500	1.215	1.737	2.070	2.721	3.475	1.243	1.777	2.117	2.783	3.555
600	1.209	1.729	2.060	2.707	3.458	1.234	1.764	2.102	2.763	3.530
700	1.204	1.722	2.052	2.697	3.445	1.227	1.755	2.091	2.748	3.511
800	1.201	1.717	2.046	2.688	3.434	1.222	1.747	2.082	2.736	3.495
900	1.198	1.712	2.040	2.682	3.426	1.218	1.741	2.075	2.726	3.483
1000	1.195	1.709	2.036	2.676	3.418	1.214	1.736	2.068	2.718	3.472
∞	1.150	1.645	1.960	2.576	3.291	1.150	1.645	1.960	2.576	3.291

of sizes n_1 and n_2 are independently drawn from normal populations with means μ_1 and μ_2, respectively, and common unknown variances $\sigma_1^2 = \sigma_2^2 = \sigma^2$, then the t distribution with $n_1 + n_2 - 2$ degrees of freedom can be applied in obtaining confidence intervals for and testing hypotheses of the difference $\delta = \mu_1 - \mu_2$. If in Theorem 8.4 we let

$$\bar{d} = \bar{x}_1 - \bar{x}_2 \quad \text{and} \quad s_{\bar{d}} = \sqrt{s_p^2 \left(\frac{1}{n_1} + \frac{1}{n_2} \right)}$$

the t statistic in Eq. (8.2) becomes

$$t = \frac{\bar{d} - \delta}{s_{\bar{d}}} \tag{8.16}$$

where $\delta = \mu_1 - \mu_2$. From Eq. (8.16) we can obtain confidence intervals for the difference in means δ in exactly the same way as we did for the single mean μ with unknown variance by replacing $\bar{x}$ by $\bar{d}$, μ by δ, s^2 by s_p^2, $1/n$ by $(1/n_1 + 1/n_2)$, and the degrees of freedom $n - 1$ by $n_1 + n_2 - 2$. Thus, the symmetric $100(1 - \alpha)$ per cent confidence limits for δ are given by

$$\bar{d} \pm t_{\alpha/2}(n_1 + n_2 - 2) \cdot s_p \sqrt{\frac{1}{n_1} + \frac{1}{n_2}} \tag{8.17}$$

The test of the hypothesis $\delta = \delta_0$ against $\delta \neq \delta_0$ is similar to that given in Example 8.1. In order to bring out certain differences in the test procedures with one and two means we illustrate with a one-sided t test.

Example 8.5. The mean breaking strength of a product made under standard conditions is to be compared to that of the same product made under different (new) conditions. (The reader may think in terms of compressive strength of bricks made in a laboratory, the tensile strength of yarn made by A, or the resistance of a sheet of metal to a certain force.) Assume that random samples of size ten had the following (coded) means and variances: $\bar{x}_1 = 8.37$ lb, $\bar{x}_2 = 9.62$ lb, $s_1^2 = 1.32$, and $s_2^2 = 1.18$. The standard is denoted by the subscript 1. Is the mean breaking strength under new conditions greater than that under standard conditions?

Using the general procedure for testing hypotheses, we have:

1. $H_0: \mu_1 = \mu_2$ (or $\delta = \mu_1 - \mu_2 = 0$) and $H_a: \mu_1 < \mu_2$ (or $\delta = \mu_1 - \mu_2 < 0$), since we are interested in knowing if the new conditions lead to greater breaking strength.
2. Assume that samples are independently drawn from two normal populations with common unknown variance σ^2. Let the significance level be $\alpha = 0.01$, $n_1 = n_2 = 10$.
3. The test statistic is

$$t = \frac{(\bar{x}_1 - \bar{x}_2) - (\mu_1 - \mu_2)}{s_p\sqrt{\dfrac{1}{n_1} + \dfrac{1}{n_2}}} = \frac{\bar{x}_1 - \bar{x}_2}{s_p\sqrt{\dfrac{1}{5}}} = \frac{\bar{d}}{s_{\bar{d}}}$$

with 18 degrees of freedom.

4. Since the experimenter is interested in detecting increased breaking strength, that is, negative difference δ, we take as the critical region all those values of t for which $t < -2.55$.

5. Since $\bar{d} = \bar{x}_1 - \bar{x}_2 = 8.37 - 9.62 = -1.25$ lb and

$$s_{\bar{d}} = \sqrt{\frac{s_p^2}{5}} = \sqrt{\frac{1}{5} \frac{9(1.32) + 9(1.18)}{10 + 10 - 2}} = 0.5$$

the calculated t statistic is

$$t_c = \frac{-1.25}{0.5} = -2.5$$

6. The sample statistic t_c falls in the noncritical region. Therefore, we do not have enough evidence to say that the breaking strength under the new conditions is greater than that under standard conditions.

When the population variances are unequal and unknown, the methods just described for finding confidence intervals and testing hypotheses involving differences of two means are not appropriate. An approximate solution to the problem of unequal variances, known as the Behrens-Fisher problem, is described below and is given by Welch [12] and Aspin [2].

If we let $s_{\bar{x}_1}^2 = s_1^2/n_1$ and $s_{\bar{x}_2}^2 = s_2^2/n_2$, it can be shown that

$$t' = \frac{\bar{d} - \delta}{\sqrt{s_{\bar{x}_1}^2 + s_{\bar{x}_2}^2}} \qquad (8.18)$$

is approximately distributed as t with ν degrees of freedom, where

$$\nu = \frac{(s_{\bar{x}_1}^2 + s_{\bar{x}_2}^2)^2}{\dfrac{(s_{\bar{x}_1}^2)^2}{n_1 - 1} + \dfrac{(s_{\bar{x}_2}^2)^2}{n_2 - 1}} \qquad (8.19)$$

The number of degrees of freedom computed by Eq. (8.19) is likely not to be a whole number. In this case it should be rounded off to the nearest integer when used in determining confidence limits or critical regions. Even if the computed ν is an integer, the t' statistic is not distributed as any t statistic. In some cases it may be better to replace the t' statistic by some other, more appropriate, approximate statistic (as illustrated below).

The ratio in Eq. (8.18) can be used as the t ratio to find approximate confidence limits and to make approximate tests of hypotheses. For example, symmetric $100(1 - \alpha)$ per cent confidence limits are given by

$$\bar{d} \pm t_{\alpha/2}(\nu)\sqrt{s_{\bar{x}_1}^2 + s_{\bar{x}_2}^2} \tag{8.20}$$

where $t_{\alpha/2}(\nu)$ is the value of t obtained from Table VI for the integral degrees of freedom given by Eq. (8.19), and the null hypothesis $H_0 : \delta = \delta_0$ can be tested against the alternative hypothesis $H_a : \delta \neq \delta_0$ by rejecting H_0 if the computed statistic t'_c is greater than $|t_{\alpha/2}(\nu)|$. Actually, exact values of the five per cent and one per cent points of t' are given by Aspin [3], but the approximate corresponding t values are generally satisfactory where the t' statistic applies.

If the population variances are unknown, an experimenter seldom knows whether they are equal or unequal. Thus, the methods described for equal variances are usually applied without knowing for certain that the variances are equal. However, this is not necessarily bad, for Box [4] found that no serious consequence will result if the population variances are only moderately different and the two sample sizes are equal. This means that the methods described for equal variances may be used as approximations in place of the approximate t' in some cases where t' is considered appropriate.

It is informative to note how Theorem 8.4, which involves two means, is related to earlier theorems. We may write Eq. (8.3) as

$$s_p^2 = \frac{(n_1 - 1)s_1^2 + (n_2 - 1)s_2^2}{n_1 + n_2 - 2}$$

Hence, from Eq. (7.8) we see that

$$\frac{(n_1 + n_2 - 2)s_p^2}{\sigma^2} \tag{8.21}$$

is distributed as χ^2 with $n_1 + n_2 - 2$ degrees of freedom. Further, from Theorem 6.6, we know that $\bar{d} = \bar{x}_1 - \bar{x}_2$ is normally distributed with mean $\delta = \mu_1 - \mu_2$ and variance

$$\sigma_{\bar{d}}^2 = \frac{\sigma^2}{n_1} + \frac{\sigma^2}{n_2}$$

Thus

$$\frac{\bar{x}_1 - \bar{x}_2 - (\mu_1 - \mu_2)}{\sqrt{\sigma^2\left(\dfrac{1}{n_1} + \dfrac{1}{n_2}\right)}} = \frac{\bar{d} - \delta}{\sigma_{\bar{d}}} \tag{8.22}$$

is normally distributed with zero mean and unit variance. According to Theorem 7.3, $\bar{x}_1$ and s_1^2 as well as $\bar{x}_2$ and s_2^2 are independent random variables. Thus, it follows that $\bar{d}$ and s_p^2 are independent random variables, and that

$$\frac{\bar{d} - \delta}{\sigma_{\bar{d}}} \quad \text{and} \quad \frac{(n_1 + n_2 - 2)s_p^2}{\sigma^2} \tag{8.23}$$

are also independent random variables. Hence, according to Theorem 8.1, the ratio

$$\frac{\dfrac{\bar{d} - \delta}{\sigma_{\bar{d}}}}{\sqrt{\dfrac{\left[\dfrac{(n_1 + n_2 - 2)s_p^2}{\sigma^2}\right]}{n_1 + n_2 - 2}}} = \frac{\bar{d} - \delta}{s_p \sqrt{\dfrac{1}{n_1} + \dfrac{1}{n_2}}} \tag{8.24}$$

is distributed as the Student t with $n_1 + n_2 - 2$ degrees of freedom.

If there is evidence that two population variances σ_1^2 and σ_2^2 are not equal, it is possible in some cases, by pairing observations, to use exact methods to examine the difference in the population means. This very important but special case is described in the next section.

8.7. PAIRED OBSERVATIONS

Due to extraneous causes, it may happen that two means are declared different when there is actually no difference in the means. The reverse may also occur; that is, one may fail to recognize real differences because of the appearance of factors other than those of interest. For example, we may wish to compare two methods of determining starch content of potatoes. If several potatoes have very different starch content, and Method 1 is used on potatoes with low starch content, whereas Method 2 is used on potatoes with high starch content, we could conclude that Method 2 is superior to Method 1 when this is not the case. In another example, we may wish to compare two analysts in their ability to measure the percentage of ammonia in a gas used in a certain manufacturing process. Unless the analysts measure the percentage of ammonia at the same time and place on the same number of days, any comparison of the abilities of the analysts is likely to be meaningless.

In each of these examples, it is clear that paired observations should be made. In the starch experiment each potato should be cut into two parts, forming a pair. The decision as to which method to apply to one part of each pair could be made by tossing a coin. In the ammonia experiment the analysts should take samples at the same time and in roughly the same place, the exact place being determined by the toss of a coin, say. In each case we try to make sure that two members of a pair are alike in all respects. Then when the observations resulting from the experiment are made, we may ascribe any difference, except for random variation, to the factor we are trying to measure.

Let x_{1j} and x_{2j} denote the observed values of the jth pair of a set of n paired values. Assume that the difference $d_j = x_{1j} - x_{2j}$ represents a sample of n random observations from a normal population with mean δ and variance σ_d^2. Denote the sample mean and variance by $\bar{d}$ and s_d^2, respectively. If there are extraneous factors affecting the jth pair, we assume that they affect both observations of the pair in exactly the same way and that subtraction removes the effect. With these assumptions it follows that

$$\frac{\bar{d} - \delta}{\dfrac{s_d}{\sqrt{n}}} \tag{8.25}$$

is distributed as t with $n - 1$ degrees of freedom. An illustration of the test procedure for the hypothesis $\delta = 0$ is given in the following example.

Example 8.6. Two methods of measuring the percentage of starch in potatoes are to be compared; the data given by C. von Scheele, G. Svensson, and J. Rasmusson in "Om Bestämning av Potatisens Stärkelse och Torrsubstanshalt med Tillhjälp av dess specifika Vikt," *Nordisk Jordbrugsforskning*, 1935, p. 22, are to be used. Sixteen potatoes with very different starch content were taken, and the two methods of measurement were applied to each potato.

Using the general procedure for testing hypotheses, we have:

1. H_0: $\delta = 0$ against H_a: $\delta \neq 0$ is used, since we wish to know if the two methods are different. Note that the mean difference δ is also the difference in the means, μ_1 and μ_2, of the two populations of measurements, i.e., $\delta = \mu_1 - \mu_2$.

2. Assume that the samples of differences are randomly drawn from a random population with mean $\delta = 0$ and unknown variance σ_d^2. Let the significance level be $\alpha = 0.05$.

3. The test statistic is

$$t = \frac{\bar{d} - \delta}{\dfrac{s_d}{\sqrt{n}}} = \frac{\bar{d}\sqrt{16}}{s_d}$$

with 15 degrees of freedom.

4. The critical region is made up of all those values of t for which $t < -2.13$ or $t > 2.13$.

5. The data in the percentage of starch experiment are given in Table 8.4, along with the differences $d_j = x_{1j} - x_{2j}$. Since $\sum d_i = 1.2$ and $\sum d_i^2 = 0.52$, $\bar{d} = 0.075$, $s_d = 0.17$, and the computed t value is

$$t_c = \frac{(0.075)4}{0.17} = 1.7$$

6. But t_c fails to fall in the critical region. Thus, we fail to reject the null hypothesis that there is no difference in the two methods of measuring the percentage of starch in potatoes.

Table 8.4

Potato Number	Percentage Starch		Difference (d)
	Method 1 (x_1)	Method 2 (x_2)	
1	21.7	21.5	0.2
2	18.7	18.7	0.0
3	18.3	18.3	0.0
4	17.5	17.4	0.1
5	18.5	18.3	0.2
6	15.6	15.4	0.2
7	17.0	16.7	0.3
8	16.6	16.9	−0.3
9	14.0	13.9	0.1
10	17.2	17.0	0.2
11	21.7	21.4	0.3
12	18.6	18.6	0.0
13	17.9	18.0	−0.1
14	17.7	17.6	0.1
15	18.3	18.5	−0.2
16	15.6	15.5	0.1

The paired-comparisons test is often used when one is comparing two measurements on the same individual or object. Thus, the experimental design consists in taking n individuals, making a measurement on each one, and then after some treatment making a second measurement in the same unit as the first. The point is to determine if any real difference, on the average, occurs as a result of the treatment. Thus, the difference in measurements for each individual is obtained, and these differences constitute a sample which is used to test the hypothesis that there is no real difference. Clearly, the test is designed to measure any effect the treatment might have. There are often obvious differences from individual to individual which do not affect the test.

It should be observed that the set of differences resulting from pairing may be treated in the same way as a set of observations in Sects. 8.2, 8.3, and 8.4. That is, confidence limits, tests of hypotheses, size of the type 2 error, power, and size of sample may be established in exactly the same way, since the assumptions concerning the d's are the same as those about the x's.

However, it is not always possible to think only in terms of the sample differences. For example, when testing $\mu_1 = \mu_2$ (or $\delta = \mu_1 - \mu_2 = 0$) at the α level we may wish to compare the paired t test with the two independent samples t test given in Sect. 8.6. We have already noted the most important advantage of the paired t test in case extraneous factors exist—namely, that the test for paired observations is based only on the variation in the differences, so that other variations which have exactly the same effect on both members of a pair do not affect the measure of sample standard deviation $s_{\bar{x}_1 - \bar{x}_2}$ as in the two independent samples t test.

In order to make further comparisons of the two test procedures, the usual assumptions are that x_{1j} and $x_{2j}(j = 1, 2, \ldots, n)$ are observations from normal populations with means μ_1 and μ_2 and variances σ_1^2 and σ_2^2, respectively. For the paired t test we do not assume that x_{1j} and x_{2j} are independent nor that the variances are equal, as in the case for the two independent samples t test. Actually, we do not need to assume that the variances σ_1^2 and σ_2^2 remain constant throughout the experiment. All we need assume is that the sum $\sigma_1^2 + \sigma_2^2$ remains constant for the n paired observations.

On the other hand, if there are no extraneous factors in an experiment, the two independent samples t test is to be preferred, since it is more powerful than the paired t test. This is so because the number of degrees of freedom, $2n - 2$, for the independent samples test is twice as great as the number of degrees of freedom, $n - 1$, for the paired t test. The difference in power (or probability of the type 2 error) is considerable for small n, but is small when the sample size is moderately large, say $n = 12$.

8.8. THE SIGN TEST

In light of the discussion in Sect. 2.5 concerning rounding-off errors, the reader might well be disturbed by the appearance of only one significant figure in the column of differences in Example 8.6. This practice should be avoided if possible. However, there are other, less powerful tests which can be applied in cases where there is only one significant figure and the experimenter has some reservations about using the appropriate paired t test because relative errors in magnitude can be quite large. These tests do not depend on the size of numerical values of the differences, but only on their sign or rank order. Further, the assumption of normality is not required in order that these new tests be valid. We illustrate how the *sign test* of the hypothesis $\delta = 0$ may be applied to the data of Example 8.6. This test does not even require that the variance $\sigma_1^2 + \sigma_2^2$ remain constant from pair to pair and is one of the simplest of all tests to apply.

Example 8.7. Use the sign test and the data of Example 8.6 to test the hypothesis that the two methods of measuring the percentage of starch in potatoes are the same.

The total number of minus signs is $k_c = 3$, and the total number of plus and minus signs is $n = 13$ (if we ignore the three zeros). For the five per cent two-sided test we need two or fewer minus signs, according to Table 8.5, before we can reject the hypothesis $\delta = 0$. Since $k_c = 3$ is greater than $k = 2$, we fail to reject the hypothesis $\delta = 0$ and reach the same conclusion as in Example 8.6.

Table 8.5

Critical Values of k for the Sign Test

[Table gives largest integral values of k such that $P[x \le k] < \alpha/2$, where x has the binomial distribution with $p = \frac{1}{2}$.]

n \ α	0.01	0.02	0.05	0.10	0.25
3	—	—	—	—	0
4	—	—	—	—	0
5	—	—	—	0	0
6	—	—	0	0	1
7	—	0	0	0	1
8	0	0	0	1	1
9	0	0	1	1	2
10	0	0	1	1	2
11	0	1	1	2	3
12	1	1	2	2	3
13	1	1	2	3	3
14	1	2	2	3	4
15	2	2	3	3	4
16	2	3	3	4	5
17	2	3	4	4	5
18	3	3	4	5	6
19	3	4	4	5	6
20	3	4	5	5	6
21	4	4	5	6	7
22	4	5	5	6	7
23	4	5	6	7	8
24	5	5	6	7	8
25	5	6	7	7	9
30	7	8	9	10	11
35	9	10	11	12	13
40	11	12	13	14	15
45	13	14	15	16	18
50	15	16	17	18	20

Let us take a closer look at the sign test as it applies generally and as it applies in particular to Example 8.7. In the example, if the method used makes no difference in the determination of the percentage of starch, then we expect roughly half of the differences to be negative. If a small proportion of the differences are either positive or negative, we suspect that there might be a real difference in the methods of determination of starch. We wish to determine just how small the proportion should be before we say the methods are different. Since this proportion depends on the sample size, we find the *number* of positive or negative signs, whichever is smaller, necessary for the rejection of the hypothesis $\delta = 0$ at the α level.

In general, if we assume any pair of values (x_{1j}, x_{2j}) $(j = 1, \ldots, n)$ to be randomly drawn from the same distribution, we expect $d_j = x_{1j} - x_{2j}$ to be positive half the time and to be negative half the time in repeated samples. That is, the null hypothesis is that the difference d_j has a distribution with median zero or, which is the same thing, true proportion of positive (or negative) signs equal to $p = \frac{1}{2}$. If we think only in terms of the signs, this means that the $+$ and $-$ signs have a dichotomous distribution with $p = \frac{1}{2}$. Thus, regardless of the nature of the distribution from which the jth pair is drawn, we expect the $+$ and $-$ signs to have a dichotomous distribution with $p = \frac{1}{2}$. Hence, in n independent trials in which a positive or negative sign for each pair is determined and for which the probability of a positive (negative) sign on each trial is $p = \frac{1}{2}$, the probability of x positive (negative) signs is given by the binomial density function

$$b(x) = b\left(x; n, p = \frac{1}{2}\right) = \frac{n!}{x! \, (n - x)!} \left(\frac{1}{2}\right)^n \qquad (8.26)$$

Table 8.5 gives the critical value k, an integer, such that

$$P[x \leq k] = \sum_{x=0}^{k} b(x) < \frac{\alpha}{2}$$

Thus, in testing the null hypothesis $p = \frac{1}{2}$ against $p \neq \frac{1}{2}$, the significance level of the test is actually less than the level indicated in Table 8.5. This is due to the fact that the binomial is a discrete distribution. Note that the density function $f_j(x)$ from which the pair of observations x_{1j} and x_{2j} is drawn is generally continuous.

The hypotheses $p = \frac{1}{2}$ and $\delta = \mu_1 - \mu_2 = 0$ are equivalent, and the alternative hypothesis $\delta \neq 0$ is equivalent to $p \neq \frac{1}{2}$. For a one-sided test the alternative hypothesis is $p < \frac{1}{2}$ $(p > \frac{1}{2})$ or its equivalent $\delta < 0$ (or $\delta > 0$). In this case, for the significance level α we enter Table 8.5 in the column headed by 2α, since only half of the probability indicated is in the lower tail (upper tail) of the binomial distribution.

8.9. COMPARISON OF THE t TEST AND THE SIGN TEST

The conclusions resulting from the t test and the sign test of the null hypothesis $\delta = 0$ against the alternative hypothesis H_a are not always the same. The t test is to be preferred to the sign test, provided all assumptions for the paired t test hold, since it is more powerful. However, some of the assumptions for the paired t test may not be valid. In this case, the sign test can generally be applied, since all one really need assume is that each pair of values be drawn randomly from the same population (populations may differ from pair to pair).

It is informative to compare the power of the sign test and the paired t test of the null hypothesis $\delta = 0$ against the alternative hypothesis $\delta \neq 0$ under assumptions which make both tests valid. We make the comparison in terms of the so-called *power efficiency*, $100n_t/n$, of the sign test relative to the t test. That is, under the assumption that random paired values are drawn from normal populations, the power efficiency in percentage is the ratio $100n_t/n$, where n_t is the sample size for a paired t test which gives the same power as a sign test based on a sample of size n. The power efficiency for the sign test decreases (1) with increasing sample size, (2) with increasing $|\delta|$, and (3) with increasing α. If random pairs are drawn from two normal populations with means μ_1 and μ_2 and common variance σ^2, the power efficiency of the sign test is given in Table 8.6, where $\Delta = |\delta|/(\sqrt{2}\sigma)$, and $\sqrt{2}\sigma$ is the standard deviation of a difference of two observations. Thus, according to Table 8.6, if the true means are ten and four and the variance

Table 8.6
Power Efficiency of Sign Test Relative to t Test for Normal Populations*

n	α	Δ				
		Near 0	.5	1.0	1.5	2.0
5	.0625	96	96	95	93	91
10	.0020	94	92	90	87	84
10	.0215	85	84	82	80	77
10	.1094	77	76	74	72	
20	.0118	76	75	73	70	
20	.0414	73	72	70	68	
20	.1153	70	69	67	65	
∞	α	$100(2/\pi) \doteq 63.7$				

* Reproduced from W. J. Dixon and F. J. Massey, *Introduction to Statistical Analysis*, 2nd ed. New York: McGraw-Hill, Inc., p. 285, Table 17–3, with permission of the publisher.

is eight, then for a 2.15 per cent level test with $n = 10$ we find

$$\Delta = \frac{10 - 4}{\sqrt{2} \cdot \sqrt{8}} = 1.5$$

and $100n_t/10 = 80$ or $n_t = 8$. That is, under the conditions indicated, a sign test with ten random differences has about the same power as a paired t test with eight random differences. Further, if n is large and Δ is not large, the t test requires approximately 64 per cent as many observations as the sign test in order to have the same power.

The above statements are based on tests in which no differences are zero (i.e., no ties in pairs). Actually, when data are drawn from continuous distributions, ties should not occur. However, this is not the case in practical work, due to limitations of measuring instruments and rounding off. In such cases, ties can either not be counted, thus decreasing the sample size, or can be counted as half plus or half minus without seriously affecting the significance test, provided the proportion of ties is not too large. In Example 8.7 ties were not counted.

The sign test can be modified and extended to include many other problems. We shall return to its use in relation to other nonparametric tests in Chap. 16. Some of the tests of Chap. 16 which take into account the rank order of the observations might also be applied in place of the paired t test and sign test.

8.10. EXERCISES

8.1. On the examination in a certain course 16 students had a mean grade of 79 and a standard deviation of 8. (a) Use a five per cent level test to determine whether there is reason to believe that the true mean is greater than 75. (b) Find a 95 per cent confidence interval for the true mean.

8.2. The bacteria content of a food product must be less than 62.0 to be acceptable. A sample of nine cans from a lot of the product has a mean of 62.5 and a standard deviation of 0.3. (a) Should the lot be rejected on the basis of the sample evidence? Use a five per cent level test. (b) Find a 90 per cent confidence interval for the true mean.

8.3. (a) Use a five per cent level two-sided test to determine if the random sample with measurements
$$55 \quad 42 \quad 52 \quad 61 \quad 76 \quad 50 \quad 56 \quad 56 \quad 38 \quad 71$$
could have been taken from a normal population with mean 50. (b) Find a 95 per cent confidence interval for the true mean. (c) Find the power function for the five per cent level one-sided test of the null hypothesis $H_0: \mu = 50$ against the alternative hypothesis $H_a: \mu > 50$, when ten random observations are taken from a normal population.

8.4. (a) A lot of rolls of paper is acceptable for making bags for grocery stores if its mean breaking strength is not less than 40 lb. A random sample of 20 pieces of paper from the lot had a mean breaking strength of 39 lb with a standard deviation of 2.4 lb. Should the lot be rejected if $\alpha = 0.1$? (b) Draw the power curve for the test in (a). What is the chance of nonrejection of a lot which has true mean breaking strength of 39 lb? (c) Draw on the same graph with (b) the power curve for the test in (a) when the sample size is ten rather than 20. Use these two curves to make a statement about the effect of the sample size on the test if $\beta \leq 0.05$. (d) Suppose that the specification that the true mean breaking strength be 40 lb allows the true mean breaking strength to be as small as 39.6 lb without rejection of the lot. For the one per cent level test that the mean breaking strength not be smaller than 40 lb, determine the sample size so that for a sample mean of $\bar{x} = 39.6$ the probability of the type 2 error is at most 6.10.

8.5. (a) Find the density function for the t distribution when $\nu = 2$. (b) Graph the curve for the distribution in (a). (c) Prove that the mean is zero for the distribution in (a).

8.6. Prove Eq. (8.14).

8.7. Find a 90 per cent tolerance interval for 75 per cent of the grades closest to the mean in Exercise 8.1.

8.8. Find a 99 per cent tolerance interval for 95 per cent of the pieces of paper closest to the mean in Exercise 8.4.

8.9. Random samples are drawn from two normal populations with the same variance. Twenty observations in sample one have mean $\bar{x}_1 = 46$ and variance $s_1^2 = 120$. Eighteen observations in sample two have mean $\bar{x}_2 = 39$ and variance $s_2^2 = 180$. (a) Is there a significant difference between the two sample means? Use a five per cent level test. (b) Find a 90 per cent confidence interval for the difference.

8.10. Two manufacturers, A and B, make the same gauge of copper wire. The measurements of tensile strength of random samples, after 5000 lb has been subtracted from each, are given in Table 8.7

Table 8.7

A	110	90	120	115	105	50	75	85
B	130	50	40	45	65	120	50	

Find a 95 per cent confidence interval for the difference in true means.

8.11. Ten albino rats are used to study the effectiveness of carbon tetrachloride as an antihelminthic. Each rat received an injection of 500 *Nippostrongylus muris* larvae. After eight days the rats were divided into two groups and each rat received via a stomach tube a dose of carbon tetrachloride dissolved in mineral oil. Each rat in one group received a dose of 0.032 cc and those in the other group a dose of 0.063 cc. Two days later the rats were killed, and the adult worms were recovered and counted. Table 8.8 shows the number of adults recovered from each rat

Table 8.8

0.032 cc	421	462	400	378	413
0.063 cc	207	17	412	74	116

Use Eq. (8.18) to find a 95 per cent confidence interval for the difference in effectiveness between the two doses. [Source: Whitlock and Bliss, "A Bioassay Technique for Antihelminthics," *The Journal of Parasitology*, Vol. **29** (1943), pp. 48–58.]

8.12. If a machined part of a certain sort is accurate to within ± 0.1 in. of specification, it can be used. Deviations from specification of a random sample of 12 such parts were as follows: -0.03, -0.01, $+0.02$, -0.01, $+0.06$, $+0.04$, -0.05, $+0.03$, $+0.02$, -0.06, -0.02, $+0.01$. What proportion of the population sampled can one be 90 per cent confident of being between -0.1 and $+0.1$ of specification?

8.13. Thirty-six boys in the same class in high school were divided into 18 pairs of almost equal I. Q. One member of each pair was randomly selected and assigned to group G_1. The remaining 18 members were assigned to group G_2. Both groups were taught mathematics by the same instructor, but different methods of instruction were used. At the end of the semester all students were given the same examination with the following resulting grades

Table 8.9

Pair No.	1	2	3	4	5	6	7	8	9
Method 1	78	59	56	94	84	81	66	78	59
Method 2	74	71	52	68	68	85	79	70	64

Pair No.	10	11	12	13	14	15	16	17	18
Method 1	56	88	88	75	75	72	81	84	78
Method 2	39	77	83	62	74	74	83	73	70

(a) Test the hypothesis that both methods are equally suited to the instructor. Use a five per cent level t test. (b) Test the hypothesis in (a), using the sign test. (c) Find a 90 per cent confidence interval for the difference in means of the two methods.

8.14. Each of 16 samples of a material is divided into two equal parts. A standard analysis, A_1, is applied to one half of each sample, and a new analysis, A_2, is applied to the other half in order to determine the percentage of a certain mineral. The percentages are

Table 8.10

Sample No.	1	2	3	4	5	6	7	8
Analysis A_1	24.45	31.52	34.04	24.30	26.48	23.95	27.63	25.11
Analysis A_2	24.40	31.57	34.08	24.32	26.38	23.93	27.63	25.20

Sample No.	9	10	11	12	13	14	15	16
Analysis A_1	27.90	22.20	27.62	24.44	28.22	32.52	26.12	22.83
Analysis A_2	27.99	22.23	27.66	24.47	28.31	32.45	26.27	22.83

(a) Test at the five per cent level to determine whether the new method of analysis gives higher percentages than the standard method. Use both the t test and the sign test. (b) How can the type 1 error be made in this experiment? How can the type 2 error be made? What are the consequences of each error? (c) Find a 95 per cent confidence interval for the difference in means of the two analyses. (d) Discuss the power efficiency of the sign test relative to the t test for the conditions given in (a).

8.15. Prove Eq. (8.1) in Theorem 8.1.

Hint. Since u and w are independently distributed, the joint density function of u and w is obtained by multiplying the density function of u by the density function of w. Then, using the relation $t = u/\sqrt{w/\nu}$, find the joint density function of t and w. By the methods of Sect. 5.4 and Exercise 5.33, the marginal density function of t is found. This is the density function given in Eq. (8.1)

8.16. Prove Theorem 8.3.
Hint. Use Stirling's approximation (see Exercise 3.67).

8.17. Prove Theorem 8.4.
Hint. Use theorems of Chaps. 6 and 7 to reduce Eq. (8.2) to the form $u/\sqrt{w/\nu}$, where u and w are difined as in Theorem 8.1.

8.18. Show that the variance of the t distribution with ν degrees of freedom is $\nu/(\nu - 2)$ when $\nu > 2$.

8.19. In preparing control charts of means $\bar{x}_i$, it often happens that the true mean μ and variance σ^2 are not known. In such a case μ and σ^2 are replaced by the unbiased estimates $\bar{x} = \sum n_i \bar{x}_i / \sum n_i$ and s_p^2 (see Sect. 7.4), obtained from the sample means and variances, and the chart is prepared the usual manner.

For random samples of size five drawn each hour for 24 consecutive hours the means and variances are as follows

Table 8.11

Sample	Mean ($\bar{x}_i$)	Variance (s_i^2)	Sample	Means ($\bar{x}_i$)	Variance (s_i^2)
1	11.03	0.589	13	11.64	2.093
2	11.17	2.251	14	11.16	4.017
3	11.10	1.347	15	11.36	1.556
4	11.16	2.231	16	12.19	1.585
5	10.03	0.419	17	12.00	0.330
6	10.71	0.392	18	11.10	1.623
7	11.30	0.873	19	11.24	2.018
8	10.93	2.041	20	11.14	3.173
9	11.17	0.927	21	10.89	1.322
10	10.63	1.239	22	11.23	0.766
11	10.71	1.249	23	10.43	3.345
12	10.93	2.222	24	10.87	2.898

(a) Find $\bar{x}$ and s_p. Draw the upper and lower control limits for sample means so that they are three standard deviations from the center line. (b) Plot the 24 sample means. Does the process appear to be out of control at any point?

8.20. The control limits about the mean (center line) differ with sample size. The mean, variance, and standard deviation for random samples in 15 consecutive weeks are shown in the table below. Find $\bar{x}$ and s_p. Draw the center line and the upper and lower control limits for each sample and plot the points (sample means). Does the process appear to be out of control at any point?

Table 8.12

Sample Number	Sample Size	Mean $\bar{x}_i$	Variance s_i^2	Standard Deviation(s_i)
1	11	11.03	0.589	0.767
2	7	11.17	2.251	1.500
3	11	11.10	1.347	1.161
4	8	11.16	2.231	1.494
5	5	10.03	0.419	0.647
6	6	10.71	0.392	0.626
7	12	11.30	0.873	0.934
8	9	10.93	2.041	1.428
9	10	11.17	0.927	0.963
10	8	10.63	1.239	1.113
11	6	10.71	1.249	1.118
12	12	10.93	2.222	1.491
13	10	11.64	2.093	1.446
14	9	11.16	4.017	2.004
15	7	11.36	1.556	1.247

8.11. MODEL EQUATIONS AND OBSERVATIONAL EQUATIONS

In much of statistics we are concerned with the mean and variance or standard deviation for both the population and sample. (This is especially true when the population or populations are normal.) Since extensions and generalizations of concepts already introduced are of considerable importance in later sections, we now take a closer look at assumptions and notations useful for this purpose.

The reader, no doubt, is already aware of the fact that each observation is considered to be the sum of two parts, namely, the mean and the deviate from the mean. It is the component parts of an observation which we wish to discuss in this section. We restrict our attention to cases where the components of an observation can be added.

To start with, observe that when we say that a random variable x is distributed with mean μ and variance σ^2, it is understood that x is the sum of μ and the deviate of x, where the deviate may be zero, positive, or negative. Letting x_j denote the jth observation in a sample of n objects and ϵ_j its deviate from the mean, we write

$$x_j = \mu + \epsilon_j \qquad (j = 1, 2, \ldots, n) \tag{8.27}$$

If the population is continuous, n represents a very small part of all possible observations. Even when the population is finite and of size N, the sample is likely to represent only a small portion of the population; otherwise, we might just as well study all observations in the population in order to make an exact statement about the population parameters and not take the chances involved in making inference statements based on a sample. In any case, it is generally reasonable to assume that the sample mean

$$\bar{x} = \frac{\sum\limits_{j=1}^{n} x_j}{n}$$

is not equal to μ for most samples, and that $\bar{x}$ has a distribution of values with mean $\mu_x = \mu$. Further, since $\bar{x}$ is not always equal to μ, and since μ is seldom known, we may wish to think of an observation as the sum of $\bar{x}$ and the deviate of x from $\bar{x}$. That is, if e_j denotes the amount x_j deviates from $\bar{x}$, the sample mean, we may write

$$x_j = \bar{x} + e_j \tag{8.28}$$

where $\bar{x}$ estimates μ and e_j estimates ϵ_j. Note that the estimator $\bar{x}$ of μ is based on n observations, but the estimator e_j of ϵ_j is based on only one observation. However, we are not particularly interested in e_j as an estimator or ϵ_j, but rather in using all e_j's to find an estimator of the variance or standard deviation. We know already that

$$\frac{\sum\limits_{j=1}^{n} e_j^2}{n-1}$$

is used to estimate σ^2 and

$$\sqrt{\frac{\sum\limits_{j=1}^{n} e_j^2}{n-1}}$$

to estimate σ. Equation (8.27) is called a *model equation* and Eq. (8.28) an *observation equation*. Each equation indicates that an observation is the sum of two components—in Eq. (8.27) the components are the *true mean* μ and the *true deviate* or *true random error* or *true error effect* ϵ_j of the jth observation, and in Eq. (8.28) the components are the *estimated mean* or *sample mean* $\bar{x}$ and the *estimated deviate* or *residual* or *error*.

There is another point of considerable interest. In Eq. (8.27) μ is constant, and ϵ_j is a random variable. Also, x_j is a random variable. However, in Eq. (8.28) the random variable x_j is the sum of two random variables $\bar{x}$ and e_j. For a particular sample, $\bar{x}$ has only one value, but from sample to sample $\bar{x}$ takes other values. In other words, it is possible for $\bar{x}$ to take any one of many values before a particular set of n observations are drawn. Since μ is not usually known, we use Eq. (8.28) in applications; Eq. (8.27) serves as a *model* (or ideal).

Next, we extend the above notations to two populations. Let x_1 be distributed with mean μ_1 and variance σ_1^2, and let x_2 be distributed with mean μ_2 and variance σ_2^2. Let x_{1j} denote the jth observation of a sample of size n_1 drawn from the first population, and x_{2j} the jth observation of a sample of size n_2 drawn from the second population. Then the model equations may be written as

$$x_{1j} = \mu_1 + \epsilon_{1j} \qquad (j = 1, 2, \ldots, n_1)$$

and

$$x_{2j} = \mu_2 + \epsilon_{2j} \qquad (j = 1, 2, \ldots, n_2)$$

These two equations may be written as

$$x_{ij} = \mu_i + \epsilon_{ij} \qquad (i = 1, 2; j = 1, \ldots, n_i) \tag{8.29}$$

where ϵ_{ij} denotes the amount x_{ij} deviates from μ_i. Further, the observation equation may be written as

$$x_{ij} = \bar{x}_i + e_{ij} \qquad (i = 1, 2; j = 1, 2, \ldots, n_i) \tag{8.30}$$

where

$$\bar{x}_i = \frac{\sum\limits_{j=1}^{n_i} x_{ij}}{n_i} \qquad (i = 1, 2) \tag{8.31}$$

and e_{ij} denotes the amount x_{ij} deviates from the sample mean $\bar{x}_i$. If k populations are to be considered, and x_i is distributed with mean μ_i and variance σ_i^2 ($i = 1, 2, \ldots, k$), the last three equations, Eqs. (8.29), (8.30), and (8.31), remain the same, except that $i = 1, 2, \ldots, k$ and n_i denotes the size of the ith sample.

When two or more population means are compared, it is often most convenient to think in terms of the amount any given mean μ_i deviates from an over-all population mean μ. For example, when comparing the mean verbal college board scores of all freshman at two or more schools, we might wish to know how much the mean score of college A, say, differs from the mean of k schools in a given state. In another example, we may wish to know how much each of k like machines in factory B on the average differs from a production standard in its ability to do a specific piece of work. (More examples and a fuller treatment of this topic are given in the chapters on analysis of variance.) In order to give a detailed comparison of the concepts about to be introduced with those discussed in Sect. 8.6, we consider the case where $k = 2$.

Let two populations P_1 and P_2 have means μ_1 and μ_2, respectively, and let

$$\mu = \frac{\mu_1 + \mu_2}{2} \tag{8.32}$$

be the *over-all mean* of the two populations combined. Letting α_i denote the amount the population mean μ_i deviates from the mean of the two population means, we may write

$$\mu_i = \mu + \alpha_i \qquad (i = 1, 2) \tag{8.33}$$

We call α_i the *true effect* of the ith population and note that

$$\sum_{i=1}^{2} \alpha_i = 0 \tag{8.34}$$

Thus, substituting Eq. (8.33) in Eq. (8.29) gives

$$x_{ij} = \mu + \alpha_i + \epsilon_{ij} \qquad (i = 1, 2; \ j = 1, \ldots, n_i) \tag{8.35}$$

which is the *model equation written in terms of true effects*. There is an observation equation corresponding to Eq. (8.35). Let the *over-all sample mean* $\bar{x}$ be defined by

$$\bar{x} = \frac{\sum_{j=1}^{n_1} x_{1j} + \sum_{j=1}^{n_2} x_{2j}}{n_1 + n_2} = \frac{n_1 \bar{x}_1 + n_2 \bar{x}_2}{n_1 + n_2} \tag{8.36}$$

or

$$\bar{x} = \frac{\sum_{i=1}^{2} \sum_{j=1}^{n_i} x_{ij}}{\sum_{i=1}^{2} n_i} \tag{8.36a}$$

Let the *sample effect of the ith population*, or, for short, the *ith sample effect*, be denoted by a_i and defined by

$$a_i = \bar{x}_i - \bar{x} \qquad (i = 1, 2) \tag{8.37}$$

Thus

$$\bar{x}_i = \bar{x} + a_i \qquad (i = 1, 2) \tag{8.38}$$

Substituting Eq. (8.38) in Eq. (8.30) gives

$$x_{ij} = \bar{x} + a_i + e_{ij} \qquad (i = 1, 2; \; j = 1, 2, \ldots, n_i) \qquad (8.39)$$

which is the *observation equation in terms of the estimated effects*. The over-all sample mean $\bar{x}$ estimates the over-all population mean μ, and the ith sample effect a_i estimates the true effect α_i. It should be noted that, in general

$$\sum_{i=1}^{2} a_i \neq 0$$

unless $n_1 = n_2$. However, it is always true that

$$\sum_{i=1}^{2} \sum_{j=1}^{n_i} a_i = 0 \qquad (8.40)$$

The relations given by Eqs. (8.32) through (8.40) can be extended in an obvious way to include k populations, in which case $i = 1, 2, \ldots, k$, and n_i denotes the size of the ith sample.

The deviates $e_{i1}, e_{i2}, \ldots, e_{in_i}$ about the mean of a random sample from population P_i may be used to estimate the variance σ_i^2 of this population. The variance estimator is given by

$$s_i^2 = \frac{\sum\limits_{j=1}^{n_i} e_{ij}^2}{n_i - 1} \qquad (i = 1, 2) \qquad (8.41)$$

If two populations have common variance $\sigma_1^2 = \sigma_2^2 = \sigma^2$, then

$$s_p^2 = \frac{\sum\limits_{j=1}^{n_1} e_{ij}^2 + \sum\limits_{j=1}^{n_2} e_{2j}^2}{n_1 + n_2 - 2} = \frac{\sum\limits_{i=1}^{2} \sum\limits_{j=1}^{n_i} e_{ij}^2}{\sum\limits_{i=1}^{2} n_i - 2} \qquad (8.42)$$

is an estimator of σ^2. Sometimes s_p^2 is called *error variance* and denoted by s_e^2. Actually, s_1^2, s_2^2, and s_p^2 are all estimators of σ^2 when the populations have common variance, but s_p^2 is preferred, since it has more degrees of freedom. Since, for a given sample, the e's of Eqs. (8.30) and (8.39) are the same, we may think in terms of either model equation, Eq. (8.29) or Eq. (8.35) when establishing confidence intervals or testing hypotheses. Now we use an example with unequal sample sizes to compare the methods and notation of Sect. 8.6 with those of this section.

Example 8.8. The following random samples are drawn from normal populations with common variance σ^2.

Sample from P_1: 27, 18, 10, 4, 19, 30, 11
Sample from P_2: 33, 9, 24, 46
(a) Test the hypothesis $\mu_1 = \mu_2$, using the methods of Sect. 8.6. (b)

Use the notation of this section to describe (a). (c) If, in addition to information given above, it is known that $\mu_1 = \mu_2 = \mu = 20$ and $\sigma_1^2 = \sigma_2^2 = \sigma^2 = 100$, compare estimated means and effects with true means and effects. (d) Discuss related topics.

In using a five per cent level test of the hypothesis $H_0: \mu_1 = \mu_2$ against the alternative hypothesis $H_a: \mu_1 \neq \mu_2$, we need the t statistic with nine degrees of freedom to find the symmetric two-tailed critical region defined by $|t| > 2.262$. Since $\bar{x}_1 = 17$, $\bar{x}_2 = 28$, $SS_1 = 527$, $SS_2 = 726$, and $s_p^2 = 139.22$, the calculated t statistic is

$$t_c = \frac{\bar{x}_1 - \bar{x}_2}{\sqrt{s_p^2\left(\dfrac{1}{n_1} + \dfrac{1}{n_2}\right)}} = \frac{-11}{\sqrt{139.22\left(\dfrac{1}{7} + \dfrac{1}{4}\right)}} = -1.5$$

The statistic t_c falls in the noncritical region. Therefore, we fail to reject H_0 and conclude that we do not have enough evidence to say that the means are different.

Observe that Eq. (8.29) is the model equation when the hypothesis $H_0: \mu_1 = \mu_2 = \mu$ is being tested. For if $\mu_1 = \mu_2$, then

$$\mu = \frac{\mu_1 + \mu_2}{2} = \frac{2\mu_1}{2} = \mu_1$$

Further, we may wish to write H_0 as

$$\mu_1 - \mu = \mu_2 - \mu = \mu - \mu$$

or, using Eq. (8.33), as

$$\alpha_1 = \alpha_2 = 0 \tag{8.43}$$

Thus, the hypothesis that the population means are the *same* is the same as the hypothesis that the true effects are *zero*. The alternative hypothesis $H_a: \mu_1 \neq \mu_2$ may also be written as $H_a: \alpha_1 \neq \alpha_2$. Thus, when the alternative hypothesis is $\alpha_1 \neq \alpha_2$, rejection of Eq. (8.43) leads to the acceptance of the statement, "The true effects are not equal." If the alternative hypothesis is $\alpha_1 < \alpha_2$ ($\alpha_1 > \alpha_2$), the rejection of Eq. (8.43) leads to the statement, "The true effect for population 1 is less (greater) than the true effect for population 2." The pooled error variance can be computed from the e's as

$$s_p^2 = \frac{[(27 - 17)^2 + \cdots + (11 - 17)^2] + [(33 - 28)^2 + \cdots + (46 - 28)^2]}{7 + 4 - 2}$$

$$= \frac{1254}{9} \doteq 139.33$$

For (c) we write each observed value as the sum of three components.

First, using Eq. (8.39) and the fact that $\bar{x} = 21$, $a_1 = 17 - 21 = -4$, and $a_2 = 28 - 21 = 7$, we obtain the results of Table 8.13a. Next, using Eq. (8.35) and the knowledge that $\mu_1 = \mu_2 = \mu = 20$, we obtain the values

Table 8.13a
Estimated Effects for Example 8.8

Sample 1 $x_{1i} = \bar{x} + a_1 + e_{ij}$	Sample 2 $x_{2j} = \bar{x} + a_2 + e_{2j}$
$27 = 21 + (-4) + 10$	$33 = 21 + 7 + 5$
$18 = 21 + (-4) + 1$	$9 = 21 + 7 + (-19)$
$10 = 21 + (-4) + (-7)$	$24 = 21 + 7 + (-4)$
$4 = 21 + (-4) + (-13)$	$46 = 21 + 7 + 18$
$19 = 21 + (-4) + 2$	
$30 = 21 + (-4) + 13$	
$11 = 21 + (-4) + (-6)$	
$\sum x_{ij} = 119 = 7(21) + 7(-4) + 0$	$\sum x_{2j} = 112 = 4(21) + 4(7) + 0$
$\bar{x}_1 = 17 = 21 + (-4)$	$\bar{x}_2 = 28 = 21 + 7$
$a_1 = -4$	$a_2 = 7$

Table 8.13b
True Effects for Example 8.8

Sample 1 $x_{ij} = \mu + \alpha_1 + \epsilon_{ij}$	Sample 2 $x_{2j} = \mu + \alpha_2 + \epsilon_{2j}$
$27 = 20 + 0 + 7$	$33 = 20 + 0 + 13$
$18 = 20 + 0 + (-2)$	$9 = 20 + 0 + (-11)$
$10 = 20 + 0 + (-10)$	$24 = 20 + 0 + 4$
$4 = 20 + 0 + (-16)$	$46 = 20 + 0 + 26$
$19 = 20 + 0 + (-1)$	
$30 = 20 + 0 + 10$	
$11 = 20 + 0 + (-9)$	

in Table 8.13b. The true over-all mean is 20, and the estimate is 21. The true population effects are zero and the estimated effects are -4 and 7, respectively.

For our discussion of (d) we start with

$$\sum_{j=1}^{n_i} (x_{ij} - \bar{x}_i) = \sum_{j=1}^{n_i} e_{ij} = 0 \qquad (i = 1, \ldots, k) \qquad (8.44)$$

and from Table 8.13a we observe that this relation holds for Samples 1 and 2. However

$$\sum_{j=1}^{n_i} (x_{ij} - \mu_i) = \sum_{j=1}^{n_i} \epsilon_{ij} \neq 0 \qquad (i = 1, \ldots, k) \qquad (8.45)$$

In particular, using Table 8.13b, we see that

$$\sum_{j=1}^{7} \epsilon_{1j} = 7 + (-2) + \cdots + (-9) = -21 \neq 0$$

and

$$\sum_{j=1}^{4} \epsilon_{2j} = 13 + (-11) + 4 + 26 = 32 \neq 0$$

Since ϵ is a random variable, the ϵ_{ij}'s may be used to estimate the common population variance σ^2. The estimate is

$$s_\epsilon^2 = \frac{7^2 + (-2)^2 + \cdots + (-9)^2 + 13^2 + \cdots + 26^2}{11}$$

$$= \frac{1573}{11} = 143$$

It can be shown that s_ϵ^2 has 11 degrees of freedom and that $11s_\epsilon^2/\sigma^2$ is distributed as χ^2 with 11 degrees of freedom.

8.12. SUM OF SQUARES IDENTITIES AND RELATED TOPICS

It is informative to note that for a sample of n_i values from a population with mean μ_i and variance σ_i^2 that

$$\sum_{j=1}^{n_i} (x_{ij} - \mu_i)^2 = \sum_{j=1}^{n_i} (x_{ij} - \bar{x}_i)^2 + n_i(\bar{x}_i - \mu_i)^2 \qquad (i = 1, 2, \ldots, k) \quad (8.46)$$

The proof of Eq. (8.46) follows

$$\sum_{j=1}^{n_i} (x_{ij} - \mu_i)^2 = \sum_j [(x_{ij} - \bar{x}_i) + (\bar{x}_i - \mu_i)]^2$$

$$= \sum_j (x_{ij} - \bar{x}_i)^2 + 2(\bar{x}_i - \mu_i) \sum_j (x_{ij} - \bar{x}_i) + \sum_j (\bar{x}_i - \mu_i)^2$$

$$= \sum_j (x_{ij} - \bar{x}_i)^2 + n_i(\bar{x}_i - \mu_i)^2$$

since

$$\sum_j (x_{ij} - \bar{x}_i) = 0$$

Equation (8.46) is called a *sum of squares identity* for one sample, and is an algebraic identity which in no way depends on any distribution assumptions associated with x_{ij}. If the sample is randomly drawn, we know that

$$s_i^2 = \sum_j \frac{(x_{ij} - \bar{x}_i)^2}{n_i - 1}$$

with $(n_i - 1)$ degrees of freedom is an unbiased estimator of σ_i^2, and we have just indicated that

$$s_{e_i}^2 = \frac{\sum\limits_{j} (x_{ij} - \mu_i)^2}{n_i}$$

with n_i degrees of freedom is an unbiased estimator of σ_i^2. Further, since $\bar{x}_i$ is distributed with mean μ_i and variance

$$\sigma_{\bar{x}_i}^2 = \frac{\sigma_i^2}{n_i}$$

it follows that $(\bar{x}_i - \mu_i)^2/1$ with one degree of freedom is an unbiased estimator of σ_i^2/n_i. Therefore

$$\frac{n_i(\bar{x}_i - \mu_i)^2}{1} = \frac{\sum\limits_{j} (\bar{x}_i - \mu_i)^2}{1} \tag{8.47}$$

with one degree of freedom is an unbiased estimator of σ_i^2. Associated with Eq. (8.46) we have a degree of freedom identity given by

$$n_i = (n_i - 1) + 1 \tag{8.48}$$

The two identities, Eqs. (8.46) and (8.48), and their extensions are useful in statistics in many ways. For example, the three ratios

$$\frac{\sum\limits_{j} (x_{ij} - \mu_i)^2}{n_i}, \quad \frac{\sum\limits_{j} (x_{ij} - \bar{x}_i)^2}{n_i - 1} \quad \text{and} \quad \frac{\sum\limits_{j} (\bar{x}_i - \mu_i)^2}{1} \tag{8.49}$$

obtained from these two identities are all unbiased estimators of σ_i^2, provided the sample is randomly drawn. Thus, the three ratios

$$\frac{\sum\limits_{j} (x_{ij} - \mu_i)^2}{\sigma_i^2}, \quad \frac{\sum\limits_{j} (x_{ij} - \bar{x}_i)^2}{\sigma_i^2} \quad \text{and} \quad \frac{\sum\limits_{j} (\bar{x}_i - \mu_i)^2}{\sigma_i^2} \tag{8.50}$$

obtained from Eq. (8.46) are distributed as χ^2 with n_i degrees, $n_i - 1$ degrees, and one degree of freedom, respectively, if the sample is drawn from a normal population. Further, if two of the sum of squares in Eq. (8.46) are known, the third can be computed directly by addition or subtraction.

Returning to the case of two samples drawn from like populations with the same mean μ and same variance σ^2, we can show that

$$\sum\limits_{i=1}^{2} \sum\limits_{j=1}^{n_i} (x_{ij} - \bar{x})^2 = \sum\limits_{i} \sum\limits_{j} (\bar{x}_i - \bar{x})^2 + \sum\limits_{i} \sum\limits_{j} (x_{ij} - \bar{x}_i)^2 \tag{8.51a}$$

or

$$\sum\limits_{i} \sum\limits_{j} (a_i + e_{ij})^2 = \sum\limits_{i} \sum\limits_{j} a_i^2 + \sum\limits_{i} \sum\limits_{j} e_{ij}^2 \tag{8.51b}$$

These equations are algebraic identities, called *sum of squares identities*, and in no way depend on any distribution assumptions associated with x_{ij}. To prove Eq. (8.51b), write

$$\sum_i \sum_j (a_i + e_{ij})^2 = \sum_i \sum_j a_i^2 + 2 \sum_i \sum_j a_i e_{ij} + \sum_i \sum_j e_{ij}^2$$

$$= \sum_i \sum_j a_i^2 + \sum_i \sum_j e_{ij}^2$$

since

$$\sum_i \sum_j a_i e_{ij} = \sum_i a_i \sum_j e_{ij} = \sum_i a_i \cdot 0 = 0$$

If the samples are randomly and independently drawn, we know that

$$\frac{\sum_i \sum_j (x_{ij} - \bar{x}_i)^2}{n_1 + n_2 - 2}$$

with $n_1 + n_2 - 2$ degrees of freedom is an unbiased estimator of σ^2. Thinking of the two samples as one large random sample of size $n_1 + n_2$ drawn from a population with mean μ and variance σ^2, we know that

$$\frac{\sum_i \sum_j (x_{ij} - \bar{x})^2}{n_1 + n_2 - 1}$$

with $n_1 + n_2 - 1$ degrees of freedom is an unbiased estimator of σ^2. Thus, we suspect that

$$\frac{\sum_i \sum_j (\bar{x}_i - \bar{x})}{1}$$

with 1 degree of freedom is an unbiased estimator of σ^2. If $n_1 = n_2 = n$

$$\sum_i \sum_j (\bar{x}_i - \bar{x})^2 = n \sum_{i=1}^{2} (\bar{x}_i - \bar{x})^2$$

Now, assuming $\bar{x}_1$ and $\bar{x}_2$ to be random means from a population of means with mean μ and variance $\sigma_{\bar{x}}^2 = \sigma^2/n$, we know that

$$\frac{\sum_i (\bar{x}_i - \bar{x})^2}{1}$$

is an unbiased estimator of σ^2/n. Thus,

$$\frac{\sum_i \sum_j (\bar{x}_i - \bar{x})^2}{1} = \frac{n \sum (\bar{x}_i - \bar{x})^2}{1}$$

is an unbiased estimator of σ^2. It can be shown that

$$\frac{\sum\limits_{i=1}^{2} n_i(\bar{x}_i - \bar{x})^2}{1} \tag{8.52}$$

with one degree of freedom is an unbiased estimator of σ^2. Thus, associated with (8.51) we have the *degrees of freedom identity*

$$n_1 + n_2 - 1 = (n_1 + n_2 - 2) + 1 \tag{8.53}$$

The two identities (8.51) and (8.53) and their extensions are very important in all analysis of variance (see Chaps. 10–13). For example, if the samples are randomly and independently drawn from populations with common mean μ and common variance σ^2 the three ratios

$$s_t^2 = \frac{\sum\limits_{i}\sum\limits_{j}(x_{ij} - \bar{x})^2}{n_1 + n_2 - 1}, \qquad s_e^2 = \frac{\sum\limits_{i}\sum\limits_{j}(x_{ij} - \bar{x}_i)^2}{n_1 + n_2 - 2}$$

and $\tag{8.54}$

$$s_m^2 = \frac{\sum\limits_{i}\sum\limits_{j}(\bar{x}_i - \bar{x})^2}{1}$$

obtained from these two identities are unbiased estimators of σ^2. If, in addition to the assumptions just mentioned, the populations are normal, then the three ratios

$$\frac{\sum\sum(x_{ij} - \bar{x})^2}{\sigma^2}, \qquad \frac{\sum\sum(x_{ij} - \bar{x}_i)^2}{\sigma^2}, \quad \text{and} \quad \frac{\sum\sum(\bar{x}_i - \bar{x})^2}{\sigma^2} \tag{8.55}$$

obtained from Eq. (8.51) are distributed as χ^2 with $n_1 + n_2 - 1$ degrees, $n_1 + n_2 - 2$ degrees, and one degree of freedom, respectively.

Since s_e^2 is computed using only e's and s_m^2 is computed using only a's, one would expect s_e^2 and s_m^2 to be independent estimators of σ^2; since s_t^2 is computed using both a's, and e's one would expect s_t^2 not to be independent of s_m^2 and s_e^2—each is actually the case. Further, the sum of squares identity is useful in finding the third sum of squares in terms of the other two.

In particular, from Example 8.8 we find, using Table 8.13a, that three sums of squares are

$$\sum_i\sum_j (a_i + e_{ij}^2) = \sum_i\sum_j (x_{ij} - \bar{x})^2$$

$$= (27 - 21)^2 + \cdots + (11 - 21)^2 + (33 - 21)^2$$

$$+ \cdots + (46 - 21)^2$$

$$= 1562$$

$$\sum_i\sum_j a_i^2 = \sum_i n_i(\bar{x}_i - \bar{x})^2$$

$$= 7(-4)^2 + 4(7)^2 = 308$$

and

$$\sum_i \sum_j e_{ij}^2 = \sum_i \sum_j (x_{ij} - \bar{x}_i)^2 = 1254$$

Actually

$$\sum_i \sum_j e_{ij}^2$$

is usually found by subtraction as $1562 - 308 = 1254$. From these sums of squares we find the following estimates of the variance σ^2: $s_t^2 = \frac{1562}{10} = 156.2$, $s_m^2 = 308$, and $s_e^2 = 139.33$. Since $\alpha_1 = \alpha_2 = 0$ in this example, $s_e^2 = 143$ is comparable to all these estimates. However, in the absence of any real knowledge about μ_1 and μ_2, only $s_e^2 = 139.33$ can be considered an unbiased estimate of σ^2—the other estimates can be expected to be inflated when α_1 and α_2 differ from zero.

In the remainder of this section we consider the case where all samples are of size n. The sum of squares identities, Eqs. (8.46) and (8.51a), may then be written as

$$\sum_{j=1}^n (x_{ij} - \mu_i)^2 = \sum_j (x_{ij} - \bar{x}_i)^2 + n(\bar{x}_i - \mu_i) \qquad (i = 1, \ldots, k) \qquad (8.56)$$

and

$$\sum_{i=1}^2 \sum_{j=1}^n (x_{ij} - \bar{x})^2 = \sum_i \sum_j (x_{ij} - \bar{x}_i)^2 + n \sum_i (\bar{x}_i - \bar{x})^2 \qquad (8.57)$$

respectively, and the degrees of freedom identities (8.48) and (8.53) as

$$n = (n - 1) + 1 \qquad (8.58)$$

and

$$2n - 1 = 2(n - 1) + 1 \qquad (8.59)$$

respectively. Since, in this case

$$\bar{x} = \frac{n\bar{x}_1 + n\bar{x}_2}{n + n} = \frac{\bar{x}_1 + \bar{x}_2}{2}$$

we have

$$\sum_{i=1}^2 (\bar{x}_i - \bar{x})^2 = \left(\bar{x}_1 - \frac{\bar{x}_1 + \bar{x}_2}{2}\right)^2 + \left(\bar{x}_2 - \frac{\bar{x}_1 + \bar{x}_2}{2}\right)^2$$

$$= \left(\frac{\bar{x}_1 - \bar{x}_2}{2}\right)^2 + \left(\frac{\bar{x}_2 - \bar{x}_1}{2}\right)^2 = \frac{(\bar{x}_1 - \bar{x}_2)^2}{2}$$

or

$$s_m^2 = n \sum_{i=1}^2 (\bar{x}_i - \bar{x})^2 = \frac{n(\bar{x}_1 - \bar{x}_2)^2}{2}$$

Thus, the ratio of the two independent variance estimators s_m^2 and s_e^2 becomes

$$\frac{s_m^2}{s_e^2} = \frac{\dfrac{n \sum (\bar{x}_i - \bar{x})^2}{1}}{s_e^2} = \frac{n(\bar{x}_1 - \bar{x}_2)^2}{2s_e^2}$$

or

$$\frac{s_m^2}{s_e^2} = \frac{(\bar{x}_1 - \bar{x}_2)^2}{\dfrac{2s_e^2}{n}} = \frac{(\bar{x}_1 - \bar{x}_2)^2}{s_{\bar{x}_1 - \bar{x}_2}^2} \tag{8.60}$$

which we recognize as the square of the t statistic used in testing the hypothesis $\mu_1 = \mu_2$ (or $\mu_1 - \mu_2 = 0$). That is, $t^2 = s_m^2/s_e^2$, where t is distributed as the Student t with $2n - 2$ degrees of freedom. Also, the ratio of the two independent variance estimators

$$s'^2 = \frac{n(\bar{x}_i - \mu_i)^2}{1} \quad \text{and} \quad s^2 = \frac{\sum_{j=1}^{n} (x_{ij} - \bar{x}_i)}{n - 1} \quad (i = 1, 2)$$

becomes

$$\frac{s'^2}{s^2} = \frac{\dfrac{n(\bar{x} - \mu)^2}{1}}{s^2} = \frac{(\bar{x}_i - \mu_i)^2}{\dfrac{s^2}{n}} = \frac{(\bar{x}_i - \mu_i)^2}{s_{\bar{x}}^2} \tag{8.61}$$

which we recognize as the square of the t statistic used in testing the hypothesis $\mu_i = \mu_0$. That is, $t^2 = s'^2/s^2$, where t is distributed as the Student t with $n - 1$ degrees of freedom. Thus, it is possible to use the ratio of two variance estimators in place of the Student t distribution for a test of the hypothesis $\mu_i = \mu_0$ (or $\mu_1 = \mu_2$) against the alternative hypothesis $\mu_i \neq \mu_0$ (or $\mu_1 \neq \mu_2$).

Since the t statistic involves calculating a square root, we might well prefer using the ratio of variance estimators if we only had its sampling distribution. Further, when s_m^2 is computed for more than two means, say k means, we cannot use the t distribution to test the hypothesis $\mu_1 = \mu_2 = \ldots = \mu_k$. However, the ratio s_m^2/s_e^2 may be used for such a test. In the next chapters we study the statistic s_m^2/s_e^2, which is a special case of the well-known F distribution.

8.13. EXERCISES

8.21. (a) Use the notation of Sect. 8.11 and the methods of Example 8.8 to discuss Exercise 8.10. (b) If, in addition to the data given in Exercise 8.10, it is known that $\mu_A = 90$, $\mu_B = 80$, and $\sigma_A^2 = \sigma_B^2 = \sigma^2 = 900$, compare estimated means and effects with true means and effects. (c) Using the information in (a) and (b), verify Eq. (8.46) for both sample

A and sample B. Use Eq. (8.49) to find six estimates of $\sigma^2 = 900$. Which of these estimates are unbiased estimates of $\sigma^2 = 900$? (d) Verify Eq. (8.51a). Find s_t^2, s_m^2, and s_e^2. Which of these estimates of $\sigma^2 = 900$ are unbiased?

8.22. (a) Use the notation of Sect. 8.11 to discuss Exercise 8.11. (b) If, in addition to the data given in Exercise 8.11, it is known that $\mu_1 = 400$, $\mu_2 = 150$, $\sigma_1^2 = 1000$, and $\sigma_2^2 = 15,000$, compare estimated means and effects with true means and effects. (c) Using the information in (a) and (b), verify Eq. (8.46) for sample A and for sample B. Use Eq. (8.49) to find three estimates of σ_1^2 and three estimates of σ_2^2. (d) Verify Eq. (8.51a). Find s_t^2, s_m^2, and s_e^2. Which of these estimates are unbiased? Explain.

8.23. Three random samples are drawn from normal populations with common variance 100. The sample values and population means are as follows

Table 8.14

Sample	Sample Values					True Mean
1	64	42	54	59	49	50
2	60	59	78	64	58	60
3	94	94	119	96	104	100

(a) Prepare tables like Tables 8.13a and 8.13b. (b) For each sample verify Eqs. (8.44), (8.45), and (8.46). (c) Verify Eq. (8.51b). Compute s_t^2, s_m^2, s_e^2, and s_m^2/s_e^2. Is s_m^2/s_e^2 approximately what you expected it to be? Explain.

REFERENCES

1. Anderson, R. L. and T. A. Bancroft, *Statistical Theory in Research*. New York: McGraw-Hill, Inc. 1952, Chap. 7.
2. Aspin, A. A., "An Examination and Further Development of a Formula Arising in the Problem of Comparing Two Mean Values," *Biometrika*, Vol. **35** (1948), pp. 88–96.
3. ———, "Tables for Use in Comparisons Whose Accuracy Involves Two Variances, Separately Estimated," *Biometrika*, Vol. **36** (1949), pp. 290–296.
4. Box, G. E. P., "Some Theorems on Quadratic Forms Applied in the Study of Analysis of Variance Problems," *Annals of Mathematical Statistics*, Vol. **25** (1954), pp. 290–302.
5. Dixon, W. J. and F. J. Massey, *Introduction to Statistical Analysis*, 2nd ed. New York: McGraw-Hill, Inc., 1957, Chaps. 9 and 17.
6. ——— and A. M. Mood, "The Statistical Sign-Test," *Journal of the American Statistical Association*, Vol. **41** (1946), pp. 557–566.
7. Hald, A., *Statistical Theory with Engineering Applications*. New York: John Wiley & Sons, Inc., 1952, Chaps. 11 and 15.

8.   Johnson, N. L. and B. L. Welch, "Application of the Non-Central t-Distribution," *Biometrika*, Vol. **31** (1940), pp. 362–389.

9.   Neyman, J., "Statistical Problems in Agricultural Experimentation," *Supplement to the Journal of the Royal Statistical Society*, Vol. **2** (1935), pp. 107–180.

10. —— and B. Tokarska, "Errors of the Second Kind in Testing Student's Hypothesis," *Journal of the American Statistical Association*, Vol. **31** (1936), pp. 318–326.

11. Resnikoff, G. J. and G. J. Lieberman, *Tables of the Non-Central t-Distribution*. Stanford: Stanford University Press, 1957.

12. Welch, B. L., "The Generalization of Student's Problem when Several Different Population Variances are Involved," *Biometrika*, Vol. **34** (1947), pp. 28–35.

9

SAMPLING
FROM NORMAL POPULATIONS—
F DISTRIBUTION WITH APPLICATIONS

Properties of the F distribution are explained. It is shown how the F distribution is useful in problems involving two variances. This includes discussions of confidence intervals, hypothesis testing, power functions and relation of the F distribution to the normal, Student t, and chi-square distributions.

9.1. INTRODUCTION

We have already discussed problems involving one variance (Chap. 7) and one or two means (Chap. 8). However, there are situations in which we need to compare two variances or more than two means, and the distributions of earlier chapters are not appropriate. It is fortunate that the same sampling distribution, the F distribution, can be used in both cases. The very important application to several means will be discussed in the chapter on analysis of variance. At this time we shall treat in some detail problems involving two variances and give approximate tests of the hypothesis that $\sigma_1^2 = \sigma_2^2 = \cdots = \sigma_k^2$. First, we consider properties of this very important distribution.

9.2. THE F DISTRIBUTION

Theorem 9.1. *Let the random variable* χ_1^2 *be distributed as chi-square* *with v_1 degrees of freedom and the random variable χ_2^2, which is independent of χ_1^2, be distributed as a chi-square with v_2 degrees of freedom. Then*

$$F \equiv F(v_1, v_2) = \frac{\dfrac{\chi_1^2}{v_1}}{\dfrac{\chi_2^2}{v_2}} = \frac{v_2 \chi_1^2}{v_1 \chi_2^2} \tag{9.1}$$

is distributed with density function

$$f(F) = \frac{\Gamma\!\left(\dfrac{v_1 + v_2}{2}\right)}{\Gamma\!\left(\dfrac{v_1}{2}\right)\Gamma\!\left(\dfrac{v_2}{2}\right)} \cdot \left(\frac{v_1}{v_2}\right)^{v_1/2} \cdot \frac{F^{(v_1-2)/2}}{\left(1 + \dfrac{v_1 F}{v_2}\right)^{(v_1+v_2)/2}}, \qquad F \geq 0 \tag{9.2}$$

with mean

$$\frac{v_2}{v_2 - 2} \qquad (v_2 > 2)$$

and variance

$$\frac{2v_2^2(v_1 + v_2 - 2)}{v_1(v_2 - 2)^2(v_2 - 4)} \qquad (v_2 > 4)$$

The gamma function $\Gamma(\alpha)$ is defined in Exercise 3.67.

Corollary. Let σ_1^2 and σ_2^2 be variances of two normal populations. Let s_1^2 with v_1 degrees of freedom and s_2^2 with v_2 degrees of freedom be two independent estimators of the variances σ_1^2 and σ_2^2, respectively. Then the ratio s_1^2/s_2^2 is distributed as F with v_1 and v_2 degrees of freedom when $\sigma_1^2 = \sigma_2^2 = \sigma^2$.

The outline of the proof of Theorem 9.1 is given in the exercises. The proof of the corollary is immediate. For s_1^2/σ^2 and s_2^2/σ^2 are distributed as χ^2 per degree of freedom with v_1 and v_2 degrees of freedom, respectively, and, therefore

$$\frac{s_1^2}{s_2^2} = \frac{\dfrac{s_1^2}{\sigma^2}}{\dfrac{s_2^2}{\sigma^2}} = \frac{\dfrac{\chi_1^2}{v_1}}{\dfrac{\chi_2^2}{v_2}} \tag{9.3}$$

is distributed as F with v_1 and v_2 degrees of freedom.

The statistic defined in Eq. (9.1) is said to have the F *distribution* with v_1 and v_2 degrees of freedom. Thus, just as with t and χ^2, the symbol F is used in referring to both the "F variate" (or F statistic) and the "F distribution." R. A. Fisher [11, 12] originally developed the exact distribution of $z = \frac{1}{2} \ln (s_1^2/s_2^2)$, i.e., the z distribution. Snedecor [22] studied $s_1^2/s_2^2 = e^{2z}$, a modified version of z, and published tables for its use. Snedecor called the distribution of the variance-ratio s_1^2/s_2^2 the F distribution in honor of

Fisher. Other sources of tables for F as well as t and χ^2 distributions are given by Bancroft [2].

The density function, Eq. (9.2), represents a two-parameter family of distributions. The graph of three members of this family is shown in Fig. 9.1 and illustrates the fact that the "shape" of the distribution changes with

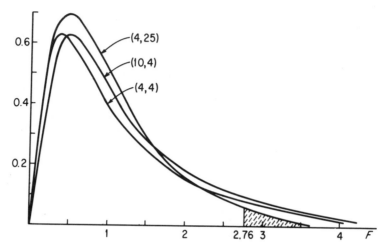

Fig. 9.1 F Distribution Curves for $(v_1, v_2) = (4, 4)$, $(10, 4)$ and $(4, 25)$. $F_{.05}$ Points are Indicated

the degrees of freedom. Let $F_\alpha = F_\alpha(v_1, v_2)$ denote that value of $F = F(v_1, v_2)$ for which

$$P[F > F_\alpha] = \alpha \tag{9.4}$$

where α is any value in the interval $0 < \alpha \le 0.5$. Values of F_α, α percentage points, corresponding to selected values of α and the most useful values of v_1 and v_2 are shown in Table VII. Usually, in applications, only percentage points on the right tail of the F distribution are required. However, if a left-tail value $F_{1-\alpha}(v_1, v_2)$ is needed, it can be found using the relation

$$F_{1-\alpha}(v_1, v_2) = \frac{1}{F_\alpha(v_2, v_1)} \tag{9.5}$$

and the right-tail value $F_\alpha(v_2, v_1)$. To prove Eq. (9.5), observe that $F_{1-\alpha}(v_1, v_2)$ satisfies the equation

$$P\left[\frac{\dfrac{\chi_1^2}{v_1}}{\dfrac{\chi_2^2}{v_2}} > F_{1-\alpha}(v_1, v_2) \right] = 1 - \alpha$$

or

$$P\left[\frac{\frac{\chi_2^2}{\nu_2}}{\frac{\chi_1^2}{\nu_1}} < \frac{1}{F_{1-\alpha}(\nu_1, \nu_2)}\right] = 1 - \alpha$$

or

$$P\left[\frac{\frac{\chi_2^2}{\nu_2}}{\frac{\chi_1^2}{\nu_1}} > \frac{1}{F_{1-\alpha}(\nu_1, \nu_2)}\right] = \alpha \tag{9.6}$$

but $F_\alpha(\nu_2, \nu_1)$ is a value such that

$$P\left[\frac{\frac{\chi_2^2}{\nu_2}}{\frac{\chi_1^2}{\nu_1}} > F_\alpha(\nu_2, \nu_1)\right] = \alpha \tag{9.7}$$

When we compare Eqs. (9.6) and (9.7), it is clear that $1/F_{1-\alpha}(\nu_1, \nu_2) = F_\alpha(\nu_2, \nu_1)$, and thus Eq. (9.5) holds.

We know that the mean of the distribution of $\chi^2/\nu = s^2/\sigma^2$ is one. Further, as the number of degrees of freedom approaches ∞, almost all of the sample statistics s^2/σ^2 will be arbitrarily close to the mean, one. In this case we say that s^2/σ^2 converges in probability to one or, symbolically, if k is an arbitrary small positive real number, then $P(|s^2/\sigma^2 - 1| < k) \to 1$ as $\nu \to \infty$. The mean of the distribution of

$$F = \frac{\frac{s_1^2}{\sigma_1^2}}{\frac{s_2^2}{\sigma_2^2}}$$

is

$$\frac{\nu_2}{\nu_2 - 2} = \frac{1}{1 - \frac{2}{\nu_2}}$$

Hence, as both ν_1 and ν_2 approach ∞, the mean of F approaches 1, and almost all of the sample statistics

$$F = \frac{\frac{s_1^2}{\sigma_1^2}}{\frac{s_2^2}{\sigma_2^2}}$$

will be arbitrarily close to 1; that is, F converges in probability to 1. This means that the distance between the two values $F_{1-\alpha}$ and F_α decreases as both ν_1 and ν_2 increase. In other words, if $\nu_1 \to \infty$ and $\nu_2 \to \infty$, then

$F_\alpha \to 1$ and $F_{1-\alpha} \to 1$. If, in particular, $\sigma_1^2 = \sigma_2^2 = \sigma^2$, F may be replaced by s_1^2/s_2^2 in the above statements. These statements may be checked by looking at the percentage points of Table VII. For a fixed α and a fixed $\nu_1(\nu_2)$, the F_α values decrease as $\nu_2(\nu_1)$ increases, and consequently the $F_{1-\alpha}$ values increase. For a fixed α the F_α values decrease toward 1, and the $F_{1-\alpha}$ values increase toward 1 as ν_1 and ν_2 increase.

Note. Using the variance of F, σ_F^2, from Theorem 9.1, we see that $\sigma_F^2 \to 0$ as ν_1 and $\nu_2 \to \infty$.

In many applications the sample variance-ratio s_1^2/s_2^2 is considered. If $\sigma_1^2 = \sigma_2^2 = \sigma^2$, then s_1^2/s_2^2 is distributed as F. However, if $\sigma_1^2 \neq \sigma_2^2$, s_1^2/s_2^2 is distributed as

$$\frac{\dfrac{\sigma_1^2 \chi_1^2}{\nu_1}}{\dfrac{\sigma_2^2 \chi_2^2}{\nu_2}} = \left(\frac{\sigma_1^2}{\sigma_2^2}\right) F(\nu_1, \nu_2) \tag{9.8}$$

Thus, it follows that any upper α percentage point

$$\frac{\sigma_1^2}{\sigma_2^2} F_\alpha$$

must satisfy the relation

$$P\left[\frac{s_1^2}{s_2^2} > \frac{\sigma_1^2}{\sigma_2^2} F_\alpha\right] = \alpha \tag{9.9}$$

If equal proportions of s_1^2/s_2^2 are to be in the tails of the distribution, then $100(1-\alpha)$ per cent are in the interval

$$\frac{\sigma_1^2}{\sigma_2^2} F_{1-\alpha/2}(\nu_1, \nu_2) < \frac{s_1^2}{s_2^2} < \frac{\sigma_1^2}{\sigma_2^2} F_{\alpha/2}(\nu_1, \nu_2) \tag{9.10}$$

The following $100(1-\alpha)$ per cent confidence interval for σ_1^2/σ_2^2 is obtained by solving the inequality (9.10) with respect to σ_1^2/σ_2^2

$$\frac{s_1^2}{s_2^2} \frac{1}{F_{\alpha/2}(\nu_1, \nu_2)} < \frac{\sigma_1^2}{\sigma_2^2} < \frac{s_1^2}{s_2^2} \frac{1}{F_{1-\alpha/2}(\nu_1, \nu_2)} \tag{9.11}$$

In practice, equal percentages are usually taken in the two tails of the $F(\nu_1, \nu_2)$ distribution. In case F_1 and F_2 are values of $F = F(\nu_1, \nu_2)$ such that $P(F < F_1) = \alpha_1$ and $P(F > F_2) = \alpha_2$, where α_1 and α_2 are both between 0 and 0.5, then Relation (9.11) may be written as

$$\frac{s_1^2}{s_2^2 F_2} < \frac{\sigma_1^2}{\sigma_2^2} < \frac{s_1^2}{s_2^2 F_1} \tag{9.12}$$

The inequality (9.12) gives a $100 (1 - \alpha_1 - \alpha_2)$ per cent confidence interval of σ_1^2/σ_2^2 .

9.3. APPLICATION OF THE F DISTRIBUTION TO PROBLEMS WITH TWO VARIANCES

The primary application of the F distribution occurs in analysis of variance (Chaps. 10 through 13) where the usual hypothesis involves several means. At this time we consider the F distribution as it applies to exactly two variances; this is a brief treatment, since these problems are so similar to one-variance problems discussed in Chap. 7. Most of our attention in this section is focused on points where the treatment of the F and χ^2 distributions differ.

It is important to know the relative sizes of two population variances (dispersions) when one is comparing two processes, temperature in two locations, traffic violations during day and night, achievement in two sections of a class, amount of foreign matter in two lots of raw material, incomes in two cities, etc. One is usually interested in knowing which of two variances is larger. Thus, the null hypothesis is likely to be $H_0: \sigma_1^2 = \sigma_2^2$ with alternative hypothesis $H_a: \sigma_1^2 > \sigma_2^2$, say. The null hypothesis in this case might also be stated as $\sigma_1^2 \leq \sigma_2^2$, since rejection of the null hypothesis leads to the acceptance of the same alternative hypothesis, namely, $H_a: \sigma_1^2 > \sigma_2^2$. In some cases, the null hypothesis is $H_0: \sigma_1^2 = \sigma_2^2$ with alternative hypothesis $H_a: \sigma_1^2 \neq \sigma_2^2$. Test procedures, confidence intervals, and power values are illustrated in Example 9.1.

Example 9.1. Let samples of sizes $n_1 = 16$ and $n_2 = 20$ be drawn from populations one and two, respectively. The variance estimates are $s_1^2 = 8.9$ and $s_2^2 = 4.6$, respectively. (The reader may wish to think of populations in his own field of interest with these variance estimates resulting from coded data.) (a) Test the hypothesis $H_0: \sigma_1^2 \leq \sigma_2^2$ against $H_a: \sigma_1^2 > \sigma_2^2$. (b) Find a 95 per cent confidence interval for the ratio σ_1^2/σ_2^2. (c) Find the probability of making the type 2 error in (a) if $\sigma_1^2 = 2\sigma_2^2$. (d) Discuss power of the test in (a).

In order for the test in (a) to be valid, we must assume that random and independent samples were drawn from two normal populations. It should be noted that no assumption is made concerning the means and that the hypothesis may be stated as $H_0: \sigma_1^2/\sigma_2^2 \leq 1$ and $H_a: \sigma_1^2/\sigma_2^2 > 1$. Under the extreme condition, $\sigma_1^2/\sigma_2^2 = 1$, of the null hypothesis we think of $s_1^2/s_2^2 = 8.9/4.6 = 1.93$ as an F statistic. For a five per cent level one-sided test, the critical region is made up of all values of F for which $F > 2.23 = F_{.05}(15, 19)$. It should be observed that for a one-sided test σ_1^2 and σ_2^2 can be selected so as always to use the upper tail of the F distribution. Since 1.93 falls in the noncritical region, we fail to reject H_0. That is, we do not have enough evidence to say that the variance of population one is actually greater than the variance of population two.

To find the 95 per cent confidence interval given by (9.11), we first find, using Table VII, $F_{.025}(15, 19) = 2.62$ and

$$F_{.975}(15, 19) = \frac{1}{F_{.025}(19, 15)} = \frac{1}{2.78}$$

by linear interpolation. Substituting these values of F along with $s_1^2/s_2^2 = 1.93$ in (9.11) gives

$$0.74 < \frac{\sigma_1^2}{\sigma_2^2} < 5.4$$

or

$$0.74\sigma_2^2 < \sigma_1^2 < 5.4\sigma_2^2$$

Let

$$\lambda^2 = \frac{\sigma_1^2}{\sigma_2^2} \tag{9.13}$$

Then the *power of the test* of $H_0: \lambda^2 \leq 1$ against $H_a: \lambda^2 > 1$ is given by

$$p(\lambda^2) = P\left[\frac{s_1^2}{s_2^2} > F_\alpha(\nu_1, \nu_2); \lambda^2 > 1\right]$$

$$= P\left[\frac{\frac{s_1^2}{\sigma_1^2}}{\frac{s_2^2}{\sigma_2^2}} > \frac{\sigma_2^2}{\sigma_1^2} F_\alpha(\nu_1, \nu_2); \lambda^2 > 1\right]$$

or

$$p(\lambda^2) = P\left[F(\nu_1, \nu_2) > \frac{F_\alpha(\nu_1, \nu_2)}{\lambda^2}\right] \tag{9.14}$$

Since $\beta(\lambda^2) = 1 - p(\lambda^2)$, it follows that

$$\beta(\lambda^2 = 2) = 1 - P\left[F(15, 19) > \frac{F_{.05}(15, 19)}{2}\right]$$

or

$$\beta(\lambda^2 = 2) = 1 - P[F(15, 19) > 1.11] \tag{9.15}$$

Using linear interpolation, we find $P[F(15, 19) > 1.11] \doteq 0.42$, since from Table VII $P[F(15, 19) > 0.99] = 0.50$ and $P[F(15, 19) > 1.38] = 0.25$. However, using the chart on p. 60 of Vogler and Norton [24], we find that $P[F(15, 19) > 1.11] \doteq 0.39$. Thus, on substituting in Eq. (9.15) we get

$$\beta(\lambda^2 = 2) = 1 - 0.39 = 0.61$$

From Eq. (9.14) it is clear that the power increases with λ^2, since $F_\alpha(\nu_1, \nu_2)/\lambda^2$ is a decreasing function of λ^2. Further, from Eq. (9.14) and the definition

$$p(\lambda^2) = 1 - \beta(\lambda^2) = P[F(\nu_1, \nu_2) > F_{1-\beta}(\nu_1, \nu_2)]$$

it follows that

$$F_{1-\beta}(\nu_1, \nu_2) = \frac{F_\alpha(\nu_1, \nu_2)}{\lambda^2}$$

or

$$\lambda^2 = \frac{F_\alpha(\nu_1, \nu_2)}{F_{1-\beta}(\nu_1, \nu_2)} = F_\alpha(\nu_1, \nu_2) \cdot F_\beta(\nu_2, \nu_1) \tag{9.16}$$

For an α level test and selected values of the power function $p(\lambda^2) = 1 - \beta(\lambda^2)$, we obtain, from Table VII, $F_\alpha(\nu_1, \nu_2)$ and $F_\beta(\nu_2, \nu_1)$ and determine values of λ^2, using Eq. (9.16). By plotting the points $(\lambda^2, p(\lambda^2))$ and connecting them by a smooth curve, we may obtain the power curve for fixed values of α, ν_1, and ν_2. The reader is probably already aware of the fact that the discussion following Eq. (9.16) is analogous to the discussion following Eq. (7.12). This being the case, we leave it to the reader to prepare a table like Table 7.1, to draw curves like those in Figs. 7.3 and 7.4, and to find power functions when the alternative hypothesis is $\sigma_1^2 < \sigma_2^2$ or $\sigma_1^2 \neq \sigma_2^2$. An extensive table of λ^2 corresponding to the five per cent and one per cent levels of significance is given by Eisenhart [23]. A table which is useful in determining the minimum equal sample sizes for fixed values of α, β, and λ^2 is given by Davies [10]. At the end of Sect. 9.4 there are statements concerning the importance of the normality assumption in testing $\sigma_1^2 = \sigma_2^2$.

9.4. TESTS FOR THE EQUALITY OF K POPULATION VARIANCES

In Sect. 9.2 we discussed an exact method for testing the equality of two normal population variances and indicated a few places in which this test might be required. Often it is not convenient to restrict the problem to two variances. Thus, we now consider tests of the equality of k normal population variances $\sigma_1^2, \sigma_2^2, \ldots, \sigma_k^2$, i.e., tests of

$$H_0: \sigma_1^2 = \sigma_2^2 = \cdots = \sigma_k^2 \tag{9.17}$$

These tests are based on variances $s_1^2, s_2^2, \ldots, s_k^2$ with $\nu_1, \nu_2, \ldots, \nu_k$ degrees of freedom, which are computed from random samples taken independently from normal populations with variances $\sigma_1^2, \sigma_2^2, \ldots, \sigma_k^2$, respectively. It should also be noted that, in testing the null hypothesis $H_0: \mu_1 = \mu_2$ by the methods of Chap. 8, we needed to know whether the population variances

σ_1^2 and σ_2^2 are equal. A preliminary test of the equality of variances described in Sect. 9.2 could be made in order to decide whether to pool the sample variances before applying the appropriate t test to $H_0: \mu_1 = \mu_2$. There are similar situations (explained in Chaps. 10 and 11) in which we need to know whether k sample variances may be pooled to obtain a single variance estimator with an increased number of degrees of freedom.

Several tests have been proposed for testing the hypothesis (9.17). For a fixed α we should select, if possible, that test for which the probability of committing a type 2 error is a minimum. This is difficult, since the size of β depends on the particular form of the alternative hypothesis, i.e., the ways in which we think Eq. (9.17) may be wrong, as well as the construction of the test. For example, if Eq. (9.17) does not hold, it may be true that the variances are more or less randomly scattered; that $k - 1$ variances are equal, and the kth variance differs from these by a large amount; that about half the variances are equal to σ_0^2 and the remainder are equal to $\sigma_0^2 + k$, where $k \neq 0$ is some real number.

When it is thought that the population variances are not equal but more or less randomly scattered, the test most often used is called *Bartlett's test* [3] and is described in many places [1, 3, 5. 15, 16]. Actually, Bartlett modified the likelihood-ratio test proposed by Neyman and Pearson [19] in constructing his test. The test statistic is

$$B = \frac{\nu \ln s^2 - \sum_{i=1}^{k} (\nu_i \ln s_i^2)}{C} = \frac{M}{C} \tag{9.18}$$

where

$$\left\{ \begin{array}{l} M = \nu \ln s^2 - \sum \nu_i \ln s_i^2 \\[2mm] \nu = \sum_{i=1}^{k} \nu_i \\[2mm] s^2 = \dfrac{\sum_{i=1}^{k} \nu_i s_i^2}{\nu} \quad \text{and} \\[4mm] C = 1 + \dfrac{\sum \left(\dfrac{1}{\nu_i}\right) - \dfrac{1}{\nu}}{3(k-1)} \end{array} \right. \tag{9.19}$$

If each $\nu_i \geq 5$, the chi-square distribution with $k - 1$ degrees of freedom serves as a satisfactory approximation to the distribution of B; otherwise, percentage points of $M = BC$ may be obtained from tables computed by Merrington and Thompson [18]. In any case, if B is sufficiently large $[B > \chi_\alpha^2(k - 1)$ or $M = CB >$ value tabled by Merrington and Thompson], we reject Eq. (9.17), the hypothesis of equal variances.

Note. Since $C > 1$, it is unnecessary to compute C when $M < \chi_\alpha^2(k - 1)$. In Example 9.2 we illustrate Bartlett's test for equal sample sizes.

Example 9.2. Suppose four manufacturers, A, B, C, and D, make the same gauge of copper wire. Suppose ten measurements of tensile strength are made on the wire produced by each manufacturer. The measurements in pounds, after subtracting 5000 pounds from each, are shown in Table 9.1, along with means, variances, and common logarithms of the variances of the samples. The problem is to determine if the manufacturers make wire of the same tensile strength, if it is assumed that all conditions of the manufacturing process are the same for this gauge of copper wire. We use Bartlett's test to compare the population variances. In Chap. 10 we consider the problem of comparing the population means.

Table 9.1
Tensile Strength in Pounds, minus 5000 Pounds,
of 10 Samples of the Same Gauge of Copper Wire

Manufacturer ↘ Sample Number	A $x_1' - 5000$	B $x_2' - 5000$	C $x_3' - 5000$	D $x_4' - 5000$	
1	110	130	100	70	
2	90	45	200	40	
3	120	50	90	100	
4	130	40	70	180	
5	115	45	90	40	
6	105	55	130	150	
7	50	65	80	200	
8	75	120	70	210	
9	85	50	80	220	
10	40	150	150	250	
$\bar{x}_i' - 5000$	92	75	106	146	
$\bar{x}_i'$	5092	5075	5106	5146	
s_i^2	895	1717	1760	6093	$s^2 = 2616.25$
$\log s_i^2$	2.95182	3.23477	3.24551	3.78483	3.41768

For the case where $n_1 = n_2 = \cdots = n_k = n$, the B statistic become

$$B = \frac{(n - 1)(k \ln s^2 - \sum_{i=1}^{k} \ln s_i^2)}{C} \qquad (9.20)$$

where

$$C = 1 + \frac{k + 1}{3k(n - 1)} \quad \text{and} \quad s^2 = \frac{\sum s_i^2}{k} \qquad (9.21)$$

In particular, since $\ln a = \log_{10} a / \log_{10} e = 2.30259 \log_{10} e$, we have

$$BC = (10 - 1)(2.30259)[4(3.41768) - (2.95182 + \cdots + 3.78483)]$$
$$= 9.4040$$

$$C = 1 + \frac{5}{3 \cdot 4 \cdot 9} = 1.0463$$

and, therefore

$$B = 8.98$$

Since $\chi^2_{.05}(3) = 7.81$ and $\chi^2_{.01}(3) = 11.34$, we reject the equality of variance hypothesis at the five per cent level, but fail to reject this hypothesis at the one per cent level. Note that for a one per cent level test we need not compute C, since $BC = 9.40$ is already less than the significant value $\chi^2_{.01}(3) = 11.34$.

If it is suspected that hypothesis (9.17) fails to hold because exactly one population variance is appreciably larger than the other $k - 1$ variances, then a test developed by Cochran [8, 23] when $\nu_1 = \cdots = \nu_k = \nu$ is more appropriate than Bartlett's test. The test statistic is

$$g = \frac{\text{largest of the } s_i^2}{\sum\limits_{i=1}^{k} s_i^2} \tag{9.22}$$

Tables of $g_\alpha(k, \nu)$ for which $P[g \geq g_\alpha(k, \nu)] \doteq \alpha$ are given in Chap. 15 of [23]. Thus, in Example 9.2, if the experimenter had suspected before making measurements that, in case Eq. (9.17) failed, the variance of manu- facturer D was greater than the variances for manufacturers A, B, and C, then Cochran's test should be applied.

Hartley [17] proposed that

$$F_{\max} = \frac{\text{largest of the } s_i^2}{\text{smallest of the } s_i^2} \tag{9.23}$$

be used to test hypothesis (9.17). Other tests, sequential tests, have been proposed by Girshick [14], Wald [25], and Cox [9].

If the populations are known to be normal or almost normal, all tests which have been mentioned involving two or more variances are appropriate. However, if the populations are not normal, these tests are very impractical, since rejection of the hypothesis (9.17) could mean that the population variances are unequal, that the populations are not normal, or both. That is, these tests are very sensitive to nonnormality (recall that the test of $\mu_1 = \mu_2$ is fairly insensitive to nonnormality). Thus, for example, one should not use any of the above-mentioned tests as preliminary tests of equality of variances before testing the equality of means whenever there is doubt concerning the normality of the populations. Concerning this point, Box

[6, p. 333] writes, "To make the preliminary test on variances is rather like putting to sea in a rowing boat to find out whether conditions are sufficiently calm for an ocean liner to leave port!" Efforts have been made, with some success, by Bartlett and Kendall [4] and Odeh and Olds [21] to find a test of hypothesis (9.17) which is not so sensitive to nonnormality.

9.5. SPECIAL CASES OF THE F DISTRIBUTION

In Sect. 9.2 we noted that χ_2^2/ν_2 converges in probability to 1 as ν_2 approaches infinity. Thus

$$F(\nu_1, \nu_2) = \frac{\dfrac{\chi_1^2}{\nu_1}}{\dfrac{\chi_2^2}{\nu_2}}$$

converges in probability to

$$F(\nu_1, \infty) = \frac{\chi_1^2}{\nu_1} = \frac{\chi^2(\nu_1)}{\nu_1} \tag{9.24}$$

as ν_2 approaches infinity. That is, the χ^2 per degree of freedom distribution with ν_1 degrees of freedom is a special case of the F distribution with ν_1 and ∞ degrees of freedom. So for any α

$$F_\alpha(\nu_1, \infty) = \frac{\chi_\alpha^2(\nu_1)}{\nu_1} \tag{9.25}$$

In particular, using Tables VII and V, we see that

$$F_{.05}(10, \infty) = 1.83 = \frac{\chi^2(10)}{10}$$

Further, as ν_1 approaches infinity, F converges in probability to

$$F(\infty, \nu_2) = \frac{1}{\dfrac{\chi_2^2}{\nu_2}} = \frac{\nu_2}{\chi^2(\nu_2)} \tag{9.26}$$

It follows that

$$F_\alpha(\infty, \nu_2) = \frac{\nu_2}{\chi_{1-\alpha}^2(\nu_2)} \tag{9.27}$$

That is, the reciprocals of lower (left) tail values of χ^2 per degree of freedom obtained from Table V are special cases of upper (right) tail values of F obtained in the extreme right-hand corner of Table VII.

From Example 7.2 we know that when $\nu_1 = 1$, $\chi^2(\nu_1)/\nu_1 = u^2$, where u is a standard normal deviate. Thus, we may write

$$F(1, \nu_2) = \frac{u^2}{\dfrac{\chi_2^2}{\nu_2}} \tag{9.28}$$

From Eq. (8.5) we know that

$$\frac{u}{\sqrt{\dfrac{\chi^2}{\nu_2}}} = t$$

with ν_2 degrees of freedom. Thus, Eq. (9.28) may be written as

$$F(1, \nu_2) = t^2(\nu_2) \tag{9.29}$$

Since the relation

$$P[F(1, \nu_2) < F_\alpha(1, \nu_2)] = 1 - \alpha$$

may be written as

$$P[-\sqrt{F_\alpha(1, \nu_2)} < t(\nu_2) < \sqrt{F_\alpha(1, \nu_2)}] = 1 - \alpha$$

we see, knowing that the t distribution is symmetric about $t = 0$, that

$$P[t(\nu_2) < -\sqrt{F_\alpha(1, \nu_2)}] = P[t(\nu_2) > \sqrt{F_\alpha(1, \nu_2)}] = \frac{\alpha}{2}$$

But

$$P[t(\nu_2) > t_{\alpha/2}(\nu_2)] = \frac{\alpha}{2}$$

Thus, it follows that

$$t_{\alpha/2}(\nu_2) = \sqrt{F_\alpha(1, \nu_2)}$$

or

$$F_\alpha(1, \nu_2) = t_{\alpha/2}^2(\nu_2) \tag{9.30}$$

Also, for $\nu_2 = 1$ it can be shown that

$$F(\nu_1, 1) = \frac{1}{t^2(\nu_1)} \tag{9.31}$$

and

$$F_\alpha(\nu_1, 1) = \frac{1}{t_{(1+\alpha)/2}^2(\nu_1)} \tag{9.32}$$

Finally, if $\nu_1 = 1$ and ν_2 approaches infinity, Eq. (9.30) becomes

$$F_\alpha(1, \infty) = u_{\alpha/2}^2 \tag{9.33}$$

and if $\nu_2 = 1$ and ν_1 approaches infinity, Eq. (9.32) becomes

$$F_\alpha(\infty, 1) = \frac{1}{u^2_{(1+\alpha)/2}} \qquad (9.34)$$

where $u_{\alpha/2}$ and $u_{(1+\alpha)/2}$ are points of the standard normal variate such that

$$P[u > u_{\alpha/2}] = \frac{\alpha}{2} \quad \text{and} \quad P[u > u_{(1+\alpha)/2}] = \frac{1 + \alpha}{2} \qquad (9.35)$$

These results are brought together in Table 9.2. It should be observed that u^2, t^2, and χ^2 percentage points are special cases of F values which appear along the borders of the α level F table.

Table 9.2

Percentage Points F_α of the $F(\nu_1, \nu_2)$ Distribution

for which $P[F > F_\alpha(\nu_1, \nu_2)] = \alpha$

ν_2 ＼ ν_1	1	$\cdots \nu_1 \cdots$	∞
1	$t^2_{\alpha/2}(1) = \dfrac{1}{t^2_{(1+\alpha)/2}(1)}$	$\cdots \dfrac{1}{t^2_{(1+\alpha)/2}(\nu_1)} \cdots$	$\dfrac{1}{u^2_{(1+\alpha)/2}}$
$\vdots$ ν_2 $\vdots$	$t^2_{\alpha/2}(\nu_2)$	$\cdots F_\alpha(\nu_1, \nu_2) \cdots$	$\dfrac{\nu_2}{\chi^2_{1-\alpha}(\nu_2)}$
∞	$u^2_{\alpha/2}$	$\cdots \chi^2_\alpha(\nu_1)/\nu_2 \cdots$	1

9.6. EXERCISES

9.1. (a) Use Eq. (9.2) to find the density functions of $F(2, 4)$, $F(4, 2)$, $F(4, 4)$, and $F(2, 6)$. (b) Graph the density function of $F(2, 4)$ and $F(4, 2)$. (c) Use the definition to find the means of the random variables $F(2, 4)$ and $F(4, 4)$. Check results, using Theorem 9.1. (d) Use the definition to find the variance of the random variable $F(2, 4)$. Check results, using Theorem 9.1. (e) Without tables, find $F_{.05}(2, 4)$ and $F_{.05}(4, 4)$ and compare with the values in Table VII.

9.2. Suppose two samples of 12 and 20 objects, respectively, have variances $s_1^2 = 20$ and $s_2^2 = 12$. (a) Is there reason to doubt that the samples are from populations having equal variances? Use a five per cent level test. (b) Find 95 per cent confidence limits for the ratio σ_1^2/σ_2^2; for the ratio σ_2^2/σ_1^2.

9.3. (a) Use the transformation

$$F = \frac{\nu_2}{\nu_1} \cdot \frac{y}{1 - y} \qquad (0 \le y \le 1) \qquad (9.36)$$

to express the density function (9.2) as a density function $f(y)$ in terms of y. Compare this new density function with Eq. (3.68) of Exercise 3.70.

We call $f(y)$ the *beta density function*. As we mentioned before, Pearson has extensively tabulated percentage points of the beta distribution. (b) Use Exercise 3.70(b) to show that the mean and variance of F are as given in the statement of Theorem 9.1.

9.4. (a) Use Eq. (9.16) and a method similar to that employed in preparing Fig. 7.3 to draw a power curve for the one-sided test of $\sigma_1^2 = \sigma_2^2$ against $\sigma_1^2 > \sigma_2^2$ when $\nu_1 = 10$, $\nu_2 = 8$, and $\alpha = 0.05$. (b) Use this curve to find the probability of making the type 2 error if $\sigma_1^2 = 3\sigma_2^2$.

9.5. Since in Theorem 9.1 χ_1^2 and χ_2^2 are independently distributed, it follows, when Eq. (7.1) is used, that their joint density function $f(\chi_1^2, \chi_2^2)$ is given by $f(\chi_1^2) \cdot f(\chi_2^2)$ and that the probability element is

$$f(\chi_1^2, \chi_2^2)\, d(\chi_1^2)\, d(\chi_2^2)$$

$$= \frac{e^{-(1/2)(\chi_1^2 + \chi_2^2)}}{4\Gamma\left(\frac{\nu_1}{2}\right)\Gamma\left(\frac{\nu_2}{2}\right)} \left(\frac{\chi_1^2}{2}\right)^{(\nu_1/2)-1} \left(\frac{\chi_2^2}{2}\right)^{(\nu_2/2)-1} d(\chi_1^2)\, d(\chi_2^2) \tag{9.37}$$

Solving for χ_1^2 in Eq. (9.1) gives

$$\chi_1^2 = \frac{\nu_1}{\nu_2} F \chi_2^2$$

which, when substituted in Eq. (9.37) yields

$$f(F, \chi_2^2) = \frac{\left(\frac{\nu_1}{\nu_2}\right)^{\nu_1/2} e^{-(1/2)[1+(\nu_1/\nu_2)F]\chi_2^2}}{2\Gamma\left(\frac{\nu_1}{2}\right)\Gamma\left(\frac{\nu_2}{2}\right)} F^{(\nu_1/2)-1} \left(\frac{\chi_2^2}{2}\right)^{(\nu_1+\nu_2-2)/2} \tag{9.38}$$

On integrating Eq. (9.38) over the domain of χ_2^2 $(0 < \chi_2^2 < \infty)$, we obtain the marginal density function given by Eq. (9.2). Prove Eq. (9.38) and then Eq. (9.2).

9.6. Prove Eqs. (9.27) and (9.32).

9.7. A new method and an old method for counting bacteria in rat feces were to be compared. Films were made by both methods, and the resulting slides were fixed and stained with crystal violet. Twenty-five random fields were examined with a microscope and the number of bacteria counted in each field for each film. The results shown in Table 9.3 were obtained:*

Table 9.3

Old Method					New Method				
12	11	4	7	0	16	28	31	20	28
5	6	0	15	7	26	20	21	27	28
13	9	15	0	7	29	23	24	17	42
11	0	13	15	49	21	21	41	27	29
0	4	44	32	10	26	27	23	28	21

* Wallace, R. H., "A Direct Method for Counting Bacteria in Feces," *Journal of Bacteriology*, Vol. **64** (1952), 593–594.

(a) Test the hypothesis that the variances are equal against the alternative hypothesis that the variance for the new method is smaller than the variance for the old method. (b) Find a 90 per cent confidence interval for the ratio of the variance of the old method to the variance of the new method. (c) Find the power function for the test in (a). Use the power function to compare the variances.

9.8. (a) The distribution of $z = \frac{1}{2} \ln_e F$, studied by R. A. Fisher, is more nearly normal than the F distribution. Show that

$$f(z)\, dz = 2Ce^{\nu_1 z}\left(1 + \frac{\nu_1}{\nu_2}e^{2z}\right)^{-(\nu_1+\nu_2)/2} dz, \qquad -\infty < z < \infty \qquad (9.39)$$

(b) Find C and show that $f(z)$ is symmetrical if $\nu_1 = \nu_2$. (c) It can be shown that z is approximately normally distributed with mean and variance given by

$$\mu_z = -\frac{1}{2}\left(\frac{1}{\nu_1} - \frac{1}{\nu_2}\right) \quad \text{and} \quad \sigma_z^2 = \frac{1}{2}\left(\frac{1}{\nu_1} + \frac{1}{\nu_2}\right) \qquad (9.40)$$

These two relations make possible a quick comparison of two variances when one computes

$$z_c = \frac{1}{2}\ln\frac{s_2^2}{s_1^2} = \ln\frac{s_1}{s_2}$$

Use this method to answer Exercise 9.7(a).

9.9. The dispersions of tensile strength of iron, as measured by testing machines at seven foundries were to be compared. Six bars were poured at each foundry under conditions as nearly alike as possible. The measurements were made in tons per square inch. The data along with the type of meehanite metal cast and the type of test machine are given in Table 9.4. (a) Test the equality of variances (measures of dispersion thought

Table 9.4*

Foundry	1	2	3	4	5	6	7
Type Meehanite	GD	GC	GA	GD	GC	GD	GB
Testing Machine	Avery	Buckton	Avery	Buckton	Buckton	Buckton	Avery
Tensile—tons/sq in.							
Bar 1	17.70	18.40	25.25	17.00	19.86	17.55	23.00
2	18.00	19.20	26.47	15.75	20.00	17.68	22.70
3	17.93	19.84	25.35	18.90	19.29	17.80	21.80
4	16.63	19.16	23.26	17.50	18.11	17.26	22.60
5	17.06	19.04	24.85	20.00	19.11	17.43	22.00
6	17.46	19.72	22.20	17.70	18.42	17.40	21.70
Mean $\bar{x}_i$	17.46	19.23	24.56	17.81	19.13	17.52	22.30
Variance s_i^2	0.2838	0.2686	2.4184	2.1984	0.5720	0.0390	0.2880
s_i	0.53	0.52	1.56	1.48	0.76	0.20	0.54

* C. A. Bennett and N. L. Franklin, *Statistical Analysis in Chemistry and the Chemical Industry*. New York: John Wiley & Sons, Inc., 1954, p. 199.

to be appropriate), using Bartlett's test. (b) Test the equality of variances, using either Cochran's or Hartley's test. (c) Test the equality of variances, using Merrington and Thompson's tables (p. 198 of Bennett and Franklin). (d) Test the equality of variances of the Avery testing machines. (e) Test the equality of variances of the Buckton testing machines. (f) In how many ways can seven variances $\sigma_1^2, \ldots, \sigma_7^2$ fail to be equal? (Only the fact that two variances are not equal is important in this question. Magnitude of difference is not to be taken into account.) Find a similar solution for k variances.

9.10. (a) What is the minimum sample size $n_1 = n_2 = n$ one would take if it is required in a five per cent level one-sided test of the null hypothesis $\sigma_1^2 = \sigma_2^2$ that β be not greater than 0.10 when $\sigma_1^2 = 3\,\sigma_2^2$ is true? (b) What should the common sample size in (a) be for a two-sided test?

9.11. An outline of the proof that B in Eq. (9.18) is approximately distributed as χ^2 with $k - 1$ degrees of freedom is given by Anderson and Bancroft [1, pp. 142–144]. Fill in the details of this proof.

It should be noted that an infinite set of constants γ_r, *called cumulants*, are used in place of moments to characterize the distribution function. The cumulants are defined by the generating function

$$C(t) = \log M(t) \tag{9.41}$$

where $M(t)$ is the generating function of μ_r'. Thus, the rth cumulant is given by

$$\gamma_r = \left. \frac{d^r C(t)}{dt_r} \right|_{t=0} \tag{9.42}$$

It can be shown that $\gamma_1 = \mu_1'$ and $\gamma_2 = \sigma^2$.

Further, Stirling's formula for approximating factorials includes one more term than the approximation given in Exercise 3.67. The extended formula is

$$\Gamma(\alpha + 1) = \sqrt{2\pi}\, e^{-\alpha} \alpha^{\alpha+1/2} \left(1 + \frac{1}{12\alpha}\right) \tag{9.43}$$

9.12. The sample sizes and variances for seven nonselected strains of guayule in the $54 \pm$ chromosome group, according to the data taken from Federer*, Table 9.5, are as follows

Table 9.5

n_i	117	119	117	115	119	116	116
s_i^2	9.28	6.80	7.26	7.43	9.99	14.02	10.80

(a) Use Bartlett's test to determine if the population variances are homogenous. (b) Test the homogeneity of variances, using any other test of variances which you consider appropriate.

---

* W. T. Federer, "Variability of Certain Seed, Seedling, and Young Plant Characters of Guayule," U. S. Dept. Agr. Tech. Bull. 919 (1946).

REFERENCES

1. Anderson, R. L. and T. A. Bancroft, *Statistical Theory in Research*. New York: McGraw-Hill, Inc., 1952, Chaps. 7 and 12.

2. Bancroft, T. A., "Probability Values for the Common Tests of Hypotheses," *Journal of the American Statistical Association*, Vol. **45** (1950), pp. 211–17.

3. Bartlett, M. S., "Properties of Sufficiency and Statistical Tests," *Proceedings of Royal Society of London*, Series A, Vol. **160,** 268–82.

4. ——— and D. G. Kendall, "The Statistical Analysis of Variance-Heterogeneity and the Logarithmic Transformation," *Journal of the Royal Statistical Society*, Supplement 8 (1948), pp. 128–38.

5. Bennett, C. A. and N. L. Franklin, *Statistical Analysis in Chemistry and the Chemical Industry*. New York: John Wiley & Sons, Inc., 1954, Chaps. 4 and 5.

6. Box, G. E. P. and S. L. Andersen, "Robust Test for Variances and Effect of Non-normality and Variance Heterogeneity on Standard Tests," *Technical Report No.* 7, Department of the Army Project No. 599–01–004, 1954.

7. ———, "Non-normality and Tests on Variances," *Biometrika*, Vol. **40** (1953), pp. 318–35.

8. Cochran, W. G., "The Distribution of the Largest of a Set of Estimated Variances as a Fraction of Their Total," *Annals of Eugenics*, London, Vol. **11** (1941), pp. 47–.

9. Cox, D. R., "Sequential Tests for Composite Hypotheses," *Proceedings of the Cambridge Philosophical Society*, Vol. **48** (1952), pp. 290–99.

10. Davies, O. L., *The Design and Analysis of Industrial Experiments*. New York: Hafner Publishing Co., Inc., 1954, pp. 614–15.

11. Fisher, R. A., "On a Distribution Yielding the Error Functions of Several Well-Known Statistics," *Proceedings of the International Mathematical Congress*, Toronto, 1924, pp. 805–13.

12. ———, *Statistical Methods for Research Workers*, 1st ed. Edinburgh: Oliver and Boyd, 1925.

13. Gayen, A. K., "The Distribution of the Variance Ratio in Random Samples of Any Size Drawn from Non-normal Universes," *Biometrika*, Vol. 37 (1950), pp. 236–55.

14. Girshick, M. A., "Contributions to the Theory of Sequential Analysis. I," *Annals of Mathematical Statistics*, Vol. **17** (1946), pp. 123–43.

15. Hald, A., *Statistical Theory with Engineering Applications*. New York: John Wiley & Sons, Inc., 1952, Chaps. 11 and 14.

16. Hartley, H. O., "Testing the Homogeneity of a Set of Variances," *Biometrika*, Vol. **31** (1944), pp. 249–55.

17. ———, "Maximum F Ratio as a Short-cut Test for Heterogeneity of Variances," *Biometrika*, Vol. 37 (1950), pp. 308–12.

18. Merrington, M. and C. M. Thompson, "Tables for Testing the Homogeneity of a Set of Estimated Variances," *Biometrika*, Vol. 33 (1946), pp. 292–304.

19. Neyman, J. and E. S. Pearson, "On the Problem of k Samples," *Bull. Acad. Polonaise Sci. Lett.*, Series A, pp. 460–81.

20. Pearson, E. S. and H. O. Hartley, *Biometrika Tables for Statisticians*, Vol. **1**. London, England: Cambridge University Press, 1954.

21. Odeh, R. E. and E. G. Olds, *Notes on the Analysis of Variance of Logarithms of Variances*, ASTIA Document No. AD211917, U. S. Department of Commerce, Washington 25, D. C., 1959.

22. Snedecor, G. W., *Analysis of Variance and Covariance*. Ames, Iowa: Collegiate Press Inc., 1934.

23. *Techniques of Statistical Analysis*, Statistical Research Group, Columbia University. New York: McGraw-Hill, Inc., 1947, Chaps. 8 and 15.

24. Vogler, L. E. and K. A. Norton, *Graphs and Tables of the Significance Levels $F(\nu_1, \nu_2, p)$ for the Fisher-Snedecor Variance Ratio*, U. S. Department of Commerce, National Bureau of Standards, Washington 25, D. C., 1957.

25. Wald, A., *Sequential Analysis*. New York: John Wiley & Sons, Inc., 1947, Chap. 4.

10

ANALYSIS OF VARIANCE— ONE-WAY CLASSIFICATION

The analysis of variance is described as it relates to the sum of squares identity. It is demonstrated how several sources of variation may be isolated, estimated, and tested. Techniques are discussed in relation to models and assumptions. The role of assumptions, sums of squares identities, and models is emphasized throughout. The concept of a single degree of freedom and its relation to the sum of squares identity, hypothesis testing, and confidence intervals is discussed. The distribution of the range of sample means is introduced and used in testing hypotheses and constructing simultaneous confidence intervals. Scheffé's method for constructing simultaneous confidence intervals is presented. Duncan's multiple test procedure is explained.

10.1. INTRODUCTION

We have already mentioned (Chaps. 8 and 9) that it is sometimes necessary to extend experiments to the comparison of several means. One might think that this can be done by testing the differences between each pair from among k means by using the t test repeatedly and, indeed, this is the case. However, there are some disadvantages to this procedure. First, it places the emphasis on pairwise comparisons, when actually the experimenter may be interested in comparisons involving $p(p = 3, \ldots, k)$ means. Second, it causes difficulty with the significance level when the experimenter is really interested in testing one or more hypotheses of the form $H_0: \mu_1 = \mu_2 = \cdots = \mu_p$. Suppose, for example, that one wishes to test

H_0 when $p = k = 6$, say. If we used the t test on pairs, it would be possible to make $\binom{6}{2} = 15$ tests. Suppose a five per cent level t test is used for each difference. In this case, the chance of saying that some one or more of the differences is significant might be as large as $1 - (1 - 0.05)^{15} \doteq 0.54$. That is, for five per cent level pairwise tests of the hypothesis $\mu_1 = \mu_2 = \cdots = \mu_6$ the significance level might actually be as large as $\alpha = 54$ per cent. This means that the hypothesis will be rejected as often as 54 per cent of the time when it is true. (This problem is discussed in some detail in Sect. 10.9.) Third, there is a loss in precision in estimating the variance if only the measurements of the two samples being compared are utilized.

If one is interested in testing a single hypothesis of the form $H_0: \mu_1 = \mu_2 = \cdots = \mu_k$ the F distribution is most often used. This requires two independent estimates of the variance σ^2 common to k normal populations from which k random and independent samples of size n_i ($i = 1, \ldots, k$), respectively, are drawn. Thus, the test involves analyzing variances. That is, the treatment of the data involves separating the variance of all observations into parts, each measuring variation which is attributed to specific causes.

The *method of analysis of variance*, an arithmetical process for splitting up a total variance into its component parts, has a much wider application than the simple generalization already noted, and is probably the most powerful procedure in the field of experimental statistics. We start our discussion of analysis of variance with the simplest case.

10.2. ANALYSIS OF VARIANCE FOR ONE-WAY CLASSIFICATION— APPLICATION

Measurements of a quantity can often be classified into categories corresponding to different conditions in an experiment. For example, the measurements resulting from 40 analyses of the concentration of iron in a standard solution may be classified into five categories of eight determinations corresponding to eight measurements made by each of five analysts. The measurements, in this example, fall into five mutually exclusive categories, where each category represents the population of all possible results which could be obtained by a particular analyst under constant conditions, and the eight specific results represent a sample from the population. If it is assumed that each analyst obtains measurements independent of the other analysts, the classification of observations is considered a one-way classification, and we say that there is a *single variable of classification*. In the example, the variable of classification is analyst, and the variable has five values. Since there is usually unexplained variation from one observation to another within a given category, we call the observations *values of a random variable*. There is one random variable for each category.

The data from an experiment are usually arranged in a rectangular array for ease of computation. A column of numbers represents measurements for one category. Following the usual convention, the different categories of measurements are often referred to as *treatments* as well as *columns of measurements*. Thus, when we refer to different treatments, we may be thinking of different schools, different analysts, different fertilizers, different concentrations of a solution, different methods, etc. Methods of estimating parameters and testing hypotheses in a one-way classification are given in Example 10.1 along with a useful notation. Certain theoretical justifications follow the example.

Example 10.1. (a) Use the coded data in Example 9.2 to find estimates of the means and effects for manufacturers A, B, and C. (b) Use a five per cent level test to determine if manufacturers A, B, and C, on the average, make copper wire of the same tensile strength. It is assumed that all conditions of the manufacturing process are as near the same as possible. (c) Find 90 per cent confidence limits for the population means μ_A, μ_B, and μ_C.

If the samples are assumed to be randomly and independently drawn, the estimates $\bar{x}_A$, $\bar{x}_B$, $\bar{x}_C$ and a_A, a_B, a_C shown in Table 10.1 are the best (in the sense of being unbiased and with minimum variance) point estimates of the true means μ_A, μ_B, μ_C and true effects α_A, α_B, α_C, respectively. The

Table 10.1
Tensile Strength of Copper Wire in Pounds

Manufacturer \ Sample No.	A $x_1 = x_1' - 5000$	B $x_2 = x_2' - 5000$	C $x_3 = x_3' - 5000$	
1	$x_{11} = 110$	$x_{21} = 130$	$x_{31} = 100$	
2	$x_{12} = 90$	$x_{22} = 45$	$x_{32} = 200$	
3	120	50	90	
4	130	40	70	
5	115	45	90	
6	105	55	130	
7	50	65	80	
8	75	120	70	
9	85	50	80	
10	$x_{1,10} = 40$	$x_{2,10} = 150$	$x_{3,10} = 150$	
$T_{i.}$	$T_A = T_1. = 920$	$T_B = T_2. = 750$	$T_C = T_3. = 1060$	$T_{..} = 2730$
$\bar{x}_{i.}$	$\bar{x}_A = \bar{x}_1. = 92$	$\bar{x}_B = \bar{x}_2. = 75$	$\bar{x}_C = \bar{x}_3. = 106$	$\bar{x} = \bar{x}_{..} = 91$
a_i	$a_A = a_1 = 1$	$a_B = a_2 = -16$	$a_C = a_3 = 15$	

notation and methods are the same as those introduced in Sect. 8.11, except that i takes values A, B, and C instead of 1, 2, and 3, and

$$T_{i.} = \sum_{j=1}^{n_i} x_{ij} \tag{10.1}$$

and

$$T.. = \sum_{i=1}^{k} \sum_{j=1}^{n_i} x_{ij} \qquad (10.2)$$

We call $T_{i.}$ the *total* of the observations in the ith column (or for the ith treatment) and $T..$ the *grand total* of all observations. For the coded data the estimate of the over-all mean is 91, and the estimates of the means of populations A, B, and C are 92, 75, and 106, respectively. Thus, for the original data the estimates of the over-all mean and the means of populations A, B, and C are, respectively, 5091, 5092, 5075, and 5106. In either case, the estimates of the effects of manufacturers A, B, and C are 1, -16 and 15, respectively. This points up one advantage of using effects, namely, that adding or subtracting a constant to all the observations does not change the estimates of the effects. Observe that

$$\sum_{i=1}^{3} a_i = 0$$

We use the general procedure, given on p. 218, to test the hypothesis in part (b) of the example.

1. The null hypothesis is $H_0: \mu_A = \mu_B = \mu_C$, and the alternative hypothesis is, "All the means are not equal." The alternative hypothesis for the F test is not $\mu_A \neq \mu_B \neq \mu_C$, for μ_A might be the same as μ_B, and both might be different from μ_C. In general, if the null hypothesis $H_0: \mu_1 = \cdots = \mu_k$ is rejected, it does not necessarily indicate that $\mu_i \, (i = 1, \ldots, k)$ is different from $\mu_j \, (j = 1, \ldots, k; i \neq j)$; it may only indicate that some linear combination of one subset of the μ's is different from a linear combination of some of the remaining μ's. That is, a significant F may indicate that $(\mu_1 + \mu_4 + \mu_5)/3$ is different from $(\mu_3 + \mu_{10})/2$, say.

2. The significance level is $\alpha = 0.05$. Assume that the three samples of common size 10 were randomly and independently drawn from three normal populations with common variance σ^2.

3. The statistic for the test is $F(2, 27) = s_m^2/s_e^2$, where, according to Eq. (8.54)

$$\begin{cases} s_m^2 = \dfrac{\displaystyle\sum_{i}^{3} \sum_{j}^{10} (\bar{x}_{i.} - \bar{x}..)^2}{3 - 1} \quad \text{and} \\[4ex] s_e^2 = \dfrac{\displaystyle\sum_{i}^{3} \sum_{j}^{10} (x_{ij} - \bar{x}_{i.})^2}{10 + 10 + 10 - 3} \end{cases} \qquad (10.3)$$

are unbiased estimators of σ^2 with two and 27 degrees of freedom, respectively. s_m^2 is an unbiased estimator of σ^2 only if H_0 is true, but s_e^2 is unbiased so long as the samples are randomly and independently drawn.

4. The critical region is made up of all those values of F for which $F > F_{.05}(2, 27) = 3.35$. The critical region is the upper tail of the F distribution, since failure of H_0 makes s_m^2 on the average larger than σ^2.

5. From Table 10.1, $a_A = 1$, $a_B = -16$, and $a_C = 15$. Thus

$$s_m^2 = \frac{10[1^2 + (-16)^2 + 15^2]}{2} = 10(241) = 2410$$

From Table 9.1, $s_A^2 = 895$, $s_B^2 = 1717$ and $s_C^2 = 1760$. Thus, the pooled or error variance is

$$s_e^2 = \frac{9(895) + 9(1717) + 9(1760)}{27} = \frac{4372}{3} = 1457.3$$

Hence, the computed F statistic F_c is

$$F_c = \frac{2410}{1457.3} = 1.65$$

6. Since $F_c = 1.65$ does not fall in the critical region, we fail to reject H_0 at the five per cent level of significance. That is, we reserve judgment about the relative strengths of wire manufacturered by A, B, and C until we have further evidence to the contrary. Hence, if we wish to take action as a consequence of the test, we act as though the three manufacturers make equally good wire.

The numerical calculations for this type of problem are often summarized in what is known as an *analysis of variance table*, shown in Table 10.2.

Table 10.2
Analysis of Variance

Source of Variation	Sum of Squares	Degrees of Freedom	Mean Square	Computed F	Critical F
Among means	4,820	2	2410	1.65	3.35
Within (error)	39,347	27	1457.3		
Total	44,167	29			

The proper conclusion can be read directly from this table, provided the assumptions hold.

To find $100(1 - \alpha)$ per cent confidence limits for mean μ of a normal population, we need independent estimates $\bar{x}_i$ and s^2 of the population mean and variance along with percentage points of the Student t distribution.

Then the symmetric $100 (1 - \alpha)$ per cent confidence limits are given by

$$\bar{x}_{i.} \pm t_{\alpha/2}(\nu) \cdot \frac{s}{\sqrt{n_i}} \tag{10.4}$$

where ν denotes the degrees of freedom of s and n_i denotes the size of the sample i. When a random sample is drawn from a single normal population with mean μ_i and variance σ^2, we know that $\bar{x}_{i.}$ and s_i^2 are independently distributed $(i = 1, 2, \ldots, k)$. Further, we know that $s_A^2, s_B^2,$ and s_C^2 are independent estimators of σ^2, the common population variance. Thus, s_e^2 is an estimator of σ^2 which is independent of $\bar{x}_{i.}$. This means that

$$\frac{(\bar{x}_{i.} - \mu_i)}{\sqrt{\dfrac{s_e^2}{n_i}}} \qquad (i = A, B, C) \tag{10.5}$$

is distributed as the Student t with degrees of freedom of s_e^2 being 27 in our problem. Now, since $t_{.05}(27) = 1.703$, the 90 per cent confidence limits of μ_i are given by

$$\bar{x}_{i.} \pm \frac{1.703 s_e}{\sqrt{10}} \tag{10.6}$$

From the analysis of variance table we get $s_e^2 = 1457.3$. Thus

$$t_{.05}(27)\sqrt{\frac{s_e^2}{n}} = 1.703\sqrt{145.73} = 20.6$$

Therefore, the 90 per cent confidence intervals for $\mu_A, \mu_B,$ and μ_C are, respectively

$$\begin{cases} 5071.4 \text{ lb} < \mu_A < 5112.6 \text{ lb} \\ 5054.4 \text{ lb} < \mu_B < 5095.6 \text{ lb} \\ 5085.4 \text{ lb} < \mu_C < 5126.6 \text{ lb} \end{cases} \tag{10.7}$$

The confidence intervals in (10.7) were found by using a pooled error variance. We could have used the individual sample variance s_i^2 with nine degrees of freedom in Eq. (10.4) to obtain

$$\begin{cases} 5074.6 \text{ lb} < \mu_A < 5109.4 \text{ lb} \\ 5051.0 \text{ lb} < \mu_B < 5099.0 \text{ lb} \\ 5081.7 \text{ lb} < \mu_C < 5130.3 \text{ lb} \end{cases} \tag{10.8}$$

since $t_{.05}(9)/\sqrt{10} = (1.833)(0.3162) = 0.5796$, $s_A = 29.92$, $s_B = 41.44$, and $s_C = 41.95$. However, both sets of intervals, (10.7) and (10.8), should not be used in the same experiment. If it is reasonable to assume the variances should be pooled, then only (10.7) should be used, since the error variance is more precise, having more degrees of freedom. On the other hand, if it is not reasonable to assume that the variances of the different populations are the same, then (10.8) should be used, even though there is a loss in the number of degrees of freedom in the variance estimator. In no case should an interval for one population mean be obtained by using the pooled variance and an interval for another population mean by using the single sample variance.

It should be noted that when the samples are randomly and independently selected the point estimates of the parameters are unbiased. If, in addition to these assumptions, the populations are normally distributed, then confidence intervals may be obtained. Further, if these three assumptions along with the equality of variance assumption hold, then we may test the hypothesis $H_0: \mu_1 = \cdots = \mu_k$ and obtain confidence intervals by using the pooled variance. All of these statements are based on the assumption that the jth observation in the ith sample can be written as

$$x_{ij} = \mu_i + \epsilon_{ij} \qquad (i = 1, \ldots, k; \, j = 1, \ldots, n_i) \qquad (10.9)$$

or

$$x_{ij} = \mu + \alpha_i + \epsilon_{ij} \qquad (i = 1, \ldots, k; \, j = 1, \ldots, n_i) \qquad (10.10)$$

where

$$\mu = \frac{\sum\limits_{i=1}^{k} \mu_i}{k} \quad \text{and} \quad \alpha_i = \mu_i - \mu \qquad (10.11)$$

These symbols were discussed and given names in Sect. 8.11.

10.3. ANALYSIS OF VARIANCE FOR ONE-WAY CLASSIFICATION— THEORY

Before we discuss in detail the one-way classification, we review and expand some of the related ideas for a single sample. In a population with mean μ and variance σ^2, we may think of any observation x as the sum of the mean μ and a deviate ϵ. Thus, in a sample $x_1, x_2, \ldots, x_n$ of size n the jth observation x_j may be written as

$$x_j = \mu + \epsilon_j \qquad (j = 1, \ldots, n) \qquad (10.12)$$

or, in case $\bar{x}$ is the sample mean, as

$$x_j = \bar{x} + e_j \qquad (j = 1, \ldots, n) \qquad (10.13)$$

where e_j is the deviate about the sample mean. If the sample is randomly selected, $\bar{x}$ and s^2, the sample variance, are unbiased estimators of μ and σ^2. It was shown in Sect. 8.12 that

$$\sum_{j=1}^{n} (x_j - \mu)^2 = \sum_{j=1}^{n} (x_j - \bar{x})^2 + n(\bar{x} - \mu)^2 \qquad (10.14)$$

is an algebraic identity, called a *sum of squares identity*. Theorem 7.3 states that $\bar{x}$ and s^2 are independently distributed, provided a random sample is drawn from a normal population. Thus, since n, μ, and σ^2 are constants, it follows that

$$\frac{n(\bar{x} - \mu)^2}{\sigma^2} = \frac{s_1^2}{\sigma^2} \quad \text{and} \quad \frac{\sum_{j=1}^{n} (x_j - \bar{x})^2}{\sigma^2} = \frac{(n-1)s^2}{\sigma^2} \qquad (10.15)$$

are independently distributed. Further, according to Theorems 7.1 and 7.4, the ratios in Eq. (10.15) are distributed as χ^2 with one degree and $n-1$ degrees of freedom. Thus, because of the additive nature of χ^2 (see Theorem 7.2), it follows that

$$\frac{\sum_{j=1}^{n} (x_j - \mu)^2}{\sigma^2} = \frac{\sum_{j=1}^{n} (x_j - \bar{x})^2}{\sigma^2} + \frac{n(\bar{x} - \mu)^2}{\sigma^2}$$

is distributed as χ^2 with n degrees of freedom. Since s_1^2/σ^2 and $(n-1)s^2/\sigma^2$ are independently distributed as χ^2 with 1 degree and $n-1$ degrees of freedom, respectively, then, according to the corollary to Theorem 9.1

$$\frac{s_1^2}{s^2} = \frac{\dfrac{n(\bar{x} - \mu)^2}{1}}{\dfrac{\sum(x_j - \bar{x})^2}{n-1}} = \frac{(\bar{x} - \mu)^2}{\dfrac{s^2}{n}} = t^2(n-1) \qquad (10.16)$$

is distributed as F with 1 degree and $n-1$ degrees of freedom.

It is clear that the variance estimator s^2 is based on deviations of the observations about the sample mean, whereas s_1^2 is based on the deviation of the sample mean from the population mean μ. Thus, for a given sample, if μ is unknown and is hypothesized to be some specified value, s^2 is fixed, but s_1^2 depends on the hypothesized value of μ, say μ_0. If the value μ_0 is used in place of μ in s_1^2, we obtain

$$s_{01}^2 = n(\bar{x} - \mu_0)^2 = n[(\bar{x} - \mu) + (\mu - \mu_0)]^2$$
$$= n(\bar{x} - \mu)^2 + n(\mu - \mu_0)^2 + 2n(\bar{x} - \mu)(\mu - \mu_0)$$

Thus

$$E(s_{01}^2) = \sigma^2 + n(\mu - \mu_0)^2 \qquad (10.17)$$

since

$$E(\bar{x} - \mu) = 0 \quad \text{and} \quad E[n(\bar{x} - \mu)^2] = \sigma^2$$

That is, the expected value of s_{01}^2 is larger than σ^2 when $\mu_0 \neq \mu$. So s_{01}^2 tends to be larger than s^2 and s_{01}^2/s^2 tends to fall in the upper tail of the distribution of $F(1, n-1)$. It is for this reason that the critical region is taken in the upper tail of the F distribution and a significantly large F is used to reject the hypothesized value μ_0 as the true mean.

Now we consider k populations. In the ith population with mean μ_i and variance σ_i^2, we may think of any observation x_i as the sum of the mean μ_i and a deviate ϵ_i ($i = 1, 2, \ldots, k$). Thus, for a sample $x_{i1}, x_{i2}, \ldots, x_{in_i}$ of size n_i, the jth observation x_{ij} in the ith population may be written as Eq. (10.9) or Eq. (10.10). Using Eqs. (10.1) and (10.2), we define the sample means and effects as follows

$$
\left\{
\begin{aligned}
\bar{x}_{i.} &= \frac{T_{i.}}{n_i} = \sum_{j=1}^{n_i} x_{ij} \bigg/ n_i \\
\bar{x}_{..} &= \frac{T_{..}}{n_.} = \sum_{i}^{k} \sum_{j}^{n_i} x_{ij} \bigg/ n. \\
a_i &= \bar{x}_{i.} - \bar{x}_{..}
\end{aligned}
\right.
\tag{10.18}
$$

where

$$
n. = \sum_{i=1}^{k} n_i
$$

Thus, as in Eqs. (8.30) and (8.39), the observation equations may be written as

$$
x_{ij} = \bar{x}_{i.} + e_{ij} \qquad (i = 1, \ldots, k; \ j = 1, \ldots, n_i) \tag{10.19}
$$

or

$$
x_{ij} = \bar{x}_{..} + a_i + e_{ij} \qquad (i = 1, \ldots, k; \ j = 1, \ldots, n_i) \tag{10.20}
$$

where ϵ_{ij} is the deviate of the jth observation about the ith sample mean. It can be shown (Exercise 10.6) that the estimated means and effects defined in Eq. (10.18) are unbiased estimators of their corresponding parameters μ_i, μ, and α_i, provided the samples are randomly and independently selected. [Actually, it can be shown that these estimators are the so-called least-squares estimators (Chap. 14) or maximum likelihood estimators. Further, the sampling distributions of these estimators have the smallest variances of all linear functions of the observations which could be used to estimate μ_i, μ, and α_i.]

The sum of squares identity for the ith sample may be written as

$$
\sum_{j=1}^{n_i} (x_{ij} - \mu_i)^2 = \sum_{j=1}^{n_i} (x_{ij} - \bar{x}_{i.})^2 + n_i(\bar{x}_{i.} - \mu_i)^2 \quad (i = 1, 2, \ldots, k) \tag{10.21}
$$

When a random sample is drawn, we may think of

$$
\sum_{j=1}^{n_i} (x_{ij} - \mu_i)^2
$$

as partitioned into two component sums of squares which lead to two independent unbiased estimators of σ^2. If the sample, in addition to being random, is drawn from a normal population, then these two component sums of squares are independently distributed as $\sigma^2 \chi^2$ with $n_i - 1$ degrees and 1 degree of freedom, respectively. Thus, if k samples, in addition to being random, are independent, then

$$\sum_{j=1}^{n_1} (x_{1j} - \mu_1)^2, \ \sum_{j=1}^{n_2} (x_{2j} - \mu_2)^2, \ \cdots, \ \sum_{j=1}^{n_k} (x_{kj} - \mu_k)^2 \qquad (10.22)$$

lead to $2k$ independent unbiased estimators of σ^2 and, if the k samples are also drawn from normal populations, the $2k$ component sum of squares are independently distributed as $\sigma^2 \chi^2$ with $n_1 - 1$, 1, $n_2 - 1$, 1, $\ldots$, $n_k - 1$, 1 degrees of freedom, respectively. By adding the k sum of squares identities in Eq. (10.21) we get

$$\sum_i^k \sum_j^{n_i} (x_{ij} - \mu_i)^2 = \sum_i^k \sum_j^{n_i} (x_{ij} - \bar{x}_{i.})^2 + \sum_i^k n_i(\bar{x}_{i.} - \mu_i)^2 \qquad (10.23)$$

Thus, due to the additive property of χ^2 (Theorem 7.2) it follows that

$$\frac{\sum_i^k \sum_j^{n_i} (x_{ij} - \bar{x}_{i.})^2}{\sigma^2} \quad \text{and} \quad \frac{\sum_i^k n_i(\bar{x}_{i.} - \mu_i)^2}{\sigma^2} \qquad (10.24)$$

are independently distributed as χ^2 with $n. - k$ and k degrees of freedom, respectively. Also

$$\frac{\sum_i^k \sum_j^{n_i} (x_{ij} - \mu_i)^2}{\sigma^2}$$

is distributed as χ^2 with $n.$ degrees of freedom, but is not independent of the distribution of either of the statistics in (10.24). Further

$$s_p^2 = \frac{\sum_i \sum_j (x_{ij} - \bar{x}_{i.})^2}{n. - k} \quad \text{and} \quad s_1^2 = \frac{\sum_i^k n_i(\bar{x}_{i.} - \mu_i)^2}{k} \qquad (10.25)$$

are independent unbiased estimators of σ^2. Since ks_1^2/σ^2 and $(n. - k)s_p^2/\sigma^2$ are independently distributed as χ^2 with k and $n. - k$ degrees of freedom, respectively

$$\frac{s_1^2}{s_p^2} = \frac{\dfrac{\sum_i n_i(\bar{x}_{i.} - \mu_i)^2}{k}}{\dfrac{\sum_i \sum_j (x_{ij} - \bar{x}_{i.})^2}{n. - k}} \qquad (10.26)$$

is distributed as F with k and $n. - k$ degrees of freedom.

The estimator s_p^2, the pooled variance, sometimes called the *within variance* or *error variance* s_e^2, is based on deviations of sample values from their sample means and, thus, does not depend on the population means μ_1, μ_2, ..., μ_k. That is, the estimator s_p^2 is independent of the population means. The estimator s_1^2 of σ^2, on the other hand, does depend on the population means. Thus, we may test a set of hypothesized means μ_{01}, μ_{02}, ..., μ_{0k}, i.e.

$$H_0 : \mu_1 = \mu_{01}, \quad \mu_2 = \mu_{02}, \quad \ldots, \quad \mu_k = \mu_{0k} \tag{10.27}$$

by substituting these values in the second part of Eq. (10.25) to obtain s_{01}^2 and then comparing this estimator of σ^2 with s_p^2, using the $F(k, n. - k)$ distribution. If H_0 is true, then s_1^2/s_p^2 is distributed as $F(k, n. - k)$. However, if some of the hypothesized values μ_{0i} are not equal to the true mean μ_i, then we can show, following the argument used in proving Eq. (10.17), that

$$E(s_{01}^2) = \sigma^2 + \sum_{i=1}^{k} n_i (\mu_i - \mu_{0i})^2 \tag{10.28}$$

That is, the expected value of s_{01}^2 is larger than σ^2 when some $\mu_{0i} \neq \mu_i$. This means that s_{01}^2 tends to be larger than s_p^2 and, thus, s_{01}^2/s_p^2 tends to fall in the upper tail of the distribution of $F(k, n. - k)$. It is for this reason that the critical region is taken in the upper tail of the F distribution and a significantly large F is used to reject H_0. It should be noted that the hypothesis (10.27) is not often tested, since its rejection leads to the unsatisfactory statement that *some* population mean is not as hypothesized. (This raises questions such as, which and how many means are not as hypothesized?) Also, it would often be difficult to decide on specific values for μ_{0i} $(i = 1, 2, \ldots, k)$.

This brings us to the problem of giving the theoretical justification for the test of the more useful null hypothesis

$$H_0 : \mu_1 = \mu_2 = \cdots = \mu_k = \mu \tag{10.29a}$$

or

$$H_0 : \alpha_1 = \alpha_2 = \cdots = \alpha_k = 0 \tag{10.29b}$$

Obviously, a test of (10.29) requires that

$$\sum_{i=1}^{k} n_i (\bar{x}_{i.} - \mu_i)^2$$

in Eq. (10.23) be expressed in terms of deviates from μ, the true over-all mean, and this suggests that the over-all sample mean $\bar{x}..$ be introduced. Thus, we write

$$\bar{x}_{i.} - \mu_i = [(\bar{x}_{i.} - \bar{x}..) - (\mu_i - \mu)] + (\bar{x}.. - \mu) \tag{10.30a}$$

or

$$\bar{x}_{i.} - \mu_i = (a_i - \alpha_i) + (\bar{x}.. - \mu) \tag{10.30b}$$

Squaring both sides of Eq. (10.30b) and summing over all $n.$ observations simplifies to

$$\sum_{i=1}^{k} n_i(\bar{x}_i - \mu_i)^2 = \sum_{i=1}^{k} n_i(a_i - \alpha_i)^2 + n.(\bar{x}.. - \mu)^2 \tag{10.31}$$

Substituting Eq. (10.31) in Eq. (10.23) gives

$$\sum_{i}^{k} \sum_{j}^{n_i} (x_{ij} - \mu_i)^2 = \sum_{i} \sum_{j} (x_{ij} - \bar{x}_{i.})^2 + \sum_{i} n_i(a_i - \alpha_i)^2 + n.(\bar{x}.. - \mu)^2 \tag{10.32}$$

From the partition theory of the χ^2 distribution it follows that the two terms on the right-hand side of Eq. (10.31) are independently distributed as $\sigma^2 \chi^2$ with $k - 1$ degrees and 1 degree of freedom, respectively. Further

$$s_2^2 = \frac{\sum_{i=1}^{k} n_i(a_i - \alpha_i)^2}{k - 1} \quad \text{and} \quad s_3^2 = \frac{n.(\bar{x}.. - \mu)^2}{1} \tag{10.33}$$

are independent and unbiased estimators of σ^2. Since s_p^2 is an unbiased estimator of σ^2 and is independent of both s_2^2 and s_3^2, it follows that

$$\frac{s_2^2}{s_p^2} = \frac{\dfrac{\sum_{i=1}^{k} n_i(a_i - \alpha_i)^2}{k - 1}}{\dfrac{\sum_{i} \sum_{j} (x_{ij} - \bar{x}_{i.})^2}{n. - k}} \tag{10.34}$$

is distributed as F with $k - 1$ and $n. - k$ degrees of freedom, and

$$\frac{s_3^2}{s_p^2} = \frac{\dfrac{n(\bar{x}.. - \mu)^2}{1}}{\dfrac{\sum_{i} \sum_{j} (x_{ij} - \bar{x}_{i.})^2}{n. - k}} \tag{10.35}$$

is distributed as F with 1 degree and $n. - k$ degrees of freedom.

We have seen that the estimator s_p^2 of σ^2 does not depend on the population means (effects). The estimator s_2^2 of σ^2, on the other hand, does depend on the population effects. Thus, we may test the hypothesis (10.29) by substituting these hypothesized values of α_i (i.e., zero) in the first part of Eq. (10.33) to obtain

$$s_{02}^2 = \frac{\sum n_i(\bar{x}_{i.} - \bar{x}..)^2}{k - 1}$$

and then comparing this estimator of σ^2 with s_p^2, using the $F(k - 1, n. - k)$ distribution. If H_0 in (10.29) is true, then s_2^2/s_p^2 is distributed as $F(k - 1, n. - k)$, since s_p^2 and s_2^2 are independent and unbiased estimators of the same variance σ^2. However, if some of the hypothesized values μ_{0i} (or α_{0i}) are not equal to the true common mean μ (or effect zero), then we can show, following the argument used in proving Eq. (10.17), that

$$E(s_{02}^2) = \sigma^2 + \frac{1}{k - 1} \sum_{i=1}^{k} n_i(\mu_i - \mu)^2 = \sigma^2 + \frac{1}{k - 1} \sum_{i=1}^{k} n_i \cdot \alpha_i^2 \qquad (10.36)$$

when

$$n_1 \alpha_1 + \cdots + n_k \alpha_k = 0$$

That is, the expected value of s_{02}^2 is larger than σ^2 when some $\mu_{0i} \neq \mu$ (or some $\alpha_i \neq 0$). It is for this reason that the critical region is taken in the upper tail of the F distribution and a significantly large F is used to reject hypothesis (10.29).

Also, one could test the null hypothesis that the over-all mean μ is some specified value μ_0 by comparing s_3^2 with s_p^2, using the $F(1, n. - k)$ distribution. If a sample ratio s_3^2/s_p^2 falls in the upper region (upper tail) of the F distribution, we reject $\mu = \mu_0$ and conclude that $\mu \neq \mu_0$. If the true mean is actually different from the hypothesized mean μ_0, then the expected value of $s_{03}^2 = n.(\bar{x}.. - \mu_0)^2$ is

$$E(s_{03}^2) = \sigma^2 + n.(\mu - \mu_0)^2 \qquad (10.37)$$

where

$$n. = \sum_{i}^{k} n_i$$

The sum of squares identity (10.32) and associated mean squares used in testing the hypothesis (10.29) and $\mu = \mu_0$ are summarized in Table 10.3.

Table 10.3

Analysis of Variance for One-Way Classification

Source of Variation	Sum of Squares	Degrees of Freedom	Mean Square	Test Statistic	Expected Mean Square
Over-all mean	$n.(\bar{x}.. - \mu_0)^2$	1	s_3^2	s_3^2/s_p^2	$\sigma^2 + n.(\mu - \mu_0)^2$
Treatment effects	$\sum_{i} n_i(\bar{x}_{i.} - \bar{x}..)^2$	$k - 1$	s_2^2	s_2^2/s_p^2	$\sigma^2 + \dfrac{1}{k-1}\sum_{i} n_i \cdot \alpha_i^2$
Within error	$\sum_{i} \sum_{j} (x_{ij} - \bar{x}_{i.})^2$	$n. - k$	s_p^2		σ^2
Total (not corrected)	$\sum_{i} \sum_{j} (x_{ij} - \mu_0)^2$	$n.$			

Following the proof given in Sect. 8.12, we can show that

$$\sum_{i}^{k} \sum_{j}^{n_i} (x_{ij} - \bar{x}..)^2 = \sum_{i} \sum_{j} (x_{ij} - \bar{x}_{i.})^2 + \sum_{i} n_i (\bar{x}_{i.} - \bar{x}..)^2 \quad (10.38)$$

Thus, the sum of the two terms on the right-hand side of Eq. (10.38) may be replaced by the left-hand side of (10.38). The relation (10.38) is called the sum of squares identity for the null hypothesis given in (10.29), and Table 10.4 is a typical analysis of variance table.

Table 10.4
Analysis of Variance for One-Way Classification

Source of Variation	Sum of Squares	Degrees of Freedom	Mean Square	Test Statistic	Expected Mean Square
Among means	$\sum_{i=1}^{k} n_i (\bar{x}_{i.} - \bar{x}..)^2$	$k - 1$	s_2^2	$F_c = \dfrac{s_2^2}{s_p^2}$	$\sigma^2 + \dfrac{\sum_{i}^{k} n_i \alpha_i^2}{k - 1}$
Within	$\sum_{i}^{k} \sum_{j}^{n_i} (x_{ij} - \bar{x}_{i.})^2$	$n. - k$	s_p^2		σ^2
Total	$\sum_{i} \sum_{j} (x_{ij} - \bar{x}..)^2$	$n. - 1$			

10.4. POWER OF ANALYSIS OF VARIANCE FOR A ONE-WAY CLASSIFICATION

If the hypothesis in (10.29) is false, then the expected value of s_2^2, the mean square for treatments, is equal to

$$\sigma^2 + \frac{1}{k-1} \sum_{i=1}^{k} n_i \alpha_i^2$$

Thus, s_2^2 is not distributed as $\sigma^2 \chi^2/(k-1)$, and the ratio s_2^2/s_p^2 is not distributed as F with $k - 1$ and $n. - k$ degrees of freedom. In fact, in this case s_2^2/s_p^2 is said to be distributed as a *noncentral F* [26, 31]. However, if we let

$$\lambda^2 = \frac{\sigma^2 + \dfrac{1}{k-1} \sum n_i \alpha_i^2}{\sigma^2} = 1 + \frac{\dfrac{\sum_{i=1}^{k} n_i \alpha_i^2}{k-1}}{\sigma^2} \quad (10.39)$$

it can be shown that s_2^2/s_p^2 is distributed approximately as $\lambda^2 F$ with ν_2' and $\nu_1 = n. - k$ degrees of freedom, where

$$\nu_2' = (k-1)\left(\frac{\lambda^4}{2\lambda^2 - 1}\right) \quad (10.40)$$

Since the power function of the test of the hypothesis in (10.29) may be expressed in terms of λ^2 as

$$p(\lambda^2) = P\left[\frac{s_2^2}{s_p^2} > F_\alpha(k-1, n.-k); \lambda^2\right]$$

it follows that

$$p(\lambda^2) \doteq P\left[F(v_2', n.-k) > \frac{F_\alpha(k-1, n.-k)}{\lambda^2}\right] \tag{10.41}$$

also gives the power of the test approximately. Clearly, the power function is an increasing function of λ^2, since $F_\alpha(k-1, n.-k)/\lambda^2$ is a decreasing function of λ^2.

To find the value of λ^2 for which $p(\lambda^2) \doteq 1 - \beta(\lambda^2)$, we solve

$$P\left[F(v_2', n.-k) > \frac{F_\alpha(k-1, n.-k)}{\lambda^2}\right] \doteq 1 - \beta(\lambda^2)$$

which leads to

$$\lambda^2 \doteq \frac{F_\alpha(k-1, n.-k)}{F_{1-\beta}(v_2', n.-k)} = F_\alpha(k-1, n.-k) \cdot F_\beta(n.-k, v_2') \tag{10.42}$$

Using this relation and Table VII, we may determine a sufficient number of points to draw a power curve in terms of λ^2.

From Eq. (10.39) it is clear that $\lambda^2 > 1$ may be used to measure an alternative hypothesis to the hypothesis (10.29b), for the more λ^2 exceeds 1, the more the alternative deviates from the null hypothesis. Since λ^2 depends on $k, \sigma^2, n_1, \ldots, n_k, \alpha_1, \ldots, \alpha_k$, we may determine power for an α level test by specifying values of these parameters. If $n_i = n \, (i = 1, \ldots, k)$, then we need specify only $n, k, \sigma^2, \alpha_1, \ldots, \alpha_{k-1}$ in order to determine power for an α level test. Actually questions concerning power are often raised after one knows α, n, and k. This means that only σ^2 and $\alpha_1, \ldots, \alpha_{k-1}$ (or σ^2 and $\sum \alpha_i^2$) must be specified in order to determine power.

For the case where $n_i = n \, (i = 1, \ldots, k)$, we may also write

$$\lambda^2 = 1 + \left(\frac{k}{k-1}\right)\phi^2 \tag{10.43}$$

where

$$\phi^2 = \frac{\dfrac{\sum\limits_{i=1}^{k} \alpha_i^2}{k}}{\dfrac{\sigma^2}{n}} \tag{10.44}$$

Clearly, λ^2 is an increasing function of ϕ^2. Thus, power of the test of the hypothesis in (10.29) may be expressed as a function of ϕ, as in the usual case. Table VIII shows eight pages of graphs with $p = 1 - \beta$ on the vertical scale corresponding to ϕ on the horizontal. Two levels of significance, $\alpha = 0.01$ and 0.05, for eight values of ν_1 and several values of ν_2 are shown. [The reader should note that ν_2 is used in place of ν_2', since the exact non-central F distribution was used in the calculations in place of the approximation $\lambda^2 F(\nu_2', \nu_1)$.] There is a different curve for each set of values α, ν_1, and ν_2. These curves may be used for purposes other than finding power, just as was the case for power curves associated with the t and χ^2 distributions.

Example 10.2. Suppose that five normal populations have common variance $\sigma^2 = 20$ with means $\mu_1 = 65$, $\mu_2 = 65$, $\mu_3 = 70$, $\mu_4 = 75$, and $\mu_5 = 75$, respectively. How many random observations n should one make on each of the five populations so that a 0.01 level analysis of variance test of the hypothesis $\mu_1 = \mu_2 = \mu_3 = \mu_4 = \mu_5 = \mu$ will have a 0.90 chance of detecting differences?

The degrees of freedom for the test are $\nu_1 = 4$ and $\nu_2 = 5(n - 1)$. Since $\mu = \sum \mu_i / 5 = 70$, $\alpha_1 = \alpha_2 = -5$, $\alpha_3 = 0$, $\alpha_4 = \alpha_5 = 5$, and

$$\phi^2 = \frac{\dfrac{[(-5)^2 + (-5)^2 + 0^2 + 5^2 + 5^2]}{5}}{\dfrac{20}{n}} = n$$

We wish to find n such that when $\alpha = 0.01$, $\nu_1 = 4$, $p = 0.90$ (or $\beta = 0.10$), then $\phi = \sqrt{n}$ approximately. Using the following selected values of n along with Table VIII, we have

n	ν_2	$\sqrt{n} = \phi$	From Table VIII ϕ
3	10	1.72	2.86
4	15	2.00	2.56
5	20	2.23	2.41
6	25	2.45	2.35

From these claculations we see that the graphic values of ϕ are greater than $\phi = \sqrt{n}$ when $n = 3$, 4, and 5 and that the graphic value of ϕ for $n = 6$ is less than $\phi = \sqrt{n}$. This indicates that the graphic value of ϕ is equal to $\sqrt{n}$ for some value of n between 5 and 6. If we take five observations from each population and run a 0.01 level analysis of variance test, the chance of making a type 2 error is greater than 0.10; if we take six observations, the chance of making a type 2 error is less than 0.10. Thus, in order to have adequate protection, we take samples of size six in this experiment.

In an actual experiment we do not usually know σ^2 and α_i ($i = 1, \ldots, k$). Thus, the above method cannot be used to determine what sample size to draw. However, it is usually possible to guess values of $\sum \alpha_i^2$ and σ^2 so that in determining a value of n the size of the type 2 error does not exceed the specified value β_0. (One should guess $\sum \alpha_i^2$ and σ^2 so as to underestimate ϕ^2. This means that one should not underestimate σ^2 or overestimate $\sum \alpha_i^2$.)

10.5. COMPUTATIONS. RELATION BETWEEN MODEL AND DATA

We have already seen how important the sum of squares identity, Eq. (10.38), is in breaking the total sum of squares into the sum of two parts and in preparing analysis of variance tables such as Tables 10.3 and 10.4. At this time we derive better computational forms for finding the total and component sum of squares in an analysis of variance table. The *total sum of squares*, denoted by *SST*, is given by

$$SST = \sum_i^k \sum_j^{n_i} (x_{ij} - \bar{x}..)^2 = \sum_i^k \sum_j^{n_i} x_{ij}^2 - \frac{\left(\sum_i^k \sum_j^{n_i} x_{ij}\right)^2}{\sum n_i}$$

or

$$SST = \sum_i^k \sum_j^{n_i} x_{ij}^2 - \frac{T_{..}^2}{n.} \tag{10.45a}$$

or

$$SST = \frac{n. \sum_i \sum_j x_{ij}^2 - T_{..}^2}{n.} = \frac{LSST}{n.} \tag{10.45b}$$

where

$$LSST = n. \sum_i \sum_j x_{ij}^2 - T_{..}^2 \tag{10.46}$$

denotes "large total sum of squares." Both Eqs. (10.45a) and (10.45b) are very useful in computing the total sum of squares. Equation (10.45b) is particularly good with desk calculators; Eq. (10.45a) is the form normally used for reasons which will soon be apparent.

The *among means sum of squares*, often called the *treatment sum of squares*, is denoted by *SSA* and given by

$$SSA = \sum_{i=1}^k n_i(\bar{x}_{i.} - \bar{x}..)^2 = \sum_i n_i \left(\frac{T_{i.}}{n_i} - \frac{T_{..}}{n.}\right)^2$$

$$= \sum_i n_i \left(\frac{T_{i.}^2}{n_i^2} - 2\frac{T_{..}}{n.} \cdot \frac{T_{i.}}{n_i} + \frac{T_{..}^2}{n_.^2}\right)$$

$$= \sum_i \left(\frac{T_{i.}^2}{n_i}\right) - \frac{2T_{..}}{n.}\left(\sum_i T_{i.}\right) + n. \frac{T_{..}^2}{n_.^2}$$

or

$$SSA = \sum_{i=1}^{k} \left(\frac{T_{i.}^2}{n_i} \right) - \frac{T_{..}^2}{n_.} \tag{10.47}$$

In case $n_i = n$ $(i = 1, \ldots, k)$, Eq. (10.47) reduces to

$$SSA = \frac{\sum_i T_{i.}^2}{n} - \frac{T_{..}^2}{nk} \tag{10.47a}$$

or

$$SSA = \frac{k \sum_i T_{i.}^2 - T_{..}^2}{nk} = \frac{LSSA}{nk} \tag{10.47b}$$

where

$$LSSA = k \sum_i T_{i.}^2 - T_{..}^2 \tag{10.48}$$

denotes "large among mean sum of squares." If the sample sizes are not equal, then Eq. (10.47) is used; otherwise, either Eq. (10.47a) or Eq. (10.47b) is applied. Equation (10.47a) is the form normally used; Eq. (10.47b) is good with the desk calculator.

The *within sample sum of squares*, often called *error* or *pooled* or *residual* sum of squares, is denoted by *SSW* and is given by

$$SSW = SST - SSA \tag{10.49}$$

For the computation of *SSW* directly from the data, use

$$SSW = \sum_i \sum_j x_{ij}^2 - \sum_i \left(\frac{T_{i.}^2}{n_i} \right) \tag{10.50a}$$

for unequal sample sizes, or

$$SSW = \sum_i \sum_j x_{ij}^2 - \frac{\sum_i T_{i.}^2}{n} \tag{10.50b}$$

for equal sample sizes.

In expressions (10.45a), (10.47), (10.47a), (10.50a), and (10.50b) note that the divisor in each case is the same as the number of observations making up the total. For example, in Eq. (10.47) $T_{i.}$ is the sum of n_i observations, and the divisor is n_i; $T_{..}$ is the sum of

$$n_. = \sum_i n_i$$

observations, and the divisor is $n_.$. This is, no doubt, the primary reason that these forms are normally preferred to the forms with large sum of squares.

In order better to understand the relation between the model and a set of data, we consider an example in which samples are drawn from known populations. The computing forms just described are used.

Example 10.3. Random samples of size ten were drawn from normal populations $n(2, 1)$, $n(2.2, 1)$, $n(2.4, 1)$ and $n(2.6, 1)$, respectively. The measurements are given in Table 10.5 along with totals and sum of squares in columns. (a) Compare the estimated means and effects with the true parameters. (b) Compare the estimates of variance with $\sigma^2 = 1$. (c) Discuss some confidence interval problems.

Table 10.5
Random Normal Deviates

	Sample 1	Sample 2	Sample 3	Sample 4	
	x_1	x_2	x_3	x_4	
	3.355	0.273	3.539	3.074	
	1.086	2.155	2.929	3.103	
	2.367	1.725	3.025	2.389	
	0.248	0.949	4.097	4.766	
	1.694	0.458	2.236	2.553	
	1.546	1.455	3.256	3.821	
	1.266	2.289	3.374	1.905	
	0.713	2.673	1.781	2.350	
	0.000	1.800	2.566	1.161	
	3.406	2.407	2.510	2.122	
$T_{i.}$	15.681	16.184	29.313	27.244	$88.422 = T_{..}$
$\sum_j x_{ij}^2$	37.071327	32.339668	90.081121	83.620342	$243.112458 = \sum_i \sum_j x_{ij}^2$

For (a), the estimated means $\bar{x}_{i.}$ are 1.57, 1.62, 2.93, and 2.72, and the population means μ_i are 2.00, 2.20, 2.40, and 2.60, respectively. Two samples underestimated their population means by ($\bar{x}_{i.} - \mu_i$ equal to) -0.44 and -0.58, and two samples overestimated their population means by 0.53 and 0.12. The over-all population mean is $\mu = 2.30$ with an estimate of $\bar{x}_{..} = T_{..}/40 = 2.21$ and a difference of $\bar{x}_{..} - \mu = -0.09$. Thus, the true effects α_i and estimated effects a_i are -0.3, -0.1, 0.1, 0.3 and -0.65, -0.59, 0.72, 0.42, respectively, with differences $a_i - \alpha_i$ of -0.35, -0.49, 0.62, and 0.21. Note that the sum of the effects is zero for both the true and estimated effects, and, therefore, the sum $\sum (a_i - \alpha_i)$ is zero. Further, note that the relation (10.30b) holds for each sample and that any one of the three differences can be found in terms of the other two. For example, in sample 1, $a_1 - \alpha_1 = (\bar{x}_{1.} - \mu_1) - (\bar{x}_{..} - \mu) = -0.44 - (-0.09) = -0.35$.

The data in Table 10.5 along with the above discussion may be used to give some idea of how much the estimates can be expected to miss the true means and effects for populations with variance 1 and sample sizes of 10 and 40. As the variance gets smaller and the sample size gets larger, we expect the point estimates on the average to get closer to the parameters they estimate; as the variance gets larger and the sample size smaller, we expect the point estimates to be farther away from the parameters on the average. For a variance of 1, we see from the above argument that for a sample of size 40 the estimated mean deviates 0.9 units from the population mean, whereas for samples of size 10 the estimated means on the average (of the absolute values of the deviates) deviate

$$\frac{|-0.44| + |-0.58| + 0.53 + 0.12}{4} = 0.42 \text{ units}$$

from their population means. Even though the above discussions are based on only 40 observations, the conclusions are fairly typical of what one expects when working with point estimates of parameters.

The good computational forms of this and other sections are used to find estimates of the common variance 1. For the four samples, the sum of squares and variances are given by

$$SS_i = \sum_{j=1}^{10} x_{ij}^2 - \frac{T_{i.}^2}{10} \tag{10.51}$$

and

$$s_i^2 = \frac{SS_i}{9} \quad (i = 1, 2, 3, 4) \tag{10.52}$$

respectively. In particular

$$SS_1 = 37.071327 - \frac{(15.681)^2}{10} = 12.4820$$

$$SS_2 = 6.1475$$
$$SS_3 = 4.1559$$
$$SS_4 = 9.3968$$

so that

$$s_1^2 = 1.3869$$
$$s_2^2 = 0.6831$$
$$s_3^2 = 0.4618$$
$$s_4^2 = 1.0441$$

Thus, the four independent samples of size 10 give independent estimates of variance with nine degrees of freedom. We see that the fourth estimate is closest to the true variance of 1. Another estimate of variance, the pooled variance s_p^2, is give by

$$s_p^2 = \frac{SS_1 + \cdots + SS_4}{(n_1 - 1) + \cdots + (n_4 - 1)} = \frac{12.4820 + \cdots + 9.3968}{9 + \cdots + 9} = \frac{32.1822}{36}$$

$$= 0.894$$

These five estimators are all unbiased, but the last one, s_p^2, is the best, since it has 36 degrees of freedom. It should be noted that s_p^2 is not closest to $\sigma^2 = 1$ in this particular problem. We select s_p^2 because it can be expected to be closer to 1 on the average than any of the others.

Further, using Eqs. (10.45a), (10.47a), and (10.49), we obtain

$$SST = 243.112458 - \frac{(88.422)^2}{40} = 47.6512$$

$$SSA = \frac{(15.681)^2 + \cdots + (27.244)^2}{10} - \frac{(88.422)^2}{40} = 15.4690$$

and

$$SSW = SST - SSA = 32.1822$$

Letting

$$s_g^2 = \frac{SST}{nk - 1}, \quad s_x^2 = \frac{SSA}{k - 1}, \quad \text{and} \quad s_p^2 = \frac{SSW}{k(n - 1)} \qquad (10.53)$$

we obtain, on substituting the above values

$$s_g^2 = \frac{47.6512}{39} = 1.222$$

$$s_x^2 = \frac{15.4690}{3} = 5.156$$

$$s_p^2 = \frac{32.1822}{36} = 0.894$$

The variance estimator s_p^2, also found above, is an unbiased estimator of σ^2, even though the population means are different. This is not the case with s_g^2 and s_x^2. In fact, according to Eq. (10.36), s_x^2 is an unbiased estimator of

$$\sigma^2 + \frac{n \sum \alpha_i^2}{k - 1} = 1 + \frac{10[(-0.3)^2 + (-0.1)^2 + (0.1)^2 + (0.3)^2]}{3} = \frac{5}{3} \doteq 1.67$$

Also, using Eqs. (10.49) and (10.53), we can write

$$(nk - 1)s_g^2 = (nk - k)s_p^2 + (k - 1)s_x^2$$

so that

$$E[(nk - 1)s_g^2] = E[(nk - k)s_p^2 + (k - 1)s_x^2]$$

or

$$E(s_g^2) = \frac{1}{nk-1}[(nk-k)E(s_p^2) + (k-1)E(s_{\bar{x}}^2)]$$

$$= \frac{1}{nk-1}\left[(nk-k)\sigma^2 + (k-1)\left(\sigma^2 + \frac{n\sum\alpha_i^2}{k-1}\right)\right]$$

or

$$E(s_g^2) = \sigma^2 + \frac{n\sum\alpha_i^2}{nk-1} \tag{10.54}$$

That is, s_g^2 is an unbiased estimator of

$$\sigma^2 + \frac{n\sum\alpha_i^2}{nk-1} = 1 + \frac{10[(-0.3)^2 + (-0.1)^2 + (0.1)^2 + (0.3)^2]}{39}$$

$$= \tfrac{41}{39} = 1.05$$

For our samples the estimates $s_{\bar{x}}^2 = 5.156$ and $s_g^2 = 1.222$ are greater than what we expect on the average. It is not particularly surprising that s_g^2 deviates less from $E(s_g^2) = 1.05$ than $s_{\bar{x}}^2$ does from $E(s_{\bar{x}}^2) = 1.67$. This is expected, since the number of degrees of freedom, 39, associated with s_g^2 is much greater than the number of degrees of freedom, 3, associated with $s_{\bar{x}}^2$.

Actually, we expect $s_{\bar{x}}^2$ to be larger than 5.156 only about 2.5 per cent of the time, since, according to the χ^2 distribution, we expect values larger than

$$\frac{(k-1)s_{\bar{x}}^2}{\sigma^2 + \frac{n\sum\alpha_i^2}{k-1}} = \frac{\dfrac{15.4690}{5}}{\dfrac{3}{}} = 9.28$$

only 2.5 per cent of the time in repeated sampling. Similarly, we find that s_g^2 would be larger than 1.222 about 25 per cent of the time and s_p^2 would be smaller than 0.894 about 25 per cent of the time in repeated sampling.

For the moment, let us pretend that we do not know that the population means are different, and let us test by analysis of variance the hypothesis that the means are equal; that is, $H_0: \mu_1 = \mu_2 = \mu_3 = \mu_4$. Under the assumption of the null hypothesis, $s_{\bar{x}}^2$ and s_p^2 are independent and unbiased estimators of the same variance $\sigma^2 = 1$. Thus, their ratio $s_{\bar{x}}^2/s_p^2$ is distributed as F with 3 and 36 degrees of freedom. For a five per cent level test, $F_{.05}(3, 36) = 2.88$, and the critical region is made up of all those values of F for which $F \geq 2.88$. For our particular samples, $F = 5.77$ falls in the critical region. Thus, we reject H_0 and conclude that the population means are not all equal. In making this conclusion, we normally take a five per cent chance of making the type 1 error, but since we know that the means are really different, we know that we do not make a type 1 error. Further, since we reject the null hypothesis, we do not have an opportunity to make the type 2 error.

Confidence intervals for means and effects may be found by the methods of Example 10.1. From Table VI and the above discussion, we find that $t_{.025}(36) = 2.03$ and $s_p^2 = 0.894$, $\bar{x}_1. = 1.57$, $\bar{x}_2. = 1.62$, $\bar{x}_3. = 2.93$, $\bar{x}_4. = 2.72$. Thus

$$t_{.025}(36) \sqrt{\frac{s_p^2}{n}} = 2.03 \sqrt{\frac{0.894}{10}} \doteq 0.61$$

so that the 95 per cent confidence intervals for μ_1, μ_2, μ_3, and μ_4 are

$$0.96 < \mu_1 < 2.18$$
$$1.01 < \mu_2 < 2.23$$
$$2.32 < \mu_3 < 3.54$$
$$2.11 < \mu_4 < 3.33$$

In each case, note that the true mean actually does fall in the indicated interval.

10.6. EXERCISES

10.1. For what purposes may we properly use the analysis of variance?

10.2. Suppose we have the analysis of variance given in Table 10.6. (a) Write the appropriate model equation. (b) What null hypothesis was the experiment probably designed to test? Give statements for both the null

Table 10.6

Source of Variation	Sum of Squares	Degrees of Freedom	Mean Squares	Expected Mean Squares
Among means	639	3	213	$\sigma^2 + 2 \sum_{i=1}^{4} \alpha_i^2$
Within	600	20	30	σ^2
Total	1239	23		

and alternative hypotheses in terms of the symbols used in (a) and in words. (c) Use a five per cent level test of the null hypothesis in (b). Under what assumptions is this test valid? (d) What are the unbiased estimates of σ^2, $\sigma^2 + 2 \sum \alpha_i^2$, and $\sum \alpha_i^2$? What is the largest value any estimator a_i can have in this experiment? (e) Find a 90 per cent confidence interval for σ^2.

10.3. An experiment was run to determine if five specific firing temperatures affect the density of bricks. The mean square for "among firing temperatures" was 1.23. Twelve bricks were used for each firing temperature, and the pooled variance "among bricks within firing temperature" was 0.64. (a) Make an analysis of variance table, showing all the things given in the analysis of variance table in Exercise 10.2. (b) Give statements for both the null and alternative hypotheses in terms of the experiment. Write the appropriate model equation and state both hypotheses in symbols. Test the null hypothesis for $\alpha = 0.05$. (c) Find a 95 per cent confidence interval for the error variance.

10.4. The following hypothetical data are for a completely randomized experiment (the student can supply his own interpretation)

Sample 1	Sample 2	Sample 3
51	53	69
47	35	59
49	43	57
68	58	51
44	60	55
51	39	72
49	49	41
43	60	54

(a) Prepare an analysis of variance table similar to Table 10.4. Use the computing formulas of Sect. 10.5. (b) Test the hypothesis that the three population means are equal, showing all steps in the general test procedure. (c) Find point estimates of the over-all mean, the three population means and effects, and the variance common to the three populations. (d) Find 95 per cent confidence intervals for the third population mean, and for the difference in the means of the first and third population. (e) Express each of the 24 observations as the sum of three estimated component parts (see illustration in Table 8.13a) and use these parts to compute the three sums of squares found in (a).

10.5. The following hypothetical data are for a completely randomized experiment (the student can supply his own interpretation)

Sample 1	Sample 2	Sample 3	Sample 4
53	53	48	62
62	45	59	75
51	56	61	59
45	59	52	50
64	56	51	55
	66	54	57
	55	61	84
		68	53
		42	
		42	

Follow the instructions given in Exercise 10.4 (a), (b), (c), (d), and (e).

10.6. Prove that the statistics defined in Eq. (10.18) are unbiased estimators of their corresponding parameters.

10.7. Prove Eq. (10.28).

10.8. Prove Eq. (10.36).

10.9. Prove Eq. (10.37).

10.10. In Exercise 10.4 Samples 1 and 2 were randomly drawn from a normal population with mean 50 and variance 100, and Sample 3 was randomly drawn from a normal population with mean 60 and variance 100. (a) Find μ, α_1, α_2, and α_3. Let these numerical values be denoted by μ_0, α_{01}, α_{02}, and α_{03}, respectively. Use this information and the data in Exercise 10.4 to verify the identity (10.32). That is, find the four sums of squares in Eq. (10.32). (b) Use the values found in (a) to test, at the five per cent level, the following two null hypotheses: $H_{01}: \alpha_i = \alpha_{0i}$ ($i = 1, 2, 3$) and $H_{02}: \mu = \mu_0$. (c) Prepare an analysis of variance table similar to Table 10.3 and test, at the five per cent level, the null hypothesis given by (10.29). (d) Express each of the 24 observations in Exercise 10.4 as the sum of three true component parts (see illustration in Table 8.13b) and compute an estimate of the variance 100, using the ϵ_{ij}. (e) Find the power of the test of the null hypothesis in (10.29) when the α effects are those found in (a). Find the power when $\alpha_1 = \alpha_2 = -5$ and $\alpha_3 = 10$.

10.11. Find power of the test in Exercise 10.6(c) when $\sum \alpha_i^2 = 75$ and $\sigma^2 = 40$.

10.12. In Exercise 10.5 Samples 1, 2, 3, and 4 were randomly drawn from normal populations with means 50, 60, 55, and 58, respectively, and common variance 100. Follow the instructions given in Exercise 10.10(a), (b), (c), (d), and (e), making necessary adjustments.

10.13. Prove Eq. (10.38).

10.14. Prove Eq. (10.40).

Hint. You may wish to check a reference.

10.15. Suppose four normal populations have common variance $\sigma^2 = 50$ with means $\mu_1 = 50$, $\mu_2 = 50$, $\mu_3 = 60$, and $\mu_4 = 80$. How many random observations n should one make on each of the populations so that a 0.05 level analysis of variance test of the hypothesis $\mu_1 = \mu_2 = \mu_3 = \mu_4$ will have a 0.90 chance of detecting differences?

10.16. In a completely randomized design with five random observations for each of seven treatments, we have the following coded data (the treatments might be such things as different temperatures, different velocities, different concentrations, different tensile strengths, different number of hours of exposure, different strains, and different methods of instruction)

			Treatments			
1	2	3	4	5	6	7
162	126	189	138	207	153	138
143	114	213	156	153	147	81
137	168	120	117	138	186	141
113	144	189	126	147	144	114
181	126	111	162	189	141	159

Follow the instructions given in Exercise 10.4(a), (b), (c), (d), and (e), making necessary adjustments.

10.7. SINGLE DEGREE OF FREEDOM

When the null hypothesis $\mu_1 = \mu_2 = \cdots = \mu_k = \mu$ is rejected, we conclude that there are differences among the means, but the particular nature of the differences is not specified by the F-test procedure described in Sect. 10.2. However, it is possbile to test specific hypotheses involving *linear combinations* of any p $(p = 2, \ldots, k)$ of the k means $\mu_1, \ldots, \mu_k$. For example, in a set of ten means we may wish to compare the mean of one population, say the control, with the average of two other population means, say the two newest. That is, we may wish to test a hypothesis of the type

$$\mu_2 = \frac{\mu_4 + \mu_7}{2}$$

or

$$2\mu_2 - \mu_4 - \mu_7 = 0 \tag{10.55}$$

which involves a linear combination of means. Actually, such a linear combination is called a *contrast* or *comparison*, since the sum of the coefficients is zero; that is, $2 + (-1) + (-1) = 0$. In general, any linear combination of k population means of the form

$$\gamma = m_1\mu_1 + m_2\mu_2 + \cdots + m_k\mu_k \tag{10.56}$$

is called a *contrast* or *comparison of true means*, provided

$$\begin{cases} m_1 + m_2 + \cdots + m_k = 0 \quad \text{and} \\ \text{some } m_i \ (i = 1, \ldots, k) \text{ is different from zero} \end{cases} \tag{10.57}$$

We know from Chap. 6 that the linear combination

$$c = m_1\bar{x}_1 + m_2\bar{x}_2 + \cdots + m_k\bar{x}_k \tag{10.58}$$

of sample means is normally distributed with mean

$$\mu_c = \gamma = m_1\mu_1 + m_2\mu_2 + \cdots + m_k\mu_k$$

and variance

$$\sigma_c^2 = m_1^2\sigma_{\bar{x}_1}^2 + \cdots + m_k^2\sigma_{\bar{x}_k}^2 = m_1^2\frac{\sigma_1^2}{n_1} + \cdots + m_k^2\frac{\sigma_k^2}{n_k} \tag{10.59}$$

provided $\bar{x}_1, \ldots, \bar{x}_k$ are means of samples of sizes $n_1, \ldots, n_k$ which are randomly and independently drawn from normal populations with means $\mu_1, \ldots, \mu_k$ and variances $\sigma_1^2, \ldots, \sigma_k^2$, respectively. In particular, the theorem holds when the linear combination (10.58) is a *contrast of sample means.* Further, if all the variances are equal to σ^2, then Eq. (10.59) becomes

$$\sigma_c^2 = \left(\frac{m_1^2}{n_1} + \cdots + \frac{m_k^2}{n_k} \right) \sigma^2 = \sigma^2 \sum_{i=1}^{k} \left(\frac{m_i^2}{n_i} \right) \tag{10.60}$$

It is clear that the statistic

$$\frac{c - \gamma}{\sigma_c} \tag{10.61}$$

is normally distributed with mean zero and variance one. Further, when all the population variances are equal to σ^2, and s_p^2 is an unbiased estimator of σ^2 so that

$$s_c^2 = s_p^2 \sum_{i=1}^{k} \left(\frac{m_i^2}{n_i} \right) \tag{10.62}$$

it follows that

$$\frac{c - \gamma}{s_c} \tag{10.63}$$

is distributed as t with

$$\left(\sum_{i=1}^{k} n_i - k \right) \text{ degrees of freedom}$$

or

$$\frac{(c - \gamma)^2}{s_c^2} \tag{10.64}$$

is distributed as F with 1 and

$$\left(\sum_{i=1}^{k} n_i - k \right) \text{ degrees of freedom}$$

Thus, in testing the null hypothesis, (10.56), that a linear combination, usually a contrast, of population means is zero, we use the statistic in Eq. (10.64). Under the assumption that γ is zero, ratio (10.64) reduces to

$$\frac{\left(\sum m_i \bar{x}_i \right)^2}{s_p^2 \sum \left(\frac{m_i^2}{n_i} \right)} = \frac{Q^2}{s_p^2} \tag{10.65}$$

where

$$Q^2 = \frac{(m_1 \bar{x}_1 + \cdots + m_k \bar{x}_k)^2}{\dfrac{m_1^2}{n_1} + \cdots + \dfrac{m_k^2}{n_k}} = \frac{\left(\sum m_i \bar{x}_i \right)^2}{\sum \left(\dfrac{m_i^2}{n_i} \right)} \tag{10.66}$$

Since $\bar{x}_i = T_{i.}/n_i$ $(i = 1, \ldots, k)$, we may write Eq. (10.66) as

$$Q^2 = \frac{\left(\dfrac{m_1 T_{1.}}{n_1} + \cdots + \dfrac{m_k T_{k.}}{n_k}\right)^2}{\dfrac{m_1^2}{n_1} + \cdots + \dfrac{m_k^2}{n_k}} = \frac{\left(\sum \dfrac{m_i T_{i.}}{n_i}\right)^2}{\sum \left(\dfrac{m_i^2}{n_i}\right)} \tag{10.67}$$

or, in case $n_1 = \ldots = n_k = n$

$$Q^2 = \frac{(m_1 T_{1.} + \cdots + m_k T_{k.})^2}{n(m_1^2 + \cdots + m_k^2)} = \frac{(\sum m_i T_{i.})^2}{n \sum m_i^2} \tag{10.68}$$

Since Q^2/s_p^2 is distributed as F with 1 and $(\sum n_i - k)$ degrees of freedom, and since s_p^2 is an unbiased estimator of σ^2 with $(\sum n_i - k)$ degrees of freedom, we know that Q^2 and s_p^2 are independently distributed and Q^2 is a variance estimator with one degree of freedom. Thus, Q^2 is called a *component of treatment sum of squares with an individual degree of freedom* or, for short, a *component with an individual degree of freedom*. (It can be shown that Q^2 is a part of the treatment sum of squares.)

Sometimes, when making inferences about k means, we are interested in two or more linear combinations, usually contrasts, in which case we need to compute the corresponding Q^2's, each with one degree of freedom. To illustrate what has already been developed in this section and to introduce the concept of orthogonal comparisons, we consider the following example.

Example 10.4. Assume that, in addition to the information in Example 10.1, we know that the manufacturer B usually supplies buyer X with wire, and that manufacturer C is a new competitor on the copper wire market. If buyer X questions that he should continue to buy wire from B, he is likely to want to know the answers to the following two questions. Does B make wire with tensile strength the same as that of A and C? Is the tensile strength of the wire made by A the same as that made by C? To answer these questions, buyer X tests the hypotheses

$$H_{01}: \mu_B = \frac{\mu_A + \mu_C}{2} \quad \text{and} \quad H_{02}: \mu_A = \mu_C$$

or, the equivalent hypotheses

$$H_{01}: \mu_A - 2\mu_B + \mu_C = 0 \quad \text{and} \quad H_{02}: \mu_A - \mu_C = 0 \tag{10.69}$$

From the solution following Example 10.1, we have $s_p^2 = 1457$ with 27 degrees of freedom, $T_A = 920$, $T_B = 750$, $T_C = 1060$, and $n = 10$. Thus, the component sums of squares with an individual degree of freedom for testing the hypotheses H_{01} and H_{02} are

$$Q_1^2 = \frac{[920 - 2(750) + 1060]^2}{10[1^2 + (-2)^2 + 1^2]} = 3840$$

and

$$Q_2^2 = \frac{[920 + 0(750) - 1060]^2}{10[1^2 + 0^2 + (-1)^2]} = 980$$

respectively. For a five per cent level test $F_{.05}(1, 27) = 4.21$, so the critical region is made up of all F for which $F > 4.21$. For hypotheses H_{01} and H_{02} we have

$$F_1 = \tfrac{3840}{1457} = 2.64 \quad \text{and} \quad F_2 = \tfrac{980}{1457} = 0.67$$

respectively. Since neither value falls in the critical region, we fail to reject both hypotheses. That is, we conclude that the tensile strength of the wire made by B is not significantly different from the tensile strength of the wire made by A and C on the average, and that the new competitor C does not produce wire that is stronger or weaker than that made by A.

The reader should note that it is possible to reject either H_{01} or H_{02} without rejecting $\mu_A = \mu_B = \mu_C$, and that the rejection of $\mu_A = \mu_B = \mu_C$ does not necessarily imply the rejection of either H_{01} or H_{02}, it being assumed that the same significance level α is used for each of the three hypotheses. However, if the hypothesis $H_0: \mu_A = \mu_B = \mu_C$ is correct, then H_{01} and H_{02} are correct, and conversely. This apparent inconsistency will be discussed in Sect. 10.9. At this time we wish to explain and illustrate what is meant by orthogonal comparisons.

In Example 10.4, the reader might have noticed that $Q_1^2 + Q_2^2 = TSS$, the treatment sum of squares. This did not just happen. It is always true, provided the two contrasts are such that the sum of the products of corresponding multipliers is equal to zero. For example, in our problem the multipliers are 1, -2, 1 and 1, 0, -1 with $1 \cdot 1 + (-2) \cdot 0 + 1(-1) = 0$. Whenever this property holds, we say that the contrasts are *orthogonal*.

In general, two or more linear combinations are said to be *orthogonal* when

> (a) Every combination is a contrast; that is, the sum of the multipliers for each combination is zero.
> (b) The sum of products of the corresponding multi-   (10.70)
> pliers of every two different linear combinations is equal to zero.

Thus, if

$$\begin{cases} C_1 = m_{11}T_1 + m_{12}T_2 + m_{13}T_3 \quad \text{and} \\ C_2 = m_{21}T_1 + m_{22}T_2 + m_{23}T_3 \end{cases} \qquad (10.71)$$

are two *linear combinations of treatment* totals such that

$$\begin{cases} m_{11} + m_{12} + m_{13} = 0 \\ m_{21} + m_{22} + m_{23} = 0 \\ m_{11}m_{21} + m_{12}m_{22} + m_{13}m_{23} = 0 \end{cases} \quad (10.72)$$

then C_1 and C_2 are said to be orthogonal or C_1 and C_2 are orthogonal contrasts. Also, C_1 and C_2 are called independent contrasts due to the third restriction of (10.72). Since the set of three equations in (10.72) involves six unknowns, we would expect that values can be assigned arbitrarily to certain m's, and that this would lead to many different pairs of sets of multipliers satisfying (10.72). For example, suppose $m_{11} = 1$, $m_{12} = 2$, and $m_{13} = -3$. Then the system (10.72) reduces to the system

$$m_{21} + m_{22} + m_{23} = 0$$

$$m_{21} + 2m_{22} - 3m_{23} = 0$$

which has the solutions

$$m_{21} = -5k, \qquad m_{22} = 4k, \qquad m_{23} = k$$

where k is any real number not equal to zero. When we use Eq. (10.68), it is clear that Q^2 corresponding to the set of multipliers $-5k$, $4k$, k does not change with k. For the k^2 factors out and cancels, leaving

$$Q_2^2 = \frac{(-5T_1 + 4T_2 + T_3)^2}{10[(-5)^2 + 4^2 + 1^2]}$$

It is for this reason that we usually (when possible) choose k so that each multiplier in a set is an integer. Thus, for three means, we see that when one set of multipliers is selected, then the second set of multipliers is uniquely determined except for a constant factor, and the corresponding components Q_1^2 and Q_2^2 are uniquely determined. Using Example 10.4 and the sets of multipliers 1, 2, -3 and -5, 4, 1, we find

$$Q_1^2 = \frac{[920 + 2(750) - 3(1060)]^2}{10[1 + 2^2 + (-3)^2]} = \frac{28880}{7} = 4126$$

and

$$Q_2^2 = \frac{[-5(920) + 4(750) + 1060]^2}{10[(-5)^2 + 4^2 + 1^2]} = \frac{4860}{7} = 694$$

so that $Q_1^2 + Q_2^2 = \frac{33740}{7} = 4820 = TSS$.

Since there are infinitely many ways in which the first of two sets of multipliers can be selected, one might ask how a particular set should be determined. The very emphatic answer is that the m_i's should be chosen before the experimenter looks at the data, and they should normally result from something in the theory underlying the experiment. However, if the data does suggest a given set of multipliers, then a second experiment could be performed in order to test the specific hypothesis involving these means. Actually, in an experiment involving three means, say, the investigator may not know in advance which two orthogonal comparisons to examine, or he may wish to test more than two comparisons using a single set of data. In either case, F or t should not be used as the test statistic. Fortunately, there does exist a test procedure for this kind of problem, and it is described in Sect. 10.8.

From the above discussion it should be clear that the treatment sum of squares for three means can be partitioned in many ways into two component sums of squares, each with a single degree of freedom, provided the associated linear combinations are orthogonal. In general, it can be shown that the treatment sum of squares for k means can be partitioned in many ways into $k - 1$ components, each with a single degree of freedom, provided the associated linear combinations are mutually orthogonal. That is, if

$$C_l = \sum_{i=1}^{k} m_{li} T_i \qquad (l = 1, \ldots, k-1) \qquad (10.73)$$

and

$$Q_l^2 = \frac{C_l^2}{n \sum_i m_{li}} \qquad (10.74)$$

then

$$TSS = Q_1^2 + Q_2^2 + \cdots + Q_{k-1}^2 \qquad (10.75)$$

provided $C_1, C_2, \ldots, C_k$ are mutually orthogonal contrasts of k treatment totals (each being the sum of n random observations). Even though orthogonal sets of multipliers $m_{l1}, m_{l2}, \ldots, m_{lk}$ may be selected in many ways, it is the responsibility of the investigator to choose only those sets which have meaning in the given experimental situation.

It should be noted that once $k - 1$ orthogonal sets of multipliers have been selected, it is impossible to select another set which is orthogonal to any of the first $k - 1$ sets. Actually, once $k - 2$ orthogonal sets are selected, the $(k - 1)$ st set is uniquely determined except for a constant multiplier. For example, when $k = 3$ and one set of multipliers m_{11}, m_{12}, m_{13} is selected, a second set m_{21}, m_{22}, m_{23} is uniquely determined except for a constant multiplier, and a third set m_{31}, m_{32}, m_{33} orthogonal to each of the first two does not exist.

The statistic $(c - \gamma)/s_c$ of (10.63) may be used to find a $100(1 - \alpha)$ per cent confidence interval for any linear combination

$$\gamma = m_1 \mu_1 + \cdots + m_k \mu_k$$

The limits are given by

$$c \pm t_{\alpha/2} s_p \sqrt{\sum_{i=1}^{k} \left(\frac{m_i^2}{n_i} \right)} \qquad (10.76)$$

where $c = m_1 \bar{x}_1 + \cdots + m_k \bar{x}_k$.

Example 10.5. To find the 95 per cent symmetric confidence limits for

$$\gamma = \mu_A - 2\mu_B + \mu_C$$

using the information of Examples 10.1 and 10.4, we have

$$c = 92 - 2(75) + 106 = 48, \quad s_p^2 = 1457,$$

$$\sum_{i=1}^{3} \frac{m_i^2}{n_i} = \frac{6}{10}$$

and $t_{.025}(27) = 2.05$. Thus, the confidence limits are

$$48 \pm 2.05 \sqrt{1457 \left(\tfrac{6}{10} \right)} = 48 \pm 60.7$$

and the confidence interval is

$$-12.7 \leq \mu_A - 2\mu_B + \mu_C \leq 108.7$$

10.8. SIMULTANEOUS CONFIDENCE INTERVALS

We have seen how the t and F distributions may be used to establish confidence intervals and to test hypotheses involving several means. In particular, we have seen how the t distribution may be used to construct confidence intervals for *specific* linear combinations (usually, linear contrasts) among several means. We learned that the number of meaningful orthogonal contrasts is limited to the number of means minus one (that is, $k - 1$). There are at least two reasons why an experimenter might wish to remove this restriction. He might want to examine linear contrasts which are not orthogonal or to examine more than $k - 1$ such comparisons. Thus, it is only natural that he ask if it is possible to give a method for constructing simultaneous confidence intervals for all possible linear contrasts among k means. The answer is in the affirmative. In fact we describe two such methods. One method depends on the distribution of the "studentized" range [24] and the other, due to Scheffé [29], requires the use of the F distribution. We also describe a method due to Dunn [7] for finding m simultaneous confidence intervals. Roy and Bose [28] have also discussed the problem of simultaneous confidence intervals. Before illustrating these methods, we describe the sampling distribution of the range of k means.

10.8.1. Distribution of the Range

It should be evident that the range can be used to measure the dispersion of k sample means and that $k + 1$ means are more dispersed on the average than k or fewer means. Thus, we expect the distribution of ranges to depend on the number of means as well as the sample sizes. We describe the sampling distribution of the range in terms of a random sample of size n and then extend the ideas to a set of k independent sample means.

Let $x_1, x_2, \ldots, x_n$ denote a random sample from a population with mean μ, variance σ^2, density function $f(x)$, and distribution function $F(x)$. If $x_{(1)}$, $x_{(2)}, \ldots, x_{(n)}$ denote the same values in increasing order of magnitude, then the sample range w is defined by

$$w = x_{(n)} - x_{(1)}$$

It can be shown that the density function $f_n(w)$ of w is given by

$$f_n(w) = n(n-1) \int_{-\infty}^{\infty} [F(x+w) - F(x)]^{n-2} f(x) f(x+w)\, dx \quad (10.77)$$

If $f(x)$ is a normal density function, then the standardized random variable W may be written as

$$W = \frac{w}{\sigma} = \frac{(x_{(n)} - \mu) - (x_{(1)} - \mu)}{\sigma} = u_{(n)} - u_{(1)} \quad (10.78)$$

where $u_{(1)}$ and $u_{(n)}$ are standardized normal variables. The density function $f_n(W)$ and distribution function $F_n(W)$ of the random variable may be obtained from Eq. (10.77). Appropriate percentage points of both $f_n(W)$ and $F_n(W)$ are tabulated in Ref. [27].

Since in most practical problems σ is unknown, we require the distribution of the so-called "studentized" range q defined by $q = w/s$. It is understood that w and s are independent variates computed from the same normal population. In this case we can write

$$q = \frac{w}{s} = \frac{W}{\sqrt{\dfrac{\chi^2}{\nu}}} = \frac{u_{(n)} - u_{(1)}}{\sqrt{\dfrac{\chi^2}{\nu}}} \quad (10.79)$$

since w is distributed as σW and s as $\sigma \sqrt{\chi^2/\nu}$. Thus, the distribution of q depends on the *sample size* n from which w is determined and ν depends on the number of degrees of freedom of s. Table IX gives values, q_α, which are exceeded with probability $\alpha = 0.05; 0.01$. That is, q_α is that value of q for which $P[q > q_\alpha] = \alpha$. We sometimes write $q(n, \nu)$ in place of q.

Let $\bar{x}_1, \ldots, \bar{x}_k$ denote the means of k independent random samples, each of size n. If $\bar{x}_{(1)}, \ldots, \bar{x}_{(k)}$ denote the same values arranged in increasing order of magnitude, then the statistic

$$q = \frac{\bar{x}_{(k)} - \bar{x}_{(1)}}{\dfrac{s}{\sqrt{n}}} \quad (10.80)$$

is distributed as the studentized range $q(k, v)$, where k is the number of means and v is the number of degrees of freedom of s^2, the independent variance estimator of σ^2. The reader should note that for $k = 2$ we have $t(v) = q(2, v)$.

Example 10.6. Use the studentized range and the information of Example 10.1 to test at the five per cent level the hypothesis $\mu_A = \mu_B = \mu_C$. Since $\bar{x}_A = 92$, $\bar{x}_B = 75$, $\bar{x}_C = 106$, $n = 10$, and $s^2 = 1457$, we have

$$q = \frac{\bar{x}_C - \bar{x}_B}{\sqrt{\dfrac{s^2}{n}}} = \frac{106 - 75}{\sqrt{145.7}} = 2.56$$

For $k = 3$, $v = 27$, and $\alpha = 0.05$, we find, using Table IX, that

$$q_{.05}(3, 27) = 3.51$$

Therefore, the critical region is made up of those values of q for which $q > 3.51$. Since the computed studentized range falls in the noncritical region, we fail to reject the null hypothesis, just as in Example 10.1.

The studentized range may also be used to test simultaneously hypotheses or to construct simultaneously confidence intervals of linear comparisons of k means. The following theorem is useful for these purposes.

Theorem 10.1. Let $\bar{x}_i$ $(i = 1, 2, \ldots, k)$ be the mean of a random sample of size n drawn from a normal population with mean μ_i and variance σ^2. Let

$$c = \sum_{i=1}^{k} m_i \bar{x}_i$$

be any comparison of k independent sample means and

$$\gamma = \sum_{i=1}^{k} m_i \mu_i$$

Let s^2 with v degrees of freedom be any unbiased estimator of σ^2 which is independent of the sample means $\bar{x}_i$. Then $1 - \alpha$ is the probability that all comparisons c simultaneously satisfy

$$-\frac{q_\alpha s}{2\sqrt{n}} \cdot \sum_{i=1}^{k} |m_i| < c - \gamma < \frac{q_\alpha s}{2\sqrt{n}} \cdot \sum_{i=1}^{k} |m_i| \qquad (10.81)$$

where q_α is the upper $100\,\alpha$ per cent studentized range value found in Table IX.

The inequality in (10.81) may be written in other ways. Perhaps the simplest is the case where we make the sum of all the positive m's equal to one. Then, since c is a contrast, the sum of all negative m's is equal to minus one, and (10.81) reduces to

$$-\frac{q_\alpha s}{\sqrt{n}} < \sum m_i \bar{x}_i - \sum m_i \mu_i < \frac{q_\alpha s}{\sqrt{n}} \qquad (10.82)$$

The reader should satisfy himself that inequality (10.82) is not a restricted form of (10.81). It should be noted that it is possible to make Theorem 10.1 more general by defining c as a linear combination or by allowing the population variances to be different or by assuming a nonzero covariance between $\bar{x}_i$ and $\bar{x}_{i'}$ $(i \neq i')$ when c is a constant.

10.8.2. Examples of Simultaneous Confidence Intervals

We now use the data of Example 10.1 to illustrate and compare the studentized range and F procedures for finding simultaneous confidence intervals. We also compare the lengths of intervals of special comparisons obtained by these procedures with those obtained by the t distribution and illustrated in Sect. 10.7.

Example 10.7. From Example 10.1 we have $\bar{x}_1 = 92$, $\bar{x}_2 = 75$, $\bar{x}_3 = 106$, $n = 10$, and $s^2 = 1457$. Use (10.82) to find simultaneously 95 per cent confidence intervals for an indefinite number of linear contrasts of μ_1, μ_2, and μ_3. Since $q_{.05} = 3.51$ when $k = 3$ and $\nu = 27$, we find

$$\frac{q_{.05} s}{\sqrt{n}} = 3.51 \sqrt{\frac{1457}{10}} = 42.5$$

Table 10.7
Confidence Limits of Contrasts of Three Means—Range Method

Contrasts	Multipliers			Value of	95 Per Cent Confidence Limits	
	m_1	m_2	m_3	$m_1\bar{x}_1 + m_2\bar{x}_2 + m_3\bar{x}_3$	Lower	Upper
(1)		(2)		(3)	(4)	(5)
$\mu_1 - \mu_2$	1	−1	0	17	−25.5	59.5
$\mu_1 - \mu_3$	1	0	−1	−14	−56.5	28.5
$\mu_2 - \mu_3$	0	1	−1	−31	−73.5	11.5
$(\mu_1 + \mu_2)/2 - \mu_3$	$\frac{1}{2}$	$\frac{1}{2}$	−1	−22.5	−65.0	20.0
$(\mu_1 + \mu_3)/2 - \mu_2$	$\frac{1}{2}$	−1	$\frac{1}{2}$	24	−18.5	66.5
$(\mu_2 + \mu_3)/2 - \mu_1$	−1	$\frac{1}{2}$	$\frac{1}{2}$	−11.5	−54.0	31.0
$\mu_1/3 + 2\mu_2/3 - \mu_3$	$\frac{1}{3}$	$\frac{2}{3}$	−1	−25.3	−67.8	17.2
$\mu_1/3 + 2\mu_3/3 - \mu_2$	$\frac{1}{3}$	−1	$\frac{2}{3}$	26.3	−16.2	68.8
$\mu_2/3 + 2\mu_1/3 - \mu_3$	$\frac{2}{3}$	$\frac{1}{3}$	−1	−19.7	−62.2	22.8
$\mu_2/3 + 2\mu_3/3 - \mu_1$	−1	$\frac{1}{3}$	$\frac{2}{3}$	3.7	−38.8	46.2
$\mu_3/3 + 2\mu_1/3 - \mu_2$	$\frac{2}{3}$	−1	$\frac{1}{3}$	21.7	−20.8	64.2
$\mu_3/3 + 2\mu_2/3 - \mu_1$	−1	$\frac{2}{3}$	$\frac{1}{3}$	− 6.7	−49.2	35.8
etc.						

Then, according to Theorem 10.1, we are 95 per cent confident of being correct in all statements of the form

$$-42.5 < \sum m_i \bar{x}_i - \sum m_i \mu_i < 42.5$$

or

$$\sum m_i \bar{x}_i - 42.5 < \sum m_i \mu_i < \sum m_i \bar{x}_i + 42.5 \qquad (10.83)$$

it being understood that the sum of the positive m's is equal to one.

Typical contrasts among three population means, along with multipliers, value of contrasts of sample means, and confidence limits, are shown in Table 10.7. We are *at least* 95 per cent confident that the contrasts listed in column (1) of Table 10.7 have values between the limits shown in columns (4) and (5). Further, we are 95 per cent confident that all the contrasts of Table 10.7, along with as many others as we wish to write, have values falling between the limits obtained by substituting in inequality (10.83).

It should be noted at this point that, when a $100(1 - \alpha)$ per cent confidence interval is required for a specific contrast, the method used in Example 10.5 gives a shorter confidence interval than the method just described. For example, applying the t distribution and Example 10.5, we obtain for the contrast $\frac{1}{2}\mu_A - \mu_B + \frac{1}{2}\mu_C$ the interval

$$-6.35 \le \frac{1}{2}(\mu_A + \mu_C) - \mu_B \le 54.35$$

and, applying the studentized range and Table 10.7, we find that the interval is

$$-18.5 \le \frac{1}{2}(\mu_A + \mu_C) - \mu_B \le 66.5$$

Clearly, the t distribution should be used if a single contrast is of interest. If more than two contrasts are of interest, the two methods are not comparable, because the confidence coefficients (when 95 per cent procedures are used) are different. This is discussed in Sect. 10.9.

The studentized range procedure for finding simultaneous confidence intervals is very useful when certain contrasts are suggested by the data. In this way an exploratory collection of data can be used to suggest contrasts which may be examined by other techniques in future experiments.

According to Scheffé [29], simultaneous intervals for any number of linear contrasts γ among k means are given by

$$c \pm \sqrt{F_\alpha' s^2 \sum \frac{m_i^2}{n_i}} \qquad (10.84)$$

where $F_\alpha' = (k - 1) F_\alpha(k - 1, \nu)$ and s^2, c, m_i, and n_i are defined as before.

Example 10.8. From Example 10.1 we have $\bar{x}_1 = 92$, $\bar{x}_2 = 75$, $\bar{x}_3 = 106$, $n = 10$, and $s^2 = 1457$. (a) Use (10.84) to find simultaneously 95 per cent confidence intervals for an indefinite number of linear contrasts of μ_1, μ_2, and μ_3. (b) Compare the results of (a) with Table 10.7.

Since $F_{.05}(2, 27) = 3.35$, we have

$$F'_{.05} = 6.70 \quad \text{and} \quad \sqrt{\frac{F'_{.05} s^2}{n}} = \sqrt{\frac{(6.70)(1457)}{10}} = 31.2$$

Thus, according to (10.84), we are 95 per cent confident of being correct in all statements of the form

$$\sum m_i \bar{x}_i - 31.2\sqrt{\sum m_i^2} < \sum m_i \mu_i < \sum m_i \bar{x}_i + 31.2\sqrt{\sum m_i^2} \quad (10.85)$$

The intervals (10.85) are to be compared with the intervals (10.83).

The contrasts of Table 10.7, as well as $\sum m_i^2$, values of $c = m_1 \bar{x}_1 + m_2 \bar{x}_2 + m_3 \bar{x}_3$, $31.2\sqrt{\sum m_i^2}$, and simultaneous 95 per cent confidence limits for Scheffé's method, are shown in Table 10.8. We are at least 95 per cent confident that the contrasts listed in column (1) of Table 10.8 have values between the limits shown in columns (5) and (6). Further, we are 95 per cent confident that all the contrasts in Table 10.8 along with as many others as we wish to write, have values falling between the limits obtained by substituting in (10.85).

Table 10.8
Confidence Limits of Contrasts of Three Means—Scheffé's Method

Contrasts	Value of c	$\sum m_i^2$	$31.2\sqrt{\sum m_i^2}$	95 Per Cent Confidence Limits	
				Lower	Upper
(1)	(2)	(3)	(4)	(5)	(6)
$\mu_1 - \mu_2$	17	2	44.1	−27.1	61.1
$\mu_1 - \mu_3$	−14	2	44.1	−58.1	30.1
$\mu_2 - \mu_3$	−31	2	44.1	−75.1	13.1
$(\mu_1 + \mu_2)/2 - \mu_3$	−22.5	$\frac{6}{4}$	38.2	−60.7	15.7
$(\mu_1 + \mu_3)/2 - \mu_2$	24	$\frac{6}{4}$	38.2	−14.2	62.2
$(\mu_2 + \mu_3)/2 - \mu_1$	−11.5	$\frac{6}{4}$	38.2	−49.7	26.7
$\mu_1/3 + 2\mu_2/3 - \mu_3$	−25.3	$\frac{14}{9}$	38.9	−64.2	13.6
$\mu_1/3 + 2\mu_3/3 - \mu_2$	26.3	$\frac{14}{9}$	38.9	−12.6	65.2
$\mu_2/3 + 2\mu_1/3 - \mu_3$	−19.7	$\frac{14}{9}$	38.9	−58.6	19.2
$\mu_2/3 + 2\mu_3/3 - \mu_1$	3.7	$\frac{14}{9}$	38.9	−35.2	42.6
$\mu_3/3 + 2\mu_1/3 - \mu_2$	21.7	$\frac{14}{9}$	38.9	−17.2	60.6
$\mu_3/3 + 2\mu_2/3 - \mu_1$	− 6.7	$\frac{14}{9}$	38.9	−45.6	32.2
etc.					

On comparing Tables 10.7 and 10.8, we observe that some intervals are shorter when the range method is applied and some are shorter when Scheffé's method is applied. But the two methods lead to the same results on the average. That is, out of all possible contrasts, 95 per cent of the intervals actually do contain the true contrast of population means when the range method is applied, and when Scheffé's method is applied.

Scheffé's method for simultaneously finding confidence intervals can be generalized to linear combinations of sample means computed from different size samples for which $\bar{x}_i$ and $\bar{x}_{i'}$ $(i \neq i')$ are correlated and have different variances. This method for finding intervals is applicable when the range method is applicable, and it is weak when the range method is weak.

It might appear that considerably more computation is required to obtain confidence intervals by Scheffé's method, since the factor $\sqrt{\sum m_i^2}$ appears in (10.85) but not in (10.83). This is misleading. We used the form (10.85) for finding limits because we wanted to compare the new limits with those for the same contrasts which are given in Table 10.7. If (10.85) had been introduced first, we could have required that $\sum m_i^2 = 1$ in order to simplify the computation.

10.8.3. A Method for Finding m Simultaneous Confidence Intervals

Sometimes an experimenter before collecting data selects a number, say m, of linear combinations (usually contrasts) among k means which he would like to estimate with confidence intervals. The range and F methods just presented could be used, but it is possible to describe a method, using the t distribution, which gives shorter intervals in some instances. Dunn [7] describes this method, pointing out in the comparison between Scheffé's intervals (or range intervals) and the t intervals that the t method is more favorable when, if all other variables except one are assumed to be held constant, (1) k is increased, or (2) ν is increased, or (3) $1 - \alpha$ is increased.

It should be emphasized that when the t intervals are used the set of linear combinations which are to be estimated must be planned in advance, whereas with Scheffé's interval (or range intervals) they may be selected after looking at the data. According to Dunn, the $100\,(1 - \alpha)$ per cent confidence intervals for a set of linear combinations selected in advance are given by

$$c \pm t'_\alpha \sqrt{s^2 \sum \frac{m_i^2}{n_i}} \tag{10.86}$$

where c, s^2, m_i, n_i are defined as before, the $\bar{x}_i$'s are independently distributed, and t' is defined by

$$\int_{-\infty}^{t'} f(t; \nu)\, dt = 1 - \frac{\alpha}{2m} \tag{10.87}$$

$f(t; \nu)$ being the density function for a Student t variate with ν degrees of freedom. Values of t' for $\alpha = 0.05$ and 0.01 and selected values of ν and m have been computed by Dunn and are reproduced in Table 10.9.

Table 10.9*

Values of t' to be Used with m Linear Combinations

$\alpha = 0.05$

m \ ν	5	7	10	12	15	20	24	30	40	∞
2	3.17	2.84	2.64	2.56	2.49	2.42	2.39	2.36	2.33	2.24
3	3.54	3.13	2.87	2.78	2.69	2.61	2.58	2.54	2.50	2.39
4	3.81	3.34	3.04	2.94	2.84	2.75	2.70	2.66	2.62	2.50
5	4.04	3.50	3.17	3.06	2.95	2.85	2.80	2.75	2.71	2.58
6	4.22	3.64	3.28	3.15	3.04	2.93	2.88	2.83	2.78	2.64
7	4.38	3.76	3.37	3.24	3.11	3.00	2.94	2.89	2.84	2.69
8	4.53	3.86	3.45	3.31	3.18	3.06	3.00	2.94	2.89	2.74
9	4.66	3.95	3.52	3.37	3.24	3.11	3.05	2.99	2.93	2.77
10	4.78	4.03	3.58	3.43	3.29	3.16	3.09	3.03	2.97	2.81
15	5.25	4.36	3.83	3.65	3.48	3.33	3.26	3.19	3.12	2.94
20	5.60	4.59	4.01	3.80	3.62	3.46	3.38	3.30	3.23	3.02
25	5.89	4.78	4.15	3.93	3.74	3.55	3.47	3.39	3.31	3.09
30	6.15	4.95	4.27	4.04	3.82	3.63	3.54	3.46	3.38	3.15
35	6.36	5.09	4.37	4.13	3.90	3.70	3.61	3.52	3.43	3.19
40	6.56	5.21	4.45	4.20	3.97	3.76	3.66	3.57	3.48	3.23
45	6.70	5.31	4.53	4.26	4.02	3.80	3.70	3.61	3.51	3.26
50	6.86	5.40	4.59	4.32	4.07	3.85	3.74	3.65	3.55	3.29
100	8.00	6.08	5.06	4.73	4.42	4.15	4.04	3.90	3.79	3.48
250	9.68	7.06	5.70	5.27	4.90	4.56	4.4†	4.2†	4.1†	3.72

$\alpha = 0.01$

m \ ν	5	7	10	12	15	20	24	30	40	∞
2	4.78	4.03	3.58	3.43	3.29	3.16	3.09	3.03	2.97	2.81
3	5.25	4.36	3.83	3.65	3.48	3.33	3.26	3.19	3.12	2.94
4	5.60	4.59	4.01	3.80	3.62	3.46	3.38	3.30	3.23	3.02
5	5.89	4.78	4.15	3.93	3.74	3.55	3.47	3.39	3.31	3.09
6	6.15	4.95	4.27	4.04	3.82	3.63	3.54	3.46	3.38	3.15
7	6.36	5.09	4.37	4.13	3.90	3.70	3.61	3.52	3.43	3.19
8	6.56	5.21	4.45	4.20	3.97	3.76	3.66	3.57	3.48	3.23
9	6.70	5.31	5.53	4.26	4.02	3.80	3.70	3.61	3.51	3.26
10	6.86	5.40	4.59	4.32	4.07	3.85	3.74	3.65	3.55	3.29
15	7.51	5.79	4.86	4.56	4.29	4.03	3.91	3.80	3.70	3.40
20	8.00	6.08	5.06	4.73	4.42	4.15	4.04	3.90	3.79	3.48
25	8.37	6.30	5.20	4.86	4.53	4.25	4.1†	3.98	3.88	3.54
30	8.68	6.49	5.33	4.95	4.61	4.33	4.2†	4.13	3.93	3.59
35	8.95	6.67	5.44	5.04	4.71	4.39	4.3†	4.26	3.97	3.63
40	9.19	6.83	5.52	5.12	4.78	4.46	4.3†	4.1†	4.01	3.66
45	9.41	6.93	5.60	5.20	4.84	4.52	4.3†	4.2†	4.1†	3.69
50	9.68	7.06	5.70	5.27	4.90	4.56	4.4†	4.2†	4.1†	3.72
100	11.04	7.80	6.20	5.70	5.20	4.80	4.7†	4.4†	4.5†	3.89
250	13.26	8.83	6.9†	6.3†	5.8†	5.2†	5.0†	4.9†	4.8†	4.11

* This table is reproduced from Olive Jean Dunn, "Multiple Comparisons Among Means," *Journal of the American Statistical Association*, Vol. **56** (1961), p. 55, Tables 1 and 2, with permission of the editor of the journal.

† Obtained by graphical interpolation.

Example 10.9. From Example 10.1 we have $\bar{x}_1 = 92$, $\bar{x}_2 = 75$, $\bar{x}_3 = 106$, $n = 10$, and $s^2 = 1457$. Use (10.86) to find simultaneously 95 per cent confidence intervals for $\mu_1 - \mu_2$, $\mu_1 - \mu_3$, and $\mu_2 - \mu_3$.

For each of these contrasts $\sum m_i^2 = 2$. With $m = 3$ and $\nu = 27$ we use Table 10.9 to find, by linear interpolation, that $t'_{.05} = 2.56$. Therefore, $t'_{.05}\sqrt{s^2 \sum m_i^2/n} = 43.8$, and the intervals are

$$-25.8 < \mu_1 - \mu_2 < 60.8$$

$$-57.8 < \mu_1 - \mu_3 < 29.8$$

$$-74.8 < \mu_2 - \mu_3 < 12.8$$

It should be observed that these intervals have roughly the same lengths as those given by the F and range methods. This is not always the case, since the relative lengths depend on the size of m as well as k, ν, and α.

In this section, Sect. 10.8, we have generally restricted our attention to confidence-interval problems. However, what has been said about simultaneous confidence intervals can easily be extended to simultaneous hypotheses involving sets of linear contrasts. References are made to such hypotheses in Sect. 10.9.

In many investigations one is not necessarily interested in estimating intervals or in testing hypotheses about some set of orthogonal linear contrasts. On the other hand, the investigator often feels, and rightly so, that methods for working with all linear combinations are too all-inclusive. For example, in an experiment involving four means, say, one often requires information concerning all pairs. That is, for means μ_1, μ_2, μ_3, μ_4, the linear contrasts $\mu_1 - \mu_2$, $\mu_1 - \mu_3$, $\mu_1 - \mu_4$, $\mu_2 - \mu_3$, $\mu_2 - \mu_4$, and $\mu_3 - \mu_4$ are the only ones of interest. For such problems we ask if it is not possible to describe procedures for testing hypotheses and constructing confidence intervals which are superior to any already introduced. The answer is in the affirmative, but the details are not completely resolved.

10.9. MULTIPLE TEST PROCEDURES FOR SEVERAL PAIRS OF MEANS

In an analysis of variance, the problem of testing for differences between pairs selected from among several means often arises. Testing the homogeneity of a set of means by the F test may result in the conclusion that they are not all alike, but it fails to signify any arrangement of distinguishable groups among the means. The problem of separating a group of heterogeneous means into subgroups of nonheterogeneous means has been approached in several different ways by a number of research workers. A common procedure, actually the oldest one, used for this purpose is the least-significant-difference test, described by Goulden [8], Davies [3], and others. There are other more recent tests which are superior to the least-significant-difference test, commonly called the l.s.d. test. We shall explain in detail how to apply one of the best multiple test procedures and show how it is superior to the least-significant-difference test by introducing a discussion on errors. Before proceeding with a comparison of the tests, we will state the assumptions generally made in applying the tests and give the example to be used in illustrating these differences.

Assume that we have a set of k means $\bar{x}_1, \bar{x}_2, \ldots, \bar{x}_k$ determined from random samples drawn independently from normal populations with means $\mu_1, \mu_2, \ldots, \mu_k$, respectively. Assume that the sampling distributions of the means have a common variance $\sigma_{\bar{x}}^2$ and that $s_{\bar{x}}^2$ is an unbiased estimator of $\sigma_{\bar{x}}^2$.

For the illustration assume that the mean response of each of eight treatments replicated four times in a one-way classification design and arranged in increasing order of magnitude are

$$\begin{array}{cccccccc} \bar{x}_4 & \bar{x}_1 & \bar{x}_8 & \bar{x}_6 & \bar{x}_5 & \bar{x}_2 & \bar{x}_7 & \bar{x}_3 \\ 1.2 & 2.7 & 3.9 & 7.2 & 8.2 & 10.3 & 10.9 & 13.1 \end{array} \qquad (10.88)$$

The error mean square is

$$s^2 = 27.66$$

with $\nu = 24$ degrees of freedom, and the standard error of means is $s_{\bar{x}} = \sqrt{27.66/4} = 2.63$. These means actually represent coded average yields in pounds per plot (two rows 25 feet long) of eight varieties of sweet potatoes. However, the reader may think of these numbers as coded means of tensile strength of eight types of wire, rainfall in August for eight years, expenditure per pupil per year in public schools in eight states, etc.

If we are interested only in testing to determine whether the difference in the response of treatments 1 and 2 is significant, the best test is the usual t test, if it is assumed that these treatments were selected before the experiment was run. We compute

$$\frac{\bar{x}_2 - \bar{x}_1}{s_{x_1 - x_2}} = \frac{\bar{x}_2 - \bar{x}_1}{\sqrt{2}\, s_{\bar{x}}} = \frac{10.3 - 2.7}{3.72} = 2.04$$

and compare this value with the two-tailed five per cent t value with 24 degrees of freedom, that is, with $t_{.05}(24) = 2.06$. Or we may compare $\bar{x}_2 - \bar{x}_1 = 7.6$ with $s_{\bar{x}}\sqrt{2}\, t_{.05}(24) = (2.63)(2.91) = 7.7$. In either case we conclude that the difference in the mean response of treatments 1 and 2 is not significantly different from zero. If the five per cent level test is satisfactory with the experimenter, there is no room for further worry, because the t test is the most powerful test, that is, "best test," in the sense that this test will accept the alternative hypothesis more often than any other test when the alternative hypothesis is actually true.

Usually, in an experiment involving eight means, we would be interested in testing several (or all) hypotheses of the form $H_0: \mu_i = \mu_j\ (i \neq j = 1, \ldots, 8)$. Suppose we want to make a statement about each difference between pairs of means, and suppose we agree that we want an over-all test corresponding to the five per cent test given above. For this purpose we will consider the multiple t test, the l.s.d. test, Newman-Keuls test [15, 25], and Duncan's multiple range test [4] in some detail. Test procedures described by Hartley [14] and Tukey [32, 33, 34] are not discussed.

A natural test to consider is provided by the joint application of α level t tests to all hypotheses H_0 using the rule which follows: if $\bar{x}_i - \bar{x}_j > s_x \sqrt{2}\, t_\alpha(\nu)$, make the decision that $\mu_i > \mu_j$; if $\bar{x}_i - \bar{x}_j \leq s_x \sqrt{2}\, t_\alpha(\nu)$, make the decision that $\mu_i = \mu_j$; that is, there is no significant difference in $\bar{x}_i$ and $\bar{x}_j$. This procedure is often referred to as an α level multiple t test. In our example, the results of the multiple t test may be shown as follows

$\bar{x}_4$	$\bar{x}_1$	$\bar{x}_8$	$\bar{x}_6$	$\bar{x}_5$	$\bar{x}_2$	$\bar{x}_7$	$\bar{x}_3$
1.2	2.7	3.9	7.2	8.2	10.3	10.9	13.1

Any two means *not underscored* by the same line are *significantly different*. Any two means *underscored* by the same line are *not significantly different*. That is, in testing each difference between two means, we determine that

$$\mu_3 > \mu_4, \quad \mu_3 > \mu_1, \quad \mu_3 > \mu_8, \quad \mu_7 > \mu_4, \quad \mu_7 > \mu_1, \quad \text{and} \quad \mu_2 > \mu_4$$

all other pairs of means being declared *not* significantly different.

The multiple t test is not recommended for use, but is introduced because of its simplicity and because some undesirable features of other multiple test procedures are easy to discuss in terms of this test. The principal disadvantage in using this test is brought out in the following discussion. In drawing a random sample of size $n\ (n > 2)$ from a normal population, we can expect the difference between the largest and the smallest observations to be greater than the difference between two randomly chosen observations on the average. (This could also be the case for the difference between less extreme observations.) Also, when $k\ (k > 2)$ independent means are computed from the same normal population, we can expect the difference between the largest and smallest mean to be greater than the difference between two randomly chosen means. That is, the dispersion of the sampling distribution of differences, d_e, between extreme means in a set of k means is greater than the dispersion of the sampling distribution of differences, d_t, between two random means. (The reader should realize that standardized values of d_e and d_t are actually studentized range variates.) Further, the larger the value of k, the larger the dispersion of d_e relative to the dispersion of d_t. For this reason it is possible in a case where two extreme means are not significantly different to make the claim that they are significantly different when the t distribution (that is, the d_t distribution) is used incorrectly. (This can also happen for differences between means which are not at the extremes.) This means that if all the means involved are homogeneous, the multiple t test has a large error probability of wrongly rejecting the null hypothesis that all the means are equal; that is, $H_k: \mu_1 = \cdots = \mu_k$. For the above five per cent level multiple t test with eight means and 24 error degrees of freedom, the error probability of wrongly rejecting the hypothesis H_8 is greater than 40 per cent. The error probability of wrongly rejecting the null hypothesis that seven of the means are equal is roughly 40 per cent; the error probability of wrongly rejecting the hypothesis that six of the means are equal is in the neighborhood of 35 per cent, and so on. These percentages were found by using Pearson and Hartley's [27] tables of the studentized range.

The least-significant-difference test was introduced to overcome the disadvantage of such a large error probability involving k means. In applying this test, the first step is to use an α level F test to determine if the variance ratio for the k means is significant. If there is no significance, then H_k is accepted; otherwise, the multiple t test is applied. In our example, the F ratio is 2.60 and the upper five per cent point for F with seven and 24 degrees of freedom is $F_{.05}(7, 24) = 2.42$. Thus, the eight means are not homogeneous, and the multiple t test is used to obtain the results given above.

As has been mentioned, the purpose of the initial F test is to remove the high error probability of the multiple t test for wrongly rejecting the hypothesis that the eight means are equal. Nevertheless, the l.s.d. test fails to insure the reduction of similarly high error probabilities for wrongly rejecting the hypothesis that p ($p = 3, \ldots, 7$) of the means are equal. However, if this principle of using a preliminary F test were carried to its logical conclusion, any test involving a group of means would need to be preceded by a preliminary F test of the group. Obviously, such a procedure becomes unwieldy. Fortunately, nearly the same results can be achieved by replacing the F tests with range tests and doing away with the t test.

Before we discuss the multiple range tests, it may be useful to give another reason why an experimenter who is interested only in testing differences in pairs of means would not choose to use an initial F test. The fact that the F test also tests the significance of all linear comparisons of the means causes trouble in connection with the significance level. For example, consider the three means 2.0, 2.1, and 3.0 which are obtained from samples of size five. The treatment mean square is 1.52, and the error mean square is $s^2 = 0.41$. The F ratio is 3.71 and the upper five per cent level F value is $F_{.05}(2, 12) = 3.89$. Thus, we fail to reject the null hypothesis that $H_3: \mu_1 = \mu_2 = \mu_3$. However, in a similar problem, when the means are

2.0, 2.0, and 3.0 with the same error mean square $s^2 = 0.41$, the treatment mean square is 1.67, and the F ratio becomes 4.17. In this case, we reject the null hypothesis and conclude that 3.0 is significantly larger than 2.0 after applying the five per cent level t test. Since the error mean square, the number of means, and the largest and smallest means are the same in both examples, it does not seem right that we should reach two different conclusions concerning the same difference. It appears that the test of the differences 3.0–2.0 in the extreme means should depend on the number of means and s. This is the situation when multiple range tests are used.

The Newman-Keuls multiple range test overcomes some of the disadvantages already encountered. This procedure, which was first suggested by "Student," developed by Newman [25], and amplified by Keuls [15], is equivalent to a multiple t test preceded by several studentized range tests. Since the t tests of which the multiple t test is composed may be regarded as range tests of subsets of two means each, the over-all procedure is composed entirely of range tests and may be usefully termed a multiple range test.

An α *level Newman-Keuls multiple range test* is given by the following rule:

> The difference between any two means in a set of k means is significant, provided the range of each subset which *contains* the given two means is significant according to an α level range test. We say that a set S of numbers *contains* a number e, provided e is not smaller than the smallest number in S and not larger than the largest number in S.

To apply this test, arrange the means in increasing order of magnitude and apply α level (studentized) range tests to all possible combinations of the k means taken p at a time ($p = 2, \ldots, k$). If any combination of p means has a nonsignificant range, then the decision is made that these p means are homogeneous, and any combination of means within these p means is homogeneous. For any combination of means which is not homogeneous, the highest mean is significantly larger than the lowest mean. In our example, the five per cent least significant ranges [$q_{0.5}(p, \nu) = q_{.05}$ denotes least significant studentized ranges and $R_p = s_{\bar{x}} \cdot q_{.05}$ denotes the *least significant range for p means*] for $\nu = 24$ degrees of freedom and subsets of size two to eight are found with the use of Table IX to be

p	2	3	4	5	6	7	8
$q_{.05}$	2.92	3.53	3.90	4.17	4.37	4.54	4.68
R_p	7.7	9.3	10.3	11.0	11.5	11.9	12.3

Since the over-all difference $13.1 - 1.2 = 11.9$ of the means in (10.88) is less

Table 10.10*

Least Significant Studentized Ranges r_p for Duncan's Multiple Range Test

$\alpha = \alpha_2 = 0.05$

ν \\ p	2	3	4	5	6	7	8	9
1	18.0	18.0	18.0	18.0	18.0	18.0	18.0	18.0
2	6.09	6.09	6.09	6.09	6.09	6.09	6.09	6.09
3	4.50	4.50	4.50	4.50	4.50	4.50	4.50	4.50
4	3.93	4.01	4.02	4.02	4.02	4.02	4.02	4.02
5	3.64	3.74	3.79	3.83	3.83	3.83	3.83	3.83
6	3.46	3.58	3.64	3.68	3.68	3.68	3.68	3.68
7	3.35	3.47	3.54	3.58	3.60	3.61	3.61	3.61
8	3.26	3.39	3.47	3.52	3.55	3.56	3.56	3.56
9	3.20	3.34	3.41	3.47	3.50	3.52	3.52	3.52
10	3.15	3.30	3.37	3.43	3.46	3.47	3.47	3.47
11	3.11	3.27	3.35	3.39	3.43	3.44	3.45	3.46
12	3.08	3.23	3.33	3.36	3.40	3.42	3.44	3.44
13	3.06	3.21	3.30	3.35	3.38	3.41	3.42	3.44
14	3.03	3.18	3.27	3.33	3.37	3.39	3.41	3.42
15	3.01	3.16	3.25	3.31	3.36	3.38	3.40	3.42
16	3.00	3.15	3.23	3.30	3.34	3.37	3.39	3.41
17	2.98	3.13	3.22	3.28	3.33	3.36	3.38	3.40
18	2.97	3.12	3.21	3.27	3.32	3.35	3.37	3.39
19	2.96	3.11	3.19	3.26	3.31	3.35	3.37	3.39
20	2.95	3.10	3.18	3.25	3.30	3.34	3.36	3.38
22	2.93	3.08	3.17	3.24	3.29	3.32	3.35	3.37
24	2.92	3.07	3.15	3.22	3.28	3.31	3.34	3.37
26	2.91	3.06	3.14	3.21	3.27	3.30	3.34	3.36
28	2.90	3.04	3.13	3.20	3.26	3.30	3.33	3.35
30	2.89	3.04	3.12	3.20	3.25	3.29	3.32	3.35
40	2.86	3.01	3.10	3.17	3.22	3.27	3.30	3.33
60	2.83	2.98	3.08	3.14	3.20	3.24	3.28	3.31
100	2.80	2.95	3.05	3.12	3.18	3.22	3.26	3.29
∞	2.77	2.92	3.02	3.09	3.15	3.19	3.23	3.26

* This table is reproduced from David B. Duncan, "Multiple Range and Multiple F. Tests," *Biometrics*, Vol. **11** (1955), pp. 3–4, with permission of the editor of the journal.

Table 10.10

Least Significant Studentized Ranges r_p for Duncan's Multiple Range Test (*cont.*)

$$\alpha = \alpha_2 = 0.05$$

ν \ p	10	12	14	16	18	20	50	100
1	18.0	18.0	18.0	18.0	18.0	18.0	18.0	18.0
2	6.09	6.09	6.09	6.09	6.09	6.09	6.09	6.09
3	4.50	4.50	4.50	4.50	4.50	4.50	4.50	4.50
4	4.02	4.02	4.02	4.02	4.02	4.02	4.02	4.02
5	3.83	3.83	3.83	3.83	3.83	3.83	3.83	3.83
6	3.68	3.68	3.68	3.68	3.68	3.68	3.68	3.68
7	3.61	3.61	3.61	3.61	3.61	3.61	3.61	3.61
8	3.56	3.56	3.56	3.56	3.56	3.56	3.56	3.56
9	3.52	3.52	3.52	3.52	3.52	3.52	3.52	3.52
10	3.47	3.47	3.47	3.47	3.47	3.48	3.48	3.48
11	3.46	3.46	3.46	3.46	3.47	3.48	3.48	3.48
12	3.46	3.46	3.46	3.46	3.47	3.48	3.48	3.48
13	3.45	3.45	3.46	3.46	3.47	3.47	3.47	3.47
14	3.44	3.45	3.46	3.46	3.47	3.47	3.47	3.47
15	3.43	3.44	3.45	3.46	3.47	3.47	3.47	3.47
16	3.43	3.44	3.45	3.46	3.47	3.47	3.47	3.47
17	3.42	3.44	3.45	3.46	3.47	3.47	3.47	3.47
18	3.41	3.43	3.45	3.46	3.47	3.47	3.47	3.47
19	3.41	3.43	3.44	3.46	3.47	3.47	3.47	3.47
20	3.40	3.43	3.44	3.46	3.46	3.47	3.47	3.47
22	3.39	3.42	3.44	3.45	3.46	3.47	3.47	3.47
24	3.38	3.41	3.44	3.45	3.46	3.47	3.47	3.47
26	3.38	3.41	3.43	3.45	3.46	3.47	3.47	3.47
28	3.37	3.40	3.43	3.45	3.46	3.47	3.47	3.47
30	3.37	3.40	3.43	3.44	3.46	3.47	3.47	3.47
40	3.35	3.39	3.42	3.44	3.46	3.47	3.47	3.47
60	3.33	3.37	3.40	3.43	3.45	3.47	3.48	3.48
100	3.32	3.36	3.40	3.42	3.45	3.47	3.53	3.53
∞	3.29	3.34	3.38	3.41	3.44	3.47	3.61	3.67

Table 10.10

Least Significant Studentized Ranges r_p for Duncan's Multiple Range Test (*cont.*)

$$\alpha = \alpha_2 = 0.01$$

ν＼p	2	3	4	5	6	7	8	9
1	90.0	90.0	90.0	90.0	90.0	90.0	90.0	90.0
2	14.0	14.0	14.0	14.0	14.0	14.0	14.0	14.0
3	8.26	8.5	8.6	8.7	8.8	8.9	8.9	9.0
4	6.51	6.8	6.9	7.0	7.1	7.1	7.2	7.2
5	5.70	5.96	6.11	6.18	6.26	6.33	6.40	6.44
6	5.24	5.51	5.65	5.73	5.81	5.88	5.95	6.00
7	4.95	5.22	5.37	5.45	5.53	5.61	5.69	5.73
8	4.74	5.00	5.14	5.23	5.32	5.40	5.47	5.51
9	4.60	4.86	4.99	5.08	5.17	5.25	5.32	5.36
10	4.48	4.73	4.88	4.96	5.06	5.13	5.20	5.24
11	4.39	4.63	4.77	4.86	4.94	5.01	5.06	5.12
12	4.32	4.55	4.68	4.76	4.84	4.92	4.96	5.02
13	4.26	4.48	4.62	4.69	4.74	4.84	4.88	4.94
14	4.21	4.42	4.55	4.63	4.70	4.78	4.83	4.87
15	4.17	4.37	4.50	4.58	4.64	4.72	4.77	4.81
16	4.13	4.34	4.45	4.54	4.60	4.67	4.72	4.76
17	4.10	4.30	4.41	4.50	4.56	4.63	4.68	4.72
18	4.07	4.27	4.38	4.46	4.53	4.59	4.64	4.68
19	4.05	4.24	4.35	4.43	4.50	4.56	4.61	4.64
20	4.02	4.22	4.33	4.40	4.47	4.53	4.58	4.61
22	3.99	4.17	4.28	4.36	4.42	4.48	4.53	4.57
24	3.96	4.14	4.24	4.33	4.39	4.44	4.49	4.53
26	3.93	4.11	4.21	4.30	4.36	4.41	4.46	4.50
28	3.91	4.08	4.18	4.28	4.34	4.39	4.43	4.47
30	3.89	4.06	4.16	4.22	4.32	4.36	4.41	4.45
40	3.82	3.99	4.10	4.17	4.24	4.30	4.34	4.37
60	3.76	3.92	4.03	4.12	4.17	4.23	4.27	4.31
100	3.71	3.86	3.98	4.06	4.11	4.17	4.21	4.25
∞	3.64	3.80	3.90	3.98	4.04	4.09	4.14	4.17

Table 10.10

Least Significant Studentized Ranges r_p for Duncan's Multiple Range Test (*cont.*)

$$\alpha = \alpha_2 = 0.01$$

ν \ p	10	12	14	16	18	20	50	100
1	90.0	90.0	90.0	90.0	90.0	90.0	90.0	90.0
2	14.0	14.0	14.0	14.0	14.0	14.0	14.0	14.0
3	9.0	9.0	9.1	9.2	9.3	9.3	9.3	9.3
4	7.3	7.3	7.4	7.4	7.5	7.5	7.5	7.5
5	6.5	6.6	6.6	6.7	6.7	6.8	6.8	6.8
6	6.0	6.1	6.2	6.2	6.3	6.3	6.3	6.3
7	5.8	5.8	5.9	5.9	6.0	6.0	6.0	6.0
8	5.5	5.6	5.7	5.7	5.8	5.8	5.8	5.8
9	5.4	5.5	5.5	5.6	5.7	5.7	5.7	5.7
10	5.28	5.36	5.42	5.48	5.54	5.55	5.55	5.55
11	5.15	5.24	5.28	5.34	5.38	5.39	5.39	5.39
12	5.07	5.13	5.17	5.22	5.24	5.26	5.26	5.26
13	4.98	5.04	5.08	5.13	5.14	5.15	5.15	5.15
14	4.91	4.96	5.00	5.04	5.06	5.07	5.07	5.07
15	4.84	4.90	4.94	4.97	4.99	5.00	5.00	5.00
16	4.79	4.84	4.88	4.91	4.93	4.94	4.94	4.94
17	4.75	4.80	4.83	4.86	4.88	4.89	4.89	4.89
18	4.71	4.76	4.79	4.82	4.84	4.85	4.85	4.85
19	4.67	4.72	4.76	4.79	4.81	4.82	4.82	4.82
20	4.65	4.69	4.73	4.76	4.78	4.79	4.79	4.79
22	4.60	4.65	4.68	4.71	4.74	4.75	4.75	4.75
24	4.57	4.62	4.64	4.67	4.70	4.72	4.74	4.74
26	4.53	4.58	4.62	4.65	4.67	4.69	4.73	4.73
28	4.51	4.56	4.60	4.62	4.65	4.67	4.72	4.72
30	4.48	4.54	4.58	4.61	4.63	4.65	4.71	4.71
40	4.41	4.46	4.51	4.54	4.57	4.59	4.69	4.69
60	4.34	4.39	4.44	4.47	4.50	4.53	4.66	4.66
100	4.29	4.35	4.38	4.42	4.45	4.48	4.64	4.65
∞	4.20	4.26	4.31	4.34	4.38	4.41	4.60	4.68

than 12.3, the least significant difference for eight means, we conclude that $\bar{x}_4$ and $\bar{x}_3$ are not significantly different and that there are no significant differences in pairs of means. The results are conviently written as follows

$\bar{x}_4$	$\bar{x}_1$	$\bar{x}_8$	$\bar{x}_6$	$\bar{x}_5$	$\bar{x}_2$	$\bar{x}_7$	$\bar{x}_3$
1.2	2.7	3.9	7.2	8.2	10.3	10.9	13.1

Duncan's multiple range test is applied like the Newman-Keuls test, but the least significant ranges r_p are different. If we let α_2 denote the significance level for two means, the significance level for p ($p = 3, \ldots, k$) means is given by

$$\alpha_p = 1 - (1 - \alpha_2)^{p-1}$$

Thus, for Duncan's five per cent level test $\alpha_2 = 0.05$, $\alpha_3 = 0.0975$, $\alpha_4 = 0.1426, \ldots, \alpha_8 = 0.3017$. That is, in Duncan's $\alpha = \alpha_2$ level test there is a set of $k - 1$ significant levels. For this reason special tables have been computed by Duncan and are reproduced in Table 10.10.

Example 10.10. Illustrate Duncan's multiple range test, using the means and standard error of (10.88).

Let $\alpha = \alpha_2 = 0.05$. Then the α_p ($p = 2, \ldots, 8$) least significant stu-dentized ranges r_p found in Table 10.10 and the least significant ranges $R_p = r_p s_{\bar{x}}$ for the data are as follows

p	2	3	4	5	6	7	8
r_p	2.92	3.07	3.15	3.22	3.28	3.31	3.34
R_p	7.7	8.1	8.3	8.5	8.6	8.7	8.8

The application of these least significant ranges R_p to the differences in ordered means of (10.88) is as follows:

1. For eight means $R_8 = 8.8$ and $\bar{x}_3 - \bar{x}_4 = 13.1 - 1.2 = 11.9$. Since $11.9 > 8.8$, we conclude that $\bar{x}_3$ is significantly larger than $\bar{x}_4$ or $\mu_3 > \mu_4$.

2. For seven means $R_7 = 8.7$ and $\bar{x}_3 - \bar{x}_1 = 13.1 - 2.7 = 10.4$. Since $10.4 > 8.7$, we conclude that $\bar{x}_3$ is significantly larger than $\bar{x}_1$ or $\mu_3 > \mu_1$.

3. For six means $R_6 = 8.6$, and $\bar{x}_3 - \bar{x}_8 = 13.1 - 3.9 = 9.2$. Since $9.2 > 8.6$, we conclude that $\bar{x}_3$ is significantly larger than $\bar{x}_8$ or $\mu_3 > \mu_8$.

4. For five means $R_5 = 8.5$ and $\bar{x}_3 - \bar{x}_6 = 13.1 - 7.2 = 5.9$. Since $5.9 < 8.5$, we conclude that $\bar{x}_3 - \bar{x}_6$ is not significantly different from zero. Further, it follows that the differences $\bar{x}_3 - \bar{x}_5$, $\bar{x}_3 - \bar{x}_2$,

$\bar{x}_3 - \bar{x}_7, \bar{x}_7 - \bar{x}_6, \bar{x}_7 - \bar{x}_5, \bar{x}_7 - \bar{x}_2, \bar{x}_2 - \bar{x}_6, \bar{x}_2 - \bar{x}_5, \bar{x}_5 - \bar{x}_6$ are not significantly different from zero. Continuing in this way, we obtain the results which follow.

5. $\bar{x}_7 - \bar{x}_4 = 9.7 > R_7 = 8.7$; thus, $\bar{x}_7$ is significantly larger than $\bar{x}_4$.

6. $\bar{x}_7 - \bar{x}_1 = 8.2 < R_6 = 8.6$; thus, $\bar{x}_7$ and $\bar{x}_1$ do not differ significantly, and it follows that the differences $\bar{x}_7 - \bar{x}_8, \bar{x}_2 - \bar{x}_1, \bar{x}_2 - \bar{x}_8, \bar{x}_5 - \bar{x}_1,$ $\bar{x}_5 - \bar{x}_8, \bar{x}_6 - \bar{x}_1, \bar{x}_6 - \bar{x}_8,$ and $\bar{x}_8 - \bar{x}_1,$ in addition to six differences listed in 4, are not significant.

7. $\bar{x}_2 - \bar{x}_4 = 9.1 > R_6 = 8.7$; thus, $\bar{x}_2$ is significantly larger than $\bar{x}_4$.

8. $\bar{x}_5 - \bar{x}_4 = 7.0 < R_5 = 8.5$; thus, $\bar{x}_5$ and $\bar{x}_4$ do not differ significantly, as well as all pairs in the set $\bar{x}_4, \bar{x}_1, \bar{x}_8, \bar{x}_6, \bar{x}_5$.

These results are conveniently shown as follows

$\bar{x}_4$	$\bar{x}_1$	$\bar{x}_8$	$\bar{x}_6$	$\bar{x}_5$	$\bar{x}_2$	$\bar{x}_7$	$\bar{x}_3$
1.2	2.7	3.8	7.2	8.2	10.3	10.9	13.1

That is, in testing each difference between two means we determine that

$$\mu_3 > \mu_4, \quad \mu_3 > \mu_1, \quad \mu_3 > \mu_8, \quad \mu_7 > \mu_4, \quad \text{and} \quad \mu_2 > \mu_4$$

all other pairs of means being declared not significantly different.

Tukey has presented two multiple range tests. His 1949 five per cent test [32] is like a multiple t test, except that he chooses as his over-all least significance range the one obtained by fixing the significance level for k means at five per cent. Theorem 10.1 is used with this method. In the above example, this fixes his "two" mean significance level around 0.5 per cent and the "three" mean significance level around one per cent. His 1953 test [34] fixes the least significant ranges halfway between those for his 1949 test and Newman-Keuls test. A comparison of the least significant studentized ranges for eight treatments with 24 degrees of freedom may be made with the aid of Table 10.11.

Other multiple test procedures are described by Hartley [13] and Tukey [33]. All multiple range tests apply to more general situations. Their appli-

Table 10.11
Least Significant Studentized Ranges for Five Per Cent Level Tests

$\nu = 24$	p						
	2	3	4	5	6	7	8
Multiple t	2.92	2.92	2.92	2.92	2.92	2.92	2.92
Duncan	2.92	3.07	3.15	3.22	3.28	3.31	3.34
Newman-Keuls	2.92	3.53	3.90	4.17	4.37	4.54	4.68
Tukey 1953	3.80	4.11	4.29	4.42	4.53	4.61	4.68
Tukey 1949	4.68	4.68	4.68	4.68	4.68	4.68	4.68

cations in other (more complicated) analyses are discussed by Duncan [5] and Kramer [18]. The effect of sample size has also been examined. Harter [11] discusses the selection of appropriate sample size, and Duncan [5] and Kramer [17] give methods for applying the multiple range tests when samples are of unequal sizes.

We have emphasized that the multiple range tests are used for testing the difference between two means. In many problems, the experimenter is interested in testing certain comparisons, in which case either Theorem 10.1 or a multiple F test should be used. A multiple F test corresponding to each multiple range test can be described, and its application would be similar to that of the range test except least significant F values would be used in place of least significant range values. Scheffé [29] has described a multiple F test corresponding to Tukey's 1949 range test, and Duncan [4] has described a multiple F test corresponding to his multiple range test. The amount of work involved in applying a multiple F test is many times greater than that required for the companion range test. This being the case, multiple F tests should not, in general, be used, unless the investigator is interested in contrasts involving more than two means.

At this point the reader no doubt wonders which multiple test procedure should be used. In general, there is no simple, clear-cut answer. If only one contrast is tested, the t test is the best, and if all contrasts are examined, either Scheffé's method, described in Sect. 10.8.2, or Theorem 10.1 should be applied. (But neither of these extremes is often considered appropriate— usually an experimenter wishes to test more than one hypothesis but fewer than 1000, say.) If one wishes to test simultaneously all hypotheses of the form $H_0: \mu_i - \mu_j = 0$ $(i \neq j; i, j = 1, \ldots, k)$ these methods should not be used. Of all the tests now available, it appears that Duncan's new multiple range test is the best for testing differences in all pairs of means. However, some recent studies by Duncan [6] indicate that there may be a better test procedure for testing differences in all pairs.

One who is interested in theory of multiple-decision problems should read what Lehmann [19, 20, 21] has to say. For a study of power for multiple tests, see Wine [35].

10.10. EXERCISES

10.17. (a) Find constants a, b, and d so that the two linear combinations

$$\gamma_1 = \mu_1 + 2\mu_2 + a\mu_3 \qquad \gamma_2 = b\mu_1 + d\mu_2 + \mu_3$$

are orthogonal contrasts. (b) The three sample totals in Exercise 10.4 are

$$T_1 = 402 \qquad T_2 = 397 \qquad T_3 = 458$$

Find the treatment sum of squares and the sums of squares for the two contrasts in (a); i. e., find SSA, Q_1^2 and Q_2^2. (c) Suppose samples 1 and 2 represent the production of old standard machines and sample 3 represents the production of a new machine. What linear contrast would you use in order to compare production of the new machine with that of the standard machines? Find a contrast orthogonal to this one.

10.18. (a) In an experiment with five treatments there are four orthogonal contrasts. Write a set of multipliers for four such contrasts. Describe an experiment in which these four contrasts are meaningful (in the experimental context). (b) Write two other different sets of multipliers for four orthogonal contrasts.

10.19. In a completely randomized design with four treatments and seven observations per treatment, the error sum of squares is 0.09 and the totals are $T_1 = 19.5$, $T_2 = 19.2$, $T_3 = 20.1$, and $T_4 = 19.2$. Treatments 1 and 2 are to be compared to treatments 3 and 4; treatments 1 and 3 to 2 and 4; treatments 1 and 4 to 2 and 3. (a) State the three hypotheses in terms of linear contrasts. Are they orthogonal? (b) Compute the error and treatment mean squares and the three components of the treatment sum of squares with a single degree of freedom. (c) Test each hypothesis at the five per cent level. What is the significance level for the experiment?

10.20. In a completely randomized experiment there are n observations for each of four treatments. Q_1^2, Q_2^2 and Q_3^2 are components with an individual degree of freedom for three orthogonal linear contrasts. Prove that $Q_1^2 + Q_2^2 + Q_3^2$ is equal to the treatment sum of squares.

10.21. Prove Eq. (10.75).

10.22. Prove Eq. (10.77).

> *Hint.* Assume that the n sample values $x_{(1)}, x_{(2)}, \ldots, x_{(n)}$ are all different. Think of $x_{(1)}$ as falling in class (interval) C_1, $x_{(2)}, \ldots, x_{(n-1)}$ as falling in class C_2, and $x_{(n)}$ as falling in class C_3, where C_1, C_2, and C_3 are mutually exclusive and exhaustive classes. Then, show that the joint distribution of the range w and smallest value $x = x_{(1)}$ in a sample is
>
> $$f(w, x) = n(n-1)f(x)f(x+w)[F(x+w) - F(x)]^{n-2}$$

10.23. Let the rectangular density function be

$$f(x) = \begin{cases} \dfrac{1}{a}, & \text{if } 0 \le x \le a \\ 0, & \text{otherwise} \end{cases}$$

Show that

$$f_n(w) = \frac{n(n-1)w^{n-2}(a-w)}{a^n}$$

10.24. Find the density function of the range w if x has the density function $f(x) = e^{-x}$, $x \ge 0$.

10.25. Find the expressions for the density function $f_n(w)$ and the distribution function $F_n(w)$ of the standardized random variable $W = w/\sigma$ if x is distributed normally with mean μ and variance σ^2.

10.26. Prove that when $k = 2$ the statistic defined by Eq. (10.80) is a t statistic.

10.27. (a) Use the studentized range and the information in Exercise 10.19 to test at the five per cent level the hypothesis $\mu_1 = \mu_2 = \mu_3 = \mu_4$. (b) Use Theorem 10.1 to find simultaneously 95 per cent confidence intervals for the three comparisons defined in Exercise 10.19. Use this information to test the three hypotheses in Exercise 10.19. (c) Use Scheffé's method to find simultaneously 95 per cent confidence intervals for the three comparisons defined in Exercise 10.19. (d) Use Dunn's method to find the 95 per cent confidence intervals of the comparisons of (b) and (c).

10.28. (a) Use the studentized range and the data in Exercise 10.16 to test at the five per cent level the hypothesis $\mu_1 = \mu_2 = \mu_3 = \mu_4 = \mu_5 = \mu_6 = \mu_7$. (b) Use Theorem 10.1 to find simultaneous 95 per cent confidence intervals for five comparisons with multipliers $\frac{1}{3}, \frac{2}{3}, -1$; five comparisons with multipliers $\frac{1}{4}, \frac{1}{4}, \frac{1}{2}, -1$; and five comparisons with multipliers $\frac{1}{5}, \frac{2}{5}, \frac{2}{5}, -\frac{2}{5}, -\frac{3}{5}$. (c) Use Scheffé's method to find simultaneous 95 per cent confidence intervals for the 15 comparisons of (b). (d) Use Dunn's method to find simultaneous 95 per cent confidence intervals for the 15 comparisons of (b).

10.29. Derive Eq. (10.82) from Eq. (10.81).

10.30. (a) Use the data of Exercise 10.5 to find simultaneous 95 per cent confidence intervals for an infinite number of linear contrasts of μ_1, μ_2, μ_3, and μ_4, applying Scheffé's method. Make a table similar to Table 10.8 and include at least 15 contrasts. (b) Use Dunn's method to find simultaneous 95 per cent confidence intervals for the contrasts listed in (a).

10.31. The means of six treatments replicated five times in a completely randomized design are $\bar{x}_1 = 70$, $\bar{x}_2 = 105$, $\bar{x}_3 = 125$, $\bar{x}_4 = 160$, $\bar{x}_5 = 100$, $\bar{x}_6 = 190$. The error sum of squares is 60,000. (The reader may think of the means as resulting from coded data. For example, the observations may be such things as weights of six groups of experimental animals, heights of cakes using six preparations, lives of six kinds of highway surface, yields of dyestuff, thrusts of rocket motors, measurements made by operators, responses after training, and concentrations of solutions.) (a) Use Duncan's multiple range test to make a pairwise ranking of these six means. Let $\alpha_2 = 0.05$. (b) Use the Newman-Keuls five per cent level multiple range test to make a pairwise ranking of these six means. (c) Use Tukey's 1953 five per cent level multiple test procedure to make a pairwise ranking of these six means. (d) Use the five per cent multiple t test to rank the six means.

10.32. Use the data of Exercise 10.5 to test at the five per cent level the significance of the difference between all pairs of means, applying (a)

Duncan's multiple range test, and (b) Newman-Keuls multiple range test.

Hint. In the multiple test procedure described in Sect. 10.9 all samples are the same size, and the least significant range R_p for p means is $R_p = r_p s_{\bar{x}}$, where r_p is the least significant studentized range and $s_{\bar{x}} = \sqrt{s^2/n}$. Since $s_{\bar{x}}$ can be written as

$$\sqrt{\frac{s^2}{2}\left(\frac{1}{n} + \frac{1}{n}\right)}$$

one would expect that, for the case of unequal sample sizes, the standard deviation of the difference of 2 means should be

$$\sqrt{\frac{s^2}{2}\left(\frac{1}{n_1} + \frac{1}{n_2}\right)}$$

Use this last expression in (a) and (b).

REFERENCES

1. Anderson, R. L. and T. A. Bancroft, *Statistical Theory in Research*. New York: McGraw-Hill, Inc., 1952, Chap. 6.

2. Bennett, C. A. and N. L. Franklin, *Statistical Analysis in Chemistry and the Chemical Industry*. New York: John Wiley & Sons, Inc., 1954, Chaps. 4 and 7.

3. Davies, Owen L., *Statistical Methods in Research and Production*, 2nd ed. London: Oliver and Boyd, 1949, pp. 71–72.

4. Duncan, D. B., "Multiple Range and Multiple F Tests," *Biometrics*, Vol. **11** (1955), pp. 1–42.

5. ———, "Multiple Range Tests for Correlated and Heteroscedastic Means," *Biometrics*, Vol. **13** (1957), pp. 164–76.

6. ———, "Bayes Rules for a Common Multiple Comparisons Problem and Related Student-t Problems," *Annals of Mathematical Statistics*, Vol. **32** (1961), pp. 1013–33.

7. Dunn, O. J., "Multiple Comparisons Among Means," *Journal of the American Statistical Association*, Vol. **56** (1961), pp. 52–64.

8. Goulden, Cyril H., *Methods of Statistical Analysis*, 1st ed. New York: John Wiley & Sons, Inc., 1939.

9. Gumbel, E. J., *Statistics of Extremes*. New York: Columbia University Press, 1958, Chap. 3.

10. Hald, A., *Statistical Theory with Engineering Applications*. New York: John Wiley & Sons, Inc., 1952, Chaps. 10, 12, and 16.

11. Harter, H. Leon, "Error Rates and Sample Sizes for Range Tests in Multiple Comparisons," *Biometrics*, Vol. **13** (1957), pp. 511–36.

12. ———, "Critical Values for Duncan's New Multiple Range Test," *Biometrics*, Vol. **16** (1960), pp. 671–85.

13. Hartley, H. O., "The Range in Random Samples," *Biometrika*, Vol. **32** (1941), pp. 334–48.

14. ———, "Some Significant Test Procedures For Multiple Comparisons," *Annals of Mathematical Statistics*, Vol. **25** (1954), Abstract, 19.

15. Keuls, M., "The Use of the 'Studentized Range' in Connection with an Analysis of Variance," *Euphytica*, Vol. 1 (1952), pp. 112–22.

16. Kolodziejczyk, S., "On an Important Test of Statistical Hypotheses," *Biometrika*, Vol. 27 (1935), pp. 161–90.

17. Kramer, Clyde Young, "Extension of Multiple Range Tests to Group Means with Unequal Numbers of Replications," *Biometrics*, Vol. 12 (1956), pp. 307–10.

18. ———, "Extension of Multiple Range Tests to Group Correlated Adjusted Means," *Biometrics*, Vol. 13 (1957), pp. 13–18.

19. Lehman, E. L., "Some Principles of the Theory of Testing Hypotheses," *Annals of Mathematical Statistics*, Vol. 21 (1950), pp. 1–26.

20. ———, "A Theory of Some Multiple Decision Problems, I," *Annals of Mathematical Statistics*, Vol. 28 (1957), pp. 1–25.

21. ———, "A Theory of Some Multiple Decision Problems, II," *Annals of Mathematical Statistics*, Vol. 28 (1957), pp. 547–72.

22. Li, J. C. R., *Introduction to Statistical Inference*. Ann Arbor, Michigan: Edwards Brothers, 1957, Chap. 15.

23. May, J. M., "Extended and Corrected Tables of the Upper Percentage Points of the 'Studentized' Range," *Biometrika*, Vol. 39 (1952), pp. 192–93.

24. ———, "Extended and Corrected Tables of the Upper Percentage Points of the Studentized Range," *Biometrika*, Vol. 40 (1953), p. 236.

25. Newman, D., "The Distribution of the Range in Samples from a Normal Population, Expressed in Terms of an Independent Estimate of Standard Deviation," *Biometrika*, Vol. 31 (1939), pp. 20–30.

26. Patnaik, P. B., "The Non-Central χ^2 and F Distributions and Their Applications," *Biometrika*, Vol. 36 (1949), pp. 202–32.

27. Pearson, E. S. and H. O. Hartley, *Biometrika Tables for Statiaticians*, Vol. 1. London, England: Cambridge University Press, 1956.

28. Roy, S. N. and R. C. Bose, "Simultaneous Confidence Interval Estimation," *Annals of Mathematical Statistics*, Vol. 24 (1953), pp. 513–36.

29. Scheffé, H., "A Method for Judging All Contrasts in the Analysis of Variance," *Biometrika*, Vol. 40 (1953), pp. 87–104.

30. Steel, R. G. D. and Torrie, J. H., *Principles and Procedures of Statistics*. New York: McGraw-Hill, Inc., 1960, Chap. 7.

31. Tang, P. C., "The Power Function of the Analysis of Variance Test with Tables and Illustrations of Their Use," *Statistical Research Memoirs*, Vol. 2 (1938), pp. 126–49.

32. Tukey, J. W., "Comparing Individual Means in the Analysis of Variance," *Biometrics*, Vol. 5 (1949), pp. 99–114.

33. ———, "Quick and Dirty Methods in Statistics," part II, Simple Analyses for Standard Designs, *Proceedings Fifth Annual Convention, American Society for Quality Control* (1951), pp. 189–97.

34. ———, "The Problem of Multiple Comparisons," unpublished notes, Princeton University, 396 pp. 1953.

35. Wine, R. L., *A Power Study of Multiple Range and Multiple F Tests*, Ph. D. thesis, Virginia Polytechnic Institute (1955), 154 pp.

11

MORE ABOUT THE ONE-WAY CLASSIFICATION

Fixed effects and random effects models for a one-way classification are compared. Concepts of the one-way classification are extended to the nested classification. Applications for both equal and unequal sample sizes are given.

11.1. INTRODUCTION

In Chap. 10 we discussed problems relating to several means (or effects) in a one-way classification. We used variances in these problems, but the emphasis was on comparisons of means. In many investigations one is primarily concerned with the estimation or testing of variances (or components of variance), means (or effects) being of secondary interest. We illustrate the difference by considering Example 10.1.

Recall that we compared the mean tensile strengths, μ_A, μ_B, and μ_C, of copper wire (of a certain gauge) of three specific manufacturers A, B, and C. Ten pieces of wire were randomly selected from A, ten from B, and ten from C. The tensile strength of each piece was measured and the sample means computed and used in making statements about the three specific population means. Other manufacturers were not mentioned; we were not interested in them. This is an example of what is called a *fixed effects experiment* or a *model* I *experiment*. However, in other situations we might be interested in using a sample of three manufacturers to make a statement about *all manufacturers* of this gauge of copper wire. For this purpose the three manufacturers would be randomly selected from among all manufacturers of copper wire, and then ten pieces of wire would be randomly selected from each of the three manufacturers. The three sample means would be computed as before, but in this case the investigator is not particularly interested in who the manufacturers are; instead, he is interested in making a statement about the variability of all population means. This is an example of what we call a *random effects experiment* or a *model* II *experiment*. The difference is that for model I a repetition of the experiment would necessarily require that the same three manufacturers be selected, but for model II a repetition of the experiment could (and is likely to) give a different set of manufacturers. To summarize, for model I we make inferences about the particular treatments selected; for model II we make inferences about the population of treatments from which a random sample of treatments was drawn.

In another illustration, if we made an inference about the mean temperature in August at six specific large cities in the United States, this would be a fixed effects experiment; if the six cities were randomly selected, it would be a random effects experiment in which an inference would be drawn about all cities. As a further illustration, if mean response to five convenient (fixed) temperatures are compared, the experiment is model I; if the five temperatures are randomly selected from some interval of values, the experiment is model II.

11.2. THE RANDOM EFFECTS EXPERIMENT IN A ONE-WAY CLASSIFICATION

The model equation for the jth observation in the ith sample of a random effects experiment is exactly the same as it is for the fixed effects experiment [see Eq. (10.10)]; that is

$$x_{ij} = \mu + \alpha_i + \epsilon_{ij} \qquad (i = 1, \ldots, k; \; j = 1, \ldots, n_i) \qquad (11.1)$$

Just as before, μ is fixed, and ϵ_{ij} represents the random (error) component of observation x_{ij}. For both model I and II it is assumed that the variance of ϵ_i is σ^2; that is, the error variances for the k populations are equal. Further, $\mu_{\epsilon_i} = E(\epsilon_i) = 0$ for each i. The difference in model I and II is in the assumption about the α_i's. For model I we assume that the α_i's are fixed and, as a consequence of Eq. (10.11), that

$$\sum_{i=1}^{k} \alpha_i = 0$$

But for model II we assume that the α_i's represent a random sample of effects from the population of α's with mean zero, that is, $\mu_{\alpha_i} = E(\alpha_i) = 0$, and variance σ_{α}^2. Also, we usually assume that α and ϵ_i are independently distributed. (Only in the rare case would we have

$$\sum_{i=1}^{k} \alpha_i = 0$$

in model II). For model I we may think of

$$\frac{\sum_{i=1}^{k} \alpha_i^2}{k-1}$$

as a type of variance of the α_i's, since we really are not interested in the population of α's.

In a fixed effects experiment we were interested in specific treatment effects α_i ($i = 1, \ldots, k$). Thus, we estimated α_i by single values (a_i) and intervals and tested the null hypothesis

$$H_0: \alpha_1 = \cdots = \alpha_k = 0 \qquad (\text{or } \sum \alpha_i^2 = 0)$$

against the alternative hypothesis

$$H_a: \text{some of the } \alpha\text{'s are not zero} \qquad (\text{or } \sum \alpha_i^2 \neq 0)$$

In the random effects experiment we are interested in the variation (or variance σ_α^2) of the population of α's. Thus, we wish to estimate σ_α^2 and test the null hypothesis

$$H_0: \sigma_\alpha^2 = 0 \tag{11.2}$$

against the alternative hypothesis

$$H_a: \sigma_\alpha^2 \neq 0 \tag{11.3}$$

The test of the null hypothesis for the fixed effects experiment was established with the aid of the sum of squares identity and the analysis of variance Table 10.4. Following arguments similar to those in Sect. 10.3,

Table 11.1

Analysis of Variance for Random Effects Experiments in One-Way Classifications

Source of Variation	Sum of Squares	Degrees of Freedom	Mean Square	Expected Mean Square
Among means	$\dfrac{\sum_i T_{i.}^2}{n} - \dfrac{T_{..}^2}{nk}$	$k-1$	s_2^2	$\sigma^2 + n\sigma_\alpha^2$
Within	$\sum_i \sum_j x_{ij}^2 - \dfrac{\sum_i T_{i.}^2}{n}$	$k(n-1)$	s_p^2	σ^2
Total	$\sum_i^k \sum_j^n x_{ij}^2 - \dfrac{T_{..}^2}{nk}$	$nk-1$		

we can justify the analysis of variance Table 11.1, which is for samples of equal size. (In Sect. 11.3 we actually prove that the expected mean squares are as indicated.)

To test the hypothesis that $\sigma_\alpha^2 = 0$, we refer to Table 11.1. It is obvious that $E(s_p^2)$ is equal to σ^2, no matter whether the hypothesis is true or false. But $E(s_2^2)$ is equal to σ^2 only when the hypothesis in (11.2) is true; otherwise, $E(s_2^2)$ is greater than σ^2 by the amount $n\sigma_\alpha^2$. Thus, just as with the fixed effects experiment, s_2^2/s_p^2 is distributed as the random variable F with $k - 1$ and $k(n - 1)$ degrees of freedom, provided that hypothesis (11.2) is true and that x_{ij} is a random normal variate. If the null hypothesis is false, we expect s_2^2/s_p^2 to be larger than unity. Thus, the hypothesis is rejected if

$$\frac{s_2^2}{s_p^2} \geq F_\alpha[k - 1, k(n - 1)]$$

where $F_\alpha[k - 1, k(n - 1)]$ is the upper α level value of F with $k - 1$ and $k(n - 1)$ degrees of freedom. Numerically, the test of $\sigma_\alpha^2 = 0$ is exactly like the test of the fixed effects hypothesis $\alpha_1 = \cdots = \alpha_k = 0$, but the interpretations are quite different.

Table 11.1 can also be used to estimate how much the treatment means (or effects) differ; that is, σ_α^2 can be estimated. The table indicates that s_2^2 is an unbiased estimator of $\sigma^2 + n\sigma_\alpha^2$, and s_p^2 is an unbiased estimator of σ^2. Thus, an unbiased point estimator, s_α^2, of σ_α^2 is given by

$$s_\alpha^2 = \frac{s_2^2 - s_p^2}{n} \tag{11.4}$$

If $(s_2^2 - s_p^2)/n$ is negative, the estimate s_α^2 is taken to be zero.

A confidence interval estimate of σ_α^2 would be more useful than a point estimate if it could be obtained. Working toward this end we note that $s_2^2/(\sigma^2 + n\sigma_\alpha^2)$ and s_p^2/σ^2 are independently distributed as χ^2 per degree of freedom with $k - 1$ and $k(n - 1)$ degrees of freedom, respectively. Thus, the ratio

$$\frac{\dfrac{s_2^2}{\sigma^2 + n\sigma_\alpha^2}}{\dfrac{s_p^2}{\sigma^2}} = \frac{s_2^2}{s_p^2}\left(\frac{1}{1 + n\dfrac{\sigma_\alpha^2}{\sigma^2}}\right) \tag{11.5}$$

is distributed as F with $k - 1$ and $k(n - 1)$ degrees of freedom; that is

$$F = \frac{s_2^2}{s_p^2}\left(1 + n\frac{\sigma_\alpha^2}{\sigma^2}\right)^{-1}$$

or

$$\frac{s_2^2}{s_p^2} = F\left(1 + n\frac{\sigma_\alpha^2}{\sigma^2}\right) \tag{11.6}$$

If F_1 and F_2 denote values of the variate F for which

$$P[F < F_1] = \frac{\alpha}{2} = P[F > F_2]$$

then $100(1 - \alpha)$ per cent of the values of F lie in the interval

$$F_1 < F < F_2 \tag{11.7}$$

From (11.5) and (11.7) it follows that

$$F_1 < \frac{s_2^2}{s_p^2}\left(\frac{1}{1 + n\dfrac{\sigma_\alpha^2}{\sigma^2}}\right) < F_2$$

and, hence, the $100\,(1 - \alpha)$ per cent confidence interval for the ratio σ_α^2/σ^2 is

$$\left(\frac{s_2^2}{s_p^2 F_2} - 1\right)\frac{1}{n} < \frac{\sigma_\alpha^2}{\sigma^2} < \left(\frac{s_2^2}{s_p^2 F_1} - 1\right)\frac{1}{n} \tag{11.8}$$

The limits of (11.8) are exact. Thus, if σ^2 is known, the limits

$$\left(\frac{s_2^2}{s_p^2 F_2} - 1\right)\frac{\sigma^2}{n} \quad \text{and} \quad \left(\frac{s_2^2}{s_p^2 F_1} - 1\right)\frac{\sigma^2}{n} \tag{11.9}$$

of σ_α^2 are also exact. Unfortunately σ^2 is not usually known. If we replace σ^2 by s_p^2, the *approximate* $100(1 - \alpha)$ *per cent confidence limits of* σ_α^2 *are*

$$\left(\frac{s_2^2}{s_p^2 F_2} - 1\right)\frac{s_p^2}{n} \quad \text{and} \quad \left(\frac{s_2^2}{s_p^2 F_1} - 1\right)\frac{s_p^2}{n} \tag{11.10}$$

The limits in (11.10) give satisfactory approximations if s_p^2 has a reasonably large number of degrees of freedom, say 15 or more. If the first value in (11.10) is negative, the lower limit is set equal to zero.

Bross [4] and other authors [2, 3, 5, 14, 16, 22] have given other methods for constructing approximate limits for components of variance. The $1 - \alpha$ limits for σ_α^2 derived by Bross are

$$\left(\frac{\dfrac{s_2^2}{s_p^2 F_2} - 1}{\dfrac{F_2' s_2^2}{s_p^2 F_2} - 1}\right) s_\alpha^2 \quad \text{and} \quad \left(\frac{\dfrac{s_2^2}{s_p^2 F_1} - 1}{\dfrac{F_1' s_2^2}{s_p^2 F_1} - 1}\right) s_\alpha^2 \tag{11.11}$$

where F' has $k - 1$ and ∞ degrees of freedom, and F_1' and F_2' are lower and upper $\alpha/2$ points, respectively, and s_α^2 is given by Eq. (11.4). The limits given by (11.11) purport to be better approximations than the limits given by (11.10), particularly for small numbers of degrees of freedom.

Example 11.1. Thinking of the data in Example 10.1 as being from a random effects experiment, estimate the components of variance σ^2 and σ_α^2.

From the analysis of variance table following Example 10.1, we find $s_2^2 = 2410$ with two degrees of freedom and $s_p^2 = 1457$ with 27 degrees of freedom. An unbiased point estimate of σ^2 is given by $s_p^2 = 1457$, and an unbiased point estimate of σ_α^2, according to Eq. (11.4), is

$$s_\alpha^2 = \frac{2410 - 1457}{10} = 95.3$$

Since $27\, s_p^2/\sigma^2$ is distributed as χ^2 with 27 degrees of freedom, and $\chi_{.95}^2 = 16.2$ and $\chi_{.05}^2 = 40.1$, the 90 per cent confidence interval for σ^2 is given by

$$\frac{39{,}347}{40.1} < \sigma^2 < \frac{39{,}347}{16.2}$$

or

$$981 < \sigma^2 < 2429$$

Further, since $F_1 = F_{.95}(2, 27) = 1/F_{.05}(27, 2) = 1/19.5 = 0.0513$ and $F_2 = F_{.05}(2, 27) = 3.35$, the 90 per cent confidence interval for σ_α^2, when (11.10) is used, is given by

$$\left[\frac{2410}{(1457)(3.35)} - 1\right]\frac{1457}{10} < \sigma_\alpha^2 < \left[\frac{2410}{(1457)(0.0513)} - 1\right]\frac{1457}{10}$$

or

$$-73.9 < \sigma_\alpha^2 < 4550$$

Replacing the negative lower limit by zero, we obtain

$$0 < \sigma_\alpha^2 < 4550$$

For the limits given by Bross we find $F_2' = F_{.05}(2, \infty) = 3.00$ and $F_1' = 1/F_{.05}(\infty, 2) = 1/19.5$. Thus, by (11.11)

$$0 < \sigma_\alpha^2 < 4550$$

The limits given by (11.10) and (11.11) are the same to four significant figures, but differ in the fifth. In this case, it is clear that the simpler Formula (11.10) should be applied.

The power of the F test of the hypothesis $\sigma_\alpha^2 = 0$ may be determined numerically as it is in Chap. 9. That is, for $n_1 = \cdots = n_k = n$ we find power as a function of λ^2, where

$$\lambda^2 = \frac{\sigma^2 + n\sigma_\alpha^2}{\sigma^2} = 1 + n\frac{\sigma_\alpha^2}{\sigma^2} \qquad (11.12)$$

Since it is fairly straightforward, we leave it to the reader to determine the power function for a particular test and to find the smallest equal-sized samples one can use for specified values of α, β, and σ_α^2.

11.3. EXPECTED MEAN SQUARES

The proofs of this section are presented primarily for two reasons. We wish to show that the expected mean squares in Table 11.1 are correct and to introduce a method of proof which is easily extended to more complicated analysis of variance experiments. It is also informative to learn where and in what order the assumptions are introduced and to see what role each assumption plays in the derivation. We use a detailed proof to show that $E(s_2^2) = \sigma^2 + n\sigma_\alpha^2$. Then we indicate how the method of proof can be used to show that

$$E(s_2^2) = \sigma^2 + \frac{n \sum_i \alpha_i^2}{k-1}$$

for model I and $E(s_p^2) = \sigma^2$ for both model I and model II experiments in a one-way classification. The method is also applied for samples of unequal size.

First, we note that the sum of squares identity is an algebraic relationship which results from the use of algebra; it in no way depends upon the distribution assumptions associated with the effects. Thus, since both the fixed effects and random effects experiments have the same model equation

$$x_{ij} = \mu + \alpha_i + \epsilon_{ij} \qquad (i = 1, \ldots, k; \, j = 1, \ldots, n_i) \qquad (11.13)$$

it follows that their sum of squares identities are identical. For the case where $n_1 = \cdots = n_k = n$, the computing form of the sum of squares identity is

$$\sum_{i=1}^{k} \sum_{j=1}^{n} x_{ij}^2 - \frac{T_{..}^2}{nk} = \left(\sum_i^k \sum_j^n x_{ij}^2 - \frac{\sum_i T_{i.}^2}{n} \right) + \left(\frac{\sum_i T_{i.}^2}{n} - \frac{T_{..}^2}{nk} \right) \qquad (11.14)$$

and the mean squares for among means and within samples are given by

$$s_2^2 = \frac{\dfrac{\sum_i T_{i.}^2}{n} - \dfrac{T_{..}^2}{nk}}{k-1} \quad \text{and} \quad s_p^2 = \frac{\displaystyle\sum_i \sum_j x_{ij}^2 - \frac{\sum_i T_{i.}^2}{n}}{k(n-1)} \qquad (11.15)$$

respectively.

In the random effects experiment we assume that k populations are randomly selected and then within each of these populations n random observations are made, the selection prodedure depending in no way upon the particular population. Thus, the effects

$$\alpha_1, \ldots, \alpha_k; \quad \epsilon_{11}, \ldots, \epsilon_{1n}; \quad \epsilon_{k1}, \ldots, \epsilon_{kn}$$

are random variables which are independently distributed. Therefore, it follows that, since we assume that α is distributed with mean $\mu_\alpha = 0$ and variance σ_α^2 and that ϵ_i is independently distributed with $\mu_{\epsilon_i} = \mu_\epsilon = 0$ and variance $\sigma_{\epsilon_i} = \sigma^2$, we can write

$$E(\alpha_i) = \mu_\alpha = 0 \quad \text{and} \quad E(\epsilon_{ij}) = \mu_{\epsilon_{ij}} = \mu_\epsilon = 0 \tag{11.16}$$

$$E[\alpha_i - E(\alpha_i)]^2 = E(\alpha_i^2) = \sigma_\alpha^2 \quad \text{(common for all } \alpha_i) \tag{11.17}$$

$$E[\epsilon_{ij} - E(\epsilon_{ij})]^2 = E(\epsilon_{ij}^2) = \sigma^2 \quad \text{(common for all } \epsilon_{ij}) \tag{11.18}$$

and

$$\begin{cases} \text{cov}\,(\alpha_i, \alpha_{i'}) = E\{[\alpha_i - E(\alpha_i)][\alpha_{i'} - E(\alpha_{i'})]\} = E(\alpha_i \alpha_{i'}) = 0 & (i = i') \\ \text{cov}\,(\epsilon_{ij}, \epsilon_{i'j'}) = E(\epsilon_{ij} \epsilon_{i'j'}) = 0 & (i \neq i' \text{ if } j = j' \text{ or } j \neq j' \text{ if } i = i') \\ \text{cov}\,(\alpha_{i'}, \epsilon_{ij}) = E(\alpha_{i'} \epsilon_{ij}) = 0 & (i = i' \text{ or } i \neq i') \end{cases} \tag{11.19}$$

The expressions in (11.19) follow from the fact that any two different effects are independently distributed. The normality assumption is not required for any of the derivations of this section. However, it is required for tests of hypotheses and the construction of confidence intervals.

From

$$E(s_2^2) = E\left(\frac{\dfrac{\sum_i T_{i.}^2}{n} - \dfrac{T_{..}^2}{nk}}{k - 1}\right)$$

we obtain, using properties of expectation

$$E(s_2^2) = \frac{1}{k-1}\left\{\frac{1}{n}[E(T_{1.}^2) + \cdots + E(T_{k.}^2)] - \frac{1}{nk}E(T_{..}^2)\right\} \tag{11.20}$$

Now, for the ith population, we may write

$$E(T_{i.}^2) = E(x_{i1} + \cdots + x_{in})^2$$
$$= E[n\mu + n\alpha_i + (\epsilon_{i1} + \cdots + \epsilon_{in})]^2$$

by (11.13).

Using properties of algebra and expectation on this last expression, we have

$$E(T_{i.}^2) = n^2\mu^2 + n^2 E(\alpha_i^2) + E(\epsilon_{i1} + \cdots + \epsilon_{in})^2 + 2n^2\mu E(\alpha_i)$$
$$+ 2n\mu E(\epsilon_{i1} + \cdots + \epsilon_{in}) + 2nE[\alpha_i(\epsilon_{i1} + \cdots + \epsilon_{in})] \tag{11.21}$$

We may write Eq. (11.21) as

$$E(T_{i.}^2) = n^2\mu^2 + n^2\sigma_\alpha^2 + n\sigma^2 \qquad (i = 1, \ldots, k) \qquad (11.22)$$

since, by Eqs. (11.16), (11.17), (11.18), and (11.19) we have

$$E(\epsilon_{i1} + \cdots + \epsilon_{in})^2 = E(\epsilon_{i1}^2) + \cdots + E(\epsilon_{in}^2)$$
$$+ 2E(\epsilon_{i1}\epsilon_{i2}) + \cdots + 2E(\epsilon_{i,n-1}\epsilon_{in})$$
$$= \sigma^2 + \cdots + \sigma^2 + 2\cdot 0 + \cdots + 2\cdot 0 = n\sigma^2$$

$$E(\alpha_i) = 0, \qquad E(\alpha_i^2) = \sigma_\alpha^2$$

$$E(\epsilon_{i1} + \cdots + \epsilon_{in}) = E(\epsilon_{i1}) + \cdots + E(\epsilon_{in}) = 0 + \cdots + 0 = 0$$

and

$$E[\alpha_i(\epsilon_{i1} + \cdots + \epsilon_{in})] = E(\alpha_i\epsilon_{i1}) + \cdots + E(\alpha_i\epsilon_{in}) = 0$$

Further, since the right-hand side of Eq. (11.22) does not depend on i, we have

$$E(T_{1.}^2) + \cdots + E(T_{k.}^2) = kn^2\mu^2 + kn^2\sigma_\alpha^2 + kn\sigma^2 \qquad (11.23)$$

In order to evaluate Eq. (11.20), it only remains to obtain $E(T_{..}^2)$. Thus

$$E(T_{..}^2) = E(T_{1.} + \cdots + T_{k.})^2$$
$$= E\{[n\mu + n\alpha_1 + (\epsilon_{11} + \cdots + \epsilon_{1n})] + \cdots + [n\mu$$
$$+ n\alpha_k + (\epsilon_{k1} + \cdots + \epsilon_{kn})]\}^2$$
$$= E[kn\mu + n(\alpha_1 + \cdots + \alpha_k) + (\epsilon_{11} + \cdots + \epsilon_{kn})]^2$$

or

$$E(T_{..}^2) = k^2 n^2 \mu^2 + n^2 E(\alpha_1 + \cdots + \alpha_k)^2 + E(\epsilon_{11} + \cdots + \epsilon_{kn})^2$$
$$+ 2kn^2\mu E(\alpha_1 + \cdots + \alpha_k) + 2kn\mu E(\epsilon_{11} + \cdots + \epsilon_{kn})$$
$$+ 2nE(\alpha_1 + \cdots + \alpha_k)(\epsilon_{11} + \cdots + \epsilon_{kn})$$

$$(11.24)$$

So

$$E(T_{..}^2) = k^2 n^2 \mu^2 + kn^2\sigma_\alpha^2 + kn\sigma^2 \qquad (11.25)$$

since

$$E(\alpha_1 + \cdots + \alpha_k)^2 = E(\alpha_1^2) + \cdots + E(\alpha_k^2) + 2E(\alpha_1\alpha_2)$$
$$+ \cdots + 2E(\alpha_{k-1}\alpha_k)$$
$$= \sigma_\alpha^2 + \cdots + \sigma_\alpha^2 + 2\cdot 0 + \cdots + 2\cdot 0 = k\sigma_\alpha^2$$

$$E(\epsilon_{11} + \cdots + \epsilon_{kn})^2 = kn\sigma^2$$

$$E(\alpha_1 + \cdots + \alpha_k) = 0$$

$$E(\epsilon_{11} + \cdots + \epsilon_{kn}) = 0$$

and

$$E(\alpha_1 + \cdots + \alpha_k)(\epsilon_{11} + \cdots + \epsilon_{kn}) = E(\alpha_1 \epsilon_{11}) + \cdots + E(\alpha_1 \epsilon_{kn})$$
$$+ \cdots + E(\alpha_k \epsilon_{11}) + \cdots + E(\alpha_k \epsilon_{kn}) = 0$$

Substituting Eqs. (11.23) and (11.25) in Eq. (11.20) gives

$$E(s_2^2) = \frac{1}{k-1}[(kn\mu^2 + kn\sigma_\alpha^2 + k\sigma^2) - (kn\mu^2 + n\sigma_\alpha^2 + \sigma^2)]$$

$$= \frac{1}{k-1}[(kn - n)\sigma_\alpha^2 + (k-1)\sigma^2]$$

or

$$E(s_2^2) = n\sigma_\alpha^2 + \sigma^2 \tag{11.26}$$

Next, we give an outline of the derivation of $E(s_2^2)$ for the fixed effects experiment. The only properties different from those used in proving Eq. (11.26) are as follows

Change from	*To*
α_i random	α_i fixed
$E(\alpha_i) = 0$	$E(\alpha_i) = \alpha_i$
$E(\alpha_i^2) = \sigma_\alpha^2$	$E(\alpha_i^2) = \alpha_i^2$
$E(c\alpha_i) = 0$	$E(c\alpha_i) = c\alpha_i$

where c is a constant. Then Eq. (11.21) becomes

$$E(T_i^2.) = n^2\mu^2 + n^2\alpha_i^2 + n\sigma^2 + 2n^2\mu\alpha_i \qquad (i = 1, \ldots, k) \tag{11.27}$$

and since $\sum \alpha_i = 0$, Eq. (11.24) becomes

$$E(T_.^2.) = k^2 n^2 \mu^2 + kn\sigma^2 \tag{11.28}$$

Substituting Eqs. (11.27) and (11.28) in Eq. (11.20) gives

$$E(s_2^2) = \frac{1}{k-1}[(kn\mu^2 + n\sum \alpha_i^2 + k\sigma^2 + 2n\mu \sum \alpha_i) - (kn\mu^2 + \sigma^2)]$$

or

$$E(s_2^2) = \sigma^2 + \frac{n\sum \alpha_i^2}{k-1} \quad \text{since} \quad \sum \alpha_i = 0 \tag{11.29}$$

To find the expected mean square for s_p^2 we first prove that

$$E\left(\sum_i^k \sum_j^n x_{ij}^2\right) = \begin{cases} nk\mu^2 + n\sum \alpha_i^2 + nk\sigma^2, & \text{for model I} \\ nk\mu^2 + nk\sigma_\alpha^2 + nk\sigma^2, & \text{for model II} \end{cases} \tag{11.30}$$

Then we substitute Eq. (11.30) and Eq. (11.23) or Eq. (11.27) in

$$E(s_p^2) = \frac{1}{k-1} E\left(\sum_i^k \sum_j^n x_{ij}^2 - \frac{\sum_i T_{i.}^2}{n}\right)$$

to obtain

$$E(s_p^2) = \sigma^2$$

When the samples are of unequal size, it is easy to show for model II that

$$E\left(\frac{T_{i.}^2}{n_i}\right) = n_i \mu^2 + n_i \sigma_\alpha^2 + \sigma^2$$

and

$$E\left(\frac{T_{..}^2}{n.}\right) = n. \, \mu^2 + \frac{\sum_{i=1}^k n_i^2}{n.} \sigma_\alpha^2 + \sigma^2$$

and, therefore, on substitution that

$$E(s_2^2) = \sigma^2 + \frac{1}{k-1}\left(n. - \frac{\sum n_i^2}{n.}\right)\sigma_\alpha^2 \tag{11.31}$$

where

$$n. = \sum_{i=1}^k n_i$$

For model I the expected mean square for treatments depends on the restriction among the α_i's. It can be shown that

$$E(s_2^2) = \begin{cases} \sigma^2 + \dfrac{\sum n_i \alpha_i^2}{k-1}, & \text{when } \sum n_i \alpha_i = 0 \\[3mm] \sigma^2 + \dfrac{\sum n_i \alpha_i^2 - \dfrac{(\sum n_i \alpha_i)^2}{n.}}{k-1}, & \text{when } \sum \alpha_i = 0 \end{cases} \tag{11.32}$$

Note that the latter expression for $E(s_2^2)$ can be written as

$$E(s_2^2) = \sigma^2 + \frac{\sum n_i(\alpha_i - \bar\alpha)^2}{k-1}$$

where

$$\bar\alpha = \frac{\sum n_i \alpha_i.}{n.} \tag{11.33}$$

With a little practice the reader should discover certain shortcuts to the above method. For example, notice that only the square terms μ^2, α_i^2, ϵ_{ij}^2 contribute nonzero values to the expected mean square. Thus, from an expression in μ, α_i and ϵ_{ij} we can read off the expected mean square almost immediately—bearing in mind, of course, all the assumptions of the particular model being considered. In the future, we give the expected mean squares in the analysis of variance table without presenting the derivation.

11.4. SUBGROUPS WITHIN A ONE-WAY CLASSIFICATION

The simplest design for an experiment is the one-way classification described in some detail in this chapter and Chap. 10. This design is also referred to as a *completely randomized design*. Sometimes it is desirable to subsample in a completely randomized design, and a more refined analysis becomes necessary. For an illustration (Example 11.2) of this new design, consider an experiment described by Davies [11, 12].

Example 11.2. Large batches of a chemical paste, regularly produced, are placed in casks for deliveries. Three casks in each batch are selected at random, and a sample is taken for analysis. Two independent analytical tests are carried out on a part of each sample to determine the paste strength. The data resulting from ten random batches are given in Table 11.2. The problem is to estimate the batch-to-batch variance σ_α^2, the cask-to-cask variance σ_β^2, and the variance σ^2 resulting from the analytical tests, or to test hypotheses about σ_α^2 or σ_β^2. Before describing the general approach, we give calculations for this equal-size sample variance components experiment.

Table 11.2
Percentage of Paste Strength of Samples

Batch	Cask 1			Cask 2			Cask 3			Total per Batch
	Observations		*Total*	*Observations*		*Total*	*Observations*		*Total*	
1	62.8	62.6	125.4	60.1	62.3	122.4	62.7	63.1	125.8	373.6
2	60.0	61.4	121.4	57.5	56.9	114.4	61.1	58.9	120.0	355.8
3	58.7	57.5	116.2	63.9	63.1	127.0	65.4	63.7	129.1	372.3
4	57.1	56.4	113.5	56.9	58.6	115.5	64.7	64.5	129.2	358.2
5	55.1	55.1	110.2	54.7	54.2	108.9	58.8	57.5	116.3	335.4
6	63.4	64.9	128.3	59.3	58.1	117.4	60.5	60.0	120.5	366.2
7	62.5	62.6	125.1	61.0	58.7	119.7	56.9	57.7	114.6	359.4
8	59.2	59.4	118.6	65.2	66.0	131.2	64.8	64.1	128.9	378.7
9	54.8	54.8	109.6	64.0	64.0	128.0	57.7	56.8	114.5	352.1
10	58.3	59.3	117.6	59.2	59.2	118.4	58.9	56.6	115.5	351.5

Grand Total   3603.2

The sums of squares (corrected) and degrees of freedom for the analysis of variance are shown in Table 11.3. They are computed from the raw data and totals by methods similar to those described in Sect. 10.5. Thus, the uncorrected sums of squares are given by

Observation $SS = (62.8)^2 + (62.6)^2 + \cdots + (56.6)^2 = 217,002.82$

Casks $SS = [(125.4)^2 + (122.4)^2 + \cdots + (115.5)^2]/2 = 216,982.48$

Batch $SS = [(373.6)^2 + (355.8)^2 + \cdots + (351.5)^2]/6 = 216{,}631.57$

Grand total $SS = (3603.2)^2/60 = 216{,}384.17$

and the corrected sums of squares by

Between batches $SS = 216{,}631.57 - 216{,}384.17 = 247.40$

Between casks within batches $SS = 216{,}982.48 - 216{,}631.57 = 350.91$

Between observations within casks $SS = 217{,}002.82 - 216{,}982.48 = 20.34$

Total $SS = 217{,}002.82 - 216{,}384.17 = 618.65$

The expected mean squares may be obtained by the methods of Sect. 11.3. The general expressions are given later in this section, and it is left as an exercise for the reader to justify these results.

Table 11.3
Analysis of Variance for Paste Strength

Source of Variation	Sum of Squares	Degrees of Freedom	Mean Square	Expected Mean Square
Batches	247.40	9	27.49	$\sigma^2 + 2\sigma_\delta^2 + 6\sigma_\alpha^2$
Casks within batches	350.91	20	17.55	$\sigma^2 + 2\sigma_\delta^2$
Observations within casks	20.34	30	0.68	σ^2
Total	618.65	59		

From Table 11.3 it is clear that point estimates s_e^2, s_d^2, and s_a^2, of the components of variance σ^2, σ_δ^2 and σ_α^2 are, respectively

$$\begin{cases} s_e^2 = 0.68 & \text{(error variance)} \\[2mm] s_d^2 = \dfrac{17.55 - 0.68}{2} = 8.44 & \text{(variance among casks)} \\[2mm] s_a^2 = \dfrac{27.49 - 17.55}{6} = 1.66 & \text{(variance among batches)} \end{cases} \qquad (11.34)$$

The standard deviations are $s_e = 0.82$, $s_d = 2.91$, and $s_a = 1.29$. Note that the cask-to-cask variation is by far the largest.

If confidence intervals are required, we use the chi-square distribution or the methods introduced in Sect. 11.2. The 0.975 and 0.025 percentage points of the χ^2 distribution for 30 degrees of freedom are given by $\chi_1^2 = 16.8$ and $\chi_2^2 = 74.0$, respectively. Thus, the 95 per cent confidence interval of σ^2 is given by

$$\frac{20.34}{74.0} < \sigma^2 < \frac{20.34}{16.8}$$

or

$$0.433 < \sigma^2 < 1.21$$

and the 95 per cent confidence interval of σ by

$$0.66 < \sigma < 1.10 \tag{11.35}$$

To calculate approximate confidence limits of σ_δ^2 we use Eq. (11.10). Since the lower and upper percentage points for F with 20 and 30 degrees of freedom are $F_1 = 1/2.35$ and $F_2 = 2.20$, the approximate 95 per cent confidence limits of σ_δ^2 are given by

$$\frac{\dfrac{17.55}{2.20} - 0.68}{2} = 3.65$$

and

$$\frac{(17.55)(2.35) - 0.68}{2} = 20.3$$

Thus the approximate 95 per cent confidence interval of σ_δ is

$$1.91 < \sigma_\delta < 4.51 \tag{11.36}$$

A similar method is used to derive approximate confidence limits of σ_α^2 (or σ_α) from the confidence limits for $\sigma_\alpha^2/\sigma_\delta^2$. Since the lower and upper percentage points for F with nine and 20 degrees of freedom are $F_3 = 1/3.67$ and $F_4 = 2.84$, the approximate 95 per cent confidence limits of σ_α^2 are given by

$$\frac{\dfrac{27.49}{2.84} - 17.55}{6} = -1.31$$

and

$$\frac{(27.49)(3.67) - 17.55}{6} = 13.9$$

Thus the approximate 95 per cent confidence interval of σ_α is

$$0 \le \sigma_\alpha < 3.73 \tag{11.37}$$

since a negative variance or standard deviation is meaningless in this context. Note that the interval which estimates σ_α is much longer than the intervals which estimate σ_δ and σ.

To test the hypothesis that there is no variation from cask to cask except that due to analytical tests, we compare the ratio $17.55/0.68 = 25.81$ with 1.93, the upper five per cent F value based on 20 and 30 degrees of freedom. This leads us to reject the hypothesis and conclude that there is real cask-to-cask variation. That is, we conclude that the variation from cask-to-cask is actually due to cask variance σ_δ^2 as well as to error variance σ^2. To test the null hypothesis that $\sigma_\alpha^2 = 0$, we compare the ratio $27.49/17.55 = 1.57$ with 2.39, the upper five per cent F value based on nine and 20 degrees of freedom. Since $1.57 < 2.39$, we fail to reject $\sigma_\alpha^2 = 0$. That is, we do not have enough evidence to say that the apparent variation from batch to batch is due to components other than cask variance σ_δ^2 and error variance σ^2.

The data of Table 11.2 were used to illustrate the various techniques. The reader should realize that normally not all the above procedures would be used with a single experiment. Further, it should be pointed out that the variance component estimates may be used in the future to determine what sampling plan to apply when the cost of sampling and testing are known. The student is referred to Cochran [6], Hansen [15], and others [1, 17, 21] for more details on selecting the optimum sampling plan.

The magnitude of the observations shown in Table 11.2 results from the following three sources of variation: (1) batch, (2) cask within batch, and (3) observation in percentage paste strength within casks. In other illustrations the three terms "batch," "cask," and "observation in percentage paste strength" may be replaced, respectively, by such terms as:

1. Chalks, laboratories, determinations of bulk density.
2. Storage times, treatments, determinations of ascorbic acid in frozen strawberries.
3. Growers, bales, determinations of percentage of foreign matter in cotton.
4. Days, sheets of building material, determinations of permeabilities.
5. Cities, blocks, determinations of response to a question.

Each of the above may serve as an illustration of an experimental design known as a *nested classification* or a *hierarchical classification* or *subsampling within a one-way classification* which has the *model equation*

$$x_{iju} = \mu + \alpha_i + \delta_{ij} + \epsilon_{iju}$$
$$i = 1, \ldots, k \quad j = 1, \ldots, n_i \quad u = 1, \ldots, n_{ij} \tag{11.38}$$

where x_{iju} denotes the uth determination within the jth sample within the ith population, μ denotes a constant over-all mean, α_i denotes the effect (constant or random) of the ith population, δ_{ij} denotes the effect (constant or random) of the jth sample within the ith population, and ϵ_{iju} denotes the uth random effect within the jth sample within the ith population. We often term α_i's the *treatment* effects, δ_{ij}'s the *experimental* effects, and ϵ_{iju}'s the *subsample* effects in order to distinguish among the effects.

If we are interested only in finding unbiased point estimates of fixed effects, then all we need assume is that the observations are randomly and independently distributed. In order to establish confidence intervals for the effects and variance components or to make tests of hypotheses in the usual way, we need the following additional assumptions. If the α_i are fixed, then we measure them as deviations from μ such that

$$\sum_{i=1}^{k} \sum_{j=1}^{n_i} n_{ij}\alpha_i = 0$$

If the α_i are random, they are assumed to be from a normal population with mean zero and variance σ_α^2. If the δ_{ij} are fixed, then we measure them as deviations from $\mu + \alpha_i$ such that

$$\sum_{j=1}^{n_i} n_{ij}\delta_{ij} = 0$$

if the δ_{ij} are random, they are assumed to be normally and independently distributed with means zero and common variance σ_δ^2. The ϵ_{iju} are assumed to be normally and independently distributed with means zero and common variance σ^2. Further, all random effects are independently distributed. For example, if the α_i are fixed, whereas the δ_{ij} and ϵ_{iju} are random, then drawing a particular value of δ does not affect the probability of drawing a particular value of ϵ. This design for subsampling can be extended indefinitely, and the model equation (11.38) and associated assumptions generalize in a straightforward fashion. The reader should note that there are

$$1 + k + \sum_{i=1}^{k} n_i$$

populations associated with the model equation (11.38).

For a collection of data with model equation (11.38), it can be shown that the sum of squares identity is

$$\sum_{i}^{k} \sum_{j}^{n_i} \sum_{u}^{n_{ij}} (x_{iju} - \bar{x})^2 = \sum_{i} \sum_{j} \sum_{u} (\bar{x}_{i..} - \bar{x})^2 + \sum_{i} \sum_{j} \sum_{u} (\bar{x}_{ij.} - \bar{x}_{i..})^2$$
$$+ \sum_{i} \sum_{j} \sum_{u} (x_{iju} - \bar{x}_{ij.})^2$$

$$(11.39)$$

where

$$\begin{cases} n_{i.} = \sum_{j} n_{ij} \qquad n_{..} = \sum_{i} \sum_{j} n_{ij} \qquad T_{ij.} = \sum_{u} x_{iju} \\[2mm] T_{i..} = \sum_{j} \sum_{u} x_{iju} \qquad T = \sum_{i} \sum_{j} \sum_{u} x_{iju} \qquad \bar{x}_{ij.} = \dfrac{T_{ij.}}{n_{ij}} \qquad (11.40) \\[2mm] \bar{x}_{i..} = \dfrac{T_{i..}}{n_{i.}} \quad \text{and} \quad \bar{x} = \dfrac{T}{n_{..}} \end{cases}$$

The computing form of the identity (11.39) is

$$\sum_i \sum_j \sum_u x_{iju}^2 - \frac{T^2}{n_{..}} = \left(\sum_i \frac{T_{i..}^2}{n_{i.}} - \frac{T^2}{n_{..}}\right) + \left(\sum_i \sum_j \frac{T_{ij.}^2}{n_{ij}} - \sum_i \frac{T_{i..}^2}{n_{i.}}\right)$$

$$+ \left(\sum_i \sum_j \sum_u x_{iju}^2 - \sum_i \sum_j \frac{T_{ij.}^2}{n_{ij}}\right) \tag{11.41}$$

The degrees of freedom identity associated with Eq. (11.39) or Eq. (11.41) is

$$n_{..} - 1 = (k - 1) + \left(\sum_{i=1}^{k} n_i - k\right) + \left(n_{..} - \sum_{i=1}^{k} n_i\right) \tag{11.42}$$

Note that $n_i \neq n_{i.}$. The symbol n_i denotes the number of experimental units in the ith population, but $n_{i.}$ denotes the total number of subsample units in the ith population. Further

$$\sum_{i=1}^{k} n_i$$

denotes the total number of experimental units in the experiment.

As with the one-way classification, the model equation, along with the assumptions which follow, may be used to determine the expectations of the three component sum of squares on the right-hand side of Eq. (11.41). These expectations, together with the degrees of freedom identity and theorems in Chaps. 7 and 9, may be applied to determine the nature of the distributions of the component sum of squares as well as the distributions of their ratios. The expected mean squares for four experimental models in a nested classification in an analysis of variance are given in Table 11.4 (for unequal samples) and Table 11.5 (for equal samples). In all models μ is fixed. In addition to this, the α_i and δ_{ij} are fixed in the *fixed model* (α, δ); only the α_i are fixed in the *mixed model* (α); only the δ_{ij} are fixed in the *mixed model* (δ), and no effects are fixed in the *random model*.

In the two models where the α_i are fixed, the unbiased estimator of the population mean

$$\mu_i = \mu + \alpha_i \qquad (i = 1, \ldots, k)$$

is given by $\bar{x}_{i..}$, and the unbiased estimator of the effect α_i is given by

$$a_i = \bar{x}_{i..} - \bar{x}$$

The procedure for testing the hypothesis

$$H_0: \mu_1 = \cdots = \mu_k = \mu \qquad (\text{or } \alpha_1 = \cdots = \alpha_k = 0) \tag{11.43}$$

when α_i are fixed, the hypothesis

$$H_0: \sigma_\alpha^2 = 0 \tag{11.44}$$

when α_i are random, and the method of estimating the components of variance may be explained in terms of the expected mean squares shown in Tables 11.4 and 11.5. (It is left as an exercise for the student to verify these expected mean squares.)

Table 11.4

Analysis of Variance and Expected Mean Squares
for Nested Classifications with *Unequal* Samples

Source of Variation	Sum of Squares	Degrees of Freedom	Mean Square	Expected Mean Square for Fixed Model (α, δ)
Among population means	$\sum_i \dfrac{T_{i..}^2}{n_{i.}} - \dfrac{T^2}{n_{..}}$	$k-1$	s_1^2	$\sigma^2 + \dfrac{\sum_i n_{i.}\alpha_i^2}{k-1}$
Subsamples within population (experimental error)	$\sum_i \sum_j \dfrac{T_{ij.}^2}{n_{ij}} - \sum_i \dfrac{T_{i..}^2}{n_{i.}}$	$\sum_i n_i - k$	s_2^2	$\sigma^2 + \dfrac{\sum_i \sum_j n_{ij}\delta_{ij}^2}{\sum n_i - k}$
Observations within subsamples (sampling error)	$\sum_i \sum_j \sum_u x_{iju}^2 - \sum_i \sum_j \dfrac{T_{ij.}^2}{n_{ij}}$	$n.. - \sum n_i$	s_p^2	σ^2
Total	$\sum_i \sum_j \sum_u x_{iju}^2 - \dfrac{T^2}{n_{..}}$	$n.. - 1$		

Source of Variation	Degrees of Freedom	Mean Square	Expected Mean Square for		
			Mixed Model (α)	Mixed Model (δ)	Random Model
Population means	$k-1$	s_1^2	$\sigma^2 + b\sigma_\delta^2 + \dfrac{\sum_i n_{i.}\alpha_i^2}{k-1}$	$\sigma^2 + c\sigma_\alpha^2$	$\sigma^2 + b\sigma_\delta^2 + c\sigma_\alpha^2$
Experimental error	$\sum_i n_i - k$	s_2^2	$\sigma^2 + a\sigma_\delta^2$	$\sigma^2 + \dfrac{\sum_i \sum_j n_{ij}\delta_{ij}^2}{\sum n_i - k}$	$\sigma^2 + a\sigma_\delta^2$
Sampling error	$n.. - \sum n_i$	s_p^2	σ^2	σ^2	σ^2
Total	$n.. - 1$				

where

$$\begin{cases} a = \dfrac{n.. - \sum_i \dfrac{\sum_j n_{ij}^2}{n_{i.}}}{\sum_i n_i - k} \\[4ex] b = \dfrac{\sum_{i'} \dfrac{\sum_j n_{ij}^2}{n_{i.}} - \dfrac{\sum_i \sum_j n_{ij}^2}{n..}}{k-1} \\[4ex] c = \dfrac{n.. - \dfrac{\sum_i n_{i.}^2}{n..}}{k-1} \end{cases} \qquad (11.45)$$

In the very important special case where $n_{ij} = r$ for every i and j and $n_i = n$ for every i, the constants in Eq. (11.45) reduce to

$$a = b = r \quad \text{and} \quad c = nr \tag{11.46}$$

Thus, the analysis of variance and expected mean squares may be written as shown in Table 11.5. From this table it is clear that, when the δ_{ij}'s are random, an exact test of either Eq. (11.43) or Eq. (11.44) may be applied by comparing the ratio s_1^2/s_2^2 to an upper α level F value with $k - 1$ and $k(n - 1)$ degrees of freedom. When the δ_{ij}'s are fixed, an exact test can be made by comparing the ratio s_1^2/s_p^2 to an upper α level F value with $k - 1$ and $kn(r - 1)$ degrees of freedom. Further, when the δ_{ij}'s are random,

Table 11.5

Analysis of Variance and Expected Mean Square
for Nested Classifications with *Equal* Samples

Source of Variation	Sum of Squares	Degrees of Freedom	Mean Square	Expected Mean Square for Fixed Model (α, δ)
Population means	$\dfrac{\sum_i T_{i..}^2}{nr} - \dfrac{T^2}{knr}$	$k - 1$	s_1^2	$\sigma^2 + nr\left(\dfrac{\sum \alpha_i^2}{k - 1}\right)$
Experimental error	$\dfrac{\sum_i \sum_j T_{ij.}^2}{r} - \dfrac{\sum_i T_{i.}^2}{nr}$	$k(n - 1)$	s_2^2	$\sigma^2 + r\dfrac{\sum_i \sum_j \delta_{ij}^2}{k(n - 1)}$
Sampling error	$\sum_i \sum_j \sum_u x_{iju}^2 - \dfrac{\sum_i \sum_j T_{ij.}^2}{r}$	$kn(r - 1)$	s_p^2	σ^2
Total	$\sum \sum \sum x_{iju}^2 - \dfrac{T^2}{knr}$	$knr - 1$		

Source of Variation	Degrees of Freedom	Mean Square	Expected Mean Square for Mixed Model (α)	Mixed Model (δ)	Random Model
Population means	$k - 1$	s_1^2	$\sigma^2 + r\sigma_\delta^2 + nr\left(\dfrac{\sum \alpha_i^2}{k - 1}\right)$	$\sigma^2 + nr\sigma_\alpha^2$	$\sigma^2 + r\sigma_\delta^2 + nr\sigma_\alpha^2$
Experimental error	$k(n - 1)$	s_2^2	$\sigma^2 + r\sigma_\delta^2$	$\sigma^2 + \dfrac{r\sum_i \sum_j \delta_{ij}}{k(n-1)}$	$\sigma^2 + r\sigma_\delta^2$
Sampling error	$kn(r - 1)$	s_p^2	σ^2	σ^2	σ^2
Total	$knr - 1$				

unbiased point estimators s^2, s_d^2, and s_a^2 of the components σ^2, σ_δ^2, and σ_α^2, respectively, are given by

$$\begin{cases} s^2 = s_p^2 \\[2mm] s_d^2 = \dfrac{s_2^2 - s_p^2}{r} \\[2mm] s_\alpha^2 = \dfrac{s_1^2 - s_2^2}{nr} \end{cases} \qquad (11.47)$$

In the rare case where the δ_{ij}'s are fixed and the α_i's are random, the estimators of σ^2 and σ_α^2 are given by

$$\begin{cases} s^2 = s_p^2 \\[2mm] s_\alpha^2 = \dfrac{s_1^2 - s_p^2}{nr} \end{cases} \qquad (11.48)$$

Confidence limits for these components may be found by the methods described in Sect. 11.2 and illustrated earlier in this section. Point and interval estimates of the population means may be found in a straightforward way.

Now we consider the general case where $a \neq b$. If the δ_{ij}'s are fixed, exact tests of Eqs. (11.43) and (11.44) apply, but when the δ_{ij}'s are random, approximate tests must be found. The reader who is interested in details of such tests is referred to [8, 9, 10, 13, 17, 19, 20]. Some of the complications in computations, tests, and estimation in the case where the sample sizes are unequal are considered in the following example.

Example 11.3. Think of the tensile strength experiment first described in Example 9.2, in which ten random observations were made for each of three manufacturers of copper wire. Now suppose that rolls of wire are randomly selected from the manufacturers so that three come from A, three from B, and two from C. Further, suppose that the number of random determinations

Table 11.6
Tensile Strength of Copper Wire in Pounds (Coded)

Manufacturer	A			B			C	
Roll	1	2	3	1	2	3	1	2
Measurement	110	130	50	130	45	120	100	130
	90	115	75	45	55	50	200	80
	120	105	85	50	65	150	90	70
			40	40			70	80
							90	150
Totals (subsamples)	320	350	250	265	165	320	550	510
Totals (samples)		920			750			1060

Grand Total   2730

of tensile strength made on each roll of wire for a given manufacturer is proportional to the size of the roll. In this case, the data may be classified as shown in Table 11.6.

The reader, no doubt, is already asking such questions as, "Isn't it going to be difficult to get random measurements from a roll of wire?" "Isn't this wasteful and time-consuming?" The answer to each question is "yes." But if such an experiment is desirable and the above suggested analysis is considered appropriate, the observations must be made in such a way that each of the eight subsamples can be considered random. We do not dwell on the practical problem of selecting random subsamples, since the primary purpose in introducing this example was to illustrate the techniques of analysis for the nested classifications when the samples and subsamples are unequal in size. The reader might just as well supply his own variables of classification. [Some of the cases listed immediately before model equation (11.38) may seem more appropriate.]

The numbers associated with the rolls have no special significance. They are used only to distinguish between rolls. We might just as well have numbered the rolls 1, 2, 3, 4, 5, 6, 7, 8 or have used any other notation which differentiates the eight rolls.

Calculations of the sums of squares are complicated by the presence of different divisors, and the coefficients for the expected mean squares require extra calculations. For the data in Table 11.6 we obtain

$$\sum_i \sum_j \sum_u x_{iju}^2 = 110^2 + 90^2 + \cdots + 150^2 = 292{,}600$$

$$\sum_i \sum_j \frac{T_{ij.}^2}{n_{ij}} = \frac{320^2}{3} + \frac{350^2}{3} + \cdots + \frac{510^2}{5} = 263{,}876$$

$$\sum_i \frac{T_{i..}^2}{n_{i.}} = \frac{920^2}{10} + \frac{750^2}{10} + \frac{1060^2}{10} = 253{,}250$$

$$\frac{T^2}{n..} = \frac{2730^2}{30} = 248{,}430$$

$$a = \frac{30 - \left(\dfrac{3^2}{10} + \dfrac{3^2}{10} + \cdots + \dfrac{5^2}{10} \right)}{8 - 3} = 3.64$$

$$b = \frac{\left(\dfrac{3^2}{10} + \cdots + \dfrac{5^2}{10} \right) - \dfrac{3^2 + \cdots + 5^2}{30}}{3 - 1} = 3.93$$

and the analysis of variance shown in Table 11.7.

Clearly, there is no difficulty in finding point and interval estimates of the α's, in finding point estimates of the components σ^2 and σ_δ, and in test-

Table 11.7
Analysis of Variance for the Nested Data of Table 11.6

Source of Variation	Sum of Squares	Degrees of Freedom	Mean Square	Expected Mean Square for Mixed Model (α)
Manufacturers	4,820	2	2410	$\sigma^2 + 3.93\sigma_\delta^2 + 5\sum_i \alpha_i^2$
Rolls within manufacturers	10,626	5	2125	$\sigma^2 + 3.64\sigma_\delta^2$
Sampling within rolls	28,724	22	1306	σ^2
Total	44,170	29		

ing the hypothesis $\sigma_\delta^2 = 0$. However, in order to test the hypothesis $\alpha_1 = \alpha_2 = \alpha_3 = 0$, we must use an approximate test, since $a \neq b$. We now present an approximate test due to Satterthwaite [18].

First note that

$$\frac{b}{a}s_2^2 - \left(\frac{b}{a} - 1\right)s_p^2 \tag{11.49}$$

is an unbiased estimator of $\sigma^2 + b\sigma_\delta^2$, which is the expected value of s_1^2 when the null hypothesis is true. Thus, if we knew the number of degrees of freedom to associate with (11.49), we would expect

$$\frac{s_1^2}{\frac{b}{a}s_2^2 - \left(\frac{b}{a} - 1\right)s_p^2} \tag{11.50}$$

to be approximately distributed as some F ratio when the null hypothesis is true. Actually, if we let

$$\hat{\nu} = \frac{\left[\frac{b}{a}s_2^2 - \left(\frac{b}{a} - 1\right)s_p^2\right]^2}{\frac{\left(\frac{b}{a}s_2^2\right)^2}{\sum n_i - k} + \frac{\left[\left(\frac{b}{a} - 1\right)s_p^2\right]^2}{n.. - \sum n_i}}$$

it can be shown that (11.50) is approximately distributed as F with $\nu_1 = 1$ and $\nu_2 = \hat{\nu}$ degrees of freedom.

In our example, $b/a = 3.93/3.64 = 1.08$ and $b/a - 1 = -0.08$. Thus

$$\hat{\nu} = \frac{[(1.08)(2125) - (0.08)(1306)]^2}{\frac{[(1.08)(2125)]^2}{5} + \frac{[(0.08)(1306)]^2}{22}}$$

$$= \frac{(2,190.52)^2}{1,053,901.18} = 4.55$$

and (11.50) becomes

$$\frac{2410}{(1.02)(2125) - (0.08)(1306)} = 1.10$$

Since the five per cent upper F value with one and five degrees of freedom is 6.61, we fail to reject the null hypothesis that the manufacturers make the same strength of copper wire. (Of course, this is obvious on examination of Table 11.7. We made the calculations only for illustrative purposes.)

The calculations in Example 11.3 should make clear the need for equal-size samples and equal-size subsamples. Thus, in planning an experiment every effort should be made to keep the sample sizes equal.

11.5. EXERCISES

11.1. The analysis of variance for a random model experiment in a one-way classification is as shown in Table 11.8. (a) What variance is used as a measure of dispersion of population means? Find a point estimate of this variance. Find a 95 per cent confidence interval estimate of this

Table 11.8

Source of Variation	Sum of Squares	Degrees of Freedom	Mean Square	Expected Mean Square
Among means	392	4	98	$\sigma^2 + 4\sigma_\alpha^2$
Within	330	15	22	σ^2
Total	722	19		

variance. What assumptions are required for each of these estimates? (b) What null hypothesis can one test with the above analysis of variance? What is the alternative hypothesis? Test the null hypothesis, stating assumptions and conclusion. (c) Describe an experiment which could lead to the above analysis of variance. State the conclusion for your experiment, using the results of (b).

11.2. An experiment was run to determine the effect of firing temperature on the density of bricks. Five firing temperatures were selected at random, and 12 bricks were used for each temperature. The sum of squares for "among firing temperatures" was 6.15, and the pooled variance among bricks within firing temperature" was 0.64. (a) Estimate, in an appropriate manner, the effect of firing temperature on the density of bricks. Discuss your results. (b) Use two methods to find approximate 95 per cent confidence intervals for the standard deviation among temperatures. (c) In addition to the above information, suppose it is known that the mean density for the 60 bricks is 2.3 (a coded value). Approximately what proportion of bricks would have density between 1.9 and 2.7?

11.3. The following hypothetical data are for a random effects experiment in a one-way classification

		Treatment			
1	2	3	4	5	6
51	44	53	60	69	55
47	51	35	39	59	72
49	49	43	49	57	41
68	43	58	60	51	54

(a) Prepare an analysis of variance table with expected mean squares. (b) Test the null hypothesis $\sigma_\alpha^2 = 0$ against the alternative hypothesis $\sigma_\alpha^2 > 0$, giving the assumptions and conclusions. (c) Graph the power curve for the test in (b). Discuss the use of the curve as it relates to this problem. (d) Find 90 per cent confidence limits for σ_α^2 by any method.

11.4. Prove Eq. (11.30) and then $E(s_p^2) = \sigma^2$.

11.5. Prove Eq. (11.31).

11.6. Prove Eq. (11.32).

11.7. Use the information in Exercise 10.5 to write the expected mean squares for both the fixed and random effects experiments in a one-way classification.

11.8. In a random effects experiment in a one-way classification, assume $n_1 = 3$, $n_2 = 6$, $n_3 = 5$, and $n_4 = 5$ with the analysis of variance given in Table 11.9. (a) Complete the analysis of variance table. (b) Find a point estimate of σ^2. (c) Find, if possible, a 95 per cent confidence interval for σ_α.

Table 11.9

Source of Variation	Sum of Squares	Degrees of Freedom	Mean Square
Treatments			53
Within			
Total	434		

11.9. Derive the expected mean squares for batches in Table 11.3.

11.10. The coded data for two determinations of ascorbic acid in frozen strawberries resulting from two storage times S_1 and S_2 and three treatments T_1, T_2, and T_3 within each storage time are as shown in Table 11.10. (a) Prepare an analysis of variance table similar to Table 11.3, assuming a fixed effects model. (b) Find unbiased estimates of effects of storage time and treatments within storage time. (c) State and test hypotheses about storage time and treatments. (d) Find 90 per cent confidence

Table 11.10

Storage time	S_1			S_2		
Treatments	T_1	T_2	T_3	T_1	T_2	T_3
Determinations	11	10	13	12	14	16
	12	9	11	15	12	15

intervals for the two storage time means and for the six treatments within storage time means. (e) Assuming the data to be for a random effects model, state and test hypotheses about storage time and treatments within storage time. (f) Find 90 per cent confidence intervals for the three variance components involved in (e).

11.11. From each of five mixes four samples are randomly drawn with three random subsamples being drawn from each of the 20 samples. Hypothetical data for such an experiment are given in Table 11.11. (a) Prepare an analysis of variance table similar to Table 11.3. Assume a random effects model. (b) Find point estimates for the variance among mixes, the variance among samples, and the error variance. Find 90 per cent confidence intervals for the standard deviations among sub-

Table 11.11

Mix Number	Sample within Mix Number	Subsample 1	Observations 2	3
1	1	51	47	49
	2	68	44	51
	3	49	43	53
	4	35	43	58
2	1	60	69	55
	2	39	59	72
	3	49	57	41
	4	60	51	54
3	1	53	62	51
	2	45	64	53
	3	45	56	59
	4	56	66	55
4	1	53	64	66
	2	57	56	59
	3	66	73	47
	4	47	59	54
5	1	62	75	59
	2	60	65	67
	3	84	63	72
	4	76	72	56

samples, samples, and mixes. (c) Assuming the data to be for a mixed effects model (δ), state and test hypotheses about mixes and samples within mixes. (d) For mix number 2, test at the five per cent level to determine if the mean for samples 1 and 2 is significantly different from the mean for samples 3 and 4. Make the same test for mix number 5. Assume a mixed effects model (δ). (e) Assuming the data to be for a mixed effects model (α), state and test hypotheses about mixes and samples within mixes. (f) For a mixed effects model (α), test at the five per cent level to determine if the mean for mix 5 is significantly different from the mean of all the other mixes. Determine if the mean for mix 4 is significantly different from the mean of mixes 1, 2, and 3.

11.12. The model equation for sub-subsampling within a one-way classification may be written as

$$x_{ijku} = \mu + \alpha_i + \delta_{ij} + \eta_{ijk} + \epsilon_{ijku}, \qquad i = 1, \ldots, p$$
$$j = 1, \ldots, r$$
$$k = 1, \ldots, t$$
$$u = 1, \ldots, n$$

(a) Write the corresponding observation equation, expressing each component part in terms of totals on x_{ijku}. Write the sum of squares identity and the analysis of variance table for such a model equation. (b) Write the expected mean squares for both the random effects and the fixed effects experiment. Find the expected mean squares for the mixed model (δ), the mixed model (α, δ) and the mixed model (α, η).

11.13. Prove Eq. (11.46).

11.14. Derive the expected mean squares for s_1^2 for each of the four models shown in Table 11.4.

11.15. Prove that $E(s_p^2) = \sigma^2$ for each of the four models shown in Table 11.4.

11.16. Prove that the expression given in (11.49) is an unbiased estimator of $\sigma^2 + b\sigma_\delta^2$.

11.17. Table 11.12 gives hypothetical data in a nested classification (the reader

Table 11.12

Experimenter	A		B			C			
Batch	1	2	1	2	3	1	2	3	4
Measurement	21	19	23	15	29	32	35	33	26
	17	13	32	34	26	45	37	42	40
	19	23	21	23	36		44	46	
	14	5		15	25			42	
		13		26					

can give his own interpretation to the measurements). (a) Prepare an analysis of variance, including the error mean squares, similar to Table

11.6. Assume the mixed model (α). (b) Find 95 per cent confidence interval estimates of σ^2 and σ_δ^2. (c) Test the hypothesis $\alpha_A = \alpha_B = \alpha_C = 0$.

REFERENCES

1. Anderson, R. L. and T. A. Bancroft, *Statistical Theory in Research*. New York: McGraw-Hill, Inc., 1952, Chap. 22.

2. Bartlett, M. S., "Approximate Confidence Intervals, II. More than One Unknown Parameter," *Biometrika*, Vol. **40** (1953), pp. 306–17.

3. Birnbaum, A. and W. C. Healy, "Estimates with Prescribed Variance Based on Two-stage Sampling," *The Annals of Mathematical Statistics*, Vol. **31** (1960), pp. 662–76.

4. Bross, Irwin, "Fiducial Intervals for F Variance Components," *Biometrics*, Vol. **6** (1950), pp. 136–44.

5. Bulmer, M. G., "Approximate Confidence Limits for Components of Variance," *Biometrika*, Vol. **44** (1957), pp. 159–67.

6. Cochran, W. G., *Sampling Techniques*. New York: John Wiley & Sons, Inc., 1953.

7. ———, "Testing a Linear Relation Among Variances," *Biometrics*, Vol. **7** (1951), pp. 17–32.

8. ——— and G. M. Cox, *Experimental Designs*. New York: John Wiley & Sons, Inc., 1957.

9. Crump, S. L., "The Estimation of Variance Components in Analysis of Variance," *Biometrics*, Vol. **2** (1946), pp. 7–11.

10. ———, "The Present Status of Variance Component Analysis," *Biometrics*, Vol. **7** (1951), pp. 1–16.

11. Davies, Owen L., *The Design and Analysis of Industrial Experiments*. New York: Hafner Publishing Co., Inc., 1954, Chap. 4.

12. ———, *Statistical Methods in Research and Production*. London: Oliver and Boyd, 1958, pp. 109–15.

13. Ganguli, M., "A Note on Nested Sampling," *Sankhyā*, Vol. **5** (1941), pp. 449–52.

14. Green, J. R., "A Confidence Interval for Variance Components," *The Annals of Mathematical Statistics*, Vol. **25** (1954), pp. 671–86.

15. Hansen, M. H., W. N. Hurwitz, and W. G. Madow, *Sample Survey Methods and Theory*, Vols. **1** and **2**. New York: John Wiley & Sons, Inc., 1953.

16. Healy, W. C., Jr., "Limits for a Variance Component with an Exact Confidence Coefficient," *The Annals of Mathematical Statistics*, Vol. **32** (1961), pp. 466–76.

17. Ostle, Bernard, *Statistics in Research*, 2nd ed. Ames, Iowa: The Iowa State College Press, 1963, Chap. 11.

18. Satterthwaite, F. E., "An Approximate Distribution of Estimates of Variance Components," *Biometrics*, Vol. **2** (1946), pp. 110–14.

19. Snedecor, G.W. and W.G. Cochran, *Statistical Methods*, 5th ed. Ames, Iowa: The Iowa State College Press, 1957, Chap. 10.

20. Steel, R. G. O. and J. H. Torrie, *Principles and Procedures of Statistics*. New York: McGraw-Hill, Inc., 1960, Chap. 7.

21. Sukhatme, P. V., *Sampling Theory of Surveys with Applications*. Ames, Iowa: The Iowa State College Press, 1953.

22. Welch, B. L., "On Linear Combinations of Several Variances," *Journal of the American Statistical Association*, Vol. **51** (1956), pp. 132–48.

12

ANALYSIS OF VARIANCE—
MULTIWAY CLASSIFICATIONS

Concepts of the last two chapters are extended, and one-, two-, and three-way classifications are compared. The uses of single and repeated observations in a two-way classification with fixed, mixed, and random models are studied. The importance of equal sample sizes is explained, and the problem of missing data is discussed. The Latin square design is introduced and treated briefly. Relative efficiency is introduced as an aid in comparing designs.

12.1. INTRODUCTION

Anyone working with data knows that in most experimental situations more than one factor (variable) affects the outcome of the experiment. Indeed, more than two or three factors often noticeably affect the magnitude of the observations. However, extensions of topics in analysis of variance to many factors may be satisfactorily accomplished through a thorough understanding of the one-, two-, and three-factor designs. Hence, we now introduce the next most simple design—the *randomized block design*, a simple two-way crossed classification design.

There are many situations in which different treatments change noticeably from experimental unit to experimental unit. For example, the amount learned by each of three teaching methods (one factor) may vary considerably for students of six different intelligence quotient (I.Q.) groups (second factor). As another example, the water absorbed by activated alumina, measured as percentage weight gain, at three different temperatures might vary with particle size. Further, the responses of different treatments may be affected remarkably by different batches of raw material; the rate of gain of weight of animals of the same age may depend on the initial weights of the animals as well as on the diets.

In each of the above examples we think of the observations as being classified according to two criteria at once. In each case one variable may be considered primary and the other secondary to the purposes of the experiment. We call the primary variable the *treatment factor* and the secondary variable the *block factor*. In the first example, the teaching methods represent different treatments and the I.Q.'s represent different blocks or experimental units. In the second example, the temperatures are treatments, and the particle size ranges are blocks. In the last example, the diets are treatments, and the weights of animals at the start of the experiment are blocks.

The simplest randomized block design is one in which each treatment is applied exactly once in each block. In the alumina example, suppose that the three treatments are 50°, 60°, and 70°F and the four blocks are for particle size ranges of 1–2 mesh, 2–4 mesh, 4–8 mesh, and 8–14 mesh. Suppose one measurement is made of the percentage of water absorbed by particles in the 1–2 mesh range at 50°F, one measurement is made of the percentage of water absorbed by particles in the 1–2 mesh range at 60°F, etc. In this way 12 measurements are made in an experiment which has a randomized block design with one observation per cross classification or, for short, per *cell*. The name "randomized block" is used because the treatments for any block are applied in random order. Thus, in the block with particle size in the 4–8 mesh range, the order of application of the temperatures to quantities of about the same initial weight could be 70°, 50°, and 60°F. An example of the work order (reading from left to right) of the whole experiment is shown in Fig. 12.1.

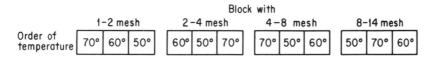

Fig. 12.1 Layout for an Experiment

Much of the analysis for the randomized block design with a single observation per cell is a simple extension of the analysis for the completely randomized design. Thus, we give an example to illustrate methods of estimation and testing before we present the general theory, models, and assumptions required for such methods.

12.2. AN APPLICATION OF THE RANDOMIZED BLOCK DESIGN

As an illustration of the nature of the randomized block design as compared with a completely randomized design, consider the following. Suppose it is required to determine whether the acidity level of the soil in a garden is the same at three different depths, say one, seven, and 13 in. Further, suppose that four recordings of the acidity level in terms of pH units are to be made at each depth of soil. Now 12 core samples of soil could be taken at random, four at 1 in., four at 7 in., and four at 13 in. depth. The results of such an experiment would then be analyzed according to the one-way classification design. On the other hand, four core samples could be randomly selected, and then the acidity level of each could be tested at each of the three depths of soil, the order being random. The data collected in this way should be analyzed as a randomized block design.

If the manual labor involved in obtaining the 12 measurements of acidity is assumed to be negligible, which design is preferred? This question cannot be answered fully at this time, but some light can be thrown on the problem. It can be shown that the paired observations experiment described in Sect. 8.7 is a special case of the randomized block design and the independent observations experiment described in Sect. 8.6 is a special case of the completely randomized design. Thus, the randomized block design is better (worse) than the completely randomized design in situations like those where the paired observation experiment is better (worse) than the independent observations experiment. Since the pH value in the acidity experiment is likely to be influenced by the particular location in the garden from which the soil sample is taken, it would appear that the randomized block design is better than the completely randomized design. That is due to the fact that one pH value is determined for each depth at any location, and hence the effect of the location can be removed in any comparison among the treatment means. But in the completely randomized design we cannot be sure to what extent the location of the soil samples influences our conclusions regarding the treatments. Further comparisons in the two designs will be made later. Now we give methods of estimating parameters and testing hypotheses relating to a randomized block design. A useful notation is introduced in the discussion of methods.

Example 12.1. Data on the acidity level of 12 soil samples in the randomized block design described above are arranged in Table 12.1 for compu-

Table 12.1
Acidity Level, in pH Units, of Soil

Core Sample Number	Depth of Soil		
	1 in.	7 in.	13 in.
1	6.7	6.5	6.3
2	6.4	6.6	6.2
3	6.9	6.8	6.4
4	6.4	6.5	6.3

tational purposes. (a) Use a five per cent level test to determine if the acidity level is the same at different depths. (b) Find estimates of treatment means and variance components.

To estimate means and effects and to prepare an analysis of variance table, we need the totals, means, and effects along with the notation shown in Table 12.2. The sum of squares identity is given by

Table 12.2
Acidity Level, in pH Units, of 12 Soil Samples

Core Sample	Depth of Soil in Inches			Core Total	Core Means
	1	7	13		
1	$x_{11} = 6.7$	$x_{21} = 6.5$	$x_{31} = 6.3$	$T_{.1} = 19.5$	$\bar{x}_{.1} = 6.5$
2	$x_{12} = 6.4$	$x_{22} = 6.6$	$x_{32} = 6.2$	$T_{.2} = 19.2$	$\bar{x}_{.2} = 6.4$
3	$x_{13} = 6.9$	$x_{23} = 6.8$	$x_{33} = 6.4$	$T_{.3} = 20.1$	$\bar{x}_{.3} = 6.7$
4	$x_{14} = 6.4$	$x_{24} = 6.5$	$x_{34} = 6.3$	$T_{.4} = 19.2$	$\bar{x}_{.4} = 6.4$
Depth totals	$T_{1.} = 26.4$	$T_{2.} = 26.4$	$T_{3.} = 25.2$	$T = 78.0$	
Depth means	$\bar{x}_{1.} = 6.6$	$\bar{x}_{2.} = 6.6$	$\bar{x}_{3.} = 6.3$		$\bar{x} = 6.5$
Depth effects	$a_1 = 0.1$	$a_2 = 0.1$	$a_3 = -0.2$		

$$\text{Total } SS = \text{Treatment } SS + \text{Block } SS + \text{Error } SS \qquad (12.1)$$

where, in our example, the sums of squares are given by

$$\text{Treatment } SS = \frac{\sum_{i=1}^{3} T_{i.}^2}{4} - \frac{T^2}{12} = \frac{(26.4)^2 + (26.4)^2 + (25.2)^2}{4} - \frac{(78.0)^2}{12}$$
$$= 507.24 - 507.00 = 0.24$$

$$\text{Block } SS = \frac{\sum_{j=1}^{4} T_{.j}^2}{3} - \frac{T^2}{12} = \frac{(19.5)^2 + \cdots + (19.2)^2}{3} - \frac{(78.0)^2}{12} = 0.18$$

$$\text{Total } SS = \sum_{i=1}^{3} \sum_{j=1}^{4} x_{ij}^2 - \frac{T^2}{12} = (6.7)^2 + \cdots + (6.3)^2 - \frac{(78.0)^2}{12} = 0.50$$

and, by subtraction

$$\text{Error } SS = 0.50 - 0.24 - 0.18 = 0.08$$

Associated with the sum of squares identity there is a degrees of freedom identity, which is illustrated in Table 12.3. For the present the student may determine the degrees of freedom for the treatment, block and total sum of squares by the principle already explained and then find the degrees of freedom for error by subtraction. A justification for the degrees of freedom for the error sum of squares will be given later. In our example, the treatment effects α_i are fixed, and the blocks are randomly selected from a population of blocks with variance σ_β^2. Further, we assume that the single observation in each cell is randomly and independently drawn from a population with variance $\sigma_{ij}^2 = \sigma^2$. That is, each of the twelve populations has the same variance. If we assume the effects α_i to be defined so that

$$\sum_{i=1}^{3} \alpha_i = 0$$

it can be shown by the methods of Chap. 11 that the expected values for the mean squares are those given in Table 12.3.

Table 12.3
Analysis of Variance for Soil Acidity

Source of Variation	Sum of Squares	Degrees of Freedom	Mean Square	Expected Mean Squares
Blocks	0.18	3	$0.06 = s_1^2$	$\sigma^2 + 3\sigma_\beta^2$
Treatments	0.24	2	$0.12 = s_2^2$	$\sigma^2 + 4 \sum_{i=1}^{3} \alpha_i^2 / (3-1)$
Error	0.08	6	$0.04/3 = s_p^2$	σ^2
Total	0.50	11		

Letting the treatment means for depths of 1, 7, and 13 in. be $\mu_1.$, $\mu_2.$, and $\mu_3.$, respectively, we wish to test the hypothesis

$$H_0: \mu_1. = \mu_2. = \mu_3. = \mu \quad (\text{or } \alpha_1 = \alpha_2 = \alpha_3 = 0 \text{ or } \sum \alpha_i^2 = 0) \quad (12.2)$$

where μ is the over-all mean. The alternative hypothesis is that at least one of the treatment means is different from the over-all population mean or, for short, $\sum \alpha_i^2 > 0$. If, in addition to the assumptions already made, we assume that the 12 populations for cells are normally and independently distributed, then it can be shown that the ratio s_2^2/s_p^2 is distributed as F with two and six degrees of freedom when the null hypothesis is true. When the alternative hypothesis is true, the numerator, s_2^2, of the ratio s_2^2/s_p^2 on the average is larger than σ^2. Hence, the critical region for the test of (12.2) is made up of all those values of F for which $F > F_{.05}(2, 6) = 5.14$. Since the particular experimental ratio

$$\frac{s_2^2}{s_p^2} = \frac{0.12}{\dfrac{0.04}{3}} = 9$$

is greater than 5.14, we reject the hypothesis (12.2) and conclude that the treatment means are different. (Actually, on looking at Table 12.2, we would say that the acidity level at 13 in. is less than the average acidity level at one and seven in.

Unbiased point estimates of the treatment means and effects, $a_i = \bar{x}_i - \bar{x}$, are shown in Table 12.2. We are not interested in the core mean estimates. However, it is useful to know the unbiased point estimator s_b^2 of the core variance component given by

$$s_b^2 = \frac{s_1^2 - s_p^2}{3} = \frac{0.06 - \dfrac{0.04}{3}}{3} = 0.0156$$

Further, an interval estimate of means may be found by the method explained in Example 10.1, and an approximate interval estimate of σ_β^2 may be found by the methods given in Sect. 11.2.

To test the hypothesis that $\sigma_\beta^2 = 0$ against the alternative hypothesis that $\sigma_\beta^2 > 0$, compute s_1^2/s_p^2 and compare with the upper α level F value with three and six degrees of freedom. If s_1^2/s_p^2 is greater than this value, reject the hypothesis $\sigma_\beta^2 = 0$ and conclude that the block component of variance exists; otherwise, fail to reject the hypothesis.

12.3. TWO-WAY CLASSIFICATIONS, SINGLE OBSERVATION PER CELL— FIXED MODEL

Think of x_{iju} as being the uth ($u = 1, \ldots, n$) random observation in the ith ($i = 1, \ldots, c$) group of factor one and the jth ($j = 1, \ldots, r$) group of factor two. Assume that x_{iju} is distributed with mean μ_{ij} and variance $\sigma_{ij}^2 = \sigma^2$; that is, each of the cr populations have the same variance. Thus, if ϵ_{iju} denotes the amount the random observation x_{iju} deviates from its mean μ_{ij}, we may write the model equation

$$x_{iju} = \mu_{ij} + \epsilon_{iju} \qquad (i = 1, \ldots, c; \; j = 1, \ldots, r; \; u = 1, \ldots, n) \qquad (12.3)$$

In the special case where $n = 1$, that is, where there is a single random observation per cell, we drop the third subscript and write

$$x_{ij} = \mu_{ij} + \epsilon_{ij} \qquad (i = 1, \ldots, c; \; j = 1, \ldots, r) \qquad (12.4)$$

understanding that μ_{ij} is a fixed parameter and that ϵ_{ij} is a single random deviate about the mean μ_{ij}.

In Example 12.1 we arranged the data in a rectangular table for ease of computation. Such a table is useful in understanding the general two-way classification. Use c columns to represent the c groups of factor *one* (treatments) and r rows to represent the r groups of factor *two* (blocks). Then the means of the cr populations along with c column means and r row means may be arranged as in Table 12.4. The *i*th *column mean* $\mu_{i.}$ and *j*th *row mean* $\mu_{.j}$ are defined by

Table 12.4
Population Means in a Two-Way Classification

Row Number	Column Numbers					Row Means	Row Effects
	1	· · ·	*i*	· · ·	*c*		
1	μ_{11}	· · ·	μ_{i1}	· · ·	μ_{c1}	$\mu_{.1}$	$\beta_1 = \mu_{.1} - \mu$
j	μ_{1j}	· · ·	μ_{ij}	· · ·	μ_{cj}	$\mu_{.j}$	$\beta_j = \mu_{.j} - \mu$
r	μ_{1r}	· · ·	μ_{ir}	· · ·	μ_{cr}	$\mu_{.r}$	$\beta_r = \mu_{.r} - \mu$
Column means	$\mu_{1.}$	· · ·	$\mu_{i.}$	· · ·	$\mu_{c.}$	μ	
Column effects	$\alpha_1 = \mu_{1.} - \mu \cdots \alpha_i = \mu_{i.} - \mu \cdots \alpha_c = \mu_{c.} - \mu$						

$$\mu_{i.} = \frac{\sum_{j=1}^{r} \mu_{ij}}{r} \quad (i = 1, \ldots, c) \tag{12.5}$$

and

$$\mu_{.j} = \frac{\sum_{i=1}^{c} \mu_{ij}}{c} \quad (j = 1, \ldots, r) \tag{12.6}$$

respectively, and the *over-all mean* μ of the *cr* populations by

$$\mu = \frac{\sum_i \sum_j \mu_{ij}}{cr} \tag{12.7}$$

Further

$$\alpha_i = \mu_{i.} - \mu \quad (i = 1, \ldots, c) \tag{12.8}$$

and

$$\beta_j = \mu_{.j} - \mu \quad (j = 1, \ldots, r) \tag{12.9}$$

denote the deviations of the column and row means from the over-all mean, and they are called the *column* and *row effects*, respectively. Thus, the cell mean μ_{ij} may be written as

$$\mu_{ij} = \mu + (\mu_{i.} - \mu) + (\mu_{.j} - \mu) + (\mu_{ij} - \mu_{i.} - \mu_{.j} + \mu) \tag{12.10}$$

or, in the important particular case, as

$$\mu_{ij} = \mu + \alpha_i + \beta_j \qquad (i = 1, \ldots, c; \ j = 1, \ldots, r) \qquad (12.11)$$

when

$$\mu_{ij} - \mu_{i.} - \mu_{.j} + \mu = 0 \qquad (12.12)$$

In case Eq. (12.12) holds, we see that the cell mean μ_{ij} can be expressed as a constant μ plus a column effect α_i which is the same for all cells in column i plus a row effect β_j which is the same for all cells in row j. This is made clear in Table 12.5. Further, we see from Eq. (12.11) that the cr parameters μ_{ij} can be expressed in terms of $1 + c + r$ parameters $\mu; \alpha_1, \ldots, \alpha_c;$ $\beta_1, \ldots, \beta_r$.

Table 12.5
Population Means in Terms of Effects

Row Number	Column Number					Row Means
	1	$\cdots$	i	$\cdots$	c	
1	$\mu_{11} =$ $\mu + \alpha_1 + \beta_1$	$\cdots$	$\mu_{i1} =$ $\mu + \alpha_i + \beta_1$	$\cdots$	$\mu_{c1} =$ $\mu + \alpha_c + \beta_1$	$\mu_{.1} = \mu + \beta_1$
j	$\mu_{1j} =$ $\mu + \alpha_1 + \beta_j$	$\cdots$	$\mu_{ij} =$ $\mu + \alpha_i + \beta_j$	$\cdots$	$\mu_{cj} =$ $\mu + \alpha_c + \beta_j$	$\mu_{.j} = \mu + \beta_j$
r	$\mu_{1r} =$ $\mu + \alpha_1 + \beta_r$	$\cdots$	$\mu_{ir} =$ $\mu + \alpha_i + \beta_r$	$\cdots$	$\mu_{cr} =$ $\mu + \alpha_c + \beta_r$	$\mu_{.r} = \mu + \beta_r$
Column means	$\mu_{1.} = \mu + \alpha_1$	$\cdots$	$\mu_{i.} = \mu + \alpha_i$	$\cdots$	$\mu_{c.} = \mu + \alpha_c$	μ

Because of the way α_i and β_j are defined, we obtain the following two linear restrictions involving the newly introduced parameters

$$\sum_{i=1}^{c} \alpha_i = 0 \quad \text{and} \quad \sum_{j=1}^{r} \beta_j = 0 \qquad (12.13)$$

That is, only $(c - 1)$ α_i's and $(r - 1)$ β_j's are independent. Thus, the cr parameters μ_{ij} in Eq. (12.11) can actually be expressed in terms of $1 + (c - 1) + (r - 1) = c + r - 1$ independent parameters $\mu;$ $\alpha_1, \ldots, \alpha_{c-1}; \beta_1, \ldots, \beta_{r-1}$.

Substituting Eq. (12.11) in Eq. (12.4) gives the *model equation*

$$x_{ij} = \mu + \alpha_i + \beta_j + \epsilon_{ij} \qquad (i = 1, \ldots, c; \ j = 1, \ldots, r) \qquad (12.14)$$

This is the form we use most of the time, since our main concern is in making statements about either the column effects or the row effects or both. For example, typical hypotheses tested by analysis of variance are

$$H_0: \alpha_1 = \cdots = \alpha_c = 0 \quad (\text{or } \mu_{1.} = \cdots = \mu_{c.} = \mu) \quad (12.15)$$

and

$$H_0: \beta_1 = \cdots = \beta_r = 0 \quad (\text{or } \mu_{.1} = \cdots = \mu_{.r} = \mu) \quad (12.16)$$

The model equation for the two-way classification given by Eq. (12.14) is said to be *additive*. This name is used because the mean of any cell is obtained by adding a row effect to a column effect to an over-all (constant) effect.

Now consider the cr populations with means μ_{ij} which are shown in Table 12.4. Select at random one observation from each population, and let each selection be independent of all others. Arrange these observations as shown in Table 12.6 and compute the means indicated in the margin. That is, compute the means $\bar{x}_{i.}$, $\bar{x}_{.j}$, and $\bar{x}$ by the formulas

$$
\begin{cases}
\bar{x}_{i.} = \dfrac{\sum\limits_{j=1}^{r} x_{ij}}{r}, \qquad \bar{x}_{.j} = \dfrac{\sum\limits_{i=1}^{c} x_{ij}}{c} \quad \text{and} \\[3ex]
\bar{x} = \dfrac{\sum\limits_{i=1}^{c}\sum\limits_{j=1}^{r} x_{ij}}{cr}
\end{cases}
\quad (12.17)
$$

respectively. The estimator effects of the columns and rows are given by

$$a_i = \bar{x}_{i.} - \bar{x} \quad \text{and} \quad b_j = \bar{x}_{.j} - \bar{x} \quad (12.18)$$

respectively.

Table 12.6
Two-way Classification with One Observation per Cell

Row Number	Column Numbers 1		i		c	Row Means	Row Effects
1	x_{11}	$\cdots$	x_{i1}	$\cdots$	x_{c1}	$\bar{x}_{.1}$	$b_1 = \bar{x}_{.1} - \bar{x}$
j	x_{1j}	$\cdots$	x_{ij}	$\cdots$	x_{cj}	$\bar{x}_{.j}$	$b_j = \bar{x}_{.j} - \bar{x}$
r	x_{1r}	$\cdots$	x_{ir}	$\cdots$	x_{cr}	$\bar{x}_{.r}$	$b_r = \bar{x}_{.r} - \bar{x}$
Column means	$\bar{x}_{1.}$	$\cdots$	$\bar{x}_{i.}$	$\cdots$	$\bar{x}_{c.}$	$\bar{x}$	
Column effects	$a_1 = \bar{x}_{1.} - \bar{x} \cdots$		$a_i = \bar{x}_{i.} - \bar{x}$	$\cdots$	$a_c = \bar{x}_{c.} - \bar{x}$		

It is easy to show that $\bar{x}$, $\bar{x}_{i.}$, $\bar{x}_{.j}$, a_i, and b_j are unbiased estimators of the parameters μ, $\mu_{i.}$, $\mu_{.j}$, α_i, and β_j, respectively. Further, if we let

$$\bar{x}_{ij} = \bar{x} + a_i + b_j \tag{12.19}$$

it follows that $\bar{x}_{ij}$ is an unbiased estimator of the cell mean μ_{ij}, since

$$E(\bar{x}_{ij}) = E(\bar{x} + a_i + b_j) = \mu + \alpha_i + \beta_j = \mu_{ij}$$

It should be noted that the unbiased property follows, in each case, from the definitions of the model equation (12.14) and the assumptions of random observations. (We do not need the assumptions of equal variances and normal populations.)

If e_{ij} denotes the amount a random observation x_{ij} deviates from cell mean $\bar{x}_{ij}$, then we may write the *observation equation* as

$$x_{ij} = \bar{x}_{ij} + e_{ij} \qquad (i = 1, \ldots, c; \ j = 1, \ldots, r) \tag{12.20}$$

or, substituting Eq. (12.19) in Eq. (12.20), as

$$x_{ij} = \bar{x} + a_i + b_j + e_{ij} \qquad (i = 1, \ldots, c; \ j = 1, \ldots, r) \tag{12.21}$$

Clearly, e_{ij} is an unbiased estimator of ϵ_{ij}.

In Sect. 10.3 we developed two sum of squares identities for the one-way classification. The first identity involved parameters of the model, and the second identity did not. (We learned in Sect. 10.3 the advantages of the second identity.) Now we could develop similar sum of squares identities for the two-way classification, but, since we wish to explain test procedures for the hypotheses in (12.15) and (12.16), we discuss only the second type of identity. The derivation of the first sum of squares identity is left to the student as an exercise.

We can partition the deviate of the observation value x_{ij} about the over-all mean $\bar{x}$ into the sum of three deviates as follows

$$x_{ij} - \bar{x} = (\bar{x}_{i.} - \bar{x}) + (\bar{x}_{.j} - \bar{x}) + (x_{ij} - \bar{x}_{i.} - \bar{x}_{.j} + \bar{x}) \tag{12.22}$$

Squaring both sides of Eq. (12.22) and summing over-all observations gives

$$\begin{aligned}
\sum_i \sum_j (x_{ij} - \bar{x})^2 &= \sum_i \sum_j (\bar{x}_{i.} - \bar{x})^2 + \sum_i \sum_j (\bar{x}_{.j} - \bar{x})^2 \\
&+ \sum_i \sum_j (x_{ij} - \bar{x}_{i.} - \bar{x}_{.j} + \bar{x})^2 \\
&+ 2 \sum_i \sum_j (\bar{x}_{i.} - \bar{x})(\bar{x}_{.j} - \bar{x}) \\
&+ 2 \sum_i \sum_j (\bar{x}_{i.} - \bar{x})(x_{ij} - \bar{x}_{i.} - \bar{x}_{.j} + \bar{x}) \\
&+ 2 \sum_i \sum_j (\bar{x}_{.j} - \bar{x})(x_{ij} - \bar{x}_{i.} - \bar{x}_{.j} + \bar{x})
\end{aligned} \tag{12.23}$$

Since the last three terms on the right-hand side of Eq. (12.23) are zero, we may write the desired *sum of squares identity* as

$$SST = SSC + SSR + SSE \tag{12.24}$$

where

$$
\begin{cases}
SST = \sum_i \sum_j (x_{ij} - \bar{x})^2, \qquad SSC = \sum_i \sum_j (\bar{x}_{i.} - \bar{x})^2 = r \sum_i a_i^2 \\
SSR = \sum_i \sum_j (\bar{x}_{.j} - \bar{x})^2 = c \sum_j b_j^2 \\
SSE - \sum_i \sum_j (x_{ij} - \bar{x}_{i.} - \bar{x}_{.j} + \bar{x})^2 = \sum_i \sum_j e_{ij}^2
\end{cases}
\tag{12.25}
$$

Note that SST denotes "total sum of squares," SSC "sum of squares for column means," SSR "sum of squares for row means," and SSE "sum of squares for error (or residual or remainder)." (The word "means" could be replaced by the word "effects" in the definition of SSC and SSR.) As an example of how the three terms on the right-hand side of Eq. (12.23) reduce to zero, we use Eq. (12.17) to write

$$\sum_i \sum_j (\bar{x}_{i.} - \bar{x})(x_{ij} - \bar{x}_{i.} - \bar{x}_{.j} + \bar{x})$$

$$= \sum_i (\bar{x}_{i.} - \bar{x}) \sum_j (x_{ij} - \bar{x}_{i.} - \bar{x}_{.j} + \bar{x})$$

$$= \sum_i (\bar{x}_{i.} - \bar{x})(\sum_j x_{ij} - \sum_j \bar{x}_{.j} - \sum_j \bar{x}_{i.} + \sum_j \bar{x})$$

$$= \sum_i (\bar{x}_{i.} - \bar{x})(r\bar{x}_{i.} - r\bar{x}_{i.} - r\bar{x} + r\bar{x})$$

$$= \sum_i (\bar{x}_{i.} - \bar{x})(0) = 0$$

In Example 12.1 the sums of squares were computed by using totals rather than means. Now, if we denote the totals by

$$T_{i.} = \sum_j x_{ij}, \qquad T_{.j} = \sum_i x_{ij} \quad \text{and} \quad T = \sum_i \sum_j x_{ij} \tag{12.26}$$

so that the means in Eq. (12.17) may be written as

$$\bar{x}_{i.} = \frac{T_{i.}}{r}, \qquad \bar{x}_{.j} = \frac{T_{.j}}{c} \quad \text{and} \quad \bar{x} = \frac{T}{cr}$$

it is easy to show that the sums of squares in Eq. (12.25) reduce to

$$
\begin{cases}
SST = \sum_i \sum_j x_{ij}^2 - \dfrac{T^2}{cr} \\[2ex]
SSC = r \sum_i (\bar{x}_{i.} - \bar{x})^2 = \dfrac{\sum_i T_{i.}^2}{r} - \dfrac{T^2}{cr} \\[2ex]
SSR = c \sum_j (\bar{x}_{.j} - \bar{x})^2 = \dfrac{\sum_j T_{.j}^2}{c} - \dfrac{T^2}{cr} \\[2ex]
SSE = \sum_i \sum_j x_{ij}^2 - \dfrac{\sum_i T_{i.}^2}{r} - \dfrac{\sum_j T_{.j}^2}{c} + \dfrac{T^2}{cr}
\end{cases}
\tag{12.27}
$$

Note that the divisor in each quotient in Eq. (12.27) is the number of observations used to obtain the total in the dividend. Also, note that SSE is usually found by subtraction as follows

$$SSE = SST - SSC - SSR$$

The expression (12.24) is an algebraic identity. It is not dependent upon any assumption relating to the model. That is, given *any* rectangular array of numbers, we can partition the sum of squares of the deviates of the observations about the over-all mean into three component sum of squares as we did in (12.24). Of course, we might not be able to give any meaningful interpretation to such a partition in many problems, but the partition can still be made.

If we assume that the additive model equation (12.14) has fixed effects α_i $(i = 1, \ldots, c)$ and β_j $(j = 1, \ldots, r)$ and random effects ϵ_{ij} which are independently distributed with common variance σ^2 and zero means, then it can be shown by the methods of Sect. 11.3 that the expectations of the sum of squares in Eq. (12.27) are

$$\left\{ \begin{aligned} E(SSC) &= (c-1)\sigma^2 + r \sum_i \alpha_i^2 \\ E(SSR) &= (r-1)\sigma^2 + c \sum_j \beta_j^2 \\ E(SSE) &= (c-1)(r-1)\sigma^2 \\ E(SST) &= (cr-1)\sigma^2 + r \sum_i \alpha_i^2 + c \sum_j \beta_j^2 \end{aligned} \right. \tag{12.28}$$

From the partition theory of the χ^2 distribution of Chap. 10 it can be shown that the three sum of squares on the right-hand side of Eq. (12.24) are independently distributed. If, in addition to the assumptions listed at the beginning of this paragraph, we assume that the cr populations are normally distributed, then the ratios

$$\left\{ \begin{aligned} &\frac{SSC}{\sigma^2 + r\left(\dfrac{\sum_i \alpha_i^2}{c-1}\right)} \\[4ex] &\frac{SSR}{\sigma^2 + c\left(\dfrac{\sum_j \beta_j^2}{r-1}\right)} \\[4ex] &\frac{SSE}{\sigma^2} \end{aligned} \right. \tag{12.29}$$

are independently distributed as chi-squares with $c - 1$, $r - 1$, and $(c - 1)(r - 1)$ degrees of freedom, respectively. Further, note that if the null hypotheses (12.15) and (12.16) hold, that is, if

$$\sum_i \alpha_i^2 = \sum_j \beta_j^2 = 0$$

then the ratios

$$
\begin{cases}
s_1^2 = \dfrac{SSC}{c-1} \\[2ex]
s_2^2 = \dfrac{SSR}{r-1} \\[2ex]
s_3^2 = \dfrac{SSE}{(c-1)(r-1)}
\end{cases}
\tag{12.30}
$$

are independent unbiased estimators of σ^2. We bring together in Table 12.7 the analysis of variance and expected mean squares for a fixed model experiment in a two-way classification design with one observation per cell. Again, we see that the degrees of freedom identity

$$
cr - 1 = (c-1) + (r-1) + (c-1)(r-1) \tag{12.31}
$$

can be used with the sum of squares identity to obtain the appropriate mean squares in the analysis of variance.

Table 12.7
Analysis of Variance for Two-Way Classification

Source of Variation	Sum of Squares	Degrees of Freedom	Mean Square	Expected Mean Square for Fixed Model
Column effects	$\displaystyle\sum_i \frac{T_{i\cdot}^2}{r} - \frac{T^2}{cr}$	$c-1$	s_1^2	$\sigma^2 + r\left(\dfrac{\sum_i \alpha_i^2}{c-1}\right)$
Row effects	$\displaystyle\sum_j \frac{T_{\cdot j}^2}{c} - \frac{T^2}{cr}$	$r-1$	s_2^2	$\sigma^2 + c\left(\dfrac{\sum_j \beta_j^2}{r-1}\right)$
Error	by subtraction	$(c-1)(r-1)$	s_3^2	σ^2
Total	$\displaystyle\sum_i \sum_j x_{ij} - \frac{T^2}{cr}$	$cr-1$		

If the null hypothesis (12.15) that the column effects are equal to zero holds, the ratio s_1^2/s_3^2 has the F distribution with $c-1$ and $(c-1)(r-1)$ degrees of freedom. Also, if the null hypothesis (12.16) that the row effects are equal to zero holds, the ratio s_2^2/s_3^2 has the F distribution with $r-1$ and $(c-1)(r-1)$ degrees of freedom. Thus, the F distribution may be used in the usual way to test the hypotheses (12.15) and (12.16), if Eq. (12.14) is a fixed effects model equation with random variables $\epsilon_{11}, \ldots, \epsilon_{cr}$ normally and independently distributed with common means of zero and common variances of σ^2.

Confidence intervals for any linear combination of row means or any linear combination of column means may be found by the method of Sect. 10.6. The error variance estimator s_3^2 with $(c-1)(r-1)$ degrees of freedom obtained in the analysis of variance is used for all confidence problems. Thus, if

$$\gamma_c = m_1 \mu_1. + \cdots + m_c \mu_c.$$ (12.32)

is any linear combination of true column means and

$$C_c = m_1 \bar{x}_1. + \cdots + m_c \bar{x}_c.$$

is the corresponding linear combination of estimated column means, then the statistic

$$\frac{C_c - \gamma_c}{\sqrt{\dfrac{s_3^2 \sum_i m_i^2}{r}}}$$ (12.33)

has the t distribution with $(c - 1)(r - 1)$ degrees of freedom. Thus, the $100\,(1 - \alpha)$ per cent confidence limits for Eq. (12.32) are given by

$$C_c \pm t_{\alpha/2} \sqrt{\dfrac{s_3^2 \sum_{i=1}^{c} m_i^2}{r}}$$ (12.34)

where $t_{\alpha/2}$ is the upper $\alpha/2$ percentage point of the t distribution with $(c - 1)(r - 1)$ degrees of freedom. The expression in (12.34) may be used to find confidence limits of single means $\mu_i.$ or differences in means $\mu_i. - \mu_{i'}.$ where $i' \neq i$. Likewise, the $100\,(1 - \alpha)$ per cent confidence limits of any linear combination of row means

$$\gamma_r = m_1' \mu._1 + \cdots + m_r' \mu._r$$

are given by

$$\sum_{j=1}^{r} m_j' \bar{x}._j \pm t_{\alpha/2} \sqrt{\dfrac{s_3^2 \sum_{j=1}^{r} m_j'^2}{c}}$$ (12.35)

where $t_{\alpha/2}$ is defined as in (12.34). The reader should remember that (12.34) and (12.35) give valid confidence limits when Eq. (12.14) is a fixed effects model equation with random variables $\epsilon_{11}, \ldots, \epsilon_{cr}$ normally and independently distributed with common means of zero and common variances of σ^2.

The methods of Sect. 10.8 for obtaining simultaneous confidence intervals (or for testing simultaneous hypotheses) may be used with either the column means or the row means, provided that the error variance s_3^2 with $(c - 1)(r - 1)$ degrees of freedom replaces s^2 with $k(n - 1)$ degrees of freedom, and that the error variance for the means is obtained by dividing s_3^2 by the number of observations used in computing an individual mean. Also, the multiple test procedures for means described in Sect. 10.9 may be applied, provided that s_3^2 of the two-way classification replaces s^2 of the one-way classification. (The reader will not have any trouble with applications of these methods if he thinks of doing two one-way classification problems in which the same error term s_3^2 is used and the divisor depends on the means.) Usually, simultaneous or multiple methods are applied either to the column means or to the row means, but not to both.

Power of the tests of hypotheses (12.15) and (12.16) may be obtained by the methods of Sect. 10.4. Also, Table VIII may be used to find power and the size of the type 2 error in a two-way classification.

The reader has probably wondered why the *randomized block design* was discussed in Example 12.1 and then was followed with a theoretical development of the *two-way classification design*. Actually, the randomized block design is a special case of the two-way classification design, and the term *randomized block* is used when one category of classification is of primary importance and the other is of secondary importance. However, in some experiments both factors may be of equal importance, in which case both factors are analyzed. In this case we may think of the experiment as a *two-factor factorial experiment with no interaction*. Factorial experiments are discussed in Chap. 13.

12.4. USE OF EFFECTS IN UNDERSTANDING ANALYSIS OF VARIANCE

In Example 12.1 of Sect. 12.2 we used the raw data and totals to prepare an analysis of variance table, and, in general, we discussed the problem in terms of means. The method given is perhaps clear, but a better understanding of the analysis may be obtained by using a more lengthy analysis in terms of effects. (This section may be omitted without destroying the continuity of the subject.)

We first note that the block and treatment sum of squares of Table 12.3 may be found directly from the estimated effects and means in Table 12.2. Using Eq. (12.25), we find that the treatment and block sum of squares are, respectively

$$SSC = r \sum_i^3 a_i^2 = 4[(0.1)^2 + (0.1)^2 + (-0.2)^2] = 0.24$$

and

$$SSR = c \sum_j^4 b_j^2 = 3[(6.5 - 6.5)^2 + (6.4 - 6.5)^2 + (6.7 - 6.5)^2$$
$$+ (6.4 - 6.5)^2] = 0.18$$

These are shortcut formulas for finding the sum of squares of the estimated treatment and block effects of each observation in the analysis.

The residual (error) sum of squares may be obtained by computing the sum of squares of all the observations after the differences in row and column means have been removed, that is, after the observations have been adjusted so that the column means are the same and all the row means are the same. First, we change the observations in Table 12.2 so that all row (block) means are the same, and are equal to the grand mean $\bar{x} = 6.5$, Thus, we leave the observations in the first row alone, add 0.1 to each value in the second row, subtract 0.2 from each value in the third row, and add 0.1 to each value in the fourth row. The resulting observations are recorded in Table 12.8 along with the new totals and adjusted row means. Note that the column

Table 12.8
Acidity Level in Table 12.2 Adjusted so that Row Means Are Constant

Core Sample	Depth of Soil in Inches			Totals	Adjusted Means
	1	7	13		
1	6.7	6.5	6.3	19.5	6.5
2	6.5	6.7	6.3	19.5	6.5
3	6.7	6.6	6.2	19.5	6.5
4	6.5	6.6	6.4	19.5	6.5
Totals	26.4	26.4	25.2	78.0	
Means	6.6	6.6	6.3		6.5

and grand means remain unchanged. That is, the block effect can be removed without changing the treatment means and grand mean. Now the variance of the adjusted sample means for the blocks is zero, but the variance of the treatment means remains the same.

Table 12.9
Acidity Level Adjusted so that Row and Column Means Are Constant

Core Sample	Depth of Soil in Inches			Totals	Adjusted Means
	1	7	13		
1	6.6	6.4	6.5	19.5	6.5
2	6.4	6.6	6.5	19.5	6.5
3	6.6	6.5	6.4	19.5	6.5
4	6.4	6.5	6.6	19.5	6.5
Totals	26.0	26.0	26.0	78.0	
Adjusted means	6.5	6.5	6.5		6.5

In a similar manner, change the values in Table 12.8 so that the column means, as well as the row and over-all means, are all equal. Thus, subtracting 0.1 from each value in the first two columns and adding 0.2 to each value in the third column gives the values recorded in Table 12.9. The remaining variation of the values in Table 12.9 is not due to differences in row or column means, since all these means are now the same. It is clearly due to experimental errors (random sampling).

In order to discuss the experimental variations in terms of deviates, e_{ij}, we subtract 6.5 from each value in Table 12.9 to obtain Table 12.10. Now, the residual (error) sum of squares is obtained directly from Table 12.10 as

$$SSE = \sum_i^3 \sum_j^4 e_{ij}^2 = (0.1)^2 + (-0.1)^2 + \cdots + (0.1)^2 = 0.08$$

This is the value found in Table 12.3.

Table 12.10
Residuals for the Acidity Level Experiment

Core Sample	Depth of Soil in Inches			Totals	Adjusted Means
	1	7	13		
1	0.1	−0.1	0.0	0.0	0.0
2	−0.1	0.1	0.0	0.0	0.0
3	0.1	0.0	−0.1	0.0	0.0
4	−0.1	0.0	0.1	0.0	0.0
Totals	0.0	0.0	0.0	0.0	
Adjusted means	0.0	0.0	0.0	0.0	0.0

Something which is easily verified for the general case is obvious in this problem, namely that

$$\begin{cases} \sum_{i=1}^{3} a_i = 0.1 + 0.1 + (-0.2) = 0 \\[2mm] \sum_{j=1}^{4} b_j = 0.0 + (-0.1) + 0.2 + (-0.1) = 0 \\[2mm] \sum_{i=1}^{3} e_{ij} = 0 \quad (j = 1, \ldots, 4) \quad \text{and} \\[2mm] \sum_{j=1}^{4} e_{ij} = 0 \quad (i = 1, 2, 3) \end{cases} \quad (12.36)$$

Thus, there is one linear restriction among the a's, one linear restriction among the b's, and $4 + 3$ (in general, $r + c$) linear restrictions among the e's. Only, $4 + 3 - 1$ (in general, $r + c - 1$) of the linear restrictions among the e's are *independent*. That is, if six of the restrictions among the e's are given, the seventh follows.

In determining the divisor of SSE so as to make the resulting mean square an unbiased estimator of σ^2, we may argue as follows. The 12 residual values in Table 12.10 have been corrected for three column means which must average 6.5, for four row means which must average 6.5, and for one overall mean which is 6.5. The degrees of freedom for the sum of squares SSE is thus $12 - [(3 - 1) + (4 - 1) + 1] = 6$ {in general, $cr - [(c - 1) + (r - 1) + 1] = (c - 1)(r - 1)$}. Thus, due to the additive nature of our model, the randomness assumption, and the equal variance assumption, it is possible to find an unbiased estimator of the common variance σ^2 when only one observation is made on each population; this is a most important property.

The residuals may be found directly from

$$e_{ij} = x_{ij} - \bar{x}_{ij} \qquad (12.37)$$

Substituting the estimated effects $\bar{x}$, a_1, a_2, a_3, b_1, b_2, b_3, b_4 in Formula (12.19), we obtain the estimated cell means shown in Table 12.11. Then the means in Table 12.11 may be subtracted term by term from the observations in Table 12.2 to give the residuals of Table 12.10. It should be noted that we may also write

$$e_{ij} = x_{ij} - \bar{x} - a_i - b_j$$

or

$$e_{ij} = x_{ij} - \bar{x}_{i.} - \bar{x}_{.j} + \bar{x} \qquad (12.38)$$

Also, note that the unbiased estimators of the cell means shown in Table 12.11 may be obtained even though a single random observation is made in each population.

Table 12.11
Estimates of Population Means for the Acidity Experiment

Core Sample	Depth of Soil in Inches		
	1	7	13
1	$6.6 = 6.5 + 0.1 + 0.0$	$6.6 = 6.5 + 0.1 + 0.0$	$6.3 = 6.5 - 0.2 + 0.0$
2	$6.5 = 6.5 + 0.1 - 0.1$	$6.5 = 6.5 + 0.1 - 0.1$	$6.2 = 6.5 - 0.2 - 0.1$
3	$6.8 = 6.5 + 0.1 + 0.2$	$6.8 = 6.5 + 0.1 + 0.2$	$6.5 = 6.5 - 0.2 + 0.2$
4	$6.5 = 6.5 + 0.1 - 0.1$	$6.5 = 6.5 + 0.1 - 0.1$	$6.2 = 6.5 - 0.2 - 0.1$

12.5. CALCULATION OF THE STANDARD ERROR FOR A COMPARISON OF MEANS

At the end of Sect. 12.3 it was stated that the error mean square found in the analysis of variance of the two-way classification could be used to establish confidence intervals for and test hypotheses about linear combinations of means. If the linear combination is a contrast, a special standard error which is appropriate for the particular contrast may be computed. Such a standard error should be applied when the variances for treatments are definitely not equal. To illustrate the techniques we analyze the data in the following example.

Example 12.2. In an experiment (hypothetical) each of four observers made counts of bacteria in milk on each of six plates. The plates were placed in a fixed position along a table illuminated by daylight so that each plate would be counted under nearly the same conditions. The observers moved from plate to plate in a random order. The bacterial counts along with totals and means are recorded in Table 12.12.

Table 12.12
Bacterial Counts in Milk

Plate Number	Observer				Totals	Plate Mean
	A	B	C	D		
1	340	250	248	282	1120	280
2	229	180	153	198	760	190
3	129	69	79	123	400	100
4	247	215	205	213	880	220
5	341	323	301	315	1280	320
6	124	73	64	99	360	90
Totals	1410	1110	1050	1230	4800	
Observation mean	235	185	175	205		200

We wish to compare the observers in their ability to count bacteria. Suppose it is known that A is a regular counter and that B, C, and D are new counters. Further, suppose that D is a woman and B and C are boys. Then, three hypotheses of interest, stated in terms of orthogonal comparisons, are as follows

$$\begin{cases} H_{01}: 3\mu_A - \mu_B - \mu_C - \mu_D = 0 \\ H_{02}: -\mu_B - \mu_C + 2\mu_D = 0 \\ H_{03}: \mu_B - \mu_C = 0 \end{cases} \qquad (12.39)$$

Test the three hypotheses in (12.39), using the standard error appropriate to each.

First, we construct the usual analysis of variance table (Table 12.13) for a fixed effects randomized blocks design. Then we use Eq. (10.68) to compute the following three components of treatment sum of squares with an individual degree of freedom

$$Q_1^2 = \frac{(\sum m_i T_i.)^2}{6 \sum m_i^2} = \frac{[3(1410) - 1(1110) - 1(1050) - 1(1230)]^2}{6[3^2 + (-1)^2 + (-1)^2 + (-1)^2]}$$

$$= \frac{840^2}{6 \cdot 12} = 9800 = \text{``} A \text{ vs. } B, C, \text{ and } D\text{''} \ SS$$

$$Q_2^2 = \frac{300^2}{6 \cdot 6} = 2500$$

$$Q_3^2 = \frac{60^2}{2 \cdot 6} = 300$$

Note that the treatment $SS = Q_1^2 + Q_2^2 + Q_3^2$; that is

$$9800 + 2500 + 300 = 12,600$$

Table 12.13

Preliminary Analysis of Variance for Data of Table 12.12

Source of Variation	Sum of Squares	Degrees of Freedom	Mean Square	F_c	$F_{.05}$ (3, 15)	$F_{.01}$ (3, 15)
Plates	173,600	5	34,720			
Observer	12,600	3	4,200	22.34**	3.59	5.42
Error	2,820	15	188			
Total	189,020	23				

** denotes "highly significant." This means that the usual hypothesis of equality of observer means is rejected at the one per cent level.

Next, in order to partition the error sum of squares, use the coefficients (see Table 12.14) which define the three linear comparisons. Then compute

Table 12.14

Coefficients of Three Orthogonal Comparisons

Comparison	Coefficients for Observer				Sum of Squares
	A	B	C	D	
1	3	−1	−1	−1	12
2	0	−1	−1	2	6
3	0	1	−1	0	2

the three comparison totals (see Table 12.15) within each of the six blocks in the same way as the numerator total for each Q^2. For example, find for the first comparison in plate 1 that

$$3(340) - 1(250) - 1(248) - 1(282) = 240$$

and for the third comparison in plate 6 that

$$1(73) - 1(64) = 9$$

Table 12.15
Comparison Totals for Plates in Table 12.12

Plate Number	Comparison		
	1	2	3
1	240	66	2
2	156	63	27
3	116	98	−10
4	108	6	10
5	84	6	22
6	136	61	9
Totals	840	300	60

Note that the totals in Table 12.15 are the same as the totals in the numerators of the Q^2's—this serves as a check.

Now, the experimental error sum of squares associated with the first comparison is given by

$$\frac{240^2 + 156^2 + \cdots + 136^2 - \dfrac{840^2}{6}}{12} = 1250.67$$

where the divisor 12 is the sum of squares of the coefficients of the comparison. In a similar way, we find the error sum of squares associated with the second and third comparisons to be 1120.33 and 449.00, respectively. The sum of these three components is equal to the error sum of squares, since

Table 12.16
Complete Analysis of Variance for the Hypotheses in (12.39)

Source of Variation	Sum of Squares	Degrees of Freedom	Mean Square	Expected Mean Square for Fixed Model (Equal Variances)
Plates	173,600	5	34,720	$\sigma^2 + 4\sum_{i=1}^{6} \beta_i^2/5$
Observers	12,600	3		
Comparison 1	9,800	1	9,800	$\sigma^2 + 36(3\alpha_1 - \alpha_2 - \alpha_3 - \alpha_4)^2/72$
Comparison 2	2,500	1	2,500	$\sigma^2 + 36(-\alpha_2 - \alpha_3 + 2\alpha_4)^2/36$
Comparison 3	300	1	300	$\sigma^2 + 36(\alpha_2 - \alpha_3)^2/12$
Error for	2,820	15	188	
Comparison 1	1,251	5	250	σ^2
Comparison 2	1,120	5	224	σ^2
Comparison 3	449	5	90	σ^2
Total	189,020	23		

the comparisons are orthogonal. The number of degrees of freedom associated with each error component is five, and the complete analysis of variance, along with the expected mean squares, is shown in Table 12.16.

It is clear from Table 12.16 that the F distribution with one and five degrees of freedom should be used in testing each of the hypotheses in (12.39). The five per cent and one per cent upper F values are $F_{.05} = 6.61$ and $F_{.01} = 16.3$. For the first comparison the computed F is $\frac{9800}{250} = 39.2$. Since $39.2 > 16.3$, comparison one is highly significant, and we conclude that observer A counts significantly more bacteria than the average of the other three observers. Since the computed F for the second and third comparisons are $\frac{2500}{224} = 11.2$ and $\frac{300}{90} = 3.3$, respectively, we say that comparison two is significant and that comparison three is not significant. Hence, we conclude that the new woman counts significantly more bacteria than the two boys on the average, but the two boys do not differ significantly in their counts. Note that had the error mean square of 188 with 15 degrees of freedom been used in each case, the conclusions would be very much the same, except comparison two would be declared highly significant rather than significant.

The tests in the last paragraph were made with Table 12.16 as a guide, even though the expected mean squares were computed on the assumption that the variances were equal. It turns out that the same tests are appropriate when the cr populations have variances which are not all equal. For, if we let the variance of the population in the ith column and jth row be σ_{ij}^2 $(i = 1, \ldots, c; j = 1, \ldots, r)$, it can be shown, under the assumption that the null hypothesis is true, that the expected values of the two mean squares used in the F test of any linear comparison of means are the same linear combinations of the variances σ_{ij}^2. For example, for comparison two it follows that

$$
\begin{aligned}
E(Q_2^2) &= E\left\{\frac{(-T_{2.} - T_{3.} + 2T_{4.})^2}{6[(-1)^2 + (-1)^2 + 2^2]}\right\} \\
&= \tfrac{1}{36} E\{[-(6\mu + 6\alpha_2 + \sum_j \beta_j + \sum_j \epsilon_{2j}) \\
&\qquad\quad - (6\mu + 6\alpha_3 + \sum_j \beta_j + \sum_j \epsilon_{3j}) \\
&\qquad\quad + 2(6\mu + 6\alpha_4 + \sum_j \beta_j + \sum_j \epsilon_{4j})]^2\} \\
&= \tfrac{1}{36} E\{[6(-\alpha_2 - \alpha_3 + 2\alpha_4) + (-\sum_j \epsilon_{2j} - \sum_j \epsilon_{3j} + 2\sum_j \epsilon_{4j})]^2\} \\
&= \tfrac{1}{36}[6^2(-\alpha_2 - \alpha_3 + 2\alpha_4)^2 + \sum_j \sigma_{2j}^2 + \sum_j \sigma_{3j}^2 + 4\sum_j \sigma_{4j}^2]
\end{aligned}
$$

and

E(error mean square for comparison 2)

$$= E \left\{ \frac{(-x_{21} - x_{31} + 2x_{41})^2 + \cdots + (-x_{26} - x_{36} + 2x_{46})^2 - \dfrac{(-T_{2.} - T_{3.} + 2T_{4.})^2}{6}}{5[(-1)^2 + (-1)^2 + 2^2]} \right\}$$

$$= \tfrac{1}{5} \cdot \tfrac{1}{36} \big[6E \{ [(-\alpha_2 - \alpha_3 + 2\alpha_4) + (-\epsilon_{21} - \epsilon_{31} + 2\epsilon_{41})]^2 + \cdots$$
$$+ [(-\alpha_2 - \alpha_3 + 2\alpha_4) + (-\epsilon_{26} - \epsilon_{36} + 2\epsilon_{46})]^2 \}$$
$$- E[(-T_{2.} - T_{3.} + 2T_{4.})^2]\big]$$

$$= \tfrac{1}{5} \cdot \tfrac{1}{36} \{ 6[(-\alpha_2 - \alpha_3 + 2\alpha_4)^2 + \sigma_{21}^2 + \sigma_{31}^2 + 4\sigma_{41}^2 + \cdots + (-\alpha_2$$
$$- \alpha_3 + 2\alpha_4)^2 + \sigma_{26}^2 + \sigma_{36}^2 + 4\sigma_{46}^2]$$
$$- [6^2(-\alpha_2 - \alpha_3 + 2\alpha_4)^2 + \sum_j (\sigma_{2j}^2 + \sigma_{3j}^2 + 4\sigma_{4j}^2)]\}$$

$$= \frac{\sum_j (\sigma_{2j}^2 + \sigma_{3j}^2 + 4\sigma_{4j}^2)}{36}$$

Thus, when the null hypothesis $-\mu_{1.} - \mu_{2.} + 2\mu_{3.} = 0$ is true, we have

$$E(Q_2^2) = E(\text{error mean square for comparison 2}) = \sigma_2^2$$

where

$$\sigma_2^2 \equiv \sum_j^6 \frac{\sigma_{2j}^2 + \sigma_{3j}^2 + 4\sigma_{4j}^2}{36}$$

In general, for any comparison

$$\gamma = m_1 \mu_{1.} + \cdots + m_c \mu_{c.}$$

the component of the treatment mean square is

$$Q^2 = \frac{\left(\sum_i^c m_i T_{i.} \right)^2}{r \sum_i^c m_i^2}$$

and the associated error mean square is

$$s_Q^2 = \frac{\dfrac{\sum_j^r \left(\sum_i^c m_i x_{ij} \right)^2}{\sum_i^c m_i^2} - \dfrac{\left(\sum_i^c m_i T_{i.} \right)^2}{r \sum_i^c m_i^2}}{r - 1} \tag{12.40}$$

It is easy to show, by the methods of Sect. 11.3, that

$$E \left(\sum_{i=1}^c m_i T_{i.} \right)^2 = r^2 \left(\sum_i m_i \alpha_i \right)^2 + \sum_i \sum_j m_i^2 \sigma_{ij}^2 \tag{12.41}$$

and

$$E\left[\sum_{j}^{r}\left(\sum_{i}^{c} m_i x_{ij}\right)^2\right] = r\left(\sum_{i} m_i \alpha_i\right)^2 + \sum_{i}\sum_{j} m_i^2 \sigma_{ij}^2 \tag{12.42}$$

Thus, it follows that

$$E(Q^2) = \frac{\sum_{i}\sum_{j} m_i^2 \sigma_{ij}^2}{r\sum_{i} m_i^2} + \frac{r\left(\sum_{i} m_i \alpha_i\right)^2}{\sum_{i} m_i^2} \tag{12.43}$$

and

$$E(s_Q^2) = \frac{\sum_{i}\sum_{j} m_i^2 \sigma_{ij}^2}{r\sum_{i} m_i^2} \tag{12.44}$$

If $\gamma = 0$, then

$$\sum_{i} m_i \alpha_i = 0$$

and Q^2/s_Q^2 has the F distribution with 1 and $r-1$ degrees of freedom, provided the ϵ_{ij} in the fixed model equation (12.14) are normally and independently distributed.

In practice, the method of partitioning the error sum of squares should not be applied unless the population variances σ_{ij}^2 can be assumed to be quite different. A small degree of heterogeneity among the variances σ_{ij}^2 does not usually noticeably disturb a test in which the regular error mean square is used. Thus, when one is deciding whether to use the regular error mean square in preference to a component of the error mean square, it should be realized that a sizeable loss in degrees of freedom in the error term is probably worse than small heterogeneity among population variances.

12.6. MISSING DATA IN A TWO-WAY CLASSIFICATION DESIGN, SINGLE OBSERVATION

Sometimes the original design of an experiment is destroyed because one or more of the observations is missing or unreliable, due to accident. Fortunately, it is possible to describe methods which give adequate (either fairly short approximations or longer exact) analyses of experiments with missing data. In the procedure we describe, the available observations are used to find estimates (dummy variables) of the missing observations so that the regular analysis can be performed on cr values in the resulting augmented table. Allen and Wishart [1] first presented the method and Yates [40, 41] and others [4, 5, 18] developed it. Now, using the data of Example 12.1, we explain the method for the case where only one value is missing.

Example 12.3. Illustrate the procedure for missing values if it is supposed that the observation in the first column and third row is missing from Table 12.1. Assume that the data are now for a fixed model.

Denoting the missing observation by y, we compute the totals and means shown in Table 12.17. It is reasonable to suppose that y should be a value with error component e_{13} equal to zero. This selection of y is an unbiased estimate of the cell mean, and, furthermore, the error sum of squares of the available data is not affected by the addition of this dummy value of y.

Table 12.17

Data of Table 12.1 with y Substituted for Missing Value

Core Number	Depth of Soil in Inches 1	7	13	Totals	Means
1	6.7	6.5	6.3	19.5	6.5
2	6.4	6.6	6.2	19.2	6.4
3	y	6.8	6.4	$13.2 + y$	$(13.2 + y)/3$
4	6.4	6.5	6.3	19.2	6.4
Totals	$19.5 + y$	26.4	25.2	$71.1 + y$	
Means	$(19.5 + y)/4$	6.6	6.3		$(71.1 + y)/12$

Thus, we write

$$y = \bar{x} + a_1 + b_3$$

or

$$y = \bar{x} + (\bar{x}_{1.} - \bar{x}) + (\bar{x}_{.3} - \bar{x}) = \bar{x}_{1.} + \bar{x}_{.3} - \bar{x} \qquad (12.45)$$

Now, substituting the values of $\bar{x}_{1.}$, $\bar{x}_{.3}$ and $\bar{x}$ from Table 12.17 in Eq. (12.45) gives

$$y = \frac{19.5 + y}{4} + \frac{13.2 + y}{3} - \frac{71.1 + y}{12}$$

or

$$y = 6.7$$

This value of y may now be substituted in Table 12.17 and the estimates of means and effects computed in the usual way. Also, this augmented table may be used to compute, in the customary manner, the components SSC, SSR, SST, and SSE in the sum of squares identity.

In order for the error mean square to be an unbiased estimate of σ^2 we divide the error sum of squares by $(3 - 1)(4 - 1) - 1 = 5$, since only five of the e_{ij}'s are now independent, because of the additional independent restriction that $e_{13} = 0$. Now to test the null hypothesis $\alpha_1 = \alpha_2 = \alpha_3 = 0$ against the alternative hypothesis

$$\sum_{i=1}^{3} \alpha_i^2 \neq 0$$

we could use the augmented data and apply the regular analysis of variance with the degrees of freedom associated with the error term reduced by one. The results of such an analysis are shown in Table 12.18, and the null hypothesis is rejected at the five per cent level. But the test indicated is an approximate test, since the expectation of the treatment mean square as calculated from the augmented data is greater than σ^2 when the null hypothesis is true. Thus, the test tends to reject a true null hypothesis too often. (In practice, if this test fails to reject the null hypothesis, there is no need to apply an exact test.) If the number of degrees of freedom for the error mean square is large, the approximate test just described might be considered satisfactory in most cases. But for a small number of degrees of freedom the test described next should be applied.

Table 12.18
Approximate Test for H_0: $\alpha_1 = \alpha_2 = \alpha_3 = 0$ when One Observation is Missing

Source of Variation	Sum of Squares	Degrees of Freedom	Mean Square	F_c	$F_{.05}$
Blocks (cores)	0.110	3			
Columns (depths)	0.207	2	0.103	8.6	5.79
Error	0.060	5	0.012		
Total	0.377	10			

As mentioned in the last paragraph, the treatment mean square computed in the usual way from the augmented data is biased upwards. In order to correct for this bias, we compute a new treatment sum of squares SSC' by the formula

$$SSC' = SSC - \text{correction for bias}$$

where

$$\text{correction for bias} = \frac{(x_{23} + x_{33} - 2y)^2}{3 \cdot 2}$$

In our example the correction for bias is

$$\frac{[13.2 - 2(6.7)]^2}{6} = 0.007$$

The ratio

$$\frac{\dfrac{SSC'}{3-1}}{\dfrac{SSE}{(3-1)(4-1)-1}}$$

is distributed as F with two and five degrees of freedom. The appropriate test of the null hypothesis $\alpha_1 = \alpha_2 = \alpha_3 = 0$, given in Table 12.19, shows that the treatment effects are different from zero at the five per cent level.

Table 12.19

Test for H_0: $\alpha_1 = \alpha_2 = \alpha_3 = 0$ when One Observation is Missing

Source of Variation	Sum of Squares	Degrees of Freedom	Mean Square	F_c	$F_{.05}$
Blocks (cores)	0.110	3			
Columns (depths)	0.200	2	0.100	8.3	5.79
Error	0.060	5	0.012		
Total	0.377	10			

Suppose, in the general case with c columns and r rows, that a single observation is missing from the lth column and mth row. Denote the dummy value by y_{lm}. Then it can be shown that the proper missing value x_{lm} can be estimated by

$$y_{lm} = \frac{cT'_{l.} + rT'_{.m} - T'}{(c-1)(r-1)} \tag{12.46}$$

where

$$\begin{cases} T'_{l.} \text{ denotes the sum of the recorded observations in column } l \\ T'_{.m} \text{ denotes the sum of the recorded observations in row } m \\ T' \text{ denotes the sum of all recorded observations} \end{cases} \tag{12.47}$$

Now, if y_{lm} is entered in the missing cell, the augmented data can be analyzed in the usual manner except for the two differences discussed above. That is, find the new error mean square s^2, using the divisor $(c-1)(r-1) - 1$, and subtract from the treatment sum of squares SSC the following correction for bias

$$\text{correction for bias} \equiv B = \frac{[T'_{.m} - (c-1)y_{lm}]^2}{c(c-1)} \tag{12.48}$$

Then the F ratio

$$\frac{\dfrac{SSC - B}{c - 1}}{\dfrac{SSE}{(c-1)(r-1) - 1}} \tag{12.49}$$

with $c - 1$ and $(c-1)(r-1) - 1$ degrees of freedom, is used to test the null hypothesis $\alpha_1 = \cdots = \alpha_c = 0$ at the desired level of significance. The standard error of the difference in the mean of the treatment with a missing value $\bar{x}_{l.}$ and the mean of any other treatment $\bar{x}_{i.}$ $(i \neq l)$ is

$$\sqrt{s^2 \left[\frac{2}{r} + \frac{c}{r(c-1)(r-1)} \right]} \tag{12.50}$$

We now consider the cases where more than one observation is missing. If one or more complete rows (columns) are missing, we carry out the usual analysis on the reduced design, provided that at least two whole rows (columns) still remain. If the missing observations are in different rows or different columns or both, we follow the same general procedure given above. Appropriate formulas for filling in the missing cells are given by Yates [40, 42] and Glenn [23]. But rather than give these formulas we describe an iterative method which gives the same values.

To illustrate, suppose three values are missing. First, place "guessed" values in two vacant cells—the mean of all recorded observations may be used. Then use Formula (12.46) with the resulting table to find an estimated value for the third missing cell. Now place this value in the third cell, remove one of the "guessed" values from a cell designated as the "first missing cell," and use Formula (12.46) to find an estimated value for the first missing cell. Now place this value in the first cell, remove the "guessed" value from the "second missing cell," and use Formula (12.46) to find an estimated value for the second missing cell. Next, place this value in the second cell, remove the estimated value from the third missing cell, and use Formula (12.46) to find a second estimated value for the third missing cell. Repeat this cycle until there is very little or no change in successive estimates in each cell. The resulting values are the ones that would have been obtained from cumbersome formulas.

In order to test the null hypothesis $\alpha_1 = \cdots = \alpha_c = 0$ with the usual F test, we compute the component sum of squares, using the table of observations augmented with k missing values, say. Next, find the error mean square, using a divisor of $(c-1)(r-1) - k$, and reduce the treatment sum of squares SSC by a factor B to correct for bias. Then the F ratio

$$\frac{\dfrac{SSC - B}{c - 1}}{\dfrac{SSE}{(c-1)(r-1) - k}} \tag{12.51}$$

with $c - 1$ and $(c-1)(r-1) - k$ degrees of freedom, should be used to avoid a biased test procedure.

In most cases where the number of degrees of freedom of the error term is large, the approximate test, in which the correction for bias is ignored, is satisfactory. However, if a correction seems desirable and more than one observation is missing, the reader is referred to Yates [40, 42] for the appropriate correction formula. Some special formulas occur in the exercises of Sect. 12.8.

12.7. RANDOM EFFECTS AND MIXED EFFECTS MODELS IN TWO-WAY CLASSIFICATION DESIGNS WITH A SINGLE OBSERVATION PER CELL

In Sect. 12.2 we gave an example of a mixed model in which rows (blocks) were randomly selected and columns (treatments) were fixed. In Sect. 12.3 we developed the fixed model (both rows and columns fixed) in some detail. The theory and methods presented in Chaps. 10, 11, and 12 (Sects. 12.1 through 12.6) are easily extended to the *random effects model* and two types of *mixed effects models* discussed in this section. Assuming that the reader will have little difficulty in making the necessary extensions, we do little more than present the analysis of variance models and analysis of variance tables including the expected mean squares.

The *model equation*, in each case, is the same as that defined in Eq. (12.14); that is

$$x_{ij} = \mu + \alpha_i + \beta_j + \epsilon_{ij} \qquad (i = 1, \ldots, c; \, j = 1, \ldots, r)$$

and the component parts are given the same designations (names). In each analysis of variance model μ is fixed and $\epsilon_{11}, \ldots, \epsilon_{cr}$ are normally and independently distributed with zero means and common variances. The effects α_i and β_j may be either fixed or random, depending on the analysis of variance model. If the $\alpha_i(\beta_j)$ are fixed, then they are measured as deviates from μ such that

$$\sum_{i=1}^{c} \alpha_i = 0 \qquad \left(\sum_{j=1}^{r} \beta_j = 0 \right)$$

If the α_i (β_j) are random, they are assumed to be from a normal population with zero mean and variance σ_α^2 (σ_β^2). Further, it is assumed that all random effects in any model are independently distributed. Both the α_i and β_j effects are fixed in the fixed effects model, which is denoted by *fixed model*. In the mixed effects model either α_i is fixed and β_j is random, or β_j is fixed and α_i is random. They are denoted by *mixed model* (α) and *mixed model* (β), respectively. The random effects model in which both the α_i and β_j are random is denoted by *random model*.

As with the fixed effects model in the two-way classification design, the model equation, along with the assumptions which follow, may be used to determine the expectations of the three component sum of squares on the right-hand side of Eq. (12.24). These expectations, together with the degrees of freedom identity and theorems in Chaps. 7 and 9, may be applied to determine the nature of the distributions of the component sum of squares as well as the distributions of their ratios. The expected mean squares for three of the four experimental models in a two-way classification design are given in Table 12.7 (one, for fixed model) and Table 12.20 [random model and mixed model (α)]. Since the form for mixed model (β) is like that for mixed model (α), we give only the latter in Table 12.20.

Table 12.20
Analysis of Variance and Expected Mean Squares for the Two-way Classification with One Observation per Cell

Source of Variation	Sum of Squares	Degrees of Freedom	Mean Square	Expected Mean Squares for	
				Mixed Model (α)	Random Model
Columns (treatments)	SSC	$c-1$	s_1^2	$\sigma^2 + r\left(\dfrac{\sum_i \alpha_i^2}{c-1}\right)$	$\sigma^2 + r\sigma_\alpha^2$
Rows (blocks)	SSR	$r-1$	s_2^2	$\sigma^2 + c\sigma_\beta^2$	$\sigma^2 + c\sigma_\beta^2$
Error (residual)	SSE	$(c-1)(r-1)$	s_3^2	σ^2	σ^2
Total	SST	$cr-1$			

No matter which experimental model is used, the fixed effects are estimated as in the fixed model, and the components of variance are estimated as in Example 12.1, Tables 12.7 and 12.20 being used as guides. The procedure for testing the hypothesis

$$H_0: \alpha_1 = \cdots = \alpha_c = 0$$

when the α_i are fixed, the hypothesis

$$H_0: \beta_1 = \cdots = \beta_r = 0$$

when the β_j are fixed, the hypothesis

$$H_0: \sigma_\alpha^2 = 0$$

when the α_i are random, or the hypothesis

$$H_0: \sigma_\beta^2 = 0$$

when the β_j are random, may be explained in terms of the expected mean squares shown in Tables 12.7 and 12.20 The power of any of these tests is obtained in the usual manner. The methods for testing or establishing confidence intervals for a linear comparison of $\alpha_1, \ldots, \alpha_c$ $(\beta_1, \ldots, \beta_r)$ when the effects are fixed are like those already described. When the effects α_i (β_j) are random, approximate confidence intervals for σ_α^2 (σ_β^2) may be obtained in the usual fashion.

12.8. EXERCISES

12.1. (a) Complete Table 12.21 for the analysis of variance and expected

Table 12.21

Source of Variation	Sum of Squares	Degrees of Freedom	Mean Square	Expected Mean Square
Blocks	26.8	4		
Treatments		3		
Error			2.5	
Total	85.3			

mean squares of a fixed effects randomized block design. (b) Test the hypothesis that the treatment effects are equal to zero, showing all steps in the general test procedure. (c) Identify the treatments and blocks with variables in your own field of study; write a short statement of an experiment which could lead to the above analysis of variance, and write your conclusion in terms of the variables introduced.

12.2. For an experiment with the randomized block design given in Fig. 12.1, suppose that the temperature mean square is 0.1076, the mesh mean square is 0.0793, and the total mean square is 0.0490. (a) Test the hypothesis that the temperature effects are all zero, showing all steps in the general test procedure. (b) Find a 95 per cent confidence interval for the common standard deviation σ. What assumptions are required for doing so?

12.3. For a fixed effects experiment in a randomized block design, the observations in Table 12.22 were made. (a) Find unbiased estimates of the

Table 12.22

Blocks	Treatments			
	1	2	3	4
1	27	24	18	23
2	19	20	17	16
3	20	22	16	18

block and treatment effects. What assumptions are required? (b) Prepare an analysis of variance table for this experiment, showing the expected mean squares. (c) Test the hypothesis that the treatment effects are equal, showing all steps in the general test procedure. (d) Use Duncan's multiple range test to make a pairwise ranking of the four treatment means. (e) Find a 95 per cent confidence interval for the difference in the means of blocks 1 and 2. (f) Use five per cent level tests on each of the null hypotheses

$$H_{01}: \mu_1. + \mu_2. - \mu_3. - \mu_4. = 0 \quad \text{and} \quad H_{02}: \mu_3. - \mu_4. = 0$$

(g) Assuming the analysis of variance table found in (b) to be for a random effects experiment, find point and 95 per cent confidence interval estimates of the variance components for block means and for treatment means. (h) Use the method of Sect. 12.5 to compute the error variance for each of the comparisons in (f). Test each of the hypotheses in (f), using the corresponding error variance with two degrees of freedom. Comment on these tests.

12.4. Table 12.23 for a randomized block design shows the length of life in years (measurements are hypothetical) of an outside paint applied under the conditions indicated (treatments fixed, mixes random). (a)

Table 12.23

Mixes	Treatments			
	1 *Hardwood, dry climate*	2 *Hardwood, damp climate*	3 *Softwood, dry climate*	4 *Softwood, wet climate*
1	4.2	3.6	4.9	3.7
2	4.6	3.6	4.9	3.7
3	4.5	3.9	4.5	3.5
4	3.7	2.7	4.1	3.5
5	3.5	3.2	4.1	3.6

Find unbiased estimates of the treatment effects. (b) Prepare the usual analysis of variance table, showing the expected mean squares. (c) Find a 99 per cent confidence interval for the mix-to-mix standard deviation. Use this to make a statement about variation in mixes. (d) Use 5 per cent level tests on each of the null hypotheses $H_{01}: \mu_{1.} + \mu_{2.} - \mu_{3.} - \mu_{4.} = 0, H_{02}: \mu_{1.} - \mu_{2.} \mid \mu_{3.} - \mu_{4.} = 0$, and $H_{03}: \mu_{1.} - \mu_{2.} - \mu_{3.} + \mu_{4.} = 0$. In each case, state your conclusion in terms of type of wood or type of climate or both. Are the three tests independent? Why? (e) Find a 99 per cent confidence interval for $\mu_{1.} - \mu_{2.} + \mu_{3.} - \mu_{4.}$; for $\mu_{1.} - \mu_{2.}$; for $\mu_{3.}$. Indicate how each of these intervals might be used. (f) Use the method of Sect. 12.5 to compute the error variance for each of the comparisons in (d). Test each of the hypotheses in (d), using the corresponding error variance with four degrees of freedom. Comment on these tests. (g) Make a table of unbiased cell means. Find a 90 per cent confidence interval for the mean for mix 2 and treatment 3; for mix 4 and treatment 2; for the difference in the means in these two cells. Comment on the use of each of these intervals.

12.5. Derive the expected mean squares in Table 12.3.

12.6. For the two-way classification, state and prove a sum of squares identity similar to (10.32). That is, find a sum of squares identity involving μ, α_i, and β_j and prove that it holds.

12.7. In Eq. (12.23) prove that (a)

$$\sum_i \sum_j (\bar{x}_{i.} - \bar{x})(\bar{x}_{.j} - \bar{x}) = 0$$

and (b)

$$\sum_i \sum_j (\bar{x}_{.j} - \bar{x})(x_{ij} - \bar{x}_{i.} - \bar{x}_{.j} + \bar{x}) = 0$$

12.8. Prove Eq. (12.27).

12.9. Prove Eq. (12.28).

12.10. (a) Table 12.24 shows the true means in the cells (populations) of a two-way classification. Find the true over-all mean, treatment effects,

Table 12.24

Blocks	Treatment				
	1	2	3	4	5
1	42	44	44	45	47
2	43	45	45	46	48
3	44	46	46	47	49
4	47	49	49	50	52˙

and block effects. What would the treatment and block effects be if the true over-all mean were 40? (b) Table 12.25 gives a single random observation made in each of twenty normal populations with means shown in (a) and common standard deviation $\sigma = 1$. Find unbiased

Table 12.25

Blocks	Treatments				
	1	2	3	4	5
1	40.7	45.3	44.6	45.7	47.1
2	41.6	44.4	45.1	47.4	49.2
3	44.8	45.1	45.2	47.0	51.6
4	46.6	50.0	50.0	50.2	52.4

estimates of all treatment means and effects. Compare the estimates with the true values. (c) Use (b) to find an unbiased estimate of σ^2 with 12 degrees of freedom. Use (a) and (b) to find an unbiased estimate of σ^2 with 20 degrees of freedom. Compare these two estimates. (d) Use (b) to prepare an analysis of variance table. (e) Test at the five per cent level the hypothesis that the treatment means are all equal. Is either a type 1 or a type 2 error made? (f) Use a five per cent level test to determine whether the values in Table 12.25 could have been drawn from the five treatment populations in (a). (g) Use (b) to find 90 per cent confidence intervals for each of the treatment means and each of the block means. Check to see how many of these nine intervals contain the true means. (h) Find simultaneous 90 per cent confidence intervals for linear combinations of the treatment means. Find the limits of 20 specific linear combinations, and use Table 12.24 to determine how many of the corresponding true linear combinations fall between these limits.

12.11. In Exercise 12.3(h), find the expected mean squares for each of the comparisons when (a) the 12 population variances are unequal, (b) four variances within a block are equal, but the three block variances are unequal, and (c) all variances are equal. (d) Find the expected mean squares for the error variances in (a).

12.12. In Exercise 12.4(d), find the expected mean squares for each comparison when (a) the 20 population variances are unequal, (b) the five variances for a treatment are equal, but the four treatment variances are unequal, and (c) all variances are equal.

12.13. Prove Eqs. (12.41), (12.42), (12.43), and (12.44).

12.14. Suppose the observation in the first block and for the first treatment is missing from the table in Exercise 12.3. Test the hypothesis $H_0: \alpha_1 = \alpha_2 = \alpha_3 = \alpha_4 = 0$ by (a) the approximate test procedure illustrated in Table 12.18, (b) the test procedure illustrated in Table 12.19. (c) Find the 95 per cent confidence interval for $\mu_{1.} - \mu_{3.}$. (d) Test the hypothesis $H_0: \alpha_1 = \alpha_2 = \alpha_3 = \alpha_4 = 0$ in the case where the observations are missing in row 1 and column 3 and in row 2 and column 1. (e) Test the hypothesis of (d) if all observations are missing in block 3. (f) Test the hypothesis $H_0: \alpha_1 = \alpha_2 = \alpha_4 = 0$ if all observations are missing from column 3 along with the observation in row 1 and column 1.

12.15. Prove Eq. (12.46).

12.16. Derive a formula similar to Eq. (12.46) for the case of two missing observations.

Hint. Consider two parts: one, when both missing observations are in the same row (or column); the other, when the observations are from different rows and different columns.

12.9. *RELATIVE EFFICIENCY OF DESIGNS*

On at least three occasions the reader has had an opportunity to ask which of two designs should be selected. In Chap. 8 the problem arises when a decision must be made on whether to compare two treatment means by pairing observations or by obtaining two independent sets of observations. In Sect. 11.4 we again found it necessary to decide on which of two nested designs to choose. Now, in this chapter we might well ask for some rule for deciding whether to use the randomized block design or the completely randomized design. To introduce a general method of comparing designs, let us first consider an example.

Example 12.4. (a) Assuming no appreciable variation from core to core in Example 12.1 test to determine if the acidity level is the same at different depths of soil. (b) Compare the experiments in Examples 12.1(a) and 12.4(a).

First, note that when there is no core-to-core variation we analyze the data in Table 12.1 as a one-way classification. This means that the sum of squares for cores (blocks) is not partitioned out of the total sum of squares. Now, since the treatment totals T_i. and grand total T for the two designs are the same, it follows that the within sum of squares for the completely randomized design is equal to the sum of the residual and block sums of squares for the randomized block design. That is

$$\text{Within } SS = \text{Residual } SS + \text{Block } SS \qquad (12.52)$$

Substituting the sum of squares from Table 12.3 in Eq. (12.52) leads to the analysis of variance shown in Table 12.26.

Table 12.26
Analysis of Variance for Soil Acidity in a One-way Classification

Source of Variation	Sum of Squares	Degrees of Freedom	Mean Square	F_c	$F_{.05}$	Expected Mean Squares
Treatments	0.24	2	0.12	4.15	4.26	$\sigma^2 + 2 \sum \alpha_i^2$
Within	0.26	9	0.03			σ^2
Totals	0.50	11				

Since the computed F is less than the tabled F, we fail to reject the null hypothesis and conclude that we do not have enough evidence to say that the acidity level differs with soil depths.

On comparing the conclusions resulting from the two designs, observe that the equality of acidity levels is rejected for the randomized block design and not rejected for the completely randomized design. Since the block mean square is 4.5 times as large as the error mean square, we expect quite different results when the two are pooled to obtain a new error (within) mean square. In this particular experiment we do not need a rule to know that the randomized block design is better. But if the two experimental error variances had nearly the same values, a decision would not be so obvious. Thus, we now give a general procedure for selecting the better of two designs.

In comparing two treatment means we need to know the error variances associated with them. Thus, if $\bar{x}_1$ and $\bar{x}_2$ are means of samples of sizes n_1 and n_2, respectively, with corresponding error variances

$$\sigma_{\bar{x}_1}^2 = \frac{\sigma_1^2}{n_1} \quad \text{and} \quad \sigma_{\bar{x}_2}^2 = \frac{\sigma_2^2}{n_2}$$

we say that mean is better which has a smaller error variance, that is, which has better precision. This can be expressed in terms of the ratio $\sigma^2_{\bar{x}_2}/\sigma^2_{\bar{x}_1}$ which is called the *relative efficiency* of mean $\bar{x}_1$ with respect to mean $\bar{x}_2$. Denoting the relative efficiency by R. E. ($\bar{x}_1$ relative to $\bar{x}_2$), we can write

$$\text{R.E.}(\bar{x}_1 \text{ relative to } \bar{x}_2) = \frac{n_1}{n_2} \frac{\sigma^2_2}{\sigma^2_1} \qquad (12.53)$$

or, if the means are based on equal sample sizes

$$\text{R.E.}(\bar{x}_1 \text{ relative to } \bar{x}_2) = \frac{\sigma^2_2}{\sigma^2_1} \qquad (12.54)$$

If the ratio is larger than one, we say that $\bar{x}_1$ is more efficient (or precise) than $\bar{x}_2$; if the ratio is smaller than one, we say that $\bar{x}_1$ is less efficient than $\bar{x}_2$.

We wish to define the relative efficiency of *two designs* which contain the same treatments. Even though this can be done in many ways, it is customary to use the approach indicated above. But, since the true error variances of designs are seldom known, it is desirable to define an estimate of relative efficiency in terms of the experimental error mean squares. If s^2_1 and s^2_2 are unbiased estimators of σ^2_1 and σ^2_2, we could define the estimate of efficiency of design 1 relative to design 2 by

$$\text{est. R.E.}(\text{design 1 relative to design 2}) = \frac{n_1 s^2_2}{n_2 s^2_1} \qquad (12.55)$$

Generally, it is understood that s^2_2 is the estimated value of the error variance of the second design as computed from the sums of squares and degrees of freedom in the first design.

The estimate given by Eq. (12.55) is useful unless the number of degrees of freedom associated with s^2_1 or s^2_2 is small. For this case we use the definition (12.56) given by Fisher [20] and described by Cochran and Cox [10]. Fisher calculated the "amount of information" which the difference between two estimated treatment means gives about the difference between the true treatment means. The amount of information is $(\nu + 1)/(\nu + 3)s^2$, where s^2 is the experimental error variance with ν degrees of freedom.

Note. If the variance σ^2 were known, the amount of information would be $1/\sigma^2$.

Thus, when the number of observations in the treatment of two designs are the same, the estimate of the efficiency of design 1 relative to design 2 is given by

$$\text{est. R.E.} = \text{est. R.E. (design 1 relative to design 2)} = \frac{\dfrac{\nu_1 + 1}{(\nu_1 + 3)s^2_1}}{\dfrac{\nu_2 + 1}{(\nu_2 + 3)s^2_2}}$$

or

$$\text{est. R. E.} = \frac{(v_1 + 1)(v_2 + 3)s_2^2}{(v_2 + 1)(v_1 + 3)s_1^2} \tag{12.56}$$

where s_1^2 and s_2^2 denote the experimental error mean squares of designs 1 and 2, respectively, and v_1 and v_2 denote their corresponding degrees of freedom. So when treatment means are based on samples of different sizes, the estimate of the efficiency of design 1 relative to design 2 is defined by

$$\text{est. R. E.} = \frac{n_1}{n_2} \cdot \frac{(v_1 + 1)(v_2 + 3)s_2^2}{(v_2 + 1)(v_1 + 3)s_1^2} \tag{12.57}$$

Note. We might also refer to Eqs. (12.56) and (12.57) as giving *relative information.*

Now we find the estimate of the efficiency of the randomized block design, *RBD*, relative to the completely randomized design, *CRD*. This requires that we obtain an *estimator* of the error variance, s_2^2, for the *CRD*, using the mean squares and degrees of freedom in the *RBD*. If s_1^2 is the error variance of the *RBD*, then it can be shown (see Ref. [10] for a proof) that

$$s_2^2 = \frac{(r - 1)s_b^2 + r(c - 1)s_1^2}{cr - 1} \tag{12.58}$$

where s_b^2 is the block mean square and c and r denote the number of columns and rows in the design.

For Example 12.1 we find that

$$s_2^2 = \frac{3(0.06) + 8\left(\frac{0.04}{3}\right)}{11} = \frac{0.86}{33}$$

Since the sample sizes for treatments are equal ($n_1 = n_2 = 4$) and the degrees of freedom for error variance small ($v_1 = 6$ and $v_2 = 9$), we use Eq. (12.56) to find

$$\text{est. R. E. } (\textit{RBD} \text{ relative to } \textit{CRD}) = \frac{(7)(12)\left(\frac{0.86}{33}\right)}{(10)(9)\left(\frac{0.04}{3}\right)} = \frac{602}{330} \doteq 1.82$$

Thus, the efficiency of the *RBD* relative to the *CRD* is 1.82 or 182 per cent. We may also say that the relative gain in efficiency is 82 per cent, or 0.82.

Note. If the error variance from Table 12.26 had been used, then the est. R. E. would be

$$\frac{(7)(12)\left(\frac{0.26}{9}\right)}{(10)(9)\left(\frac{0.04}{3}\right)} = \frac{91}{45} \doteq 2.02$$

and this overestimates the efficiency of the randomized block design in this example.

As indicated in Sect. 11.4, the concept of relative efficiency may also be used to compare a given design to some other design of the *same type* having different sample and subsample sizes. We use Eq. (12.55) and Example 11.2 to show how this may be done. We would like to know if the nested design in Example 11.2 having ten treatments with samples of size $n = 3$ and subsamples of size $r = 2$ is more or less efficient than the nested design having ten treatments with $n = 2$ and $r = 3$. The experimental error variance $s_1^2 = 17.55$ of the design in Example 11.2 is equal to $s_e^2 + rs_d^2 = 0.68 + 2(8.44)$. Assuming that the estimates of σ^2 and σ_δ^2 would remain unchanged even though r and n change, we find the estimate of the experimental error variance s_2^2 of the new design to be

$$s_2^2 = 0.68 + 3(8.44) = 26.00$$

The number of replications for each treatment in the old design is $n_1 = rn = 2 \cdot 3 = 6$ and in the new design is $n_2 = 3 \cdot 2 = 6$. Thus, the estimated efficiency of the old design relative to the new design is

$$\frac{6(26.00)}{6(17.55)} = 1.48, \text{ or } 148 \text{ per cent}$$

The reader should verify (Exercise 12.21) that the paired-observation experiment in Example 8.6 can be analyzed as a randomized block design with 16 blocks and two treatments. Then a special formula should be found which gives an estimate of the efficiency of a paired-observation experiment relative to an experiment with two independent sets of observations.

12.10. SUBSAMPLING IN A RANDOMIZED BLOCK DESIGN

Sometimes it is desirable to subsample in a randomized block design just as in a completely randomized design. For example, in the acidity experiment (Example 12.1) let the depths of the soil be *top* (depth of zero to two in.), *middle* (depth of six to eight in.), and *bottom* (depth of 11 to 13 in.), instead of one, seven, and 13 inches. Remove the top two inches of soil in a core sample, mix the soil thoroughly, and select two samples at random to analyze for *pH* level. In the same way select two samples at random from each of the other 11 treatment and block combinations.

Since the amount of information per observation is greatest, and since the calculations are less involved for an equal number of observations per cell, we give the analysis for this case only. The general case may be found in other references [2, 26, 36]. We write the *model equation* for subsampling in a randomized block design as

$$x_{iju} = \mu + \alpha_i + \beta_j + \delta_{ij} + \epsilon_{iju}$$
$$i = 1, \ldots, c \quad j = 1, \ldots, r \quad u = 1, \ldots, n \tag{12.59}$$

where x_{iju} denotes the uth random observation in the jth block for the ith treatment, μ denotes a constant over-all mean, α_i denotes the effect (constant or random) of the ith treatment, β_j denotes the effect (constant or random) of the jth block, δ_{ij} denotes the effect (constant or random) of the jth block on the ith treatment, and ϵ_{iju} denotes the uth random effect in the jth block for the ith treatment. We call δ_{ij} an experimental effect and ϵ_{iju} a sampling effect.

Clearly, there are eight experimental models which have the model equation in (12.59). At this time, we discuss only the four with random experimental effects. (Actually, we list only three, since the two mixed models are "symmetric" in α and β.) The case where δ_{ij} is fixed is discussed in Chapter 13 on *factorial experiments*.

Since the assumptions and definitions associated with the effects μ, α_i, β_j are the same as those in Sect. 12.7, we do not restate them here. The δ_{ij} takes the place of ϵ_{ij} in the earlier sections of this chapter, and ϵ_{iju} is a new random component which is assumed to be normally distributed with mean zero and mean σ^2. Any effects which are random are assumed to be independently distributed.

The cell, column, row, and grand totals are given by

$$\begin{cases} T_{ij.} = \sum_u x_{iju}, \qquad T_{i..} = \sum_j \sum_u x_{iju} \\ T_{.j.} = \sum_i \sum_u x_{iju} \quad \text{and} \quad T = \sum_i \sum_j \sum_u x_{iju} \end{cases} \tag{12.60}$$

respectively. The cell, column, row, and grand means are given by

$$\begin{cases} \bar{x}_{ij.} = \dfrac{T_{ij.}}{n}, \qquad \bar{x}_{i..} = \dfrac{T_{i..}}{rn} \\ \bar{x}_{.j.} = \dfrac{T_{.j.}}{cn}, \qquad \bar{x} = \dfrac{T}{crn} \end{cases} \tag{12.61}$$

respectively. If the α_i are fixed, they are estimated by

$$a_i = \bar{x}_{i..} - \bar{x} \qquad (i = 1, \ldots, c)$$

If the β_j are fixed, they are estimated by

$$b_j = \bar{x}_{.j.} - \bar{x} \qquad (j = 1, \ldots, r)$$

For any of the four experimental models under consideration, the sum of squares identity is given by

$$SST = SSC + SSR + SSE + SSS \tag{12.62}$$

where

$$SST = \sum_i \sum_j \sum_u (x_{iju} - \bar{x})^2 = \sum_i \sum_j \sum_u x_{iju}^2 - \frac{T^2}{crn} \qquad (12.63)$$

$$SSS = \sum_i \sum_j \sum_u (x_{iju} - \bar{x}_{ij.})^2 = \sum_i \sum_j \sum_u x_{iju}^2 - \frac{\sum_i \sum_j T_{ij.}^2}{n} \qquad (12.64)$$

$$SSC = rn \sum_i (\bar{x}_{i..} - \bar{x})^2 = \frac{\sum_i T_{i..}^2}{rn} - \frac{T^2}{crn} \qquad (12.65)$$

$$SSR = cn \sum_j (\bar{x}_{.j.} - \bar{x})^2 = \frac{\sum_j T_{.j.}^2}{cn} - \frac{T^2}{crn} \qquad (12.66)$$

$$SSE = n \sum_i \sum_j (\bar{x}_{ij.} - \bar{x}_{i..} - \bar{x}_{.j.} + \bar{x})^2$$

$$= SS(ST) - SSC - SSR \qquad (12.67)$$

$$SS(ST) = \frac{\sum_i \sum_j T_{ij.}^2}{n} - \frac{T^2}{crn} \qquad (12.68)$$

Note that Eq. (12.64) may also be written as

$$SSS = SST - SS(ST) \qquad (12.64a)$$

We use SST, SSC, SSR, SSE, SSS, and $SS(ST)$ to denote total sum of squares, column sum of squares, row sum of squares, experimental error sum of squares, sample error sum of squares, and sum of squares for subtotals (cell totals), respectively.

The analysis of variance is given in Table 12.27, and expected mean squares for three experimental models are shown in Table 12.28. The justifications for the entries in these tables are similar to those already given for other designs. As in other cases already discussed, the expected mean squares

Table 12.27
Analysis of Variance for Sampling in a Randomized Block Design

Source of Variation	Sum of Squares	Degrees of Freedom	Mean Square
Subtotals	$SS(ST)$		
Blocks	SSR	$r - 1$	$s_1^2 = SSR/(r - 1)$
Treatments	SSC	$c - 1$	$s_2^2 = SSC/(c - 1)$
Experimental error	$SS(ST) - SSR - SSC$	$(c - 1)(r - 1)$	$s_3^2 = SSE/(c - 1)(r - 1)$
Sampling error	$SST - SS(ST)$	$cr(n - 1)$	$s_4^2 = SSS/cr(n - 1)$
Total	SST	$crn - 1$	

can be used as guides in estimating variance components, testing hypotheses, in establishing confidence limits, and in finding power for a test.

Table 12.28
Expected Mean Square for Sampling in a Randomized Block Design

Source of Variation	Degrees of Freedom	Mean Square	Random Model	Fixed Model (α, β)	Mixed Model (α)
			Expected Mean Squares for		
Blocks (rows)	$r - 1$	s_1^2	$\sigma^2 + n\sigma_\delta^2$ $+ cn\sigma_\beta^2$	$\sigma^2 + n\sigma_\delta^2$ $+ \dfrac{cn \sum\limits_{j} \beta_j^2}{r - 1}$	$\sigma^2 + cn\sigma_\beta^2$
Treatments (columns)	$c - 1$	s_2^2	$\sigma^2 + n\sigma_\delta^2$ $+ rn\sigma_\alpha^2$	$\sigma^2 + n\sigma_\delta^2$ $+ \dfrac{rn \sum\limits_{i} \alpha_i^2}{c - 1}$	$\sigma^2 + n\sigma_\delta^2$ $+ \dfrac{rn \sum\limits_{i} \alpha_i^2}{c - 1}$
Experimental error	$(c - 1)(r - 1)$	s_3^2	$\sigma^2 + n\sigma_\delta^2$	$\sigma^2 + n\sigma_\delta^2$	$\sigma^2 + n\sigma_\delta^2$
Sampling error	$cr(n - 1)$	s_4^2	σ^2	σ^2	σ^2
Total	$crn - 1$				

12.11. THE LATIN SQUARE DESIGN

The randomized block design was introduced to eliminate from the analysis of treatment means the larger part of the variation due to heterogeneity of the experimental material. That is, in using the randomized block design instead of the completely randomized design, we impose one restriction (the blocks) on the experimental units so as to decrease the error mean square. Actually, it may be desirable in some experiments to impose two or more restrictions on the experimental units. The Latin square design is introduced for the case where a two-way classification (two restrictions) is made on the experimental units.

The plan for the Latin square design with t^2 experimental units is to form a square with t rows and t columns and then to apply t treatments in the cells

Table 12.29
A Plan for a Latin Square Design

		Columns			
		1	2	3	4
Rows	1	A	B	C	D
	2	B	A	D	C
	3	C	D	B	A
	4	D	C	A	B

in such a way that each treatment occurs only once in each column and once in each row. In this way each row and each column contains a complete replication of the treatments. Table 12.29 illustrates a Latin square design with four treatments A, B, C, and D. Such a design eliminates from the errors differences among columns as well as differences among rows, and allows a better opportunity to reduce the errors than a randomized block design does. As an illustration consider the following example.

Example 12.5. A continuous sheet of paper eight feet wide is manufactured and placed on rolls for distribution. Later these rolls of paper are to be made into bags for grocery stores. It is claimed that the application of a certain chemical solution will increase the tensile strength of the paper. Four different solutions are available. The problem is to (a) estimate the effect of each solution on tensile strength, (b) test the hypothesis that they are all equal in their ability to increase tensile strength, and (c) test the hypothesis that the special chemical solution is equally as effective as the average of the other three.

It is known that tensile strength varies with *batches of raw material* (pulpwood, etc.) and with *distance from the edge* of the sheet. (Tensile strength is least near the edges and greatest in the middle of the sheet.) Thus, a Latin square design is considered appropriate for the analysis.

We obtain a particular experimental plan by the method explained below. For each of four batches, cut across the sheet from edge to edge to obtain rectangular *sections* eight feet long and one foot wide. Then cut and label four pieces from each section, as illustrated in Fig. 12.2. Denoting the solutions by A, B, C, D, the sections by row 1, row 2, row 3, row 4, and the four pieces numbered 1 by column 1, the four pieces numbered 2 by column 2, etc., we can use the plan in Table 12.29 for applying solutions to pieces. For an experiment conducted according to this plan, the tensile strength (in pounds) of pieces of paper and the appropriate totals are given in Table 12.30.

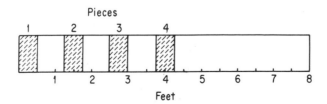

Fig. 12.2 Plan for Cutting Pieces from a Typical Section of Paper in the Tensile Strength Experiment

Before giving the solutions to the three problems of Example 12.5, we present the general model and analysis for the Latin square design.

Table 12.30

Tensile Strength of Paper in a Latin Square Design with Random Assignment
of Solutions Shown by the Letters in Parentheses

Batch Number	Piece Number (distance from edge in inches)				Row Totals
	1 (3″)	2 (18″)	3 (33″)	4 (48″)	
1	41.3(A)	38.8(B)	39.3(C)	42.2(D)	161.6
2	37.7(B)	41.0(A)	41.2(D)	39.3(C)	159.2
3	38.2(C)	40.4(D)	39.3(B)	43.3(A)	161.2
4	38.8(D)	38.2(C)	42.6(A)	38.4(B)	158.0
Column totals	156.0	158.4	162.4	163.2	640.0

$$\text{Total for treatment } A = 41.3 + 41.0 + 42.6 + 43.3 = 168.2$$
$$\text{Total for treatment } B = 37.7 + 38.8 + 39.3 + 38.4 = 154.2$$
$$\text{Total for treatment } C = 38.2 + 38.2 + 39.3 + 39.3 = 155.0$$
$$\text{Total for treatment } D = 38.8 + 40.4 + 41.2 + 42.2 = 162.6$$
$$\text{Grand total} = 640.0$$

The *model equation* is

$$x_{ij(k)} = \mu + \alpha_i + \beta_j + \gamma_k + \epsilon_{ij(k)}$$
$$i = 1, \ldots, t \quad j = 1, \ldots, t \quad k = 1, \ldots, t \tag{12.69}$$

where $x_{ij(k)}$ denotes the observation for the kth treatment falling in the ith column and jth row, μ, α_i, and β_j are defined as in the two-way classification, γ_k is the effect of the kth treatment, the $\epsilon_{ij(k)}$ are independently and normally distributed with means zero and common variance σ^2, and

$$\sum_{i=1}^{t} \alpha_i = \sum_{j=1}^{t} \beta_j = \sum_{k=1}^{t} \gamma_k = 0 \tag{12.70}$$

Note that this is a fixed model design. The k is placed in parentheses to indicate that it is not independent of i and j, and to emphasize the fact that there are *not* $t \cdot t \cdot t$ observations in the experiment. This model is the correct one to use in an experiment where column and row effects are additive, and where the errors can be made more homogeneous by imposing two restrictions on the experimental units.

The column, row, grand, and treatment totals are given by

$$\begin{cases} T_{i.} = \sum_j x_{ij(k)}, \quad T_{.j} = \sum_i x_{ij(k)} \\ T = \sum_i \sum_j x_{ij(k)} = \sum_i \sum_k x_{ij(k)} = \sum_j \sum_k x_{ij(k)} \quad \text{and} \\ T_{..(k)} = \text{total of all observations associated with the} \\ \qquad \qquad k\text{th treatment} \end{cases} \tag{12.71}$$

respectively. The column, row, grand, and treatment means are given by

$$\begin{cases} \bar{x}_{i.} = \dfrac{T_{i.}}{t}, \qquad \bar{x}_{.j} = \dfrac{T_{.j}}{t}, \qquad \bar{x} = \dfrac{T}{t^2} \quad \text{and} \\[2mm] \bar{x}_{..(k)} = \dfrac{T_{..(k)}}{t} \end{cases} \tag{12.72}$$

respectively. Unbiased estimates a_i, b_j, and c_k of the effects α_i, β_j, and γ_k, respectively, are given by

$$\begin{cases} a_i = \bar{x}_{i.} - \bar{x} & i = 1, \dots, t \\ b_j = \bar{x}_{.j} - \bar{x} & j = 1, \dots, t \\ c_k = \bar{x}_{..(k)} - \bar{x} & k = 1, \dots, t \end{cases} \tag{12.73}$$

The treatment effects are the only ones which are usually estimated. The sum of squares identity is

$$SST = SSC + SSR + SSTr + SSE \tag{12.74}$$

where

$$\begin{cases} SST = \displaystyle\sum_i \sum_j x^2_{ij(k)} - \frac{T^2}{t^2} \\[4mm] SSC = \dfrac{\displaystyle\sum_i T^2_{i.}}{t} - \dfrac{T^2}{t^2} \\[4mm] SSR = \dfrac{\displaystyle\sum_j T^2_{.j}}{t} - \dfrac{T^2}{t^2} \\[4mm] SSTr = \dfrac{\displaystyle\sum_k T^2_{..(k)}}{t} - \dfrac{T^2}{t^2} \\[4mm] SSE = SST - SSC - SSR - SSTr \end{cases} \tag{12.75}$$

The analysis of variance and expected mean squares are derived in the usual way and are shown in Table 12.31.

Now, returning to Example 12.5, we find the unbiased estimates of solution effects on tensile strength to be

$$c_1 = c_A = \frac{168.2}{4} - \frac{640.0}{16} = 42.05 - 40.00 = 2.05$$

$$c_2 = c_B = 38.55 - 40.00 = -1.45$$

$$c_3 = c_C = 38.75 - 40.00 = -1.25$$

$$c_4 = c_D = 40.65 - 40.00 = 0.65$$

The special solution is treatment A, and the point estimate indicates that this solution gives largest tensile strength of any of the solutions. In order to

Table 12.31

Analysis of Variance and Expected Mean Squares for a Latin Square Design

Source of Variation	Sum of Squares	Degrees of Freedom	Mean Square	Expected Mean Square for Fixed Model
Columns	SSC	$t-1$	$s_1^2 = \dfrac{SSC}{t-1}$	$\sigma^2 + t\dfrac{\sum\limits_i \alpha_i^2}{t-1}$
Rows	SSR	$t-1$	$s_2^2 = \dfrac{SSR}{t-1}$	$\sigma^2 + t\dfrac{\sum\limits_j \beta_j^2}{t-1}$
Treatments	SSTr	$t-1$	$s_3^2 = \dfrac{SSTr}{t-1}$	$\sigma^2 + t\dfrac{\sum\limits_k \gamma_k^2}{t-1}$
Error	SSE (by subtraction)	$(t-1)(t-2)$	$s_4^2 = \dfrac{SSE}{(t-1)(t-2)}$	σ^2
Total	SST	t^2-1		

determine if the solutions are significantly different from each other, make the analysis of variance shown in Table 12.32. It is clear, on looking at the table, that the solutions have significantly different effects. (Note that the

Table 12.32

Analysis of Variance for the Data in Table 12.30

Source of Variation	Sum of Squares	Degrees of Freedom	Mean Square	F_c	$F_{.05}$	Expected Mean Square
Distances from edge	8.64	3	2.88	11.5		$\sigma^2 + 4\sum\limits_i \alpha_i^2/3$
Batches	2.16	3	0.72	2.9		$\sigma^2 + 4\sum\limits_j \beta_j^2/3$
Solutions	33.16	3	11.04	44.2	4.76	$\sigma^2 + 4\sum\limits_k \gamma_k^2/3$
Error	1.50	6	0.25			σ^2
Total	45.46	15				

number of degrees of freedom for the error term is very small. Generally, in practice, squares of size 4×4 or less should be repeated so as to increase the degrees of freedom for the error variance.) In fact, the solutions are significantly different at the 0.001 level. To determine whether solution A is significantly better than the average of the other three, compute

$$Q^2 = \frac{(\sum m_i T_{i.})^2}{4\sum m_i^2} = \frac{[3(168.2) - 1(154.2) - 1(155.0) - 1(162.6)]^2}{4 \cdot 12}$$

$$= 22.4$$

and

$$\frac{Q^2}{s_4^2} = \frac{22.4}{0.25} = 89.6$$

and compare with the upper five per cent value of the F distribution with one and six degrees of freedom. Since $F_{.05}(1, 6) = 5.99$ and $F_{.0001}(1, 6) = 82.5$, we conclude that solution A gives paper greater tensile strength than the other three solutions, understanding that there is less than a 0.0001 chance of making a type 1 error. Further, since

$$t_{.025}(6) = 2.447, \qquad s_{\bar{x}..(k)} = \tfrac{1}{4} \quad \text{and} \quad t_{.025} s_{\bar{x}..(k)} = 0.61$$

the 95 per cent confidence intervals for the treatment means are

$$41.44 < \mu_A < 42.66$$
$$37.94 < \mu_B < 39.16$$
$$38.41 < \mu_C < 39.36$$
$$40.04 < \mu_D < 41.26$$

The number of possible arrangements of treatments in Latin squares increases rapidly as t increases. For a 2×2 square, treatments may be arranged in only two ways, namely

A	B
B	A

and

B	A
A	B

There are $1 \cdot 3!2! = 12$ different arrangements in a 3×3 square, $4 \cdot 4!3! = 576$ ways to arrange a 4×4 square, $56 \cdot 5!4! = 161,280$ ways to arrange a 5×5 square, and $9408 \cdot 6!5! = 812,851,200$ ways to arrange a 6×6 square. There are four standard 4×4 squares, 56 standard 5×5 squares, and 9408 standard 6×6 squares, and each standard square can be permuted in $t!(t - 1)!$ ways. A standard square is one in which the letters in the first row and first column are arranged in alphabetical order. The four standard 4×4 Latin squares are given in Table 12.33. Fisher and Yates

Table 12.33
The Standard 4×4 Latin Squares

A B C D	A B C D	A B C D	A B C D
B A D C	B C D A	B D A C	B A D C
C D B A	C D A B	C A D B	C D A B
D C A B	D A B C	D C B A	D C B A

[22] give all 4×4, 5×5, and 6×6 standard squares, along with $7 \times 7, \ldots, 12 \times 12$ sample squares. Cochran and Cox [10] give sample squares from 3×3 through 12×12. Norton [32] and Sade [35] give 562 7×7 squares from which it is possible to generate the 16,942,080 standard squares. Not all standard squares of higher-order Latin squares have been

tabulated. Other discussion on the formation of Latin squares is found in Refs. [21, 22, 26, 31, 32, 35].

A random Latin square design may be obtained by selecting a standard square at random, arranging the columns in random order, and then arranging the last $t - 1$ rows in random order. If standard squares are not available, it is usually adequate to construct a Latin square, then randomly to arrange columns and rows, and, finally, to assign treatments to the letters at random.

The requirement that a Latin square design must have the same number of rows, columns, and treatments is its principal disadvantage. If the number of treatments is small, the number of degrees of freedom for the error mean square is small or does not exist. For a 2×2 square there are no degrees of freedom and for the 3×3 square there are only two degrees of freedom for the error mean square. However, the degrees of freedom may be increased by using more than one square in the same experiment. On the other hand, if the number of treatments is large, the number of experimental units soon becomes too large for practical purposes. Thus, 5×5 through 8×8 squares are the most useful. Numerous examples of Latin square designs occur in the literature. These designs are particularly useful in research in social studies, agriculture, industry, medicine, and marketing studies.

12.12. MISSING PLOTS IN LATIN SQUARES

The procedures for analysis of a Latin square design with one or more missing values are about the same as for the randomized block design. If a single observation is missing from the cell in the lth column and mth row, denote the dummy value by $y_{lm(p)}$ and estimate the missing value $x_{lm(p)}$ by

$$y_{lm(p)} = \frac{t(T'_{l.} + T'_{.m} + T'_{..(p)}) - 2T'}{(t - 1)(t - 2)} \qquad (12.76)$$

where

$$\begin{cases} T'_{l.} \text{ denotes the sum of the observed values in column } l \\ T'_{.m} \text{ denotes the sum of the observed values in row } m \\ T'_{..(p)} \text{ denotes the sum of the observed values for treatment } p \quad (12.77) \\ T' \text{ denotes the sum of all observations} \end{cases}$$

Then substitute the dummy value $y_{lm(p)}$ for the missing value and compute the sums of squares in the usual way. The degrees of freedom for error mean square is now $(t - 1)(t - 2) - 1$, and in testing the hypothesis

$$H_0: \gamma_1 = \cdots = \gamma_t = 0 \qquad (12.78)$$

a correction for bias in the treatment sum of squares should be made. The correction is made by computing

$$SSC' = SSC - B \tag{12.79}$$

where

$$B = \frac{[T' - T'_{l.} - T'_{.m} - (t - 1)T'_{..(p)}]^2}{[(t - 1)(t - 2)]^2} \tag{12.80}$$

Then the F ratio

$$\frac{\dfrac{SSC'}{t - 1}}{\dfrac{SSE}{(t - 1)(t - 2) - 1}} \tag{12.81}$$

with $t - 1$ and $(t - 1)(t - 2) - 1$ degrees of freedom is used to test (12.78). Kramer and Glass [29] give a direct method of analysis which does not require a correction for bias in the treatment sum of squares. The standard error of the difference in the mean of the treatment with a missing value $\bar{x}_{..(p)}$ and the mean of any other treatment $\bar{x}_{..(k)}$ $(k \neq p)$ is

$$\sqrt{s^2\left[\frac{2}{t} + \frac{1}{(t - 1)(t - 2)}\right]} \tag{12.82}$$

If more than one value is missing, the reader is referred to articles by Yates [43], Yates and Hale [45], DeLury [15], and Kramer and Glass [29]. The articles by Yates and Hale give methods for analyzing the Latin squares when a single observation is missing; when one or more columns, rows, or treatments are missing; and when one column and one or more other observations is missing. DeLury also gives many of these methods. Kramer and Glass give explicit formulas for each missing value for many special cases, a procedure for analyzing the general case, and a direct method of analysis of variance which does not require correction for bias in the treatment sum of squares.

12.13. EFFICIENCY OF THE LATIN SQUARE RELATIVE TO OTHER DESIGNS

The definitions and methods of Sect. 12.9 may be applied to find the efficiency of the Latin square design relative to the randomized block and completely randomized designs. If we use the definitions for mean squares given in Table 12.31, the estimated error mean square, s_w^2, for the completely randomized design is

$$s_w^2 = \frac{s_1^2 + s_2^2 + (t - 1)s_4^2}{t + 1} \tag{12.83}$$

with $v_2 = t(t - 1)$ degrees of freedom. An estimate of the efficiency of the Latin square design relative to the completely randomized design may then be obtained by substituting $(t-1)(t-2)$, $t(t-1)$, s_w^2, and s_4^2, respectively, in place of v_2, v_1, s_2^2, and s_1^2 in Eq. (12.56).

When one is comparing the Latin square design to the randomized block design, two estimates of relative efficiency can be made, one when columns in the Latin square are used as blocks in the randomized block design, and the other when rows are used as blocks. If the columns are used as blocks, the estimated error mean square s_e^2 for the randomized block design is

$$s_e^2 = \frac{s_2^2 + (t-1)s_4^2}{t} \tag{12.84}$$

with $(t-1)(t-1)$ degrees of freedom. If rows are used as blocks, replace s_2^2 in Eq. (12.84) by s_1^2 from Table 12.31 to obtain the estimated error mean square. On substituting Eq. (12.84) in Eq. (12.56), we have an estimate of the efficiency of the Latin square design relative to the randomized block design.

12.14. OTHER EXPERIMENTAL MODELS IN AND EXTENSIONS OF THE LATIN SQUARE DESIGN

The analysis for the Latin square design given in Sect. 12.11 was for the fixed model. That is, the column, row, and treatment effects were considered fixed. However, in some experiments any one or all of the effects might be considered random. For any set of effects which are fixed, make the assumptions following Eq. (12.69). When the column effects are random, assume that they are normally distributed with mean zero and variance σ_α^2; when the row effects are random, assume that they are normally distributed with mean zero and variance σ_β^2; when the treatment effects are random, assume that they are normally distributed with mean zero and variance σ_γ^2. Assume all random effects to be independently distributed. The sums of squares

Table 12.34
Analysis of Variance and Expected Mean Squares for a $t \times t$ Latin Square

Source of Variation	Mean Square	Expected Mean Square for		
		Random Model	Mixed Model (γ)	Mixed Model (α, γ)
Columns	$s_1^2 = \dfrac{SSC}{t-1}$	$\sigma^2 + t\sigma_\alpha^2$	$\sigma^2 + t\sigma_\alpha^2$	$\sigma^2 + t\dfrac{\sum_i \alpha_i^2}{t-1}$
Rows	$s_2^2 = \dfrac{SSR}{t-1}$	$\sigma^2 + t\sigma_\beta^2$	$\sigma^2 + t\sigma_\beta^2$	$\sigma^2 + t\sigma_\beta^2$
Treatments	$s_3^2 = \dfrac{SSTr}{t-1}$	$\sigma^2 + t\sigma_\gamma^2$	$\sigma^2 + t\dfrac{\sum_k \gamma_k^2}{t-1}$	$\sigma^2 + t\dfrac{\sum_k \gamma_k^2}{t-1}$
Error	$s_4^2 = \dfrac{SSE}{(t-1)(t-2)}$	σ^2	σ^2	σ^2

identity and analysis of variance are the same for each experimental model. However, as shown in Table 12.34, the expected mean squares depend on the model. The mixed models not shown are given by writing the second component as σ_l^2 or $\sum l^2/(t-1)$ $(l = \alpha, \beta,$ or $\gamma)$ according as the effect is random or fixed. Methods already explained may be used to give testing and estimating procedures, with the expected mean squares as guides.

In some experiments with Latin square designs it is desirable that samples be taken in each of the experimental units. If n samples are randomly taken in each of the t^2 experimental units, the analysis of variance is a straightforward extension of the analysis for a single observation per cell. The sums of squares due to columns, rows, and treatments are computed in the usual way, with the understanding that there are nt observations in each total for columns, rows, and treatments. The subtotal sum of squares $SS(ST)$, showing the variation among cells, is computed from the totals in the t^2 cells. The experimental error and sampling error sum of squares are given by

$$SSE = SS(ST) - SSC - SSR - SSTr \qquad (12.85)$$

and

$$SSS = SST - SS(ST) \qquad (12.86)$$

respectively. The analysis of variance is shown in Table 12.35. The usual

Table 12.35
Analysis of Variance for a $t \times t$ Latin Square with n Samples per Experimental Unit

Source of Variation	Sum of Squares	Degrees of Freedom	Mean Squares
Columns	SSC	$t - 1$	$s_1^2 = SSC/(t-1)$
Rows	SSR	$t - 1$	$s_2^2 = SSR/(t-1)$
Treatments	SSTr	$t - 1$	$s_3^2 = SSTr/(t-1)$
Experimental error	SSE	$(t-1)(t-2)$	$s_4^2 = SSE/(t-1)(t-2)$
Sampling error	SSS	$t^2(n-1)$	$s_5^2 = SSS/t^2(n-1)$
Total	SST	$t^2n - 1$	

techniques may be applied to find the expected mean squares for the various experimental models. This is left as a exercise for the reader. In any case, if we remember that the samples are random, the null hypothesis

$$H_0: \gamma_1 = \cdots = \gamma_t = 0$$

may be tested by comparing the ratio

$$\frac{\dfrac{SSTr}{t-1}}{\dfrac{SSE}{(t-1)(t-2)}} \qquad (12.87)$$

with the F distribution with $t - 1$ and $(t - 1)(t - 2)$ degrees of freedom.

Designs which impose more than two types of restrictions on the experimental units are occasionally applied. We illustrate designs with three restrictions. Just as Latin letters representing treatments were used in a two-way classification to define a Latin square design, Greek letters may be used in a Latin square to define a *Graeco-Latin* square design. The rule is to place t Greek letters in a square with t^2 cells in such a way that each treatment (Greek letter) occurs only once in each column, once in each row, and once with each Latin letter. Examples of 3×3 and 4×4 Graeco-Latin squares are given in Table 12.36. Squares of all sizes greater than 3×3 do not exist. For example, a 6×6 Graeco-Latin square is impossible to construct. However, squares of size t do exist when t is a positive odd integer greater than 1 or a power of a prime or a number satisfying the relation $t = 4k + 2\,(k = 2, 3, \ldots)$. [In fact, Parker, Bose, and Shirkhande disproved

Table 12.36
Sample Graeco-Latin Square Designs

$A\beta$	$B\alpha$	$C\gamma$
$C\alpha$	$A\gamma$	$B\beta$
$B\gamma$	$C\beta$	$A\alpha$

$B\beta$	$D\delta$	$A\alpha$	$C\gamma$
$A\delta$	$C\beta$	$B\gamma$	$D\alpha$
$D\gamma$	$B\alpha$	$C\delta$	$A\beta$
$C\alpha$	$A\gamma$	$D\beta$	$B\delta$

a 177-year-old conjecture when in 1958 they proved that squares of size $t = 4k + 2\,(k = 2, 3, \ldots)$ can be constructed. See Gardner, M., "How Three Modern Mathematicians Disproved a Celebrated Conjecture of Leonhard Euler," *Scientific American*, Nov., 1959.] These designs tend to be useful in areas where the Latin square design has proved useful. For further reading see Refs. [6, 7, 14, 17, 19, 20, 27, 28, 30].

12.15. OTHER TOPICS RELATING TO ANALYSIS OF VARIANCES

In general, only methods of analysis of variance and their applications have been presented. Many other considerations enter in experimentation, and there are good books [2, 6, 10, 19, 20, 26, 31, 34, 36, 37, 38] on experimental design which should be consulted. The choice of the appropriate design, the selection of the particular treatments, the number of replications, and the particular information required of the data are all very important topics. In all the designs discussed we have assumed that the

$$\begin{cases} \text{(a) Samples are random} \\ \text{(b) Effects are additive} \\ \text{(c) Populations are normal} \\ \text{(d) Error variances are equal} \end{cases} \qquad (12.88)$$

What should an experimenter do in case some of these assumptions do not hold?

The points just raised and many others need to be considered in experimentation. The student should seek for answers to his particular problems. Many answers are given in detail in the literature. Some questions still remain unanswered, but there are rules and approximations which often are satisfactory. Some of the above topics are now discussed very briefly.

In Sects. 6.2 and 8.4, for example, we have already discussed for one and two treatments the problem of determining the appropriate sample size in an experiment when certain conditions are required. We have also pointed out that samples of equal sizes are desirable when two or more treatment means are compared. When the total number of observations is fixed, we know that equal-size samples make the computations simpler and the error variance for treatment means smaller. However, we have not discussed the problem of determining the best size of an experiment or the size of a treatment replication. The number of treatment replications may be found in terms of specified values of α (size of type 1 error) and β (size of type 2 error), the error variance σ^2 (or its estimate s^2) and the difference to be detected. Cochran and Cox [10], Harris, Howitz, and Mood [24], and others [25, 39] give detailed discussions along with tables which are appropriate for finding the sample size.

If the four assumptions in (12.88) hold, the procedures already explained are valid; otherwise, the procedures may lead to approximate results or downright false results. It should be remembered that all inferences in statistics require the randomness assumption, but certain adjustments may be made so that failure of any of the other assumptions does not invalidate the analysis. Often a transformation can be made so that the usual analysis can be carried out on the transformed data. Some common transformations are the arc sine, square root, and logarithm transformations. The reader who is interested in transformations and related assumptions is referred to articles [3, 8, 9, 11, 12, 13, 33], to mention only a few. Applications may be found in these and numerous other references.

12.16. EXERCISES

12.17. Assuming no block-to-block variation in the randomized block design *RBD* in Exercise 12.1, we could analyze the experiment as a completely randomized design *CRD*. (a) Use Eq. (12.57) to find an estimate of the efficiency of the *RBD* relative to the *CRD*. (b) Pool the block and error sum of squares in Exercise 12.1 to find the within mean square of a one-way classification. Test the hypothesis that the treatment effects are equal to zero. (c) Discuss the desirability of separating the block-to-block variation from the within variation.

12.18. Answer (a), (b), and (c) of Exercise 12.17, replacing "Exercise 12.1" by "Exercise 12.2."

12.19. Answer (a), (b), and (c) of Exercise 12.17, replacing "Exercise 12.1" by "Exercise 12.3."

12.20. Prove Eq. (12.58).

12.21. (a) Analyze Example 8.6 as a randomized block design with 16 blocks and two experiments. (b) Estimate the efficiency of the paired-observation experiment relative to the experiment with two independent sets of observations.

12.22. In the nested design in Example 11.2 there were ten treatments with samples of size three and subsamples of size two. A new nested design is to have the same ten treatments, but with both samples and subsamples of size two. Estimate the efficiency of the old design relative to the new design.

12.23. Prove Eq. (12.62).

12.24. In a randomized block design with three random blocks (car loads, say), three fixed treatments (densities, say), and two random subsamples (determinations of similar kind) the coded determinations in Table 12.37

Table 12.37

Blocks	Treatments					
	1		2		3	
1	26	28	23	26	18	18
2	20	18	20	20	16	19
3	19	21	23	21	14	18

were obtained. (a) Prepare the usual analysis of variance table, showing the expected mean squares. (b) Test the equality of the treatment effects at the five per cent level. (c) Find a 95 per cent confidence interval for the difference in the means of treatments 1 and 2. (d) Find point and 95 per cent confidence interval estimates of the variance components for blocks and experimental error, that is, for σ_β^2 and σ_δ^2. (e) Assuming a random effects experiment, find 95 per cent confidence limits of σ_β^2. (f) For the purpose of comparing treatments, estimate the efficiency of the design in (a) and (b) relative to a similar design with two blocks and three subsamples.

12.25. Derive the three expected mean squares for blocks in Table 12.23.

12.26. Prove that a_i, b_j, and c_k as defined in (12.73) are unbiased estimators of α_i, β_j, and γ_k, respectively.

12.27. Derive Eq. (12.74).

12.28. (a) Complete Table 12.38 for the analysis of variance and expected mean squares of a fixed effects Latin square design. (The reader may think of

Table 12.38

Source of Variation	Sum of Squares	Degrees of Freedom	Mean Square	Expected Mean Square
Columns		5		
Rows	4.20			
Treatments			2.43	
Error			0.65	
Total	39.65			

the columns as representing schools, the rows as classes, the treatments as methods of teaching spelling, and the observations as grades based on 100 points.) (b) Test the hypothesis that the treatment effects are equal to zero, showing all steps in the general test procedure.

12.29. In a marketing experiment the price of a staple item, potatoes, was studied. There were five cities and five types of stores in the region of study. Since only the five most popular kinds of potatoes were regularly sold in each store in each city, the Latin square design was used in the experiment. The particular random Latin square design used and the mean price (in cents) per ten lb (for a selected month) are shown in Table 12.39. (Unless otherwise indicated, the reader should consider this a fixed effects experiment. The size of city, type of store, and kind of potato are

Table 12.39

City	Type of Store				
	1	2	3	4	5
1	69.2(C)	69.0(A)	63.2(B)	61.6(D)	64.5(E)
2	65.1(E)	64.4(D)	68.4(C)	67.5(A)	62.6(B)
3	63.9(B)	63.9(E)	66.7(A)	65.7(C)	62.8(D)
4	62.7(D)	68.2(C)	62.4(E)	62.4(B)	67.3(A)
5	68.1(A)	64.5(B)	61.8(D)	63.3(E)	65.8(C)

indicated when appropriate.) (a) Prepare the usual analysis of variance table, showing the expected mean squares. (The reader should check to see if the sum of squares for columns, rows, treatments, and totals are 14.30, 5.10, 115.47, and 141.24, respectively.) (b) Use a five per cent level test to determine whether (1) "kind of potato" effects are equal; (2) "type of store" effects are equal; (3) "city" effects are equal. Write a summary statement. (c) Find unbiased estimates of each of the kind of potato effects and each of the type of store effects. Establish 95 per cent confidence intervals for each of the treatment means. (d) The "kinds of potatoes" A, B, C, D, and E are baking potato of packer X, all-purpose potato of packer X, baking potato of packer Y, all-purpose potato of packer Y, and all-purpose potato of packer Z, respectively. Knowing this, we wish to test more specific hypotheses than (1) of (b). Use a five per cent level test to determine whether (1) baking potatoes bring a significantly better price than all-purpose potatoes; (2) packer X and packer Y receive the same price on the average; (3) packer Z receives a better price for all-purpose potatoes than packers X and Y on the average. (e) The store types 1, 2, 3, 4, and 5 are large chain store downtown, small private store downtown, large chain store in residential shopping center, large private store in residential shopping center, and small private store in residential shopping center, respectively. With this added information available, state and test three hypotheses involving linear contrasts of type of store. (f) Use Duncan's multiple range test to rank the five treatment means. (g) Test hypothesis (1) in (b), assuming that the observation in the first row and first column is missing. Find a 90 per cent confidence interval for the difference in the means of treatments A and C. (h) Estimate the efficiency of this experiment relative to the completely randomized design. Also, estimate the efficiency of this experiment relative to the randomized block design if rows are used as blocks; if columns are used as blocks. Write a summary statement.

12.30. In an experiment with only three treatments a single Latin square design does not allow enough degrees of freedom for the error mean square. The following data is for two replications of the same 3×3 Latin square design

54(C)	56(A)	53(B)		50(C)	60(A)	53(B)
52(A)	47(B)	46(C)		56(A)	44(B)	45(C)
50(B)	41(C)	54(A)		47(B)	40(C)	58(A)

(a) Prepare an analysis of variance table. (b) Test the equality of the three treatment means using a five per cent level test. (c) Establish a 95 per cent confidence interval for the difference in the means of treatments A and B.

12.31. Derive the expected mean squares for the mixed model (γ) of Table 12.34.

12.32. (a) Construct an experimental layout for a 5×5 Graeco-Latin square design. (b) Give an analysis of variance table for (a), indicating how the sum of squares are computed. (c) Write the model equation for the general Graeco-Latin square design, defining all technical terms. (d) Write the expected mean squares for all experiments in Graeco-Latin square designs in which the errors are random.

REFERENCES

1. Allan, F. E. and J. Wishart, "A Method of Estimating Yield of a Missing Plot in Field Experimental Work," *Journal of Agricultural Science*, Vol. **20** (1930), pp. 399–406.

2. Anderson, R. L. and T. A. Bancroft, *Statistical Theory in Research*. New York: McGraw-Hill, Inc., 1952, Chaps. 18, 20, and 23.

3. Bartlett, M. S., "The Use of Transformations," *Biometrics*, Vol. **3** (1947), pp. 39–52.

4. Baten, W. D., "Formulas for Finding Estimates for Two and Three Missing Plots in Randomized Layouts," *Mich. Agr. Exp. Sta. Tech. Bul.*, Vol. **165** (1939).

5. ———, "Variances of Differences Between Means when there are two Missing Values in Randomized Block Designs," *Biometrics*, Vol. **8** (1952), pp. 42–50.

6. Bennett, C. A. and N. L. Franklin, *Statistical Analysis in Chemistry and the Chemical Industry*. New York: John Wiley & Sons, Inc., 1954, Chaps. 7 and 8.

7. Bose, R. C., "On the Application of Galois Fields to the Problem of the Construction of Hyper-Graeco-Latin Squares," *Sankhyā*, Vol. **3** (1938), pp. 323–38.

8. Box, G. E. P., "Non-normality and Tests on Variances," *Biometrika*, Vol. **40** (1953), pp. 318–35.

9. ——— and S. L. Anderson, "Robust Tests for Variances and Effect of Non-normality and Variance Heterogeneity on Standard Tests," *Technical Report Number* 7, Department of the Army, Project Number 599–01–004, 1954.

10. Cochran, W. G. and G. M. Cox, *Experimental Designs*, 2nd ed. New York: John Wiley & Sons, Inc., 1957, Chaps. 3, 4, and 5.

11. ———, "Some Difficulties in the Statistical Analysis of Replicated Experiments," *Empire Journal of Experimental Agriculture*, Vol. **6** (1938), pp. 157–75.

12. ———, "Some Consequences when the Assumptions for the Analysis of Variance are Not Satisfied," *Biometrics*, Vol. **3** (1947), pp. 22–38.

13. David, F. N. and N. L. Johnson, "The Effect of Non-normality on the Power Function of the F-test in the Analysis of Variance," *Biometrika*, Vol. **38** (1951), pp. 43–57.

14. Davies, H. M., "The Application of Variance Analysis to Some Problems of Petroleum Technology," *Journal of Inst. Petroleum*, Vol. **32** (1946), p. 465.

15. DeLury, D. B., "The Analysis of Latin Squares when Some Observations are Missing, "*Journal of the American Statistical Association*, Vol. **41** (1946), pp. 370–89.

16. Dixon, W. J. and F. J. Massey, *Introduction to Statistical Analysis*, 2nd. ed. New York: McGraw-Hill, Inc., 1957, Chap. 10.

17. Dunlop, G., "Methods of Experimentation in Animal Nutrition," *Journal of Agricultural Science*, Vol. **23** (1933), pp. 580–614.

18. Federer, W. T., "Evaluation of Variance Components from a Group of Experiments with Multiple Classification," *Iowa Agr. Exp. Bul.*, Vol. **380** (1951), pp. 241–310.

19. ———, *Experimental Design*. New York: The Macmillan Company, 1955, Chaps. 5, 6, and 15.

20. Fisher, R. A., *The Design of Experiments*, 6th. ed. New York: Hafner Publishing Co., Inc., 1951.

21. —— and F. Yates, "The 6 × 6 Latin Squares," *Cambridge Phil. Soc. Proc.*, Vol. **30** (1934), pp. 492–507.

22. —— and F. Yates, *Statistical Tables for Biological, Agricultural, and Medical Research*. London: Oliver and Boyd, 1957.

23. Glenn, W. A. and C. Y. Kramer, "Analysis of Variance of a Randomized Block Design with Missing Observations," *Applied Statistics*, Vol. **7** (1958), p. 173.

24. Harris, M., D. G. Horwitz, and A. M. Mood, "On the Determination of Sample Sizes in Designing Experiments," *Journal of the American Statistical Association*, Vol. **43** (1948), pp. 391–402.

25. Harter, H. L., "Error Rates and Sample Sizes for Range Tests in Multiple Comparisons," *Biometrics*, Vol. **13** (1957), pp. 511–36.

26. Kempthorne, Oscar, *The Design and Analysis of Experiments*. New York: John Wiley & Sons, Inc., 1952, Chaps. 6, 9, 10, and 12.

27. Kishen, K., "On Latin and Hyper-Greaco-Latin Cubes and Hyper-Cubes," *Current Science*, Vol. **2** (1942), pp. 98–99.

28. ——, "On the Construction of Latin and Hyper-Graeco-Latin Cubes and Hyper-Cubes," *J. Indian Soc. Agr. Stat.*, Vol. **2** (1949), pp. 20–48.

29. Kramer, C. Y. and S. Glass, "Analysis of Variance of a Latin Square Design with Missing Observations," *Applied Statistics*, Vol. **9** (1960), pp. 43–50.

30. Main, W. R. and L. H. C. Tippett, "The Design of Weaving Experiments," *Shirley Inst. Mem.*, Vol. **18** (1941), pp. 109–20.

31. Mann, H. B., *Analysis and Design of Experiments*. New York: Dover Publications, Inc., 1949.

32. Norton, H. W., "The 7 × 7 Squares," *Annals of Eugenics*, Vol. **9** (1939), pp. 269–307.

33. Odeh, R. E. and E. G. Olds, *Notes on the Analysis of Variance of Logarithims of Variance*, ASTIA Document No. AD211917, U. S. Department of Commerce, Washington 25, D. C., 1959.

34. Ostle, Bernard, *Statistics in Research*, Ames, Iowa: The Iowa State College Press, 1954, Chaps. 10 and 11.

35. Sade, A., "An Omission in Norton's List of 7 × 7 Squares," *Annals of Mathematical Statistics*, Vol. **22** (1951), pp. 306–7.

36. Scheffé, Henry, *The Analysis of Variance*. New York: John Wiley & Sons, Inc., 1959, Chap. 4.

37. Snedecor, G. W. and W. G. Cochran, *Statistical Methods*, 5th ed. Ames, Iowa: The Iowa State College Press, 1957, Chaps. 11 and 12.

38. Steel, R. D. G. and J. H. Torrie, *Principles and Procedures of Statistics*. New York: McGraw-Hill, Inc., 1960, Chap. 8.

39. Tang, P. G., "The Power Function of the Analysis of Variance Tests with Tables and Illustrations of their Use, "*Statistical Research Memoirs*, Vol. **2** (1938), pp. 126–57.

40. Yates, F., "The Analysis of Replicated Experiments when the Field Results are Incomplete," *Empire Journal of Agriculture*, Vol. **1** (1933), pp. 129–42.

41. ———, "The Principles of Orthogonality and Confounding in Replicated Experiments," *Journal of Agricultural Science*, Vol. **23** (1933), pp. 108–45.

42. ———, "Incomplete Randomized Blocks, "*Annals of Eugenics*, Vol. **7** (1936), pp. 121–40.

43. ———, "Incomplete Latin Squares," *Journal of Agricultural Science*, Vol. **26** (1936), pp. 301–15.

44. ———, "Orthogonal Functions and Tests of Significance in the Analysis of Variance," *Journal of the Royal Statistical Society*, Supplement 5 (1938), p. 177.

45. ———, and R. W. Hale, "The Analysis of Latin Squares when Two or More Rows, Columns, or Treatments are Missing," *Supplement of the Journal of the Royal Statistical Society*, Vol. **6** (1939), pp. 67–79.

13

AN INTRODUCTION TO FACTORIALS

Factorials are described, related to designs, and analyzed according to the procedures given in earlier chapters on analysis of variance. Factorial experiments are compared to the so-called classical experiments. Interaction is illustrated and defined. Different experimental models are introduced and compared. The topics of fractional factorials, confounding, and split plots are treated in summary fashion.

13.1. INTRODUCTION

With the introduction of each new design emphasis was given to the refinements which reduce the error mean square. In these designs we thought of the responses, or observations, as being affected by only one *treatment factor*, the other factors being for the control of the experimental error. However, two or more treatment factors may cause variation in the observations. Thus, in this chapter we think of variation in observations as resulting from two or more treatment factors in designs with one or more error-control factors. Experiments, or studies, with two or more treatment factors are abundant and can be found almost anywhere data are collected. Illustrations are found in the effect that

1. Different temperatures and different fabrics have on the percentage of shrinkage during dyeing.
2. Time of shift and amount of music have on the amount of absenteeism in a large plant.
3. Nitrogen, phosphorous, and potassium have on the yield of a crop.

4. Size of an egg, type of diet, and sex of a chick have on its weight at ten weeks of age.
5. Different breakers and different gaugers have on the compressive strength of a hardened cement cube.
6. Concentration of detergent, sodium carbonate, and sodium carboxymethyl cellulose have on the cleaning ability of a solution.
7. Baking temperature and recipes have on the size of a cake.
8. Weight and sexes of subjects and rate of stimulus have on the amount of physical response.

We use illustration 1 to introduce some of the terminology associated with factorials. Let T denote temperature and F fabric. Then three different temperatures may be designated by T_1, T_2, T_3 and four different fabrics by F_1, F_2, F_3, F_4. We refer to T_1, T_2, T_3 as *levels* of factor T, T_1 being termed the first level. Likewise, F_1, F_2, F_3, F_4 are termed the first, second, third, and fourth levels of F. The combination of the first level of T and the fourth level of F is denoted by T_1F_4 and called "the treatment combination of T_1 and F_4."

If in an experiment there are only three levels of T and four levels of F, then there are 12 possible treatment combinations. If each of the 12 possible treatment combinations is applied, the experiment is termed a *factorial experiment*, or, for short, a *factorial*. We also refer to this as a 3×4 *factorial experiment*, meaning that there are three levels of the first factor and four levels of the second factor in the experiment. (We do not say "factorial design," since the term "factorial" has to do with the combination of treatment levels in an experiment.) The treatment combinations may be applied to experimental units in a completely randomized design, randomized block design, Latin square design, and many other designs which have not been described.

13.2. AN APPLICATION OF A 3×2 FACTORIAL

In Example 13.1 we illustrate these definitions and principles as well as the method of analysis for a hypothetical 3×2 factorial experiment in a completely randomized design.

Example 13.1. Three levels of factor A and two levels of factor B are fixed in an experiment with 18 experimental units in which each of the six treatment combinations A_1B_1, A_1B_2, A_2B_1, A_2B_2, A_3B_1, A_3B_2 is to be randomly assigned to three units. Hypothetical data and notation for such an experiment are shown in Table 13.1. (The student may assume the data to be coded data for any two-factor factorial experiment he chooses to consider.) (a) Prepare an analysis of variance table, showing the variation due to factor

A, factor B, and interaction of A and B. (b) Find estimates of effects and test appropriate hypotheses.

Table 13.1

Hypothetical Data for a 3×2 Factorial in a Completely Randomized Design

Level of factor A		1		2		3	
Level of factor B		1	2	1	2	1	2
Treatment combination		$A_1 B_1$	$A_1 B_2$	$A_2 B_1$	$A_2 B_2$	$A_3 B_1$	$A_3 B_2$
Replication	1	$x_{111} = 24$	$x_{121} = 23$	$x_{211} = 16$	$x_{221} = 21$	$x_{311} = 19$	$x_{321} = 24$
	2	$x_{112} = 29$	$x_{122} = 19$	$x_{212} = 11$	$x_{222} = 21$	$x_{312} = 16$	$x_{322} = 21$
	3	$x_{113} = 25$	$x_{123} = 24$	$x_{213} = 15$	$x_{223} = 18$	$x_{313} = 16$	$x_{323} = 18$
Total		$T_{11.} = 78$	$T_{12.} = 66$	$T_{21.} = 42$	$T_{22.} = 60$	$T_{31.} = 51$	$T_{32.} = 63$

Grand total $T = 360$

Thinking of the data in Table 13.1 as a one-way classification, the treatment and total sums of squares are given by

$$SSTr = \frac{\sum\limits_{i}^{3} \sum\limits_{j}^{2} T_{ij.}^2}{3} - \frac{T^2}{18} = \frac{78^2 + \cdots + 63^2}{3} - \frac{360^2}{18} = 258$$

and

$$SST = \sum\limits_{i}^{3} \sum\limits_{j}^{2} \sum\limits_{u}^{3} x_{iju}^2 - \frac{T^2}{18} = 24^2 + \cdots + 18^2 - \frac{360^2}{18} = 330$$

respectively. The within sum of squares is then found to be

$$SSW = SST - SSTr = 72$$

These sums of squares, along with their corresponding degrees of freedom and mean squares, are shown in Table 13.2, an intermediate analysis of variance table.

Table 13.2

An Intermediate Analysis of Variance for the Data of Table 13.1

Source of Variation	Sum of Squares	Degrees of Freedom	Mean Square	Expected Mean Square for Fixed Model*
Treatment combinations	258	5	51.6	$\sigma^2 + 3 \sum\limits_{i}^{3} \sum\limits_{j}^{2} (\mu_{ij} - \mu)^2 / 5$
Within	72	12	6	σ^2
Total	330	17		

* See Eq. (12.10) for definition of $\mu_{ij} - \mu$.

In order to separate the treatment combination sum of squares into component sum of squares for factor A, factor B, and the "interaction of factors A and B," arrange the totals of Table 13.1 in a two-way classification, as shown in Table 13.3 (see Sect. 13.3 for a definition of interaction and Sect. 13.4 for a discussion of interaction). Observe that the total sum of

Table 13.3
Treatment Combination Totals in a Two-way Classification

Levels for Factor B	Levels for Factor A			Totals for Factor B
	1	2	3	
1	$T_{11.} = 78$	$T_{21.} = 42$	$T_{31.} = 51$	$T_{.1.} = 171$
2	$T_{12.} = 66$	$T_{22.} = 60$	$T_{32.} = 63$	$T_{.2.} = 189$
Totals for Factor A	$T_{1..} = 144$	$T_{2..} = 102$	$T_{3..} = 114$	$T = 360$

squares for the two-way classification of Table 13.3 is the same as the treatment sum of squares computed from Table 13.1. The sums of squares for factors A and B are, respectively

$$SSA = \frac{\sum_i^3 T_{i..}^2}{6} - \frac{T^2}{18} = \frac{144^2 + 102^2 + 114^2}{6} - \frac{360^2}{18} = 156$$

and

$$SSB = \frac{\sum_j^2 T_{.j.}^2}{9} - \frac{T^2}{18} = \frac{171^2 + 189^2}{9} - \frac{360^2}{18} = 18$$

Table 13.4
Analysis of Variance and Expected Mean Squares for the Data of Table 13.1

Source of Variation	Sum of Squares	Degrees of Freedom	Mean Square	Computed F	Expected Mean Squares for Fixed Model (α, β)*
Treatment combinations	258	5			$\sigma^2 + 3 \sum_i^3 \sum_j^2 (\mu_{ij} - \mu)^2/5$
Factor A	156	2	88	14.7	$\sigma^2 + 6 \sum_i^3 \alpha_i^2/2$
Factor B	18	1	18	3.0	$\sigma^2 + 9 \sum_j^2 \beta_j^2/1$
Interaction AB	84	2	42	7.0	$\sigma^2 + 3 \sum_i^3 \sum_j^2 (\mu_{ij} - \mu_{i.} - \mu_{.j} + \mu)^2/2$
Within	72	12	6		σ^2
Total	330	17			

* See Sect. 12.3 for definitions of $\mu_{ij} - \mu$, α_i, β_j, and $\mu_{ij} - \mu_{i.} - \mu_{.j} + \mu$.

Denoting the sum of squares due to the interaction of A and B by $SSAB$, we find, on subtraction, that

$$SSAB = SSTr - SSA - SSB = 84$$

Bringing all these sums of squares together, we have the analysis of variance shown in Table 13.4. That is, for the complete analysis of variance we first obtain the within, or error, sum of squares by making a one-way classification analysis, and then we partition the treatment combination sum of squares by doing a two-way classification analysis. Combining the two analyses, we see that the sum of squares identity is

$$SST = SSA + SSB + SSAB + SSW \qquad (13.1)$$

On comparing Eq. (13.1) with Eq. (12.62), we see that the sums of squares for factor A, factor B, interaction of A and B, and within are the same as the sums of squares for columns, rows, experimental error, and sampling error, respectively. It is clear from the above discussion that this analysis is the same as that for a two-way classification with the same number of observations per cell (see Sect. 12.8). But we shall soon see that the test procedures and interpretations are different.

Dividing the cell totals, factor A totals, factor B totals, and grand total in Table 13.3 by 3, 6, 9, and 18, respectively, gives the means which are shown in Table 13.5. Estimates of effects are found in terms of these means.

Table 13.5
Means for a 3×2 Factorial with Three Replications (Computed from Table 13.3)

Levels for Factor B	Levels for Factor A			Means for Factor B
	1	2	3	
1	$\bar{x}_{11.} = 26$	$\bar{x}_{21.} = 14$	$\bar{x}_{31.} = 17$	$\bar{x}_{.1.} = 19$
2	$\bar{x}_{12.} = 22$	$\bar{x}_{22.} = 20$	$\bar{x}_{32.} = 21$	$\bar{x}_{.2.} = 21$
Means for Factor A	$\bar{x}_{1..} = 24$	$\bar{x}_{2..} = 17$	$\bar{x}_{3..} = 19$	$\bar{x} = 20$

The estimated effects for the different levels of factors A and B are

$$a_1 = \bar{x}_{1..} - \bar{x} = 4$$
$$a_2 = \bar{x}_{2..} - \bar{x} = -3$$
$$a_3 = \bar{x}_{3..} - \bar{x} = -1$$
$$b_1 = \bar{x}_{.1.} - \bar{x} = -1$$
$$b_2 = \bar{x}_{.2.} - \bar{x} = 1$$

These effects estimate how much the means for levels A_1, A_2, A_3, B_1, B_2

deviate from the over-all mean. In a somewhat similar way we estimate the
effect of the cells by subtracting out the sum of the over-all effect (mean),
the factor A effect and the factor B effect from the cell means. These estimates
are shown in Table 13.6 and are termed the effects due to the interaction of
A and B, or, for short, the *interaction effects*. The notation $(ab)_{ij}$ is used to
denote the estimated interaction effect of the treatment combination with
the ith level of factor A and the jth level of factor B after the over-all A_i and
B_j effects have been removed. [The expression (ab) is to be read as one

Table 13.6

Estimated Interaction Effects for Data of Table 13.1

Levels for Factor B	Levels for Factor A		
	1	2	3
1	$(ab)_{11} = 3$ $[26 - 20 - 4 - (-1)]$	$(ab)_{21} = -2$ $[14 - 20 - (-3) - (-1)]$	$(ab)_{31} = -1$ $[17 - 20 - (-1) - (-1)]$
2	$(ab)_{12} = -3$ $(22 - 20 - 4 - 1)$	$(ab)_{22} = 2$ $[20 - 20 - (-3) - 1]$	$(ab)_{32} = 1$ $[21 - 20 - (-1) - 1]$

symbol denoting estimated interaction effect and is not to be taken as a
product.] The reader should notice that the sum of the estimated interaction
effects in any column or any row is zero. That is

$$\begin{cases} \sum_{i=1}^{3} (ab)_{ij} = 0 & (j = 1, 2) \\ \sum_{j=1}^{2} (ab)_{ij} = 0 & (i = 1, 2, 3) \end{cases} \qquad (13.2)$$

Appropriate hypotheses and corresponding test procedures are indicated
in Table 13.4. The three hypotheses are

$$\begin{cases} H_{01}: \text{the true effects for factor } A \text{ are equal to zero} \\ H_{02}: \text{the true effects for factor } B \text{ are equal to zero} \\ H_{03}: \text{the true interaction effects are equal to zero} \end{cases} \qquad (13.3)$$

To test H_{01}, find the ratio of the factor A mean square to the within mean
square and compare with the upper α level value of the F distribution with
two and 12 degrees of freedom. Since the computed F, 14.7, is larger than
the upper five per cent F value, 3.89, we reject H_{01} and conclude that the true
effects for factor A are different. Following similar procedures, we fail to
reject H_{02}, but we do reject H_{03}. That is, we conclude that the estimated
effects in Table 13.6 deviate too much from zero to attribute them to chance
fluctuations.

We now compare the factorial experiment for Example 13.1 with the so-called "classical" experiment, that is, the experiment in which only one factor is examined, all others being held constant, or nearly constant. In the factorial experiment each of three hypotheses was independently tested, using the same error variance with 12 degrees of freedom. In the classical experiment we could test hypothesis H_{01}, using an F distribution with two and 12 degrees of freedom, provided each of the three levels of A was replicated five times; we could test hypothesis H_{02}, using an F distribution with one and 12 degrees of freedom, if each of the two levels of B was replicated seven times. Thus, by making 29 observations we could test the two hypotheses H_{01} and H_{02} independently with 12 degrees of freedom, just as in the factorial experiment. Clearly, the factorial experiment is superior to the experiment just described on at least two counts; namely, fewer observations are required, and interaction of two factors may be estimated and tested.

It is interesting to note that the factorial experiment is frequently applied by investigators who have no special interest in a statistical analysis. Then interest lies in obtaining a "picture" of a large field of investigation rather than a detailed "picture" of selected topics. Thus, by applying a statistical analysis on a factorial experiment it would be possible, in addition to getting a broad "picture," to obtain unbiased estimators of effects, to make significant tests involving these effects, and to establish confidence intervals for single means or comparisons. In these ways a statistical analysis is likely to give much wider applications to experimental results.

13.3. TWO-FACTOR FACTORIAL EXPERIMENTS IN ONE-WAY CLASSIFICATION DESIGNS—FIXED MODEL

Before writing model equation (12.14), that is

$$x_{ij} = \mu + \alpha_i + \beta_j + \epsilon_{ij} \qquad (i = 1, \ldots, c; j = 1, \ldots, r)$$

for the two-way classification with one observation per cell, we made the assumption that $\mu_{ij} - \mu_{i.} - \mu_{.j} + \mu = 0$. Letting

$$(\alpha\beta)_{ij} = \mu_{ij} - \mu_{i.} - \mu_{.j} + \mu \qquad (i = 1, \ldots, c; j = 1, \ldots, r) \quad (13.4)$$

we now consider the analysis of variance for the case where $(\alpha\beta)_{ij}$ is not zero. A complete analysis for such a model usually requires that more than one observation per cell be made. Actually, the calculations are easiest and the information per observation greatest when an equal number of observations are made in each cell.

Thus, we consider the two-factor factorial experiment with n observations per treatment combination, that is, the particular two-way classification with n observations per cell which has the *model equation*

$$x_{iju} = \mu + \alpha_i + \beta_j + (\alpha\beta)_{ij} + \epsilon_{iju} \qquad \begin{aligned} i &= 1, \ldots, c \\ j &= 1, \ldots, r \\ u &= 1, \ldots, n \end{aligned} \qquad (13.5)$$

where x_{iju} denotes the uth random observation for the ith level of factor A and the jth level of factor B, μ denotes a constant over-all mean, α_i denotes the effect of the ith level of factor A, β_j denotes the effect of the jth level of factor B, $(\alpha\beta)_{ij}$ denotes the effect of the ith level of factor A and the jth level of factor B, and ϵ_{iju} denotes the uth random effect for the ith level of factor A and the jth level of factor B. Expression (13.5) has the same form as the model equation (12.59) for subsampling in a randomized block design. In Eq. (12.59) three of the effects are for the control of the error variance, whereas only one is for the description of treatment effects, but in Eq. (13.5) only one effect is for control of the error variance, whereas three are for the description of treatments. Replacing δ_{ij} of Eq. (12.59) by $(\alpha\beta)_{ij}$ indicates at a glance the change in design and emphasis. For a *fixed model* experiment the α_i, β_j, and $(\alpha\beta)_{ij}$ are assumed to be fixed, so that

$$\begin{cases} \sum_{i=1}^{c} \alpha_i = \sum_{j=1}^{r} \beta_j = 0 \\[2mm] \sum_{i=1}^{c} (\alpha\beta)_{ij} = 0 \qquad j = 1, \ldots, r \\[2mm] \sum_{j=1}^{r} (\alpha\beta)_{ij} = 0 \qquad i = 1, \ldots, c \end{cases} \qquad (13.6)$$

and the ϵ_{iju} are normally and independently distributed with zero means and common variance σ^2.

The sum of squares identity, not depending on the assumptions regarding effects, the same as Eq. (12.62), is illustrated in Example 13.1 and indicated in Table 13.7. Under the assumptions for the fixed model experiment, *fixed model* $[\alpha, \beta, (\alpha\beta)]$, it is easy to derive the expected mean squares shown in Table 13.7.

The effects of factor A and the effects of factor B are estimated in the usual way, and the interaction effect $(\alpha\beta)_{ij}$ is estimated by

$$(ab)_{ij} = \bar{x}_{ij} - \bar{x}_{i.} - \bar{x}_{.j} + \bar{x} \qquad (13.7a)$$

or

$$(ab)_{ij} = \bar{x}_{ij} - \bar{x} - a_i - b_j \qquad (13.7b)$$

For the analysis of variance of Table 13.7 the three hypotheses usually tested are given in (13.3). They may also be stated as

Table 13.7

Analysis of Variance for a Fixed Model Two-Factor Factorial Experiment
in a One-Way Classification Design

Source of Variation	Sum of Squares	Degrees of Freedom	Mean Squares	Expected Mean Squares for Fixed Model $[\alpha, \beta, (\alpha\beta)]$
Treatment combinations	$SSTr$	$cr - 1$	$s_0^2 = \dfrac{SSTr}{cr - 1}$	$\sigma^2 + n \dfrac{\sum\limits_{i}^{c}\sum\limits_{j}^{r}(\mu_{ij} - \mu)^2}{cr - 1}$
Factor A	SSA	$c - 1$	$s_1^2 = \dfrac{SSA}{c - 1}$	$\sigma^2 + rn \dfrac{\sum\limits_{i}^{c}\alpha_i^2}{c - 1}$
Factor B	SSB	$r - 1$	$s_2^2 = \dfrac{SSB}{r - 1}$	$\sigma^2 + cn \dfrac{\sum\limits_{j}^{r}\beta_j^2}{r - 1}$
Interaction	$SSAB = SSTr$ $- SSA - SSB$	$(c - 1)(r - 1)$	$s_3^2 = \dfrac{SSAB}{(c - 1)(r - 1)}$	$\sigma^2 + n \dfrac{\sum\limits_{i}^{c}\sum\limits_{j}^{r}(\alpha\beta)_{ij}^2}{(c - 1)(r - 1)}$
Within (error)	$SSW = SST$ $- SSTr$	$cr(n - 1)$	$s_4^2 = \dfrac{SSW}{cr(n - 1)}$	σ^2
Total	SST	$crn - 1$		

where

$$SSTr = \frac{\sum\limits_{i}^{c}\sum\limits_{j}^{r} T_{ij.}^2}{n} - \frac{T^2}{crn} \qquad SSA = \frac{\sum\limits_{i}^{c} T_{i..}^2}{rn} - \frac{T^2}{crn}$$

$$SSB = \frac{\sum\limits_{j}^{r} T_{.j.}^2}{cn} - \frac{T^2}{crn} \qquad SST = \sum\limits_{i}^{c}\sum\limits_{j}^{r}\sum\limits_{u}^{n} x_{iju}^2 - \frac{T^2}{crn}$$

$$T_{ij.} = \sum\limits_{u}^{n} x_{iju} \qquad T_{i..} = \sum\limits_{j}^{r}\sum\limits_{u}^{n} x_{iju} \qquad T_{.j.} = \sum\limits_{i}^{c}\sum\limits_{u}^{n} x_{iju}$$

$$
\begin{cases}
H_{01}: \alpha_1 = \cdots = \alpha_c = 0 \\
H_{02}: \beta_1 = \cdots = \beta_r = 0 \\
H_{03}: (\alpha\beta)_{11} = \cdots = (\alpha\beta)_{cr} = 0 \quad \text{or} \quad (\alpha\beta)_{ij} = 0 \quad (i = 1, \ldots, c \\
\hphantom{H_{03}: (\alpha\beta)_{11} = \cdots = (\alpha\beta)_{cr} = 0 \quad \text{or} \quad (\alpha\beta)_{ij} = 0 \quad} j = 1, \ldots, r)
\end{cases}
\tag{13.8}
$$

To test H_{01}, compute the ratio s_1^2/s_4^2 and compare with the upper α level value of the F distribution with $c - 1$ and $cr(n - 1)$ degrees of freedom. To test H_{02}, compare the ratio s_2^2/s_4^2 with the upper α level value of the F distribution with $r - 1$ and $cr(n - 1)$ degrees of freedom. To test H_{03}, compare the ratio s_3^2/s_4^2 with the upper α level value of the F distribution with $(c - 1)(r - 1)$ and $cr(n - 1)$ degrees of freedom. In a fixed model experiment all three hypotheses in (13.8) are usually tested at the same significance level α. Since the tests are independent, the over-all significance level for the experiment is then $1 - (1 - \alpha)^3$; that is, the probability that at least one of the hypotheses will be falsely rejected is $1 - (1 - \alpha)^3$.

It is possible to test other hypotheses in a factorial experiment. The null hypothesis that all the treatment combination means are equal, that is, $\mu_{ij} = \mu$ ($i = 1, \ldots, c; j = 1, \ldots, r$), can be tested by comparing s_0^2/s_4^2 with the upper α level of the F distribution with $cr - 1$ and $cr(n - 1)$ degrees of freedom. However, such a test is likely to be of no interest. Tests of linear comparisons are likely to be of much more interest. Usually, the linear comparisons involve the means for the levels of only one factor. The test procedure is illustrated in the following example.

Example 13.2. Use the data of Example 13.1 to test the following two hypotheses about linear comparisons among the three levels of factor A

$$H_{04}: \gamma_1 = 2\mu_{1.} - \mu_{2.} - \mu_{3.} = 0$$

$$H_{05}: \gamma_2 = \mu_{2.} - \mu_{3.} = 0$$

Using the totals of Table 13.3, we find that components of the treatment sum of squares are

$$Q_1^2 = \frac{\left(\sum\limits_{i=1}^{3} m_i T_{i.} \right)^2}{rn \sum\limits_{i=1}^{3} m_i^2} = \frac{[2(144) - 1(102) - 1(114)]^2}{6[2^2 + (-1)^2 + (-1)^2]} = 144$$

and

$$Q_2^2 = \frac{[1(102) - 1(114)]^2}{6[1^2 + (-1)^2]} = 12$$

The error variance is $s_4^2 = 6$ with 12 degrees of freedom. Thus, the computed F ratios are

$$\frac{Q_1^2}{s_4^2} = 24 \quad \text{and} \quad \frac{Q_2^2}{s_4^2} = 2$$

Since the upper five per cent value of the F distribution with one and 12 degrees of freedom is $F_{.05}(1, 12) = 4.75$, we reject H_{04} and fail to reject H_{05}. That is, we conclude that the true mean response for the first level of factor A is greater than the mean of the other two levels, and the true mean responses at the second and third levels are not different.

Duncan's multiple test procedure may also be applied in ranking the means for the levels of factor A or in ranking the means for the levels of factor B. Confidence intervals for means and comparisons may be obtained by the methods already described. Power of the tests may also be found by the usual methods.

13.4. INTERACTION IN A FACTORIAL EXPERIMENT

Since interaction is the source of variation that makes factorial experiments different from those already described, we now give special consideration to this very important concept. The term "interaction" has already been used several times as we referred to "interaction effects," "the interaction of factors A and B," "the sum of squares for the interaction of A and B," and "the test for interaction." It is clear that the interaction effect of the ith level of factor A and the jth level of factor B is defined by Eq. (13.4); that is, $(\alpha\beta)_{ij}$ is defined by

$$(\alpha\beta)_{ij} = \mu_{ij} - \mu_{i.} - \mu_{.j} + \mu$$

Expressed in terms of factor A and factor B effects, $(\alpha\beta)_{ij}$ may be written as

$$(\alpha\beta)_{ij} = \mu_{ij} - \alpha_i - \beta_j - \mu \tag{13.9}$$

It is understood that μ is an effect (mean) common to all cr populations, that α_i is an effect common to all levels of factor B, that β_j is an effect common to all levels of factor A, and $(\alpha\beta)_{ij}$ is an effect due to the combination of the ith level of A and the jth level of B, that is, a cell effect. But interaction may be explained in other, perhaps simpler, ways.

Consider a 2×3 factorial. There are six populations with means μ_{ij} ($i = 1, 2; j = 1, 2, 3$) and common variance σ^2. If interaction does not exist, $\mu_{ij} = \mu + \alpha_i + \beta_j$; if it does exist, then $\mu_{ij} = \mu + \alpha_i + \beta_j + (\alpha\beta)_{ij}$. Two-way tables of population means of treatment combinations, expressed in terms of effects, are given in Table 13.8. If the means in the A_1 column are subtracted term for term from the means in the A_2 column, observe that the differences are always the same for the case of no interaction, but that the differences are not the same for the case of interaction. Further, if the means in the $B_1(B_2)$ row are subtracted term for term from the means in the $B_2(B_1)$ row, a similar thing happens.

Table 13.8
Population Means in Terms of Effects

	No Interaction		Interaction	
	A_1	A_2	A_1	A_2
B_1	$\mu + \alpha_1 + \beta_1$	$\mu + \alpha_2 + \beta_1$	$\mu + \alpha_1 + \beta_1 + (\alpha\beta)_{11}$	$\mu + \alpha_2 + \beta_1 + (\alpha\beta)_{21}$
B_2	$\mu + \alpha_1 + \beta_2$	$\mu + \alpha_2 + \beta_2$	$\mu + \alpha_1 + \beta_2 + (\alpha\beta)_{12}$	$\mu + \alpha_2 + \beta_2 + (\alpha\beta)_{22}$
B_3	$\mu + \alpha_1 + \beta_3$	$\mu + \alpha_2 + \beta_3$	$\mu + \alpha_1 + \beta_3 + (\alpha\beta)_{13}$	$\mu + \alpha_2 + \beta_3 + (\alpha\beta)_{23}$

As an example, consider the particular means of two 2×3 factorials shown in Table 13.9. By taking the differences between responses at varying levels of $A(B)$ for each level of $B(A)$, we find that the pattern of differences is the same for each level of $B(A)$ in Table 13.9a. In fact, at each level of B the difference in responses at the two levels of A is 2 and at each level of A the differences are 1 and 4. That is, there is no interaction between factors A and B in Table 13.9a. In Table 13.9b the pattern of differences at each level of $B(A)$ changes. That is, at levels B_1 and B_2 the differences are 2, but at level B_3 the difference is -1. Further, at level A_1 the differences are 1 and 4 and at level A_2 the differences are 1 and 1. Thus, there is interaction between factors A and B in Table 13.9b.

Table 13.9

Population Means of Responses for Treatment Combinations in a 2×3 Factorial

	(a) *No Interaction*		*Factor B means*		(b) *Interaction*		*Factor B means*
	A_1	A_2			A_1	A_2	
B_1	12	14	13	B_1	12	14	13
B_2	13	15	14	B_2	13	15	14
B_3	17	19	18	B_3	17	16	16.5
Factor A means	14	16	15	Factor A means	14	15	14.5

Interaction may also be illustrated graphically by plotting the means $\mu_{i.}$ for the levels of factor A (means $\mu_{.j}$ for the levels of factor B) as abscissas and the several corresponding cell means μ_{ij} for each level of factor A (for each level of factor B) as ordinates and connecting the points for each level of factor B (factor A) by straight-line segments. If the resulting broken-line polygons are parallel, there is no interaction; otherwise, there is interaction. Two graphs for each of the 2×3 factorials given in Table 13.9 are shown in Fig. 13.1. When there is no interaction, the broken-line polygon is actually a straight-line segment with slope of one. Both a and b of Fig. 13.1 illustrate this point. Further, c and d of Fig. 13.1 indicate the presence of interaction. In most cases the investigator has a choice as to which of two graphs to construct in illustrating the presence or absence of interaction. [It is left as an exercise for the reader to determine what to do in cases where two or more of the means $\mu_{i.}$ ($\mu_{.j}$) are the same.] After trends are discussed in Chap. 14, we give another geometric representation of interaction.

Thus, in every way we have looked at interaction, we see that interaction of A and B exists if and only if the magnitude of the difference in response in changing from one level of $A(B)$ to another depends on the level of $B(A)$ at which the difference is determined. This means that two factors combine to produce an effect not due to either one of them alone.

In working with sample means $\bar{x}_{ij.}$, $\bar{x}_{i..}$, and $\bar{x}_{.j.}$, constructions similar to those in Fig. 13.1 can be made, but, since these means are random variables, the absence of parallel lines no longer necessarily indicates the

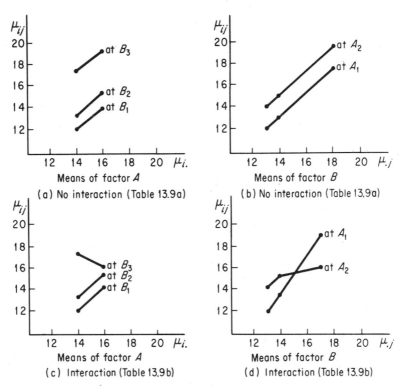

Fig. 13.1 Graphs Illustrating Absence and Presence of Interaction

presence of interaction. Actually, when there is no interaction, the lines may be far from parallel, particularly when the samples are small. The apparent interaction is declared significant when it is too large to explain on the basis of chance. To illustrate, Fig. 13.2 shows the significant interaction of Example 13.1.

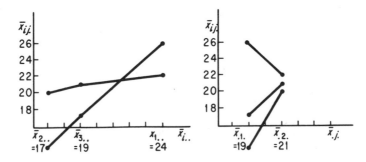

Fig. 13.2 Interaction of A and B in Example 13.1

13.5. EXTENSIONS OF THE FIXED-MODEL FACTORIAL EXPERIMENT

Two extensions of the factorial are considered. First, we discuss the two-factor factorial in the randomized block design. Then we give the analysis for factorials with three or more factors in either the completely randomized, randomized block, or Latin square design.

For the two-factor factorial in a randomized block design, the model equation is

$$x_{ijk} = \mu + \alpha_i + \beta_j + (\alpha\beta)_{ij} + \rho_k + \epsilon_{ijk} \qquad \begin{aligned} i &= 1, \ldots, c \\ j &= 1, \ldots, r \qquad (13.10) \\ k &= 1, \ldots, b \end{aligned}$$

where μ, α_i, β_j, $(\alpha\beta)_{ij}$, and ϵ_{ijk} are defined as in Sect. 13.3, and ρ_k denotes the effect of the kth block. We define the effects such that

$$\begin{cases} \sum_i \alpha_i = \sum_j \beta_j = \sum_k \rho_k = 0 \\ \sum_i (\alpha\beta)_{ij} = \sum_j (\alpha\beta)_{ij} = 0 \end{cases} \qquad (13.11)$$

and, for the analysis, assume that the ϵ_{ijk} are normally and independently distributed with mean zero and common variance σ^2. It is to be understood that each of the cr treatment combinations is randomly assigned to the cr experimental units in each block.

From these statements it follows that the sum of squares for factor A, factor B, interaction of A and B, and total are computed as in Table 13.7, and that the block and error sum of squares are given by

$$SSBl = \frac{\sum_k T_{..k}^2}{cr} - \frac{T^2}{crn} \qquad (13.12)$$

and

$$SSE = SST - SSTr - SSBl \qquad (13.13)$$

respectively. That is, the within sum of squares, SSW, in Table 13.7 is partitioned so that

$$SSW = SSBl + SSE \qquad (13.14)$$

It is easy to derive the expected mean squares shown in Table 13.10, and to see that the estimation and test procedures are similar to those for a fixed-model factorial experiment in a completely randomized design. Since the error sum of squares is obtained by taking the block sum of squares out of the within sum of squares, the number of degrees of freedom for the error variance is reduced from $cr(b - 1)$ to $(cr - 1)(b - 1)$.

Table 13.10

Analysis of Variance for a Fixed-Model Two-Factor Factorial Experiment in a Randomized Block Design

Source of Variation	Sum of Squares	Degrees of Freedom	Mean Squares	Expected Mean Squares for Fixed Model $[\alpha, \beta, (\alpha\beta), \rho]$
Block (replicates)	$SSBl$	$b - 1$	$s_0^2 = \dfrac{SSBl}{b - 1}$	$\sigma^2 + cr\dfrac{\sum\limits_k \rho_k^2}{b - 1}$
Factor A	SSA	$c - 1$	$s_1^2 = \dfrac{SSA}{c - 1}$	$\sigma^2 + rb\dfrac{\sum\limits_i \alpha_i^2}{c - 1}$
Factor B	SSB	$r - 1$	$s_2^2 = \dfrac{SSB}{r - 1}$	$\sigma^2 + cb\dfrac{\sum\limits_j \beta_j^2}{r - 1}$
Interaction	$SSAB$	$(c - 1)(r - 1)$	$s_3^2 = \dfrac{SSAB}{(c - 1)(r - 1)}$	$\sigma^2 + b\dfrac{\sum\limits_i \sum\limits_j (\alpha\beta)_{ij}^2}{(c - 1)(r - 1)}$
Error	SSE	$(cr - 1)(b - 1)$	$s_4^2 = \dfrac{SSE}{(cr - 1)(b - 1)}$	σ^2
Total	SST	$crb - 1$		

A two-factor factorial experiment in a Latin square design has restricted application due to the fact that the number of treatment combinations, cr, must equal the size of the square, b. Certain factorials, particularly the 2×2, 2×4, 2×5, 2×6, and 3×4 factorials, may be used to advantage in some instances. The model equation, sum of squares identity, analysis of variance, and expected mean squares are easy to write as simple extensions of the arguments for the two-factor factorial in the completely randomized and randomized block designs.

Extensions of the factorial to three or more factors is straightforward. We discuss the case for three factors with each treatment combination replicated an equal number of times in a completely randomized design. (The analysis of factorials with more than two factors in a randomized block design or Latin square design is left as an exercise for the reader.) The model equation is given by

$$x_{ijku} = \mu + \alpha_i + \beta_j + \gamma_k + (\alpha\beta)_{ij} + (\alpha\gamma)_{ik} + (\beta\gamma)_{jk} + (\alpha\beta\gamma)_{ijk} + \epsilon_{ijku}$$

$$i = 1, \ldots, c \quad j = 1, \ldots, r \quad k = 1, \ldots, l \quad u = 1, \ldots, n \qquad (13.15)$$

where the terms x_{ijku}, μ, α_i, β_j, $(\alpha\beta)_{ij}$, and ϵ_{ijku} are like those given in Sect. 13.3. Also, γ_k denotes the effect of the kth level of the factor C, $(\alpha\gamma)_{ik}$ denotes the effect of the one-way interaction of the ith level of A and the kth level of C, $(\beta\gamma)_{jk}$ denotes the effect of the one-way interaction of the jth level of B and the kth level of C, and $(\alpha\beta\gamma)_{ijk}$ denotes the effect of the two-way interaction of the ith level of A, jth level of B, and kth level of C. These effects are to have the restrictions given in Eq. (13.6) along with other similar restrictions required because of the additional factor C.

The sum of squares identity has nine terms. In order to give computing formulas for these sums of squares, first let

$$\begin{cases} T_{ijk.} = \sum_u x_{ijku} & T_{ij..} = \sum_k \sum_u x_{ijku} \\[2mm] T_{i.k.} = \sum_j \sum_u x_{ijku} & T_{.jk.} = \sum_i \sum_u x_{ijku} \\[2mm] T_{i...} = \sum_j \sum_k \sum_u x_{ijku} & T_{.j..} = \sum_i \sum_k \sum_u x_{ijku} \\[2mm] T_{..k.} = \sum_i \sum_j \sum_u x_{ijku} & T = \sum_i \sum_j \sum_k \sum_u x_{ijku} \\[2mm] CT = \dfrac{T^2}{crln} \end{cases} \quad (13.16)$$

The necessary sums of squares are given by

$$SST = \sum_i \sum_j \sum_k \sum_u x_{ijku}^2 - CT \qquad (13.17)$$

$$SSTr(ABC) = \frac{\sum_i \sum_j \sum_k T_{ijk.}^2}{n} - CT \qquad (13.18)$$

$$SSTr(AB) = \frac{\sum_i \sum_j T_{ij..}^2}{ln} - CT \qquad (13.19)$$

$$SSTr(AC) = \frac{\sum_i \sum_k T_{i.k.}^2}{rn} - CT \qquad (13.20)$$

$$SSTr(BC) = \frac{\sum_j \sum_k T_{.jk.}^2}{cn} - CT \qquad (13.21)$$

$$SSA = \frac{\sum_i T_{i...}^2}{rln} - CT \qquad (13.22)$$

$$SSB = \frac{\sum_j T_{.j..}^2}{cln} - CT \qquad (13.23)$$

$$SSC = \frac{\sum_k T_{..k.}^2}{crn} - CT \qquad (13.24)$$

$$SSAB = SSTr(AB) - SSA - SSB \qquad (13.25)$$

$$SSAC = SSTr(AC) - SSA - SSC \tag{13.26}$$

$$SSBC = SSTr(BC) - SSB - SSC \tag{13.27}$$

$$\begin{aligned} SSABC = SSTr(ABC) &- SSA - SSB - SSC \\ &- SSAB - SSAC - SSBC \end{aligned} \tag{13.28}$$

$$SSE = SST - SSTr(ABC) \tag{13.29}$$

The analysis of variance and expected mean squares are shown in Table 13.11.

Table 13.11
Analysis of Variance for a Fixed-Model Three-Factor Factorial Experiment in a
Completely Randomized Design

Source of Variation	Sum of Squares	Degrees of Freedom	Mean Squares	Expected Mean Square for Fixed Model
Factor A	SSA	$c - 1$	s_1^2	$\sigma^2 + rln\dfrac{\sum_i \alpha_i^2}{c - 1}$
Factor B	SSB	$r - 1$	s_2^2	$\sigma^2 + cln\dfrac{\sum_j \beta_j^2}{r - 1}$
Factor C	SSC	$l - 1$	s_3^2	$\sigma^2 + crn\dfrac{\sum_k \gamma_k^2}{l - 1}$
Interaction $A \times B$	$SSAB$	$(c - 1)(r - 1)$	s_4^2	$\sigma^2 + ln\dfrac{\sum_i \sum_j (\alpha\beta)_{ij}^2}{(c - 1)(r - 1)}$
Interaction $A \times C$	$SSAC$	$(c - 1)(l - 1)$	s_5^2	$\sigma^2 + rn\dfrac{\sum_i \sum_k (\alpha\gamma)_{ik}^2}{(c - 1)(l - 1)}$
Interaction $B \times C$	$SSBC$	$(r - 1)(l - 1)$	s_6^2	$\sigma^2 + cn\dfrac{\sum_j \sum_k (\beta\gamma)_{jk}^2}{(r - 1)(l - 1)}$
Interaction $A \times B \times C$	$SSABC$	$(c - 1)(r - 1)(l - 1)$	s_7^2	$\sigma^2 + n\dfrac{\sum_i \sum_j \sum_k (\alpha\beta\gamma)_{ijk}^2}{(c - 1)(r - 1)(l - 1)}$
Within error	SSE	$crl(n - 1)$	s_8^2	σ^2
Total	SST	$crln - 1$		

For a three-factor factorial experiment in a randomized block design, the block sum of squares (degrees of freedom) is taken out of the within sum of squares (degrees of freedom) of Table 13.11, all other sums of squares (degrees of freedom) remaining the same. For a factorial in a Latin square design, the row and column sums of squares are taken out of the within sum of squares of Table 13.11. If there are more than three factors in either a completely randomized, randomized block, or Latin square design, the methods already given can be used to obtain the analysis of variance. In order to illustrate a good computational technique for the analysis of a four-factor factorial in a randomized block design, we give Example 13.3.

Example 13.3. The primary objective of this experiment was to gain quantitative information about the ability of various concrete mixes to protect reinforcing steel against corrosion over long periods of time. The electrical-resistance method was used to evaluate the protection given the reinforcing steel by the surrounding concrete. In this method a thin steel ribbon was cast in the thin concrete test sections to simulate the reinforcing steel. When corrosion of the steel occurred, the cross-sectional area of the steel ribbon decreased, and the electrical resistance of the ribbon increased. Hence, by periodic measurements of the electrical resistance of the ribbon, the progress of the corrosion was recorded and plotted.

Test panels were 2.75 in. wide and 12 in. long. Their thickness T was either 0.750, 1.125, or 1.500 in. A steel ribbon 0.008 by 0.250 by 10 in. was soldered to two 0.25 in. square copper bars, each 2 in. long. These copper bars served as electrode terminals and protruded about one inch from the ends of the concrete test panel. Prior to embedment in the fresh concrete, the ribbon assembly was cleaned with a solvent, and the soldered joint and terminal were covered with three coats of Seal-Glo. The solvent was applied to de-oil the ribbon, and the Seal-Glo was applied to prevent electrolytic action between the copper, the solder, and the steel. The ribbon was centered in the panel with its width parallel to the panel depth. The concrete used in molding the test panels was made from six or seven sacks S of the same type of cement, 6 or 6.5 gallons G of water per sack of cement, and tap water W_1 or sea water W_2 having a specific gravity of 1.028. The sand and limestone were the same for all test panels. The experiment was replicated three times.

The resistance R at any time divided by the initial resistance R_0 was taken as the measure of corrosion ρ. A graph of ln ρ against time (number of cycles under the accelerated test conditions) usually plotted as a straight line. Its "least squares" slope, b, measured the relative protection given the reinforcing, large numerical values of b indicating shorter life for the reinforcing steel.

The above is a $3 \times 2 \times 2 \times 2$, or 3×2^3, factorial experiment replicated three times in a randomized block design. The slopes, b, of the regression lines multiplied by 10^5 are given in Table 13.12 for the 72 test specimens.

The replicate (block) totals are 104,901, 88,760, and 112,572, respectively. Other totals required for the analysis of variance are shown in Tables 13.13 through 13.23. The border totals are useful for checking purposes, and those in the two-way tables are required for the sums of squares of factors G, W, S, and T.

Table 13.12

Slope of Regression Line Times 10^5 for a $3 \times 2 \times 2 \times 2$ Factorial Experiment in a Randomized Block Design* (Protection of Concrete Mixes against Corrosion of Steel)

Replicate				G_1		G_2	
				W_1	W_2	W_1	W_2
1	S_1	T_1		434	1,995	28	20,240
		T_2		196	10,410	94	9,738
		T_3		1,233	14,310	-705	5,808
	S_2	T_1		13	2,308	-19	13,310
		T_2		641	3,026	27	4,125
		T_3		-1	16,940	113	627
2	S_1	T_1		341	4,661	62	16,740
		T_2		174	12,620	289	5,078
		T_3		3,466	9,105	216	1,306
	S_2	T_1		48	4,224	60	3,942
		T_2		$-7,131$	3,669	184	2,915
		T_3		$-2,998$	13,350	49	16,390
3	S_1	T_1		224	8,661	115	5,561
		T_2		78	17,730	85	13,090
		T_3		$-4,159$	6,982	182	7,951
	S_2	T_1		$-1,001$	4,374	17	5,041
		T_2		55	1,903	-98	9,772
		T_3		37	21,620	52	14,300

* Source: Applied Mechanics Department, Virginia Polytechnic Institute, 1956.

Table 13.13

Four-way ($G \times W \times S \times T$) Table Obtained by Summing over Replicates in Table 13.12

		G_1		G_2		Totals
		W_1	W_2	W_1	W_2	
S_1	T_1	999	15,317	205	42,541	59,062
	T_2	448	40,760	468	27,906	69,582
	T_3	540	30,397	-307	15,065	45,695
S_2	T_1	-940	10,906	58	22,293	32,317
	T_2	$-6,435$	8,598	113	16,812	19,088
	T_3	$-2,962$	51,920	214	31,317	80,489
Totals		$-8,350$	157,898	751	155,934	306,233

Table 13.14

Three-way ($W \times S \times T$) Table Obtained by Summing over Factor G
in Table 13.13

		W_1	W_2	*Totals*
	T_1	1,204	57,858	59,062
S_1	T_2	916	68,666	69,582
	T_3	233	45,462	45,695
	T_1	−882	33,199	32,317
S_2	T_2	−6,322	25,410	19,088
	T_3	−2,748	83,237	80,489
Totals		−7,599	313,832	306,233

Table 13.15

Three-way ($G \times S \times T$) Table Obtained by Summing over Factor W
in Table 13.13

		G_1	G_2	*Totals*
	T_1	16,316	42,746	59,062
S_1	T_2	41,208	28,374	69,582
	T_3	30,937	14,758	45,695
	T_1	9,966	22,351	32,317
S_2	T_2	2,163	16,925	19,088
	T_3	48,958	31,531	80,489
Totals		149,548	156,685	306,233

Table 13.16

Three-way ($G \times W \times T$) Table Obtained by Summing over Factor S
in Table 13.13

	G_1		G_2		*Totals*
	W_1	W_2	W_1	W_2	
T_1	59	26,223	263	64,834	91,379
T_2	−5,987	49,358	581	44,718	88,670
T_3	−2,422	82,317	−93	46,382	126,184
Totals	−8,350	157,898	751	155,934	306,233

Table 13.17

Three-way ($G \times W \times S$) Table Obtained by Summing over Factor T in Table 13.13

	G_1		G_2		Totals
	W_1	W_2	W_1	W_2	
S_1	1,987	86,474	366	85,512	174,339
S_2	−10,337	71,424	385	70,422	131,894
Totals	−8,350	157,898	751	155,934	306,233

Table 13.18

$G \times T$ Table from Table 13.16

	G_1	G_2	Totals
T_1	26,282	65,097	91,379
T_2	43,371	45,299	88,670
T_3	79,895	46,289	126,184
Totals	149,548	156,685	306,233

Table 13.19

$W \times T$ Table from Table 13.14

	W_1	W_2	Totals
T_1	322	91,057	91,379
T_2	−5,406	94,076	88,670
T_3	−2,515	128,699	126,184
Totals	−7,599	313,832	306,233

Table 13.20

$S \times T$ Table from Table 13.15

	S_1	S_2	Totals
T_1	59,062	32,317	91,379
T_2	69,582	19,088	88,670
T_3	45,695	80,789	126,184
Totals	174,339	131,894	306,233

Table 13.21

$G \times W$ Table from Table 13.17

	G_1	G_2	Totals
W_1	−8,350	751	−7,599
W_2	157,898	155,934	313,832
Totals	149,548	156,685	306,233

Table 13.22

$W \times S$ Table from Table 13.14

	W_1	W_2	Totals
S_1	2,353	171,986	174,339
S_2	−9,952	141,846	131,894
Totals	−7,599	313,832	306,233

Table 13.23

$G \times S$ Table from Table 13.15

	G_1	G_2	Totals
S_1	88,461	85,878	174,339
S_2	61,087	70,807	131,894
Totals	149,546	156,685	306,233

Once the sum of squares of factors G, W, S, and T are found, the six two-factor interaction sums of squares are computed in the usual way from Tables 13.18 through 13.23. Then the four three-factor interaction sums of squares are found by using formulas like Eq. (13.28). That is, any three-factor interaction sum of squares is obtained by subtracting from the three-factor total sum of squares three single-factor sums of squares and three two-factor interaction sums of squares. The four-factor sum of squares is obtained in an analogous manner by subtracting from the four-factor total sum of squares the four single-factor, the six two-factor, and the four three-factor sums of squares. This rule can be extended to any number of factors. The replicate, total, and error sums of squares are computed in the usual way. Table 13.24 shows the resulting analysis of variance.

Table 13.24

Analysis of Variance for the Fixed-Effects Four-Factor Factorial Experiment in the Randomized Block Design of Example 13.3

Source of Variation	Sum of Squares	Degrees of Freedom	Mean Square	Calculated F
Replication	12,310,937	2	6,155,469	
G (gal/sack)	707,455	1	707,455	
W (type water)	1,434,970,666	1	1,434,970,666	111.49
S (sacks/yd^3)	25,021,917	1	25,021,917	
T (thickness)	36,472,596	2	18,236,298	
$G \times T$	109,279,406	2	54,639,703	4.25
$W \times T$	37,805,268	2	18,902,634	
$S \times T$	161,459,812	2	80,729,906	6.27
$G \times W$	1,700,475	1	1,700,475	
$W \times S$	4,417,878	1	4,417,878	
$G \times S$	2,102,275	1	2,102,275	
$W \times S \times T$	140,077,459	2	70,038,730	5.44
$G \times S \times T$	37,912,665	2	18,956,333	
$G \times W \times T$	126,001,634	2	63,000,817	4.89
$G \times W \times S$	2,129,704	1	2,129,704	
$G \times W \times S \times T$	23,227,457	2	11,613,719	
Error	592,187,398	46	12,871,030	
Total	2,735,474,065	71		

Since all treatment and interaction effects are fixed in this experiment, each hypothesis is tested by comparing the appropriate mean square with the error mean square. Those F ratios which are significant at the five per cent level are shown in Table 13.24. (Actually, since $F_{.05}(2, 46) = 3.13$, $F_{.01}(2, 46) = 4.92$, and $F_{.01}(1, 46) = 7.00$, we see that three of the five effects which are significant at the five per cent level are also significant at the one per cent level, and a fourth, the three-way interaction of T, G, and W is almost significant at the one per cent level.) The four significant interaction effects all involve thickness of concrete panels, but the thickness effects are not significant. The two-factor interaction effects can be interpreted by by the methods of Sect. 13.4. In studying the three-factor interaction $T \times S \times W$, say, it should be noted that three factors in combination work in such a way that the average response at certain combinations of levels is significantly different from the average response at other combinations of levels. For example, levels W_1, S_2, and T_2 in combination give the best response; that is, concrete panels 1.125 in. thick made with tap water, 7 sacks of cement per cubic yard, and either 6 or 6.5 gal of water give the least corrosion to embedded steel. (For a better understanding of the inter-actions in this experiment the interested reader can study Tables 13.14, 13.16, 13.18, and 13.20 in some detail.) It is obvious, on looking at Table 13.13, that tap water is better than salt water for all treatment combinations considered. Since this is a fixed-level experiment, statistical statements can be made only about the levels included in the experiment.

The above computational procedure was given on account of the many checks, and because the data can be studied as the tables of totals are constructed in preparation for the analysis of variance. Then, after certain effects are declared significant, these same tables of totals may be used in a detailed examination of differences in the experiment. Further, these tables are useful in finding estimates of effects and in presenting final conclusions resulting from the experiment.

Special important cases of factorials, such as 2^n and 3^n factorials, may be analyzed by shorter, more appropriate techniques. Yates [56, 57] has presented a special notation and a special method for computing mean squares and interpreting results, and others [5, 12, 15, 19, 22, 23, 30, 46] give important details for analysis of certain regular factorials. The interested reader can consult these and other references along with exercises in this chapter for an understanding of the principles.

The response to a quantitative factor may show a recognizable trend over the various levels. If this is the case, the trend should be specified and taken into account in the analysis. So far, we have assumed that no such trend could be specified in advance of the experiment. Thus, in a sense the methods presented so far are incomplete. After discussing trends in Chap. 14, we return to the problem of trends in analysis of variance. The reader should study Refs. [14, 41, 54] for further information on quantitative and quali-tative factors and interactions.

13.6. RANDOM-EFFECTS AND MIXED-EFFECTS TWO-FACTOR FACTORIAL EXPERIMENTS IN A ONE-WAY CLASSIFICATION

The model equation and sum of squares identity for the random- and mixed-effects two-factor factorials are the same as those given in Sect. 13.3 for the fixed-effects model. But the expected mean squares, and, thus, tests of hypotheses and estimation of variance components depend on the assump-tions regarding the effects.

For the random-model factorial experiment with n observations per treatment combination, we assume that the effects α_i and β_j are inde-pendently selected from normal populations with zero means and variances

σ_α^2 and σ_β^2, respectively. The interaction effects $(\alpha\beta)_{ij}$ are also assumed to be independently and normally distributed with mean zero and variance $\sigma_{\alpha\beta}^2$. The restrictions in Eq. (13.6) do not hold for the random model.

Two mixed models are possible, but, since they are symmetric in A and B, we consider only one. Assume that the α_i are fixed so that

$$\begin{cases} \sum_i \alpha_i = 0 \\ \sum_i (\alpha\beta)_{ij} = 0 \qquad j = 1, \ldots, r \end{cases} \qquad (13.30)$$

Note that we do *not* assume that

$$\sum_j (\alpha\beta)_{ij} = 0 \qquad i = 1, \ldots, c$$

The β_j are assumed to be randomly and normally distributed with mean zero and varaince σ_β^2. Since the levels of A are fixed and the levels of B are randomly selected, the effects $(\alpha\beta)_{ij}$ are assumed to be independently and normally distributed with mean zero and variance $\sigma_{\alpha\beta}^2$. All random effects in all models are independently distributed.

Under the assumptions given above, the analysis of variance and expected mean squares shown in Table 13.25 along with the distribution of variance ratios are easy to derive. As in other cases, the expected mean squares can be used as guides in estimating variance components, testing hypotheses, establishing confidence limits, and finding power for the various tests.

The test procedures for the random and mixed models are clearly indicated by the expected mean squares. Nevertheless, it is important to note that for the mixed model the mean square for fixed effects α_i is compared with the interaction mean square, whereas the mean square for the random effects β_j is compared with the within mean square.

Table 13.25

Analysis of Variance and Expected Mean Squares for Random- and Mixed-Model Two-factor Factorial Experiments in a One-way Classification

Source of Variation	Mean Squares	Expected Mean Squares for	
		Random Model	Mixed Model (α)
Factor A	$s_1^2 = \dfrac{SSA}{c-1}$	$\sigma^2 + n\sigma_{\alpha\beta}^2 + rn\sigma_\alpha^2$	$\sigma^2 + n\sigma_{\alpha\beta}^2 + rn\dfrac{\sum_i \alpha_i^2}{c-1}$
Factor B	$s_2^2 = \dfrac{SSB}{r-1}$	$\sigma^2 + n\sigma_{\alpha\beta}^2 + cn\sigma_\beta^2$	$\sigma^2 + cn\sigma_\beta^2$
Interaction	$s_3^2 = \dfrac{SSAB}{(c-1)(r-1)}$	$\sigma^2 + n\sigma_{\alpha\beta}^2$	$\sigma^2 + n\sigma_{\alpha\beta}^2$
Within (error)	$s_4^2 = \dfrac{SSW}{cr(n-1)}$	σ^2	σ^2
Total	$s_5^2 = \dfrac{SST}{crn-1}$		

13.7. EXTENSIONS OF THE RANDOM- AND MIXED-MODEL FACTORIAL EXPERIMENT

We consider in turn the analysis of variance for three-factor factorial experiments in a completely randomized design in which three factors are random, two factors are random, and one factor is random. The notation, model equation (13.15), and sum of squares identity are the same as in Sect. 13.5.

For the random model assume the effects α_i, β_j, γ_k, $(\alpha\beta)_{ij}$, $(\alpha\gamma)_{ik}$, $(\beta\gamma)_{jk}$, $(\alpha\beta\gamma)_{ijk}$ and ϵ_{ijku} to be independently and randomly distributed with means zero and variances σ_α^2, σ_β^2, σ_γ^2, $\sigma_{\alpha\beta}^2$, $\sigma_{\alpha\gamma}^2$, $\sigma_{\beta\gamma}^2$, $\sigma_{\alpha\beta\gamma}^2$, and σ^2, respectively. The analysis of variance and expected mean squares are shown in Table 13.26. From the expected mean squares of Table 13.26 it is clear that there are exact procedures for testing all two-factor and three-factor interaction variance components. However, exact tests for the hypotheses

$$H_{01}: \sigma_\alpha^2 = 0$$

$$H_{02}: \sigma_\beta^2 = 0 \qquad\qquad (13.31)$$

$$H_{03}: \sigma_\gamma^2 = 0$$

are not possible if the mean squares of Table 13.26 are used. An approximate

Table 13.26

Analysis of Variance for a Random-Model Three-Factor Factorial Experiment in a Completely Randomized Design

Source of Variation	Mean Squares	Expected Mean Squares for the Random Model
Factor A	$s_1^2 = \dfrac{SSA}{c-1}$	$\sigma^2 + n\sigma_{\alpha\beta\gamma}^2 + ln\sigma_{\alpha\beta}^2 + rn\sigma_{\alpha\gamma}^2 + rln\sigma_\alpha^2$
Factor B	$s_2^2 = \dfrac{SSB}{r-1}$	$\sigma^2 + n\sigma_{\alpha\beta\gamma}^2 + ln\sigma_{\alpha\beta}^2 + cn\sigma_{\beta\gamma}^2 + cln\sigma_\beta^2$
Factor C	$s_3^2 = \dfrac{SSC}{l-1}$	$\sigma^2 + n\sigma_{\alpha\beta\gamma}^2 + rn\sigma_{\alpha\gamma}^2 + cn\sigma_{\beta\gamma}^2 + crn\sigma_\gamma^2$
$A \times B$	$s_4^2 = \dfrac{SSAB}{(c-1)(r-1)}$	$\sigma^2 + n\sigma_{\alpha\beta\gamma}^2 + ln\sigma_{\alpha\beta}^2$
$A \times C$	$s_5^2 = \dfrac{SSAC}{(c-1)(l-1)}$	$\sigma^2 + n\sigma_{\alpha\beta\gamma}^2 + rn\sigma_{\alpha\gamma}^2$
$B \times C$	$s_6^2 = \dfrac{SSBC}{(r-1)(l-1)}$	$\sigma^2 + n\sigma_{\alpha\beta\gamma}^2 + cn\sigma_{\beta\gamma}^2$
$A \times B \times C$	$s_7^2 = \dfrac{SSABC}{(c-1)(r-1)(l-1)}$	$\sigma^2 + n\sigma_{\alpha\beta\gamma}^2$
Within error	$s_8^2 = \dfrac{SSE}{crl(n-1)}$	σ^2
Total	$s_9^2 = \dfrac{SST}{crln-1}$	

test (see Sect. 11.4) due to Satterthwaite [43] may be applied. In the approximate test of H_{01}, compute

$$\frac{s_1^2}{s_4^2 + s_5^2 - s_7^2} \tag{13.32}$$

and compare with the upper α level value of the F distribution with $\nu_1 = c - 1$ and $\nu_2 = \hat{\nu}$ degrees of freedom where

$$\hat{\nu} = \frac{(s_4^2 + s_5^2 - s_7^2)^2}{\dfrac{(s_4^2)^2}{(c-1)(r-1)} + \dfrac{(s_5^2)^2}{(c-1)(l-1)} - \dfrac{(s_7^2)^2}{(c-1)(r-1)(l-1)}} \tag{13.33}$$

Note that when the variance components σ^2, $\sigma_{\alpha\beta\gamma}^2$, $\sigma_{\alpha\beta}^2$, $\sigma_{\alpha\gamma}^2$, and σ_α^2 are replaced by their estimates s^2, $s_{\alpha\beta\gamma}^2$, $s_{\alpha\beta}^2$, $s_{\alpha\gamma}^2$, and s_α^2, respectively, the expression in (13.32) may be written as

$$\frac{s^2 + ns_{\alpha\beta\gamma}^2 + lns_{\alpha\beta}^2 + rns_{\alpha\gamma}^2 + rlns_\alpha^2}{(s^2 + ns_{\alpha\beta\gamma}^2 + lns_{\alpha\beta}^2) + (s^2 + ns_{\alpha\beta\gamma}^2 + rns_{\alpha\gamma}^2) - (s^2 + ns_{\alpha\beta\gamma}^2)}$$

or

$$\frac{(s^2 + ns_{\alpha\beta\gamma}^2 + lns_{\alpha\beta}^2 + rns_{\alpha\gamma}^2) + rlns_\alpha^2}{s^2 + ns_{\alpha\beta\gamma}^2 + lns_{\alpha\beta}^2 + rns_{\alpha\gamma}^2}$$

The hypotheses H_{02} and H_{03} may be tested by a similar method.

For the analysis of the mixed-model three-factor factorial experiment,

Table 13.27

Expected Mean Squares for Two Mixed-Model Three-Factor Factorial Experiments in a Completely Randomized Design

Source of Variation	Mean Squares	Expected Mean Squares for the	
		Mixed Model (α)	Mixed Model (α, β)
Factor A	s_1^2	$\sigma^2 + n\sigma_{\alpha\beta\gamma}^2 + ln\sigma_{\alpha\beta}^2 + rn\sigma_{\alpha\gamma}^2 + rln\dfrac{\sum \alpha_i^2}{c-1}$	$\sigma^2 + rn\sigma_{\alpha\gamma}^2 + rln\dfrac{\sum_i \alpha_i^2}{c-1}$
Factor B	s_2^2	$\sigma^2 + cn\sigma_{\beta\gamma}^2 + cln\sigma_\beta^2$	$\sigma^2 + cn\sigma_{\beta\gamma}^2 + cln\dfrac{\sum_j \beta_j^2}{r-1}$
Factor C	s_3^2	$\sigma^2 + cn\sigma_{\beta\gamma}^2 + crn\sigma_\gamma^2$	$\sigma^2 + crn\sigma_\gamma^2$
$A \times B$	s_4^2	$\sigma^2 + n\sigma_{\alpha\beta\gamma}^2 + ln\sigma_{\alpha\beta}^2$	$\sigma^2 + n\sigma_{\alpha\beta\gamma}^2 + ln\dfrac{\sum_i \sum_j (\alpha\beta)_{ij}^2}{(c-1)(r-1)}$
$A \times C$	s_5^2	$\sigma^2 + n\sigma_{\alpha\beta\gamma}^2 + rn\sigma_{\alpha\gamma}^2$	$\sigma^2 + rn\sigma_{\alpha\gamma}^2$
$B \times C$	s_6^2	$\sigma^2 + cn\sigma_{\beta\gamma}^2$	$\sigma^2 + cn\sigma_{\beta\gamma}^2$
$A \times B \times C$	s_7^2	$\sigma^2 + n\sigma_{\alpha\beta\gamma}^2$	$\sigma^2 + n\sigma_{\alpha\beta\gamma}^2$
Within error	s_8^2	σ^2	σ^2

consider (1) the case where the effects α_i are fixed and all others are random, and (2) the case where the effects α_i and β_j are fixed, the effects $(\alpha\beta)_{ij}$ also being fixed, and all others are random. The usual assumptions regarding the independence and normality of random components are made, and the resulting expected mean squares are shown in Table 13.27. On close examination of these expected mean squares, it is observed that exact tests exist for all hypotheses except

$$H_{01}: \alpha_1 = \cdots = \alpha_c = 0 \qquad (13.34)$$

in the mixed-model (α) experiment. An approximate test of (13.34) can be made by computing (13.32) and comparing it with the upper α level value of the F distribution with $\nu_1 = c - 1$ and $\nu_2 = \hat{\nu}$ degrees of freedom, where $\hat{\nu}$ is defined by Eq. (13.33). That is, the test of (13.34) is exactly like the test of the hypothesis that $\sigma_\alpha^2 = 0$.

The analysis of variance and expected mean squares for factorials with more than three factors follows the same general pattern already discussed. Also, subsampling in completely randomized (or randomized block or Latin square) designs in which the treatments are combinations of levels of factors leads to the same type of analysis given in Sects. 11.4 and 12.7.

13.8. SUMMARY REMARKS ON FACTORIALS

Methods for partitioning sums of squares, ranking factor means, testing hypotheses with a single degree of freedom, and establishing confidence intervals for levels of a single factor in higher-order factorials are similar to those already explained for one-, two-, and three-factor experiments. Fixed effects and components of variance may be estimated with the aid of tables of means and expected mean squares. Power of the various tests may be determined along the lines given in earlier chapters.

The question of pooling certain mean squares to increase the number of degrees of freedom of the error mean square often arises in analysis of variance. For example, on looking at the two models in Table 13.25, one might argue that failure to reject the hypothesis that $\sigma_{\alpha\beta}^2 = 0$ is evidence that $\sigma_{\alpha\beta}^2$ is so near zero that the mean squares s_3^2 and s_4^2 both estimate the error variance σ^2 and, thus, should be pooled to obtain an estimate of error variance with more degrees of freedom. That is, one might propose

$$s_p^2 = \frac{(c-1)(r-1)s_3^2 + cr(n-1)s_4^2}{(c-1)(r-1) + cr(n-1)} \qquad (13.35)$$

as an estimator of σ^2 with $cr(n-1) + (c-1)(r-1)$ degrees of freedom. Thus, the hypothesis for factor A could be tested by comparing s_1^2 with the residual mean square s_p^2 with more degrees of freedom, making the test more powerful. Such a procedure is not generally recommended, but there is evidence [2, 3, 4, 10, 27, 37, 38] to indicate that such pooling is justified in certain cases.

Suppose we consider the test of $\sigma^2_{\alpha\beta} = 0$ a "preliminary" test and the test of either $\sigma^2_{\alpha} = 0$ or $\sum \alpha^2_i = 0$ the "final" test, and suppose an α level final test is required. Then, in no case should an α level preliminary test be applied. Instead, the level for the preliminary test should be much greater. For example, for a five per cent level final test the preliminary test should be in the neighborhood of 50 per cent. The rationale for this is reasonably clear. If one fails to reject the hypothesis that $\sigma^2_{\alpha\beta} = 0$ at such a high level as 50 per cent, there is a good chance that $\sigma^2_{\alpha\beta}$ is very near zero; with a lower significance level, one is not as sure that $\sigma^2_{\alpha\beta}$ is near zero. Examples of the pooling procedures are given by Paull [37, 38] and Bozivich, Bancroft, Hartley, and Huntsberger [10].

It the above treatment of factorials we were concerned with presenting the model, analysis, and certain applications; we did not discuss such problems as the choice of levels, the selection of treatment combinations, giving the largest (or smallest or specified) response, or the design necessary to get a "picture" of the total "response surface" over the region of interest. The choice of factors and levels of factors depends on the objectives and the cost of the experiment. These topics are discussed in Refs. [14, 48]. The factorial experiment may be preliminary to the main objective. For example, the factorial may be used to estimate the particular combinations of levels of factors which give the maximum (or minimum or specified) response, and then a further experiment may be made to determine a better estimate of the desired response. Sometimes two such experiments are combined into one. These and similar problems are discussed in Refs. [6, 16, 17, 22, 42, 55]. Special procedures for studying the response surface are given by Box [8, 9]. Further references on fixed, mixed, and random models are cited in Refs. [44, 45, 52, 53], and special techniques for finding expected mean squares are given in Refs. [25, 47].

Certain advantages and disadvantages of the factorial have already become apparent. We now list these and others for quick reference. Some advantages are

1. The interactions of two or more factors may be examined
2. The most efficient use is made of resources, in that all responses are used in estimating effects and mean squares
3. The results of the experiment may be applied over a wider range of conditions
4. There are usually more degrees of freedom for the residual mean square, this being particularly true in the fixed model

Some disadvantages are

1. The number of treatment combinations required to study several factors at several levels is large and often prohibitive
2. There may be no interest in certain treatment combinations
3. The experiment and computations are more complicated

Some of the disadvantages, particularly the first one, may be overcome by applying one of the experiments of Sect. 13.9. For elaboration on these and other advantages and disadvantages, the reader is referred to Refs. [12, 15, 19, 30].

13.9. CONFOUNDING IN FACTORIAL EXPERIMENTS

The number of treatment combinations in a factorial experiment can become quite large. If every treatment combination appears in every block of a randomized block design, the large number of experimental units within each block may cause considerable increase in the error variance. In order to reduce the error variance, Fisher [22] developed a method in which all treatment combinations are placed in two or more blocks and the error variance is still dependent only on the variation within blocks. This is done at the expense of losing total or partial information on one or more of the treatment comparisons; that is, factors or interactions. Since in such a design the variation of certain treatment comparisons are linked together (mixed up) with block-to-block variation, these comparisons are said to be *confounded* with blocks. The blocks which contain only a fraction of the treatment combinations are called *incomplete blocks* [7, 12, 13, 19, 24, 30].

We illustrate the principle of confounding with a 2^3 factorial. Actually, eight treatments should usually be placed in each block of a randomized block design, but we use the small 2^3 factorial simply to get across the concepts with minimum writing. Before we explain the details of confounding, it is desirable that we look at seven particular orthogonal treatment comparisons with a single degree of freedom, which come from the eight treatment combinations in the usual randomized block design replicated n times.

For treatment combinations

$$A_1B_1C_1, \ A_1B_1C_2, \ A_1B_2C_1, \ A_1B_2C_2, \ A_2B_1C_1, \ A_2B_1C_2, \ A_2B_2C_1, \ A_2B_2C_2$$

the corresponding totals may be denoted by

$$T_{111}, \ T_{112}, \ T_{121}, \ T_{122}, \ T_{211}, \ T_{212}, \ T_{221}, \ T_{222} \qquad (13.36)$$

respectively, where the dots indicating summation are omitted. Comparisons of means or totals are determined by multipliers. For example, the multipliers for the comparison of the lower level with the higher level of A are

$$-1, \ -1, \ -1, \ -1, \ 1, \ 1, \ 1, \ 1 \qquad (13.37)$$

It is easy to show that the mean square for factor A in a regular analysis of variance table (see Table 13.11) is the same as

$$Q_A^2 = \frac{(-T_{111} - T_{112} - T_{121} - T_{122} + T_{211} + T_{212} + T_{221} + T_{222})^2}{n[(-1)^2 + (-1)^2 + \cdots + 1^2]} \quad (13.38)$$

the mean square for the comparison with the multipliers in (13.37). According to Eq. (13.22) the sum of squares for factor A is

$$SSA = \frac{T_{1..}^2 + T_{2..}^2}{2 \cdot 2n} - \frac{(T_{111} + T_{112} + \cdots + T_{222})^2}{2 \cdot 2 \cdot 2n} \quad (13.39)$$

Since

$$T_{i..} = T_{i11} + T_{i12} + T_{i21} + T_{i22} \quad (i = 1, 2)$$

it follows that

$$Q_A^2 = \frac{(-T_{1..} + T_{2..})^2}{8n}$$

and

$$SSA = \frac{1}{8n}[2T_{1..}^2 + 2T_{2..}^2 - (T_{1..} + T_{2..})^2]$$

$$= \frac{1}{8n}(T_{1..}^2 + T_{2..}^2 - 2T_{1..} T_{2..}) = Q_A^2$$

Therefore, the mean square for factor A, s_1^2, is equal to Q_A^2, since A has only two levels.

Seven sets of multipliers for orthogonal comparisons along with corresponding source of variation are shown in Table 13.28. Using a technique

Table 13.28
Multipliers for Orthogonal Treatment Comparisons

Source of Variation	Treatment Comparisons							
	$A_1B_1C_1$	$A_1B_1C_2$	$A_1B_2C_1$	$A_1B_2C_2$	$A_2B_1C_1$	$A_2B_1C_2$	$A_2B_2C_1$	$A_2B_2C_2$
A	-1	-1	-1	-1	1	1	1	1
B	-1	-1	1	1	-1	-1	1	1
C	-1	1	-1	1	-1	1	-1	1
AB	1	1	-1	-1	-1	-1	1	1
AC	1	-1	1	-1	-1	1	-1	1
BC	1	-1	-1	1	1	-1	-1	1
ABC	-1	1	1	-1	1	-1	-1	1

similar to the above, we can show that the other mean squares for treatment in Table 13.11 are equal to the Q^2's for six other orthogonal treatment comparisons. That is, it can be shown that s_2^2, s_3^2, s_4^2, s_5^2, s_6^2, s_7^2 are the same as Q_B^2, Q_C^2, Q_{AB}^2, Q_{AC}^2, Q_{BC}^2, Q_{ABC}^2, respectively. The analysis of variance for the eight treatment comparisons randomized in three blocks with eight units each is a particular case of that given in Table 13.11. To test the seven hypotheses indicated, each treatment comparison mean square is compared with an error mean square with 16 degrees of freedom.

If, in a particular experiment, it is not possible to obtain blocks with eight nonheterogeneous experimental units, then an effort should be made to use blocks with fewer units. Suppose blocks of four fairly homogeneous experimental units may be obtained. Then four selected treatment combinations could be randomly placed in one block, and the remaining four placed in a second block to make a complete replicate of the experiment. For example, the four treatment combinations with lower levels of A could be placed in one block and those with upper levels of A in a second block. Then, if one attempted to obtain the mean square for factor A, the variation due to the two blocks would be mixed with the variation due to A. That is, the comparison for factor A is computed in the same way as the comparison between blocks. In such a case we say that factor A is confounded with blocks and, thus, cannot be analyzed (or separated out) in the experiment. If one of the factors A, B, C, sometimes called *main effect*, is confounded with blocks, we say that the experiment is a *split plot*. Clearly, any one of the sources of variation in Table 13.28 could be confounded with blocks. Often one of the two-factor or three-factor interaction comparisons which is considered unimportant is confounded with blocks.

Figure 13.3 gives an experimental layout for eight treatment combinations

Replicate 1		Replicate 2		Replicate 3	
$A_2B_1C_2$	$A_1B_2C_1$	$A_1B_2C_2$	$A_1B_1C_2$	$A_2B_2C_1$	$A_2B_2C_2$
$A_1B_1C_1$	$A_2B_2C_2$	$A_2B_1C_2$	$A_2B_1C_1$	$A_1B_2C_2$	$A_1B_2C_1$
$A_2B_2C_1$	$A_1B_1C_2$	$A_2B_2C_1$	$A_2B_2C_2$	$A_1B_1C_1$	$A_2B_1C_1$
$A_1B_2C_2$	$A_2B_1C_1$	$A_1B_1C_1$	$A_1B_2C_1$	$A_2B_1C_2$	$A_1B_1C_2$
Block 1	Block 2	Block 1	Block 2	Block 1	Block 2

Fig. 13.3 Layout for a 2^3 Factorial Experiment in which the Three-factor Interaction Comparison is Completely Confounded with Blocks

replicated three times in six blocks with four experimental units in which the three-factor interaction comparison is confounded with blocks. When the same comparison is confounded in every replication, we say that there is *complete confounding*. Thus, for the incomplete block design in Fig. 13.3, the ABC interaction comparison is completely confounded with blocks. Note that all treatment combinations with positive multipliers are randomized in one incomplete block and that those with negative multipliers are randomized in the other incomplete block of each replication.

If all main and interaction effects are important, but some are of less importance than others, then it is possible to design the experiment so that full information may be obtained on some and partial information on those of least importance. Figure 13.4 gives an experimental layout for eight treatment combinations replicated three times in six blocks with four experimental units in which the AB interaction is confounded with blocks in the first replicate, AC interaction is confounded with blocks in the second replicate, and BC interaction is confounded with blocks in the third replicate. For the incomplete block design in Fig. 13.4 we say there is *partial confounding*. Actually, since each two-factor interaction comparison is confounded

Replicate 1		Replicate 2		Replicate 3	
$A_1B_1C_2$	$A_1B_2C_2$	$A_1B_2C_1$	$A_2B_1C_1$	$A_2B_1C_1$	$A_1B_2C_1$
$A_2B_2C_2$	$A_2B_1C_1$	$A_2B_1C_2$	$A_1B_2C_2$	$A_1B_1C_1$	$A_2B_1C_2$
$A_2B_2C_1$	$A_1B_2C_1$	$A_2B_2C_2$	$A_2B_2C_1$	$A_1B_2C_2$	$A_1B_1C_2$
$A_1B_1C_1$	$A_2B_1C_2$	$A_1B_1C_1$	$A_1B_1C_2$	$A_2B_2C_2$	$A_2B_2C_1$
Block 1	Block 2	Block 1	Block 2	Block 1	Block 2
AB Confounded		AC Confounded		BC Confounded	

Fig. 13.4 Layout for a 2^3 Factorial Experiment in which the Three Two-factor Interaction Comparisons are Partially Confounded with Blocks

and is confounded the same number of times, the term *balanced partial confounding* is often applied. Also when the four interaction comparisons or the three main effects comparisons are partially confounded the same number of times in incomplete blocks, we say that there is balanced partial confounding.

It should be noted that in each replicate of Figs. 13.3 and 13.4 only one treatment comparison is confounded with incomplete blocks. For example, in the second replicate of Fig. 13.4 the AC interaction comparison is con-

Table 13.29
Analysis of the Experiment of
Figure 13.3

Source of Variation	Degrees of Freedom	
Replicate	2	
ABC (or block)	1	
Replicate $\times$ ABC	2	
Treatments	6	
A		1
B		1
C		1
AB		1
AC		1
BC		1
Residual	12	
Total	23	

Table 13.30
Analysis of the Experiment of
Figure 13.4

Source of Variation	Degrees of Freedom
Blocks	5
A	1
B	1
C	1
ABC	1
AB	1
AC	1
BC	1
Residual	11
Total	23

founded, but the other six treatment comparisons are not confounded, since of the four treatment combinations in each of two incomplete blocks exactly two have positive and two negative multipliers.

The source of variation and partition of degrees of freedom for the experiments of Figs. 13.3 and 13.4 are shown in Tables 13.29 and 13.30, respectively. All mean squares for treatment comparisons which are either not confounded or partially confounded are tested against the residual mean square. The sum of squares (or mean squares) in Table 13.29 are computed by the usual technique. In Table 13.30 all treatment sum of squares except the three two-factor interaction comparisons which are partially confounded are found in the usual way. The sum of squares for the AB comparison is found by using only totals from replicates 2 and 3. In computing the sum of squares for the treatment comparisons which are not confounded, the divisor is 24; for those which are partially confounded, the divisor is 16. Thus, the treatment comparisons which are not confounded are obtained with higher precision than those which are partially confounded, the ratio of precision being 3 to 2. Since the different divisors take care of the different precision, no further correction is required in the analysis of variance and test procedures.

We have illustrated the principle of confounding in a very simple case. It is possible to confound 3^k factorials in replicates with three blocks, 2^k factorials in replicates with four blocks, etc. Different experimental models, relative efficiency, missing values, and many other topics could be discussed. The reader is referred to Refs. [5, 12, 15, 19, 28, 30, 33, 34, 35, 39, 40, 55, 56, 57] for further study of these and other topics.

The split-plot design is particularly important in experiments in which the levels of one factor, say A, require larger experimental units than other factors, say B, C, ... For example, in the study of some property of alloys, the differences in furnaces, A, may require large amounts of material and the differences in molds and specific gravity, B and C, relatively small amounts. In another example, in the stimulus-response of plants, one factor might be for differences in whole plants, A, and another for differences in leaves, B. Two factors which affect the response in a laboratory experiment with students might be solution A, in a large quantity, and an additive, B, to a relatively small sample. The analysis for split plots is similar to those already given in Tables 13.29 and 13.30, except that there are two error terms, one for the "whole plot" factor A and another for the "split-plot" factor (or factors). If more than two sizes of experimental units are required, the plots are split accordingly. Thus, split-split plots and higher-order split plots are also used. For details on these and other topics the reader is referred to Refs. [1, 5, 11, 12, 18, 19, 30, 31, 49, 50, 51, 55].

In experiments treated so far in this chapter *all* treatment combinations are arranged in some design. For a large number of factors, say more than four, the number of treatment combinations as well as the number of possible tests becomes quite large. By confounding we reduced both the number of experimental units in a block and the number of test procedures. But when there are five or more factors, there is often a need for a reduction in the number of treatment combinations in the experiment. If only the main effects and certain interactions are of interest, it is possible to analyze these effects when only a fraction of the treatment combinations are selected. For example, in an experiment involving only the first block for each replicate in Fig. 13.3, we could still obtain information on all treatment comparisons except the ABC interaction comparison. Such an experiment is called a fractional factorial. That is, an experiment in which the treatment combinations of a block are equivalent to those in one block of a replicate in a system of confounding is called a *fractional factorial*. Details of fractional factorials are discussed in Refs. [15, 20, 21, 29, 30, 32, 39].

13.10. EXERCISES

13.1. The data in Table 13.31 are for a fixed-effects 2×3 factorial experiment in a completely randomized design. (a) Express each of the six observa-

Table 13.31

Levels of factor A		1			2		
Levels of factor B		1	2	3	1	2	3
Treatment combination		A_1B_1	A_1B_2	A_1B_3	A_2B_1	A_2B_2	A_2B_3
Replication	1	25	23	18	26	10	18
	2	28	22	23	20	13	17
	3	25	21	22	20	13	16

tions for replication 1 as the sum of the over-all mean, an A effect, a B effect, an AB interaction effect, and an error. (b) Prepare an analysis of variance table showing the expected mean squares. (c) Test the three hypotheses in (13.3), showing all steps in the general test procedure. (d) Identify the two factors with variables in your own field of study, write a short statement of an experiment which could lead to the above analysis of variance, and write your conclusion in terms of the variables introduced. (e) Find a 95 per cent confidence interval for the difference in the means of the two levels of factor A. (f) Find a 95 per cent confidence interval for each of the three means for the levels of factor B.

13.2. The data in Table 13.32 are for a fixed-effects 2×3 factorial experiment in a randomized block design. (a) Express each of the six observations for treatment combinations A_1B_3 and A_2B_1 as the sum of the over-all mean, a block effect, an A effect, a B effect, an AB interaction effect, and an error. (b) Prepare an analysis of variance table, showing the expected mean squares. (c) Test the three hypotheses for the equality of

Table 13.32

Level of factor A		1			2		
Level of factor B		1	2	3	1	2	3
Treatment combination		A_1B_1	A_1B_2	A_1B_3	A_2B_1	A_2B_2	A_2B_3
Block	1	28	13	22	26	23	18
	2	25	10·	18	20	22	17
	3	25	13	23	20	21	16

factor A effects, factor B effects, and interaction effects, (d) Find a 90 per cent confidence interval for each of the following: (1) means for levels of factor A, (2) differences in means for levels of B, and (3) mean of the treatment combination A_1B_2 in block 2.

13.3. The true population means for 15 treatment combinations are given in Table 13.33. (a) Find the true effects for factor A, factor B, and interaction of A and B. (b) Find the value of

$$\frac{r \sum_i \alpha_i^2}{(c-1)}, \quad \frac{c \sum_j \beta_j^2}{(r-1)}, \quad \frac{\sum_i \sum_j (\alpha\beta)_{ij}^2}{(c-1)(r-1)} \quad \text{and} \quad \frac{\sum_i \sum_j (\mu_{ij} - \mu)^2}{(cr-1)}$$

Table 13.33

B \ A	1	2	3	4	5
1	66	62	58	47	42
2	54	54	53	42	37
3	51	49	45	46	44

Use this information to compare the effects of the two factors.

13.4. Prove the sum of squares identity (13.1).

13.5. Prove that a_i, b_i, and $(ab)_{ij}$ are unbiased estimators of α_i, β_j, and $(\alpha\beta)_{ij}$, respectively. Assume a fixed-effects c × r factorial experiment in a completely randomized design.

13.6. In Table 13.7, derive the expected mean square for s_3^2.

13.7. The data in Table 13.34 are for a fixed-effects 3 × 2 factorial experiment in which six random samples, each of size two, were drawn from a normal population with mean 50 and variance 100. (a) Prepare an

Table 13.34

B \ A	1		2		3	
1	62	63	53	59	61	46
2	45	34	46	39	59	46

analysis of variance table and test each of the hypotheses in (13.3) at the five per cent level. What is the significance level for the whole experiment? Since you know the source of the six samples, indicate whether a type 1 or type 2 error was committed. (b) Compute the true and estimated effects of factor A and of the interaction of A and B. Compare the true and estimated effects. Draw graphs to compare the true and estimated interaction effects. (c) Add 50 to each observation of the A_2B_1 treatment combination, 100 to each observation of the A_3B_1 treatment combination, and leave the other eight observations unchanged. Prepare an analysis of variance table and test each of the hypotheses in (13.3) at the five per cent level. Since you know the sources of the six samples, indicate whether a type 1 or type 2 error was committed in each test. (d) Use the data in (c) to do (b). (e) Add 50 to each observation of the A_1B_2, A_2B_1, and A_3B_1 treatment combinations and 100 to each observation of the A_1B_1 treatment combination, leaving the other four observations unchanged. Prepare an analysis of variance table and test each of the hypotheses in (13.3) at the five per cent level. Indicate whether a type 1 or type 2 error was committed in each test. (f) Use the data in (e) to do (b). (g) Find the numerical value of the expected mean squares for (a), (c), and (e). Use these expected mean squares to explain any differences in the actual mean squares found in (a), (c), and (e).

13.8. The data in Table 13.35 are for a fixed-effects 5×3 factorial experiment in a completely randomized design (the reader may think of the levels of A as different temperatures, the levels of B as different fabrics, and the observations as coded values for percentage of shrinkage during

Table 13.35

Factor B	Replication	Factor A				
		1	2	3	4	5
1	1	50	89	70	50	34
	2	64	59	53	56	49
2	1	39	57	44	41	26
	2	57	82	60	32	38
3	1	36	53	42	51	44
	2	44	36	31	47	40

dying). (a) Prepare an analysis of variance table. (b) Find 90 per cent confidence intervals for each of the 15 cell means. Since the observations were taken from populations with means shown in Exercise 13.3, determine how many intervals include the true mean. (c) Use Duncan's range test to rank the means for the five levels of factor A. Also, use Duncan's range test to rank the means for the 15 cells. Compare both rankings with the true rankings obtained from Exercise 13.3. (d) Use the true means for the five levels of factor A to write four orthogonal comparisons. Use the data of this exercise to determine 95 per cent confidence intervals for these comparisons. In each case, check to see if the true value for the comparison falls in the interval. (e) Find a 95 per cent confidence interval for the difference in the mean of the third level of factor A and the mean of the combination of the second level of factor A and the third level of factor B. Indicate how such information might be useful. (f) Discuss the interaction effects in terms of a graph. Assume that a small percentage of shrinkage is desirable. (g) Assuming the levels of B to be random and those of A to be fixed, test the hypothesis that the A effects are equal. Also, find a 95 per ecnt confidence interval for the mean of the fifth level of factor A. (h) Assuming the first replications form one block and the second replications a second block, test the hypotheses in (13.3). Assume a fixed-effects experiment.

13.9. Derive the expected mean squares for s_3^2 and s_4^2 shown in Table 13.10.

13.10. The data in Table 13.36 are for a fixed-effects 2×2 factorial in a Latin square design. (a) Prepare an analysis of variance table and test the three hypotheses of (13.3). (b) Identify the two factors, rows, and columns with variables in your own field of study, write a short statement of an

Table 13.36

		Columns			
		1	2	3	4
Rows	1	$18(A_2B_1)$	$29(A_1B_1)$	$10(A_2B_2)$	$19(A_1B_2)$
	2	$16(A_2B_2)$	$20(A_1B_2)$	$14(A_2B_1)$	$22(A_1B_1)$
	3	$30(A_1B_2)$	$20(A_2B_1)$	$25(A_1B_1)$	$17(A_2B_2)$
	4	$28(A_1B_1)$	$15(A_2B_2)$	$23(A_1B_2)$	$14(A_2B_1)$

experiment which could lead to the above analysis, and write your conclusions in terms of the variables introduced.

13.11. Write the model equation for an $A \times B$ factorial experiment in a Latin square design. Derive the sum of squares identity for the usual analysis of variance. Make an analysis of variance table showing the expected mean squares.

13.12. (a) Complete Table 13.37 for the analysis of variance and expected mean squares of a mixed-model (β) factorial experiment in a randomized block design (the reader may assume that the computations are for any three-factor factorial experiment he chooses). (b) Use the analysis of variance table, Table 13.37, to test all hypotheses concerning main effects, two-factor interaction effects, and three-factor interaction effects. (c) Find unbiased point estimates of all the variance components in (a). Find 90 per cent confidence intervals for $\sigma_{\alpha\beta\gamma}$, $\sigma_{\alpha\gamma}$, and σ_α. (d) Test all hypotheses in (b), assuming a mixed model (β, γ). Compare the results in (b) and (d).

Table 13.37

Source of Variation	Sum of Squares	Degrees of Freedom	Mean Squares	Expected Mean Squares
Blocks		2	28	
A			160	
B		2	210	
C				
AB	170	2		
AC	84			
BC		6	180	
ABC	228			
Error			15	
Total	3338			

13.13. In a study to determine the effect of age and type of soil on the unconfined compressive strength of mixtures of cement with soil, the data in Table 13.38 were obtained. (The compressive strength was in pounds per square inch, and ten per cent of each mixture was cement. Soils Leighton Buzzard sand, Tunstall Common A, Tunstall Common B, and Whitchurch are denoted by I, II, III, and IV, respectively.)

Table 13.38

Age	Soil	Batch*							
		A	B	C	D	E	F	G	H
7 days	I	490	380	345	295	380	350	410	335
	II	520	460	45	75	60	220	80	265
	III	590	400	20	250	20	430	175	500
	IV	810	620	395	550	90	540	700	905
28 days	I	830	760	720	660	670	640	860	750
	II	1020	870	750	770	940	800	870	980
	III	970	680	860	670	620	845	770	830
	IV	1500	1210	1330	1040	1000	930	1130	1270

* Data from P. T. Sherwood, "Rapid Method for Detecting the Presence of Deleterious Organic Matter in Soil-Cement," *Journal of Applied Chemistry*, Vol. **12** (1962), pp. 279–88.

Prepare an analysis of variance table to use in discussing this experiment. Write a detailed report, bringing out any information which seems pertinent to you. You should note such things as possible differences in error variance for 7 and 28 days, variability in batches, comparison of sand with the other three soils, and size of confidence intervals. Assume that the batches are random.

13.14. An investigation was made to determine the sources of variation in the wool content of nominally ten per cent wool blankets. Table 13.39 gives the percentage of wool in 16 pieces of blanket, eight for each color, four for each batch, and two for each loom. (The batches, loom, and pieces were randomly selected.) Estimate and discuss the sources of variation.

Hint. This type of problem was discussed in an earlier chapter, and should be carefully distinguished from a factorial experiment.

Table 13.39*

Color of Material	Batch Number	Loom Number	Piece Number	Percentage of Wool
Green	1	1	1	10.1
			2	9.7
		2	3	14.2
			4	15.1
	2	3	5	14.7
			6	14.9
		4	7	10.1
			8	14.4
Yellow	3	5	9	10.6
			10	10.3
		6	11	12.5
			12	12.4
	4	7	13	10.3
			14	12.8
		8	15	12.8
			16	12.4

* Data from W. S. Connor, "Locating Important Sources of Variation," *Industrial and Engineering Chemistry*, Vol. **53**, No. 12 (1961), pp. 73A–74A.

13.15. Give graphic interpretations of all the interactions in Example 13.3.

13.16. Let A denote breakers, B gaugers, C curing time, and D mixes. Coded

Table 13.40

		A_1		A_2	
		B_1	B_2	B_1	B_2
C_1	D_1	42.5	34.0	26.0	24.0
		42.0	33.0	26.0	25.0
		41.5	32.0	23.0	26.0
	D_2	40.5	21.0	24.0	7.0
		41.2	21.0	23.5	6.0
		41.3	18.0	24.5	5.0
C_2	D_1	38.5	36.5	27.5	32.5
		39.0	35.7	28.0	32.0
		39.5	35.8	28.5	31.5
	D_2	44.0	32.5	29.5	25.0
		45.0	31.8	29.0	24.0
		46.0	31.7	28.5	23.0

data for the compressive strength of hardened cement cubes for such a 2^4 fixed-effects factorial experiment in a completely randomized design given in Table 13.40. (a) Prepare the analysis of variance table for this data applying the usual techniques. (b) Prepare the analysis of variance table for this data, using the systematic method due to Yates [56, 57].

Hint. List the 16 cell totals in the order indicated in Table 13.41. (The treatment combination $A_1 B_2 C_2 D_1$ is written as 1221, etc.) The first eight entries in column (3) are the sums $126 + 75$, $99 + 75$, ..., $96 + 72$, and the second eight entries are the differences $75 - 126$, $75 - 99$, ..., $72 - 96$. The entries in column (4) are found by operating on column (3) in the same way. The process is continued until all entries through column (6) are found. Finally, an entry in column (7) is found by squaring the corresponding entry in column (6) and dividing by (3)(16). The reader can now finish this table and justify the computations. For example, show that 1692.2 is the sum of squares for the main effect A. (c) Test the indicated hypotheses and write a careful summary statement for the experiment, explaining any significant main effects and interactions in terms of breakers, gaugers, curing time, and mixes. (d) After pooling all treatment mean squares with the error mean square which you consider appropriate, discuss (c) in terms of the new error mean square.

Table 13.41

Treatment Combination (1)	Totals in 16 Cells (2)	Sums and Differences of Pairs				$Q^2 = \dfrac{[(6)]^2}{48}$
		(3)	(4)	(5)	(6)	(7)
1111	126	201	375	780	1443	
2111	75	174	405	663	-285	1692.2(A)
1211	99	201	273	-120	-195	792.2(B)
2211	75	204				(AB)
1121	117	195				(C)
2121	84	78				(AC)
1221	108	222				(BC)
2221	96	168				(ABC)
1112	123	-51				(D)
2112	72	-24				(AD)
1212	60	-33				(BD)
2212	18	-12				(ABD)
1122	135	-51				(CD)
2122	87	-42				(ACD)
1222	96	-48				(BCD)
2222	72	-24	24	15	21	($ABCD$)

13.17. Let A denote concentration of a detergent, B concentration of sodium carbonate, and C concentration of sodium carboxy-methyl cellulose. Table 13.42 gives data for the cleaning ability of a solution in washing tests for a single replication of a fixed-effects 3^3 factorial design (large numbers indicate improved cleaning ability). (a) Prepare an

Table 13.42*

Detergent		A_1			A_2			A_3		
Sodium Carbonate		B_1	B_2	B_3	B_1	B_2	B_3	B_1	B_2	B_3
Sodium CMC	C_1	106	197	223	198	329	320	270	361	321
	C_2	149	255	294	243	364	410	315	390	415
	C_3	182	259	297	232	389	416	340	406	387

* Data from Fuell and Wagg, "Statistical Methods in Detergency Investigations," *Research*, Vol. 2 (1949), p. 334.

analysis of variance table in which the sums of squares for interactions *AC* and *ABC* are pooled and used for the error mean square in testing all other two-factor interaction effects and all main effects. Write a summary report of your findings. (b) For each main effect sum of squares it is possible to define two orthogonal linear comparisons, each with a single degree of freedom. The comparisons with multipliers 1, 0, -1 and 1, -2, 1 are called *linear* and *quadratic* components, and, for factor *A*, are designated by A_L and A_Q, respectively. Compute and interpret the variation due to A_L, A_Q, B_L, B_Q, C_L, C_Q, $A_L B_L$, $A_L B_Q$, $A_Q B_L$, $A_Q B_Q$, $B_L C_L$, $B_L C_Q$, $B_Q C_L$, and $B_Q C_Q$. (c) Use the systematic method due to Yates to compute the sums of squares in (a). It would be very informative if the reader determined a good computational technique, using Exercise 13.16(b) as a guide.

13.18. Derive the expected mean squares in Table 13.25.

13.19. Use an appropriate pooling technique to increase the degrees of freedom in the error mean squre of Exercise 13.17 and to discuss the significance of the comparisons in Exercise 13.17(b).

13.20. Determine an experimental layout for a 2^4 factorial experiment in which the three-factor interaction comparison *BCD* is completely confounded with blocks. Give two replicates. Write the appropriate analysis of variance, indicating only the sources of variation and degrees of freedom. Show how typical sums of squares are computed.

13.21. (a) Determine an experimental layout for a 2^4 factorial experiment in which the three-factor interaction comparisons *ABC*, *ABD*, and *ACD* are partially confounded with blocks. (b) Write the appropriate analysis of variance and show how typical sums of squares are computed.

13.22. The analysis of variance for a split-plot experiment with *c* whole plots *A*, *r* split plots *B*, and *b* blocks (or replicates) is given in Table 13.43. An experimental layout and the observations for two subplots within each of four whole plots replicated three times is given in Table 13.44. (a) Write the model equation and prepare an analysis of variance table for this data. (b) State and test hypotheses about factor *A*, factor *B*, and interaction of *A* and *B*. Write a summary statement for this experiment in terms of factors in your own field. (c) State and prove the sum of squares identity for a split-plot experiment.

Table 13.43

Source of Variation	Degrees of Freedom	Expected Mean Squares for Fixed-Effects Model
Blocks	$b - 1$	
A	$c - 1$	$\sigma^2 + r\sigma_2^2 + br \sum \alpha_i^2/(c - 1)$
Whole-plot error	$(b - 1)(c - 1)$	$\sigma^2 + r\sigma_2^2$
B	$r - 1$	$\sigma^2 + bc \sum \beta_j^2/(r - 1)$
AB	$(c - 1)(r - 1)$	$\sigma^2 + b \sum\sum (\alpha\beta)_{ij}^2/(c - 1)(r - 1)$
Split-plot error	$c(b - 1)(r - 1)$	σ^2
Total	$bcr - 1$	

Table 13.44

Block 1	$A_2 B_1$ (21)	$A_4 B_2$ (8)	$A_1 B_2$ (21)	$A_3 B_1$ (21)
	$A_2 B_2$ (13)	$A_4 B_1$ (16)	$A_1 B_1$ (23)	$A_3 B_1$ (13)

Block 2	$A_3 B_1$ (26)	$A_1 B_1$ (26)	$A_4 B_2$ (13)	$A_2 B_2$ (19)
	$A_3 B_2$ (16)	$A_1 B_2$ (24)	$A_4 B_1$ (19)	$A_2 B_1$ (25)

Block 3	$A_1 B_2$ (26)	$A_4 B_2$ (17)	$A_2 B_1$ (28)	$A_3 B_2$ (13)
	$A_1 B_1$ (26)	$A_4 B_1$ (23)	$A_2 B_2$ (20)	$A_3 B_1$ (23)

13.23. Spherical particles $\frac{1}{8}$ in. in diameter were dropped into a glass tube three in. in diameter, and the percentage of porosity was measured. Data for an experiment with four kinds of material, two heights of drop, and eight distances from the wall of the container are given in Table 13.45.

Table 13.45
Percentages of Porosities of Successive Annuli or Layers

Material	Deposition	Over-all Porosity	Distance from Wall (inches)							
			0	$\frac{1}{4}$		$\frac{1}{2}$		$\frac{3}{4}$		1
Lead	Cascaded	39.3	47.3	40.3	36.6	37.5	37.0	38.9	37.3	37.7
	36 in. drop	38.0	44.0	35.3	34.2	37.7	38.4	36.4	36.6	38.7
Phosphor-bronze	Cascaded	40.9	49.0	39.3	37.8	38.1	40.1	39.3	40.2	40.3
	36 in. drop	38.2	45.7	37.3	33.1	34.0	38.3	37.4	37.3	37.9
Polystyrene	Cascaded	40.9	47.8	38.8	39.2	38.7	40.3	41.1	40.6	40.8
	36 in. drop	36.3	45.0	34.0	29.1	33.8	37.0	37.1	37.5	35.5
Glass	Cascaded	40.1	47.9	39.3	39.0	36.4	37.8	38.8	38.5	37.3
	36 in. drop	35.6	42.4	37.9	29.8	31.5	34.5	36.1	34.9	36.6

* Data from J. C. Macrae and W. A. Gray, "Significance of the Properties of Materials in the Packing of Real Spherical Particles," *British Journal of Applied Physics*, Vol. **12** (1961), pp. 164–72, Table 6.

Analyze the data to determine the effects of material, deposition, and distance on porosity.

REFERENCES

1. Anderson, R. L., "Missing-plot Techniques," *Biometrics*, Vol. **2** (1946), pp. 41–47.
2. Bancroft, T. A., "On Biases in Estimation Due to the Use of Preliminary Tests of Significance," *Annals of Mathematical Statistics*, Vol. **15** (1944), pp. 190–204.
3. Bechhofer, R. E., "The Effect of Preliminary Tests of Significance on the Size and Power of Certain Tests of Univariate Linear Hypotheses," unpublished Ph. D. thesis, Columbia University Library.
4. Bennett, B. M., "Estimation of Means on the Basis of Preliminary Tests of Significance, "*Annals of Mathematical Statistics*, Vol. **23** (1952), pp. 31–43.
5. Bennett, C. A. and N. L. Franklin, *Statistical Analysis in Chemistry and the Chemical Industry*. New York: John Wiley & Sons, Inc., 1954, Chap. 8.
6. Bliss, C. I., "The Comparison of Dosage-mortality Data, "*Annals of Applied Biology*, Vol. **22** (1935), pp. 307–33.
7. Bose, R. C., "On the Construction of Balanced Incomplete Block Designs," *Annals of Eugenics*, Vol. **9** (1939), pp. 353–99.
8. Box, G. E. P., "The Exploration and Exploitation of Response Surfaces: Some General Considerations and Examples," *Biometrics*, Vol. **10** (1954), pp. 16–60.
9. ——— and J. S. Hunter, "Experimental Designs for Exploring Response Surfaces, "*Experimental Designs in Industry*, V. Chew, ed. New York: John Wiley & Sons, Inc., 1958, pp. 138–90.
10. Bozivich, Bancroft, Hartley, and Huntsberger, "Analysis of Variance: Preliminary Tests, Pooling, and Linear Models," *WADC Technical Report 55–244*, Vol. **1**, U. S. Dept. of Commerce, Washington, D. C., 1956, pp. 1–137.
11. Cochran, W. G., "A Survey of Experimental Design," *Mimeo*, U. S. D. A., 1940.
12. ——— and G. M. Cox, *Experimental Design*, 2nd ed. New York: John Wiley & Sons, Inc., 1957, Chaps. 5, 6, 6A, and 7.
13. Connor, W. S., Jr., "On the Structure of Balanced Incomplete Block Designs," *Annals of Mathematical Statistics*, Vol. **23** (1952), pp. 57–71.
14. Daniel, C. and N. Heerema, "Design of Experiments for Most Precise Slope Estimation on Linear Extrapolation," *Journal of the American Statistical Association*, Vol. **45** (1950), pp. 546–56.
15. Davies, O. L., *The Design and Analysis of Industrial Experiments*. New York: Hafner Publishing Co., Inc., 1954, Chaps. 7, 8, 9, and 10.
16. Dixon, W. J. and A. M. Mood, "A Method for Obtaining and Analyzing Sensitivity Data," *Journal of the American Statistical Association*, Vol. **43** (1948), pp. 109–26.
17. Eisenhart, C., M. W., Hastay, and W. A. Wallis, *Selected Techniques of Statistical Analysis*. New York: McGraw-Hill, Inc., 1947.

18. Federer, W. T., "A Note on Error (b) in the Split-plot Design," *Mimeo BU-19-M*, Cornell University, 1951.

19. ———, *Experimental Design*. New York: The Macmillan Company, 1955, Chaps. 7, 8, 9, and 10.

20. Finney, D. J., "The Fractional Replication of Factorial Arrangements," *Annals of Eugenics*, Vol. 12 (1945), pp. 291–301.

21. ———, "Recent Developments in the Design of Filed Experiments, III. Fractional Replication," *Journal of Agricultural Science*, Vol. 36 (1946), pp. 184–91.

22. Fisher, R. A., *The Design of Experiments*, 6th ed. New York: Hafner Publishing Co., Inc., 1951.

23. Graybill, F. A., *An Introduction to Linear Statistical Models*, Vol. 1. New York: McGraw-Hill, Inc., 1961, Chaps. 12, 16, 17, and 18.

24. Hanani, Haim, "The Existence and Construction of Balanced Incomplete Block Designs," *Annals of Mathematical Statistics*, Vol. 32 (1961), pp. 361–86.

25. Henderson, C. R., "Estimation of Variance and Covariance Components," *Biometrics*, Vol. 9 (1953), pp. 226–52.

26. Hotelling, H., "Experimental Determination of the Maximum of a Function," *Annals of Mathematical Statistics*, Vol. 12 (1941), pp. 20–45.

27. Huntsberger, D. V., "An Extension of Preliminary Tests for Pooling Data," Abstract, *Journal of the American Statistical Association*, Vol. 49 (1954), p. 348.

28. Kempthorne, O., "A Note on Differential Response in Blocks," *Journal of Agricultural Science*, Vol. 37 (1947), pp. 245–48.

29. ———, "A Simple Approach to Confounding and Fractional Replication in Factorial Experiments," *Biometrika*, Vol. 34 (1947), pp. 255–72.

30. ———, *The Design and Analysis of Experiments*. New York: John Wiley & Sons, Inc., 1952, Chaps. 13–20.

31. Khargonkar, S. A., "The Estimation of Missing Plot Value in Split-plot and Strip Trials," *J. Indian Soc. Agri. Stat.*, Vol. 1 (1948), pp. 147–61.

32. Kitagawa, T. and M. Mitome, *Tables for the Design of Factorial Experiments*. Tokyo: Baifukan Co., Ltd., 1953.

33. Li, J. C. R., "Design and Statistical Analysis of Some Confounded Factorial Experiments," *Iowa Agri. Exp. Station Res. Bul.*, Vol. 333 (1944), pp. 449–92.

34. Nair, K. R., "On a Method of Getting Confounded Arrangements in the General Symmetrical Type of Experiment," *Sankyhyā*, Vol. 4 (1938), pp. 121–38.

35. ——— and C. R. Rao, "Confounding in Asymmetrical Factorial Experiments," *Journal of the Royal Statistical Society, Supplement*, Vol. 10 (1948), pp. 109–31.

36. Ostle, B., *Statistics in Research*, 2nd ed. Ames, Iowa: The Iowa State College Press, 1963, Chaps. 12 and 13.

37. Paull, A. E., "On a Preliminary Test for Pooling Mean Squares in the Analysis of Variance," unpublished Ph. D. thesis, University of North Carolina, 1948.

38. ———, "On a Preliminary Test for Pooling Mean Squares in the Analysis

of Variance," *Annals of Mathematical Statistics*, Vol. **21** (1950), pp. 539–56.

39. Plackett, R. L. and J. P. Burman, "The Design of Optimum Multifactorial Experiments," *Biometrika*, Vol. **33** (1946), pp. 305–25.

40. Rao, C. R., "General Methods of Analysis for Incomplete Block Designs," *Journal of the American Statistical Association*, Vol. **42** (1947), pp. 541–61.

41. Rayner, A. A., "Quality × Quantity Interaction," *Biometrics*, Vol. **9** (1953), pp. 387–411.

42. Robbins, H. and S. Munro, "A Stochastic Approximation Method," *Annals of Mathematical Statistics*, Vol. **22** (1951), pp. 400–407.

43. Satterthwaite, F. E., "An Approximate Distribution of Estimates of Variance Components," *Biometrics*, Vol. **2** (1946), pp. 110–14.

44. Scheffé, H., "A Mixed Model for the Analysis of Variance," *Annals of Mathematical Statistics*, Vol. **27** (1956), pp. 23–36.

45. ———, "Alternative Models for Analysis of Variance," *Annals of Mathematical Statistics*, Vol. **27** (1956), pp. 251–71.

46. ———, *The Analysis of Variance*. New York: John Wiley & Sons, Inc., 1959.

47. Schultz, E. F., Jr., "Rules of Thumb for Determining Expectations of Mean Squares in Analysis of Variance," *Biometrics*, Vol. **11** (1955), pp. 123–35.

48. Smith, H. F. "Simplified Calculation of a Linear Regression," *Nature*, Vol. **167** (1951), pp. 367–68.

49. Steel, R. G. D. and J. H. Torrie, *Principles and Procedures of Statistics*. New York: McGraw-Hill, Inc., 1960, Chaps. 11 and 12.

50. Taylor, J., "The Comparison of Pairs of Treatments in Split-plot Experiments," *Biometrika*, Vol. **37** (1950), pp. 443–44.

51. Tukey, J. W., "Diadic Anova, an Analysis of Variance of Vectors," *Human Biology*, Vol. **21** (1949), pp. 65–110.

52. Wilk, M. B. and O. Kempthorne, "Fixed, Mixed, and Random Models," *Journal of the American Statistical Association*, Vol. **50** (1955), pp. 1144–67.

53. ——— and O. Kempthorne, "Some Aspects of the Analysis of Factorial Experiments in a Completely Randomized Design," *Annals of Mathematical Statistics*, Vol. **27** (1956), pp. 950–84.

54. Williams, E. J., "The Interpretation of Interactions in Factorial Experiments," *Biometrika*, Vol. **39** (1952), pp. 65–81.

55. Yates, F., "The Principles of Orthogonality and Confounding in Replicated Experiments," *Journal of Agricultural Science*, Vol. **23** (1933), pp. 108–45.

56. ———, "Complex Experiments," *Journal of the Royal Statistical Society*, *Supplement* 2, Vol. **2** (1935), pp. 181–247.

57. ———, "The Design and Analysis of Factorial Experiments," *Tech. Comm. No.* 35, Imperial Bureau of Soil Science, 1937, pp. 1–95.

58. ——— and W. G. Cochran, "The Analysis of Groups of Experiments," *Journal of Agricultural Science*, Vol. **28** (1938), pp. 556–88.

14

REGRESSION AND RELATED TOPICS

The dependence of the mean response on increasing levels of a quantitative factor (or quantitative factors) is studied. It is shown how the sum of squares identity and procedures for estimation and hypothesis testing may be extended to problems in which observations are made on more than one characteristic of an object. Linear and polynomial regression are developed in some detail. It is shown how orthogonal polynomials may be applied in determining the nature of a trend. Correlation is related to regression. Simple covariance analysis is used in comparing slopes of lines.

14.1. INTRODUCTION

So far in our studies we have analyzed differences in responses at different levels of a treatment factor or at different treatment combinations. We have not studied any trend that might exist between response and levels of a factor (or combination of levels of several factors). That is, we have not attempted to relate changes in responses to changes in levels of the treatment factor. For example, in the soil-sample experiment (Example 12.1) we were concerned with the acidity level at different depths, but we were not concerned with a trend of acidity level with depth of soil. In many problems the relationship between the levels of a quantitative variable x_u and the corresponding average response $\eta_u = \mu_{x_u}$ should be taken into account. Some examples are

1. The average tensile strength of cement increases with curing time
2. The amount of β-erythroidine in an aqueous solution changes in a regular fashion with the colorimeter reading of the turbidity

3. Students' grades on a subject might be related to a college entrance examination

4. The yield of a crop is related to the amount of water and amount of fertilizer

5. The maximum temperature during each day is related to the time of year

6. The price of a certain type of item is related to time, cost of raw material, and cost of labor

In some areas of study relationships are said to be exact and are expressed as mathematical functions of the form

$$y = G(x_1, \ldots, x_k)$$

This is particularly true in physics and some areas of chemistry. Examples of such relations are found in Boyle's gas law, Newton's law of force and acceleration, Ohm's law in electricity, etc. Such relations may be used to predict y with good accuracy when values of $x_1, \ldots, x_k$ are specified. There may be problems in finding the "correct" relationship or in making accurate measurements, but there is seldom little doubt that a single "correct" underlying *functional relationship* exists. We do not consider such "exact" functional relations.

In most investigations y cannot be determined exactly from a set of values $x_1, \ldots, x_k$ even when the x's are known without error. For example, there is no functional relationship which gives the exact weight of a person whose exact height is known. Still, it is well-known that a tall man is likely to weigh more than a short man. In fact, we are told by health experts what weight is "best" for a man of a given age, body build, and height. In this type of problem it is often assumed that a functional relationship exists between the variables $x_1, \ldots, x_k$ and the average response η associated with these variables. That is, it is assumed that

$$\eta = g(x_1, \ldots, x_k; \theta_1, \ldots, \theta_p) \qquad (14.1)$$

where $\theta_1, \ldots, \theta_p$ denote parameters. The functional relation given in Eq. (14.1) is known as a *regression equation*. Usually the problem is to specify the best functional form (family), and then to determine values of the parameters which give the most appropriate equation (particular member of the family) in a given experimental situation.

How does one set about selecting a particular regression equation for the population being studied? The *functional form* may be determined by the experimenter by (1) taking into account his knowledge of the theoretical structure existing or seeming to exist among the variables involved, or (2) by plotting sample points in a *scatter diagram*. Both methods are important in experimentation. Method (1) should get first consideration, as it is also useful in selecting the appropriate variables before the observations are made. Method (2) should be used to supplement method (1), or, if little is known about the underlying theoretical structure, to make a decision about the form of the relationship. The scatter diagram has limited use, since it consists in plotting sample points in a rectangular co-ordinate system with axes corresponding to the variables involved. The scatter diagram is illustrated in Example 14.1.

In this chapter we usually assume the functional form to be known and present a method, the method of *least squares*, for choosing estimates of particular parameters which specify a unique relationship connecting the average response η with the variables $x_1, \ldots, x_k$. The method of least squares gives estimators of the parameters $\theta_1, \ldots, \theta_p$ which are unbiased and have minimum variance of all estimators which are linear functions of the observations. In fact, if we assume the response variable to be normally distributed, it can be shown that the method of least squares is the same as the method of *maximum likelihood*, the most widely accepted general procedure for estimation at this time. The method of least squares is simply a procedure for finding estimators $\hat{\theta}_1, \ldots, \hat{\theta}_p$ of parameters $\theta_1, \ldots, \theta_p$ which minimize the function

$$Q = \sum_{u=1}^{n} (y_u - \hat{\eta}_u)^2 \qquad (14.2)$$

where

$$\hat{\eta}_u = g(x_{1u}, \ldots, x_{ku}; \hat{\theta}_1, \ldots, \hat{\theta}_p) \qquad (u = 1, \ldots, n) \qquad (14.3)$$

and $y_u, x_{1u}, \ldots, x_{ku}$ denotes the uth set of observations on $k + 1$ variables, only y_u being random. The relation

$$\hat{\eta} = g(x_1, \ldots, x_k; \hat{\theta}_1, \ldots, \hat{\theta}_p) \qquad (14.4)$$

is called the equation of the *regression curve of best fit*. The above procedures, along with some notation, are illustrated in the following example.

Example 14.1. Use the following hypothetical data to (a) plot a scatter diagram, (b) find the least-squares estimates of the parameters in the line of best fit, (c) find the error mean square of the deviates about the line of best fit and (d) test the hypothesis that the slope of the true regression line is zero

x	1	4	4	6	6	6
y	7	5	3	3	2	1

The scatter diagram is shown in Fig. 14.1, and it is obvious that a straight line should be used to fit the data. In fact, it appears that the line through the points with co-ordinates (1, 7) and (6, 2) is the appropriate line.

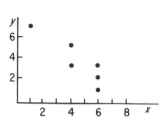

Fig. 14.1 Scatter Diagram for the Data in Example 14.1

For every pair of values (x_u, y_u) it is assumed that x_u is fixed and y_u is a random variable. That is, for each fixed value of x there is a corresponding array of y's. We assume, as is usually the case, that each array of y's corresponding to a fixed x is randomly and normally distributed with a common variance σ^2. It is to be understood that x is a quantitative variable which can be observed without error. (In particular, all one need assume is that the error in x is negligible as compared to that in y.) In this example, the mean η of an array of y's is assumed to fall along a straight line; that is, the mean η is given by

$$\eta = \alpha + \beta x \tag{14.5}$$

where α and β are parameters. The *model equation* for the uth pair of observations (x_u, y_u) is given by

$$y_u = \eta_u + \epsilon_u \qquad\qquad (u = 1, \ldots, n)$$

or

$$y_u = \alpha + \beta x_u + \epsilon_u \qquad (u = 1, \ldots, n)$$

Our problem is to find least-squares estimators a and b of α and β, respectively, and hence an estimator

$$\hat{\eta}_u = a + bx_u \tag{14.6}$$

of

$$\eta_u = \alpha + \beta x_u$$

for each u. Sometimes we write Eq. (14.6) as

$$\bar{y}_x = a + bx \tag{14.7}$$

indicating that the regression mean $\bar{y}_x$ is dependent on x. We refer to Eq. (14.5) as the *true regression of y on x* and to Eq. (14.6) or Eq. (14.7) as the *estimated regression of y on x*.

By the definition of the method of least squares, a and b are values which make

$$Q = \sum_u (y_u - a - bx_u)^2 \tag{14.8}$$

a minimum. Since x_u and y_u are data values, Q is a function of a and b. Thus, by the methods of the calculus (see Sect. 14.2), it can be shown that Eq. (14.8) is a minimum when

$$b = \frac{n \sum x_u y_u - (\sum x_u)(\sum y_u)}{n \sum x_u^2 - (\sum x_u)^2} \qquad (14.9)$$

and

$$a = \bar{y} - b\bar{x} \qquad (14.10)$$

After computing the totals shown in Table 14.1, we substitute them in Eqs. (14.9) and (14.10) to obtain

$$b = \frac{(6)(75) - (27)(21)}{(6)(141) - (27)(27)} = -1 \quad \text{and} \quad a = \frac{21}{6} - (-1)\left(\frac{27}{6}\right) = 8$$

Therefore, the regression line of best fit is

$$\bar{y}_x = 8 - x \qquad (14.11)$$

Table 14.1
Computations for Example 14.1

	x	y	x^2	y^2	xy
	1	7	1	49	7
	4	5	16	25	20
	4	3	16	9	12
	6	3	36	9	18
	6	2	36	4	12
	6	1	36	1	6
Totals	27	21	141	97	75

Equation (14.11) may be used to find $\bar{y}_x$ for any value of x for which the true regression equation (14.5) is the model. For example, if it is assumed that Eq. (14.5) is the proper model for only those values of x in the interval $1 \le x \le 6$, then estimates of η should not be obtained for values of x outside this interval. In most problems the line of best fit is assumed to hold for *all* values between the two extreme values of x in the experiment. For example, in our problem, when $x = 2$ the estimated mean of the array is 6. But, in many cases, values of x outside the two extreme values should not be used to find estimates of the means of the y arrays. This is usually the case in problems in which the x-axis is a time axis. The point is that estimates of array means may usually be found from the line of best fit for all those values of x between the two extreme x values of the experiment, but caution should be applied in other cases.

It should be noted that the estimated mean $\hat{\eta}_u$ of an array is also a y value and that it is obtained by taking into account all y values, not just those in the array above x_u. Further, note that the mean of all y's corresponding to x_u may or may not be the same as the estimated regression mean for the array corresponding to x_u. In our example, they happen to be the same for the arrays above $x = 1, 4$, and 6.

The fitted regression line passes through point $(4, 4)$ as well as the points $(1, 7)$ and $(6, 2)$, obtained by inspection in part (a). Thus, the sum of squares of deviates of the y values about the fitted line is given by

$$Q = (7 - 7)^2 + (5 - 4)^2 + (3 - 4)^2 + (3 - 2)^2 + (2 - 2)^2 + (1 - 2)^2 = 4$$

This is also called the *residual sum of squares* and is denoted by SSres. It can be shown (see Sect. 14.2) that the residual mean square $s_{y/x}^2$ with $n - 2$ degrees of freedom defined by

$$s_{y/x}^2 = \frac{SS\text{res}}{n - 2}$$

is an unbiased estimator of σ^2. In our problem

$$s_{y/x}^2 = \frac{4}{6 - 2} = 1$$

has four degrees of freedom and is an unbiased estimate of the common unknown variance σ^2.

The value $b = -1$ gives an unbiased point estimate of the true slope β. In many experiments it is not enough to find the point estimate of the slope. For example, it might be very desirable that we determine whether $b = -1$ is significantly different from zero. For this reason, we often test the null hypothesis that $\beta = 0$ against the alternative hypothesis that $\beta \neq 0$. This may be done by comparing the ratio of the regression mean square (see Sect. 14.2)

$$s_r^2 = \frac{[n \sum x_u y_u - (\sum x_u)(\sum y_u)]^2}{n[n \sum x_u^2 - (\sum x_u)^2]}$$

to the residual mean square $s_{y/x}^2$ with the upper α level value of the F distribution with 1 and $n - 2$ degrees of freedom. For the data of Example 14.1 we find that

$$s_r^2 = \frac{[(6)(75) - (27)(21)]^2}{6[6(141) - (27)^2]} = 19.5$$

and

$$\frac{s_r^2}{s_{y/x}^2} = \frac{19.5}{1} = 19.5$$

Since 19.5 is greater than 7.71, the upper five per cent value of F with one and four degrees of freedom, we reject the null hypothesis and conclude that $\beta \neq 0$. Since the sample slope is negative, we actually conclude that the true slope is negative. Thus, as x increases, y decreases.

In this section we have given an indication of some of the types of problems in regression analysis. In the next section the theory for linear regression of y on x is developed.

14.2. SIMPLE LINEAR REGRESSION

Linear regression of y on x is sometimes called *simple linear regression*. The key formulas of Sect. 14.1 are now brought together for quick reference and comparison.

The *model equation* for the uth pair of observations (x_u, y_u) is given by

$$y_u = \eta_u + \epsilon_u \qquad (u = 1, \ldots, n) \tag{14.12}$$

where

$$\eta_u = \alpha + \beta x_u \qquad (u = 1, \ldots, n) \tag{14.13}$$

denotes the *true regression of y on x*. The estimating equation is given by

$$y_u = \hat{\eta}_u + e_u \tag{14.14}$$

where

$$\hat{\eta}_u = \bar{y}_{x_u} = a + b x_u \tag{14.15}$$

denotes the *estimated regression of y on x*, and

$$e_u = y_u - \hat{\eta}_u \tag{14.16}$$

denotes the amount the uth random observation y_u deviates from the corresponding estimated regression ordinate. By studying Fig. 14.2, relations

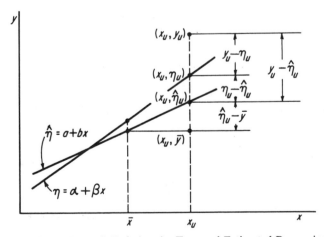

Fig. 14.2 An Example Relating the True and Estimated Regression of y on x to a Typical Pair of Observations (x_u, y_u)

among the true and estimated regression and observation (x_u, y_u) should become clearer. Note that $\bar{y} = a + b\bar{x}$, where

$$\bar{x} = \frac{\sum\limits_{u} x_u}{n} \quad \text{and} \quad \bar{y} = \frac{\sum\limits_{u} y_u}{n}$$

The estimators a and b are selected so as to make

$$Q = \sum_{u=1}^{n} e_u^2 = \sum_{u=1}^{n} (y_u - a - bx_u)^2 = SS\text{res} \qquad (14.17)$$

a minimum. If we think of Q as a function of a and b, those values of a and b which make Q a minimum, if it exists, are obtained by solving simultaneously the equations

$$\frac{\partial Q}{\partial a} = 0 \quad \text{and} \quad \frac{\partial Q}{\partial b} = 0$$

The partial derivatives of Q with respect to a and b are given by

$$\frac{\partial Q}{\partial a} = -2 \sum_{u} (y_u - a - bx_u)$$

and

$$\frac{\partial Q}{\partial b} = -2 \sum_{u} (y_u - a - bx_u) x_u$$

respectively. Setting these partials equal to zero and rearranging terms gives

$$\begin{cases} na + \left(\sum\limits_{u} x_u\right) b = \sum\limits_{u} y_u \\ \left(\sum\limits_{u} x_u\right) a + \left(\sum\limits_{u} x_u^2\right) b = \sum\limits_{u} x_u y_u \end{cases} \qquad (14.18)$$

which are referred to as the *normal equations* for estimating α and β. Any time Eq. (14.18) has a unique solution for a and b, these values make Q a minimum, since Q is bounded below (i.e., cannot be less than zero). The solution may be written as

$$b = \frac{SP}{SSx} \quad \text{and} \quad a = \bar{y} - b\bar{x} \qquad (14.19)$$

where

$$SP = \sum_{u} x_u y_u - \frac{\left(\sum\limits_{u} x_u\right)\left(\sum\limits_{u} y_u\right)}{n} = \sum_{u} (x_u - \bar{x})(y_u - \bar{y}) \qquad (14.20)$$

and

$$SSx = \sum_{u} x_u^2 - \frac{\left(\sum\limits_{u} x_u\right)^2}{n} = \sum_{u} (x_u - \bar{x})^2 \qquad (14.21)$$

When y is a random variable, it follows that b, a, and $\hat{\eta}$ are unbiased estimators of β, α, and η, respectively. Since

$$SP = \sum_u (x_u - \bar{x})(y_u - \bar{y})$$

$$= \sum_u (x_u - \bar{x}) y_u - \bar{y} \sum (x_u - \bar{x})$$

$$= \sum (x_u - \bar{x}) y_u$$

we may express b as the following linear function of $y_1, \ldots, y_n$

$$b = \left(\frac{x_1 - \bar{x}}{SSx}\right) y_1 + \cdots + \left(\frac{x_n - \bar{x}}{SSx}\right) y_n \qquad (14.22)$$

Thus

$$E(b) = \left(\frac{x_1 - \bar{x}}{SSx}\right) E(y_1) + \cdots + \left(\frac{x_n - \bar{x}}{SSx}\right) E(y_n)$$

$$= \left(\frac{x_1 - \bar{x}}{SSx}\right)(\alpha + \beta x_1) + \cdots + \left(\frac{x_n - \bar{x}}{SSx}\right)(\alpha + \beta x_n)$$

$$- \frac{\alpha}{SSx} \sum_u (x_u - \bar{x}) + \frac{\beta}{SSx} \sum_u (x_u - \bar{x})x_u$$

or

$$E(b) = \beta \qquad (14.23)$$

since

$$\sum_u (x_u - \bar{x})x_u = \sum_u (x_u - \bar{x})x_u - \bar{x} \sum_u (x_u - \bar{x})$$

$$= \sum_u (x_u - \bar{x})(x_u - \bar{x}) = SSx$$

Further

$$E(a) = E(\bar{y}) - \bar{x}E(b)$$

$$= \frac{1}{n}[E(y_1) + \cdots + E(y_n)] - \bar{x}E(b)$$

$$= \frac{1}{n}[(\alpha + \beta x_1) + \cdots + (\alpha + \beta x_n)] - \bar{x}\beta$$

$$= \alpha + \beta \left(\frac{\sum_u x_u}{n}\right) - \bar{x}\beta$$

or

$$E(a) = \alpha \qquad (14.24)$$

Also

$$E(\hat{\eta}) = E(a) + xE(b) = \alpha + \beta x = \eta \qquad (14.25)$$

Note that it is to be understood that the same set of x's occur in repeated samples of size n.

In order to test hypotheses about and establish confidence intervals for the parameters β, α, and η, it is necessary to know the nature of the distributions of b, a, and $\hat{\eta}$, respectively. For this purpose we assume that

(a) The model equation is given by Eq. (14.12)
(b) x_u is an exact observation and y_u is a random variable
(c) All arrays of y's have the same variance σ^2 (14.26)
(d) For every x there is an array of y's which has the normal distribution

These assumptions are pictured in Fig. 14.3. It should be clear that the assumptions, with the exception of (a), are the same as those made for the fixed-effects model in analysis of variance. In fact, when $\beta = 0$ the two are identical. Thus, much of our attention will be given to β and its estimator b. In this way we emphasize those topics in regression which are different from the usual analysis of variance.

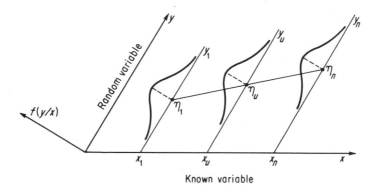

Fig. 14.3 Diagram Illustrating the Assumptions in (14.26)

To find unbiased points estimators of β, α, and η, only assumptions (a) and (b) were required, but to obtain confidence intervals for these parameters we use all assumptions in (14.26). Applying assumptions (a) and (b) along with linear relation (14.22) and Theorem 6.1, we see that the variance of b is given by

$$\sigma_b^2 = \frac{(x_1 - \bar{x})^2}{(SSx)^2}\sigma_{y_1}^2 + \cdots + \frac{(x_n - \bar{x})^2}{(SSx)^2}\sigma_{y_n}^2 \qquad (14.27)$$

If, in addition, we assume that assumption (c) holds, then Eq. (14.27) becomes

$$\sigma_b^2 = \frac{\displaystyle\sum_u (x_u - \bar{x})^2}{(SSx)^2}\sigma^2$$

or

$$\sigma_b^2 = \frac{\sigma^2}{SSx} \tag{14.28}$$

Finally, if assumption (d) is used in addition to assumptions (a), (b), and (c), it follows that b is normally distributed with mean β and variance σ_b^2. Therefore

$$\frac{(b - \beta)\sqrt{SSx}}{\sigma} \tag{14.29}$$

has the standard normal distribution. When σ^2 is known, we can use (14.29) to test the null hypothesis $\beta = \beta_0$, where β_0 is a constant, and to find $100(1 - \alpha)$ per cent confidence intervals for β. The symmetric confidence interval is given by

$$b - u_{\alpha/2}\frac{\sigma}{\sqrt{SSx}} < \beta < b + u_{\alpha/2}\frac{\sigma}{\sqrt{SSx}} \tag{14.30}$$

The variance σ^2 is not usually known. Hence, a sample variance must be used. It can be shown by the methods of Chap. 11 that

$$s_{y/x}^2 = \frac{SSres}{n - 2} = \frac{\sum_u (y_u - a - bx_u)^2}{n - 2} \tag{14.31}$$

is an unbiased estimator of σ^2. (The divisor $n - 2$ makes $s_{y/x}^2$ an unbiased estimator of σ^2. This seems reasonable, since the deviates about $a + bx$ are used, and a and b are two linear restrictions among the observations y_1, ..., y_n.) Thus

$$\frac{(n - 2)s_{y/x}^2}{\sigma^2} = \frac{SSres}{\sigma^2} \tag{14.32}$$

is distributed as χ^2 with $n - 2$ degrees of freedom and

$$\frac{(b - \beta)\sqrt{SSx}}{s_{y/x}} \tag{14.33}$$

is distributed as t with $n - 2$ degrees of freedom. The standardized statistic in (14.33) may be used to test the null hypothesis $\beta = \beta_0$ in the same way that

$$\frac{(\bar{x} - \mu)\sqrt{n}}{s}$$

is used to test the null hypothesis $\mu = \mu_0$. Further, a $100(1 - \alpha)$ per cent confidence interval for β is given by

$$b - t_{\alpha/2}(n - 2)\frac{s_{y/x}}{\sqrt{SSx}} < \beta < b + t_{\alpha/2}(n - 2)\frac{s_{y/x}}{\sqrt{SSx}} \qquad (14.34)$$

In Example 14.1 we used an F distribution to illustrate a test of the hypothesis $\beta = 0$. This was done mainly to indicate that regression analysis is tied in with analysis of variance. Actually, in testing the null hypothesis $\beta = 0$ against the alternative hypothesis $\beta \neq 0$, the statistic (14.33) becomes

$$\frac{b\sqrt{SSx}}{s_{y/x}}$$

and may be written as

$$\frac{b^2\,SSx}{s_{y/x}^2} = \frac{SS\text{reg}}{s_{y/x}^2} = \frac{s_r^2}{s_{y/x}^2} = F(1, n - 2) \qquad (14.35)$$

since b is a linear combination with one degree of freedom. That is, the two-sided t test is identical to the one-sided F test of the hypothesis $\beta = 0$. However, it should be noted at this time that the F statistic used in the example cannot be used to test the null hypothesis $\beta = \beta_0$ when $\beta_0 \neq 0$ or to establish a confidence interval for β. For these reasons the t distribution is generally preferred to the F distribution in working with simple linear regression.

It should be noted that the variance of the estimator b, given in Eq. (14.28), decreases with increasing values of SSx. Thus, values of x should be as widely scattered as possible in order to make SSx as large as possible in an experiment. For example, if there is strong evidence that the regression of y on x is linear from x_l to x_m but may not be linear outside this interval, then half of the observations should be made at x_l and half at x_m in order to obtain minimum variance. If, for some of the values intermediate to x_l and x_m, there is doubt that the regression is linear, then, of course, some observations should be made at the intermediate values.

Sometimes it is important to know the nature of the distribution of the estimated regression mean $\hat{\eta}_0$ of an array of y's corresponding to a particular x_0. Substituting $a = \bar{y} - b\bar{x}$ in Eq. (14.15) gives

$$\hat{\eta}_0 = \bar{y} + b(x_0 - \bar{x}) \qquad (14.36)$$

where $(x_0, \hat{\eta}_0)$ is any point on the admissible fitted regression line. Since $\bar{y}$ and b are normally distributed random variables, the linear combination in (14.36) is also normally distributed with mean

$$\mu_{\hat{\eta}_0} = \eta_0 = \alpha + \beta x_0 \qquad (14.37)$$

and variance

$$\sigma_{\hat{\eta}_0}^2 = \left[\frac{1}{n} + \frac{(x_0 - \bar{x})^2}{SSx}\right]\sigma^2 \qquad (14.38)$$

The mean was found in Eq. (14.25). We obtain the variance as follows, using the notation $\sigma_q^2 = \text{Var}(q)$, where q is any random variable

$$
\begin{aligned}
\sigma_{\hat{\eta}_0}^2 = \text{Var}(\hat{\eta}_0) &= \text{Var}[\bar{y} + (x_0 - \bar{x})b] \\
&= \text{Var}(\bar{y}) + \text{Var}[(x_0 - \bar{x})b] + 2\,\text{cov}[\bar{y}, (x_0 - \bar{x})b] \\
&= \frac{\text{Var}(y)}{n} + (x_0 - \bar{x})^2\,\text{Var}(b) + 2\cdot 0 \\
&= \frac{\sigma^2}{n} + (x_0 - \bar{x})^2\,\frac{\sigma^2}{SSx}
\end{aligned}
$$

since

$$
\begin{aligned}
\text{cov}[\bar{y}, (x_0 - \bar{x})b] &= \left(\frac{x_0 - \bar{x}}{nSSx}\right)\text{cov}[y_1 + \cdots + y_n, (x_1 - \bar{x})y_1 + \cdots \\
&\qquad\qquad + (x_n - \bar{x})y_n] \\
&= \left(\frac{x_0 - \bar{x}}{nSSx}\right)\{\text{cov}[y_1, (x_1 - \bar{x})y_1] + \cdots \\
&\qquad + \text{cov}[y_1, (x_n \quad \bar{x})y_n] + \cdots + \text{cov}[y_n, (x_1 - \bar{x})y_1] \\
&\qquad + \cdots + \text{cov}[y_n, (x_n - \bar{x})y_n]\} \\
&= \left(\frac{x_0 - \bar{x}}{nSSx}\right)[(x_1 - \bar{x})\,\text{Var}(y_1) + \cdots + 0 + \cdots \\
&\qquad + 0 + \cdots + (x_n - \bar{x})\,\text{Var}(y_n)] \\
&= \left(\frac{x_0 - \bar{x}}{nSSx}\right)[\sigma^2\sum_u(x_u - \bar{x})] = 0
\end{aligned}
$$

If σ^2 is unknown, it follows that

$$
\frac{\hat{\eta}_0 - \eta_0}{s_{\hat{\eta}_0}} = \frac{\hat{\eta}_0 - (\alpha + \beta x_0)}{s_{y/x}\sqrt{\dfrac{1}{n} + \dfrac{(x_0 - \bar{x})^2}{SSx}}} \tag{14.39}
$$

has the t distribution with $n - 2$ degrees of freedom. There are two cases of particualr interest. When $x_0 = \bar{x}$, then $\hat{\eta}_0 = \bar{y}$, and Eq. (14.39) reduces to

$$
\frac{\bar{y} - \mu_{\bar{y}}}{s_{\bar{y}}} = \frac{\bar{y} - (\alpha + \beta\bar{x})}{s_{y/x}|\sqrt{n}} \tag{14.40}
$$

where $s_{\bar{y}} = s_{y/x}/\sqrt{n}$ has $n - 2$ degrees of freedom. When $x_0 = 0$, $\eta_0 = \bar{y} - b\bar{x} = a$, and Eq. (14.39) reduces to

$$
\frac{a - \alpha}{s_{y/x}\sqrt{\dfrac{1}{n} + \dfrac{\bar{x}^2}{SSx}}} \tag{14.41}
$$

which is also distributed as t with $n - 2$ degrees of freedom. The symmetric

$100 (1 - \alpha)$ per cent confidence limits of $\eta_0 = \alpha + \beta x_0$ at $x = x_0$ are given by

$$\bar{y} + b(x_0 - \bar{x}) \pm t_{\alpha/2}(n - 2)s_{y/x} \sqrt{\frac{1}{n} + \frac{(x_0 - \bar{x})^2}{SSx}} \qquad (14.42)$$

The limits are functions of x_0; the greater the distance between x_0 and $\bar{x}$, the more extreme the limits of η_0. The symmetric $100 (1 - \alpha)$ per cent confidence limits of α, the parameter, are

$$a \pm t_{\alpha/2}(n - 2)s_{y/x} \sqrt{\frac{1}{n} + \frac{\bar{x}^2}{SSx}} \qquad (14.43)$$

These principles are illustrated in the following example.

Example 14.2. Use the data of Example 14.1 to find (a) a symmetric 90 per cent confidence interval for β and (b) symmetric 90 per cent confidence intervals for η_0 when $x_0 = 3$ and 4.5.

From Example 14.1 we know that $b = -1$, $SSx = 19.5$, and $s_{y/x}^2 = 1$ with four degrees of freedom. Since $t_{.05}(4) = 2.13$, we find, using (14.34), that the 90 per cent confidence interval of β is

$$-1.48 \leq \beta \leq -0.52$$

Since both bounds of the slope are negative, one has a high degree of confidence that the response decreases with increasing x. From this information we could also reject the null hypothesis $\beta = 0$ and conclude that β is negative, understanding that there is at most a ten per cent chance of making an error.

We know also that $\bar{x} = 27/6$ and $\bar{y} = 21/6$. Thus, when $x_0 = 3$, $\hat{\eta}_0 = \bar{y} + b(x_0 - \bar{x}) = 5$, and when $x_0 = 4.5 = \bar{x}$, $\hat{\eta}_0 = y = 3.5$. Now substituting in Eq. (14.42) gives confidence intervals

$$3.87 \leq (\alpha + 3\beta) \leq 6.13 \qquad (14.44)$$

and

$$2.63 \leq \mu_{\bar{y}} \leq 4.37 \qquad (14.45)$$

Note that the length of the interval in (14.44) is 2.26 and that in (14.45) it is 1.74. Since $x_0 = 6$ is 1.5 units larger than $\bar{x}$, we know that the length of the interval about $\eta = \alpha + 6\beta$ is also 2.26. In other words, we can expect our estimate of η to be best in the vicinity of the mean $\bar{x}$.

Sometimes it is reasonable to assume that $\alpha = 0$. The true regression of y on x then becomes

$$\eta = \beta x \qquad (14.46)$$

It can be shown by a method analogous to the above that

$$b = \frac{\sum_u x_u y_u}{\sum_u x_u^2} \tag{14.47}$$

and

$$s_{y/x}^2 = \frac{\sum_u (y_u - bx_u)^2}{n - 1} \tag{14.48}$$

is an unbiased estimator of σ^2 with $n - 1$ degrees of freedom. It then follows that

$$\frac{(b - \beta)\sqrt{\sum x_u^2}}{s_{y/x}} \tag{14.49}$$

is distributed as t with $n - 1$ degrees of freedom.

Analysis of variance in a simple linear regression is best explained in terms of a sum of squares identity. Such an identity is derived by substituting Eqs. (14.19), (14.20), and (14.21) in the expression of SSres defined in Eq. (14.17). Then

$$\begin{aligned} SS\text{res} &= \sum_u (y_u - a - bx_u)^2 \\ &= \sum_u [(y_u - \bar{y}) - b(x_u - \bar{x})]^2 \\ &= \sum_u (y_u - \bar{y})^2 - 2b \sum_u (x_u - \bar{x})(y_u - \bar{y}) + b^2 \sum_u (x_u - \bar{x})^2 \\ &= \sum_u (y_u - \bar{y})^2 - 2b \cdot bSSx + b^2 SSx \end{aligned}$$

or

$$SS\text{res} = SSy - b^2 SSx \tag{14.50}$$

where

$$SSy = \sum_u (y_u - \bar{y})^2 \tag{14.51}$$

Using Eq. (14.50), we may write the *sum of squares identity for simple linear regression* as

$$SSy = SS\text{res} + SS\text{reg} \tag{14.52}$$

where

$$SS\text{reg} = b^2 SSx = \frac{SP^2}{SSx} = bSP \tag{14.53}$$

Note that the sum of squares identity (14.52) can be obtained directly from the identity

$$y_u - \bar{y} = (y_u - \bar{y}_{x_u}) + (\bar{y}_{x_u} - \bar{y}) \tag{14.54}$$

where

$$\bar{y}_{x_u} - \bar{y} = (a + bx_u) - (a + b\bar{x}) = b(x_u - \bar{x})$$

The deviates on the right-hand side of Eq. (14.54) are shown in Fig. 14.2. The analysis of variance and expected mean squares for simple linear regression are given in Table 14.2. The test of the null hypothesis $\beta = 0$ has already been indicated.

Table 14.2

Analysis of Variance for Simple Linear Regression

Source of Variation	Sum of Squares	Degrees of Freedom	Mean Squares	Expected Mean Squares
Due to regression	$SSreg = \dfrac{SP^2}{SSx}$	1	$s_r^2 = \dfrac{SSreg}{1}$	$\sigma^2 + \beta^2 SSx$
Deviation about regression	$SSres = SSy - SSreg$	$n - 2$	$s_{y/x}^2 = \dfrac{SSres}{n-2}$	σ^2
Total	SSy	$n - 1$		

14.3. A TEST FOR LINEARITY OF REGRESSION

In Sect. 14.2 we assumed the regression to be linear. A method for checking the validity of this assumption is often desirable. For example, one may suspect that data relating the curing time to tensile strength of a given size of cement cube are not linear. Hypothetical data for such an experiment are given in Table 14.3, and the scatter diagram and line of best fit are shown in Fig. 14.4. The data points are indicated by dots and the array means by circles.

The method for testing linearity links regression analysis and analysis

Table 14.3

Data for Tensile Strength of Cement Cubes

Cube Number	Curing Time in Days x	Tensile Strength in Pounds y
1	2	19
2	2	21
3	3	24
4	3	27
5	3	27
6	4	29
7	4	31
8	6	35
9	6	36
10	6	37

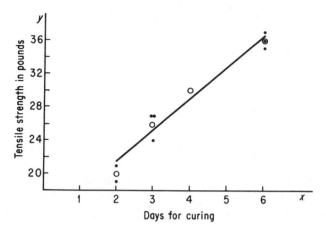

Fig. 14.4 Scatter Diagram and Fitted Regression Line
for the Data of Table 14.3

of variance of a one-way classification. Perhaps the scatter diagram indicated how this is possible. The four-array means do not fall on a straight line. In fact, due to sampling variation, it would be very unusual to have four-array means fall on a straight line. The problem is to determine if these means deviate too much from linearity to be explained by chance causes. Such a decision is made after comparing the mean square for deviation from regression with a valid error mean square. The required variance estimate is obtained by pooling the error variances of the arrays, that is, by finding the error variance of a one-way classification. It is for this reason that the notation of analysis of variance is used.

Let data points be denoted by (x_i, y_{iu}). The regression line of best fit is still written as

$$\bar{y}_{x_i} = a + bx_i$$

but the estimating equation is written as

$$y_{iu} = a + bx_i + e_{iu} \qquad (i = 1, \ldots, k; u = 1, \ldots, n_i) \qquad (14.55)$$

The residual sum of squares, SSres, then becomes

$$SS\text{res} = \sum_i \sum_u (y_{iu} - a - bx_i)^2$$

Now, denoting the mean of the ith array by

$$\bar{y}_i = \sum_u \frac{y_{iu}}{n_i}$$

we may write

$$SS\text{res} = \sum_i \sum_u [(y_{iu} - \bar{y}_i) + (\bar{y}_i - \bar{y}_{x_i})]^2$$

or

$$SS\text{res} = \sum_i \sum_u (y_{iu} - \bar{y}_i)^2 + \sum_i n_i(\bar{y}_i - \bar{y}_{x_i})^2 \qquad (14.56)$$

since

$$\sum_i \sum_u 2(y_{iu} - \bar{y}_i)(\bar{y}_i - \bar{y}_{x_i}) = 2 \sum_i (\bar{y}_i - \bar{y}_{x_i}) \sum_u (y_{iu} - \bar{y}_i) = 0$$

The first sum of squares on the right-hand side of Eq. (14.56) is the within sum of squares, SSW, in the one-way classification. The second sum of squares represents the sum of squares of deviates of array means about the estimated regression array means and is denoted by SSD. The degrees of freedom identity associated with the sum of squares identity (14.56) is

$$n - 2 = (n - k) + (k - 2)$$

where

$$n = \sum_i n_i$$

From the partition theory of Chap. 10 we know that SSW and SSD are independently distributed. Further, we know that $SSW/(n - k)$ is an unbiased estimator of σ^2. It can be shown by the methods of Chap. 11 that $SSD/(k - 2)$ is an unbiased estimator of σ^2, provided that the true regression is linear. Therefore, the ratio

$$\frac{\dfrac{SSD}{(k-2)}}{\dfrac{SSW}{(n-k)}} \qquad (14.57)$$

is distributed as F with $k - 2$ and $n - k$ degrees of freedom when the hypothesis of linearity is true; that is, when

$$H_0 : \quad \eta_i = \alpha + \beta x_i \qquad (i = 1, \ldots, k) \qquad (14.58)$$

is true. Note that the hypothesis specifies only that the k array means fall on a straight line.

If the regression is not linear, the expected value of $SSD/(k - 2)$ is greater than σ^2. Thus, the critical region for the α level test of (14.58) is made up of all values of F greater than $F_\alpha = F_\alpha(k - 2, n - k)$. To test (14.58), use the experimental data and (14.57) to compute a sample statistic F_c, and then compare F_c with F_α. If $F_c > F_\alpha$, reject (14.58) and conclude that the regression is not linear; if $F_c \leq F_\alpha$, fail to reject (14.58), understanding that there is not enough evidence to conclude that the true regression is different from $\eta = \alpha + \beta x$. (It should be noted that the test cannot be made unless more than one y observation is made for at least one x. Otherwise, there would be zero degrees of freedom for the within mean square.)

The sum of squares, SSD, is not usually computed by

$$\sum_i n_i (\bar{y}_i - \bar{y}_{x_i})^2$$

The following derivation leads to a better computational formula. According to Eqs. (10.38) and (14.52), the total sum of squares can be partitioned in two ways. Equating these two expressions gives

$$SS\text{res} + SS\text{reg} = SSW + SSTr$$

where $SSTr$ denotes the sum of squares among array means. Solving for SSres gives

$$SS\text{res} = SSW + (SSTr - SS\text{reg}) \qquad (14.59)$$

On comparing Eq. (14.56) with Eq. (14.59), we see that

$$SSD = SSTr - SS\text{reg} \qquad (14.60)$$

Thus, for the test of linearity we make the usual one-way classification analysis, compute the regression sum of squares, and then apply Eq. (14.60). Such an analysis of variance is shown in Table 14.5.

Example 14.3. Use the data in Table 14.3 to test the hypothesis of linearity, that is

$$H_0: \eta_i = \alpha + \beta x_i \qquad (i = 1, \ldots, 4)$$

Table 14.4 shows the data arranged in a one-way classification. The total and treatment sum of squares are found to be 348.4 and 336.4, respectively. From Table 14.3 we find $SP = 86.6$ and $SSx = 22.9$. Therefore, $SS\text{reg} = 327.5$. The resulting analysis of variance is shown in Table 14.5. Since the computed F is less than $F_{.05}(2, 6) = 5.14$, we fail to reject the hypothesis of linearity. That is, we do not have enough evidence to say that the regression is not linear.

Table 14.4

Data of Table 14.3 Arranged in a One-way Classification

	Tensile Strength in Pounds for Day				
	2	3	4	6	
	19	24	29	35	
	21	27	31	36	
		27		37	
Totals	40	78	60	108	286

Table 14.5
Analysis of Variance for Linearity of Regression

Source of Variation	Sum of Squares	Degrees of Freedom	Mean Square	Computed F	Tabled F
Among means	336.4	3			
Regression	327.5	1			
Deviation from regression	8.9	2	4.45	2.22	5.14
Within	12.0	6	2.00		
Total	348.4	9			

If our primary objective is to determine whether the regression is non-linear or not, we should probably stop with the statement that the data fit the regression equation

$$\bar{y}_x = 13.9 + 3.78x \qquad (14.61)$$

satisfactorily. [Eq. (14.61) was obtained from the data of Table 14.3.] However, if our main objective is to establish a confidence interval or to test a hypothesis about β, the test of linearity should be considered a preliminary test. In this case, it is doubtful whether the significance level should be as low as five per cent; perhaps it should be in the neighborhood of 25 or 40 per cent.

If the hypothesis is rejected, it is necessary to look further for the best-fitting regression curve. This topic is discussed in Sect. 14.5. For further reading in simple linear regression, see Refs. [1, 5, 9, 16, 32, 33, 36, 44, 46, 52, 59].

14.4. MULTIPLE LINEAR REGRESSION

There are numerous examples in many areas of study where variation in two or more variables affect the random response. Two illustrations, 4 and 6 on page 498 are mentioned in Sect. 14.1. Others may be found almost anywhere regression analysis is presented. We use data taken from Gingrich and Meyer [31] and shown in Table 14.6 to illustrate principles of multiple regression.

Consider now the simplest and most common model for multiple regression, namely, the model where the random response, y, is *linearly* dependent on two or more variables, $x_1, x_2, \ldots, x_p$. We discuss the cases where $p = 2$ or 3, extensions to more than three dependent variables being obvious.

When $p = 2$, we assume each observation y_u to be randomly drawn from a normal distribution with mean

$$\eta_u = \mu_{y_u/x_1 x_2} = \alpha + \beta_1 x_{1u} + \beta_2 x_{2u} \qquad (u = 1, \ldots, n) \qquad (14.62)$$

Table 14.6
Aerial Stand Volume Data for Upland Oak

Plot Number*	Stand Height in Feet x_1	Crown Cover in Per Cent x_2	Volume per Plot in Cubic Feet y
1	90	84	460
2	74	58	433
3	70	68	365
4	75	88	376
5	68	88	419
6	58	87	330
7	80	48	362
8	87	53	381
9	80	72	431
10	120	88	1038
11	72	96	482
12	60	59	322
13	72	77	379
14	72	62	381
15	80	86	526
16	80	84	466
17	90	78	558
18	55	85	281
19	60	72	335
20	58	92	329
21	90	85	433
22	85	93	508
23	74	94	521
24	110	72	655
25	86	73	531

* Data for only the first 25 of 93 plots examined by Gingrich and Meyer.

and variance σ^2. Estimators a, b_1, and b_2, of α, β_1, and β_2 are found so as to minimize

$$Q = \sum_u (y_u - a - b_1 x_{1u} - b_2 x_{2u})^2 \tag{14.63}$$

Setting the partial derivatives with respect to a, b_1, and b_2 equal to zero leads to the *normal equations*

$$\begin{cases} na + (\sum x_{1u})b_1 + (\sum x_{2u})b_2 = \sum y_u \\ (\sum x_{1u})a + (\sum x_{1u}^2)b_1 + (\sum x_{1u}x_{2u})b_2 = \sum x_{1u}y_u \\ (\sum x_{2u})a + (\sum x_{1u}x_{2u})b_1 + (\sum x_{2u}^2)b_2 = \sum x_{2u}y_u \end{cases} \tag{14.64}$$

The solution of these equations gives unbiased estimators of the parameters α, β_1, and β_2, and, furthermore, it can be shown (see Gauss-Markoff theorem [15, 32, 49]) that these estimators have minimum variance of all estimators which are linear combinations of the observations. The estimated regression of y on x_1 and x_2 is given by

$$\hat{\eta}_u = a + b_1 x_{1u} + b_2 x_{2u} \tag{14.65}$$

By the methods of Chap. 11 it can be shown that an unbiased estimator of σ^2 is given by

$$s^2_{y/x_1 x_2} = \frac{\sum\limits_{u}(y_u - \hat{\eta})^2}{n - 3} \tag{14.66}$$

The solution of the system of equations in (14.64) can be found by any of the methods of solving simultaneous linear equations. We solve by first eliminating a and then applying Cramer's rule with determinants. Solving for a in the first equation of (14.64) gives

$$a = \bar{y} - \bar{x}_1 b_1 - \bar{x}_2 b_2 \tag{14.67}$$

where

$$\bar{y} = \frac{\sum y_u}{n}, \quad \bar{x}_1 = \frac{\sum x_{1u}}{n} \quad \text{and} \quad \bar{x}_2 = \frac{\sum x_{2u}}{n}$$

Substituting Eq. (14.67) in the second and third equations of (14.64) gives

$$(\sum x_{1u})(\bar{y} - \bar{x}_1 b_1 - \bar{x}_2 b_2) + (\sum x_{1u}^2)b_1 + (\sum x_{1u}x_{2u})b_2 = \sum x_{1u}y_u$$
$$(\sum x_{2u})(\bar{y} - \bar{x}_1 b_1 - \bar{x}_2 b_2) + (\sum x_{1u}x_{2u})b_1 + (\sum x_{2u}^2)b_2 = \sum x_{2u}y_u$$

and, on further reduction, this leads to

$$\begin{cases} a_{11} b_1 + a_{12} b_2 = g_1 \\ a_{21} b_1 + a_{22} b_2 = g_2 \end{cases} \tag{14.68}$$

where

$$\begin{cases} a_{ii} = SSx_i = \sum x_{iu}^2 - \dfrac{(\sum x_{iu})^2}{n} = \sum (x_{iu} - \bar{x}_i)^2 & (i = 1, 2) \\[3mm] a_{12} = a_{21} = SP_{12} = \sum x_{1u}x_{2u} - \dfrac{(\sum x_{1u})(\sum x_{2u})}{n} = \sum (x_{1u} - \bar{x}_1)(x_{2u} - \bar{x}_2) \\[3mm] g_i = SP_{iy} = \sum x_{iu}y_u - \dfrac{(\sum x_{iu})(\sum y_u)}{n} = \sum (x_{iu} - \bar{x}_i)(y_u - \bar{y}) & (i = 1, 2) \end{cases}$$

Systems (14.64) and (14.68) are similar, except that in system (14.68) the sum of squares and products are for deviates instead of original observations. Note that if the true regression of y on x_1 and x_2 had been written as

$$\eta_u = \bar{y} + \beta_1(x_{1u} - \bar{x}_1) + \beta_2(x_{2u} - \bar{x}_2) \tag{14.69}$$

and the estimators b_1 and b_2 of β_1 and β_2, respectively, had been found so as to minimize

$$Q = \sum_u [(y_u - \bar{y}) - b_1(x_{1u} - \bar{x}_1) - b_2(x_{2u} - \bar{x}_2)]^2$$

the resulting normal equations would be those given in (14.68). Further, Q may be expressed as

$$Q = \sum_u (y_u - \hat{\eta}_u)^2 = SSy - g_1 b_1 - g_2 b_2 \qquad (14.70)$$

where

$$\hat{\eta}_u = \bar{y} + b_1(x_{1u} - \bar{x}_1) + b_2(x_{2u} - \bar{x}_2)$$

The residual sum of squares is usually computed by (14.70).

The regression coefficients are found by solving (14.68). There are many compact methods [8, 17, 19, 21, 22, 23, 29, 56] of solving normal equations. But since our primary concern is in finding interval estimates and in testing hypotheses about the β's, we use a method which is also very useful in determining the distribution of b_i.

In order to see the generality of the method, consider the case where $p = 3$. Assume that y_u is a random normal variable with mean

$$\eta_u = \alpha + \beta_1 x_{1u} + \beta_2 x_{2u} + \beta_3 x_{3u}$$

and variance σ^2. The normal equations are

$$\begin{cases} a_{11}b_1 + a_{12}b_2 + a_{13}b_3 = g_1 \\ a_{21}b_1 + a_{22}b_2 + a_{23}b_3 = g_2 \\ a_{31}b_1 + a_{32}b_2 + a_{33}b_3 = g_3 \end{cases} \qquad (14.71)$$

where b_i estimates β_i $(i = 1, 2, 3)$ and

$$\begin{cases} a_{ii} = \sum (x_{iu} - \bar{x})^2 \\ a_{ij} = a_{ji} = \sum (x_{iu} - \bar{x}_i)(x_{ju} - \bar{x}_j) \qquad (i \neq j) \\ g_i = \sum (x_{iu} - \bar{x}_i)(y_u - \bar{y}) \end{cases} \qquad (14.72)$$

Let Δ denote the determinant of the matrix, A, of coefficients on the right-hand side of (14.71), that is

$$\Delta = \begin{vmatrix} a_{11} & a_{12} & a_{13} \\ a_{21} & a_{22} & a_{23} \\ a_{31} & a_{32} & a_{33} \end{vmatrix}$$

When $\Delta \neq 0$, the solution of (14.71), when Cramer's rule is applied, is

$$b_1 = \frac{\Delta_1}{\Delta}, \qquad b_2 = \frac{\Delta_2}{\Delta}, \qquad b_3 = \frac{\Delta_3}{\Delta} \qquad (14.73)$$

where Δ_i denotes the determinant which results by replacing the ith column of Δ by the column of g's in (14.71).

In order to determine the distribution of b_i we express (14.73) as a linear combination of g's which are linear combinations of y's. To illustrate, note that by expanding Δ_1 by the first column we may write

$$b_1 = \frac{1}{\Delta}\left(g_1 \begin{vmatrix} a_{22} & a_{23} \\ a_{32} & a_{33} \end{vmatrix} - g_2 \begin{vmatrix} a_{12} & a_{13} \\ a_{32} & a_{33} \end{vmatrix} + g_3 \begin{vmatrix} a_{12} & a_{13} \\ a_{22} & a_{23} \end{vmatrix} \right)$$

or

$$b_1 = c_{11}g_1 + c_{12}g_2 + c_{13}g_3 \qquad (14.74)$$

where

$$c_{11} = \frac{1}{\Delta} \begin{vmatrix} a_{22} & a_{23} \\ a_{32} & a_{33} \end{vmatrix}, \qquad c_{12} = -\frac{1}{\Delta} \begin{vmatrix} a_{12} & a_{13} \\ a_{32} & a_{33} \end{vmatrix}, \qquad c_{13} = \frac{1}{\Delta} \begin{vmatrix} a_{12} & a_{13} \\ a_{22} & a_{23} \end{vmatrix}$$

In a similar way, we may write

$$\begin{cases} b_2 = c_{21}g_1 + c_{22}g_2 + c_{23}g_3 \\ b_3 = c_{31}g_1 + c_{32}g_2 + c_{33}g_3 \end{cases} \qquad (14.75)$$

where

$$c_{ji} = \frac{(-1)^{i+j} A_{ij}}{\Delta} \qquad (i = 1, 2, 3; \; j = 1, 2, 3)$$

A_{ij} being the two-by-two determinant obtained by deleting the ith column and jth row of Δ. Actually, $c_{11}, c_{12}, \ldots, c_{33}$ are numbers in the matrix, C, which is the inverse of the matrix A. That is, the c's may be found by solving the matrix equation

$$\begin{bmatrix} a_{11} & a_{12} & a_{13} \\ a_{21} & a_{22} & a_{23} \\ a_{31} & a_{32} & a_{33} \end{bmatrix} \cdot \begin{bmatrix} c_{11} & c_{21} & c_{31} \\ c_{12} & c_{22} & c_{32} \\ c_{13} & c_{23} & c_{33} \end{bmatrix} = \begin{bmatrix} 1 & 0 & 0 \\ 0 & 1 & 0 \\ 0 & 0 & 1 \end{bmatrix} \qquad (14.76)$$

In particular, c_{11}, c_{12}, c_{13} are found by solving the system of linear equations

$$\begin{cases} a_{11}c_{11} + a_{12}c_{12} + a_{13}c_{13} = 1 \\ a_{21}c_{11} + a_{22}c_{12} + a_{23}c_{13} = 0 \\ a_{31}c_{11} + a_{32}c_{12} + a_{33}c_{13} = 0 \end{cases} \qquad (14.77)$$

The numbers c_{21}, c_{22}, c_{23} and c_{31}, c_{32}, c_{33} are obtained by solving two similar systems of linear equations with the right-hand-side coefficients 1, 0, 0 replaced by 0, 1, 0 and 0, 0, 1, respectively. Now substituting $c_{11}, c_{12}, \ldots, c_{33}$

in Eqs. (14.74) and (14.75) gives b_1, b_2, and b_3, which may be used to write the regression equation of y on x_1, x_2, x_3 as

$$\hat{\eta} = a + b_1 x_1 + b_2 x_2 + b_3 x_3 \tag{14.78}$$

or

$$\hat{\eta} = \bar{y} + b_1(x_1 - \bar{x}_1) + b_2(x_2 - \bar{x}_2) + b_3(x_3 - \bar{x}_3) \tag{14.78a}$$

where

$$a = \bar{y} - b_1 \bar{x}_1 - b_2 \bar{x}_2 - b_3 \bar{x}_3$$

In order to find the distribution of b_1, we first use Eq. (14.74) and the definition of g_i ($i = 1, 2, 3$) in Eq. (14.72) to write

$$b_1 = c_{11} \sum_u (x_{1u} - \bar{x}_1)(y_u - \bar{y}) + c_{12} \sum_u (x_{2u} - \bar{x}_2)(y_u - \bar{y})$$
$$+ c_{13} \sum_u (x_{3u} - \bar{x}_3)(y_u - \bar{y})$$

or

$$b_1 = \sum_u [c_{11}(x_{1u} - \bar{x}) + c_{12}(x_{2u} - \bar{x}_2) + c_{13}(x_{3u} - \bar{x})] y_u \tag{14.79}$$

since

$$\sum_u [c_{11}(x_{1u} - \bar{x}_1) + c_{12}(x_{2u} - \bar{x}_2) + c_{13}(x_{3u} - \bar{x}_3)] \bar{y} = 0 \tag{14.80}$$

In Eq. (14.79) we see that b_1 is a linear combination of $y_1, \ldots, y_n$ which are normally and independently distributed. Thus, when the x's are fixed, b_1 is also normally distributed. The mean of b_1 may be written as

$$\mu_{b_1} = \sum_u [c_{11}(x_{1u} - \bar{x}_1) + c_{12}(x_{2u} - \bar{x}_2) + c_{13}(x_{3u} - \bar{x}_3)] \cdot (\alpha + \beta_1 x_{1u}$$
$$+ \beta_2 x_{2u} + \beta_3 x_{3u})$$
$$= \alpha \sum_u [c_{11}(x_{1u} - \bar{x}_1) + c_{12}(x_{2u} - \bar{x}_2) + c_{13}(x_{3u} - \bar{x}_3)]$$
$$+ \beta_1(c_{11}a_{11} + c_{12}a_{12} + c_{13}a_{13})$$
$$+ \beta_2(c_{11}a_{21} + c_{12}a_{22} + c_{13}a_{23})$$
$$+ \beta_3(c_{11}a_{31} + c_{12}a_{32} + c_{13}a_{33})$$

or

$$\mu_{b_1} = \beta_1$$

by applying Eqs. (14.80) and (14.77). Since the y's are independent, the variance of b_1, if we use Theorem 6.1, is

$$\sigma_{b_1}^2 = \sum_u [c_{11}(x_{1u} - \bar{x}_1) + c_{12}(x_{2u} - \bar{x}_2) + c_{13}(x_{3u} - \bar{x}_3)]^2 \sigma_{y_u}^2$$

Using the assumption of equal variances and properties of algebra, we may write the variance of b_1 as

$$\sigma_{b_1}^2 = \sigma^2[c_{11}(c_{11}a_{11} + c_{12}a_{12} + c_{13}a_{13})$$
$$+ c_{12}(c_{11}a_{21} + c_{12}a_{22} + c_{13}a_{23})$$
$$+ c_{13}(c_{11}a_{31} + c_{12}a_{32} + c_{13}a_{33})]$$

or, on substituting the relations of (14.77), as

$$\sigma_{b_1}^2 = c_{11}\sigma^2$$

In a similar way we can show that b_i is distributed normally with mean β_i and variance $c_{ii}\sigma^2$ ($i = 2, 3$). But b_1, b_2, and b_3 are not generally independently distributed, since it can be shown that

$$\text{cov}(b_1, b_2) = c_{12}\sigma^2, \qquad \text{cov}(b_1, b_3) = c_{13}\sigma^2, \qquad \text{cov}(b_2, b_3) = c_{23}\sigma^2$$

If, as is ordinarily the case, σ^2 is unknown, one uses the unbiased estimator

$$s_{y123}^2 = s_{y/x_1x_2x_3}^2 = \frac{\sum_{u}(y_u - \hat{\eta}_u)^2}{n - 3 - 1}$$

where

$$\sum_{u}(y_u - \hat{\eta}_u)^2 = SSy - g_1b_1 - g_2b_2 - g_3b_3$$

Then it follows that

$$\frac{(n - 4)\, s_{y123}^2}{\sigma^2} \tag{14.81}$$

is distributed as χ^2 with $n - 4$ degrees of freedom, and

$$\frac{b_i - \beta_i}{s_{y_{123}}\sqrt{c_{ii}}} \qquad (i = 1, 2, 3) \tag{14.82}$$

is distributed as the Student t with $n - 4$ degrees of freedom. Then (14.82) may be used to test any hypothesis of the form $\beta_i = \beta_{i0}$, where β_{i0} is any constant. The $100(1 - \alpha)$ per cent symmetric confidence interval of β_i is given by

$$b_i - t_{\alpha/2}(n - 4)\, s_{y_{123}}\sqrt{c_{ii}} < \beta_i < b_i + t_{\alpha/2}(n - 4)\, s_{y_{123}}\sqrt{c_{ii}} \tag{14.83}$$

It can easily be shown that

$$\frac{\dfrac{\sum_{i=1}^{3} g_i b_i}{3}}{s_{y_{123}}^2} \tag{14.84}$$

is distributed as F with three and $n - 4$ degrees of freedom, and that (14.84) is the appropriate statistic to use in testing the common hypothesis

$$H_0: \quad \beta_1 = \beta_2 = \beta_3 = 0$$

However, a test involving two β parameters is not straightforward. For more information on this topic, see Exercises 14.14 and 14.15 and Refs. [5, 6, 9, 32, 36, 59].

Example 14.4. Use the data in Table 14.6 to illustrate applications of multiple linear regression. (a) Find the linear equation for regression of y on x_1 and x_2. (b) Test the hypothesis $\beta_1 = \beta_2 = 0$ at the five per cent level. (c) Find the inverse matrix. (d) Test the hypothesis $\beta_1 = 0$ at the five per cent level, and find a 95 per cent confidence interval for β_2.

Table 14.6 shows only a fraction of the observations which were made. According to Gingrich and Meyer [31, p. 140], "A total of 93 one-fifth acre plots were randomly located on $1 : 12{,}000$-scale photographs of upland oak stands in Centre County, Pennsylvania. The photographs were taken on infrared film with a minus-blue filter. Contact prints were made on semi-matte paper." After carefully locating each plot, a two-man crew measured at chest height the diameter of every tree equal to or greater than five inches. Standard tables were then used to estimate the gross cubic-foot volume, y_u, of the uth plot. The *stand height*, x_{1u}, is the mean of the three tallest trees in plot u. The *relative crown cover*, x_{2u}, is the mean of six measurements made by three different observers using a standard crown diameter scale. Several other ground and aerial measurements were made, but we study only these three in order to determine how gross volume may be estimated from stand height and relative crown cover obtained from aerial photographs. (Since only a fraction of the available data is used to illustrate techniques in multiple regression, the reader should not expect the results of the example necessarily to be the same as those reported by the authors [31].)

The sums, sums of squares, and sums of products are

$$\sum x_{1u} = 1946 \qquad \sum x_{2u} = 1942 \qquad \sum y_u = 11{,}302$$

$$\sum x_{1u}^2 = 157{,}196 \qquad \sum x_{2u}^2 = 155{,}244 \qquad \sum y_u^2 = 5{,}654{,}974$$

$$\sum x_{1u}x_{2u} = 151{,}133 \qquad \sum x_{1u}y_u = 927{,}352 \qquad \sum x_{2u}y_u = 890{,}387$$

The sums of squares and sums of products about the means are

$$a_{11} = SSx_1 = 5719.36 \qquad a_{12} = SP_{12} = -32.28$$

$$a_{22} = SSx_2 = 4389.44 \qquad g_1 = SP_{1y} = 47{,}604.32$$

$$SSy = 545{,}565.84 \qquad g_2 = SP_{2y} = 12{,}447.64$$

Thus, the normal equations are

$$5719.36b_1 + (-32.28)b_2 = 47{,}604.32$$

$$-32.28 b_1 + 4389.44 b_2 = 12{,}447.64$$

The determinant of the coefficients is

$$\Delta = (5719.36)(4389.44) - (-32.28)^2 = 25{,}103{,}745.56$$

Therefore

$$b_1 = \frac{\Delta_1}{\Delta} = 8.340 \quad \text{and} \quad b_2 - \frac{\Delta_2}{\Delta} = 2.897$$

since

$$\Delta_1 = \begin{vmatrix} 47{,}604.32 & -32.28 \\ 12{,}447.64 & 4{,}389.44 \end{vmatrix} = 209{,}358{,}116.20$$

and

$$\Delta_2 = \begin{vmatrix} 5{,}719.36 & 47{,}604.32 \\ -32.28 & 12{,}447.64 \end{vmatrix} = 72{,}729{,}201.76$$

Applying Eq. (14.67) gives

$$a = 452.08 - 77.84(8.340) - 77.68(2.897) = -422.1$$

Finally, substituting a, b_1, and b_2 in Eq. (14.65), we find the estimated regression of y on x_1 and x_2 to be

$$\hat{\eta} = -422.1 + 8.340 x_1 + 2.897 x_2 \tag{14.85}$$

Equation (14.85) may be very useful in obtaining estimates of means of arrays of y's corresponding to any pair of values x_{1u}, x_{2u}. But one does not usually stop here, since it is important to know whether the linear trend is significant, and, if so, which variables contribute most to the trend. For example, if there is no need to take into account the regression of some of the variables x_i, then considerably less computation is involved.

In order to test the null hypothesis

$$H_0: \quad \beta_1 = \beta_2 = 0$$

we find

$$SS\text{reg} = \sum g_i b_i = 433{,}069$$
$$SS\text{res} = SSy - SS\text{reg} = 112{,}497$$

and, therefore

$$\frac{\dfrac{SS\text{reg}}{2}}{\dfrac{SS\text{res}}{22}} = 42.3$$

Since $F_{.005}(2, 22) = 6.81$, we reject the null hypothesis at the 0.5 per cent level and conclude that either β_1 or β_2 or both β_1 and β_2 are significantly different from zero. That is, we conclude that there is a linear trend which depends on stand height or on crown cover or on both stand height and crown cover.

Often we are interested in the contribution of a specific variable x_i. This is determined by testing a hypothesis or establishing a confidence interval for a specific coefficient β_i. For such purposes it is convenient to have the inverse matrix, C, of the coefficient matrix

$$A = \begin{bmatrix} a_{11} & a_{12} \\ a_{21} & a_{22} \end{bmatrix} = \begin{bmatrix} 5719.36 & -32.28 \\ -32.28 & 4389.44 \end{bmatrix}$$

The inverse matrix of A is particularly easy to compute, since

$$c_{11} = \frac{A_{11}}{\Delta} = \frac{4389.44}{25,103,746} = 0.000174852$$

$$c_{22} = \frac{5719.36}{25,103,746} = 0.000227829$$

$$c_{12} = \frac{(-1)(-32.28)}{25,103,746} = 0.000001286$$

Therefore

$$C = 10^{-6} \begin{bmatrix} 174.852 & 1.286 \\ 1.286 & 227.829 \end{bmatrix}$$

Since this method of finding an inverse is not very good for higher-order matrices, we describe a more compact method in Sect. 14.6.

The error variance estimate is

$$s_{y_{12}}^2 = 5113.49$$

Thus

$$s_{b_1} = \sqrt{c_{11} s_{y_{12}}^2} = \sqrt{0.894103} = 0.9456$$

To test the null hypothesis $\beta_1 = 0$, we compare

$$\frac{b_1 - \beta_1}{s_{b_1}} = \frac{8.340}{0.9456} = 8.82$$

with $t_{.025}(22) = 2.074$. Since $8.82 > 2.07$, we reject the null hypothesis and conclude that there is a positive slope in the x_1 direction. Further, the 95 per cent confidence interval for β_1 is

$$6.38 < \beta_1 < 10.30$$

since

$$t_{.025}(22) \cdot s_{b_1} = 1.96$$

Actually, since both limits are positive, we could use this information to reject the null hypothesis and conclude that β_1 is positive.

Also, note that the null hypothesis $\beta_2 = 0$ is rejected at the five per cent level, since the statistic

$$\frac{b_2 - \beta_2}{s_{b_2}} = \frac{2.897}{1.079} = 2.68$$

is larger than $t_{.025}(22) = 2.074$. However, if $|b_2|$ had been in the neighborhood of 1 or less, say, we might argue that gross volume per plot depends only on stand height. This being the case, we would require a regression equation of y on x_1. Since

$$\text{cov}(b_1, b_2) = c_{12} s_{y_{12}}^2 = 0.006575 \neq 0$$

we suspect that b_1 and b_2 are not independently distributed. (When $c_{12}\sigma^2 \neq 0$, we know that b_1 and b_2 are not independently distributed.) In case b_1 and b_2 are not independently distributed, we cannot find the required regression equation by simply dropping the last term in Eq. (14.85) and writing

$$\hat{\eta} = -422.1 + 8.340 x_1$$

The proper regression line is found by ignoring the x_2 values and fitting y on x_1 in the usual manner. This leads to estimates

$$b = b_1 = \frac{g_1}{a_{11}} = \frac{47{,}604}{5{,}719} = 8.324$$

and

$$a = 452.08 - (77.84)(8.324) = -195.86$$

and regression equation

$$\hat{\eta} = -195.86 + 8.324 x_1$$

14.5. POLYNOMIAL REGRESSION

In many applications the regression of y on x is assumed to be a polynomial function. Thus, for our model we assume that each observation y_u is randomly drawn from a normal distribution with mean

$$\eta_u = \mu_{y_u/x} = \alpha + \beta_1 x_u + \cdots + \beta_q x_u^q \qquad (u = 1, \ldots, n) \qquad (14.86)$$

and variance σ^2, where q is the degree of the polynomial to be fitted. Equation (14.86) can be viewed as a special case of multiple linear regression. For if

$$p_{iu} = x_u^i \qquad (i = 1, \ldots, q)$$

is substituted in Eq. (14.86), the regression equation becomes

$$\eta_u = \alpha + \beta_1 p_{1u} + \cdots + \beta_q p_{qu} \qquad (14.87)$$

which is *linear* in $p_1, \ldots, p_q$. Hence, the methods of analysis of multiple linear regression may be applied. A sum of products becomes a sum of a power of x_u. For example

$$\sum_u p_{1u} p_{2u} = \sum_u x_u^1 x_u^2 = \sum_u x_u^3$$

Thus, in addition to solving normal equations, one must also find sums of powers of x_u from 1 to $2q$ when fitting a qth degree polynomial. This means that the amount of computation increases rapidly as the degree of the polynomial increases. For this reason, along with others, it is desirable that an experiment be planned so that shortcut techniques can be employed. The *method of orthogonal polynomials*, described below, is very useful in this regard.

On finding the regression of y on x_1 at the end of Example 14.4, we did not use the regression of y on x_1 and x_2. Instead, we started again with the raw data $(x_{11}, y_1), (x_{12}, y_2), \ldots, (x_{1n}, y_n)$ and solved a new set of normal equations. In general, when we use the methods already presented, it is necessary that we start with the raw data each time we fit a regression equation on some of the variables $x_1, x_2, \ldots, x_q$. However, there are methods which allow us, in certain cases, to use an equation already fitted in determining the equation of best fit on fewer or more variables. We now outline and illustrate such a procedure, the method of orthogonal polynomials, for polynomial regression when the x_u are equally spaced and only one y_u is associated with each x_u. Further details, along with applications, may be found in Refs. [2, 3, 5, 9, 10, 37] and the exercises.

If the values of x are at unit intervals, it is always possible to express Eq. (14.87) in the form

$$\eta_u = \alpha' + \beta_1' p_{1u}' + \cdots + \beta_q' p_{qu}' \qquad (u = 1, \ldots, n) \qquad (14.88)$$

where $\alpha', \beta_1', \ldots, \beta_q'$ are parameters, and $p_1', \ldots, p_q'$ are orthogonal polynomials in x of degrees $1, \ldots, q$, respectively. Two polynomials p_i' and $p_j' \, (i \neq j)$ are *orthogonal* when

$$\sum_u p_{iu}' p_{ju}' = 0 \qquad (14.89)$$

that is, when the products of the polynomials evaluated at $x_1, \ldots, x_n$ sum to zero. The first three orthogonal polynomials are

$$\begin{cases} p_1' = \lambda_1 (x - \bar{x}) \\[2mm] p_2' = \lambda_2 \left[(x - \bar{x})^2 - \dfrac{n^2 - 1}{12} \right] \\[2mm] p_3' = \lambda_3 \left[(x - \bar{x})^3 - (x - \bar{x}) \dfrac{3n^2 - 7}{20} \right] \end{cases} \qquad (14.90)$$

where for fixed n the λ_i are constants chosen so that values of p'_{iu} are integers reduced to their lowest terms. If p''_i ($i = 1, 2, 3$) denotes the polynomial on the right-hand side of (14.90), other polynomials may be obtained from the recursion formula

$$p''_{i+1} = p''_i \cdot p''_1 - \frac{i^2(n^2 - i^2)}{4(4i^2 - 1)} p''_{i-1} \quad (i = 2, 3, \ldots, q - 1) \quad (14.91)$$

If x_0 is any integer so that x_0, $x_0 + 1$, $x_0 + 2$, $x_0 + 3$ are four consecutive integers, it is easy to compute the values in Table 14.7. Other values of orthogonal polynomials are shown in Table 14.8. Values of p'_{iu} for $n = 3$ to 75 and $q = 1$ to 5 are given by Fisher and Yates [27]. These tables have been extended to $n = 104$ by Anderson and Houseman [4].

Table 14.7
Values of Polynomials when $n = 4$

Consecutive Integers	$x - \bar{x}$ p''_1	$(x - \bar{x})^2 - \frac{5}{4}$ p''_2	$(x - \bar{x})^3 - \frac{41}{20}(x - \bar{x})$ p''_3	$2p''_1$ p'_1	$1p''_2$ p'_2	$\frac{10}{3}p''_3$ p'_3
x_0	$-\frac{3}{2}$	1	$-\frac{3}{10}$	-3	1	-1
$x_0 + 1$	$-\frac{1}{2}$	-1	$\frac{9}{10}$	-1	-1	3
$x_0 + 2$	$\frac{1}{2}$	-1	$-\frac{9}{10}$	1	-1	-3
$x_0 + 3$	$\frac{3}{2}$	1	$\frac{3}{10}$	3	1	1

Table 14.8
Values of Orthogonal Polynomials for $n = 3, 4, 5, 6$

Consecutive Integers	$n = 3$ p'_1 p'_2	$n = 4$ p'_1 p'_2 p'_3	$n = 5$ p'_1 p'_2 p'_3 p'_4	$n = 6$ p'_1 p'_2 p'_3 p'_4 p'_5
x	-1 1	-3 1 -1	-2 2 -1 1	-5 5 -5 1 -1
$x + 1$	0 -2	-1 -1 3	-1 -1 2 -4	-3 -1 7 -3 5
$x + 2$	1 1	1 -1 -3	0 -2 0 6	-1 -4 4 2 -10
$x + 3$		3 1 1	1 -1 -2 -4	1 -4 -4 2 10
$x + 4$			2 2 1 1	3 -1 -7 -3 -5
$x + 5$				5 5 5 1 1
$k_i = \sum_u p'^2_{iu}$	2 6	20 4 20	10 14 10 70	70 84 180 28 252
λ_i	1 3	2 1 $\frac{10}{3}$	1 1 $\frac{5}{6}$ $\frac{35}{12}$	2 $\frac{3}{2}$ $\frac{5}{3}$ $\frac{7}{12}$ $\frac{21}{10}$

The least-squares estimators a', b'_1, $\ldots$, b'_q of α', β'_1, $\ldots$, β'_q, respectively, are found by solving the normal equations

$$\begin{cases} na' + \quad 0 \quad + \cdots + \quad 0 \quad = \sum y_u \\ 0 + (\sum p'^2_{1u})b'_1 + \cdots + \quad 0 \quad = \sum p'_{1u} y_u \\ \quad \cdots \\ 0 + \quad 0 \quad + \cdots + (\sum_u p'^2_{qu})b'_q = \sum p'_{qu} y_u \end{cases} \quad (14.92)$$

The ease with which the solutions are found is one of the advantages of orthogonal polynomials. The solutions are

$$\begin{cases} a' = \dfrac{\sum y_u}{n} = \bar{y} \\[4mm] b'_i = \dfrac{\sum\limits_u p'_{iu} y_u}{\sum p'^2_{iu}} = \dfrac{\sum p'_{iu} y_u}{k_i} \qquad (i = 1, \ldots, q) \end{cases} \tag{14.93}$$

A great advantage is due to the fact that the coefficients $a', b'_1, \ldots, b'_q$ are independently distributed. Thus, any regression coefficient b'_i and the corresponding regression sum of squares

$$b'_i \sum_u p'_{iu} y_u = \dfrac{\left(\sum\limits_u p'_{iu} y_u\right)^2}{k_i} = b'^2_i k_i \qquad (i = 1, \ldots, q) \tag{14.94}$$

may be found directly. Also, the cubic regression equation, for example, can be found directly from the quadratic regression equation by simply adding on the term $b'_3 p'_3$. Further, the significance of b'_3 can be assessed by comparing $b'^2_3 k_3$ with the independently distributed residual mean square

$$\dfrac{SSy - b'^2_1 k_1 - b'^2_2 k_2 - b'^2_3 k_3}{n - 1 - 3}$$

In general, the mean squares

$$b'^2_i k_i$$

and

$$\dfrac{SSy - b'^2_1 k_1 - \cdots - b'^2_i k_i}{n - 1 - i} \tag{14.95}$$

are independently distributed.

If the degree of the polynomial is known to be (or there is strong evidence that it is) q, the procedures explained in multiple linear regression may be applied. However, if there is doubt as to the degree of the polynomial which fits a set of data (with equally spaced x's), the procedure described in Example 14.5 should be used.

Example 14.5. Find the simplest polynomial which adequately represents the hypothetical data below

x	1	2	3	4	5	6	7	8
y	2	7	11	14	16	12	13	11

This example is given to illustrate the technique. The reader can supply his own application. For example, one might think of this as coded data for

a. Average rainfall in eight consecutive months
b. Density of a glass in terms of annealing temperature
c. Yield of a crop in terms of distance in which rows are spaced
d. Cost of living index over time
e. Amount of assimilation of nitrogen in terms of amount of Chile saltpeter administered

The values of the orthogonal polynomials, along with the sums of squares k_i, sums of products

$$\sum_u p'_{iu} y_u$$

and regression sum of squares due to b'_i are shown in Table 14.9. The analysis of variance is given in Table 14.10. When we look at the sequence of tests indicated in Table 14.10, it is clear that the quadratic polynomial adequately describes the data at the five per cent level. Actually, the fit is significant at the 0.5 per cent level.

Table 14.9
Computations for Orthogonal Polynomials

x	p'_1	p'_2	p'_3	p'_4	y	$p'_1 y$	$p'_2 y$	$p'_3 y$	$p'_4 y$
1	-7	7	-7	7	2	-14	14	-14	14
2	-5	1	5	-13	7	-35	7	35	-91
3	-3	-3	7	-3	11	-33	-33	77	-33
4	-1	-5	3	9	14	-14	-70	42	126
5	1	-5	-3	9	16	16	-80	-48	144
6	3	-3	-7	-3	12	36	-36	-84	-36
7	5	1	-5	-13	13	65	13	-65	-169
8	7	7	7	7	11	77	77	77	77
k_i	168	168	264	616	$\sum_u p'_{iu} y'_u$	98	-108	20	32
λ_i	2	1	$\frac{2}{3}$	$\frac{7}{12}$	$\dfrac{(\sum_u p'_{iu} y_u)^2}{k_i}$	$98^2/168$	$108^2/168$	$20^2/264$	$32^2/616$
						57.17	69.43	1.52	1.66

Two features of the test procedure should be noted. First, two consecutive nonsignificant results were obtained after the last significant result. This was done because odd-degree polynomials are very likely to be nonsignificant when an even-degree polynomial is significant. Thus, if the sequence of the tests had stopped after testing the cubic fit, we might have missed a significant result. Second, the residual mean square changes with each test. However, once a decision is made regarding the degree of the polynomial, the residual mean square which gives the last significant result can be used to estimate confidence intervals or to make other tests. For example, the residual mean

Table 14.10
Analysis of Variance for Example 14.5

Source of Variation	Sum of Squares	Degrees of Freedom	Mean Square	F	$F_{.05}$
Total	$\sum y_u^2 =$ 1060.00	8			
Mean ($a' = \bar{y}$)	$(\sum y_u)^2/8 = 924.50$	1			
Residual from mean	$SSy = 135.50$	7			
Linear	57.17	1	57.17	4.4	5.99
Residual from linear	78.33	6	13.06		
Quadratic	69.43	1	69.43	39.0	6.61
Residual from quadratic	8.90	5	1.78		
Cubic	1.52	1	1.52	0.82	7.71
Residual from cubic	7.38	4	1.85		
Quartic	1.66	1	1.66	0.87	10.13
Residual from quartic	5.72	3	1.91		

square for the quadratic term should be used to estimate a confidence interval for β_1'.

The reader should note that the procedure described in Example 14.5 is for the case where the degree of the polynomial is uncertain. Otherwise, the proper residual mean square can be found directly and the usual procedure for testing and estimating applied.

Using Eq. (14.93), we find regression coefficients to be

$$a' = \bar{y} = 10.750$$
$$b_1' = \frac{98}{168} = 0.583$$
$$b_2' = -\frac{108}{168} = -0.643$$

and the quadratic regression equation is

$$\hat{\eta} = 10.750 + 0.583p_1' - 0.643p_2' \tag{14.96}$$

where

$$\begin{cases} p_1' = 2(x - 4.5) = 2x - 9 \\ p_2' = (x - 4.5)^2 - \frac{21}{4} = x^2 - 9x + 15 \end{cases} \tag{14.97}$$

Substituting Eqs. (14.97) in Eq. (14.96) and simplifying gives the following function in terms of x

$$\hat{\eta} = -4.143 + 6.953x - 0.643x^2$$

Note that by dropping the p_2' term in Eq. (14.96) the resulting linear regression equation is

$$\hat{\eta} = 10.750 + 0.583(2x - 9)$$

or

$$\hat{\eta} = 5.503 + 1.166x$$

This is the same equation which would be obtained by fitting a simple linear regression directly.

14.6. CALCULATIONS IN MULTIPLE LINEAR REGRESSION

As long as we fit y on one or two x variables, the computations are relatively simple. But this situation changes rapidly as the number of variables increases. For this reason we present a method which applies to all multiple linear regression and which reduces the required calculations to a minimum. The method, due to Doolittle [17] and Gauss [29], is particularly useful with a desk calculator.

In regression problems there are lengthy calculations associated with (1) finding the equation of best fit and (2) making statements about parameters. For (1) this requires that we solve a system of linear normal equations [see Eq. (14.71)] and for (2) we must usually find the inverse of the coefficient matrix of (1). We explain how the *Doolittle-Gauss method* [17, 22, 29, 57] may be used for both (1) and (2). For a theoretical developement of this and many other methods available for solving simultaneous linear equations the reader should study Dwyer [20].

To simplify the explanation, we describe the case of three equations in three unknowns. The normal equations are

$$\begin{cases} a_{11}b_1 + a_{12}b_2 + a_{13}b_3 = g_1 \\ a_{21}b_1 + a_{22}b_2 + a_{23}b_3 = g_2 \\ a_{31}b_1 + a_{32}b_2 + a_{33}b_3 = g_3 \end{cases} \tag{14.98}$$

The coefficient matrix A is symmetric; that is, $a_{ij} = a_{ji}\ (i,j = 1, 2, 3)$. The Doolittle-Gauss procedure applies only to *symmetric matrices*.

The method involves finding another system of three equations in three unknowns which has the same solution as (14.98) and is easy to solve. If it is assumed that a unique solution exists, it can be shown that such a system is given by

$$\begin{cases} a_{11}b_1 + a_{12}b_2 + a_{13}b_3 = g_1 \\ \phantom{a_{11}b_1 +\ } a_{22\cdot 1}b_2 + a_{23\cdot 1}b_3 = g_{2\cdot 1} \\ \phantom{a_{11}b_1 + a_{22\cdot 1}b_2 +\ } a_{33\cdot 12}b_3 = g_{3\cdot 12} \end{cases} \tag{14.99}$$

where

$$
\begin{cases}
a_{2j \cdot 1} = a_{2j} - d_{12}a_{1j} \quad \text{and} \quad d_{1j} = a_{1j}/a_{11} \qquad (j = 2,3) \\
g_{2 \cdot 1} = g_2 - d_{12}g_1 \\
a_{33 \cdot 12} = a_{33} - d_{13}a_{13} - d_{23 \cdot 1}a_{23 \cdot 1} \text{ and } d_{23 \cdot 1} = a_{23 \cdot 1}/a_{22 \cdot 1} \\
g_{3 \cdot 12} = g_3 - d_{13}g_1 - d_{23 \cdot 1}g_{2 \cdot 1}
\end{cases}
\qquad (14.100)
$$

If the equations in (14.99) are divided by a_{11}, $a_{22 \cdot 1}$, $a_{33 \cdot 12}$, respectively, the resulting system has the same solution as (14.98). That is, the system

$$
\begin{cases}
b_1 + d_{12}b_2 + d_{13}b_3 = h_1 \\
 b_2 + d_{23 \cdot 1}b_3 = h_{2 \cdot 1} \\
\phantom{b_1 + d_{12}b_2 + } b_3 = h_{3 \cdot 12}
\end{cases}
\qquad (14.101)
$$

where

$$
h_1 = \frac{g_1}{a_{11}}, \qquad h_{2 \cdot 1} = \frac{g_{2 \cdot 1}}{a_{22 \cdot 1}}, \qquad h_{3 \cdot 12} = \frac{g_{3 \cdot 12}}{a_{33 \cdot 12}}
$$

gives the solution of (14.98). An illustration is given in Example 14.6.

Solving a symmetric system of linear equations by the Doolittle-Gauss method is particularly attractive because it lends itself to a compact computational form. Since the solution depends on the coefficients, we start by writing (14.98) in synthetic form as

$$
\begin{array}{ccc|c}
a_{11} & a_{12} & a_{13} & g_1 \\
a_{21} & a_{22} & a_{23} & g_2 \\
a_{31} & a_{32} & a_{33} & g_3
\end{array}
\qquad (14.102)
$$

where the unknowns b_1, b_2, b_3 are omitted and the equality marks are replaced by a vertical line segment. Next, the coefficients of the two linear systems (14.99) and (14.101) are meshed and arranged below the coefficients of (14.102), as shown in Table 14.11. As is illustrated in Example 14.6, this allows a convenient computational pattern to be utilized.

Example 14.6. Use the Doolittle-Gauss method to solve the simple linear system

$$
\begin{aligned}
5b_1 + 3b_2 + 2b_3 &= 5 \\
3b_1 + 4b_2 + b_3 &= -2 \\
2b_1 + b_2 + 3b_3 &= 9
\end{aligned}
$$

The solution is given in Table 14.11. The first three rows are the coefficients in the system. The fourth is the same as the first row. The fifth row is obtained by dividing each number in the fourth by the leading number 5.

Table 14.11

Synthetic Doolittle-Gauss Method for Solving Linear Equations

General				Particular			
a_{11}	a_{12}	a_{13}	g_1	5	3	2	5
a_{21}	a_{22}	a_{23}	g_2	3	4	1	-2
a_{31}	a_{32}	a_{33}	g_3	2	1	3	9
a_{11}	a_{12}	a_{13}	g_1	5	3	2	5
1	d_{12}	d_{13}	h_1	1	0.600	0.400	1.000
	$a_{22 \cdot 1}$	$a_{23 \cdot 1}$	$g_{2 \cdot 1}$		2.200	-0.200	-5.000
	1	$d_{23 \cdot 1}$	$h_{2 \cdot 1}$		1.000	-0.091	-2.273
		$a_{33 \cdot 12}$	$g_{3 \cdot 12}$			2.182	6.546
		1	$h_{3 \cdot 12}$			1.000	3.000
b_1	b_2	b_3		1.000	-2.000	3.000	

The numbers in the sixth row are given by

$$4 - 3(0.600) = 2.200$$
$$1 - 2(0.600) = -0.200$$
$$-2 - 5(0.600) = -5.000$$

The seventh row is found on dividing each of these numbers by the leading number 2.200. The numbers in the eighth row are computed as follows

$$3 - 2(0.400) - (-0.200)(-0.091) = 2.182$$
$$9 - 5(0.400) - (-5.000)(-0.091) = 6.546$$

Dividing these numbers by 2.182 gives the numbers in the ninth row. Using the ninth, seventh, and fifth rows, respectively, we find the solutions to be

$$b_3 = h_{3 \cdot 12} = 3.000$$
$$b_2 = h_{2 \cdot 1} - d_{23 \cdot 1} b_3 = -2.273 - (-0.091)(3.000) = -2.000$$
$$b_1 = h_1 - d_{12} b_2 - d_{13} b_3 = 1.000 - (0.600)(-2.000) - (0.400)(3.000)$$
$$= 1.000$$

Since the values are found in reverse order, b_3 being used to find b_2 and b_3 and b_2 being used to find b_1, we term this the *back solution* for a set of linear equations. With a little practice the reader should become very proficient with this pivotal method.

Example 14.7. Find the inverse matrix of the coefficient matrix in Example 14.6. That is, find the inverse of

$$A = \begin{bmatrix} 5 & 3 & 2 \\ 3 & 4 & 1 \\ 2 & 1 & 3 \end{bmatrix}$$

The columns of the inverse matrix [see (14.76)] are found by solving three systems of linear equations like (14.77). In each case A is the coefficient matrix of the unknowns. Thus, the calculations of three synthetic Doolittle-Gauss solutions can be meshed into one as shown in Table 14.12.

Table 14.12
Doolittle-Gauss Method for Finding the Inverse of a Symmetric Matrix

Coefficient Matrix			Identity Matrix			Row Sum Check
5	3	2	1	0	0	11
3	4	1	0	1	0	9
2	1	3	0	0	1	7
5	3	2	1	0	0	11
1	0.600	0.400	0.200	0	0	2.200
	2.200	−0.200	−0.600	1.000	0	2.400
	1.000	−0.091	−0.273	0.455	0	1.091
		2.182	−0.455	−0.091	1.000	2.636
		1.000	−0.208	−0.042	0.458	1.208
0.458	−0.292	−0.208				
−0.292	0.459	0.042				
−0.208	0.042	0.458				

The computations in the first nine rows of Table 14.12 are carried out just as in Table 14.11. The entries in the row sum check column after the fifth are obtained in two ways. For example, the sixth entry is given by

$$2.200 + (-0.200) + (-0.600) + 1.000 = 2.400$$

or

$$9 - 11(0.600) = 2.400$$

The seventh entry is

$$1.000 + (-0.091) + (-0.273) + 0.455 = 1.091$$

or

$$\frac{2.400}{2.200} = 1.090$$

Since a pair of answers are the same to three significant figures, we consider the *check* indicates that no errors have been made in computations through the seventh row. This simple check is one of the good features of the Doolittle-Gauss method. It protects against most errors made in computation.

The inverse matrix C, rows 10, 11, and 12 of Table 14.12, is symmetric, since A is symmetric. Thus, the first row and first column of C can be found by using the first four columns in Table 14.12 in exactly the same way that the four columns were used in Table 14.11. That is

$$c_{13} = c_{31} = -0.208$$

$$c_{12} = c_{21} = -0.273 - (-0.091)(-0.208) = -0.292$$

$$c_{11} = 0.200 - (0.400)(-0.208) - (0.600)(-0.292) = 0.458$$

Using the same method, replacing the fourth column by the fifth and sixth, respectively, we find that

$$c_{23} = c_{32} = -0.042$$

$$c_{22} = 0.455 - (-0.091)(0.042) = 0.459$$

$$c_{33} = 0.458$$

This is sometimes called the *back solution* method [15] for finding the inverse. The student should practice the procedure until it becomes thoroughly familiar to him.

The back solution method has the obvious disadvantage of requiring that the numbers in the inverse be found sequentially. However, the numbers in columns 4, 5, and 6 of Table 14.12 can be used in a simple direct method (see Exercise 14.26 and Ref. [20]) to find any number of the inverse without using others. For large matrices either the abbreviated Doolittle method [20, 57] or the square root method [8, 19, 23] should be applied. Of course, where electronic machines are avaliable, the complexities of computation may be avoided.

14.7. REGRESSION FOR OTHER CASES

In the earlier section of this chapter we discussed regression only for cases in which η_u, the array mean, is a linear function of the parameters, regression coefficients. We assumed the random variables $y_1, \ldots, y_n$ to be normally distributed with means η_u and homogeneous variances $\sigma^2_{y/x_u} = \sigma^2$. Now, we look briefly at some other cases.

We first note that the usual procedures of linear regression may be carried out in cases where the uth array variance σ^2_u is given by

$$\sigma^2_u = \frac{\sigma^2}{w_u} \tag{14.103}$$

σ^2 being unknown and w_u known constants. Actually, in simple linear regression w_u is usually taken to be a known function of x. It is left as an exercise for the student to find the expression for and distribution of a, b, and s^2_{y/x_u} for a simple regression model.

In cases where $\eta = E(y|x)_{\backslash}$ is not a linear function of x, but is a linear function of the parameters, we may still use the methods of linear regesssion. For example, in the two-parameter case, if $k(y)$ is normally distributed with mean

$$\kappa(\eta) = E[k(y)|g(x)] = \alpha'' + \beta'' g(x) \qquad (14.104)$$

and variance

$$V[k(y)|g(x)] = \sigma^2 \qquad (14.105)$$

the usual theory applies. That is, the scatter diagram of points $(g(x_u), k(y_u))$ in the rectangular co-ordinate system with horizontal axis $g(x)$ and vertical axis $k(y)$ fall along the true straight line whose equation is given in Eq. (14.104). Thus, to find a simple regression curve by the *usual indirect procedure*, one transforms each data point (x_u, y_u) to $(g(x_u), k(y_u))$, finds estimates a'' and b'' of α'' and β'' in the usual way, and then solves the estimating regression equation

$$\hat{\kappa}(\hat{\eta}) = a'' + b'' g(x)$$

for $\hat{\eta}$. The resulting equation is not generally identical with the one that would have been obtained had we proceeded in the direct and more complicated way of Sect. 14.2. The estimators of the regression coefficients obtained in this way do not generally have the desirable properties of estimators obtained in the direct way. But since the difference is usually not large, the indirect approximate method is normally preferred due to simplicity of application. The student should recognize that such a device allows an investigator to obtain fairly good estimates of scatter diagrams of many shapes. (Of course, the investigator's theoretical knowledge of the underlying structure is of primary importance in making a decision about the model to use.) A few functional relationships which satisfy condition (14.104) are given in the next paragraph. The reader who is interested in applications may work the exercises or consult the literature for illustrations. A few references [33, 40, 41, 46, 48, 53, 59] are given at the end of this chapter.

Putting $g(x)$ equal to x, x^{-1}, $\log_b x$, and b^x, respectively, and $k(y)$ equal to y^{-1} and $\log_b y$, respectively, leads to the eight equations given in Table 14.13. It is to be understood that b is a constant. On solving each of these equations for η we obtain the equations of Table 14.14, which have horizontal or vertical asymptotes in most cases and varying degrees of curvature. The symbol γ denotes $b^{\alpha''}$.

Table 14.13
Particular Cases of $\kappa(\eta) = \alpha'' + \beta'' g(x)$

$g(x)$	$\kappa(\eta) = \eta^{-1}$	$\kappa(\eta) = \log_b \eta$
x	$\dfrac{1}{\eta} = \alpha'' + \beta'' x$	$\log_b \eta = \alpha'' + \beta'' x$
$\dfrac{1}{x}$	$\dfrac{1}{\eta} = \alpha'' + \dfrac{\beta''}{x}$	$\log_b \eta = \alpha'' + \dfrac{\beta''}{x}$
$\log_b x$	$\dfrac{1}{\eta} = \alpha'' + \beta'' \log_b x$	$\log_b \eta = \alpha'' + \beta'' \log_b x$
b^{-x}	$\dfrac{1}{\eta} = \alpha'' + \beta'' b^{-x}$	$\log_b \eta = \alpha'' + \beta'' b^{-x}$

Equations (1) and (2) in Table 14.14 represent hyperbolas; Eqs. (5) and (7) represent exponential and power curves, respectively. When $b = e$, Eq. (4) is a special case of the very important "logistic equation." All these equations serve as useful models, and there are many illustrations on their use. The interested student may first look to Refs. [12, 33, 41, 46, 53, 59], which give other references. In particular, the logistic curve, a special case of a growth curve, has frequently been used in studies of such things as human and insect populations and growth of cells and telephone subscribers. There are also other growth curves [33, 48, 59] which have proved to be very useful in applied work.

Table 14.14
Solution of the Equations in Table 14.13 with Respect to η

$g(x)$	Equation Number	$\kappa(\eta) = \dfrac{1}{\eta}$	Equation Number	$\kappa(\eta) = \log_b \eta$
x	1	$\eta = \dfrac{1}{\beta''\left(x + \dfrac{\alpha''}{\beta''}\right)}$	5	$\eta = \gamma b^{\beta'' x}$
$\dfrac{1}{x}$	2	$\eta = \dfrac{1}{\alpha''} - \dfrac{\beta}{\alpha''^2} \cdot \dfrac{1}{x + \dfrac{\beta''}{\alpha''}}$	6	$\eta = \gamma b^{\beta''/x}$
$\log_b x$	3	$\eta = \dfrac{1}{\beta''\left(\log_b x + \dfrac{\alpha''}{\beta''}\right)}$	7	$\eta = \gamma x^{\beta''}$
b^{-x}	4	$\eta = \dfrac{1}{\alpha'' + \beta'' b^{-x}}$	8	$\eta = \gamma b^{\beta'' b^{-x}}$

There are important cases where the parameters of the regression model should be estimated by an iterative procedure. For a detailed discussion of such methods, see Garwood [28] and others [30, 34, 35]. Illustrations are found in Refs. [28, 42, 43].

Other procedures are also used to estimate parameters in regression models. Cornell [12, 13] gives four methods of estimating parameters along with illustrations for linear combinations of exponentials. Hald [33] gives references to studies in time series. Askovitz [7] gives a graphic method for fitting a line. For other studies in regression, read Refs. [12, 30, 40, 55].

14.8. EXERCISES

14.1. (a) Plot the scatter diagram for the following data

x	1	3	3	4	6	6	8
y	8	7	6	5	5	2	2

(b) Use the data in (a) to find the estimated linear regression of y on x.
(c) Test the hypothesis that the slope of the true regression line is zero.

14.2. Give three examples from your area of study of the use of regression. Carefully define all variables, suggest the functional form of the regression equation, and indicate some uses of each example.

14.3. (a) At $x = 1$ it is known that $y \sim n$ (2, $\sigma^2 = 16$) and at $x = 7$, $y \sim n(5, \sigma^2 = 16)$. Find the equation for the true linear regression of y on x. Making the usual assumptions of linear regression, what can you say about the distribution of the array at $x = 2$? (b) The following data were obtained for the regression in (a)

x	1	4	4	7	8	8
y	3	4	3	4	5	4

Find the estimated linear regression of y on x. At $x = 8$, find the true and estimated regression mean, the array mean, and then compare the true and estimated error effects for both observed values. (c) Find a 90 per cent confidence interval for the true slope. Does the true slope fall in this interval? (d) Find a 90 per cent confidence interval for the true array mean at $x = 2$. Does the true array mean fall in this interval?

14.4. Prove that $s^2_{y/x}$ is an unbiased estimator of σ^2.

14.5. For a sample of 11 pairs (x, y) it was found that $\sum x = 34$, $\sum y = 85$, $\sum x^2 = 676$, $\sum y^2 = 815$, and $\sum xy = 326$. (a) Find the three sums of squares in (14.52). (b) Use a five per cent level test to determine if the true slope is significantly different from 2. (c) Find a 95 per cent confidence interval for the common variance. (d) Test the linearity of regression.

14.6. Prove Eq. (14.47) and show that $s^2_{y/x}$ as defined in Eq. (14.48) is an unbiased estimator of σ^2.

14.7. Derive Eq. (14.52) starting with Eq. (14.54).

14.8. Show that $SSW/(n - k)$ is an unbiased estimator of σ^2, and that

$SSD/(k-2)$ is an unbiased estimator of σ^2, provided that the true regression is linear.

14.9. Prove that $s_{y/x_1x_2}^2$ as defined in Eq. (14.66) is an unbiased estimator of σ^2.

14.10. Prove the Gauss-Markoff theorem.

14.11. The data in Table 14.15 on multiple linear regression are chosen so as to restrict the amount of computation. (The reader may think of these as coded data which relate price of an item to cost of raw material and cost of labor, death rate of adult males to percentage of fat calories in diet and value of home, grades on a particular course to I. Q. and amount of time spent in study, reciprocal of tar content of a gas stream to gas inlet temperature and rotor speed, etc.) (a) Find the linear regression of

Table 14.15

Object	x_1	x_2	y
1	1	1	3
2	1	2	4
3	2	2	3
4	2	3	5
5	2	4	6
6	3	2	5
7	3	3	4
8	3	4	7
9	4	4	8
10	4	5	9
11	5	3	7
12	5	5	10

y on x_1 and x_2. (b) Test the hypothesis $\beta_1 = \beta_2 = 0$ at the five per cent level. (c) Find the inverse matrix. (d) Test the hypothesis $\beta_2 = 0$ at the five per cent level. (e) Find a 95 per cent confidence interval for β_1.

14.12. Derive Eq. (14.70).

14.13. Show that cov $(b_1, b_2) = c_{12}\sigma^2$ in multiple linear regression of y on x_1, x_2, and x_3.

14.14. In a study with 25 data points of the form (y, x_1, x_2, x_3) it was found that

$$\sum x_1 = 2189 \qquad \sum y^2 = 89{,}341$$
$$\sum x_2 = 758 \qquad \sum x_1 x_2 = 78{,}225$$
$$\sum x_3 = 141 \qquad \sum x_1 x_3 = 15{,}030$$
$$\sum y = 1418 \qquad \sum x_1 y = 140{,}444$$
$$\sum x_1^2 = 327{,}350 \qquad \sum x_2 x_3 = 4593$$
$$\sum x_2^2 = 24{,}781 \qquad \sum x_2 y = 44{,}783$$
$$\sum x_3^2 = 877 \qquad \sum x_3 y = 8527$$

(a) Find the linear regression of y on x_1, x_2, and x_3. (b) Find the linear regression of y on x_1 and x_2; of y on x_1 and x_3; of y on x_2 and x_3. (c) Find the linear regression of y on x_1; of y on x_2; of y on x_3. (d) Find the inverse matrix for part (a). (e) Test the hypothesis $\beta_1 = \beta_2 = \beta_3 = 0$ at the five per cent level. (f) Use (a) and (d) to establish 90 per cent confidence intervals for β_1, β_2, and β_3, respectively. (g) Test the hypothesis $\beta_2 = \beta_3 = 0$ at the five per cent level.

Hint. To find the regression sum of squares due to b_2 and b_3, $RSS(b_2, b_3)$, subtract the regression sum of squares due to fitting y on x_1, $RSS(b_1')$, from the regression sum of squares due to fitting b_1, b_2, and b_3, $RSS(b_1, b_2, b_3)$. It can be shown that the expected value of

$$RSS(b_1, b_2, b_3) - RSS(b_1') = RSS(b_2, b_3)$$

is σ^2 plus a function of β_2 and β_3 only. Thus, under the null hypothesis

$$\frac{\dfrac{RSS(b_2, b_3)}{3-1}}{s^2_{y123}}$$

Table 14.16*

Round Number	x_1	x_2	x_3	y
1	55	65	177	−188
2	74	45	139	−113
3	97	60	141	−59
4	168	75	60	276
5	126	98	71	232
6	111	82	113	114
7	113	75	137	47
8	45	70	131	−155
9	79	43	136	−101
10	81	77	89	4
11	92	45	178	−147
12	114	74	104	108
13	77	55	144	−76
14	64	69	100	−22
15	127	73	115	−11
16	159	61	112	113
17	85	78	135	−45
18	96	61	97	69
19	103	74	134	−61
20	158	63	77	85
21	111	62	141	72
22	104	81	200	116
23	83	51	143	−67
24	95	52	113	−69
25	91	66	141	−148

* Data by Courtesy of Paul C. Cox.

is distributed as F with two and 21 degrees of freedom.

(h) Test each of the following hypotheses at the five per cent level

$$H_{01}: \quad \beta_1 = \beta_3 = 0 \qquad H_{03}: \quad \beta_1 = 0$$
$$H_{02}: \quad \beta_1 = \beta_2 = 0 \qquad H_{04}: \quad \beta_3 = 0$$

(i) Letting y, x_1, x_2, and x_3 be particular variables in your area of research, write a summary statement for the experiment of this exercise.

14.15. After the propellant weight, propellant temperature, and total weight were recorded, 25 rockets were fired at a fixed target. The range error, that is, the component of miss distance parallel to the straight line determined by the target and the launcher, was then measured. Table Table 14.16 gives the propellent weight minus 2000 lb, the propellant temperature in centigrade units, the total weight minus 5700 lb, and the range error in yards and they are designated by x_1, x_2, x_3, and y, respectively. Use these data to find solutions required in Exercise 14.14(a), (b), (c), (d), (e), (f), (g), (h), and (i).

14.16. Find the polynomial which adequately represents the following hypothetical data

x	1	2	3	4	5	6	7	8
y	11	13	12	16	14	11	7	2

14.17. Derive the orthogonal polynomials in Eq. (14.90).

Hint. Let

$$p'_{iu} = c_{i0} + c_{i1}x_u + c_{i2}x_u^2 + \cdots + c_{ii}x_u^i$$

where $i = 1, \ldots, q$ and $u = 1, \ldots, n$. Find c's so that

$$\sum_u p'_{iu} p'_{ju} = 0$$

for every pair of polynomials where $i \neq j$.

14.18. Use Eq. (14.91) to find p''_{i+1} for $i = 1, 2, \ldots, 6$. Also, find λ_{i+1} and p'_{i+1}.

14.19. Verify the values in Table 14.8.

14.20. Derive the normal equations in Eq. (14.92).

14.21. Find 95 per cent level confidence intervals for each of the nonzero parameters in Exercise 14.16.

14.22. Find general expressions for

$$\sum_u^n p_{iu}^2 \qquad (i = 1, 2, 3, 4, 5)$$

14.23. Management programs for the conservation of salmon fisheries depend on investigations and records of the past. Among the many variables studied are total runs in thousands, escapements in thousands, and percentage of escapement. The percentage of escapement in the years 1894–1945 for the Fraser River Sockeye Salmon in each of the four-year age cycles are shown in Table 14.17. (a) Use the 52 percentages to find

Table 14.17*

CYCLE A		CYCLE B		CYCLE C		CYCLE D	
Year	*Percentage Escapement*	*Year*	*Percentage Escapement*	*Year*	*Percentage Escapement*	*Year*	*Percentage Escapement*
1894	44.5	1895	40.5	1896	21.8	1897	24.3
1898	13.1	1899	11.2	1900	7.9	1901	8.4
1902	14.5	1903	12.5	1904	14.3	1905	14.4
1906	23.4	1907	19.5	1908	19.7	1909	7.3
1910	20.6	1911	23.2	1912	24.5	1913	18.6
1914	10.8	1915	16.0	1916	8.3	1917	5.9
1918	14.3	1919	20.3	1920	26.3	1921	17.4
1922	29.4	1923	34.0	1924	31.1	1925	25.3
1926	47.8	1927	31.1	1928	25.0	1929	23.1
1930	17.1	1931	25.9	1932	29.9	1933	16.1
1934	16.2	1935	33.5	1936	42.6	1937	27.1
1938	25.4	1939	25.0	1940	28.4	1941	30.1
1942	22.4	1943	23.3	1944	26.4	1945	22.2

* Data taken from Table 8 of G. A. Rounsefell, "Methods of Estimating Total Runs and Escapements of Salmon," *Biometrics*, Vol. 5 (1949), pp. 115–126.

the polynomial which adequately represents the data. (b) For each of the four cycles find the polynomial which adequately represents the data. (c) Write a summary statement of your findings in (a) and (b) and indicate how this information might be used. (d) Discuss the assumptions of independence and normality as they relate to the data.

14.24. Prove Eqs. (14.99) and (14.101).

14.25. (a) Use the Doolittle-Gauss method to solve the simple linear system

$$5a + 2b + 3c = 5$$
$$2a + 4b + c = 12$$
$$3a + b + 3c = 0$$

(b) Find the inverse matrix of the coefficient matrix in (a).

14.26. (a) Use the numbers in columns 4, 5, and 6 of Table 14.12 to find the inverse matrix.

Hint. An element in the ith row and jth column of the inverse matrix may be found, after the elements of the identity matrix have been omitted, by multiplying term for term the top (bottom) elements in the $i + 3$ column by the bottom (top) elements in the $j + 3$ column and adding. For example, the element in the first row and third column of the inverse matrix is given by

$$(1)(0) + (-0.600)(0) + (-0.455)(0.458) = -0.208$$

or by

$$(0.200)(0) + (-0.273)(0) + (-0.208)(1.000) = -0.208$$

(b) Use the method of (a) to find the inverse matrix in Exercise 14.25(b).
(c) For a general symmetric 3×3 matrix, prove that the inverse matrix
is obtained by the method of (a).

14.27. In simple linear regression let all the usual assumptions hold except that
of homogenous variances. Let the variance of the uth array be given by

$$\sigma_u^2 = \frac{\sigma^2}{w_u}$$

σ^2 being unknown and w_u a known constant. Find the expression for and
the distribution of a, b, and s_{y/x_u}^2 for a simple regression model.

14.28. (a) Show that each of the equations in Table 14.13 can be solved for
η to give the corresponding equations in Table 14.14. (b) Graph and
discuss from the mathematical point of view four families of curves in
Table 14.14.

14.29. The following data were drawn from a family of regression curves of
the type $\eta = e^{\alpha'' + \beta''/x}$

x	1.0	1.1	1.2	1.5	2.0	3.0	4.0	5.0	6.0	7.0
y	.382	.303	.260	.195	.135	.096	.082	.072	.070	.065

(a) Make a transformation so that the regression is linear in the parameters α'' and β'', and then find the fitted linear regression equation for
the transformed data. (b) Use the linear regression equation found in
(a) to obtain a regression equation of the form

$$\hat{\eta} = e^{a'' + b''/x}$$

Check to see how well this equation fits the data. (c) Without transforming the data, find, if possible, the best-fitting curve, using the method of
least squares.

14.30. Use two methods to fit the following data to a model of the type
$\eta = e^{\alpha''} x^{\beta''}$

x	1	2	3	4	5	6	7
y	.06	.21	.45	.78	1.22	1.80	2.40

14.31. For standard pieces of equipment (for example, electronic) the model
for the proportion η surviving x hours of operation is $\eta = e^{\alpha + \beta x}$. Use
the data in Table 14.18 to find estimates of α and β. Disucss some pos-

sible applications of the fitted equation. What is the fitted equation in case $\alpha = 0$?

Table **14.18**

x	y	x	y
0	1.000	320	0.612
40	0.950	360	0.583
80	0.875	400	0.550
120	0.840	440	0.523
160	0.778	480	0.488
200	0.735	520	0.458
240	0.690	560	0.430
280	0.657	600	0.405

14.32. (a) Use the data in Table 14.19 to find an appropriate polynomial regression equation of density on depth of rock. (b) The densities are

Table **14.19***

Depth in Feet	Mean Density in g/cm³	Depth in Feet	Mean Density in g/cm³
851–951	2.378	2261–2361	2.577
951–1051	2.367	2361–2461	2.506
1051–1151	2.423	2461–2561	2.400
1151–1251	2.337	2561–2661	2.345
1251–1351	2.435	2661–2761	2.535
1351–1451	2.391	2761–2861	2.612
1451–1561	2.557	2861–2961	2.569
1561–1661	2.441	2961–3061	2.615
1661–1761	2.462	3061–3161	2.648
1761–1861	2.425	3161–3261	2.637
1861–1961	2.456	3261–3361	2.659
1961–2061	2.507	3361–3461	2.623
2061–2161	2.508	3461–3491	2.624
2161–2261	2.548		

  * Data taken from Table 2b of M. J. S. Innes, "The Use of Gravity Methods to Study the Underground Structure and Impact Energy of Meteorite Craters," *Journal of Geophysical Research*, Vol. **66** (1961), pp. 2225–39.

for fragmental rocks located at the Brent Crater in the Canadian Shield. The crater has characteristics which strongly suggest meteoric origin. Discuss the data and regression analysis in terms of this incomplete information.

14.9. CORRELATION AND ITS RELATION TO REGRESSION

In certain investigations both values of a pair (x, y) are random—the x value is not controlled as in simple linear regression. For example, both the grade in history and the grade in algebra of college students may be assumed to be random, or both the temperature and the humidity on a given day may be assumed to be random. In such cases, we think of sampling from a bivariate distribution. When the random variables x and y are not independently distributed, we think of the degree of association as being measured by the correlation coefficient ρ, which is defined in Eq. (3.58) as

$$\rho = \frac{\sigma_{xy}}{\sigma_x \sigma_y}$$

In order to relate correlation to regression, we require the denstiy function of y for a given x. According to Definition (5.62) the conditional density function of y for a given x is

$$f(y\,|\,x) = \frac{f(x, y)}{f(x)} \tag{14.106}$$

where $f(x, y)$ denotes a bivariate density function of x and y and $f(x)$ the marginal density function of x.

Since we assumed the arrays in regression to be normal, we consider the case where $f(x, y)$ is the bivariate normal density function given in Eq. (3.57). It can be shown (see Exercise 14.34) that y for a given value of x is normally distributed with mean

$$\mu_{y/x} = \mu_y + \rho\frac{\sigma_y}{\sigma_x}(x - \mu_x) \tag{14.107}$$

and variance

$$\sigma^2_{y/x} = \sigma^2_y(1 - \rho^2) \tag{14.108}$$

From Eq. (14.107) it is clear that the means of the conditional distributions (array distributions) fall on a straight line. From Eq. (14.108) we observe that the variance is constant and does not depend on x, and ρ^2 lies between 0 and 1. Thus, the assumptions of simple linear regression of y on x are satisfied. That is, the straight line on which the means of the conditional distributions fall is a regression line. Hence, on comparing Eq. (14.107) with Eq. (14.69) when $\beta_2 = 0$, we see that

$$\beta_{y/x} = \rho\frac{\sigma_y}{\sigma_x} \tag{14.109}$$

where $\beta_{y/x}$ denotes β_1. Since in a bivariate normal distribution x is a random normal variable when y is fixed, we can show, in a similar way, that the slope, $\beta_{x/y}$, of the regression of x on y is related to the correlation coefficient by

$$\beta_{x/y} = \rho\frac{\sigma_x}{\sigma_y} \tag{14.110}$$

Further, the correlation coefficient is the geometric mean of the slopes of the two regression lines; that is

$$\rho = \sqrt{\beta_{y/x}\beta_{x/y}}$$

Now, we observe [see Eq. (14.108)] that when $\rho = 1$, the variation about the regression line (14.107) is zero. But this is also true of the variation about the regression of x on y. This means that all points fall on the regression line. Thus, when $\rho = 1$, it follows that the two regression lines are the same, and the two-variate distribution actually is a one-variate distribution, since x and y are linearly dependent variables. When $\rho = 0$, the variables are independent and the regression lines are parallel to the co-ordinate axes. Finally, when $0 < \rho < 1$, both regression lines have positive slopes; when $-1 < \rho < 0$, both have negative slopes. Further, using Eq. (14.108), we observe that

$$1 - \frac{\sigma^2_{y/x}}{\sigma^2_y} \tag{14.111}$$

can be used as a measure of the degree of dependence of x and y. In a similar way, we also observe that

$$1 - \frac{\sigma^2_{x/y}}{\sigma^2_x} \tag{14.112}$$

can be used as a measure of the degree of dependence of x and y. That is, the degree of dependence may be measured by the correlation coefficient squared, by using the ratio of the variance about the regression of y on x to the variance of the marginal distribution of y, or by using the ratio of the variance about the regression of x on y to the variance of the marginal distribution of x.

As an estimator of ρ, we define the sample correlation coefficient of x and y by

$$r = \frac{\dfrac{\sum\limits_{u}^{n} (x_u - \bar{x})(y_u - \bar{y})}{(n-1)}}{\sqrt{\dfrac{\sum\limits_{u}^{n} (x_u - \bar{x})^2 \sum\limits_{u}^{n} (y_u - \bar{y})^2}{(n-1)^2}}} \tag{14.113}$$

which may also be written as

$$r = \frac{\sum (x_u - \bar{x})(y_u - \bar{y})}{\sqrt{\sum (x_u - \bar{x})^2 \sum (y_u - \bar{y})^2}}$$

or

$$r_{xy} = \frac{SP}{\sqrt{SSx\, SSy}} \tag{14.114}$$

where

$$r_{xy} = r, \qquad SP = \sum x_u y_u - \frac{(\sum x_u)(\sum y_u)}{n}$$

$$SSx = \sum x_u^2 - \frac{(\sum x_u)^2}{n}, \qquad SSy = \sum y_u^2 - \frac{(\sum y_u)^2}{n}$$

The sample correlation coefficient r defined in this way is an unbiased estimator of the population correlation coefficient ρ only when $\rho = 0$. Still, for reasons beyond the scope of this book, r is usually considered the best estimator of ρ.

It is easy to see that the numerical value of the sample correlation coefficient does not depend on the unit of measure of either x or y. For, if we let

$$\begin{cases} v = c_1 x + c_2 \\ w = c_3 y + c_4 \end{cases} \qquad (14.115)$$

where c_1, c_2, c_3, c_4 are any constants except $c_1 \neq 0$, $c_3 \neq 0$, it follows that the correlation coefficient in terms of v and w is

$$r_{vw} = \frac{\sum (v - \bar{v})(w - \bar{w})}{\sqrt{\sum (v - \bar{v})^2 \sum (w - \bar{w})^2}}$$

$$= \frac{\sum (c_1 x - c_1 \bar{x})(c_3 y - c_3 \bar{y})}{\sqrt{\sum (c_1 x - c_1 \bar{x})^2 \sum (c_3 y - c_3 \bar{y})^2}}$$

or

$$r_{vw} = r_{xy} \qquad (14.116)$$

Since the correlation coefficient does not depend on either scale of measure-

Table 14.20
Data from Table 14.3 Transformed

x	y	$v = x - 4$	$w = y - 28$	v^2	w^2	vw
2	19	-2	-9	4	81	18
2	21	-2	-7	4	49	14
3	24	-1	-4	1	16	4
3	27	-1	-1	1	1	1
3	27	-1	-1	1	1	1
4	29	0	1	0	1	0
4	31	0	3	0	9	0
6	35	2	7	4	49	14
6	36	2	8	4	64	16
6	37	2	9	4	81	18
Totals		-1	6	23	352	86

ment, we may use it in much the same way in which we applied the coefficient of variation when we were discussing a single variate. Also, as can be observed in Example 14.8, the computations are reduced considerably.

Example 14.8. Code the data in Table 14.3 and find the sample correlation coefficient. Also, discuss the relation of r to the sample regression coefficient $b_{y/x} = b$.

Table 14.20 gives the coded data along with the required products and totals. Thus, we obtain directly

$$SP_{vw} = 86 - \frac{(-1)(6)}{10} = 86.5$$

$$SS_{v} = 23 - \frac{(-1)^2}{10} = 22.9$$

$$SS_{w} = 352 - \frac{6^2}{10} = 348.4$$

$$r^2 = \frac{(86.6)^2}{(22.9)(348.4)} = 0.9400$$

$$r = 0.884$$

Since the sample correlation coefficient is a relatively large positive value, we conclude that there is a strong degree of association between x and y, or, for the regression of y on x, there is a strong dependence of y on a given value of x. Furthermore, we conclude that the slope for the regression of y on x is positive. Thus, as x increases, y increases and does not deviate much from the regression line. But knowing r alone is not enough to determine the mean value of y corresponding to any x. For this we need the regression equation. (It should be understood that the statements about r require that both x and y be random variables.)

Using property (14.53) for the regression sum of squares, the sum of squares identity (14.52), and relation (14.114), we find that it follows that

$$r^2 = \frac{SP^2}{SSx\ SSy} = \frac{SSreg}{SSy} \tag{14.117}$$

or

$$r^2 = \frac{SSy - SSres}{SSy} = 1 - \frac{SSres}{SSy} \tag{14.118}$$

From Eq. (14.117) we see that the square of the correlation coefficient is equal to the ratio of the regression sum of squares to the total sum of squares. It is sometimes called the *coefficient of determination* and is very useful in explaining what proportion of the total variation is due to regression.

In connection with Example 14.8, we observe that 94 per cent of the variation in the y observations is due to the regression of y on x. Since

$$0 \le SS\text{res} \le SSy$$

we conclude, using Eq. (14.118), that

$$0 \le r^2 \le 1$$

or

$$-1 \le r \le 1$$

Thus, the sample correlation ranges over the same set of values as ρ. Furthermore, for any fixed value of ρ except -1 or 1, the sample values of r fall in the interval $-1 \le r \le 1$.

It should be realized that the correlation coefficient r is a measure of *linear* association or dependence. If all points fall on a straight line not parallel to one of the coordinate axes, r is 1 or -1. The fact that two variables x and y are functionally related is not enough to insure that $r = 1$ or -1. Consider, for example, the three functions

(a) $y = x + 1$

(b) $y = x^2 - 4x + 4$

(c) $y = x^2 + 2x - 3$

Compute r for each function when $x = 0, 1, 2, 3, 4$. The corresponding y values are shown in Table 14.21 along with sums of squares and products and the required totals. By using Eq. (14.114), it is easy to show that the correlation coefficients for (a), (b), and (c) are 1, 0, 0.98, respectively. This should illustrate why it is so important that the trend be linear or nearly linear when the correlation coefficient r is used.

Table 14.21
Values for Three Functions

x	Functional Values			x^2	y^2 Values for			xy Values for		
	(a)	(b)	(c)		(a)	(b)	(c)	(a)	(b)	(c)
0	1	4	−3	0	1	16	9	0	0	0
1	2	1	0	1	4	1	0	2	1	0
2	3	0	5	4	9	0	25	6	0	10
3	4	1	12	9	16	1	144	12	3	36
4	5	4	21	16	25	16	441	20	16	84
Totals 10	15	10	35	30	55	34	619	40	20	130

If $\rho = 0$, the sampling distribution of r is symmetric, but as ρ approaches either -1 or 1, asymmetry increases. Also, the distribution of r depends on

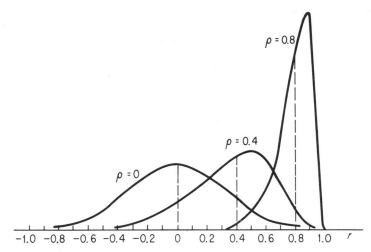

Fig. 14.5 Distribution of r for $\rho = 0$, 0.4 and 0.8 when $n = 10$

the sample size n. For small samples, r is quite variable, particularly when ρ is in the neighborhood of zero. Three graphs are shown in Fig. 14.5.

Since the sampling distribution of r is complicated, the percentage points are quite difficult to compute directly. Fortunately, due to the work of David [14] and Fisher [26], we can determine confidence intervals and test hypotheses directly with the use of charts and simple transformations. For selected values of n smaller than 400, David computed the distribution of r. From these extensive calculations David obtained charts reproduced in Table X which are useful in finding 95 and 99 per cent confidence intervals or in testing hypotheses about ρ. For example, suppose a sample of size 20 has a correlation coefficient of 0.6 and we wish to find the 95 per cent confidence interval for ρ. Using Table X, we draw a vertical line through $r = 0.6$ until it cuts the two curves corresponding to $n = 20$. At the points of intersection, draw horizontal lines until they cut the vertical axis ρ. The limits are found to be 0.20 and 0.82 so that the 95 per cent confidence interval for ρ based on $n = 20$ and $r = 0.6$ is

$$0.20 \leq \rho \leq 0.82$$

Note that this same chart can be used to test the hypothesis $\rho = \rho_0$ against $\rho < \rho_0$ at the 2.5 per cent level or to test the hypothesis $\rho = \rho_0$ against $\rho \neq \rho_0$ at the five per cent level, ρ_0 being any constant in the interval $-1 < \rho_0 < 1$.

For values of n greater than 49, say, we can use a transformation given by Fisher [26]. He showed that the random variable

$$z = \frac{1}{2} \log_e \left(\frac{1 + r}{1 - r} \right) = \text{arc tanh } r, \qquad -1 < r < 1 \qquad (14.119)$$

is approximately normally distributed with mean

$$\mu_z = \frac{1}{2} \log_e \left(\frac{1 + \rho}{1 - \rho} \right) + \frac{\rho}{2(n - 1)} \tag{14.120}$$

and variance

$$\sigma_z^2 = \frac{1}{n - 3}$$

To find the $100 (1 - \alpha)$ per cent confidence interval for ρ, first use the normal distribution to find a $100 (1 - \alpha)$ per cent confidence interval for μ_z and then apply the inverse transformation to obtain the interval for ρ. That is, use Eq. (14.119) to find z, the normal tables to find $u_{\alpha/2}$ and then compute

$$z \pm \frac{u_{\alpha/2}}{\sqrt{n - 3}} \tag{14.121}$$

Letting z_1 and z_2 denote the lower and upper limits, respectively, in (14.121), and assuming that $\rho/[2(n - 1)]$ in Eq. (14.120) is small enough to be ignored, we then transform, by the use of

$$r = \frac{e^{2z} - 1}{e^{2z} + 1} = \tanh z \tag{14.122}$$

to find r_1 and r_2, which are approximate $100 (1 - \alpha)$ per cent confidence limits of ρ. Any standard mathematics table may be used in the transformations (14.119) and (14.122). To test the hypothesis $\rho = \rho_0$, we may use the confidence limits as guides or we may compute

$$\frac{z - \mu_z}{\sqrt{n - 3}} \tag{14.123}$$

and compare it with the appropriate standard normal u value.

In the particular important case where $\rho = 0$, it can be shown that the random variable

$$\frac{r\sqrt{n - 2}}{\sqrt{1 - r^2}}$$

has the Student t distribution with $n - 2$ degrees of freedom. Thus, we can make an *exact* test of the hypothesis $\rho = 0$. Since $t^2 = F$, we may also use the F distribution with one degree and $n - 2$ degrees of freedom for the test of the hypothesis $\rho = 0$. In either case, when ρ actually is zero, the y arrays are independent (not dependent) of x. But this is the same situation that exists in the linear regression problem when $\beta = 0$. Indeed, if we test $\beta = 0$ and $\rho = 0$, using the t distribution, the particular values of the respective statistics always turn out to be the same value for

$$\frac{b}{s_b} = \frac{\dfrac{SP}{SSx}}{\sqrt{\dfrac{SS\text{res}}{SSx\,(n-2)}}} = \frac{\dfrac{SP}{\sqrt{SSx}}\sqrt{n-2}}{\sqrt{SSy\,(1-r^2)}} = \frac{r\sqrt{n-2}}{\sqrt{1-r^2}} \qquad (14.124)$$

We have only introduced the subject of correlation. It has been expanded in many directions. For example, Snedecor [51] describes a method for testing the hypothesis that several independent sample correlation coefficients are estimates of the same population correlation coefficient. We have studied only one measure of correlation. There are other measures which have important specialized applications. The *biserial* and *tetrachoric* measures are useful in such areas as public health, education, and psychology. The interested reader may start his studies by referring to Refs. [18, 45, 47, 54] for detailed discussion. The *rank* correlation is also very useful when one wishes to know whether two rankings are in substantial agreement. The reader is referred to Kendall's books [38, 39] as a starting point in studies of this type of correlation. The degree of association among more than two variables is studied with the use of *partial* and *multiple* correlation. References in this area are [6, 18, 24, 38]. References to these and other topics in correlation may be found in the bibliography and reference lists of the references already cited in this paragraph.

14.10. TWO OR MORE SIMPLE REGRESSIONS (COVARIANCE ANALYSIS)

In Chaps. 6, 8, 10, 11, 12, and 13 we discussed problems associated with comparing two or more means. There are times when we need to treat regression in a similar way. In this section we restrict our attention to some of the simplest problems involving the comparisons of two or more simple linear regressions. The statement of Example 14.9 should serve as an illustration of the type of problem which one might wish to examine.

Example 14.9. Suppose the penetration of different kinds of steel plates by 50-caliber projectiles is being studied. Suppose that five projectiles are fired at each of three plates. Let y'_{iu} denote the depth to which the uth projectile penetrates the ith plate, and let x'_{iu} denote the initial velocity of the uth projectile which is fired toward the ith plate. Let y and x denote the coded values of y' and x', respectively. The measurements are given in Table 14.22. (a) Compare the penetration into the first two plates, taking into account the initial velocity. (b) Relate the regression analysis to the usual analysis of variance. (c) Compare the penetration into the three plates. (The table of data, calculations, and conclusions are given after the theory.)

Other examples in which similar questions arise are in the comparison of

1. Achievements in three algebra classes when the I.Q. of each student is taken into account

2. Weight gained by a certain type of animal using four diets when the initial weight is known

3. Tensile strength of paper tested by different processes where strength depends on thickness

4. Cost of living index in different cities where the index changes with time

5. Yield of a type of grain in different regions where yield depends on production cost

6. Volumes of two or more gases where volume is influenced by temperature and pressure

7. Logarithm of sieve residue for two tube mills for the production of cement related to the production in tons per hour

8. Number of a certain type of insect at different altitudes where number emerging depends on temperature

Let the uth observation for the ith line be denoted by (x_{iu}, y_{iu}) with $i = 1, \ldots, k$; $u = 1, \ldots, n_i$. Assume, as usual, that for a fixed x_{iu} the corresponding y_{iu} is normally distributed with mean

$$\eta_i = \mu_{y_i/x_i} = \alpha_i + \beta_i x_i \qquad (i = 1, \ldots, k) \qquad (14.125)$$

and variance

$$\sigma^2_{y_i/x_i} = \sigma^2_i \qquad (i = 1, \ldots, k)$$

Then, according to Sect. 14.2, the least-squares estimator b_i of β_i is normally distributed with mean

$$\mu_{b_i} = \beta_i$$

and variance

$$\sigma^2_{b_i} = \frac{\sigma^2_i}{SSx_i}$$

where

$$SSx_i = \sum_u (x_{iu} - \bar{x}_i)^2 = \sum_u x^2_{iu} - \frac{(\sum x_{iu})^2}{n_i} \qquad (i = 1, \ldots, k)$$

Further, the least-squares estimator a_i of α_i is normally distributed with mean

$$\mu_{a_i} = \alpha_i$$

and variance

$$\sigma^2_{a_i} = \sigma^2_i \left(\frac{1}{n_i} + \frac{\bar{x}^2_i}{SSx_i} \right) = \frac{\sigma^2_i \sum_u x^2_{iu}}{n_i \, SSx_i} \qquad (14.126)$$

Two hypotheses which we are interested in testing are

$$H_{01}: \quad \beta_1 = \cdots = \beta_k = \beta$$

and

$$H_{02}: \quad \alpha_1 = \cdots = \alpha_k = \alpha$$

We first consider the special case where $k = 2$. For this we require a knowledge of the sampling distributions of $b_1 - b_2$ and $a_1 - a_2$. We assume that the variances σ_1^2 and σ_2^2 are equal to σ^2. When b_1 and b_2 are normally and independently distributed, it follows that $b_1 - b_2$ is normally distributed with mean

$$\beta_1 - \beta_2$$

and variance

$$\sigma_{b_1 - b_2}^2 = \frac{\sigma_1^2}{SSx_1} + \frac{\sigma_2^2}{SSx_2} = \sigma^2 \left(\frac{1}{SSx_1} + \frac{1}{SSx_2} \right)$$

Thus, under the null hypothesis that $\beta_1 = \beta_2$ the statistic

$$\frac{b_1 - b_2}{\sigma \sqrt{\dfrac{1}{SSx_1} + \dfrac{1}{SSx_2}}} \tag{14.127}$$

is normally distributed with mean zero and variance 1. If the common population variance is unknown, we estimate it by pooling the two error variances

$$s_{y/x_1}^2 = \frac{SS\mathrm{res}_1}{(n_1 - 2)} \quad \text{and} \quad s_{y/x_2}^2 = \frac{SS\mathrm{res}_2}{(n_2 - 2)}$$

The resulting pooled estimator is given by

$$s^2 = \frac{(n_1 - 2)\, s_{y/x_1}^2 + (n_2 - 2)\, s_{y/x_2}^2}{(n_1 - 2) + (n_2 - 2)} = \frac{SS\mathrm{res}_1 + SS\mathrm{res}_2}{n_1 + n_2 - 4} \tag{14.128}$$

or

$$s^2 = \frac{SSy_1 + SSy_2 - b_1^2 SSx_1 - b_2^2 SSx_2}{n_1 + n_2 - 4} \tag{14.128a}$$

where

$$SSy_i = \sum_u y_{iu}^2 - \frac{\left(\sum\limits_u y_{iu} \right)^2}{n_i} \qquad (i = 1, 2)$$

Therefore, when $\beta_1 = \beta_2$ the statistic

$$\frac{b_1 - b_2}{s \sqrt{\dfrac{1}{SSx_1} + \dfrac{1}{SSx_2}}} \tag{14.129}$$

is distributed as t with $n_1 + n_2 - 4$ degrees of freedom.

Following a similar argument, we can show that, under the null hypothesis that $\alpha_1 = \alpha_2$,

$$\frac{a_1 - a_2}{s\sqrt{\dfrac{\sum\limits_u x_{1u}^2}{n_1\, SSx_1} + \dfrac{\sum\limits_u x_{2u}^2}{n_2\, SSx_2}}} \tag{14.130}$$

is distributed as t with $n_1 + n_2 - 4$ degrees of freedom. In the particular case where $\beta_1 = \beta_2 = \beta$, we find the pooled (weighted) estimator b of β given by

$$\bar{b} = \frac{SSx_1 b_1 + SSx_2 b_2}{SSx_1 + SSx_2} = \frac{SP_1 + SP_2}{SSx_1 + SSx_2} \tag{14.131}$$

where

$$SP_i = \sum_u x_{iu} y_{iu} - \frac{\left(\sum\limits_u x_{iu}\right)\left(\sum\limits_u y_{iu}\right)}{n_i} \qquad (i = 1, 2)$$

We then estimate α_1 and α_2 by

$$a_1 = \bar{y}_1 - \bar{b}\bar{x}_1 \quad \text{and} \quad a_2 = \bar{y}_2 - \bar{b}\bar{x}_2$$

respectively. The pooled variance estimator is given by

Table 14.22
Velocity (x) and Penetration (y) Data for Example 14.9

Item Number	Plate I		Plate II		Plate III		Totals	
	x	y	x	y	x	y		
1	19	24	17	33	16	12		
2	28	24	11	35	31	8		
3	13	22	3	29	26	13		
4	20	26	8	28	35	25		
5	5	14	21	40	12	7		
Totals	85	110	60	165	120	65	265	340

General		Particular						
$\sum x_{iu}^2 \quad \sum y_{iu}^2$	1739	2508	924	5539	3262	1051	5925	9098
$\sum x_{iu} y_{iu}$	2004		2097		1737		5838	
							Calculated from above totals	
$SSx_i \quad SSy_i$	294	88	204	94	382	206	1243.3	1391.3
SP_i	134		117		177		−168.7	
b_i		0.4558		0.5735		0.4634		
$SSreg_i$		61.07		67.10		82.01	22.88	
$SSres_i$		26.93		26.90		123.99	1368.45	

$$s'^2 = \frac{SSy_1 + SSy_2 - \bar{b}^2(SSx_1 + SSx_2)}{n_1 + n_2 - 3} \qquad (14.132)$$

Thus, when $\beta_1 = \beta_2 = \beta$ we test the null hypothesis $\alpha_1 = \alpha_2 = \alpha$, using the statistic

$$\frac{a_1 - a_2}{s'\sqrt{\dfrac{\sum x_{1u}^2}{n_1\,SSx_1} + \dfrac{\sum x_{2u}^2}{n_2\,SSx_2}}} \qquad (14.133)$$

which is distributed as t with $n_1 + n_2 - 3$ degrees of freedom.

Now we return to Example 14.9. The coded data along with the totals are shown in Table 14.22, and the sums of squares, sums of products, slopes and sums of squares for regression, and residual for each plate are shown below the data.

To illustrate the techniques for the comparison of two regression lines, we first use numerical information on plates I and II. To test the hypothesis $\beta_1 = \beta_2$ against the alternative $\beta_1 \neq \beta_2$ at the five per cent level, we use the t statistic in Eq. (14.129). Since $t_{.025}(6) = 2.45$, the critical region is made up of all values of t for which $t < -2.45$ or $t > 2.45$. Using the formulas of this section and numbers of Table 14.22, we find

$$s^2 = \frac{26.93 + 26.90}{6} = 8.970$$

$$s_{b_1 - b_2}^2 = 8.970(\tfrac{1}{294} + \tfrac{1}{204}) = 0.07448$$

$$s_{b_1 - b_2} = 0.2729$$

$$t = \frac{0.4558 - 0.5735}{0.2729} = -0.4315$$

Since the sample t statistic does not fall in the critical region, we fail to reject the hypothesis of equal slopes.

Since we have two statistics, (14.130) and (14.133), for testing the hypothesis $\alpha_1 = \alpha_2 = \alpha$ against the alternative $\alpha_1 \neq \alpha_2$, we should make a decision, if possible, about the relative sizes of slopes β_1 and β_2. Sometimes we know whether or not β_1 is approximately equal to β_2; other times we may wish to run a preliminary test of the hypothesis $\beta_1 = \beta_2$ to determine if b_1 and b_2 are close enough to compute a weighted common slope b. In the above procedure the value of the sample statistic in absolute value is actually less than the two-sided 50 per cent t value. Thus, we feel safe in pooling the sample slopes and using (14.133) to test the hypothesis $\alpha_1 = \alpha_2$. For this purpose we find

$$\bar{b} = \frac{134 + 117}{294 + 204} = 0.5040$$

$$s'^2 = \frac{88 + 94 - 0.5040(134 + 117)}{10 - 3} = 7.927$$

$$s^2_{a_1-a_2} = 7.927\left(\tfrac{1739}{1470} + \tfrac{924}{1020}\right) = 16.56$$

$$s_{a_1-a_2} = 4.07$$

$$a_1 = \frac{110 - (0.5040)85}{5} = 13.43$$

$$a_2 = \frac{165 - (0.5040)60}{5} = 26.97$$

$$t = \frac{13.43 - 26.97}{4.07} = -3.33$$

The lower 0.025 t value with $5 + 5 - 3 = 7$ degrees of freedom is -2.36. Since -3.33 is less than -2.36, we reject the hypothesis and conclude that the true intercept α_1 of the regression equation for plate I is actually smaller than the true intercept α_2.

Putting the information of the two tests together, we conclude that velocity appears to follow the same law, but that there is a marked difference in the depth of penetration on the two plates. That is, in terms of the fitted regression lines, we may say that the slopes are equal but the parallel lines are significantly far apart. This is shown in Fig. 14.6. The estimated equations for the two plates are

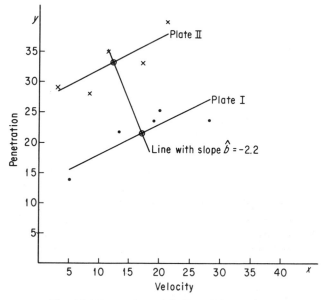

Fig. 14.6 Regression Lines for Plates I and II

$$\hat{\eta}_1 = 13.43 + 0.5040x$$

$$\hat{\eta}_2 = 26.97 + 0.5040x$$

The difference

$$\hat{\eta}_2 - \hat{\eta}_1 = 13.54$$

which does not depend on x, is the best estimate of the difference in penetration for any velocity in the range from 5 to 21. This is due to the fact that the pooled regression slope was used. If we had used the slopes shown in Table 14.22, the regression equations would be

$$\hat{\eta}_1 = 14.25 + 0.4558x$$

$$\hat{\eta}_2 = 26.12 + 0.5735x$$

The difference

$$\hat{\eta}_2 - \hat{\eta}_1 = 11.87 + 0.1177x$$

is a function of x. This brings out one of the advantages of pooling the slopes when possible.

In the simple case where the value of the estimated difference, $d_{y/x} = \hat{\eta}_1 - \hat{\eta}_2$, does not depend on x, one may not be satisfied with a point estimate; one may require a $100(1 - \alpha)$ per cent confidence interval. In order to find the distribution of $d_{y/x}$, we replace a_i by $\bar{y}_i - \bar{b}\bar{x}_i$ $(i = 1, 2)$ so that

$$d_{y/x} = \hat{\eta}_1 - \hat{\eta}_2 = [(\bar{y}_1 - \bar{b}\bar{x}_1) + \bar{b}x] - [(\bar{y}_2 - \bar{b}\bar{x}_2) + \bar{b}x]$$

or

$$d_{y/x} = \bar{y}_1 - \bar{y}_2 - \bar{b}(\bar{x}_1 - \bar{x}_2) \tag{14.134}$$

Since $\bar{y}_1$, $\bar{y}_2$, and $\bar{b}$ are normally and independently distributed, it follows that $d_{y/x}$ is normally distributed with mean $\delta_{y/x} = E(d_{y/x}) = \eta_1 - \eta_2$ and variance

$$\sigma^2 \left[\frac{1}{n_1} + \frac{1}{n_2} + \frac{(\bar{x}_1 - \bar{x}_2)^2}{SSx_1 + SSx_2} \right] \tag{14.135}$$

In case σ^2 is unknown, we may replace it by s'^2 and use the t distribution with $n_1 + n_2 - 3$ degrees of freedom to establish a confidence interval or test a hypothesis about $\delta_{y/x}$. The reader should note that Eq. (14.133) may be used to find a confidence interval for the intercept difference $\alpha_1 - \alpha$, but that Eq. (14.134) should be used when one is interested in finding an interval for $\eta_1 - \eta_2$ for any value of x. In a similar way we could use Eq. (14.39) to find the distribution of $\hat{\eta}_1 - \hat{\eta}_2$ when the true slopes β_1 and β_2 are different. But any difference, $\hat{\eta}_{10} - \hat{\eta}_{20}$, would be a function of x_0. That is, the difference would change with x, and thus $\hat{\eta}_{10} - \hat{\eta}_{20}$ would not generally be a very useful statistic.

When $x_{iu} = \bar{x}_i$, the true regression equation (14.125) becomes

$$\mu_{y_i/x_i} = \alpha_i + \beta_i \bar{x}_i$$

Hence, we may write the true regression equation as

$$\eta_i = \mu_{y_i/x_i} + \beta_i(x_i - \bar{x}_i) \qquad (i = 1, \ldots, k) \qquad (14.136)$$

If two lines are identical, the slopes β_1 and β_2 are equal to a common slope β, and $\eta_1 = \eta_2$. Therefore

$$\mu_{y_1/x_1} + \beta x - \beta \bar{x}_1 = \mu_{y_2/x_2} + \beta x - \beta \bar{x}_2$$

or

$$\beta = \frac{\mu_{y_1/x_1} - \mu_{y_2/x_2}}{\bar{x}_1 - \bar{x}_2} \qquad (14.137)$$

As an estimator of β we calculate

$$\hat{b} = \frac{\bar{y}_1 - \bar{y}_2}{\bar{x}_1 - \bar{x}_2} \qquad (14.138)$$

Therefore, the estimated regression equation with slope $\hat{b}$ passes through the points $(\bar{x}_1, \bar{y}_1)$ and $(\bar{x}_2, \bar{y}_2)$. If the two theoretical regression lines are actually the same, the slope $\hat{b}$ will only deviate at random from the slope, $\bar{b}$, common to the two estimated regression lines. Thus, the distribution of $\hat{b} - \bar{b}$ could be used to determine if $\hat{b}$ is significantly different from $\bar{b}$, that is, to determine if two theoretical regression lines which have the same slope are identical.

Now $\bar{b}$ is calculated from the variation within sets and $\hat{b}$ from variation between sets. Therefore, $\hat{b}$ and $\bar{b}$ are independently distributed, and the variance of $\hat{b} - \bar{b}$ is

$$\sigma^2_{\hat{b}-\bar{b}} = \sigma^2 \left[\frac{1}{(\bar{x}_1 - \bar{x}_2)^2} \left(\frac{1}{n_1} + \frac{1}{n_2} \right) + \frac{1}{SSx_1 + SSx_2} \right] \qquad (14.139)$$

Thus, under the assumption that the two true regression lines are identical, the statistic

$$\frac{\hat{b} - \bar{b}}{\sigma_{\hat{b}-\bar{b}}} \qquad (14.140)$$

is normally distributed with mean zero and variance one. If σ^2 in Eq. (14.135) is replaced by s'^2 and $\sigma^2_{\hat{b}-\bar{b}}$ by $s^2_{\hat{b}-\bar{b}}$, the resulting statistic

$$\frac{\hat{b} - \bar{b}}{s_{\hat{b}-\bar{b}}} \qquad (14.141)$$

is distributed as the Student t with $n_1 + n_2 - 3$ degrees of freedom.

In Example 14.9 we find

$$s^2_{\hat{b}-\bar{b}} = 7.927 \left[\frac{1}{(17 - 12)^2} \left(\frac{1}{5} + \frac{1}{5} \right) + \frac{1}{294 + 204} \right] = 0.1427$$

$$s_{\hat{b}-\bar{b}} = 0.378$$

$$\hat{b} = \frac{22 - 33}{17 - 12} = -2.200$$

$$t = \frac{-2.200 - 0.5040}{0.378} = -7.15$$

Since -7.15 is less than $t_{.025}(7) = -2.36$, we reject the hypothesis of identical true regression lines and conclude that there are two true regression lines. This is indicated in Fig. 14.6. Note that this test does not indicate whether the lines differ with respect to slopes, with respect to intercepts, or with respect to both slopes and intercepts. It simply indicates that the true lines are different.

So far we have considered special cases of *covariance* problems. Perhaps all of the discussion in this section is not necessary, but it was given in the hope that the reader could be led from the usual simple linear regression to the place where the covariance analysis can be easily generalized to any number of simple linear regressions.

Now we consider the comparison of the three plates in Example 14.9 with data and calculations in Table 14.22. If we wished to compare the mean penetration of the plates, ignoring velocity, we would use a one-way classification analysis. That is, to test the hypothesis of equality of means, we would find a *mean square for within* and a *mean square for among plates*. Under the assumption of the hypothesis these mean squares are independently distributed. Thus, we would form a ratio of these mean squares and use the F table to compare with the computed F ratio. If we wish to compare the

Table 14.23

Analysis of Covariance for Example 14.9

	SSx	SP	SSy	Degrees of Freedom	SSreg	SSres	Degrees of Freedom
Plate I	294	134	88	4	61.07	26.93	3
Plate II	204	117	94	4	67.10	26.90	3
Plate III	382	177	206	4	82.01	123.99	3
Among	363.33	−596.67	1003.33	2	979.87	23.46	1
Within	880.00	428.00	388.00	12	208.16	179.84	11
Total	1243.33	−168.67	1391.33	14	22.88	1368.45	13

mean penetration of plates, *taking into account the velocity*, there is an analogous test procedure (which we leave to the student to justify). In this procedure independent estimates of variation about the regression line are found under the assumption that the lines are the same. Hence, we find estimates of the error variation about regression from *within plates* and from *among plates* after taking into account the dependence of y on x. The necessary calculations are shown in Table 14.23. (Since we wish to give the procedure for testing several hypotheses, a compact table which is useful for each hypothesis is given.)

If we did not make use of the possible effect that the x values (velocities) might have on the mean penetrations of the plates, the F value of the usual analysis of variance test would be

$$\frac{\dfrac{1391.33 - 388}{2}}{\dfrac{388}{12}} = 15.52$$

Since $F_{.05}(2, 12) = 3.89$, we reject the hypothesis of equal means, and conclude that we can recognize significant differences in the mean penetration in different plates when the initial velocity is ignored. Taking into account the initial velocity, we compute in an analogous way, using the SSres and adjacent degrees of freedom columns, the value

$$\frac{\dfrac{1368.45 - 179.84}{2}}{\dfrac{179.84}{11}} = 36.35 \qquad (14.142)$$

Since $F_{.05}(2, 11) = 3.98$, we reject the hypothesis of equal means, and conclude that the mean depths of penetration in the plates differ, after making allowances for the differences in initial velocities.

Before describing other tests in covariance analysis, we explain the computations in Table 14.23. Each degrees of freedom column is to be used with the adjacent column to the left. The entries, with the exception of degrees of freedom, in the rows labeled Plate I, Plate II, Plate III, and Total, are brought forward from Table 14.22. The first three entries in the row for within groups are obtained by adding over plates. For example, the first entry is $294 + 204 + 382 = 880$. The first three entries in the row for among means are obtained by subtracting within values from corresponding total values. The SSreg and SSres for both within groups and among means are found in the usual way.

If k simple linear regressions have a common variance σ^2, we may, by generalizing Eq. (14.128), write the pooled estimator of σ^2 as

$$s_1^2 = \frac{\sum_i SS\,res_i}{\sum_i n_i - 2k} \tag{14.143}$$

This variance estimator represents the variation about the k fitted regression lines. Since it is known that

$$u_i = \frac{b_i - \beta_i}{\dfrac{\sigma}{\sqrt{SSx_i}}} \qquad (i = 1, \ldots, k)$$

is normally distributed with mean zero and variance one, then when

$$\beta_1 = \cdots = \beta_k = \beta$$

$$u_i = \frac{(b_i - \beta)\sqrt{SSx_i}}{\sigma} \qquad (i = 1, \ldots, k)$$

is normally distributed with mean zero and variance one. Thus, for independent b_i it follows that

$$\sum u_i^2 = \frac{\sum (b_i - \beta)^2\, SSx_i}{\sigma^2} \tag{14.144}$$

is distributed as χ^2 with k degrees of freedom. Generalizing Eq. (14.131), we find that the least-squares estimator of β is given by

$$\bar{b} = \frac{\sum SP_i}{\sum SSx_i}$$

Since

$$b_i - \beta = (b_i - \bar{b}) + (\bar{b} - \beta)$$

we may, after substituting in the right-hand side of Eq. (14.144), write

$$\sum (b_i - \beta)^2\, SSx_i = \sum (b_i - \bar{b})^2\, SSx_i + (\bar{b} - \beta)^2 \sum SSx_i$$

Thus, from the partition theory of the χ^2 distribution, it follows that

$$\sum (b_i - \bar{b})^2\, SSx_i \tag{14.145}$$

is distributed as $\sigma^2 \chi^2$ with $k - 1$ degrees of freedom

$$\frac{\bar{b} - \beta}{\dfrac{\sigma}{\sqrt{\sum SSx_i}}} \tag{14.146}$$

is normally distributed with mean zero and variance one, and that (14.145) and $\bar{b}$ are independently distributed. Letting

$$s_2^2 = \frac{\sum (b_i - \bar{b})^2 SSx_i}{k - 1} \tag{14.147}$$

we have an estimator of σ^2 which is independent of s_1^2 when the hypothesis

$$\beta_1 = \cdots = \beta_k = \beta \tag{14.148}$$

is true. Thus, the ratio

$$\frac{s_2^2}{s_1^2} \tag{14.149}$$

is distributed as F with $k - 1$ and $\sum n_i - 2k$ degrees of freedom. We may use (14.149) to test the hypothesis in (14.148), that is, the hypothesis of parallel regression lines. The reader should realize that this is the generalization of the test of the hypothesis $\beta_1 = \beta_2$ in which the t distribution was applied.

To illustrate this test, we use the SSres column of Table 14.23 and note that

$$\sum_{i=1}^{3} (b_i - \bar{b}) SSx_i = SS\text{res (within)} - (SS\text{res}_1 + SS\text{res}_2 + SS\text{res}_3) \tag{14.150}$$

The proof of Eq. (14.150) is left as an exercise for the reader. Now the estimates s_1^2 and s_2^2 are

$$s_1^2 = \frac{26.93 + 26.90 + 123.99}{9} = \frac{177.82}{9} = 19.76$$

and

$$s_2^2 = \frac{179.84 - 177.82}{2} = 1.01$$

respectively. Therefore

$$F = \frac{1.01}{19.76} = 0.0511$$

Since 0.0511 is less than $F_{.05} (2, 9) = 4.26$, we fail to reject the hypothesis of parallelism of the true regression lines.

If the true regression lines are parallel, we may combine the two estimators s_1^2 and s_2^2 to obtain the estimator

$$\frac{(\sum n_i - 2k)s_1^2 + (k - 1)s_2^2}{(\sum n_i - 2k) + (k - 1)} \tag{14.151}$$

with $(\sum n_i - 2k) + (k - 1) = \sum n_i - (k + 1)$ degrees of freedom. Further, if the true regression lines are identical, a straight line with the common slope β passes through the k points $(\bar{x}_1, \mu_{y_1/x_1}), \ldots, (\bar{x}_k, \mu_{y_k/x_k})$. The slope of this line is estimated by

$$\hat{b} = \frac{\sum_i n_i (\bar{x}_i - \bar{x})(\bar{y}_i - \bar{y})}{\sum_i n_i (\bar{x}_i - \bar{x})^2} \qquad (14.152)$$

where

$$\bar{x} = \frac{\sum_i \sum_u x_{iu}}{\sum_i n_i} \quad \text{and} \quad \bar{y} = \frac{\sum_i \sum_u y_{iu}}{\sum_i n_i}$$

The regression equation of best fit for the k points $(\bar{x}_1, \bar{y}_1), \ldots, (\bar{x}_k, \bar{y}_k)$ is

$$\hat{\mu}_{y_i/x_i} = \bar{y} + \hat{b}(\bar{x}_i - \bar{x}) \qquad (14.153)$$

and the variation of the k sample pair of means about this line is given by the estimator

$$s_3^2 = \frac{\sum_i n_i [\bar{y}_i - \bar{y} - \hat{b}(\bar{x}_i - \bar{x})]^2}{k - 2} \qquad (14.154)$$

Thus, the *linearity* of the true regression line through points $(\bar{x}_1, \mu_{y_1/x_1}), \ldots,$ $(\bar{x}_k, \mu_{y_k/x_k})$ may be tested by comparing the variance estimator s_3^2 with s_1^2 or the variance estimator given in (14.151), whichever seems appropriate.

In Example 14.9

$$s_3^2 = \frac{23.46}{3 - 2} = 23.46$$

since it can be shown that

$$SS\mathrm{res} \,(\mathrm{among}) = \sum_i n_i [\bar{y}_i - \bar{y} - \hat{b}(\bar{x}_i - \bar{x})]^2 \qquad (14.155)$$

Further, the variance estimator given by (14.151), which is also the one used in the denominator of (14.142), is found to be

$$\frac{9(19.76) + 2(1.01)}{9 + 2} = \frac{179.84}{11} = 16.35$$

Since the ratio

$$\frac{23.46}{16.35} = 1.43$$

is smaller than $F_{.05}\,(1, 11) = 4.84$, we fail to reject the hypothesis. In fact, since 1.43 is roughly the upper 25 per cent point of the F distribution, we feel that any deviation for linearity in the true regression equation is not sizeable.

If the hypothesis that *the k true regression lines are identical* is correct, the slope $\hat{b}$ is a random normal variable with mean β and variance

$$\frac{\sigma^2}{\sum_i n_i(\bar{x}_i - \bar{x})^2}$$

Thus, it follows that an estimator of the variation of $\hat{b} - \bar{b}$ about the weighted mean b of $\hat{b}$ and $\bar{b}$ is given by

$$s_4^2 = (\hat{b} - \bar{b})^2 \left[\frac{1}{\sum n_i(\bar{x} - \bar{x})^2} + \frac{1}{\sum SSx_i} \right] \qquad (14.156)$$

Hence, a test of the identity of k regression lines is given by comparing s_4^2 with the variance in (14.151), or with the variance obtained by combining s_1^2, s_2^2, and s_3^2. It can be shown that

$$SS\text{res (total)} - SS\text{res (among)} - SS\text{res (within)} = s_4^2 \qquad (14.157)$$

Thus, for Example 14.9 we have

$$s_4^2 = 1368.45 - 179.84 - 23.46 = 1165.15$$

so that the computed F ratio is

$$\frac{1165.15}{16.35} = 71.26$$

Since 71.26 is larger than $F_{.05}(1, 11) = 4.84$, we reject the hypothesis and conclude that the three regression lines are not identical. This test is made under the assumptions that the regression means lie on a straight line and that the slopes of the regression lines are all the same. If these assumptions cannot be made, than the appropriate test to use is the one given in Eq. (14.142).

In the above tests we used $k = 3$ individual regression lines, $k = 3$ parallel lines with slope $\bar{b}$, one regression line of means with slope $\hat{b}$ and one regression line for all observations with slope obtained by getting a weighted mean, b, of $\hat{b}$ and $\bar{b}$. The reader should plot scatter diagrams and regression lines until he is thoroughly familiar with the interrelations of all these lines and the corresponding tests. To "see" the residual in a scatter diagram, it is sometimes helpful to transform the data so that the means for x and y in the regression equations fall at the same point on the graph. It might also be informative to study the partition of the component parts which are indicated in the following identity

$$y_{iu} - \bar{y} - b(x_{iu} - \bar{x}) = [y_{iu} - \bar{y}_i - b_i(x_{iu} - \bar{x}_i)]$$
$$+ [(b_i - \bar{b})(x_{iu} - \bar{x}_i)] + [\bar{y}_i - \bar{y} - \hat{b}(\bar{x}_i - \bar{x})] \qquad (14.158)$$
$$+ [(\bar{b} - \hat{b})(x_{iu} - \bar{x}_i) + (\hat{b} - b)(x_{iu} - \bar{x})]$$

The identity indicates that the variation about a single over-all regression line is equal to the sum of the variation of the observations about k individual regression lines plus the variation among the slopes of the k lines plus the variation of the means about the regression line of means plus the variation for the difference between the mean slope $\bar{b}$ and the slope $\hat{b}$ of the regression line of means. This identity leads to a sum of squares identity which contains the components used in the above tests.

There are other tests which can be made in a covariance analysis of a one-way classification, but those given in this section illustrate the kinds of tests which are usually made. Analysis of covariance can also be extended to include two-way and multiway classifications with a single linear control variable x, to include one-, two-, and multiway classifications with more than one linear control variable, and to include cases where the regression is not linear.

For further study the reader is referred to Refs. [5, 9, 25, 33, 36, 37, 51]. A special issue on the analysis of covariance was published by the Biometric Society in 1957. Articles in this issue which might be of particular interest are Refs. [11, 50, 58]. Other references to covariance analysis can be found in the references already cited.

14.11. EXERCISES

14.33. Table 14.24 gives the height (inches) and weight (pounds) of a random sample of 24 college freshmen (male). (a) Find the linear regression of

Table 14.24

Height	Weight	Height	Weight	Height	Weight
73	190	69	157	70	166
71	179	67	150	64	124
66	146	71	170	67	149
67	145	71	172	73	197
66	144	65	132	70	173
70	164	73	187	69	159
72	183	74	205	73	195
74	200	71	175	74	195

height on weight. (b) Find the linear regression of weight on height. (c) Find the correlation coefficient. (d) Determine a 90 per cent confidence interval for the true correlation coefficient. (e) Test the hypothesis $\rho = \frac{1}{2}$ at the five per cent level. (f) Write a summary statement in terms of your findings in (a), (b), (c), and (d).

14.34. Let $f(x, y)$ be the bivariate normal density function given in (3.57). Prove that y for a given value of x is normally distributed with mean

$$\mu_{y/x} = \mu_y + \rho \frac{\sigma_y}{\sigma_x}(x - \mu_x)$$

and variance

$$\sigma^2_{y/x} = \sigma^2_y (1 - \rho^2)$$

14.35. Prove that $E(r) \neq \rho$ when $\rho \neq 0$.

14.36. (a) Give the co-ordinates (x, y) of five points which have a correlation coefficient of -0.8. (b) Leaving four of the points found in (a) fixed, determine the fifth point so that the correlation coefficient is -0.4.

14.37. For a sample of size 18 the correlation coefficient is -0.35. Find a 95 per cent confidence interval for ρ.

14.38. A random sample of size 52 has a correlation coefficient of 0.23. Find a 95 per cent confidence interval for ρ, using Table X; using the arc tanh transformation.

14.39. Let $(x_1, y_1), \ldots, (x_n, y_n)$ denote n pairs of observations. Let $x'_1, \ldots, x'_n$ and $y'_1, \ldots, y'_n$ denote the ranks corresponding to observations $x_1, \ldots, x_n$ and $y_1, \ldots, y_n$, respectively. That is, $x'_1, \ldots, x'_n$ (and $y'_1, \ldots, y'_n$) is some arrangement of the positive integers $1, \ldots, n$. Prove that the correlation between ranks is

$$r' = 1 - \frac{6 \sum (x'_i - y'_i)^2}{n(n^2 - 1)}$$

Assume that no two x(or y) values are the same.

14.40. Prove that the ratio in (14.130) is distributed as t with $n_1 + n_2 - 4$ degrees of freedom.

14.41. Prove Eq. (14.150)

14.42. Prove Eq. (14.157).

14.43. Starting with Eq. (14.158), derive the sum of squares identity used in the test procedures of Sect. 14.10.

14.44. Plot all regression lines discussed in Sect. 14.10.

14.45. The data in Table 14.25 give the average number of parts manufactured per hour, x, and the production cost per part, y, for five factories in

Table 14.25

City A		City B		City C	
x	y	x	y	x	y
11	$1.60	19	$1.00	17	$0.90
13	1.20	19	0.50	13	1.40
10	1.80	21	0.70	16	0.70
11	1.30	17	0.70	18	0.60
13	1.50	10	2.00	10	2.00

three cities. Use the methods of Sect. 14.10 to analyze these data. Write a summary report of your findings, constructing regression lines for illustrative purposes.

14.46. In a study of grades, y, in three algebra classes the effect of I.Q., x, is taken into account. Use the methods of Sect. 14.10 to analyze the data

Table 14.26

Class I		Class II		Class III	
x	y	x	y	x	y
99	81	101	85	108	90
103	84	95	78	99	75
108	81	105	93	126	99
109	79	94	80	119	97
96	78	101	83	109	93
104	79	126	95	110	91
96	81	107	90	105	88
105	85	104	89	119	93
94	72	120	97	128	90
91	79	95	75	94	78
		103	79	103	82
		89	68		

in Table 14.26. Write a summary report of your findings, constructing regression lines for illustrative purposes.

REFERENCES

1. Acton, F. S., *Analysis of Straight-Line Data.* New York: John Wiley & Sons, Inc., 1959, Chaps. 1–10.

2. Aitken, A. C., "On the Graduation of Data by the Orthogonal Polynomials of Least Squares, "*Proceedings of the Royal Society of Edinburgh*, Vol. **53** (1933), pp. 54–78.

3. Allen, F. E., "The General Form of the Orthogonal Polynomials for Simple Series with Proofs of Their Simple Properties," *Proceedings of the Royal Society of Edinburgh*, Vol. **50** (1930), pp. 310–320.

4. Anderson, R. L. and E. E. Houseman, "Tables of Orthogonal Polynomial Values Extended to $N = 104$, "*Iowa State College Agr. Expt. Station Research Bull.* 297, 1942.

5. ——— and T. A. Bancroft, *Statistical Theory in Research.* New York: McGraw-Hill, Inc., 1952, Chaps. 13, 14, 15, and 16.

6. Anderson, T. W., *An Introduction to Multivariate Statistical Analysis.* New York: John Wiley & Sons, Inc., 1958, Chaps. 8, 9, and 12.

7. Askovitz, S. I., "A Short-cut Graphic Method for Fitting the Best Straight Line to a Series of Points According to the Criterion of Least Squares," *Journal of the American Statistical Association*, Vol. **52** (1957), pp. 13–17.

8. Banachiewicz, T., "Principes d'une Nouvelle Technique de la Méthode des Moindres Carrés; Méthode de Résolution Numérique des Équations Linéaires, du Calcul des Déterminants et des Inverses, et de Réduction des Formes Quadratiques, "*Wydzial Matematyczno-Przyrodniczy*, Bull. Intern. Sci. Math., (1938), Akademija Umiejetnosci, Krakow, pp. 134–35, 393–404.

9. Bennett, C. A. and N. L. Franklin, *Statistical Analysis in Chemistry and the Chemical Industry*. New York: John Wiley & Sons, Inc., 1954, Chaps. 6 and 6A.

10. Birge, R. T. and J. W. Weinberg, "Least Squares Fitting of Data by Means of Polynomials, "*Reviews of Modern Physics*, Vol. 19 (1947), pp. 298–360.

11. Cochran, W. G., "Analysis of Covariance: Its Nature and Uses," *Biometrics*, Vol. 13 (1957), pp. 261–81.

12. Cornell, R. G., *A New Estimation Procedure for Linear Combinations of Exponentials*, U. S. Department of Commerce, Washington 25, D. C., 1956, 160 pp.

13. ———, "Non-Linear Estimation for Linear Combinations of Exponentials," Abstracts, *Biometrics*, Vol. 14 (1958), p. 567.

14. David, F. N., *Tables of the Ordinates and Probability Integral of the Distribution of the Correlation Coefficient in Small Samples*. London: Cambridge University Press, 1938.

15. ——— and J. Neyman, "Extension of the Markoff Theorem on Least Squares, "*Statistical Research Memoirs*, Vol. 2 (1938), pp. 105–16.

16. Deming, W. E., *Statistical Adjustment of Data*. New York: John Wiley & Sons, Inc., 1943, Chaps. 1–11.

17. Doolittle, M. H., "Method Employed in the Solution of Normal Equations and the Adjustment of a Triangulation," *U. S. Coast and Geodetic Survey Report*, Washington, D. C.,1878, pp. 115–20.

18. DuBois, P. H., *Multivariate Correlation Analysis*. New York: Harper & Row, Publishers, 1957.

19. Duncan, D. B. and J. F. Kenney, *On the Solution of Normal Equations and Related Topics*. Ann Arbor, Mich.: Edwards Brothers, 1946.

20. Dwyer, P. S., *Linear Computations*. New York: John Wiley & Sons, Inc., 1951, Chaps. 4, 5, 6, and 13.

21. ———, "The Solution of Simultaneous Equations," *Psychometrika*, Vol. 6 (1941), pp. 101–29.

22. ———, "The Doolittle Technique," *Annals of Mathematical Statistics*, Vol. 12 (1941), pp. 449–58.

23. ———, "The Square Root Method and its Use in Correlation and Regression," *Journal of the American Statistical Association*, Vol. 40 (1945), pp. 493–503.

24. Ezekiel, Mordecai, *Methods of Correlation Analysis*, 2nd ed. New York: John Wiley & Sons, Inc., 1941, Chaps. 1–8.

25. Federer, W. T., *Experimental Designs*. New York: The Macmillan Company, 1955, Chap. 16.

26. Fisher, R. A. "On the 'Probable Error' of a Coefficient of Correlation Deduced from a Small Sample," *Metron*, Vol. 1 (1921), pp. 3–32.

27. ——— and F. Yates, *Statistical Tables for Biological, Agricultural, and Medical Research*. London: Oliver and Boyd, 1957.

28. Garwood, F., "The Application of Maximum Likelihood to Dosage-Mortality Curves," *Biometrika*, Vol. 32 (1941), pp. 46–58.

29. Gauss, K. F., "Supplementum Theoriae Combinationis Observationum Erroribus Minimis Obnoxiae," *Werke*, Vol. **15** (1873).

30. Geary, R. C., "Non-Linear Functional Relationship Between Two Variables when One Variable is Controlled," *Journal of the American Statistical Association*, Vol. **48** (1953), pp. 94–103.

31. Gingrich, S. F. and H. A. Meyer, "Construction of an Aerial Stand Volume Table for Upland Oak," *Forest Service*, Vol. **1** (1955), pp. 140–47.

32. Graybill, F. A., *An Introduction to Linear Statistical Models*, Vol. **1**. New York: McGraw-Hill, Inc., 1961, Chaps. 6, 7, 8, 9, and 10.

33. Hald, A., *Statistical Theory with Engineering Applications*. New York: John Wiley & Sons, Inc., 1952, Chaps. 18, 19, and 20.

34. Hartley, H. O., "The Estimation of Non-Linear Parameters by Internal Least Squares, "*Biometrika*, Vol. **35** (1948), pp. 32–45.

35. Hotelling, H., "Tubes and Spheres in *n*-spaces and a Class of Statistical Problems," *American Journal of Mathematics*, Vol. **61** (1939), pp. 440–60.

36. Kempthorne, Oscar, *The Design and Analysis of Experiments*. New York: John Wiley & Sons, Inc., 1952, Chap. 5.

37. Kendall, M. G., *The Advanced Theory of Statistics*, Vol. **2**, 3rd ed. New York: Hafner Publishing Co., Inc., 1951, Chaps. 22, 29, and 30.

38. ———, *The Advanced Theory of Statistics*, Vol. **1**, 5th ed. New York: Hafner Publishing Co., Inc., 1952, Chaps. 14, 15, and 16.

39. ———, *Rank Correlation Methods*, 2nd ed. New York: Hafner Publishing Co., Inc., 1955.

40. ———, "Regression, Structure and Functional Relationship," *Biometrika*, Vol. **39** (1952), pp. 96–108.

41. Keeping, E. S., "A Significance Test for Exponential Regression," *Annals of Mathematical Statistics*, Vol. **22** (1951), pp. 180–98.

42. Koshal, R. S., "Application of the Method of Maximum Likelihood to the Improvement of Curves Fitted by the Method of Moments, "*Journal of the Royal Statistical Society*, Vol. **96** (1933), pp. 303–13.

43. ———, "Maximum Likelihood and Minimal χ^2 in Relation to Frequency Curves," *Annals of Eugenics*, Vol. **9** (1939), pp. 209–31.

44. Li, J. C. R., *Introduction to Statistical Inference*. Ann Arbor, Mich.: Edwards Brothers, 1957, Chaps. 16, 17, and 19.

45. McNemar, O., *Psychological Statistics*. New York: John Wiley & Sons, Inc., 1955, Chap. 12.

46. Ostle, B., *Statistics in Research*, 2nd ed. Ames, Iowa: The Iowa State College Press, 1963, Chaps. 7, 8, 9, and 14.

47. Pearson, K., "On a New Method of Determining Correlation when One Variable is Given an Alternative and the Other by Multiple Categories," *Biometrika*, Vol. **7** (1910), pp. 248–57.

48. Pimentel-Gomes, F., "The Use of Mitscherlich's Regression Law in the Analysis of Experiments with Fertilizers, "*Biometrics*, Vol. **9** (1953), pp. 498–516.

49. Rao, C. R., "Generalization of Markoff's Theorem and Tests of Linear Hypotheses," *Sankhyā*, Vol. **7** (1945–46.), pp. 9–19.

50. Smith, H. F., "Interpretation of Adjusted Treatment Means and Regressions in Analysis of Covariance," *Biometrics*, Vol. **13** (1957), pp. 282–308.

51. Snedecor, G. W., *Statistical Methods*, 5th ed. Ames, Iowa: The Iowa State College Press, 1956, Chaps. 6, and 7.

52. Steel, R. G. D. and J. H. Torrie, *Principles and Procedures of Statistics.* New York: McGraw-Hill, Inc., 1960, Chaps. 9, 10, 14, 15, and 16.

53. Stevens, W. L., "Asymptotic Regression," *Biometrics*, Vol. **7** (1951), pp. 247–67.

54. Treloar, A. E., *Correlation Analysis.* Minneapolis: Burgess Publishing Co., 1942.

55. Wald, A., "The Fitting of Straight Lines if Both Variables are Subject to Error," *Annals of Mathematical Statistics*, Vol. **11** (1940), pp. 284–300.

56. Waugh, F. V., "A Simplified Method of Determining Multiple Regression Constants," *Journal of the American Statistical Association*, Vol. **30** (1935), pp. 694–700.

57. ——— and P. S. Dwyer, "Compact Computation of the Inverse of a Matrix," *Annals of Mathematical Statistics*, Vol. **16** (1945), pp. 259–71.

58. Wilkinson, G. N., "The Analysis of Covariance with Incomplete Data," *Biometrics*, Vol. **13** (1957), pp. 363–72.

59. Williams, E. J., *Regression Analysis.* New York: John Wiley & Sons, Inc., 1956, Chaps. 1–11.

15

ANALYSIS OF COUNTED DATA

In the preceding chapters most of the discussion was devoted to the treatment of quantitative data measured on a continuous scale. This chapter is concerned with problems relating to data which occur as frequencies, or counts, in categories. A category may or may not fall on a continuous scale, and it may be either quantitative or qualitative. The binomial, Poisson, hypergeometric, and multinomial distributions serve as models. It is shown how the chi-square distribution is useful in obtaining good approximations to probabilities applied in *goodness-of-fit* and *contingency-table* problems. Simplified computing formulas are given for important cases.

15.1. INTRODUCTION

We have already considered the nature of models, along with a few of the problems, associated with counting the number of objects in each of several categories. In Chap. 3 the dichotomous and Poisson distributions were introduced; in Chap. 4 a derivation of the binomial density function was given, and in the exercises of Chap. 5 the density functions and some characteristics of the hypergeometric and multinomial distributions were presented. Several applications of the binomial distribution were given in Chap. 6, and it was shown how the computations can be reduced considerably by using the normal approximation in many cases. We have not yet considered problems associated with counting objects which may fall in more than two categories. In order to focus our attention on the nature of a typical problem, we first consider the familiar case with only two categories which are mutually exclusive.

Example 15.1. A librarian at a college claims that 20 per cent of the books are catalogued as science books. Suppose that during this academic year 750 of 3500 new books bought are science books. We wish to know if the proportion of science books bought this year is significantly different from the proportion bought during all the preceding years.

We assume that the "science" and "nonscience" categories are mutually exclusive. Let $p = 0.20$ denote the true proportion of science books catalogued before this year, and $1 - p = q = 0.80$ the true proportion of nonscience books. If books during the present year were bought in the same proportion as in former years, then $3500(0.20) = 700$ would be science and $3500(0.80) = 2800$ nonscience books. We wish to know if the numbers of books actually bought, 750 and $3500 - 750 = 2750$, are significantly different from the numbers the librarian might expect to be brought, 700 and 2800. Clearly, this is a dichotomous situation in which the binomial distribution with $p = 0.20$ and $n = 3500$ should be applied. In this example we wish to know the probability of a random sample of size 3500 being as extreme or more extreme than the one selected. If the probability is less than 0.05, say, we conclude that the sample proportion of science (and nonscience) books is significantly different from 0.20 (0.80); otherwise, we reserve judgment; that is, we conclude that we do not have enough evidence to say that the true proportion is different from 0.20.

Under the above assumptions, the probability of buying 750 or more science books is given by

$$P[x = 750, 751, \ldots, 3500; \text{ binomial with } p = 0.20 \text{ and } n = 3500]$$

$$\doteq P[x \geq 749.5; \text{ normal with } \mu = np = 700 \quad \text{and}$$

$$\sigma = \sqrt{npq} = 23.66] \qquad (15.1)$$

$$= P\left[u \geq \frac{749.5 - 700}{23.66}\right] = P[u \geq 2.09] = 0.0183$$

Since this is a two-sided test, we must also find the probability of buying 650 or fewer science books. Due to symmetry of the normal distribution, this probability is the same as that given by Eq. (15.1). Thus, the required probability is $2(0.0183) = 0.0366$. With such a small probability we conclude that the true proportion must be different from 0.20. Actually, in this case, we conclude that the proportion of science books bought in the last year is significantly larger than 0.20, understanding that there is a 2.5 per cent chance of making a type 1 error.

To make the test more general, suppose n random observations are classified into two mutually exclusive categories C_1 and C_2 with corresponding true proportions $p_1 = p$ and $p_2 = 1 - p = q$. Let o_1 and $o_2 = n - o_1$ denote the frequencies of the sample values, also called *observed frequencies*, in categories C_1 and C_2, respectively, and let e_1 and $e_2 = n - e_1$ denote the *theoretical* or *expected* frequencies. Now, since o_1 is a random variable with a binomial distribution, we know

$$\frac{o_1 - np_1}{\sqrt{np_1 p_2}} \tag{15.2}$$

is approximately normally distributed with mean zero and variance one, provided both $np_1 > 5$ and $np_2 > 5$ when n is large. [If n is not large, we replace o_1 (in 15.2) by $|o_1 - 1/2|$.] Thus

$$\frac{(o_1 - np_1)^2}{np_1(1 - p_1)} \tag{15.3}$$

is approximately distributed as χ^2 with one degree of freedom, provided $np_1 > 5$ and $np_2 > 5$. Further, since

$$\frac{(o_1 - e_1)^2}{e_1} + \frac{(o_2 - e_2)^2}{e_2} = \frac{(o_1 - np_1)^2}{np_1} + \frac{[n - o_1 - n(1 - p_1)]^2}{n(1 - p_1)}$$

$$= (o_1 - np_1)^2 \left[\frac{1 - p_1 + p_1}{np_1(1 - p_1)} \right]$$

$$= \frac{(o_1 - np_1)^2}{np_1 p_2}$$

it follows that

$$\sum_{i=1}^{2} \frac{(o_i - e_i)^2}{e_i} = \chi'^2 \tag{15.4}$$

is approximately distributed as χ^2 with one degree of freedom, provided $np_i > 5$ ($i = 1, 2$). The expression on the left-hand side of Eq. (15.4) is denoted by χ^2 in most places, but we use χ'^2 (chi prime square) to indicate that it is only approximately distributed as χ^2. The reader will recognize that Eq. (15.4) has certain computational advantages over (15.2), and that it is preferred to (15.3) because of its symmetry in terms of frequencies. But the primary reason for introducing Eq. (15.4) is that it can be easily generalized to any number of categories.

Suppose that all values of a population (discrete or continuous, quantitative or qualitative) fall in k mutually exclusive categories $C_1, C_2, \ldots, C_k$. Let p_i denote the true proportion of values falling in category C_i ($i = 1, \ldots, k$), where

$$\sum_{i}^{k} p_i = 1$$

In a random sample of n observations let o_i and $e_i = np_i$ denote the *observed* and *expected* frequency in category C_i, where

$$\sum_{i} o_i = \sum_{i} e_i = n$$

Then it can be shown (see Exercise 15.18) that

$$\chi'^2 = \sum_{i=1}^{k} \frac{(o_i - e_i)^2}{e_i} \tag{15.5}$$

is approximately distributed as χ^2 with $k - 1$ degrees of freedom. This statistic is used in testing the hypothesis that the theoretical (or expected) frequencies are $e_i, \ldots, e_k$ against the alternative that at least one is not as specified. The test procedure is illustrated in the following example.

Example 15.2. Suppose that in addition to the information given in Example 15.1 we are told that 30 per cent are humanities, 35 per cent social science, and 15 per cent general books. Further, suppose that 1000 new humanities, 1200 new social science, and 550 new general books are bought. We wish to determine if the proportion of new books in the four categories is significantly different from the expected proportions of $p_1 = 0.20$, $p_2 = 0.30$, $p_3 = 0.35$, and $p_4 = 0.15$, where the subscripts 1, 2, 3, 4, denote science, humanities, social science, and general, respectively.

If the null hypothesis is true, the expected number of new books in the categories are

$$e_1 = 700, \qquad e_2 = 1050, \qquad e_3 = 1225, \qquad e_4 = 525$$

The observed frequencies are

$$o_1 = 750, \qquad o_2 = 1000, \qquad o_3 = 1200, \qquad o_4 = 550$$

Thus, we find that the computed value of χ'^2, using Eq. (15.5) is

$$\chi'^2 = \frac{(750 - 700)^2}{700} + \frac{(1000 - 1050)^2}{1050} + \frac{(1200 - 1225)^2}{1225} + \frac{(550 - 525)^2}{525}$$
$$= 3.57 + 2.38 + 0.52 + 1.19 = 7.66$$

Now, when $\chi'^2 = 0$ the observed frequencies are identical to the expected frequencies, and as χ'^2 increases, the amount the observed frequencies deviate from the expected frequencies also increases. Thus, the upper tail of the χ^2 distribution with $k - 1$ degrees of freedom is used to test the hypothesis. Since $\chi^2_{.05}(3) = 7.82$, $\chi^2 = 7.76$ barely falls in the noncritical region. Thus, we do not have enough evidence to say the proportion of new books in the four categories differs significantly from those catalogued.

In both these examples the χ^2 distribution has been used to determine whether the observed frequencies match or *fit* the theoretical frequencies. Thus, we call this the χ^2 *goodness-of-fit test*. It should be clear that applications are almost as extensive as problems in which counting occurs. For example, the χ^2 goodness-of-fit test may be used in

1. Almost any opinion poll
2. Checking characteristics of insects against theoretical values
3. Comparing defective parts produced by different machines
4. Studies of almost any human characteristic—amount of education, income, color of hair, occupation
5. Basic research when only a relatively small finite number of categories are meaningful or when rapid results are needed
6. Studies that affect us almost every day—traffic, weather, amount of sleep, health, news, working conditions

If counts are made in two or more categories, the χ^2 goodness-of-fit test may be applied, provided n is sufficiently large. Rules controlling the minimum size of o_i vary. It appears [4, 5, 7, 14, 28, 35, 37] that as many as 20 per cent of the categories may have expected frequencies between one and five, but not less than one, and still have χ'^2 values which are closely approximated by the proper χ^2 distribution.

It should be noted that the only assumption relating to the categories is that they be mutually exclusive. Otherwise, the selection of categories is completely arbitrary. As a matter of fact, a goodness-of-fit technique for one or more variables could have been applied in most of the studies already made. As an illustration, in Example 12.1 we could have arbitrarily defined the three categories to be "very acid," "average acid," and "low acid," and have given a rule for deciding in which category a given sample value falls. Of course, in this particular experiment we would lose some precision, since acidity can be more accurately measured, but we might gain something in time and simplicity. One of the great advantages is that we do not need to make normality assumptions in the χ^2 goodness-of-fit test. Of course, care should be taken to avoid the areas in which mistakes in application are made [3, 7, 21].

Great caution should be used in stating conclusions resulting from the goodness-of-fit test. One should never accept the null hypothesis solely on the evidence of the test. This is illustrated in Example 15.3, and further discussion is given at the end of the example.

15.2. NATURE OF THE GOODNESS-OF-FIT STATISTIC

In Exercise 5.24 we gave the multinomial density function

$$\frac{n!}{x_1! \cdots x_k!} p_1^{x_1} \cdots p_k^{x_k}$$

where x_i denotes the frequency of observations in the ith category in which the true proportion of observations is p_i ($i = 1, \ldots, k$). It was assumed that the categories are mutually exclusive, that the observations are randomly selected, and that $p_i + \ldots + p_k = 1$. Clearly, the multinomial density function gives exact probabilities for any set of observed frequencies

$$x_1 = o_1, \; x_2 = o_2, \ldots, x_k = o_k$$

Thus, the reader, given enough time, could compute the exact probability of a pattern of observations' being as extreme or more extreme from those hypothesized than those observed. Since the computations involved would be so lengthy, an approximation is very desirable.

If one repeatedly draws random samples of size n from a population divided into k categories with proportions $p_1, \ldots, p_k$, the observed frequency o_i in the ith category is a random variable with mean np_i and variance $np_i(1 - p_i)$. Thus, the standardized random variable

$$\frac{o_i - np_i}{\sqrt{np_i(1 - p_i)}} \qquad (i = 1, \ldots, k) \tag{15.6}$$

is approximately normally distributed with mean zero and variance one, provided both np_i and $n(1 - p_i)$ are not less than five and n is sufficiently large. Now, if $o_i, \ldots, o_k$ were independently distributed, the statistic

$$\sum_{i=1}^{k} \frac{(o_i - np_i)^2}{np_i(1 - p_i)} \tag{15.7}$$

would be approximately distributed as χ^2 with k degrees of freedom. The fact that the o_i are correlated and are restricted by the relation $\sum o_i = n$ makes it necessary that (15.7) be replaced by another statistic. It can be shown [6, 9, 10, 11, 13, 16, 26, 30] that the appropriate statistic is

$$\chi'^2 = \sum_{i=1}^{k} \frac{(o_i - np_i)^2}{np_i} \tag{15.8}$$

which is approximately distributed as χ^2 with $k - 1$ degrees of freedom, provided $np_i > 5$ for every i.

The use of Eq. (15.8), as we have seen in two examples, requires that $e_i = np_i$ be known or be specified by the hypothesis, that is, requires that the theoretical distribution be known. Usually this is not the case. Instead, the observed frequencies are used to find maximum likelihood estimators, p_i, of the parameters p_i which replace the parameters in Eq. (15.8). Further, it can be shown [8, 10, 11, 12, 25, 27, 29, 31, 32] that the statistic

$$\chi''^2 = \sum_{i=1}^{k} \frac{(o_i - n\hat{p}_i)^2}{n\hat{p}_i} \tag{15.9}$$

is distributed approximately as χ^2 with $k - 1 - c$ degrees of freedom, where c denotes the number of independent parameters of a distribution which must be estimated in order to determine the estimators $\hat{p}_i$. This is illustrated in the following example.

Example 15.3. Use Eq. (15.9) to test how well the teak tree data of Example 2.10 fit a normal curve.

The type of family of distributions is assumed to be known, but two parameters, μ and σ^2, must be estimated in order to determine which member of this family should be used as a fit for the data. In Example 3.10 the estimates are given as

$$\hat{\mu} = 21.69 \quad \text{and} \quad \hat{\sigma}^2 = 34.5156$$

and the theoretical frequencies for each interval are computed. The observed and theoretical frequencies for the intervals specified in Exercise 2.10 are given in Table 15.1, along with other values which are useful in the computation of χ''^2. (If a fully automatic desk calculator is available, χ''^2 can be computed directly from o_i and e_i, and other recordings are then unnecessary.)

Table 15.1
Calculations for the Teak Tree Data of Exercise 2.10

Diameter of Tree in Inches	Observed Frequency	Theoretical Frequency	Computations	
			$o_i - e_i$	$(o_i - e_i)^2/e_i$
4.5– 7.5	8	8.5*	−0.5	0.029
7.5–10.5	26	22.7	3.3	0.480
10.5–13.5	50	58.3	−8.3	1.182
13.5–16.5	120	116.5	3.5	0.105
16.5–19.5	181	180.9	0.1	0.000
19.5–22.5	215	217.6	−2.6	0.031
22.5–25.5	213	202.9	10.1	0.503
25.5–28.5	145	146.7	−1.7	0.020
28.5–31.5	76	82.1	−6.1	0.453
31.5–34.5	36	35.8	0.2	0.001
34.5–37.5	18	15.9†	2.1	0.227
			Total	3.08

* Includes area to the left of 4.5.
† Includes area to the right of 37.5.

In any case

$$\chi''^2 = 3.08$$

From the χ^2 distribution with $11 - 1 - 2 = 8$ degrees of freedom, we find $\chi^2_{.05}(8) = 15.5$. Since 3.08 is less than 15.5, we fail to reject the null hypothesis that the data were drawn from a normal population with mean $\mu = 21.69$ and variance $\sigma^2 = 34.5156$. However, it would be a gross error to conclude that the data came from this distribution. Indeed, as indicated in Fig. 15.1, it could have come from any of the numerous distributions which have

roughly the same proportions in the various categories. But we may conclude that the data came from a distribution not greatly different from the
particular normal distribution specified by the null hypothesis.

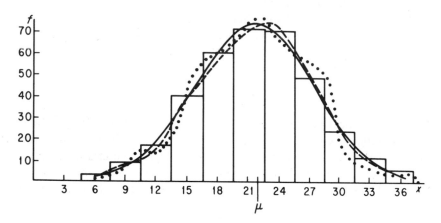

Fig. 15.1 Three Curves which Could Fit the Teak Tree Data

Since the intervals may be selected in any fashion so long as they are
mutually exclusive and exhaustive, the experimenter (investigator) has
unlimited choices in selecting them. The areas under the curve and above
the intervals represent the true proportions $p_1, \ldots, p_k$ in the continuous
case. (In the discrete case the proportions are the lengths of line segments.)
Figure 15.2 illustrates these statements for the continuous case. Note that
the χ^2 test may be used if two or more categories are combined to form a new

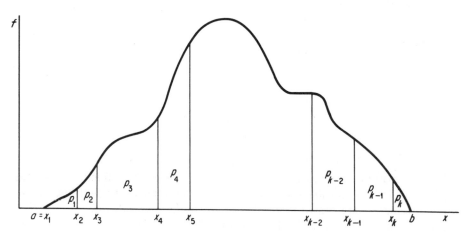

Fig. 15.2 Illustration of the Arbitrary Way in which
Intervals may be Selected

category, provided the proper adjustment is made in the number of degrees of freedom.

The properties for the single variable of classification generalize to more than one variable of classification. A very simple illustration is found in the Mendelian theory of inheritance [23]. Suppose that a characteristic is inherited in the ratio 3 to 1, written 3: 1. That is, in the long run, the characteristic is dominant in three-fourths of the progeny and is recessive in the remaining one-fourth. Then when two characteristics, both having the ratio 3: 1, are crossed, the progeny should have, according to the Mendelian theory, frequencies in the ratios 9: 3: 3: 1. That is, $\frac{9}{16}$ should have both dominant characteristics, $\frac{3}{16}$ a dominant first characteristic and recessive second characteristic, $\frac{3}{16}$ a recessive first characteristic and a dominant second characteristic, and $\frac{1}{16}$ both recessive characteristics. The χ^2 goodness-of-fit test for such a two-way classification is illustrated in the following example.

Example 15.4. In a certain hypothetical experiment peas were crossed with the result that 220 were round and yellow, 78 round and green, 71 wrinkled and yellow, and 31 wrinkled and green. Do these frequencies conform to the theoretical assumption that "round" is the dominant shape and "yellow" is the dominant color such that the true ratios are 9: 3: 3: 1?

The calculations are shown in Table 15.2. Since the theory gives the theoretical frequencies in each of the categories, there are $4 - 1 = 3$ degrees of freedom for the χ^2 statistic. The critical χ^2 value is $\chi^2_{.05}(3) = 7.81$. Since 1.88 is less than 7.81, we fail to reject the hypothesis and conclude that we do not have enough evidence to say that the Mendelian theory fails.

Table 15.2
Computations for the Pea Experiment in Example 15.4

Categories	Frequency		$o_i - e_i$	$\frac{(o_i - e_i)^2}{e_i}$
	Observed	Theoretical		
Round and yellow	220	225	-5	0.111
Round and green	78	75	3	0.120
Wrinkled and yellow	71	75	-4	0.213
Wrinkled and green	31	25	6	1.440
Totals	400	400		1.884

The number of characteristics can be extended indefinitely, and we can still use the χ'^2 statistic as an approximation to the χ^2 distribution to test the goodness of fit of observed frequencies against those hypothesized by the theory. Further, this procedure can be extended to any experiment (or investigation) in which theory gives the expected number of observations in each category.

15.3. TEST OF INDEPENDENCE IN TWO-WAY TABLES

When n observations are tabulated according to two variables of classification, rows and columns, say, we are often interested in determining whether the variables are associated or not. Such two-way tables of frequencies are often called *contingency tables*. The null hypothesis subject to test is that the two classifications are *independent*, that is, the probability that an observation falls in a particular row (column) is not affected by the particular column (row) to which it belongs. If the null hypothesis is rejected, the two variables of classification are said to be *dependent* or correlated. The test of independence is illustrated in the following example.

Example 15.5. A random sample of 200 students in a college were asked the question, "Do you think scientists are slightly unbalanced people?" The results are shown in Table 15.3. We wish to know if the proportion of students saying "yes" is independent of class in school, that is, is the same for each class.

Table 15.3
Number of Students—Classified According to Question Response and
Class in School

	Freshmen	*Sophomore*	*Junior*	*Senior*	*Total*
Yes	15	8	5	2	30
No	58	46	34	32	170
Total	73	54	39	34	200

A statistic, χ''^2, of the type defined by Eq. (15.9) is used. Thus, estimators of cell proportions (probabilities) must be found.

If the number of "yes" responses to the question is independent of class in school, then the proportion of freshmen saying "yes" should be the same as the proportion in any other class. That is, the true proportion of freshmen saying "yes," p_{11}, should be the same as the true proportion of yes answers in all classes, $p_{.1}$. Thus, an estimator of the expected number of freshmen saying "yes," $\hat{e}_{11}$, is the product of the number of freshmen, $n_{1.}$, times an estimator of the proportion in all classes saying "yes," $p_{.1}$; that is

$$\hat{e}_{11} = n_{1.}\,\hat{p}_{.1} = 73\left(\frac{30}{200}\right) = 11.0$$

Estimators of the other expected cell frequencies are found in a similar way and shown in Table 15.4. For example

$$\hat{e}_{24} = \frac{(34)(170)}{200} = 28.9$$

Table 15.4

Estimates of Expected Frequencies for Table 15.3 when Variables of Classification
are Independent

	Freshmen	Sophomore	Junior	Senior	Total
Yes	11.0	8.1	5.9	5.1	30.1
No	62.0	45.9	33.1	28.9	169.9
Total	73	54	39	34	200

The totals in Table 15.4 should be the same as those in Table 15.3 except for rounding-off errors. This being the case, it is clear that, once three selected estimates, $\hat{e}_{11}, \hat{e}_{12}, \hat{e}_{13}$, say, are determined by the above method, all others can be found by subtraction. Now we find

$$\chi''^2 = \frac{(15 - 11.0)^2}{11.0} + \frac{(8 - 8.1)^2}{8.1} + \cdots + \frac{(32 - 28.9)^2}{28.9}$$

$$= 4.09$$

and $\chi_{.05}^2(3) = 7.81$. Hence, we fail to reject the hypothesis of independence. That is, there is not enough evidence to say that the number of "yes" responses depends on the class in school.

In general, let c mutually exclusive categories of one variable (attribute) be arranged in columns and r mutually exclusive categories of a second variable (attribute) be arranged in rows so that the resulting cells represent the cr mutually exclusive categories of *joint attributes*. The cell in the ith column and jth row represents the joint attribute of the ith category of variable one and the jth category of variable two ($i = 1, \ldots, c; j = 1, \ldots, r$). Let p_{ij} denote the probability of an object selected at random falling in the cell in the ith column and jth row

$$p_{i.} = \sum_j p_{ij}$$

denote the probability of an object's falling in the ith column, and

$$p_{.j} = \sum_i p_{ij}$$

denote the probability of an object's falling in the jth row. The null hypothesis that the two variables are independent may then be written as

$$H_0: p_{ij} = p_{i.} p_{.j} \qquad (i = 1, \ldots, c; \ j = 1, \ldots, r) \qquad (15.10)$$

If n objects are randomly selected and n_{ij} fall in the cell in the ith column and jth row, then, according to Eq. (15.8), the statistic

$$\chi'^2 = \sum_i^c \sum_j^r \frac{(n_{ij} - np_{ij})^2}{np_{ij}} \qquad (15.11)$$

is approximately distributed as χ^2 with $cr - 1$ degrees of freedom, provided $np_{ij} > 5$ for every pair of i and j. Under the assumption of the null hypothesis (15.10) the statistic in Eq. (15.11) may be written as

$$\chi'^2 = \sum_i^c \sum_j^r \frac{(n_{ij} - np_{i.}p_{.j})^2}{np_{i.}p_{.j}} \qquad (15.12)$$

But Eq. (15.12) is not very useful, since the probabilities $p_{i.}$ and $p_{.j}$ are seldom known.

If we let

$$n_{i.} = \sum_j n_{ij} \quad \text{and} \quad n_{.j} = \sum_i n_{ij}$$

it is easy to show that

$$\begin{cases} \hat{p}_{i.} = \dfrac{n_{i.}}{n} & (i = 1, \ldots, c) \\[2mm] \hat{p}_{.j} = \dfrac{n_{.j}}{n} & (j = 1, \ldots, r) \end{cases} \qquad (15.13)$$

are maximum likelihood estimators of $p_{i.}$ and $p_{.j}$, respectively. Thus, an estimator of the expected frequency in the cell in the ith column and jth row is given by

$$\hat{e}_{ij} = n\hat{p}_{i.}\hat{p}_{.j}$$

or

$$\hat{e}_{ij} = \frac{(n_{i.})(n_{.j})}{n} \qquad (i = 1, \ldots, c; j = 1, \ldots, r) \qquad (15.14)$$

Since

$$\sum_i p_{i.} = \sum_j p_{.j} = 1$$

there are only $(c - 1) + (r - 1) = c + r - 2$ independent parameters, and only $c + r - 2$ independent estimators used in finding the cr estimators $\hat{e}_{11}, \hat{e}_{12}, \ldots, \hat{e}_{cr}$. Thus, on substituting the estimator (15.14) for $np_{i.}p_{.j}$ in Eq. (15.12), we obtain the statistic

$$\chi''^2 = \sum_i^c \sum_j^r \frac{(n_{ij} - \hat{e}_{ij})^2}{\hat{e}_{ij}} \qquad (15.15)$$

or

$$\chi''^2 = \sum_i^c \sum_j^r \frac{\left[n_{ij} - \dfrac{(n_{i.})(n_{.j})}{n}\right]^2}{\dfrac{(n_{i.})(n_{.j})}{n}} \qquad (15.15a)$$

which is distributed approximately as χ^2 with $cr - 1 - (c + r - 2) =$

$(c - 1)(r - 1)$ degrees of freedom. For further study the student may check Refs. [6, 9, 10, 12, 16, 18, 34]. The application of this statistic has already been given in Example 15.5.

Table 15.5

Observed and Theoretical (Estimated) Frequencies in a Contingency Table

		Variable One Categories					Totals
		1	· · ·	i	· · ·	c	
Variable Two Categories	1	$n_{11}(\hat{e}_{11})$	· · ·	$n_{i1}(\hat{e}_{i1})$	· · ·	$n_{c1}(\hat{e}_{c1})$	$n_{.1}$
	j	$n_{1j}(\hat{e}_{1j})$	· · ·	$n_{ij}(\hat{e}_{ij})$	· · ·	$n_{cj}(\hat{e}_{cj})$	$n_{.j}$
	r	$n_{1r}(\hat{e}_{1r})$	· · ·	$n_{ir}(\hat{e}_{ir})$	· · ·	$n_{cr}(\hat{e}_{cr})$	$n_{.r}$
Totals		$n_{1.}$	· · ·	$n_{i.}$	· · ·	$n_{c.}$	n

Table 15.5 is given as an aid in computing χ''^2 and in understanding the relations among the observed and expected frequencies. The entries and computations in a problem are made in the following order. Record all observed frequencies and then find marginal totals. Then use the marginal frequencies and Eq. (15.14) to find the estimated theoretical frequencies. If a fully automatic desk calculator is available, χ''^2 can be found without any further intermediate recording. If such a machine is not available, intermediate recordings like those in Table 15.2 should be made. In any case, Table 15.5 is very useful in making a rapid estimate of the significance of the two variables of classification.

The test of mutual independence of more than two variables (attributes) is similar to the one explained for the two-variable case. We indicate the procedure with three variables. Suppose variables U, V, and W are classified into c columns, r rows, and l layers, respectively, so that there are crl cells in a three-way table. Let n_{ijk} denote the number of objects (from among n random objects) falling in the cell in the ith column, jth row, and kth layer. Then it can be shown that the statistic

$$\chi''^2 = \sum_i^c \sum_j^r \sum_k^l \frac{(n_{ijk} - \hat{e}_{ijk})^2}{\hat{e}_{ijk}} \qquad (15.16)$$

is approximately distributed as χ^2 with $(c - 1)(r - 1)\ (l - 1)$ degrees of freedom, where

$$\hat{e}_{ijk} = \frac{n_{i..}\, n_{.j.}\, n_{..k}}{n^2}$$

$$n_{i..} = \sum_j \sum_k n_{ijk}$$

$$n_{.j.} = \sum_i \sum_k n_{ijk}$$

$$n_{..k} = \sum_i \sum_j n_{ijk}$$

$$n = \sum_i \sum_j \sum_k n_{ijk}$$

The statistic (15.16) is applied in the usual way to test the mutual independence of the variables U, V, and W. Other hypotheses [1, 2, 17, 24, 36] may also be of interest. For example, one may wish to test whether variable U is independent of V and W.

15.4. SPECIAL CASES OF TESTS OF INDEPENDENCE

If there are only two categories for one variable of classification, the computation of the statistic χ''^2 may be simplified considerably. If $c = r = 2$, then Eq. (15.15) reduces to

$$\chi''^2 = \frac{n(n_{11}n_{22} - n_{12}n_{21})^2}{n_{1.}n_{2.}n_{.1}n_{.2}} \tag{15.17}$$

which is approximately distributed as χ^2 with one degree of freedom, provided each estimated theoretical frequency $\hat{e}_{ij}(i, j = 1, 2)$ is large. If some $\hat{e}_{ij}$ is small, the approximation is improved considerably by applying Yates' correction [9, 37]. This is done by adding $\frac{1}{2}$ to the smallest observed frequency and keeping the marginal totals the same. Thus, the corrected statistic, $\chi_c''^2$ is computed by

$$\chi_c''^2 = \frac{n\left(|n_{11}n_{22} - n_{12}n_{21}| - \dfrac{n}{2}\right)^2}{n_{1.}n_{2.}n_{.1}n_{.2}} \tag{15.18}$$

At best, the application of the above statistic still leads to approximate results. Hence, in some doubtful cases the exact procedure [12] should be applied, even though it requires considerably more calculation. The method is also useful in arriving at a better understanding of the inferences resulting from the use of the χ''^2 or $\chi_c''^2$ statistic. It can be shown that, for fixed marginal totals, the probability of a set of observed values $n_{11}, n_{12}, n_{21}, n_{22}$ in a 2×2 contingency table is

$$P[n_{11}, n_{12}, n_{21}, n_{22}] = \frac{n_{1.}!\, n_{2.}!\, n_{.1}!\, n_{.2}!}{n!\, n_{11}!\, n_{12}!\, n_{21}!\, n_{22}!} \tag{15.19}$$

An application of the exact method is given in the following example.

Example 15.6. On exposure to a certain disease 20 people responded according to the frequencies shown in Table 15.6. We wish to know, at the five per cent level, whether the inoculation is effective.

Table 15.6
Results of Exposure to a Disease (Observed Frequencies)

	Attacked	*Not Attacked*	*Total*
Not inoculated	10	3	13
Inoculated	2	5	7
Total	12	8	20

First, using Eq. (15.18), we find that

$$\chi_c''^2 = \frac{20(|10 \cdot 5 - 2 \cdot 3| - 10)^2}{12 \cdot 8 \cdot 13 \cdot 7} = 2.65$$

Since 2.65 is less than $\chi_{.05}^2 (1) = 3.84$, we fail to reject the hypothesis of independence. That is, we do not have enough evidence to say that the inoculation is effective.

Applying Eq. (15.19) and the method of exact probabilities, we reason as follows. If inoculation makes no difference in the proportion of people contracting the disease, then

$$\frac{(12)(7)}{20} = 4.2$$

would be attacked and

$$\frac{(8)(7)}{20} = 2.8$$

would not be attacked. In our example, only two people who were inoculated contracted the disease. Thus, only in cases with observed frequencies of zero or one would the number of people contracting the disease be more extreme. That is, only the two sets of frequencies shown in Table 15.7 would be more extreme from expectation than those of Table 15.6. Thus, the exact

Table 15.7
Sets of Frequencies with Totals Fixed

	Attacked	*Not*	*Total*	*Attacked*	*Not*	*Total*
Not inoculated	11	2	13	12	1	13
Inoculated	1	6	7	0	7	7
Total	12	8	20	12	8	20

probability of a set of frequencies' being as extreme or more extreme than those observed is given by

$P[10, 3, 2, 5] + P[11, 2, 1, 6] + P[12, 1, 0, 7]$

$$= 13! \, 7! \, 12! \, 8! \left(\frac{1}{20! \, 10! \, 3! \, 2! \, 5!} + \frac{1}{20! \, 11! \, 2! \, 1! \, 6!} + \frac{1}{20! \, 12! \, 1! \, 0! \, 4!} \right)$$

$$= \frac{101}{1938} \doteq 0.0521$$

Assuming the hypothesis of independence to be true, we conclude that 5.21 per cent of the sets (patterns) of frequencies are as extreme or more extreme than the set in Table 15.6. Thus, at the five per cent level, we barely fail to reject the hypothesis of independence.

If we had used Eq. (15.17) so that

$$\chi''^2 = \frac{20(10 \cdot 5 - 2 \cdot 3)^2}{12 \cdot 8 \cdot 13 \cdot 7} = 4.43$$

then we would have rejected the hypothesis of independence, in which case, the conclusion is different from that obtained by applying the exact procedure or Yates' correction. However, the conclusions resulting from the exact procedure and the application of $\chi_c''^2$ do not always agree. This is illustrated in the following paragraph.

Remembering that

$$P[\chi^2 > k] = 2P[u > \sqrt{k}] \qquad (15.20)$$

where k is any positive constant and $u = \sqrt{\chi^2}$ is the standard normal deviate, we have

$$P[\chi^2 > 2.65 = \chi_c''^2] = 2P[u > 1.63] = 0.104$$

and

$$P[\chi^2 > 4.43 = \chi''^2] = 2P[u > 2.10] = 0.036$$

Now, if a ten per cent level test is used, the exact test leads to the rejection of the hypothesis of independence, but application of Yates' correction leads to the conclusion that we fail to reject this hypothesis.

In the particular case where there are two columns (rows) and r rows (c columns) in a contingency table, it is easy to show that the χ''^2 statistic in Eq. (15.15) reduces to

$$\chi''^2 = \frac{1}{n_{1.} \, n_{2.}} \sum_{j=1}^{r} \frac{(n_{1j} n_{2.} - n_{2j} n_{1.})^2}{n_{.j}} \qquad (15.21)$$

$$\left[\chi''^2 = \frac{1}{n_{.1} \, n_{.2}} \sum_{i=1}^{c} \frac{(n_{i1} n_{.2} - n_{i2} n_{.1})^2}{n_{i.}} \right] \qquad (15.21a)$$

which is approximately distributed as χ^2 with $r - 1$ ($c - 1$) degrees of freedom. If we let

$$p_j = \frac{n_{1j}}{n_{.j}} \qquad (j = 1, \ldots, r)$$

and

$$p = \frac{n_{1.}}{n}$$

it can be shown that Eq. (15.21) reduces to

$$\chi''^2 = \frac{1}{p(1-p)} \left(\sum_j n_{1j} p_j - np \right) \tag{15.22}$$

This expression, due to Snedecor [34], is a good computational form.

15.5. TESTS OF HOMOGENEITY OF SEQUENCES

In the usual contingency table the n observations are free to fall in any of the cr cells. In other situations the frequency table may have the same notation and general appearance, but be quite different in interpretation. (An analogous situation is found in the one- and two-way classification in analysis of variance.)

Suppose c samples of sizes $n_{1.}, \ldots, n_{i.}, \ldots, n_{c.}$ are independently drawn from the same population and suppose the $n_{i.}$ observations in sample i are classified in r mutually exclusive categories. Let n_{ij} denote the number of observations of the ith sample falling in the jth group. The resulting observed frequency table will look exactly like Table 15.5. However, in this case we think of c independent sequences of frequencies, each classified in r cells, rather than a single sequence of frequencies classified in cr cells.

If $p_1, \ldots, p_j, \ldots, p_r$ denote the true probabilities of the population values falling in the r categories, then for the ith sample the probability of a sample value falling in the jth category is p_j. Thus, for each of the c columns in Table 15.5 we assume that

$$\sum_{j=1}^{r} p_j = 1 \tag{15.23}$$

In certain investigations one wishes to test to determine whether c samples, each classified according to the same r categories, could have been drawn from the same population, that is, to determine if each of the c samples has true probabilities $p_1, \ldots, p_j, \ldots, p_r$. It can be shown [9, 14] that the proper statistic to use in such a test is given by Eq. (15.15). If the hypothesis is rejected, we say that the samples are drawn from different populations. If we fail to reject the hypothesis, we say that there is not enough evidence to claim that the populations are different. Thus, this is sometimes called a test of *homogeneity of sequences*.

There are two cases of special interest. In the first, suppose that c samples, each of size n, are drawn from binomial populations. Let $n_{11}, \ldots, n_{i1}, \ldots, n_{c1}$ denote the number of successes in each of n independent trials. Then $n_{i2} = n - n_{i1}$ denotes the number of failures in the ith sample. The observed frequencies and marginal totals are shown in Table 15.8. Under the assump-

Table 15.8
Observed Frequencies for c Binomial Samples of Equal Size

Sample	1	· · ·	i	· · ·	c	Total
Successes	n_{11}	· · ·	n_{i1}	· · ·	n_{c1}	$n_{.1}$
Failures	$n - n_{11}$	· · ·	$n - n_{i1}$	· · ·	$n - n_{c1}$	$cn - n_{.1}$
Totals	n	· · ·	n	· · ·	n	cn

tion that each of the c samples is randomly drawn from a binomial population with probability of success p, it can be shown that the maximum likelihood estimator $\hat{e}_i$ of the theoretical frequency of success in the ith sample is

$$\hat{e}_i = n\left(\frac{n_{.1}}{cn}\right) = \bar{n}_{.1} \qquad (i = 1, \ldots, c) \tag{15.24}$$

where $\bar{n}_{.1} = n_{.1}/c$ denotes the mean number of successes over all samples. To test the null hypothesis that all samples are drawn from the same binomial population, we may use the statistic χ''^2 given in Eq. (15.15) or its equivalent form

$$\chi''^2 = \frac{\sum\limits_{i=1}^{c} (n_{i1} - \bar{n}_{.1})^2}{\bar{n}_{.1}\left(1 - \dfrac{\bar{n}_{.1}}{n}\right)} \tag{15.25}$$

The statistic in Eq. (15.25) is sometimes called the *binomial index of dispersion*.

If n is large and p is small so that $np = \mu$ is constant, we may use the Poisson distribution to approximate the binomial distribution. Then $\bar{n}_{.1}/n$ as a maximum likelihood estimator of p is small and $1 - \bar{n}_{.1}/n$ is approximately equal to one. Furthermore, the right-hand side of Eq. (15.25) reduces to

$$\frac{\sum\limits_{i=1}^{c} (n_{i1} - \bar{n}_{.1})^2}{\bar{n}_{.1}} \tag{15.26}$$

which is known as the *Poisson index of dispersion* [35].

Example 15.7. In ten consecutive weeks the number of illegal left-hand turns at a busy stoplight were as follows: 27, 21, 23, 19, 18, 25, 23, 28, 24, 22.

Use a five per cent level test to determine whether the number of left-hand turns is nonheterogeneous.

Assuming the number of left-hand turns per week to have the Poisson distribution, we use (15.26) to find

$$\frac{\sum (n_{i1} - \bar{n}_{.1})^2}{\bar{n}_{.1}} = \frac{(27 - 23)^2 + \cdots + (22 - 23)^2}{23} = 4$$

For nine degrees of freedom $\chi^2_{.05} = 16.9$. Thus, we fail to reject the hypothesis that the data came from the same Poisson distribution. That is, on the basis of these results we cannot say that the data came from different Poisson distributions.

The χ^2 test for counted data may be related to the F test for measured data. Many of the procedures in analysis of variance have counterparts in experiments with counted data. For example, instead of using the χ^2 distribution to test the homogeneity of c binomial population means, say, we may use it to test hypotheses involving a single degree of freedom [15, 18, 19, 20, 22]. For further study the reader is referred to Refs. [6, 7, 9, 12, 22, 33, 34].

15.6. EXERCISES

15.1. Two hundred sixty persons of voting age, selected at random, were asked two days before a city election which of two candidates, A and B, they favored. One hundred forty-three favored A, and 117 favored B. Use a five per cent level test to determine whether the opinion in the population may be equally divided.

15.2. In manufacturing a certain type of small unit it is claimed that no more than one per cent are defective. In a random sample of 500 units 11 were found to be defective. What conclusion do you make concerning the claim?

15.3. In 120 fair tosses of a die the following data were obtained

Number of spots	1	2	3	4	5	6
Frequency	15	27	18	12	25	23

Is there reason to believe the die is not properly balanced?

15.4. Suppose the number of defective units in a day's production was tabulated by shifts with the following results

Shift	1	2	3
Frequency	19	36	25

Is there reason to believe the true relative frequencies of defectives are the same for all shifts?

15.5. During a three-month period there were 145 machine breakdowns. Table 15.9 gives the number of breakdowns for each machine during each shift. Determine if the number of breakdowns for any machine is independent of shift.

Table 15.9

Shift	Machine			
	A	B	C	D
1	9	5	11	12
2	9	11	18	20
3	12	9	12	17

15.6. Table 15.10 shows the grade distribution of three instructors who taught the same course for the same period of time. Did the instructors give significantly different percentages of the five grades?

Table 15.10

Instructor	Grade				
	A	B	C	D	E
I	20	38	126	18	21
II	14	45	183	25	14
III	38	48	275	24	33

15.7. Prove Eq. (15.17).

15.8. (a) Write all possible patterns where the marginal totals in a two-way frequency table are those given in Table 15.11. (b) Find the probabilities

Table 15.11

		7
		9
11	5	16

associated with these patterns. (c) What is the exact probability that one would have the following observed pattern of frequencies or a pattern more extreme?

3	4
8	1

(d) What is the comparable probability if Yates' correction is used?

(e) What is the comparable probability if Yates' correction is not used?

15.9. Prove that Eq. (15.15a) reduces to

$$\chi''^2 = \frac{1}{n_{1.}n_{2.}} \sum_{j}^{r} \frac{(n_{1j}n_{2.} - n_{2j}n_{1.})^2}{n_{.j}}$$

when $c = 2$.

15.10. Table 15.12 shows the hair color distribution of boys and girls in a certain town. Use the formula in Exercise 15.9 to test the independence of hair color and sex. Make comments on your findings.

Table 15.12

	Fair	Red	Hair Color Medium	Dark	Black
Boys	450	120	1232	435	27
Girls	575	110	1104	422	35

15.11. Prove that the estimators given in Eq. (15.13) are maximum likelihood estimators of $p_{i.}$ and $p_{.j}$, respectively.

15.12. Prove that Eq. (15.19) gives the probability of a set of observed values $n_{11}, n_{12}, n_{21}, n_{22}$ in a 2×2 contingency table.

15.13. Prove Eq. (15.22).

15.14. Justify using Eqs. (15.25) and (15.26) in testing equality of proportions.

15.15. Twenty boys were randomly selected from each of five high schools in a city, and the median height was determined for the total group of 100. Table 15.13 shows the number of boys in each high school with heights greater than the median and heights equal to or less than the median. Test to determine if the five high schools are likely to have boys with the same median height.

Table 15.13

Frequency	A	B	High School C	D	E
Greater than	14	9	10	5	12
Equal to or less than	6	11	10	15	8
Total	20	20	20	20	20

15.16. The number of deaths by accidents in each of six large universities during the same academic year were as follows

10, 5, 8, 3, 2, 12

Use a five per cent level test to determine whether the number of deaths is nonheterogeneous.

15.17. Four groups of students were asked the same question. Table 15.14 gives frequencies of "yes" and "no" answers. (a) Determine whether there is a significant difference in the percentages of affirmative answers for boys and girls; for high school and college. (b) Note that this is a 2 × 2 factorial experiment. Apply the usual analysis of variance to determine the solution to part (a).

Table 15.14

Answer	High School Boys	High School Girls	College Boys	College Girls
Yes	30	15	50	25
No	70	85	50	75
Total	100	100	100	100

15.18. Prove that χ'^2 as given in Eq. (15.5) or Eq. (15.8) is approximately distributed as χ^2 with $k - 1$ degrees of freedom.

Hint. Derive the χ^2 distribution from the multinomial in much the same way that the normal (and thus χ^2 with one degree of freedom) was derived from the binomial distribution. That is, start with the multinomial distribution with k categories, replace factorials with Stirling's approximation, rearrange terms, and take logarithms, expand the logarithmic expression log $(1 + z)$, etc.

15.19. Use the literature to prepare a derivation of Eq. (15.9).

15.20. Justify Eq. (15.16) and give a 2 × 2 × 2 illustration.

REFERENCES

1. Anderson, R. L. and T. A. Bancroft, *Statistical Theory in Research.* New York: McGraw-Hill, Inc., 1952, Chap. 12.

2. Bartlett, M. S., "Contingency Table Interaction," *The Journal of the Royal Statistical Society Supplement*, Vol. 2 (1935), pp. 248–252.

3. Berkson, J., "Some Difficulties of Interpretation Encountered in the Application of the Chi-Square Test," *Journal of the American Statistical Association*, Vol. 33 (1938), pp. 526–36.

4. Cochran, W. G., "The χ^2 Distribution for the Binomial and Poisson Series, with Small Expectations," *Annals of Eugenics*, Vol. 7 (1936), pp. 207–17.

5. ———, "The χ^2 Correction for Continuity," *Iowa State College Journal of Science*, Vol. 16 (1942), pp. 421–36.

6. ———, "The χ^2 Test of Goodness of Fit," *Annals of Mathematical Statistics*, Vol. 23 (1952), pp. 315–45.

7. ———, "Some Methods for Strengthening the Common χ^2 Tests," *Biometrics*, Vol. 10 (1954), pp. 417–51.

8. Cramér, H., "On the Composition of Elementary Errors," *Skandinavisk Aktuarietidskrift*, Vol. **11** (1928), pp. 13–74, 141–80.

9. ——, *Mathematical Methods of Statistics*. Princeton, N. J.: Princeton University Press, 1946, Chap. 30.

10. Fisher, R. A., "On the Interpretation of χ^2 from Contingency Tables, and the Calculation of Probabilities," *Journal of the Royal Statistical Society*, Vol. **85** (1922), pp. 87–94.

11. ——, "The Conditions Under Which χ^2 Measures the Discrepancy Between Observation and Hypothesis," *Journal of the Royal Statistical Society*, Vol. **87** (1924), pp. 442–49.

12. ——, *Statistical Methods for Research Workers*, 7th ed. London: Oliver and Boyd, 1938, Chap. 4.

13. Greenhood, E. R., *A Detailed Proof of the Chi-Square Test of Goodness of Fit*. Cambridge, Mass.: Harvard University Press, 1940.

14. Haldane, J. B. S., "The Mean and Variance of χ^2, When Used as a Test of Homogeneity, When Expectations are Small," *Biometrika*, Vol. **31** (1940), pp. 346–55.

15. Irwin, J. O., "A Note the Subdivision of χ^2 into Components," *Biometrika*, Vol. **36** (1949), pp. 130–34.

16. Kenney, J. F., *Mathematics of Statistics*, Part 2. New York: D. Van Nostrand Co., Inc., 1939, Chap. 8.

17. Kermack, W. O. and A. G. McKendrick, "The Design and Interpretation of Experiments Based on a Four-fold Table: The Statistical Assessment of the Effects of Treatment," *Proceedings of the Royal Society of Edinburgh*, Vol. **60** (1940), pp. 362–75.

18. Kimball, A. W., "Short-Cut Formulas for the Exact Partition of χ^2 in Contingency Tables," *Biometrics*, Vol. **10** (1954), pp. 452–58.

19. Lancaster, H. O., "The Deviation and Partition of χ^2 in Certain Discrete Distributions," *Biometrika*, Vol. **36** (1949), pp. 117–29.

20. ——, "The Exact Partition of χ^2 and Its Application to the Problem of Pooling of Small Expectations," *Biometrika*, Vol. **37** (1950), pp. 267–70.

21. Lewis, D. and C. J. Burke, "The Use and Misuse of Chi-Square Test," *Psychological Bulletin*, Vol. **46** (1949), pp. 433–98.

22. Li, J. C. R., *Introduction to Statistical Inference*. Ann Arbor, Michigan: Edwards Brothers, 1957, Chaps. 21 and 22.

23. Mendel, G., "Versuche mit Pflanzenhybriden," *Verhandlung Naturforschung Verlag, Brunn*, Vol. **4** (1865).

24. Mood, A. M., *Introduction to the Theory of Statistics*. New York: McGraw-Hill, Inc., 1950, Chap. 12.

25. Neyman, J., "Smooth Test for Goodness of Fit," *Skandinavisk Aktuarietidskrift*, Vol. **20** (1937), pp. 150–99.

26. ——, "Contribution to the Theory of the χ^2 Test," *Proceedings of the Berkeley Symposium on Mathematical Statistics and Probability*, University of California Press, 1949, pp. 239–73.

27. —— and E. S. Pearson, "On the Use and Interpretation of Certain Test

Criteria for Purposes of Statistical Inference," *Biometrika*, Vol. **20A** (1928), pp. 175–240, 263–94.

28. ———, ———, "Further Notes on the χ^2 Distribution," *Biometrika*, Vol. **22** (1931), pp. 298–305.

29. Pearson, E. S., "The Probability Integral Transformation for Testing Goodness of Fit and Combining Independent Tests of Significance," *Biometrika*, Vol. **30** (1938), pp. 134–48.

30. Pearson, Karl, "On the Criterion that a Given System of Deviations from the Probable in the Case of a Correlated System of Variables Is Such That It Can Be Reasonably Supposed to Have Arisen from Random Sampling," *Philosophical Magazine*, Series 5, Vol. **50** (1900), pp. 157–72.

31. ———, "On the Probability that Two Independent Distributions of Frequency Are Really Samples from the Same Population," *Biometrika*, Vol. **8** (1911), pp. 250–254.

32. Sheppard, W. F., "The Fit of a Formula for Discrepant Observations," *Philosophical Transactions of the Royal Society*, Vol. **228** (1929).

33. Smirnoff, N., "Sur la Distribution de ω^2," *Comptes Rendus de l'Academie des Sciences*, Vol. **202** (1936), pp. 449–52.

34. Snedecor, G. W., *Statistical Methods*, 5th ed. Ames, Iowa: The Iowa State College Press, 1956, Chap. 9.

35. Sukhatme, P. V., "On the Distribution of χ^2 in Small Samples of the Poisson Series," *Journal of the Royal Statistical Society, Supplement*, Vol. **5** (1938), pp. 75–79.

36. Winsor, C. P., "Factorial Analysis of a Multiple Dichotomy," *Human Biology*, Vol. **20** (1948), pp. 195–204.

37. Yates, F., "Contingency Tables Involving Small Numbers and the χ^2 Test," *Journal of the Royal Statistical Society, Supplement*, Vol. **1** (1934), pp. 217–35.

16

DISTRIBUTION-FREE METHODS

Some methods are presented for testing and comparing properties of distributions when their functional forms (or shapes) are unknown. An extension of the sign test, Wilcoxon's signed test, the Mann-Whitney U test, and a median test are described. The problems of testing the randomness assumption are discussed, and one procedure is described.

16.1. INTRODUCTION

In the great majority of the estimation and test procedures described in the other chapters, we assumed the observations to be *randomly* drawn from populations with *known functional form*. Even though the populations were usually assumed to be normal, this is not a crucial requirement so long as the functional form is known. For, according to Chap. 3 and Theorem 3.4, any continuous distribution can be transformed to a normal distribution. (Actually, in practice the true nature of the population is seldom known for sure, but there is much evidence that mild departures from normality do not usually seriously affect the conclusions.)

When the *functional form is unknown*, any statistical statements made as a result of an experiment cannot depend upon the form of the distribution; that is, the statistical statements are *distribution-free*. For example, the sign test of Chap. 8 and the χ^2 tests of Chap. 15 are distribution-free tests. Since distribution-free procedures do not depend upon knowing the forms of populations, one does not usually deal with parameters. That is, distributions are compared without the use of parameters. Hence, distribution-free procedures are also called *nonparametric procedures*, even though neither is wholly appropriate in some cases.

In all procedures the assumption of randomness is basic. But sometimes there is reason to question this assumption. Thus, tests have been developed to determine whether the sample values are nonrandom. Two such tests are described in this chapter.

16.2. EXTENSIONS OF THE SIGN TEST

In Sect. 8.8 we used the sign test to compare two methods of measuring the percentage of starch in potatoes. That is, we tested the hypothesis $\mu_1 = \mu_2$ or $\delta \equiv \mu_1 - \mu_2 = 0$. However, in some situations one might be interested in hypothesizing that one method is a fixed number of units, δ_0, better than the other. In this case, we would require a test of the hypothesis $\mu_1 = \mu_2 + \delta_0$ or $\delta = \delta_0$, where δ_0 is a real number. The procedure is simply to apply the sign test to the signs of the differences

$$x_{11} - (x_{21} + \delta_0), x_{12} - (x_{22} + \delta_0), \ldots, x_{1n} - (x_{2n} + \delta_0) \qquad (16.1)$$

If the number of positive signs, say, is significantly different from $n/2$ at the α level, we reject the null hypothesis that

$$\mu_1 = \mu_2 + \delta_0 \qquad (16.2)$$

Otherwise, we fail to reject this hypothesis. Usually there is an interval of values of δ_0 for which we fail to reject Eq. (16.2). Such an interval is a confidence interval for the difference $\delta = \mu_1 - \mu_2$.

16.2.1. Wilcoxon's Signed Rank Test

In the paired-t test we used both the *order* and *magnitude* properties of the observations; in the sign test we applied only the *order* properties. Further, the sign test does not require the assumption of normality as does the paired-t test. Wilcoxon [52, 53, 54] described a test for paired values which uses both the order and magnitude (of ranks) properties but does not require the normality assumption.

Example 16.1. Illustrate Wilcoxon's signed rank test with the following differences taken from the potato problem of Example 8.6

$$0.2, 0.0, 0.0, 0.1, 0.2, 0.2, 0.3, -0.3, 0.1, 0.2, 0.3, 0.0, -0.1, 0.1, -0.2, 0.1$$

After discarding differences of zero, arrange the remaining differences according to increasing order of the absolute value of the differences, that is

$$0.1, 0.1, +0.1, 0.1, 0.1, 0.2, 0.2, 0.2, 0.2, +0.2, 0.3, +0.3, 0.3 \qquad (16.3)$$

Then assign ranks and attach the sign of the differences. (When the absolute value of two or more differences is the same, assign to each the mean of the ranks they would have had if all were different. Since, in our example, the absolute value of the first five differences is 0.1, we assign the rank

$$\frac{1 + 2 + 3 + 4 + 5}{5} = 3$$

to each, attaching a minus sign to the rank corresponding to -0.1.) It follows that the signed ranks corresponding to the differences in (16.3) are

$$3, 3, -3, 3, 3, 8, 8, 8, 8, -8, 12, -12, 12$$

respectively. Finally, find the sum, T, of the positive or negative ranks, whichever is smaller, and compare with the critical value shown in Table 16.1. If T is smaller than the critical value, reject the null hypothesis $\delta = 0$ and accept the alternative hypothesis, $\delta \neq 0$, that the methods are different;

Table 16.1

Critical Values of T for Wilcoxon's Signed Rank Two-Sided Test* (Absolute values of T less than the tabulated values occur with indicated probability. In a one-sided test the probabilities are 0.025, 0.01, and 0.005, respectively.)

Pairs	Probability			Pairs	Probability		
n	.05	.02	.01	n	.05	.02	.01
6	0	—	—	16	30	24	20
7	2	0	—	17	35	28	23
8	4	2	0	18	40	33	28
9	6	3	2	19	46	38	32
10	8	5	3	20	52	43	38
11	11	7	5	21	59	49	43
12	14	10	7	22	66	56	49
13	17	13	10	23	73	62	55
14	21	16	13	24	81	69	61
15	25	20	16	25	89	77	68

* This table is reproduced from F. Wilcoxon, *Some Rapid Approximate Statistical Procedures*, American Cyanamid Company, Stamford, Connecticut, 1949, Table I, p. 13, with the permission of the American Cyanamid Company.

otherwise, fail to reject the null hypothesis. Since the sum of the absolute value of the negative ranks is

$$T = 3 + 8 + 12 = 23$$

and the five per cent critical value of T when $n = 13$ is 17, we fail to reject the null hypothesis. Note that the same conclusion was made when the sign and paired-t tests were used.

Application of the paired t, Wilcoxon's signed rank, and the sign tests to the same experimental data will not always lead to the same conclusion.

In the case where the assumptions of the paired-t test hold, all three tests are valid. If an α-level two-sided (or one-sided) procedure is applied, the paired-t test is the most powerful and the sign test the least powerful of the three. In fact, it can be shown that the power efficiency of the Wilcoxon's signed rank test relative to the paired-t test is near 0.95 for small samples and near $3/\pi = 0.955$ for large samples when the true values of δ are near zero.

The signed rank test may also be applied to test the hypothesis that δ is some specified value δ_0. (In fact, in a single sample, the same test procedure may be used to test the hypothesis that the median of a group of observations is equal to some specified value, say, μ_0.) The general procedure for Wilcoxon's signed rank test is as follows

1. Subtract δ_0 (or μ_0) from each difference $d = x_1 - x_2$ (or value x)
2. Rank the resulting adjusted differences (values) in order of size, ignoring sign. In case of ties in adjusted differences (or values), assign the mean rank to each tied adjusted difference (value)
3. Attach the sign of the adjusted difference (or value) to the corresponding rank
4. Find the sum, T, of the positive or negative ranks, whichever is smaller in absolute value
5. Compare T with the critical value T_c in Table 16.1. If $|T| < T_c$, reject the null hypothesis that the mean difference (value) is δ_0 (μ_0); otherwise, fail to reject the null hypothesis

Table 16.1 is not adequate for more than 25 pairs. In such a case it can be shown that T is approximately normally distributed with mean

$$\mu_T = \frac{n(n + 1)}{4} \tag{16.4}$$

and variance

$$\sigma_T^2 = \frac{n(n + 1)(2n + 1)}{24} \tag{16.5}$$

Therefore

$$u' = \frac{T - \mu_T}{\sigma_T}$$

is approximately normally distributed with zero mean and unit variance. Thus, in experiments with more than 25 pairs of observations, the normal distribution may be used as an approximate test of the significance of the difference in means.

16.2.2. The Walsh Test

Walsh [45, 46] describes another very powerful nonparametric test which is based on ranking the differences $d_i = x_{1i} - x_{2i}$ ($i = 1, \ldots, n$). In order for the test to be valid, it is assumed that the n differences are independently drawn from n symmetrical populations, each having the same mean (median) μ.

16.3. TESTS FOR TWO INDEPENDENT SAMPLES

Comparisons of similar characteristics of two or more independent samples are given in earlier sections. In Sect. 8.6 the two-sample t test was used to test the null hypothesis that two population means, μ_1 and μ_2, are equal. For this purpose it was assumed that random samples of sizes n_1 and n_2 are independently drawn from two normal populations with equal variances. In Chaps. 10 and 11 we described extensions of this problem to k normal populations with equal variances. Sometimes one or more of the assumptions for the two-sample t test or k-sample F test are not valid. Thus, in Chap. 15 we described methods for comparing two or more populations with unknown distributions. For example, Fisher's exact probability test (Sect. 15.4) and the χ^2 test of homogeneity of sequences (Sect. 15.5) were discussed.

In this section we describe three more useful distribution-free tests which require that the sample values be *ranked* (ordered). In these tests it is assumed that random and independent samples are drawn from two continuous distributions which have the same form but possibly different values of the location parameter (e.g., mean or median). Thus, under the usual null hypothesis, the random and independent samples

$$x_1, \ldots, x_{n_1} \quad \text{and} \quad y_1, \ldots, y_{n_2} \tag{16.6}$$

are assumed to come from a single population. The alternative hypothesis, expressed in terms of a difference in location parameters, may be either *one-sided* or *two-sided*.

16.3.1. The Rank-Sum Test

Under the null hypothesis that there is a single population, the two samples in (16.6) may be combined to give a single random sample of size $n = n_1 + n_2$. Arrange these n observations in order of increasing size, assigning score 1 to the smallest value, score 2 to the second smallest, $\ldots$, n to the largest, preserving the identity of the samples. There is no loss in generality if we assume $n_1 \le n_2$. Let T denote the total of the ranks of the sample with n_1 observations. Then the smallest value of T is

$$1 + \cdots + n_1 = \frac{n_1(n_1 + 1)}{2} \tag{16.7}$$

and the largest is

$$n + (n - 1) + \cdots + (n - n_1 + 1) = \frac{n_1(n_1 + 2n_2 + 1)}{2} \quad (16.8)$$

The sampling distribution of T is used to test the significance of the difference in location parameters. Either one-sided or two-sided tests may be described. Extremely large or extremely small values indicate that the location parameters of two populations are different. In order better to understand the nature of the sampling distribution of T and the use of Table 16.3, consider the following example.

Example 16.2. Find the sampling distribution of T for samples of sizes four and five.

In the example $n_1 = 4$, $n_2 = 5$, and $n = 9$. The ranks of the combined sample of nine observations are

$$1, 2, 3, 4, 5, 6, 7, 8, 9 \quad (16.9)$$

Since the extreme values of the sum of four ranks, T, are

$$\frac{4 \cdot 5}{2} = 10 \quad \text{and} \quad \frac{4 \cdot 15}{2} = 30$$

Table 16.2
Sampling Distribution of Rank Totals T of Samples of Size Four in Combination with Samples of Size Five

Rank Total (T)	Frequency f	Relative Frequency	Cumulative Rel. Freq.
10	1	0.008	0.008
11	1	0.008	0.016
12	2	0.016	0.032
13	3	0.024	0.056
14	5	0.040	0.096
15	6	0.048	0.144
16	8	0.063	0.207
17	9	0.071	0.278
18	11	0.087	0.365
19	11	0.087	0.452
20	12	0.095	0.548
21	11	0.087	0.635
22	11	0.087	0.722
23	9	0.071	0.793
24	8	0.063	0.856
25	6	0.048	0.904
26	5	0.040	0.944
27	3	0.024	0.968
28	2	0.016	0.984
29	1	0.008	0.992
30	1	0.008	1.000

the possible values of T are 10, 11, ... , 30. We require the number of ways in which a specified T can be obtained.

There are $\binom{n}{n_1} = \binom{9}{4} = 126$ different ways to select four scores from among the nine in (16.9); that is, the total number of ways (frequency) in which T is found is 126. The frequencies of specific rank sums shown in Table 16.2 may be obtained by an exhaustive listing of sums of four ranks. For example, the frequencies for $T = 10, 11, 12, 13$ are 1, 1, 2, 3, respectively, since $1 + 2 + 3 + 4 = 10$; $1 + 2 + 3 + 5 = 11$; $1 + 2 + 3 + 6 = 12$; $1 + 2 + 4 + 5 = 12$; $1 + 2 + 3 + 7 = 13$; $1 + 2 + 4 + 6 = 13$; $1 + 3 + 4 + 5 = 13$. This method is satisfactory for small samples, but as the samples become larger a better counting technique becomes more desirable. Such a procedure is now described.

If we let $f(T; n_1, n_2)$ denote the number of ways of obtaining the specific rank total T when n_1 and n_2 are the number of observations in two groups, it can be shown that

$$f(T; n_1, n_2) = f(T; n_1, n_2 - 1) + f(T - n; n_1 - 1, n_2) \qquad (16.10)$$

For the highest rank $n = n_1 + n_2$ is either used in finding the rank total T, or it is not. When n is not used, the frequency of the rank sum T is given by $f(T; n_1, n_2 - 1)$. When n is used, the frequency of the rank sum T is given by $f(T - n; n_1 - 1, n_2)$. Formula (16.10) is particularly useful in computing a sequence of frequency tables of T. That is, after tables have been computed for small n_1 and n by an exhaustive listing, (16.10) may be applied to obtain frequency tables for larger n_1 and n.

Table 16.2 is typical of sampling distributions of T in at least two respects. The relative frequencies may be considered to be probabilities, since each of the $\binom{n}{n_1}$ ways in which T is computed is considered to be equally likely. Every distribution of T is symmetric. Thus, the mean of the sampling distribution of T is given by

$$\mu_T = \frac{\left[\dfrac{n_1(n_1 + 1)}{2} + \dfrac{n_1(n_1 + 2n_2 + 1)}{2}\right]}{2}$$

or

$$\mu_T = \frac{n_1(n_1 + n_2 + 1)}{2} \qquad (16.11)$$

and the probability of obtaining a particular value of T is the same as the probability of obtaining

$$T' = n_1(n_1 + n_2 + 1) - T \qquad (16.12)$$

We call T' the *conjugate* of T and illustrate how it is used in the following example.

Example 16.3. One might expect that the speed with which turtles swim across a tank of water to a platform on the opposite side depends on the temperature of the water. In an experiment (hypothetical) four turtles took 85, 210, 432, 183 sec. in 33°C water and five turtles took 72, 89, 13, 56, 145 sec. in 42°C water. Use the T statistic to test the hypothesis that there is no difference in swimming time at 33°C and 42°C temperatures against the alternative hypothesis that turtles swim faster in 42°C water.

First, rank the nine observations according to increasing number of seconds and find that the smaller sample has ranks 4, 7, 8, 9. Thus $T = 28$. According to Table 16.2, the probability of the ranks' being this large or larger is $0.016 + 0.008 + 0.008 = 0.032$. Hence, at the five per cent level, we reject the null hypothesis and conclude that turtles swim faster in 42°C water than in 33°C water. We might conclude from this that the cold-blooded turtle finds hot water noxious.

Note that $T' = 4(10) - 28 = 12$ and that the probability that T' is equal to or less than 12 is 0.032. Further, note that T' is the value obtained by summing the ranks of the observations arranged so that score 1 is assigned to the largest value, score 2 to the second largest, etc. That is, $T' = 1 + 2 + 3 + 6 = 12$, since the ranks of the four values 432, 210, 183, 85 are 1, 2, 3, 6, respectively. Thus, the frequencies of only those values of T from the smallest to the mean of the T distribution need be computed and recorded. In our example this includes $T = 10, 11, \ldots, 20$.

Due to the symmetry property the lower part of the T distribution may be used in either one-sided or two-sided test procedures. In order to apply a two-sided test, we require a critical point, T_0, for which $\alpha/2$ of the rank sums in the appropriate sampling distribution lie below. The samples are declared significantly different at the α level if either T or T' lies below this critical point T_0. In a one-sided test procedure T_0 is that rank sum below which α of the rank sums lie. In the one-sided test, use either the statistic T or the statistic T', whichever is appropriate.

From Table 16.2 we find that 5.6 per cent of the T values are equal to or less than 13, and 3.2 per cent are equal to or less than 12. Since there is no value of T below which *exactly* five per cent of the rank sums lie, we take as the critical point T_0 that value below which *five per cent or less* of the rank sums lie; that is, in this case $T_0 = 12$. The five per cent critical value defined in this way is not usually true to its name. However, the discrepancy is not usually considered serious. The value $T_0 = 12$ may be used as a five per cent point in a one-sided test or a ten per cent point in a two-sided test. For a one-sided 2.5 per cent test or a two-sided five per cent test $T_0 = 11$, the true significance levels being 1.6 per cent and 3.2 per cent, respectively.

Table 16.3 has been constructed in this manner for all meaningful values of n_1 and n_2 for which $n_1 + n_2 \leq 30$ and $n_1 \leq n_2$.

In all the above discussion of the rank-sum test we have ignored the case of tied values of the observations. In case of ties, the usual procedure, although there are good alternatives, is to find the mean of ranks that the tied observations would have if they were distinguishable, and then to assign each this mean rank. When the tied observations fall in one sample, the test is not affected; when observations in different samples are tied, the test is not seriously affected.

It is not difficult to show that the variance of the sampling distribution of T is

Table 16.3
Critical Values for Rank Sums*
For a two-sided test they are 5 per cent (or less) values.
For a one-sided test they are 2.5 per cent (or less) values.

n_2 \ n_1	2	3	4	5	6	7	8	9	10	11	12	13	14	15
4			10											
5		6	11	17										
6		7	12	18	26									
7		7	13	20	27	36								
8	3	8	14	21	29	38	49							
9	3	8	15	22	31	40	51	63						
10	3	9	15	23	32	42	53	65	78					
11	4	9	16	24	34	44	55	68	81	96				
12	4	10	17	26	35	46	58	71	85	99	115			
13	4	10	18	27	37	48	60	73	88	103	119	137		
14	4	11	19	28	38	50	63	76	91	106	123	141	160	
15	4	11	20	29	40	52	65	79	94	110	127	145	164	185
16	4	12	21	31	42	54	67	82	97	114	131	150	169	
17	5	12	21	32	43	56	70	84	100	117	135	154		
18	5	13	22	33	45	58	72	87	103	121	139			
19	5	13	23	34	46	60	74	90	107	124				
20	5	14	24	35	48	62	77	93	110					
21	6	14	25	37	50	64	79	95						
22	6	15	26	38	51	66	82							
23	6	15	27	39	53	68								
24	6	16	28	40	55									
25	6	16	28	42										
26	7	17	29											
27	7	17												
28	7													

* This table is reproduced from C. White, "The Use of Ranks in a Test of Significance for Comparing Two Treatments," *Biometrics*, Vol. **8** (1950), Tables for .05 and .01, Critical Points of Rank Sums, pp. 37, 38, with permission of the editor of the journal.

Table 16.3

Critical Values for Rank Sums (*cont.*)

For a two-sided test they are 1 per cent (or less) values.

For a one-sided test they are 0.5 per cent (or less) values.

n_2 \ n_1	2	3	4	5	6	7	8	9	10	11	12	13	14	15
5				15										
6			10	16	23									
7			10	17	24	32								
8			11	17	25	34	43							
9		6	11	18	26	35	45	56						
10		6	12	19	27	37	47	58	71					
11		6	12	20	28	38	49	61	74	87				
12		7	13	21	30	40	51	63	76	90	106			
13		7	14	22	31	41	53	65	79	93	109	125		
14		7	14	22	32	43	54	67	81	96	112	129	147	
15		8	15	23	33	44	56	70	84	99	115	133	151	171
16		8	15	24	34	46	58	72	86	102	119	137	155	
17		8	16	25	36	47	60	74	89	105	122	140		
18		8	16	26	37	49	62	76	92	108	125			
19	3	9	17	27	38	50	64	78	94	111				
20	3	9	18	28	39	52	66	81	97					
21	3	9	18	29	40	53	68	83						
22	3	10	19	29	42	55	70							
23	3	10	19	30	43	57								
24	3	10	20	31	44									
25	3	11	20	32										
26	3	11	21											
27	4	11												
28	4													

$$\sigma_T^2 = \frac{n_1 n_2 (n_1 + n_2 + 1)}{12} \tag{16.13}$$

and that the distribution of T approaches the normal as n_1 and n_2 increase. Therefore, approximate critical points, T_0'', may be found from

$$u = \frac{T - \dfrac{n_1(n_1 + n_2 + 1)}{2}}{\sqrt{\dfrac{n_1 n_2 (n_1 + n_2 + 1)}{12}}} \tag{16.14}$$

with the use of the standard normal. For values of n_1 and n_2 outside the range of Table 16.3 the approximation is very good when $\alpha = 5$ per cent, and, also, very good when $\alpha = 1$ per cent unless n_1 is small. In fact, when $n_1 + n_2 = 30$ and $\alpha = 5$ per cent, the approximate values T_0'' are the same as T_0 of Table 16.3 in 11 of 14 cases, and only one unit higher in the remaining three.

If all the assumptions of the two-sample t test hold, the rank-sum test is valid. It can be shown [13] that the power efficiency of the rank-sum test relative to the two-sample t test is near 0.95 when the true difference in location parameter is near zero. Thus, in order to provide the same power, approximately five per cent more observations are required for the rank-sum test than for the two-sample t test. However, for nonnormal populations the rank-sum test may be more powerful [13, 41] than the two-sample t test (in which case it should be noted that the two-sample t test is not valid and the rank-sum test is).

There are possible advantages as well as disadvantages to using the rank-sum test and other ranking procedures. Some of the advantages are

1. The calculations are simpler. The sample statistic is very easy to compute. Most of the work is in ranking the observations, and this can be made easier with shortcut techniques

2. The assumptions are few and not very restrictive. The form of the distributions need not be known. Knowledge of the population mean and variance are not required

3. The data may be in ordinal form only

4. Real differences in location parameters may be more easily detected. This is particularly true when the populations deviate considerably from normality or the location parameters are not means

Some of the disadvantages are

1. There may be a loss in efficiency. Some information may be sacrificed when the quantitative nature of the data is replaced by ranks

2. It is not possible to establish confidence limits for the difference in location parameters of two populations. The hypothesis of equality of location parameters may be tested against either a one-sided or a two-sided alternative, but an interval estimate for this difference is not possible

The rank-sum statistic was described for equal sample sizes by Wilcoxon [52, 54]. The relation of the rank-sum test to other ranking procedures is discussed in Refs. [41, 44]. Mann and Whitney [29] proposed an alternative test procedure for unequal sample sizes. This test is described in the next section.

16.3.2. The U Test

The test procedure described by Mann and Whitney [29] requires that the observations in the two independent samples of (16.6) be arranged in ascending order with the identity of the samples preserved. *They define a statistic, U, to be equal to the number of times a y precedes an x.*

As an illustration, suppose the nine observations of Example 16.3 are arranged in increasing order with "times" associated with 33°C water designated by y and those associated with 42°C water designated by x. Thus, the order and identity of the sample values are as follows

Order of values	13	56	72	85	89	145	183	210	432
Identity of sample	x	x	x	y	x	x	y	y	y

Since the y score 85 precedes each of the x scores 89 and 145, and since the y scores 183, 210, and 432 do not precede any x scores, the sample value of the U statistic is

$$U = 2 + 0 + 0 + 0 = 2$$

The test, in its original form, was designed to test the null hypothesis

H_0: the x and y values have the same distribution   (16.15)
against the alternative hypothesis

H_a: the location parameter of y is larger than the location parameter of x; that is, the bulk of the   (16.16) distribution of y's is to the right of the bulk of the distribution of x's.

If H_a is true, we expect U to be small. Mann and Whitney [29] computed tables which give probabilities associated with small (lower tail) values of U, and Auble [3] gives tables of critical values of U for significant levels of 0.001, 0.01, 0.025, and 0.05 for a one-sided test. For the one-sided alternative hypothesis that the location parameter of y is smaller than the parameter of x, we compute the statistic U', defined to be the number of times an x precedes a y, and use Auble's tables to test H_0. For a two-sided alternative hypothesis we compute U or U', whichever is smaller, and use Auble's tables, understanding that the significance levels are now 0.002, 0.02, 0.05, and 0.10.

We do not give tables of the U statistic, since it can be shown that this statistic and the T statistic of Sect. 16.3.1 give identical results. In fact, it is not difficult to show that

$$U = n_1 n_2 + \frac{n_1(n_1 + 1)}{2} - T_1 \qquad (16.17)$$

where T_1 denotes the sum of the ranks assigned to the samples of size n_1, or that

$$U = n_1 n_2 + \frac{n_2(n_2 + 1)}{2} - T_2 \qquad (16.17a)$$

where T_2 denotes the sum of the ranks assigned to the sample of size n_2. (It is to be understood that the ranks of a sample are those obtained by arranging the $n_1 + n_2$ observations of the two independent samples according to ascending order.) The U statistic is usually computed by Eq. (16.17) or Eq. (16.17a), since it is tedious to compute by using the definition of U when n_1 and n_2 become fairly large.

Under the null hypothesis (16.15) it can be shown that the mean and variance of the sampling distribution of U are given by

$$\mu_U = \frac{n_1 n_2}{2} \tag{16.18}$$

and

$$\sigma_U^2 = \frac{n_1 n_2 (n_1 + n_2 + 1)}{12} \tag{16.19}$$

In fact, when both n_1 and n_2 are larger than eight, the statistic U is approximately normally distributed, and the larger the sample sizes, the better the approximation.

Since the U statistic is equivalent to the T statistic of Sect. 16.3.1, the statements about the relative efficiency of rank-sum tests apply to the Mann-Whitney U test. Also, the rank-sum and Mann-Whitney test procedures have the same advantages and disadvantages. For a discussion of tied scores and other topics the reader is referred to Refs. [16, 48, 49].

16.3.3. The Median Test

Let the combined $n_1 + n_2$ observations in (16.6) be arranged according to increasing order of magnitude with the identity of the samples preserved. Let r_l (r_u) denote the ratio of the number of y's to the number of x's to the left (right) of the median of the combined samples. Now, if the null hypothesis (16.15) is true, r_u should not differ greatly from r_l. However, if the median of the distribution of y's is to the right of the median of the distribution of the x's, then r_u should be larger than r_l. This suggests that a statistic defined in terms of the number of x's (y's) to the right of the combined sample median would be useful in testing the null hypothesis against either the one-sided or the two-sided alternative hypothesis that the population medians are different.

Let m_1 (m_2) denote the number of x (y) values larger than the combined sample median. Observe that $m_1 + m_2$ is $(n_1 + n_2)/2$ when $n_1 + n_2$ is even and is $(n_1 + n_2 - 1)/2$ when $n_1 + n_2$ is odd. If the $n_1 + n_2$ values are assumed to be distinct, it can be shown that the probability, $P[m_1]$, of exactly m_1 x's (and, of course, exactly m_2 y's) exceeding the combined sample median is given by

$$P[m_1] = \frac{\binom{n_1}{m_1} \cdot \binom{n_2}{m_2}}{\binom{n_1 + n_2}{m_1 + m_2}} \tag{16.20}$$

The expression (16.20) is the hypergeometric density function. Thus, to test the null hypothesis H_0 against the alternative hypothesis that the median of the distribution of y values is to the right of the median of the distribution of x values, we need to know the probability that m_1 is equal to or less than the sample value, m_{10}; that is, we require

$$P[m_1 \leq m_{10}] = \sum_{m_1=0}^{m_{10}} \frac{\binom{n_1}{m_1} \cdot \binom{n_2}{m_2}}{\binom{n_1 + n_2}{m_1 + m_2}} \tag{16.21}$$

Lieberman and Owen [28] have computed tables of the hypergeometric distribution which may be used to find Eq. (16.21) so as to test H_0 against the alternative indicated. These tables may also be used for a two-sided alternative.

It is informative to note that the two independent samples may be dichotomized and arranged as shown in Table 16.4. Thus, when

Table 16.4

Grouped Data for Median Test

Number of Sample Observations	Sample of x values	Sample of y values	Total
Greater than the combined median	m_1	m_2	$m_1 + m_2$
Equal to or less than the combined median	$n_1 - m_1$	$n_2 - m_2$	$n_1 + n_2 - (m_1 + m_2)$
Total	n_1	n_2	$n_1 + n_2$

$n_1 + n_2 < 20$, Fisher's exact test may be used to test H_0; when $n_1 + n_2 \geq 20$, the χ^2 statistic corrected for continuity may be applied. Also, when n_1 and n_2 are quite large, Theorem 6.7 and the methods of Sect. 6.2 may be applied to give an approximate test of H_0. In this case the statistic

$$\frac{\dfrac{m_1}{n_1} - \dfrac{m_2}{n_2}}{\sqrt{\dfrac{m_1 + m_2}{n_1 + n_2}\left(\dfrac{1}{n_1} + \dfrac{1}{n_2}\right)}} \tag{16.22}$$

approximates the standard normal statistic u.

The median test is not as powerful as the rank-sum, Mann-Whitney or student tests. However, it has the advantage that it is easily generalized to k samples where the χ^2 test is still appropriate. Also, the median test may be extended to apply to any percentile of the grouped data instead of the fiftieth percentile (median). For a discussion of these statements, a proof of Eq. (16.20), and information on other related topics the reader is referred to Refs. [10, 14, 24, 32, 35, 38, 55].

16.4. DISTRIBUTION-FREE METHODS IN ANALYSIS OF VARIANCE

In Sect. 16.2 and 16.3 we described a few procedures which are useful in comparing two populations. These test procedures have been generalized so that several distribution-free methods are available for comparing more than two populations. Kruskel and Wallis [27] described a test based on ranks which may be used for the completely randomized design. Distribution-free methods for the randomized complete-block design have been described by Friedman [19], Mood [31], and Kendall and Smith [25]. Durbin [15] has generalized the method of Kendall and Smith to balanced incomplete-block designs, and Bradley and Terry [6], Bradley [7, 8, 9] and Abelson and Bradley [1] have described methods of paired comparisons which are useful in incomplete-block designs. Descriptions of these and other methods may be found in Refs. [11, 17, 24, 31, 36, 38, 39].

16.5. RANDOMNESS

Statistical inference is based in some way on the assumption that sample values are randomly drawn from one or more populations. So long as the sample values are obtained with the aid of a random number table or recognized probability methods, the randomness assumption is not likely to be questioned. However, in situations where the experimenter (or investigator) has little or no control over the selection of the data values, he might well question the randomness assumption. For example, a long-range prediction of death in highway accidents must be based on whatever records are available, so one may wish to test to see if the observations can be considered to be random. Also, one may wish to test the randomness assumption in an effort to detect the presence (or absence) of assignable causes in an investigation of statistical control.

The effect of time on an experiment has not been taken into account in most of the techniques already described. Still, knowing the particular time, that is, the *order in time*, in which observations are made might give valuable information on the randomness of the sample values. In fact, we describe a test procedure based upon *order in time* which is useful in testing the usual randomness assumption. (A careful study of the effect of "time" in an experiment might well require a second volume. Such topics as quality control and sampling inspection, time series analysis, sequential test procedures, and stochastic processes fall in this category. There are many articles (see the bibliography in Ref. [4]) and books [4, 12, 18, 21, 26, 43, 50, 51] written on these topics.)

It is difficult to test for randomness. However, tests have been described which detect nonrandomness. This means that it is possible to conclude that a given sequence of observations is *not random* at a specified significance level, but that it is not possible to conclude that the sequence *is random*.

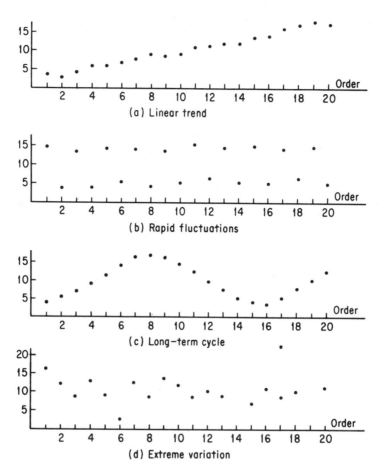

Fig. 16.1 Dot Diagrams Indicating Order in which
Observations were Drawn

Before describing a test, we consider some illustrations of nonrandomness shown in Fig. 16.1. The horizontal axis designates the order in which the observations are drawn, and the vertical axis designates the magnitude of the values. Figure 16.1a shows a linear trend in time, and Figs. 16.1b and 16.1c illustrate cyclic patterns. There are other patterns of possible nonrandomness, but these are adequate for our purposes. They certainly indicate that taking into account the order of occurrence of the observations might be useful in constructing tests of nonrandomness.

The reader should realize that the pattern sometimes depends on the nature of the population sampled. For example, the pattern in Fig. 16.1d might have resulted from randomly selecting the observations from a population with a long tail to the right.

Some common tests of nonrandomness in a sequence of observations are based on *number of runs, length of runs, runs up and down* [30, 34], and *control charts*. Other tests are based on such things as *serial correlation* [2, 42] and *mean square of successive differences* [23, 33].

16.5.1. Use of Runs in Detecting Nonrandomness

A sequence of i identical symbols which is preceded and followed by a different symbol or no symbol is called a *run of length i*. Thus, in the sequence

$$a\ a\ b\ a\ a\ a\ b\ b\ b\ b\ b\ a\ b\ b\ a \qquad (16.23)$$

consisting of seven a's and eight b's, there are four runs of a's and three runs of b's. The lengths of the runs of a's are 2, 3, 1, and 1, respectively; the lengths of the runs of b's and 1, 5, and 2, respectively. The same 15 a's and b's can be arranged in other ways. For example, the sequence

$$b\ a\ b\ a\ b\ a\ b\ a\ b\ a\ b\ a\ b\ a\ b \qquad (16.24)$$

has seven runs of a's and eight runs of b's, each of length 1, and the sequence

$$a\ a\ a\ a\ a\ a\ a\ b\ b\ b\ b\ b\ b\ b\ b \qquad (16.25)$$

has one run of a's of length 7 and one run of b's of length 8. Thus, we observe that the seven a's and eight b's can be arranged so that the total number of runs ranges from two to 15.

In the particular case where a represents "heads" and b represents "tails" on a fair toss of a well-balanced coin, we might expect sequence (16.23) to occur, but we probably would not expect sequences (16.24) and (16.25) to occur, since they indicate short- and long-term cyclic patterns, respectively. That is, sequences (16.24) and (16.25) indicate nonrandomness, and sequence (16.23) could well indicate randomness. Thus, a sequence with too many or too few runs could indicate the absence of randomness. That is, the total number of runs, r, could be used to test for randomness.

If there are many runs, each one can be expected to be short; if there are few runs, at least one can be expected to be long. Thus, the *length* of the longest run could be used to *test randomness*.

Sequences such as (16.23), (16.24), and (16.25) may result from sampling a dichotomous population, but such sequences can also be formed from continuous populations by letting the observed values above or below a given value x_0 be indicated by a or b, respectively. For example, in a routine production, b could denote nondefective and a defective parts. Further, in a sample of college students b could denote an I.Q. score below 110 and a an I.Q. score above 110. The value x_0 may be any percentile (fractile) value, but the median is frequently used.

Now, consider the sampling distribution of the total number of runs r, that is, the sampling distribution of

$$r = r_1 + r_2$$

where r_1 and r_2 denote the number of runs of a's and b's, respectively. Let n_1 and n_2 denote the number of a's and b's, respectively. Thus, the total number of observations n is given by

$$n = n_1 + n_2$$

provided x_0 is not an observation. (In case one or more of the observations have the value x_0, the usual practice is to omit them from the sample, and thus from the resulting computations and conclusions. In some cases x_0 may be selected before the data is collected, so that none of the observed values will be x_0.) Under the assumption that every arrangement of n_1 a's and n_2 b's is equally likely, it can be shown [5, 11, 22, 24, 30] by first finding the joint density function of r_1 and r_2 that the density function of r is given by

$$f(r) = \begin{cases} \dfrac{2\dbinom{n_1 - 1}{\frac{r}{2} - 1}\dbinom{n_2 - 1}{\frac{r}{2} - 1}}{\dbinom{n}{n_1}} & \text{when } r \text{ is even} \\[3em] \dfrac{\dbinom{n_1 - 1}{\frac{r-1}{2}}\dbinom{n_2 - 1}{\frac{r-3}{2}} + \dbinom{n_1 - 1}{\frac{r-3}{2}}\dbinom{n_2 - 1}{\frac{r-1}{1}}}{\dbinom{n}{n_1}} & \text{when } r \text{ is odd} \end{cases} \tag{16.26}$$

Further, it is not difficult to show that the mean and variance of the sampling distribution are given by

$$\mu_r = \frac{2n_1 n_2}{n} + 1 \tag{16.27}$$

and

$$\sigma_r^2 = \frac{2n_1 n_2 (2n_1 n_2 - n)}{n^2 (n - 1)} \tag{16.28}$$

The density function (16.26) is useful in computing lower, r', and upper, r'', critical values of r. In fact, Swed and Eisenhart [40] give, for $n_1 \leq n_2 \leq 20$, the critical values such that

$$\sum_{r=2}^{r'} f(r) = P[r \leq r'] \leq \alpha \qquad \text{for } \alpha = 0.005, 0.01, 0.025 \text{ and } 0.05$$

Table 16.5
Lower 0.025 Critical Values for Number of Runs*
(The probability is 0.025 or less that r is equal to or less than the value tabled.)

n_1 \ n_2	5	6	7	8	9	10	11	12	13	14	15	16	17	18	19	20
2								2	2	2	2	2	2	2	2	2
3		2	2	2	2	2	2	2	2	2	3	3	3	3	3	3
4	2	2	2	3	3	3	3	3	3	3	3	4	4	4	4	4
5	2	3	3	3	3	3	4	4	4	4	4	4	5	5	5	5
6		3	3	3	4	4	4	4	5	5	5	5	5	6	6	6
7			3	4	4	5	5	5	5	5	6	6	6	6	6	6
8				4	5	5	5	6	6	6	6	6	7	7	7	7
9					5	5	6	6	6	7	7	7	7	8	8	8
10						6	6	7	7	7	7	8	8	8	8	9
11							7	7	7	8	8	8	9	9	9	9
12								7	8	8	8	9	9	9	10	10
13									8	9	9	9	10	10	10	10
14										9	9	10	10	10	11	11
15											10	10	11	11	11	12
16												11	11	11	12	12
17													11	12	12	13
18														12	13	13
19															13	13
20																14

* This table is abridged from F. S. Swed and C. Eisenhart, "Tables for Testing Randomness of Grouping in a Sequence of Alternative," *Annals of Mathematical Statistics*, Vol. **14** (1943), Table II, with permission of the editor of the journal.

Upper 0.025 Critical Values for Number of Runs*
(The probability is 0.025 or less that r is equal to or greater than the value tabled.)

n_1 \ n_2	5	6	7	8	9	10	11	12	13	14	15	16	17	18	19	20
4	9	9														
5	10	10	11	11												
6		11	12	12	13	13	13	13								
7			13	13	14	14	14	14	15	15	15					
8				14	14	15	15	16	16	16	16	17	17	17	17	17
9					15	16	16	16	17	17	18	18	18	18	18	18
10						16	17	17	18	18	18	19	19	19	20	20
11							17	18	19	19	19	20	20	20	21	21
12								19	19	20	20	21	21	21	22	22
13									20	20	21	21	22	22	23	23
14										21	22	22	23	23	23	24
15											22	23	23	24	24	25
16												23	24	25	25	25
17													25	25	26	26
18														26	26	27
19															27	27
20																28

* This table is abridged from F. S. Swed and C. Eisenhart, "Tables for Testing Randomness of Grouping in a Sequence of Alternatives," *Annals of Mathematical Statistics*, Vol. **14** (1943), Table III, with permission of the editor of the journal.

and

$$\sum_{r=r''}^{n_1+n_2} f(r) = P[r \geq r''] \geq \alpha \qquad \text{for } \alpha = 0.95, 0.975, 0.99 \text{ and } 0.995$$

Parts of their Tables II and III which can be used in a 2.5 per cent one-sided test or a five per cent two-sided test are reproduced in Table 16.5. For an illustration of a two-sided test, consider the following example.

Example 16.4. On inspecting a row of 22 plants for wilt, the sequence of hardy, H, and droopy, D, plants was as follows

$$H\ H\ H\ H\ D\ H\ D\ D\ H\ H\ H\ H\ H\ H\ D\ D\ D\ H\ H\ H\ H\ H$$

Use a two-sided five per cent level test for the null hypothesis

$$H_0: \text{ the } H\text{'s and } D\text{'s occur in random order}$$

against the alternative hypothesis

$$H_a: \text{ the } H\text{'s and } D\text{'s occur in nonrandom order}$$

There are $n_1 = 6$ D's, $n_2 = 16$ H's, and $r = 7$ runs. (Note that n_1 refers to the number of plants in the smaller category.) According to Table 16.5, the lower 2.5 per cent critical value of r is $r'_{.025} = 5$. Thus, since $7 > 5$, we fail to reject H_0, and conclude that the sample of 22 hardy and droopy plants may be arranged in random order.

If n_1 and n_2 are large, then r is approximately normally distributed with mean μ_r and variance σ_r^2, as given by Eqs. (16.27) and (16.28). That is

$$\frac{r - \mu_r}{\sigma_r}$$

is approximately normally distributed with mean zero and variance one.

If both n_1 and n_2 exceed 20, the normal approximation is quite good. When $n_1 < 5$ and $n_2 > 20$, the probabilities for an extreme number of runs should be computed directly by Eq. (16.26). In other cases where Table 16.5 cannot be applied, the normal approximation is probably satisfactory.

16.5.2. Use of Control Charts in Detecting Nonrandomness

The control chart, in addition to its use in detecting and eliminating assignable causes of variation, can also be used to determine whether the sequence of observations can be considered a controlled or random sequence. Even though the control chart may detect any type of nonrandomness, it is particularly useful in detecting extreme variation of individual measurements. For details involving the use of control charts, the reader is referred to Refs. [5, 20, 22, 37].

16.6. EXERCISES

16.1. The claim was made that stimulus S increases blood pressure (systolic) by five units. Table 16.6 gives the blood pressure of 16 subjects immediately before and m minutes after the application of S. The null hypo-

Table 16.6

Order Number of Subject	Before Stimulus	After Stimulus	Order Number of Subject	Before Stimulus	After Stimulus
1	113	116	9	106	115
2	111	116	10	109	115
3	124	128	11	109	116
4	120	128	12	120	123
5	115	122	13	105	104
6	122	121	14	129	130
7	102	105	15	103	110
8	122	125	16	115	124

thesis is that the stimulus increases blood pressure five units; the alternative hypothesis is that blood pressure is not increased five units. Test the null hypothesis at the five per cent level by applying (a) the sign test, (b) Wilcoxon's sign rank procedure, and (c) the paired-t procedure. (d) Compare the conclusions of the three tests.

16.2. Use the data of Exercise 8.13 to test the hypothesis that the mean grade for method 1 is at least three units higher than the mean grade for method 2. Apply the sign, the paired-t, and Wilcoxon's sign rank test procedures and compare results.

16.3. Use the data of Exercise 8.14 and Wilcoxon's sign rank procedure to test the hypothesis in Exercise 8.14(a).

16.4. Use the data of Exercise 8.10 to test the hypothesis of no difference in tensile strength against the alternative hypothesis of unequal tensile strength. Apply the (a) rank-sum test, (b) U test, (c) median test, and (d) t test. (e) Compare the conclusions of the four test procedures.

16.5. Along two stretches of highway, 55 miles per hour speed limit signs are regularly posted. The following table gives the speeds of drivers who were caught exceeding the limit

Highway I	65, 82, 60, 59, 95, 78, 62, 63
Highway II	66, 69, 61, 67, 68, 64, 72

Determine whether there is any difference in violator mean speed on the two highways. Test using the (a) rank-sum test, (b) U test, and (c) median test. (d) Compare the conclusions of the three test procedures, and explain why the t test is not appropriate.

618 DISTRIBUTION-FREE METHODS CHAP. 16

16.6. On a given day in two widely separated counties the maximum relative humidity readings (in percentages), as recorded by 13 instruments of the same type, were as follows

County A	18, 42, 25, 26, 20, 36
County B	30, 33, 58, 12, 65, 33, 32

Use two test procedures to compare the mean relative humidity of the two counties. Discuss the pros and cons of your conclusions.

16.7. The lengths of life minus 1500 kilowatt hours of 18 of the same type of electronic tube made by two manufacturers were as follows

Manufacturer X	38, 132, 0, 159, 81, 118, 6, 78, 171, 137
Manufacturer Y	125, 78, 79, 210, 405, 79, 811, 81

Compare the mean lives of the electronic tubes made by X and Y.

16.8. Prove Eq. (16.10).

16.9. Prove Eq. (16.13).

16.10. Prove Eq. (16.17).

16.11. Prove Eqs. (16.18) and (16.19).

16.12. Prove Eq. (16.20).

16.13. (a) Define a statistic which may be used to compare the first quartiles of two distributions. (b) Give an illustration of the use of your statistic. (c) Find the mean and variance of the sampling distribution of your statistic.

16.14. On inspecting 25 consecutive units on a production line, it was found that the sequence of defective D and nondefective N units was as

$$N\ N\ N\ N\ N\ N\ N\ D\ N\ N\ N\ N\ N\ D\ D\ D\ N\ N\ N\ N\ N\ N\ N\ D$$

Determine whether the order of the sample could be considered random.

16.15. Thirty-five people waiting in line to buy tickets were arranged as follows (M denotes male and F female)

$$MMMFMMMFMFMFMFMFFFMMFMFMFMFMFMMMMFM$$

Determine whether the order of the sample could be considered random.

16.16. Eighteen cars of logs to be used as pulpwood are waiting on a track. If more than ten per cent of a load is hardwood, the car is denoted by H; otherwise, the car is denoted by S (softwood). Determine whether the order of the following sample can be considered random

$$S\ S\ S\ S\ S\ S\ S\ S\ S\ S\ H\ H\ S\ S\ S\ S\ H$$

16.17. Show that Eq. (16.26) holds.

16.18. Prove Eq. (16.27).

16.19. Prove Eq. (16.28).

REFERENCES

1. Abelson, R. M. and R. A. Bradley, "A 2×2 Factorial with Paired Comparisons," *Biometrics*, Vol. **10** (1954), pp. 487–502.

2. Anderson, R. L., "Distribution of the Serial Correlation Coefficient," *Annals of Mathematical Statistics*, Vol. **13** (1942), pp. 1–13.

3. Auble, D., "Extended Tables for the Mann-Whitney Statistic," *Bulletin of Institution of Educational Research*, Indiana University, Vol. 1, No. 2 (1953).

4. Bartlett, M. S., *An Introduction to Stochastic Processes with Special References to Methods and Applications*. London: Cambridge University Press, 1956.

5. Bennett, C. A. and N. L. Franklin, *Statistical Analysis in Chemistry and the Chemical Industry*. New York: John Wiley & Sons, Inc., 1954, Chaps. 10 and 11.

6. Bradley, R. A. and M. E. Terry, "The Rank Analysis of Incomplete Block Designs. I. The Method of Paired Comparisons," *Biometrika*, Vol. **39** (1952), pp. 324–45.

7. ———, "Incomplete Block Rank Analysis: On the Appropriateness of the Model for a Method of Paired Comparisons," *Biometrics*, Vol. **10** (1954), pp. 375–90.

8. ———, "Rank Analysis of Incomplete Block Designs. II. Additional Tables for the Method of Paired Comparisons," *Biometrika*, Vol. **41** (1954), pp. 502–37.

9. ———, "Rank Analysis of Incomplete Block Designs. III. Some Large-sample Results on Estimation and Power for a Method of Paired Comparison," *Biometrika*, Vol. **42** (1955), pp. 450–69.

10. Brown, G. W. and A. M. Mood, "On Median Tests for Linear Hypotheses," *Proceedings of the Second Berkley Symposium on Mathematical Statistics and Probability*, University of California Press, 1951, pp. 159–66.

11. Brunk, H. D., *An Introduction to Mathematical Statistics*. Boston: Ginn & Company, 1960, Chap. 17.

12. Davis, H. T., *The Analysis of Economic Time Series*. Bloomington, Indiana: Principia Press, 1941.

13. Dixon, W. J., "Power Functions of the Sign Test and Power Efficiency for Normal Alternatives," *Annals of Mathematical Statistics*, Vol. **24** (1953), pp. 467–73.

14. ——— and F. J. Massey, *Introduction to Statistical Analysis*, 2nd ed. New York: McGraw-Hill, Inc., 1957, Chap. 17.

15. Durbin, J., "Incomplete Blocks in Ranking Experiments," *British Journal of Psychology, Statistics Section*, Vol. **4** (1951), pp. 85–90.

16. Festinger, L., "The Significance of Differences Between Means Without

Reference to the Frequency Distribution Function," *Psychometrika*, Vol. **11** (1946), pp. 97–105.

17. Fraser, D. A. S., *Nonparametric Methods in Statistics*. New York: John Wiley & Sons, Inc., 1957.

18. Freund, J. E., *Modern Elementary Statistics*, 2nd ed. Englewood Cliffs, N. J.: Prentice-Hall, Inc., 1960, Chaps. 13 and 17.

19. Friedman, M., "The Use of Ranks to Avoid the Assumption of Normality Implicit in the Analysis of Variance," *Journal of the American Statistical Association*, Vol. **32** (1937), pp. 675–701.

20. Grant, E. L., *Statistical Quality Control*. New York: McGraw-Hill , Inc., 1946.

21. Grenander, U. and M. Rosenblatt, *Statistical Analysis of Stationary Time Series*. New York: John Wiley & Sons, Inc., 1957.

22. Hald, A., *Statistical Theory with Engineering Applications*. New York: John Wiley & Sons, Inc., 1952, Chaps. 13 and 20.

23. Hart, B. I., "Significance Levels for the Ratio of the Mean Square Successive Difference to the Variance," *Annals of Mathematical Statistics*, Vol. **13** (1942), pp. 445–47.

24. Hoel, P. G., *Introduction to Mathematical Statistics*, 3rd ed. New York: John Wiley & Sons, Inc., 1962, Chap. 13.

25. Kendall, M. G. and B. Smith, "The Problem of *m* Rankings," *Annals of Mathematical Statistics*, Vol. **10** (1939), pp. 275–87.

26. ———, *Contributions to the Study of Oscillatory Time Series*. London: Cambridge University Press, 1946.

27. Kruskel, W. H. and W. A. Wallis, "Use of Ranks in One-Criterion Variance Analysis," *Journal of the American Statistical Association*, Vol. **47** (1952), pp. 583–621.

28. Lieberman, G. J. and D. B. Owen, *Tables of the Hypergeometric Distribution*. Stanford, California: Stanford University Press, 1960.

29. Mann, H. B. and D. R. Whitney, "On a Test of Whether One of Two Random Variables is Stochastically Larger than the Other," *Annals of Mathematical Statistics*, Vol. **18** (1947), pp. 50–60.

30. Mood, A. M., "The Distribution Theory of Runs," *Annals of Mathematical Statistics*, Vol. **11** (1940), pp. 367–92.

31. ———, *Introduction to the Theory of Statistics*. New York: McGraw-Hill, Inc., 1950, Chap. 16.

32. Moses, L. E., "Non-Parametric Statistics for Psychological Research," *Psychological Bulletin*, Vol. **49** (1952), pp. 122–43.

33. Neumann, J. von, "Distribution of the Ratio of the Mean Square Successive Difference to the Variance," *Annals of Mathematical Statistics*, Vol. **12** (1941), pp. 367–95.

34. Olmstead, P. S., "Distribution of Sample Arrangements for Runs Up and Down," *Annals of Mathematical Statistics*, Vol. **17** (1946), pp 24–33.

35. Savage, I. R., "Bibliography of Nonparametric Statistics and Related Topics," *Journal of the American Statistical Association*, Vol. **48** (1953), pp. 884–906.

36. Scheffé, H., "Statistical Inference in the Non-parametric Case," *Annals of Mathematical Statistics*, Vol. **14** (1943), pp. 305–32.

37. Shewhart, W. A., *Economic Control of Quality of Manufactured Product*. New York: D. Van Nostrand Co., Inc., 1931.

38. Siegel, S., *Nonparametric Statistics for the Behavioral Sciences*. New York: McGraw-Hill, Inc., 1956.

39. Steel, R.G.D. and J. H. Torrie, *Principles and Procedures of Statistics*. New York: McGraw-Hill, Inc., 1960, Chap. 21.

40. Swed, F. S. and C. Eisenhart, "Tables for Testing Randomness of Grouping in a Sequence of Alternatives," *Annals of Mathematical Statistics*, Vol. **14** (1943), pp. 66–87.

41. Terry, M. E., "Some Rank Order Tests which Are Most Powerful Against Specific Parametric Alternatives," *Annals of Mathematical Statistics*, Vol. **23** (1952), pp. 346–66.

42. Wald, A. and J. Wolfowitz, "An Exact Test for Randomnesss in the Non-Parametric Case Based on Serial Correlation," *Annals of Mathematical Statistics*, Vol. **14** (1943), pp. 378–88.

43. ———, *Sequential Analysis*. New York: John Wiley & Sons, Inc., 1947.

44. Wallis, W. A., "Rough-and-Ready Statistical Tests," *Industrial Quality Control*, Vol. **8** (1952), pp. 35–40.

45. Walsh, J. E., "Some Significance Tests for the Median Which Are Valid Under Very General Conditions," *Annals of Mathematical Statistics*, Vol. **20** (1949), pp. 64–81.

46. ———, "Applications of Some Significance Tests for the Median Which Are Valid Under Very General Conditions," *Journal of The American Statistical Association*, Vol. **44** (1949), pp. 342–55.

47. White, C., "The Use of Ranks in a Test of Significance for Comparing Two Treatments," *Biometrics*, Vol. **8** (1952), pp. 33–41.

48. Whitney, D. R., *A Comparison of the Power of Non-Parametric Tests and Tests Based on the Normal Distribution Under Non-Normal Alternatives*, unpublished doctoral dissertation, Ohio State University, 1948.

49. ———, "A Bivariate Extension of the *U* Statistic," *Annals of Mathematical Statistics*, Vol. **22** (1951), pp. 274–82.

50. Whittle, P., *Hypothesis Testing in Time Series Analysis*. New York: Hafner Publishing Co., Inc., 1951.

51. Wiener, N., *Extrapolation, Interpolation and Smoothing of Stationary Time Series*. New York: John Wiley & Sons, Inc., 1949.

52. Wilcoxon, F., "Individual Comparisons by Ranking Methods," *Biometrics Bulletin*, Vol. **1** (1945), pp. 80–83.

53. ———, "Probability Tables for Individual Comparisons by Ranking Methods," *Biometrics*, Vol. **3** (1947), pp. 119–22.

54. ———, *Some Rapid Approximate Statistical Procedures*. American Cyanamid Company, Stamford, Conn., 1949.

55. Wilks, S. S., "Order Statistics," *Bulletin of the American Mathematical Society*, Vol. **54** (1948), pp. 6–50.

STATISTICAL TABLES

STATISTICAL TABLES

Table I

Ordinates of the Normal Density Function

$$n(t) = \frac{1}{\sqrt{2\pi}} e^{-t^2/2}$$

t	.00	.01	.02	.03	.04	.05	.06	.07	.08	.09
.0	.3989	.3989	.3989	.3988	.3986	.3984	.3982	.3980	.3977	.3973
.1	.3970	.3965	.3961	.3956	.3951	.3945	.3939	.3932	.3925	.3918
.2	.3910	.3902	.3894	.3885	.3876	.3867	.3857	.3847	.3836	.3825
.3	.3814	.3802	.3790	.3778	.3765	.3752	.3739	.3725	.3712	.3697
.4	.3683	.3668	.3653	.3637	.3621	.3605	.3589	.3572	.3555	.3538
.5	.3521	.3503	.3485	.3467	.3448	.3429	.3410	.3391	.3372	.3352
.6	.3332	.3312	.3292	.3271	.3251	.3230	.3209	.3187	.3166	.3144
.7	.3123	.3101	.3079	.3056	.3034	.3011	.2989	.2966	.2943	.2920
.8	.2897	.2874	.2850	.2827	.2803	.2780	.2756	.2732	.2709	.2685
.9	.2661	.2637	.2613	.2589	.2565	.2541	.2516	.2492	.2468	.2444
1.0	.2420	.2396	.2371	.2347	.2323	.2299	.2275	.2251	.2227	.2203
1.1	.2179	.2155	.2131	.2107	.2083	.2059	.2036	.2012	.1989	.1965
1.2	.1942	.1919	.1895	.1872	.1849	.1826	.1804	.1781	.1758	.1736
1.3	.1714	.1691	.1669	.1647	.1626	.1604	.1582	.1561	.1539	.1518
1.4	.1497	.1476	.1456	.1435	.1415	.1394	.1374	.1354	.1334	.1315
1.5	.1295	.1276	.1257	.1238	.1219	.1200	.1182	.1163	.1145	.1127
1.6	.1109	.1092	.1074	.1057	.1040	.1023	.1006	.0989	.0973	.0957
1.7	.0940	.0925	.0909	.0893	.0878	.0863	.0848	.0833	.0818	.0804
1.8	.0790	.0775	.0761	.0748	.0734	.0721	.0707	.0694	.0681	.0669
1.9	.0656	.0644	.0632	.0620	.0608	.0596	.0584	.0573	.0562	.0551
2.0	.0540	.0529	.0519	.0508	.0498	.0488	.0478	.0468	.0459	.0449
2.1	.0440	.0431	.0422	.0413	.0404	.0396	.0387	.0379	.0371	.0363
2.2	.0355	.0347	.0339	.0332	.0325	.0317	.0310	.0303	.0297	.0290
2.3	.0283	.0277	.0270	.0264	.0258	.0252	.0246	.0241	.0235	.0229
2.4	.0224	.0219	.0213	.0208	.0203	.0198	.0194	.0189	.0184	.0180
2.5	.0175	.0171	.0167	.0163	.0158	.0154	.0151	.0147	.0143	.0139
2.6	.0136	.0132	.0129	.0126	.0122	.0119	.0116	.0113	.0110	.0107
2.7	.0104	.0101	.0099	.0096	.0093	.0091	.0088	.0086	.0084	.0081
2.8	.0079	.0077	.0075	.0073	.0071	.0069	.0067	.0065	.0063	.0061
2.9	.0060	.0058	.0056	.0055	.0053	.0051	.0050	.0048	.0047	.0046
3.0	.0044	.0043	.0042	.0040	.0039	.0038	.0037	.0036	.0035	.0034
3.1	.0033	.0032	.0031	.0030	.0029	.0028	.0027	.0026	.0025	.0025
3.2	.0024	.0023	.0022	.0022	.0021	.0020	.0020	.0019	.0018	.0018
3.3	.0017	.0017	.0016	.0016	.0015	.0015	.0014	.0014	.0013	.0013
3.4	.0012	.0012	.0012	.0011	.0011	.0010	.0010	.0010	.0009	.0009
3.5	.0009	.0008	.0008	.0008	.0008	.0007	.0007	.0007	.0007	.0006
3.6	.0006	.0006	.0006	.0005	.0005	.0005	.0005	.0005	.0005	.0004
3.7	.0004	.0004	.0004	.0004	.0004	.0004	.0003	.0003	.0003	.0003
3.8	.0003	.0003	.0003	.0003	.0003	.0002	.0002	.0002	.0002	.0002
3.9	.0002	.0002	.0002	.0002	.0002	.0002	.0002	.0002	.0001	.0001

Table II

Cumulative Normal Distribution

$$N(t) = \int_{-\infty}^{t} \frac{1}{\sqrt{2\pi}} e^{-u^2/2} du$$

t	.00	.01	.02	.03	.04	.05	.06	.07	.08	.09
.0	.5000	.5040	.5080	.5120	.5160	.5199	.5239	.5279	.5319	.5359
.1	.5398	.5438	.5478	.5517	.5557	.5596	.5636	.5675	.5714	.5753
.2	.5793	.5832	.5871	.5910	.5948	.5987	.6026	.6064	.6103	.6141
.3	.6179	.6217	.6255	.6293	.6331	.6368	.6406	.6443	.6480	.6517
.4	.6554	.6591	.6628	.6664	.6700	.6736	.6772	.6808	.6844	.6879
.5	.6915	.6950	.6985	.7019	.7054	.7088	.7123	.7157	.7190	.7224
.6	.7257	.7291	.7324	.7357	.7389	.7422	.7454	.7486	.7517	.7549
.7	.7580	.7611	.7642	.7673	.7704	.7734	.7764	.7794	.7823	.7852
.8	.7881	.7910	.7939	.7967	.7995	.8023	.8051	.8078	.8106	.8133
.9	.8159	.8186	.8212	.8238	.8264	.8289	.8315	.8340	.8365	.8389
1.0	.8413	.8438	.8461	.8485	.8508	.8531	.8554	.8577	.8599	.8621
1.1	.8643	.8665	.8686	.8708	.8729	.8749	.8770	.8790	.8810	.8830
1.2	.8849	.8869	.8888	.8907	.8925	.8944	.8962	.8980	.8997	.9015
1.3	.9032	.9049	.9066	.9082	.9099	.9115	.9131	.9147	.9162	.9177
1.4	.9192	.9207	.9222	.9236	.9251	.9265	.9279	.9292	.9306	.9319
1.5	.9332	.9345	.9357	.9370	.9382	.9394	.9406	.9418	.9429	.9441
1.6	.9452	.9463	.9474	.9484	.9495	.9505	.9515	.9525	.9535	.9545
1.7	.9554	.9564	.9573	.9582	.9591	.9599	.9608	.9616	.9625	.9633
1.8	.9641	.9649	.9656	.9664	.9671	.9678	.9686	.9693	.9699	.9706
1.9	.9713	.9719	.9726	.9732	.9738	.9744	.9750	.9756	.9761	.9767
2.0	.9772	.9778	.9783	.9788	.9793	.9798	.9803	.9808	.9812	.9817
2.1	.9821	.9826	.9830	.9834	.9838	.9842	.9846	.9850	.9854	.9857
2.2	.9861	.9864	.9868	.9871	.9875	.9878	.9881	.9884	.9887	.9890
2.3	.9893	.9896	.9898	.9901	.9904	.9906	.9909	.9911	.9913	.9916
2.4	.9918	.9920	.9922	.9925	.9927	.9929	.9931	.9932	.9934	.9936
2.5	.9938	.9940	.9941	.9943	.9945	.9946	.9948	.9949	.9951	.9952
2.6	.9953	.9955	.9956	.9957	.9959	.9960	.9961	.9962	.9963	.9964
2.7	.9965	.9966	.9967	.9968	.9969	.9970	.9971	.9972	.9973	.9974
2.8	.9974	.9975	.9976	.9977	.9977	.9978	.9979	.9979	.9980	.9981
2.9	.9981	.9982	.9982	.9983	.9984	.9984	.9985	.9985	.9986	.9986
3.0	.9987	.9987	.9987	.9988	.9988	.9989	.9989	.9989	.9990	.9990
3.1	.9990	.9991	.9991	.9991	.9992	.9992	.9992	.9992	.9993	.9993
3.2	.9993	.9993	.9994	.9994	.9994	.9994	.9994	.9995	.9995	.9995
3.3	.9995	.9995	.9995	.9996	.9996	.9996	.9996	.9996	.9996	.9997
3.4	.9997	.9997	.9997	.9997	.9997	.9997	.9997	.9997	.9997	.9998

t	1.282	1.645	1.960	2.326	2.576	3.090	3.291	3.891	4.417
N(t)	.90	.95	.975	.99	.995	.999	.9995	.99995	.999995
2[1 − N(t)]	.20	.10	.05	.02	.01	.002	.001	.0001	.00001

Table III

Confidence Belts for Proportions*
(Confidence coefficient .95)

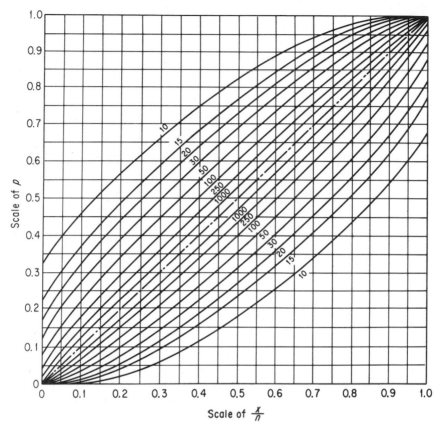

Scale of $\frac{x}{n}$

* This chart is reproduced from C. J. Clopper and E. S. Pearson, "The Use of Confidence or Fiducial Limits Illustrated in the Case of the Binomial," *Biometrika*, Vol. **26** (1934), p. 404, with the permission of Professor E. S. Pearson.

Table III

Confidence Belts for Proportions* (*cont.*)
(Confidence coefficient .99)

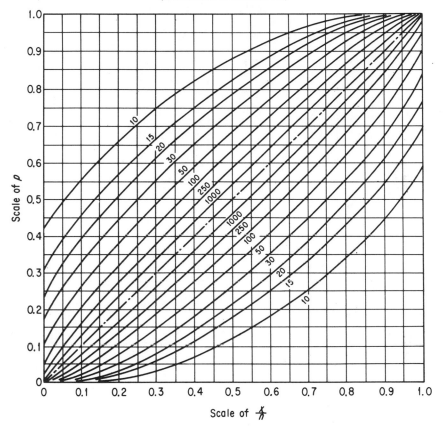

Scale of $\frac{x}{n}$

* This chart is reproduced from C. J. Clopper and E. S. Pearson, "The Use of Confidence or Fiducial Limits Illustrated in the Case of the Binomial," *Biometrika*, Vol. **26** (1934), p. 404, with the permission of Professor E. S. Pearson.

STATISTICAL TABLES

Table IV

Percentage Points of the χ^2 Distribution*

ν \ α	.995	.990	.975	.950	.900	.750
1	.0⁴393	.0³157	.0³982	.0²393	.0158	.102
2	.0100	.0201	.0506	.103	.211	.575
3	.0717	.115	.216	.352	.584	1.21
4	.207	.297	.484	.711	1.06	1.92
5	.412	.554	.831	1.15	1.61	2.67
6	.676	.872	1.24	1.64	2.20	3.45
7	.989	1.24	1.69	2.17	2.83	4.25
8	1.34	1.65	2.18	2.73	3.49	5.07
9	1.73	2.09	2.70	3.33	4.17	5.90
10	2.16	2.56	3.25	3.94	4.87	6.74
11	2.60	3.05	3.82	4.57	5.58	7.58
12	3.07	3.57	4.40	5.23	6.30	8.44
13	3.57	4.11	5.01	5.89	7.04	9.30
14	4.07	4.66	5.63	6.57	7.79	10.2
15	4.60	5.23	6.26	7.26	8.55	11.0
16	5.14	5.81	6.91	7.96	9.31	11.9
17	5.70	6.41	7.56	8.67	10.1	12.8
18	6.26	7.01	8.23	9.39	10.9	13.7
19	6.84	7.63	8.91	10.1	11.7	14.6
20	7.43	8.26	9.59	10.9	12.4	15.5
21	8.03	8.90	10.3	11.6	13.2	16.3
22	8.64	9.54	11.0	12.3	14.0	17.2
23	9.26	10.2	11.7	13.1	14.8	18.1
24	9.89	10.9	12.4	13.8	15.7	19.0
25	10.5	11.5	13.1	14.6	16.5	19.9
26	11.2	12.2	13.8	15.4	17.3	20.8
27	11.8	12.9	14.6	16.2	18.1	21.7
28	12.5	13.6	15.3	16.9	18.9	22.7
29	13.1	14.3	16.0	17.7	19.8	23.6
30	13.8	15.0	16.8	18.5	20.6	24.5
40	20.7	22.2	24.4	26.5	29.1	33.7
50	28.0	29.7	32.4	34.8	37.7	42.9
60	35.5	37.5	40.5	43.2	46.5	52.3

* This table is abridged from Catherine M. Thompson, "Tables of Percentage Points of the χ^2 Distribution," *Biometrika*, Vol. **32** (1941), pp. 188–89, with the permission of Professor E. S. Pearson.

Table IV

Percentage Points of the χ^2 Distribution (*cont.*)

.500	.250	.100	.050	.025	.010	.005
.455	1.32	2.71	3.84	5.02	6.63	7.88
1.39	2.77	4.61	5.99	7.38	9.21	10.6
2.37	4.11	6.25	7.81	9.35	11.3	12.8
3.36	5.39	7.78	9.49	11.1	13.3	14.9
4.35	6.63	9.24	11.1	12.8	15.1	16.7
5.35	7.84	10.6	12.6	14.4	16.8	18.5
6.35	9.04	12.0	14.1	16.0	18.5	20.3
7.34	10.2	13.4	15.5	17.5	20.1	22.0
8.34	11.4	14.7	16.9	19.0	21.7	23.6
9.34	12.5	16.0	18.3	20.5	23.2	25.2
10.3	13.7	17.3	19.7	21.9	24.7	26.8
11.3	14.8	18.5	21.0	23.3	26.2	28.3
12.3	16.0	19.8	22.4	24.7	27.7	29.8
13.3	17.1	21.1	23.7	26.1	29.1	31.3
14.3	18.2	22.3	25.0	27.5	30.6	32.8
15.3	19.4	23.5	26.3	28.8	32.0	34.3
16.3	20.5	24.8	27.6	30.2	33.4	35.7
17.3	21.6	26.0	28.9	31.5	34.8	37.2
18.3	22.7	27.2	30.1	32.9	36.2	38.6
19.3	23.8	28.4	31.4	34.2	37.6	40.0
20.3	24.9	29.6	32.7	35.5	38.9	41.4
21.3	26.0	30.8	33.9	36.8	40.3	42.8
22.3	27.1	32.0	35.2	38.1	41.6	44.2
23.3	28.2	33.2	36.4	39.4	43.0	45.6
24.3	29.3	34.4	37.7	40.6	44.3	46.9
25.3	30.4	35.6	38.9	41.9	45.6	48.3
26.3	31.5	36.7	40.1	43.2	47.0	49.6
27.3	32.6	37.9	41.3	44.5	48.3	51.0
28.3	33.7	39.1	42.6	45.7	49.6	52.3
29.3	34.8	40.3	43.8	47.0	50.9	53.7
39.3	45.6	51.8	55.8	59.3	63.7	66.8
49.3	56.3	63.2	67.5	71.4	76.2	79.5
59.3	67.0	74.4	79.1	83.3	88.4	92.0

STATISTICAL TABLES

Table V

Percentage Points of the χ^2/ν Distribution*

ν \ α	.995	.990	.975	.950	.900	.750	.500	.250	.100	.050	.025	.010	.005
1	$.0^439$	$.0^316$	$.0^398$	$.0^239$	.016	.102	.455	1.32	2.71	3.84	5.02	6.63	7.88
2	.005	.010	.025	.052	.106	.288	.693	1.39	2.30	3.00	3.69	4.61	5.30
3	.024	.038	.072	.117	.195	.404	.789	1.37	2.08	2.60	3.12	3.78	4.28
4	.052	.074	.121	.178	.266	.481	.839	1.35	1.94	2.37	2.79	3.32	3.72
5	.082	.111	.166	.229	.322	.535	.870	1.33	1.85	2.21	2.57	3.02	3.35
6	.113	.145	.206	.272	.367	.576	.891	1.31	1.77	2.10	2.41	2.80	3.09
7	.141	.177	.241	.310	.405	.608	.907	1.29	1.72	2.01	2.29	2.64	2.90
8	.168	.206	.272	.342	.436	.634	.918	1.28	1.67	1.94	2.19	2.51	2.74
9	.193	.232	.300	.369	.463	.655	.927	1.26	1.63	1.88	2.11	2.41	2.62
10	.216	.256	.325	.394	.487	.674	.934	1.25	1.60	1.83	2.05	2.32	2.52
11	.237	.278	.347	.416	.507	.689	.940	1.25	1.57	1.79	1.99	2.25	2.43
12	.256	.298	.367	.436	.525	.703	.945	1.24	1.55	1.75	1.94	2.18	2.36
13	.274	.316	.385	.453	.542	.715	.949	1.23	1.52	1.72	1.90	2.13	2.29
14	.281	.333	.402	.469	.556	.726	.953	1.22	1.50	1.69	1.87	2.08	2.24
15	.307	.349	.418	.484	.570	.736	.956	1.22	1.49	1.67	1.83	2.04	2.19
16	.321	.363	.432	.498	.582	.745	.959	1.21	1.47	1.64	1.80	2.00	2.14
17	.335	.377	.445	.510	.593	.752	.961	1.21	1.46	1.62	1.78	1.97	2.10
18	.348	.390	.457	.522	.604	.760	.963	1.20	1.44	1.60	1.75	1.93	2.06
19	.360	.402	.469	.532	.613	.766	.965	1.20	1.43	1.59	1.73	1.90	2.03
20	.372	.413	.480	.543	.622	.773	.967	1.19	1.42	1.57	1.71	1.88	2.00
21	.383	.424	.490	.552	.630	.778	.968	1.19	1.41	1.56	1.69	1.85	1.97
22	.393	.434	.499	.561	.638	.784	.970	1.18	1.40	1.54	1.67	1.83	1.95
23	.403	.443	.508	.569	.646	.789	.971	1.18	1.39	1.53	1.66	1.81	1.92
24	.412	.452	.517	.577	.652	.793	.972	1.18	1.38	1.52	1.64	1.79	1.90
25	.421	.461	.525	.584	.659	.798	.973	1.17	1.38	1.51	1.63	1.77	1.88
26	.429	.469	.532	.592	.665	.802	.974	1.17	1.37	1.50	1.61	1.76	1.86
27	.437	.477	.540	.598	.671	.806	.975	1.17	1.36	1.49	1.60	1.74	1.84
28	.445	.484	.547	.605	.676	.809	.976	1.17	1.35	1.48	1.59	1.72	1.82
29	.452	.492	.553	.611	.682	.813	.977	1.16	1.35	1.47	1.58	1.71	1.80
30	.460	.498	.560	.616	.687	.816	.978	1.16	1.34	1.46	1.57	1.70	1.79
40	.518	.554	.611	.663	.726	.842	.983	1.14	1.30	1.39	1.48	1.59	1.67
50	.560	.594	.647	.695	.754	.859	.987	1.13	1.26	1.35	1.43	1.52	1.59
60	.592	.625	.675	.720	.774	.872	.989	1.12	1.24	1.32	1.39	1.47	1.53

* This table is obtained from Table IV by division.

Table VI

Percentage Points of the t Distribution*

ν \ α	.25	.20	.15	.10	.05	.025	.01	.005	.0005
1	1.000	1.376	1.963	3.078	6.314	12.706	31.821	63.657	636.619
2	.816	1.061	1.386	1.886	2.920	4.303	6.965	9.925	31.598
3	.765	.978	1.250	1.638	2.353	3.182	4.541	5.841	12.941
4	.741	.941	1.190	1.533	2.132	2.776	3.747	4.604	8.610
5	.727	.920	1.156	1.476	2.015	2.571	3.365	4.032	6.859
6	.718	.906	1.134	1.440	1.943	2.447	3.143	3.707	5.959
7	.711	.896	1.119	1.415	1.895	2.365	2.998	3.499	5.405
8	.706	.889	1.108	1.397	1.860	2.306	2.896	3.355	5.041
9	.703	.883	1.100	1.383	1.833	2.262	2.821	3.250	4.781
10	.700	.879	1.093	1.372	1.812	2.228	2.764	3.169	4.587
11	.697	.876	1.088	1.363	1.796	2.201	2.718	3.106	4.437
12	.695	.873	1.083	1.356	1.782	2.179	2.681	3.055	4.318
13	.694	.870	1.079	1.350	1.771	2.160	2.650	3.012	4.221
14	.692	.868	1.076	1.345	1.761	2.145	2.624	2.977	4.140
15	.691	.866	1.074	1.341	1.753	2.131	2.602	2.947	4.073
16	.690	.865	1.071	1.337	1.746	2.120	2.583	2.921	4.015
17	.689	.863	1.069	1.333	1.740	2.110	2.567	2.898	3.965
18	.688	.862	1.067	1.330	1.734	2.101	2.552	2.878	3.922
19	.688	.861	1.066	1.328	1.729	2.093	2.539	2.861	3.883
20	.687	.860	1.064	1.325	1.725	2.086	2.528	2.845	3.850
21	.686	.859	1.063	1.323	1.721	2.080	2.518	2.831	3.819
22	.686	.858	1.061	1.321	1.717	2.074	2.508	2.819	3.792
23	.685	.858	1.060	1.319	1.714	2.069	2.500	2.807	3.767
24	.685	.857	1.059	1.318	1.711	2.064	2.492	2.797	3.745
25	.684	.856	1.058	1.316	1.708	2.060	2.485	2.787	3.725
26	.684	.856	1.058	1.315	1.706	2.056	2.479	2.779	3.707
27	.684	.855	1.057	1.314	1.703	2.052	2.473	2.771	3.690
28	.683	.855	1.056	1.313	1.701	2.048	2.467	2.763	3.674
29	.683	.854	1.055	1.311	1.699	2.045	2.462	2.756	3.659
30	.683	.854	1.055	1.310	1.697	2.042	2.457	2.750	3.646
40	.681	.851	1.050	1.303	1.684	2.021	2.423	2.704	3.551
60	.679	.848	1.046	1.296	1.671	2.000	2.390	2.660	3.460
120	.677	.845	1.041	1.289	1.658	1.980	2.358	2.617	3.373
∞	.674	.842	1.036	1.282	1.645	1.960	2.326	2.576	3.291

* This table is abridged from Table III of R. A. Fisher and Frank Yates, *Statistical Tables for Biological, Agricultural and Medical Research*, 5th ed., published by Oliver and Boyd, Ltd., Edinburgh. By permission of the authors and publishers.

Table VII

Percentage Points of the F Distribution*

$$\alpha = 0.50$$

ν_2 \ ν_1	1	2	3	4	5	6	7	8	9
1	1.00	1.50	1.71	1.82	1.89	1.94	1.98	2.00	2.03
2	.667	1.00	1.13	1.21	1.25	1.28	1.30	1.32	1.33
3	.585	.881	1.00	1.06	1.10	1.13	1.15	1.16	1.17
4	.549	.828	.941	1.00	1.04	1.06	1.08	1.09	1.10
5	.528	.799	.907	.965	1.00	1.02	1.04	1.05	1.06
6	.515	.780	.886	.942	.977	1.00	1.02	1.03	1.04
7	.506	.767	.871	.926	.960	.983	1.00	1.01	1.02
8	.499	.757	.860	.915	.948	.971	.988	1.00	1.01
9	.494	.749	.852	.906	.939	.962	.978	.990	1.00
10	.490	.743	.845	.899	.932	.954	.971	.983	.992
11	.486	.739	.840	.893	.926	.948	.964	.977	.986
12	.484	.735	.835	.888	.921	.943	.959	.972	.981
13	.481	.731	.832	.885	.917	.939	.955	.967	.977
14	.479	.729	.828	.881	.914	.936	.952	.964	.973
15	.478	.726	.826	.878	.911	.933	.949	.960	.970
16	.476	.724	.823	.876	.908	.930	.946	.958	.967
17	.475	.722	.821	.874	.906	.928	.943	.955	.965
18	.474	.721	.819	.872	.904	.926	.941	.953	.962
19	.473	.719	.818	.870	.902	.924	.939	.951	.961
20	.472	.718	.816	.868	.900	.922	.938	.950	.959
21	.471	.717	.815	.867	.899	.921	.936	.948	.957
22	.470	.715	.814	.866	.898	.919	.935	.947	.956
23	.470	.714	.813	.864	.896	.918	.934	.945	.955
24	.469	.714	.812	.863	.895	.917	.932	.944	.953
25	.468	.713	.811	.862	.894	.916	.931	.943	.952
26	.468	.712	.810	.861	.893	.915	.930	.942	.951
27	.467	.711	.809	.861	.892	.914	.930	.941	.950
28	.467	.711	.808	.860	.892	.913	.929	.940	.950
29	.467	.710	.808	.859	.891	.912	.928	.940	.949
30	.466	.709	.807	.858	.890	.912	.927	.939	.948
40	.463	.705	.802	.854	.885	.907	.922	.934	.943
60	.461	.701	.798	.849	.880	.901	.917	.928	.937
120	.458	.697	.793	.844	.875	.896	.912	.923	.932
∞	.455	.693	.789	.839	.870	.891	.907	.918	.927

* This table is abridged from Maxine Merrington and Catherine Thompson, "Table of Percentage Points in the Inverted Beta Distribution," *Biometrika*, Vol. 33 (1943), pp. 73–88, with permission of Professor E. S. Pearson.

Table VII

Percentage Points of the F Distribution (*cont.*)

$\alpha = 0.50$

10	12	15	20	24	30	40	60	120	∞
2.04	2.07	2.09	2.12	2.13	2.15	2.16	2.17	2.18	2.20
1.35	1.36	1.38	1.39	1.40	1.41	1.42	1.43	1.43	1.44
1.18	1.20	1.21	1.23	1.23	1.24	1.25	1.25	1.26	1.27
1.11	1.13	1.14	1.15	1.16	1.16	1.17	1.18	1.18	1.19
1.07	1.09	1.10	1.11	1.12	1.12	1.13	1.14	1.14	1.15
1.05	1.06	1.07	1.08	1.09	1.10	1.10	1.11	1.12	1.12
1.03	1.04	1.05	1.06	1.07	1.08	1.08	1.09	1.10	1.10
1.02	1.03	1.04	1.05	1.06	1.07	1.07	1.08	1.08	1.09
1.01	1.02	1.03	1.04	1.05	1.05	1.06	1.07	1.07	1.08
1.00	1.01	1.02	1.03	1.04	1.05	1.05	1.06	1.06	1.07
.994	1.01	1.02	1.03	1.03	1.04	1.05	1.05	1.06	1.06
.989	1.00	1.01	1.02	1.03	1.03	1.04	1.05	1.05	1.06
.984	.996	1.01	1.02	1.02	1.03	1.04	1.04	1.05	1.05
.981	.992	1.00	1.01	1.02	1.02	1.03	1.04	1.04	1.05
.977	.989	1.00	1.01	1.02	1.02	1.03	1.03	1.04	1.05
.975	.986	.997	1.01	1.01	1.02	1.03	1.03	1.04	1.04
.972	.983	.995	1.01	1.01	1.02	1.02	1.03	1.03	1.04
.970	.981	.992	1.00	1.01	1.02	1.02	1.03	1.03	1.04
.968	.979	.990	1.00	1.01	1.01	1.02	1.02	1.03	1.04
.966	.977	.989	1.00	1.01	1.01	1.02	1.02	1.03	1.03
.965	.976	.987	.998	1.00	1.01	1.02	1.02	1.03	1.03
.963	.974	.986	.997	1.00	1.01	1.01	1.02	1.03	1.03
.962	.973	.984	.996	1.00	1.01	1.01	1.02	1.02	1.03
.961	.972	.983	.994	1.00	1.01	1.01	1.02	1.02	1.03
.960	.971	.982	.993	1.00	1.00	1.01	1.02	1.02	1.03
.959	.970	.981	.992	1.00	1.00	1.01	1.01	1.02	1.03
.958	.969	.980	.991	1.00	1.00	1.01	1.01	1.02	1.03
.957	.968	.979	.990	1.00	1.00	1.01	1.01	1.02	1.02
.956	.967	.978	.990	1.00	1.00	1.01	1.01	1.02	1.02
.955	.966	.978	.989	.994	1.00	1.01	1.01	1.02	1.02
.950	.961	.972	.983	.989	.994	1.00	1.01	1.01	1.02
.945	.956	.967	.978	.983	.989	.994	1.00	1.01	1.01
.939	.950	.961	.972	.978	.983	.989	.994	1.00	1.01
.934	.945	.956	.967	.972	.978	.983	.989	.994	1.00

STATISTICAL TABLES

Table VII

Percentage Points of the F Distribution

$$\alpha = 0.25$$

ν_2 \ ν_1	1	2	3	4	5	6	7	8	9
1	5.83	7.50	8.20	8.58	8.82	8.98	9.10	9.19	9.26
2	2.57	3.00	3.15	3.23	3.28	3.31	3.34	3.35	3.37
3	2.02	2.28	2.36	2.39	2.41	2.42	2.43	2.44	2.44
4	1.81	2.00	2.05	2.06	2.07	2.08	2.08	2.08	2.08
5	1.69	1.85	1.88	1.89	1.89	1.89	1.89	1.89	1.89
6	1.62	1.76	1.78	1.79	1.79	1.78	1.78	1.78	1.77
7	1.57	1.70	1.72	1.72	1.71	1.71	1.70	1.70	1.69
8	1.54	1.66	1.67	1.66	1.66	1.65	1.64	1.64	1.64
9	1.51	1.62	1.63	1.63	1.62	1.61	1.60	1.60	1.59
10	1.49	1.60	1.60	1.59	1.59	1.58	1.57	1.56	1.56
11	1.47	1.58	1.58	1.57	1.56	1.55	1.54	1.53	1.53
12	1.46	1.56	1.56	1.55	1.54	1.53	1.52	1.51	1.51
13	1.45	1.55	1.55	1.53	1.52	1.51	1.50	1.49	1.49
14	1.44	1.53	1.53	1.52	1.51	1.50	1.49	1.48	1.47
15	1.43	1.52	1.52	1.51	1.49	1.48	1.47	1.46	1.46
16	1.42	1.51	1.51	1.50	1.48	1.47	1.46	1.45	1.44
17	1.42	1.51	1.50	1.49	1.47	1.46	1.45	1.44	1.43
18	1.41	1.50	1.49	1.48	1.46	1.45	1.44	1.43	1.42
19	1.41	1.49	1.49	1.47	1.46	1.44	1.43	1.42	1.41
20	1.40	1.49	1.48	1.47	1.45	1.44	1.43	1.42	1.41
21	1.40	1.48	1.48	1.46	1.44	1.43	1.42	1.41	1.40
22	1.40	1.48	1.47	1.45	1.44	1.42	1.41	1.40	1.39
23	1.39	1.47	1.47	1.45	1.43	1.42	1.41	1.40	1.39
24	1.39	1.47	1.46	1.44	1.43	1.41	1.40	1.39	1.38
25	1.39	1.47	1.46	1.44	1.42	1.41	1.40	1.39	1.38
26	1.38	1.46	1.45	1.44	1.42	1.41	1.39	1.38	1.37
27	1.38	1.46	1.45	1.43	1.42	1.40	1.39	1.38	1.37
28	1.38	1.46	1.45	1.43	1.41	1.40	1.39	1.38	1.37
29	1.38	1.45	1.45	1.43	1.41	1.40	1.38	1.37	1.36
30	1.38	1.45	1.44	1.42	1.41	1.39	1.38	1.37	1.36
40	1.36	1.44	1.42	1.40	1.39	1.37	1.36	1.35	1.34
60	1.35	1.42	1.41	1.38	1.37	1.35	1.33	1.32	1.31
120	1.34	1.40	1.39	1.37	1.35	1.33	1.31	1.30	1.29
∞	1.32	1.39	1.37	1.35	1.33	1.31	1.29	1.28	1.27

Table VII

Percentage Points of the F Distribution (*cont.*)

$\alpha = 0.25$

10	12	15	20	24	30	40	60	120	∞
9.32	9.41	9.49	9.58	9.63	9.67	9.71	9.76	9.80	9.85
3.38	3.39	3.41	3.43	3.43	3.44	3.45	3.46	3.47	3.48
2.44	2.45	2.46	2.46	2.46	2.47	2.47	2.47	2.47	2.47
2.08	2.08	2.08	2.08	2.08	2.08	2.08	2.08	2.08	2.08
1.89	1.89	1.89	1.88	1.88	1.88	1.88	1.87	1.87	1.87
1.77	1.77	1.76	1.76	1.75	1.75	1.75	1.74	1.74	1.74
1.69	1.68	1.68	1.67	1.67	1.66	1.66	1.65	1.65	1.65
1.63	1.62	1.62	1.61	1.60	1.60	1.59	1.59	1.58	1.58
1.59	1.58	1.57	1.56	1.56	1.55	1.54	1.54	1.53	1.53
1.55	1.54	1.53	1.52	1.52	1.51	1.51	1.50	1.49	1.48
1.52	1.51	1.50	1.49	1.49	1.48	1.47	1.47	1.46	1.45
1.50	1.49	1.48	1.47	1.46	1.45	1.45	1.44	1.43	1.42
1.48	1.47	1.46	1.45	1.44	1.43	1.42	1.42	1.41	1.40
1.46	1.45	1.44	1.43	1.42	1.41	1.41	1.40	1.39	1.38
1.45	1.44	1.43	1.41	1.41	1.40	1.39	1.38	1.37	1.36
1.44	1.43	1.41	1.40	1.39	1.38	1.37	1.36	1.35	1.34
1.43	1.41	1.40	1.39	1.38	1.37	1.36	1.35	1.34	1.33
1.42	1.40	1.39	1.38	1.37	1.36	1.35	1.34	1.33	1.32
1.41	1.40	1.38	1.37	1.36	1.35	1.34	1.33	1.32	1.30
1.40	1.39	1.37	1.36	1.35	1.34	1.33	1.32	1.31	1.29
1.39	1.38	1.37	1.35	1.34	1.33	1.32	1.31	1.30	1.28
1.39	1.37	1.36	1.34	1.33	1.32	1.31	1.30	1.29	1.28
1.38	1.37	1.35	1.34	1.33	1.32	1.31	1.30	1.28	1.27
1.38	1.36	1.35	1.33	1.32	1.31	1.30	1.29	1.28	1.26
1.37	1.36	1.34	1.33	1.32	1.31	1.29	1.28	1.27	1.25
1.37	1.35	1.34	1.32	1.31	1.30	1.29	1.28	1.26	1.25
1.36	1.35	1.33	1.32	1.31	1.30	1.28	1.27	1.26	1.24
1.36	1.34	1.33	1.31	1.30	1.29	1.28	1.27	1.25	1.24
1.35	1.34	1.32	1.31	1.30	1.29	1.27	1.26	1.25	1.23
1.35	1.34	1.32	1.30	1.29	1.28	1.27	1.26	1.24	1.23
1.33	1.31	1.30	1.28	1.26	1.25	1.24	1.22	1.21	1.19
1.30	1.29	1.27	1.25	1.24	1.22	1.21	1.19	1.17	1.15
1.28	1.26	1.24	1.22	1.21	1.19	1.18	1.16	1.13	1.10
1.25	1.24	1.22	1.19	1.18	1.16	1.14	1.12	1.08	1.00

STATISTICAL TABLES

Table VII

Percentage Points of the F Distribution

$\alpha = 0.10$

ν_2 \\ ν_1	1	2	3	4	5	6	7	8	9
1	39.86	49.50	53.59	55.83	57.24	58.20	58.91	59.44	59.86
2	8.53	9.00	9.16	9.24	9.29	9.33	9.35	9.37	9.38
3	5.54	5.46	5.39	5.34	5.31	5.28	5.27	5.25	5.24
4	4.54	4.32	4.19	4.11	4.05	4.01	3.98	3.95	3.94
5	4.06	3.78	3.62	3.52	3.45	3.40	3.37	3.34	3.32
6	3.78	3.46	3.29	3.18	3.11	3.05	3.01	2.98	2.96
7	3.59	3.26	3.07	2.96	2.88	2.83	2.78	2.75	2.72
8	3.46	3.11	2.92	2.81	2.73	2.67	2.62	2.59	2.56
9	3.36	3.01	2.81	2.69	2.61	2.55	2.51	2 47	2 44
10	3 29	2.92	2.73	2.61	2.52	2.46	2.41	2.38	2.35
11	3.23	2.86	2.66	2.54	2.45	2.39	2.34	2.30	2.27
12	3.18	2.81	2.61	2.48	2.39	2.33	2.28	2.24	2.21
13	3.14	2.76	2.56	2.43	2.35	2.28	2.23	2.20	2.16
14	3.10	2.73	2.52	2.39	2.31	2.24	2.19	2.15	2.12
15	3.07	2.70	2.49	2.36	2.27	2.21	2.16	2.12	2.09
16	3.05	2.67	2.46	2.33	2.24	2.18	2.13	2.09	2.06
17	3.03	2.64	2.44	2.31	2.22	2.15	2.10	2.06	2.03
18	3.01	2.62	2.42	2.29	2.20	2.13	2.08	2.04	2.00
19	2.99	2.61	2.40	2.27	2.18	2.11	2.06	2.02	1.98
20	2.97	2.59	2.38	2.25	2.16	2.09	2.04	2.00	1.96
21	2.96	2.57	2.36	2.23	2.14	2.08	2.02	1.98	1.95
22	2.95	2.56	2.35	2.22	2.13	2.06	2.01	1.97	1.93
23	2.94	2.55	2.34	2.21	2.11	2.05	1.99	1.95	1.92
24	2.93	2.54	2.33	2.19	2.10	2.04	1.98	1.94	1.91
25	2.92	2.53	2.32	2.18	2.09	2.02	1.97	1.93	1.89
26	2.91	2.52	2.31	2.17	2.08	2.01	1.96	1.92	1.88
27	2.90	2.51	2.30	2.17	2.07	2.00	1.95	1.91	1.87
28	2.89	2.50	2.29	2.16	2.06	2.00	1.94	1.90	1.87
29	2.89	2.50	2.28	2.15	2.06	1.99	1.93	1.89	1.86
30	2.88	2.49	2.28	2.14	2.05	1.98	1.93	1.88	1.85
40	2.84	2.44	2.23	2.09	2.00	1.93	1.87	1.83	1.79
60	2.79	2.39	2.18	2.04	1.95	1.87	1.82	1.77	1.74
120	2.75	2.35	2.13	1.99	1.90	1.82	1.77	1.72	1.68
∞	2.71	2.30	2.08	1.94	1.85	1.77	1.72	1.67	1.63

Table VII

Percentage Points of the F Distribution (*cont.*)

$$\alpha = 0.10$$

10	12	15	20	24	30	40	60	120	∞
60.19	60.71	61.22	61.74	62.00	62.26	62.53	62.79	63.06	63.33
9.39	9.41	9.42	9.44	9.45	9.46	9.47	9.47	9.48	9.49
5.23	5.22	5.20	5.18	5.18	5.17	5.16	5.15	5.14	5.13
3.92	3.90	3.87	3.84	3.83	3.82	3.80	3.79	3.78	3.76
3.30	3.27	3.24	3.21	3.19	3.17	3.16	3.14	3.12	3.10
2.94	2.90	2.87	2.84	2.82	2.80	2.78	2.76	2.74	2.72
2.70	2.67	2.63	2.59	2.58	2.56	2.54	2.51	2.49	2.47
2.54	2.50	2.46	2.42	2.40	2.38	2.36	2.34	2.32	2.29
2.42	2.38	2.34	2.30	2.28	2.25	2.23	2.21	2.18	2.16
2.32	2.28	2.24	2.20	2.18	2.16	2.13	2.11	2.08	2.06
2.25	2.21	2.17	2.12	2.10	2.08	2.05	2.03	2.00	1.97
2.19	2.15	2.10	2.06	2.04	2.01	1.99	1.96	1.93	1.90
2.14	2.10	2.05	2.01	1.98	1.96	1.93	1.90	1.88	1.85
2.10	2.05	2.01	1.96	1.94	1.91	1.89	1.86	1.83	1.80
2.06	2.02	1.97	1.92	1.90	1.87	1.85	1.82	1.79	1.76
2.03	1.99	1.94	1.89	1.87	1.84	1.81	1.78	1.75	1.72
2.00	1.96	1.91	1.86	1.84	1.81	1.78	1.75	1.72	1.69
1.98	1.93	1.89	1.84	1.81	1.78	1.75	1.72	1.69	1.66
1.96	1.91	1.86	1.81	1.79	1.76	1.73	1.70	1.67	1.63
1.94	1.89	1.84	1.79	1.77	1.74	1.71	1.68	1.64	1.61
1.92	1.87	1.83	1.78	1.75	1.72	1.69	1.66	1.62	1.59
1.90	1.86	1.81	1.76	1.73	1.70	1.67	1.64	1.60	1.57
1.89	1.84	1.80	1.74	1.72	1.69	1.66	1.62	1.59	1.55
1.88	1.83	1.78	1.73	1.70	1.67	1.64	1.61	1.57	1.53
1.87	1.82	1.77	1.72	1.69	1.66	1.63	1.59	1.56	1.52
1.86	1.81	1.76	1.71	1.68	1.65	1.61	1.58	1.54	1.50
1.85	1.80	1.75	1.70	1.67	1.64	1.60	1.57	1.53	1.49
1.84	1.79	1.74	1.69	1.66	1.63	1.59	1.56	1.52	1.48
1.83	1.78	1.73	1.68	1.65	1.62	1.58	1.55	1.51	1.47
1.82	1.77	1.72	1.67	1.64	1.61	1.57	1.54	1.50	1.46
1.76	1.71	1.66	1.61	1.57	1.54	1.51	1.47	1.42	1.38
1.71	1.66	1.60	1.54	1.51	1.48	1.44	1.40	1.35	1.29
1.65	1.60	1.55	1.48	1.45	1.41	1.37	1.32	1.26	1.19
1.60	1.55	1.49	1.42	1.38	1.34	1.30	1.24	1.17	1.00

STATISTICAL TABLES

Table VII

Percentage Points of the F Distribution

$\alpha = 0.05$

$\nu_2 \backslash \nu_1$	1	2	3	4	5	6	7	8	9
1	161	200	216	225	230	234	237	239	241
2	18.5	19.0	19.2	19.2	19.3	19.3	19.4	19.4	19.4
3	10.1	9.55	9.28	9.12	9.01	8.94	8.89	8.85	8.81
4	7.71	6.94	6.59	6.39	6.26	6.16	6.09	6.04	6.00
5	6.61	5.79	5.41	5.19	5.05	4.95	4.88	4.82	4.77
6	5.99	5.14	4.76	4.53	4.39	4.28	4.21	4.15	4.10
7	5.59	4.74	4.35	4.12	3.97	3.87	3.79	3.73	3.68
8	5.32	4.46	4.07	3.84	3.69	3.58	3.50	3.44	3.39
9	5.12	4.26	3.86	3.63	3.48	3.37	3.29	3.23	3.18
10	4.96	4.10	3.71	3.48	3.33	3.22	3.14	3.07	3.02
11	4.84	3.98	3.59	3.36	3.20	3.09	3.01	2.95	2.90
12	4.75	3.89	3.49	3.26	3.11	3.00	2.91	2.85	2.80
13	4.67	3.81	3.41	3.18	3.03	2.92	2.83	2.77	2.71
14	4.60	3.74	3.34	3.11	2.96	2.85	2.76	2.70	2.65
15	4.54	3.68	3.29	3.06	2.90	2.79	2.71	2.64	2.59
16	4.49	3.63	3.24	3.01	2.85	2.74	2.66	2.59	2.54
17	4.45	3.59	3.20	2.96	2.81	2.70	2.61	2.55	2.49
18	4.41	3.55	3.16	2.93	2.77	2.66	2.58	2.51	2.46
19	4.38	3.52	3.13	2.90	2.74	2.63	2.54	2.48	2.42
20	4.35	3.49	3.10	2.87	2.71	2.60	2.51	2.45	2.39
21	4.32	3.47	3.07	2.84	2.68	2.57	2.49	2.42	2.37
22	4.30	3.44	3.05	2.82	2.66	2.55	2.46	2.40	2.34
23	4.28	3.42	3.03	2.80	2.64	2.53	2.44	2.37	2.32
24	4.26	3.40	3.01	2.78	2.62	2.51	2.42	2.36	2.30
25	4.24	3.39	2.99	2.76	2.60	2.49	2.40	2.34	2.28
26	4.23	3.37	2.98	2.74	2.59	2.47	2.39	2.32	2.27
27	4.21	3.35	2.96	2.73	2.57	2.46	2.37	2.31	2.25
28	4.20	3.34	2.95	2.71	2.56	2.45	2.36	2.29	2.24
29	4.18	2.33	2.93	2.70	2.55	2.43	2.35	2.28	2.22
30	4.17	3.32	2.92	2.69	2.53	2.42	2.33	2.27	2.21
40	4.08	3.23	2.84	2.61	2.45	2.34	2.25	2.18	2.12
60	4.00	3.15	2.76	2.53	2.37	2.25	2.17	2.10	2.04
120	3.92	3.07	2.68	2.45	2.29	2.18	2.09	2.02	1.96
∞	3.84	3.00	2.60	2.37	2.21	2.10	2.01	1.94	1.88

Table VII

Percentage Points of the F Distribution (*cont.*)

$$\alpha = 0.05$$

10	12	15	20	24	30	40	60	120	∞
242	244	246	248	249	250	251	252	253	254
19.4	19.4	19.4	19.4	19.5	19.5	19.5	19.5	19.5	19.5
8.79	8.74	8.70	8.66	8.64	8.62	8.59	8.57	8.55	8.53
5.96	5.91	5.86	5.80	5.77	5.75	5.72	5.69	5.66	5.63
4.74	4.68	4.62	4.56	4.53	4.50	4.46	4.43	4.40	4.36
4.06	4.00	3.94	3.87	3.84	3.81	3.77	3.74	3.70	3.67
3.64	3.57	3.51	3.44	3.41	3.38	3.34	3.30	3.27	3.23
3.35	3.28	3.22	3.15	3.12	3.08	3.04	3.00	2.97	2.93
3.14	3.07	3.01	2.94	2.90	2.86	2.83	2.79	2.75	2.71
2.98	2.91	2.84	2.77	2.74	2.70	2.66	2.62	2.58	2.54
2.85	2.79	2.72	2.65	2.61	2.57	2.53	2.49	2.45	2.40
2.75	2.69	2.62	2.54	2.51	2.47	2.43	2.38	2.34	2.30
2.67	2.60	2.53	2.46	2.42	2.38	2.34	2.30	2.25	2.21
2.60	2.53	2.46	2.39	2.35	2.31	2.27	2.22	2.18	2.13
2.54	2.48	2.40	2.33	2.29	2.25	2.20	2.16	2.11	2.07
2.49	2.42	2.35	2.28	2.24	2.19	2.15	2.11	2.06	2.01
2.45	2.38	2.31	2.23	2.19	2.15	2.10	2.06	2.01	1.96
2.41	2.34	2.27	2.19	2.15	2.11	2.06	2.02	1.97	1.92
2.38	2.31	2.23	2.16	2.11	2.07	2.03	1.98	1.93	1.88
2.35	2.28	2.20	2.12	2.08	2.04	1.99	1.95	1.90	1.84
2.32	2.25	2.18	2.10	2.05	2.01	1.96	1.92	1.87	1.81
2.30	2.23	2.15	2.07	2.03	1.98	1.94	1.89	1.84	1.78
2.27	2.20	2.13	2.05	2.00	1.96	1.91	1.86	1.81	1.76
2.25	2.18	2.11	2.03	1.98	1.94	1.89	1.84	1.79	1.73
2.24	2.16	2.09	2.01	1.96	1.92	1.87	1.82	1.77	1.71
2.22	2.15	2.07	1.99	1.95	1.90	1.85	1.80	1.75	1.69
2.20	2.13	2.06	1.97	1.93	1.88	1.84	1.79	1.73	1.67
2.19	2.12	2.04	1.96	1.91	1.87	1.82	1.77	1.71	1.65
2.18	2.10	2.03	1.94	1.90	1.85	1.81	1.75	1.70	1.64
2.16	2.09	2.01	1.93	1.89	1.84	1.79	1.74	1.68	1.62
2.08	2.00	1.92	1.84	1.79	1.74	1.69	1.64	1.58	1.51
1.99	1.92	1.84	1.75	1.70	1.65	1.59	1.53	1.47	1.39
1.91	1.83	1.75	1.66	1.61	1.55	1.50	1.43	1.35	1.25
1.83	1.75	1.67	1.57	1.52	1.46	1.39	1.32	1.22	1.00

STATISTICAL TABLES

Table VII

Percentage Points of the F Distribution

$\alpha = 0.01$

ν_2 \ ν_1	1	2	3	4	5	6	7	8	9
1	4052	5000	5403	5625	5764	5859	5928	5982	6022
2	98.5	99.0	99.2	99.2	99.3	99.3	99.4	99.4	99.4
3	34.1	30.8	29.5	28.7	28.2	27.9	27.7	27.5	27.3
4	21.2	18.0	16.7	16.0	15.5	15.2	15.0	14.8	14.7
5	16.3	13.3	12.1	11.4	11.0	10.7	10.5	10.3	10.2
6	13.7	10.9	9.78	9.15	8.75	8.47	8.26	8.10	7.98
7	12.2	9.55	8.45	7.85	7.46	7.19	6.99	6.84	6.72
8	11.3	8.65	7.59	7.01	6.63	6.37	6.18	6.03	5.91
9	10.6	8.02	6.99	6.42	6.06	5.80	5.61	5.47	5.35
10	10.0	7.56	6.55	5.99	5.64	5.39	5.20	5.06	4.94
11	9.65	7.21	6.22	5.67	5.32	5.07	4.89	4.74	4.63
12	9.33	6.93	5.95	5.41	5.06	4.82	4.64	4.50	4.39
13	9.07	6.70	5.74	5.21	4.86	4.62	4.44	4.30	4.19
14	8.86	6.51	5.56	5.04	4.70	4.46	4.28	4.14	4.03
15	8.68	6.36	5.42	4.89	4.56	4.32	4.14	4.00	3.89
16	8.53	6.23	5.29	4.77	4.44	4.20	4.03	3.89	3.78
17	8.40	6.11	5.18	4.67	4.34	4.10	3.93	3.79	3.68
18	8.29	6.01	5.09	4.58	4.25	4.01	3.84	3.71	3.60
19	8.18	5.93	5.01	4.50	4.17	3.94	3.77	3.63	3.52
20	8.10	5.85	4.94	4.43	4.10	3.87	3.70	3.56	3.46
21	8.02	5.78	4.87	4.37	4.04	3.81	3.64	3.51	3.40
22	7.95	5.72	4.82	4.31	3.99	3.76	3.59	3.45	3.35
23	7.88	5.66	4.76	4.26	3.94	3.71	3.54	3.41	3.30
24	7.82	5.61	4.72	4.22	3.90	3.67	3.50	3.36	3.26
25	7.77	5.57	4.68	4.18	3.86	3.63	3.46	3.32	3.22
26	7.72	5.53	4.64	4.14	3.82	3.59	3.42	3.29	3.18
27	7.68	5.49	4.60	4.11	3.78	3.56	3.39	3.26	3.15
28	7.64	5.45	4.57	4.07	3.75	3.53	3.36	3.23	3.12
29	7.60	5.42	4.54	4.04	3.73	3.50	3.33	3.20	3.09
30	7.56	5.39	4.51	4.02	3.70	3.47	3.30	3.17	3.07
40	7.31	5.18	4.31	3.83	3.51	3.29	3.12	2.99	2.89
60	7.08	4.98	4.13	3.65	3.34	3.12	2.95	2.82	2.72
120	6.85	4.79	3.95	3.48	3.17	2.96	2.79	2.66	2.56
∞	6.63	4.61	3.78	3.32	3.02	2.80	2.64	2.51	2.41

Table VII

Percentage Points of the F Distribution (*cont.*)

$$\alpha = 0.01$$

10	12	15	20	24	30	40	60	120	∞
6056	6106	6157	6209	6235	6261	6287	6313	6339	6366
99.4	99.4	99.4	99.4	99.5	99.5	99.5	99.5	99.5	99.5
27.2	27.1	26.9	26.7	26.6	26.5	26.4	26.3	26.2	26.1
14.5	14.4	14.2	14.0	13.9	13.8	13.7	13.7	13.6	13.5
10.1	9.89	9.72	9.55	9.47	9.38	9.29	9.20	9.11	9.02
7.87	7.72	7.56	7.40	7.31	7.23	7.14	7.06	6.97	6.88
6.62	6.47	6.31	6.16	6.07	5.99	5.91	5.82	5.74	5.65
5.81	5.67	5.52	5.36	5.28	5.20	5.12	5.03	4.95	4.86
5.26	5.11	4.96	4.81	4.73	4.65	4.57	4.48	4.40	4.31
4.85	4.71	4.56	4.41	4.33	4.25	4.17	4.08	4.00	3.91
4.54	4.40	4.25	4.10	4.02	3.94	3.86	3.78	3.69	3.60
4.30	4.16	4.01	3.86	3.78	3.70	3.62	3.54	3.45	3.36
4.10	3.96	3.82	3.66	3.59	3.51	3.43	3.34	3.25	3.17
3.94	3.80	3.66	3.51	3.43	3.35	3.27	3.18	3.09	3.00
3.80	3.67	3.52	3.37	3.29	3.21	3.13	3.05	2.96	2.87
3.69	3.55	3.41	3.26	3.18	3.10	3.02	2.93	2.84	2.75
3.59	3.46	3.31	3.16	3.08	3.00	2.92	2.83	2.75	2.65
3.51	3.37	3.23	3.08	3.00	2.92	2.84	2.75	2.66	2.57
3.43	3.30	3.15	3.00	2.92	2.84	2.76	2.67	2.58	2.49
3.37	3.23	3.09	2.94	2.86	2.78	2.69	2.61	2.52	2.42
3.31	3.17	3.03	2.88	2.80	2.72	2.64	2.55	2.46	2.36
3.26	3.12	2.98	2.83	2.75	2.67	2.58	2.50	2.40	2.31
3.21	3.07	2.93	2.78	2.70	2.62	2.54	2.45	2.35	2.26
3.17	3.03	2.89	2.74	2.66	2.58	2.49	2.40	2.31	2.21
3.13	2.99	2.85	2.70	2.62	2.54	2.45	2.36	2.27	2.17
3.09	2.96	2.82	2.66	2.58	2.50	2.42	2.33	2.23	2.13
3.06	2.93	2.78	2.63	2.55	2.47	2.38	2.29	2.20	2.10
3.03	2.90	2.75	2.60	2.52	2.44	2.35	2.26	2.17	2.06
3.00	2.87	2.73	2.57	2.49	2.41	2.33	2.23	2.14	2.03
2.98	2.84	2.70	2.55	2.47	2.39	2.30	2.21	2.11	2.01
2.80	2.66	2.52	2.37	2.29	2.20	2.11	2.02	1.92	1.80
2.63	2.50	2.35	2.20	2.12	2.03	1.94	1.84	1.73	1.60
2.47	2.34	2.19	2.03	1.95	1.86	1.76	1.66	1.53	1.38
2.32	2.18	2.04	1.88	1.79	1.70	1.59	1.47	1.32	1.00

STATISTICAL TABLES

Table VII

Percentage Points of the F Distribution

$\alpha = 0.005$

ν_2 \ ν_1	1	2	3	4	5	6	7	8	9
1	16211	20000	21615	22500	23056	23437	23715	23925	24091
2	198	199	199	199	199	199	199	199	199
3	55.6	49.8	47.5	46.2	45.4	44.8	44.4	44.1	43.9
4	31.3	26.3	24.3	23.2	22.5	22.0	21.6	21.4	21.1
5	22.8	18.3	16.5	15.6	14.9	14.5	14.2	14.0	13.8
6	18.6	14.5	12.9	12.0	11.5	11.1	10.8	10.6	10.4
7	16.2	12.4	10.9	10.0	9.52	9.16	8.89	8.68	8.51
8	14.7	11.0	9.60	8.81	8.30	7.95	7.69	7.50	7.34
9	13.6	10.1	8.72	7.96	7.47	7.13	6.88	6.69	6.54
10	12.8	9.43	8.08	7.34	6.87	6.54	6.30	6.12	5.97
11	12.2	8.91	7.60	6.88	6.42	6.10	5.86	5.68	5.54
12	11.8	8.51	7.23	6.52	6.07	5.76	5.52	5.35	5.20
13	11.4	8.19	6.93	6.23	5.79	5.48	5.25	5.08	4.94
14	11.1	7.92	6.68	6.00	5.56	5.26	5.03	4.86	4.72
15	10.8	7.70	6.48	5.80	5.37	5.07	4.85	4.67	4.54
16	10.6	7.51	6.30	5.64	5.21	4.91	4.69	4.52	4.38
17	10.4	7.35	6.16	5.50	5.07	4.78	4.56	4.39	4.25
18	10.2	7.21	6.03	5.37	4.96	4.66	4.44	4.28	4.14
19	10.1	7.09	5.92	5.27	4.85	4.56	4.34	4.18	4.04
20	9.94	6.99	5.82	5.17	4.76	4.47	4.26	4.09	3.96
21	9.83	6.89	5.73	5.09	4.68	4.39	4.18	4.01	3.88
22	9.73	6.81	5.65	5.02	4.61	4.32	4.11	3.94	3.81
23	9.63	6.73	5.58	4.95	4.54	4.26	4.05	3.88	3.75
24	9.55	6.66	5.52	4.89	4.49	4.20	3.99	3.83	3.69
25	9.48	6.60	5.46	4.84	4.43	4.15	3.94	3.78	3.64
26	9.41	6.54	5.41	4.79	4.38	4.10	3.89	3.73	3.60
27	9.34	6.49	5.36	4.74	4.34	4.06	3.85	3.69	3.56
28	9.28	6.44	5.32	4.70	4.30	4.02	3.81	3.65	3.52
29	9.23	6.40	5.28	4.66	4.26	3.98	3.77	3.61	3.48
30	9.18	6.35	5.24	4.62	4.23	3.95	3.74	3.58	3.45
40	8.83	6.07	4.98	4.37	3.99	3.71	3.51	3.35	3.22
60	8.49	5.80	4.73	4.14	3.76	3.49	3.29	3.13	3.01
120	8.18	5.54	4.50	3.92	3.55	3.28	3.09	2.93	2.81
∞	7.88	5.30	4.28	3.72	3.35	3.09	2.90	2.74	2.62

Table VII

Percentage Points of the F Distribution (*cont.*)

$$\alpha = 0.005$$

10	12	15	20	24	30	40	60	120	∞
24224	24426	24630	24836	24940	25044	25148	25253	25359	25465
199	199	199	199	199	199	199	199	199	200
43.7	43.4	43.1	42.8	42.6	42.5	42.3	42.1	42.0	41.8
21.0	20.7	20.4	20.2	20.0	19.9	19.8	19.6	19.5	19.3
13.6	13.4	13.1	12.9	12.8	12.7	12.5	12.4	12.3	12.1
10.2	10.0	9.81	9.59	9.47	9.36	9.24	9.12	9.00	8.88
8.38	8.18	7.97	7.75	7.64	7.53	7.42	7.31	7.19	7.08
7.21	7.01	6.81	6.61	6.50	6.40	6.29	6.18	6.06	5.95
6.42	6.23	6.03	5.83	5.73	5.62	5.52	5.41	5.30	5.19
5.85	5.66	5.47	5.27	5.17	5.07	4.97	4.86	4.75	4.64
5.42	5.24	5.05	4.86	4.76	4.65	4.55	4.44	4.34	4.23
5.09	4.91	4.72	4.53	4.43	4.34	4.23	4.12	4.01	3.90
4.82	4.64	4.46	4.27	4.17	4.07	3.97	3.87	3.76	3.65
4.60	4.43	4.25	4.06	3.96	3.86	3.76	3.66	3.55	3.44
4.42	4.25	4.07	3.88	3.79	3.69	3.58	3.48	3.37	3.26
4.27	4.10	3.92	3.73	3.64	3.54	3.44	3.33	3.22	3.11
4.14	3.97	3.79	3.61	3.51	3.41	3.31	3.21	3.10	2.98
4.03	3.86	3.68	3.50	3.40	3.30	3.20	3.10	2.99	2.87
3.93	3.76	3.59	3.40	3.31	3.21	3.11	3.00	2.89	2.78
3.85	3.68	3.50	3.32	3.22	3.12	3.02	2.92	2.81	2.69
3.77	3.60	3.43	3.24	3.15	3.05	2.95	2.84	2.73	2.61
3.70	3.54	3.36	3.18	3.08	2.98	2.88	2.77	2.66	2.55
3.64	3.47	3.30	3.12	3.02	2.92	2.82	2.71	2.60	2.48
3.59	3.42	3.25	3.06	2.97	2.87	2.77	2.66	2.55	2.43
3.54	3.37	3.20	3.01	2.92	2.82	2.72	2.61	2.50	2.38
3.49	3.33	3.15	2.97	2.87	2.77	2.67	2.56	2.45	2.33
3.45	3.28	3.11	2.93	2.83	2.73	2.63	2.52	2.41	2.29
3.41	3.25	3.07	2.89	2.79	2.69	2.59	2.48	2.37	2.25
3.38	3.21	3.04	2.86	2.76	2.66	2.56	2.45	2.33	2.21
3.34	3.18	3.01	2.82	2.73	2.63	2.52	2.42	2.30	2.18
3.12	2.95	2.78	2.60	2.50	2.40	2.30	2.18	2.06	1.93
2.90	2.74	2.57	2.39	2.29	2.19	2.08	1.96	1.83	1.69
2.71	2.54	2.37	2.19	2.09	1.98	1.87	1.75	1.61	1.43
2.52	2.36	2.19	2.00	1.90	1.79	1.67	1.53	1.36	1.00

Table VIII

Power of the Analysis-of-Variance F Test*

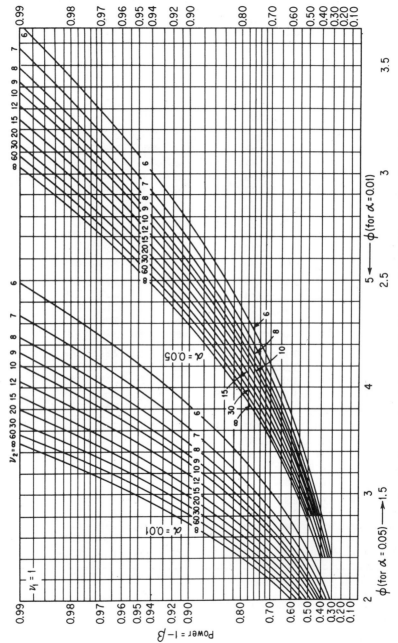

* This table is reproduced from W. J. Dixon and F. J. Massey, Jr., *Introduction to Statistical Analysis*, 2nd ed. McGraw-Hill, Inc., New York, 1957, Table A-13, pp. 426–33, with the permission of the publishers.

Table VIII

Power of the Analysis-of-Variance F Test (cont.)

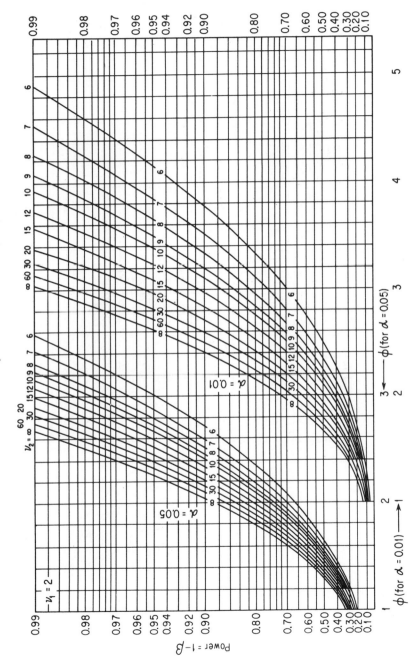

Table VIII

Power of the Analysis-of-Variance F Test (*cont.*)

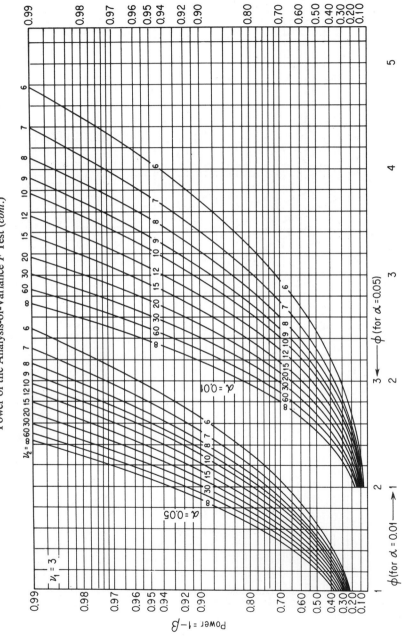

Table VIII

Power of the Analysis-of-Variance *F* Test *(cont.)*

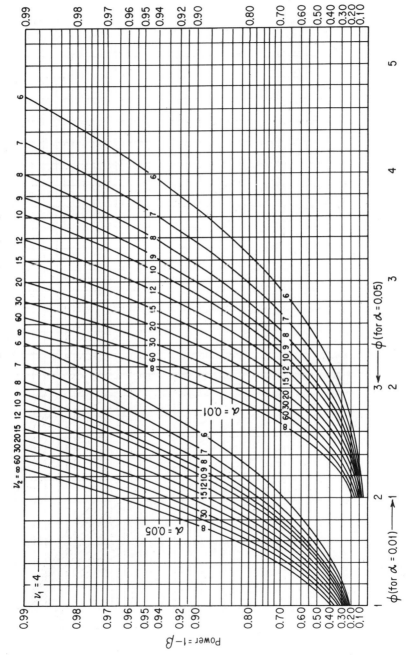

Table VIII

Power of the Analysis-of-Variance *F* Test (*cont.*)

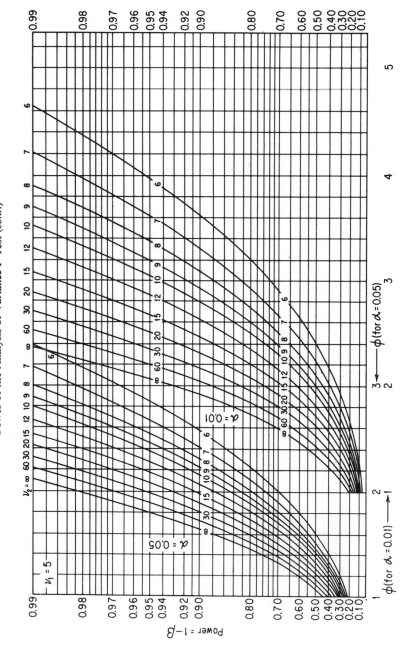

Table VIII

Power of the Analysis-of-Variance F Test (cont.)

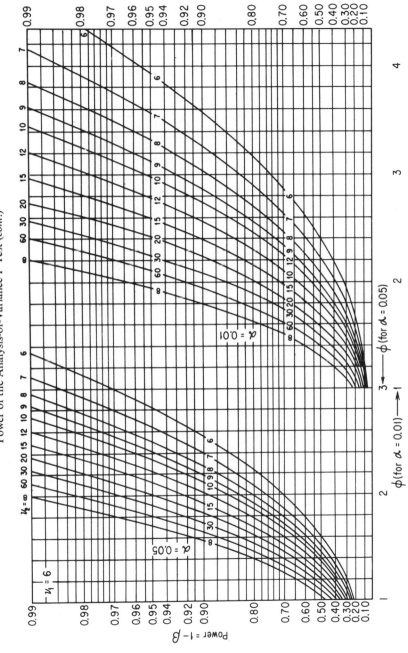

Table VIII

Power of the Analysis-of-Variance F Test (cont.)

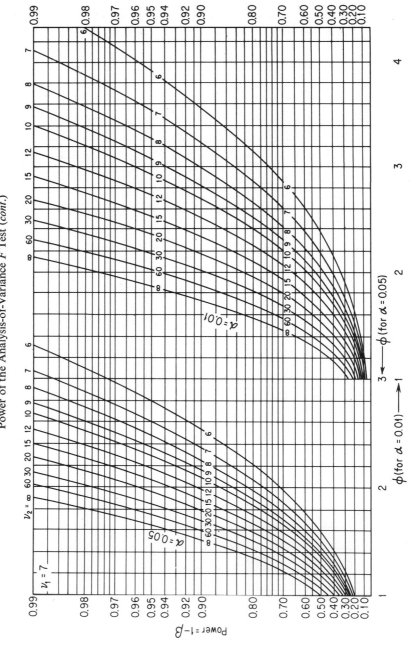

Table VIII

Power of the Analysis-of-Variance *F* Test (*cont.*)

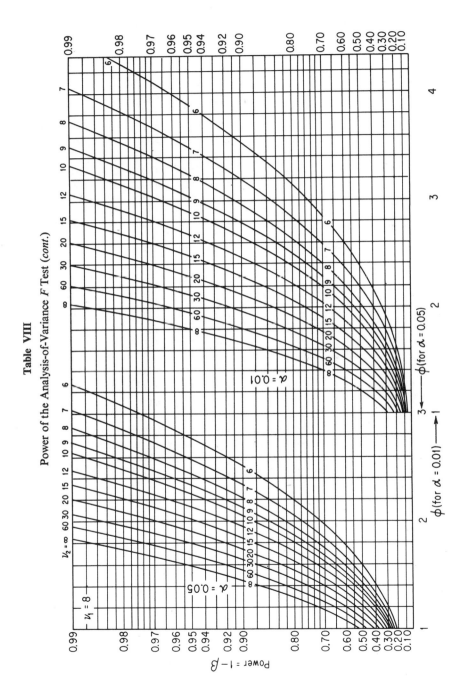

Table IX

Percentage Points of the Studentized Range*

$$\alpha = 0.05$$

ν \ k	2	3	4	5	6	7	8	9
5	3.64	4.60	5.22	5.67	6.03	6.33	6.58	6.80
6	3.46	4.34	4.90	5.31	5.63	5.89	6.12	6.32
7	3.34	4.16	4.68	5.06	5.36	5.61	5.82	6.00
8	3.26	4.04	4.53	4.89	5.17	5.40	5.60	5.77
9	3 20	3.95	4.42	4.76	5.02	5.24	5.43	5.60
10	3.15	3.88	4.33	4.65	4.91	5.12	5.30	5.46
11	3.11	3.82	4.26	4.57	4.82	5.03	5.20	5.35
12	3.08	3.77	4.20	4.51	4.75	4.95	5.12	5.27
13	3.06	3.73	4.15	4.45	4.69	4.88	5.05	5.19
14	3.03	3.70	4.11	4.41	4.64	4.83	4.99	5.13
15	3.01	3.67	4.08	4.37	4.60	4.78	4.94	5.08
16	3.00	3.65	4.05	4.33	4.56	4.74	4.90	5.03
17	2.98	3.63	4.02	4.30	4.52	4.71	4.86	4.99
18	2.97	3.61	4.00	4.28	4.49	4.67	4.82	4.96
19	2.96	3.59	3.98	4.25	4.47	4.65	4.79	4.92
20	2.95	3.58	3.96	4.23	4.45	4.62	4.77	4.90
24	2.92	3.53	3.90	4.17	4.37	4.54	4.68	4.81
30	2.89	3.49	3.84	4.10	4.30	4.46	4.60	4.72
40	2.86	3.44	3.79	4.04	4.23	4.39	4.52	4.63
60	2.83	3.40	3.74	3.98	4.16	4.31	4.44	4.55
120	2.80	3.36	3.69	3.92	4.10	4.24	4.36	4.48
∞	2.77	3.31	3.63	3.86	4.03	4.17	4.29	4.39

* This table is abridged from E. S. Pearson and H. O. Hartley, *Biometrika Tables for Statisticians*, Vol. 1 (1954) Cambridge University Press for the Biometrika Trustees, p. 176, Table 29 (upper percentage points), with permission of Professor E. S. Pearson.

Table IX

Percentage Points of the Studentized Range (*cont.*)

$$\alpha = 0.05$$

10	11	12	13	14	15	16	17	18	19	20
6.99	7.17	7.32	7.47	7.60	7.72	7.83	7.93	8.03	8.12	8.21
6.49	6.65	6.79	6.92	7.03	7.14	7.24	7.34	7.43	7.51	7.59
6.16	6.30	6.43	6.55	6.66	6.76	6.85	6.94	7.02	7.09	7.17
5.92	6.05	6.18	6.29	6.39	6.48	6.57	6.65	6.73	6.80	6.87
5.74	5.87	5.98	6.09	6.19	6.28	6.36	6.44	6.51	6.58	6.64
5.60	5.72	5.83	5.93	6.03	6.11	6.20	6.27	6.34	6.40	6.47
5.49	5.61	5.71	5.81	5.90	5.99	6.06	6.14	6.20	6.26	6.33
5.40	5.51	5.62	5.71	5.80	5.88	5.95	6.03	6.09	6.15	6.21
5.32	5.43	5.53	5.63	5.71	5.79	5.86	5.93	6.00	6.05	6.11
5.25	5.36	5.46	5.55	5.64	5.72	5.79	5.85	5.92	5.98	6.03
5.20	5.31	5.40	5.49	5.58	5.65	5.72	5.79	5.85	5.90	5.96
5.15	5.26	5.35	5.44	5.52	5.59	5.66	5.72	5.79	5.84	5.90
5.11	5.21	5.31	5.39	5.47	5.55	5.61	5.68	5.74	5.79	5.84
5.07	5.17	5.27	5.35	5.43	5.50	5.57	5.63	5.69	5.74	5.79
5.04	5.14	5.23	5.32	5.39	5.46	5.53	5.59	5.65	5.70	5.75
5.01	5.11	5.20	5.28	5.36	5.43	5.49	5.55	5.61	5.66	5.71
4.92	5.01	5.10	5.18	5.25	5.32	5.38	5.44	5.50	5.54	5.59
4.83	4.92	5.00	5.08	5.15	5.21	5.27	5.33	5.38	5.43	5.48
4.74	4.82	4.91	4.98	5.05	5.11	5.16	5.22	5.27	5.31	5.36
4.65	4.73	4.81	4.88	4.94	5.00	5.06	5.11	5.16	5.20	5.24
4.56	4.64	4.72	4.78	4.84	4.90	4.95	5.00	5.05	5.09	5.13
4.47	4.55	4.62	4.68	4.74	4.80	4.85	4.89	4.93	4.97	5.01

STATISTICAL TABLES

Table IX

Percentage Points of the Studentized Range*

$$\alpha = 0.01$$

k / ν	2	3	4	5	6	7	8	9
5	5.70	6.97	7.80	8.42	8.91	9.32	9.67	9.97
6	5.24	6.33	7.03	7.56	7.97	8.32	8.61	8.87
7	4.95	5.92	6.54	7.01	7.37	7.68	7.94	8.17
8	4.74	5.63	6.20	6.63	6.96	7.24	7.47	7.68
9	4.60	5.43	5.96	6.35	6.66	6.91	7.13	7.32
10	4.48	5.27	5.77	6.14	6.43	6.67	6.87	7.05
11	4.39	5.14	5.62	5.97	6.25	6.48	6.67	6.84
12	4.32	5.04	5.50	5.84	6.10	6.32	6.51	6.67
13	4.26	4.96	5.40	5.73	5.98	6.19	6.37	6.53
14	4.21	4.89	5.32	5.63	5.88	6.08	6.26	6.41
15	4.17	4.83	5.25	5.56	5.80	5.99	6.16	6.31
16	4.13	4.78	5.19	5.49	5.72	5.92	6.08	6.22
17	4.10	4.74	5.14	5.43	5.66	5.85	6.01	6.15
18	4.07	4.70	5.09	5.38	5.60	5.79	5.94	6.08
19	4.05	4.67	5.05	5.33	5.55	5.73	5.89	6.02
20	4.02	4.64	5.02	5.29	5.51	5.69	5.84	5.97
24	3.96	4.54	4.91	5.17	5.37	5.54	5.69	5.81
30	3.89	4.45	4.80	5.05	5.24	5.40	5.54	5.65
40	3.82	4.37	4.70	4.93	5.11	5.27	5.39	5.50
60	3.76	4.28	4.60	4.82	4.99	5.13	5.25	5.36
120	3.70	4.20	4.50	4.71	4.87	5.01	5.12	5.21
∞	3.64	4.12	4.40	4.60	4.76	4.88	4.99	5.08

* This table is abridged from E. S. Pearson and H. O. Hartley, *Biometrika Tables for Statisticians*, Vol. 1 (1954), Cambridge University Press for the Biometrika Trustees, p. 177, Table 29 (upper percentage points), with permission of Professor E. S. Pearson.

Table IX

Percentage Points of the Studentized Range (*cont.*)

$$\alpha = 0.01$$

10	11	12	13	14	15	16	17	18	19	20
10.24	10.48	10.70	10.89	11.08	11.24	11.40	11.55	11.68	11.81	11.93
9.10	9.30	9.49	9.65	9.81	9.95	10.08	10.21	10.32	10.43	10.54
8.37	8.55	8.71	8.86	9.00	9.12	9.24	9.35	9.46	9.55	9.65
7.87	8.03	8.18	8.31	8.44	8.55	8.66	8.76	8.85	8.94	9.03
7.49	7.65	7.78	7.91	8.03	8.13	8.23	8.32	8.41	8.49	8.57
7.21	7.36	7.48	7.60	7.71	7.81	7.91	7.99	8.07	8.15	8.22
6.99	7.13	7.25	7.36	7.46	7.56	7.65	7.73	7.81	7.88	7.95
6.81	6.94	7.06	7.17	7.26	7.36	7.44	7.52	7.59	7.66	7.73
6.67	6.79	6.90	7.01	7.10	7.19	7.27	7.34	7.42	7.48	7.55
6.54	6.66	6.77	6.87	6.96	7.05	7.12	7.20	7.27	7.33	7.39
6.44	6.55	6.66	6.76	6.84	6.93	7.00	7.07	7.14	7.20	7.26
6.35	6.46	6.56	6.66	6.74	6.82	6.90	6.97	7.03	7.09	7.15
6.27	6.38	6.48	6.57	6.66	6.73	6.80	6.87	6.94	7.00	7.05
6.20	6.31	6.41	6.50	6.58	6.65	6.72	6.79	6.85	6.91	6.96
6.14	6.25	6.34	6.43	6.51	6.58	6.65	6.72	6.78	6.84	6.89
6.09	6.19	6.29	6.37	6.45	6.52	6.59	6.65	6.71	6.76	6.82
5.92	6.02	6.11	6.19	6.26	6.33	6.39	6.45	6.51	6.56	6.61
5.76	5.85	5.93	6.01	6.08	6.14	6.20	6.26	6.31	6.36	6.41
5.60	5.69	5.77	5.84	5.90	5.96	6.02	6.07	6.12	6.17	6.21
5.45	5.53	5.60	5.67	5.73	5.79	5.84	5.89	5.93	5.98	6.02
5.30	5.38	5.44	5.51	5.56	5.61	5.66	5.71	5.75	5.79	5.83
5.16	5.23	5.29	5.35	5.40	5.45	5.49	5.54	5.57	5.61	5.65

Table X

Confidence Belts for the Correlation Coefficient ρ: $\rho = .95$*

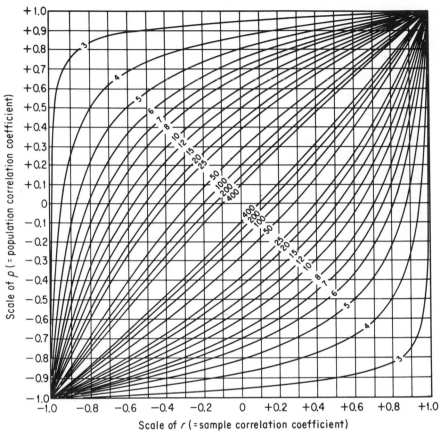

Scale of ρ (= population correlation coefficient)

Scale of r (= sample correlation coefficient)

Table X

Confidence Belts for the Correlation Coefficient ρ: $\rho = .99$ (*cont.*)

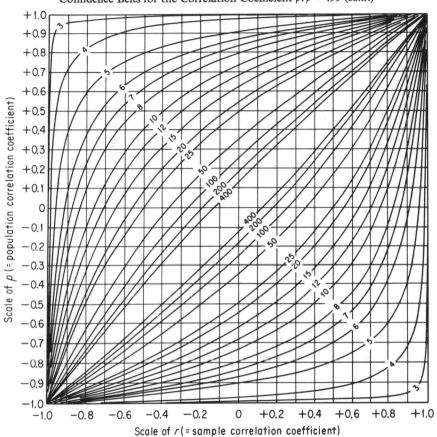

Scale of r (= sample correlation coefficient)

Scale of ρ (= population correlation coefficient)

INDEX